OCEAN

Bismarck Archipelago

Manus

Admiralty Islands

PAPUA NEW GUINEA

Wewak

BISMARCK SEA

Schleintz Range

New Ireland

Ramu River

Adelbert Range

Veron Range

Ulawun (2334 m)

Bougainville Island

Wabag

Madang

Bismarck Range

Finisterre Range

Nakanai Mountains

Mt. Wilhelm (4509 m)

Mt. Hagen

Huon Peninsula

Sarawaget Range

New Britain

Mount Balbi (2715 m)

Takuan (2243 m)

Lae

Kikori River

Purai River

Kikori

SOLOMON SEA

Kerema

Gulf of Papua

Mt. Victoria (4038 m)

Popondetta

Owen Stanley Range

Daru

Port Moresby

Alotau

Milne Bay

Louisiade Archipelago

CORAL SEA

ISLAND

PALMS OF NEW GUINEA

PALMS OF NEW GUINEA

William J. Baker, Anders S. Barfod, Rodrigo Cámara-Leret,
John L. Dowe, Charlie D. Heatubun, Peter Petoe,
Jessica H. Turner, Scott Zona and John Dransfield

Illustrated by **Lucy T. Smith**

Kew Publishing
Royal Botanic Gardens, Kew

First published in 2024 by
Royal Botanic Gardens, Kew,
Richmond, Surrey, TW9 3AB, UK
www.kew.org

ISBN 978 1 84246 810 4
eISBN 978 1 84246 811 1

Distributed on behalf of the Royal Botanic Gardens, Kew in North America
by the University of Chicago Press, 1427 East 60th St, Chicago, IL 60637, USA

British Library Cataloguing in Publication Data
A catalogue record for this book is available from the British Library

Design and page layout: Christine Beard
Production management: Georgie Hills
Copy-editing: Sharon Whitehead

FRONT COVER
Pigafetta filaris and *Metroxylon sagu* (sago palm) in coastal vegetation near Sorong. Photo: WB.

BACK COVER
Background – *Sommieria leucophylla*. Photo: GP. Top row, left to right – *Ptychococcus paradoxus, Metroxylon sagu, Hydriastele flabellata, Hydriastele biakensis*. Bottom row, left to right – *Cyrtostachys excelsa, Calamus nudus, Hydriastele pinangoides, Calamus bulubabi*.
All photos: WB (except *Calamus nudus* [SV]).

ENDPAPERS
Front and back: Political map of the New Guinea region.

FRONTISPIECE
Cyrtostachys loriae, near Timika. Photo: WB.

Printed in the United Kingdom by Short Run Press Ltd

For information or to purchase all Kew titles please visit
shop.kew.org/kewbooksonline or email publishing@kew.org

Kew's mission is to understand and protect plants and fungi, for the wellbeing of people and the future of all life on Earth.

Kew receives approximately one third of its funding from Government through the Department for Environment, Food and Rural Affairs (Defra). All other funding needed to support Kew's vital work comes from members, foundations, donors and commercial activities, including book sales.

CONTENTS

TAXONOMIC ACCOUNTS (contd.)

FOREWORD

by **Dominggus Mandacan**
for Indonesia

When we talk about palms, we immediately think of their importance to the indigenous people of New Guinea. The sago palm not only provides the staple diet of most lowlanders, but also sits at the heart of our culture and way of life. Such is its prominence in our worldview that some ethnic groups believe that their ancestors came from the sago palm. Taking together all of the properties and traditional uses of palms, it is clear that they are the most widely utilised plant family in New Guinea. This strong connection has helped to shape the unique cultural identity, customs and indigenous knowledge of the island's people.

The publication of *Palms of New Guinea* is a milestone in New Guinea biodiversity research and a precious legacy for future generations. To produce such a monumental work requires extraordinary effort – it is no surprise that this book has taken more than 20 years to complete. But what is more important than the ways and means of producing a book is the genuine collaboration that has been shown by the whole team of dedicated palm experts, who solved all challenges together and worked together to increase local capacity in the region, ensuring a sustainable cadre of palm botanists from the region who will continue this work into the future.

This book has opened our horizons in the way we think about the diversity of such a familiar group of plants in New Guinea, an island with floristic riches exceeding all other islands. I really appreciate the work of the authors who have brought such a significant contribution to our understanding of the potential of the region's palm diversity. The information in this book will be invaluable for determining the policy direction for the future use and conservation of all palm species. Personally, I am also extremely flattered to have been honoured with the naming of one of the new species of betel nut palm discovered during this project, *Areca mandacanii*, alongside *Areca jokowi*, which was named after the 7th President of the Republic of Indonesia, H.E. Joko Widodo. It is a great honour for me to be enshrined in the botanical world and to have been asked to give this foreword.

I believe this book is only the beginning. It will stimulate more research on palms and other plant groups and their conservation needs in New Guinea, creating foundations for sustainable development, the prevention of biodiversity loss and climate change on our spectacular, irreplaceable island home.

H. E. Dominggus Mandacan
Governor of West Papua Province 2017–2022
Great Chief of the Arfak Tribe
Global Conservation Hero – Conservation International & Elle Magazine 2019

FOREWORD

by **Kipiro Damas**
for Papua New Guinea

The island of New Guinea is one of the world's highest diversity areas in terms of its 13,600 vascular plant species. It is also one of the least botanically explored parts of the world, so documentation of the flora is very important. The documentation of palms is as important as that of any other vascular plant group.

In the past, the indigenous people of Papua New Guinea lived a nomadic life. Palms played very significant roles in their survival by providing food, medicine and shelter. The fronds and their sheaths were used as roofing and walls for makeshift huts. The juvenile shoots from some species were readily available as a source of food, and the seeds were eaten as stimulants (betel nut). The nomadic lifestyle has now been replaced. People no longer live that life due to the introduction of the westernised education system. This has led to completely different ways of thinking and surviving.

Traditional knowledge of plants no longer exists in the younger generations of today because they are stationed permanently in one location. Food, shelter, stimulants and medicines are now much more easily reached. There are only a few elders left who know the old ways, and the young generation of today does not see the need to seek advice from them. The behaviours of the young are influenced by the western world, and for this reason, the transfer of traditional knowledge is now almost non-existent.

It is very important to properly document the entirety of our flora, especially the most useful species, such as palms. This book, *Palms of New Guinea*, builds on two previous field guides, but is much more comprehensive than those forerunners. It is full of information on all 250 species, their distribution, their local names, their traditional uses, their identification and morphological features, and more. This book is very user friendly, with many photographs and illustrations that will help anyone, especially the layperson, to identify a palm species on sight.

I am one of the remnant, still-active, senior botanists in Papua New Guinea, home-grown after the colonial era, and I truly regard *Palms of New Guinea* as one of the best modern-day books published on our palms. This book will be treasured by many up-and-coming New Guinea plant enthusiasts. I am so grateful to William Baker and his team because this book, I believe, will really make a difference. Tenkyu tru!

Kipiro Q. Damas
Senior Botanist, PNG National Herbarium Lae
PNG Forest Research Institute Forest Biodiversity Program Manager (2023)

PREFACE

No other part of the world fires the imagination quite like New Guinea. This vast island, the largest in the tropics, still retains more than 70% of its vegetation intact, from the swamps and tall rainforest of the lowlands, through rugged mountain forests and alpine meadows thick with tree ferns, to glacier-capped summits at almost 5,000 m. It is one of the world's most important remaining wilderness areas. Its iconic fauna, renowned for its birds of paradise, tree-dwelling marsupials and spectacular butterflies, is more than matched by its flora, the richest of any tropical island (Cámara-Leret *et al.* 2020), with its countless species of orchids, spiny ant-plants and extravagant rhododendrons. New Guinea's astounding biodiversity, however, remains substantially under-explored and inadequately documented. As environmental pressures relentlessly intensify, robust scientific knowledge of the New Guinea biota has never been more urgently needed. This is especially true for species-rich groups such as the palms, which are keystone species of tropical forests that also underpin human livelihoods in innumerable ways.

New Guinea is the largest remaining gap in our understanding of the global palm flora. When the *Palms of New Guinea* project was initiated in the 1990s, we found palm taxonomy on the island in "an advanced state of chaos" (Baker 2002a). Few genera had been studied in a coherent way. It was suspected that many names were redundant and needed to be synonymised, while at the same time New Guinea palm specimens were extremely hard to name in the herbarium due to a lack of keys and taxonomic accounts, hinting at the existence of many undescribed species. The problems were compounded by the patchy nature of palm exploration in the field; much of the island was effectively unstudied for palms.

Encouraged by the success and impact of the regional monograph *The Palms of Madagascar* (Dransfield & Beentje 1995), a similar monograph was proposed for New Guinea at a workshop of palm botanists at the *3rd Flora Malesiana Symposium* at the Royal Botanic Gardens, Kew in July 1995. In 1998, the *Palms of New Guinea* project commenced (see Baker [2000, 2001a, b, 2002a] for summaries of the early history of the project). Unlike the situation for Madagascar, however, several experts were already interested in New Guinea palms at the project's inception, and therefore an international collaborative team was assembled that involved botanists from Papua New Guinea, Indonesia, Australia, Denmark, the United Kingdom and the United States. The early team included William Baker, Roy Banka, Anders Barfod, Sasha Barrow, Ross Bayton, John Dowe, John Dransfield, Michael Ferrero, Osia Gideon, Charlie Heatubun, Ary Keim, Anders Kjaer, Rudi Maturbongs, Johanis Mogea, Jack Wanggai, Scott Zona and botanical illustrator Lucy Smith. While not all of these individuals ultimately contributed to the writing of this book, all have played an integral role at different stages, and additional team members have joined at later stages (Rodrigo Cámara-Leret, Peter Petoe and Jessica Turner).

The publication of this book, the culmination of the *Palms of New Guinea* project, has been long anticipated. We underestimated the scientific challenges of the work and struggled at times to fit it around our other working commitments. But we were not idle in the intervening 25 years. We have explored extensively for palms in the field, collecting 1,100 new specimens. We have published nearly 40 preparatory papers and have described 91 species new to science, 37% of the entire native palm flora. Numerous generic monographs and two field guides (Barfod *et al.* 2001, Baker & Dransfield 2006a, b) have also been delivered. More than 250 diagnostic botanical illustrations have been prepared. We have tackled systematic problems in the family that pertain to New Guinea genera using cutting-edge DNA-based methods and have described three new genera. Several postgraduate degrees have been awarded to team members for research addressing New Guinea palm taxonomy.

We hope that this book will stimulate an entirely new phase of study of the palms of New Guinea, as has happened in Madagascar, where the documented palm flora has grown by 20% since 1995 thanks to the sound taxonomic foundations provided by *The Palms of Madagascar*. We look forward to *Palms of New Guinea* soon becoming outdated by the discovery of new species, distribution records and other palm knowledge, or by the identification of inevitable omissions. These will be the signs that the book has served its purpose.

Lower montane rainforest in the Tamrau Mountains. Photo: WB.

ACKNOWLEDGEMENTS

The origins of the *Palms of New Guinea* project reach back to the early 1990s and a wide array of colleagues, friends, institutions and funders have supported us over the ensuing decades. There are too many to mention everyone by name, but we are indebted to all of them. We would like to acknowledge our home institutions – especially the Royal Botanic Gardens, Kew, Universitas Papua (UNIPA), Aarhus University and James Cook University – for their commitment to this project.

The project team has collaborated widely in research that underpins this book. Chief among our collaborators are our colleagues in New Guinea. In particular, the late Roy Banka of the Papua New Guinea Forest Research Institute was a key player in the project until his untimely passing in 2017. Roy had an uncanny ability to make things happen, even in difficult circumstances, and is very much missed by the New Guinea palm community. Similarly, Rudi Maturbongs and Jack Wanggai of Universitas Papua were critical partners in the initiation of palm research in Indonesian New Guinea, contributing extensively to the exploration of palms in the region and building important links with Kew.

Many others collaborated closely with the authors on the taxonomy of New Guinea palms and other aspects of their biology, namely Christine Bacon, Megan Barstow, Ross Bayton, Zoe Dennehy, Fred Essig, Edwino Fernando, Jack Fisher, Lauren Gardiner, Gregori Hambali, Tiberius Jimbo, Ary Prihardyanto Keim, Anders Kjaer, Carl Lewis, Adrian Loo, Maria Norup, Melinda Trudgen and Stephanus Venter.

Numerous colleagues gave their time generously to advise and support the authors. Michael Ferrero, the late David Frodin, the late Robert Johns, Osia Gideon, Damien Hicks and Timothy Utteridge shared their many years of experience of New Guinea botany with us. Fred Essig, Jeff Marcus and Johanis Mogea offered varied insights into New Guinea palms. Rafaël Govaerts was always available to give nomenclatural guidance. Monika Shaffer-Fehre and Petra Hoffmann provided translations of difficult German taxonomic literature. Sharon Willoughby gave thoughtful feedback on drafts of the introductory chapters. Mary and Michael Lock offered encouragement and opportunities to WJB that ultimately broke the project log jam, allowing it to reach fruition at last.

Fieldwork in New Guinea has been a cornerstone of the project. We thank all national agencies that authorised and facilitated fieldwork, in particular the Indonesian National Research and Innovation Agency (Badan Riset dan Inovasi Nasional [BRIN], which now incorporates the former Indonesian Institute of Sciences (Lembaga Ilmu Pengetahuan Indonesia [LIPI], the Indonesian Ministry of Forestry, the Papua New Guinea National Research Institute and the Papua New Guinea Forest Research Institute (PNG FRI). PNG FRI, UNIPA, Herbarium Bogoriense, the Christensen Research Institute and WWF

Papua New Guinea provided invaluable institutional support for our fieldwork and collaborations. Numerous colleagues from these institutions, our home institutions and elsewhere joined us in the field to help with the sweaty business of palm collecting, including Deby Arifiani, Sasha Barrow, Billy Bau, Ziggy Boru, Marie Briggs, Mark Coode, Allan Damborg, Barnabas Desianto, Soejatmi Dransfield, Michael Ferrero, Lauren Gardiner, Olo Gebia, Pieter Gusbager, Marthinus Iwanggin, Marten Jitmau, Ary Prihardyanto Keim, Lawrence Kage, Paul Katik, Kathleen King, Max Kuduk, Krisma Lekitoo, Pilep Mambor, Silver Masbong, Onasius Piter Matani, Rudi Maturbongs, Tobias Paiki, Oliver Paul, Pratito Puradyatmika, Ransuddin, Arkilaus Rumaikewi, Himmah Rustiami, Matthias Sagisolo, Victor Simbiak, Timothy Utteridge, Elisa Wally, Jack Wanggai, Jimmy Wanma, Joko Witono and Otto Geisler Wutoi. Max Kuduk, Tanya Leary, Larry Orsak and Geoff Stocker facilitated fieldwork in Papua New Guinea. Numerous local people, too many to name here, helped us as guides and assistants in the field; their insight and expertise enriched our field experiences immeasurably.

The herbarium was a major frontier of botanical discovery for the *Palms of New Guinea* project. We are grateful to the curators of A, AAU, B, BH, BM, BO, BRI, CANB, CNS, FI, FTG, JCT, K, L, LAE, MAN, MEL, NY, SING, UPNG and WRSL (acronyms follow Thiers [2023]) for providing access to specimens through loans and herbarium visits, and for curating and housing our own field-collected specimens. Chiara Nepi at FI was especially helpful in responding to queries relating to the Beccari herbarium. Many people contributed to the building and curation of our database of New Guinea palm specimen records, especially Steven Bachman, Roy Banka, Kate Davis, Lauren Gardiner, Anders Kjaer, Helen Sanderson, Melinda Trudgen and Jovita Yesilyurt. BRI and AAU shared data directly from their own institutional specimen databases.

We have endeavoured to ensure that every species in this book is comprehensively illustrated. The authors wish to express their sincere appreciation for the dedication of artist Lucy T. Smith to the project. Not only has she prepared an extraordinary body of scientific illustrations for the book, she has also opened our eyes to numerous botanical details that we had often overlooked. Camilla Speight prepared illustrations of *Pigafetta* in the scoping phase of the project. Carrick Chambers, Phil Cribb and David Simpson helped immensely with the sourcing of funds to facilitate these illustrations. Maps were designed and rendered by Tim Wilkinson, based on the meticulous georeferencing of our specimen database by Steven Bachman, undertaken during his placement year at Kew while an undergraduate at Brunel University. Photographs were provided by Sasha Barrow, George Beccaloni, Mark Coode, Ian Cowan, Aaron Davis, Allan Damborg, Fred Essig, Zacky Ezedin, Michael Ferrero, Lauren Gardiner, Joe Gentili, Osia Gideon, Garrick Hitchcock, Charles Humfrey, Zaki Jamil, Laura Jennings, Anders Kjaer, Norbert Kilian, Penniel Lamei, Carl Lewis, Fred Lohrer, Jeff Marcus, Sujin Marcus, Andrew McRobb, Christopher Nash, Chiara Nepi, Jonah Philip, Gilles Pierson, Axel Dalberg Poulsen, Martin Sands, Andre Schuiteman, Jeff Searle, Tim Utteridge and Stephanus Venter.

Kew Publishing has remained committed to this project throughout its protracted genesis. We have received tireless support from Head of Publishing Gina Fullerlove and her successor Lydia White. Christine Beard has worked her usual magic in the elegant design and typesetting of the book. Production Manager Georgie Hills has chaperoned us through the process with considerable good humour. Pei Chu and Paul Little assisted with many loose ends and Sharon Whitehead copy-edited the text. Any errors that remain are the authors' own.

This book was made possible through generous funding from a range of bodies. Fieldwork was supported by Balai Penelitian Kehutanan Manokwari, the Bentham Moxon Trust, the Carlsberg Foundation, the Danish Natural Science Research Council, the European Union, Fairchild Tropical Botanic Garden, the International Palm Society Endowment Fund, the South Florida Palm Society, the Royal Botanic Gardens, Kew and WWF Papua New Guinea. The Australia and Pacific Science Foundation funded the production of many of the botanical illustrations. The Merwin Conservancy created a unique opportunity for WJB to write the introductory chapters for this book through a residency in the home and garden of the late William and Paula Merwin in Maui in January and February 2023.

Finally, the authors wish to thank their partners and families for their support and encouragement, and for their tolerance of long absences and the occasional sleepless night during the production of this book.

Areca macrocalyx fruit for sale as betel nut in a market in Finschhafen. Photo: WB.

ABOUT THE AUTHORS

William J. Baker PhD is a Senior Research Leader at the Royal Botanic Gardens, Kew and Honorary Professor at Aarhus University. His research focuses on the taxonomy, phylogenetics, biogeography and conservation of flowering plants. For 30 years, he has specialised in palms, especially rattans and arecoids, making many contributions to the understanding the evolution and classification of the family using evidence from DNA. He has studied palms extensively in the field in the Pacific, Africa, Madagascar, Borneo and, of course, New Guinea. He co-authored the second edition of *Genera Palmarum* and co-edits *PALMS*, the journal of the International Palm Society.

Anders S. Barfod PhD is Associate Professor of Tropical Plant Resources at Aarhus University, where he teaches a wide range of courses from introductory botanical morphology to tropical ecosystem management. His research is focused on various aspects of palms, such as systematics, ethnobotany, floral development, biogeography, conservation and reproductive ecology. He has conducted fieldwork across the tropics in South America, Africa, South-East Asia and Australasia, including Papua New Guinea.

Rodrigo Camara-Leret PhD is Assistant Professor of Botany at the University of Zurich. His research is centred on the conservation of biological and cultural heritage. He has spent 30 months doing collaborative fieldwork with 28 indigenous groups in South America, organized capacity-building workshops across the tropics, and led a team of 99 scientists from 56 institutions to build the first expert-verified checklist of the vascular plants of New Guinea.

John L. Dowe PhD is Adjunct Senior Research Fellow at the Australian Tropical Herbarium, James Cook University in Cairns. His admiration for palms was secured when appointed curator of Townsville Botanic Gardens in 1992. This included the Palmetum, the first public botanic garden devoted to palms. He completed his PhD on the genus *Livistona* in 2001. He was employed as a botanist and ecologist by James Cook University from the early 2000s. He has conducted extensive research on palms in Australia, New Guinea and the south-west Pacific archipelagos.

John Dransfield PhD is an Honorary Research Fellow at the Royal Botanic Gardens Kew where, until his retirement in 2005, he was head of palm research. He has a broad interest in palm systematics on a global scale and has specialised in the palms of Malesia, Madagascar and Africa, and in the systematics, uses and cultivation of rattans. He is co-author of the two editions of *Genera Palmarum* and co-edits *PALMS*, the journal of the International Palm Society.

Charlie D. Heatubun PhD is Professor of Forest Botany at the University of Papua, Manokwari. He works in the fields of plant systematics, tropical ecology, conservation, forest policy and sustainable development. His palm research focuses on Malesia, mainly New Guinea, where he lives. He is Head of the Regional Research and Innovation Agency of West Papua Province, Indonesia, which he has chaired since 2017. Over the years, he has devoted his time, knowledge and experience to establishing West Papua as the first conservation province in Indonesia and implementing sustainable development to protect the natural capital of New Guinea, including palms.

Peter Petoe MSc is an Ecologist and Conservation Officer at the local government authority of Kalundborg, Denmark. He is a plant taxonomist and systematist trained at Aarhus University and the Royal Botanic Gardens, Kew. He has employed alpha-taxonomic approaches, as well as high-throughput DNA sequencing methods, to study the classification of palms and to improve our understanding of their diversity. In addition to his work on New Guinea palms for this book, he has co-authored various papers about palms and has undertaken palm-collecting field work in Borneo and Madagascar.

Jessica H. Turner BSc is a Biology graduate from the University of Bath. She has a particular interest in ecology and evolution, studying these extensively during her degree. She completed a nine-month internship at the Royal Botanic Gardens, Kew in 2021, where she assisted on the *Palms of New Guinea* project, especially in the compilation and completion of conservation assessments. She has recently worked as a Research Assistant in Costa Rica for the NGO Latin American Sea Turtles.

Lucy T. Smith MA is an award-winning botanical artist who has illustrated plants professionally for 30 years, many of those at the Royal Botanic Gardens, Kew, which holds hundreds of her illustrations in its collection. Australian born, Lucy first began drawing and painting palms in Townsville, North Queensland before settling in Kew to work as a freelance botanical artist specialising in palms. She is passionate about combining art and science to document the diversity of the plant world.

Scott Zona PhD is a Research Technician in Horticultural Science at North Carolina State University and a research collaborator with the Herbarium of the University of North Carolina at Chapel Hill. His research interests are palm taxonomy, anatomy, morphology and the use of living collections in the study of palms. He is co-author of *The Encyclopedia of Cultivated Palms* and recently published *A Gardener's Guide to Botany*. He also co-edits *PALMS*, the journal of the International Palm Society.

ABOUT THIS BOOK

Scope

This book is a comprehensive taxonomic account of the palms of New Guinea that addresses classification, nomenclature, morphology, identification, distribution, habitat, ecology, local names, uses and conservation status. All species treated in this book are native to the region, with the exception of three species of uncertain origin (*Areca catechu, Arenga pinnata, Cocos nucifera*), which are included here as they are traditionally cultivated and utilised. Other non-native species, such as the oil palm (*Elaeis guineensis*) or ornamental species are not featured.

The geographical scope of this book encompasses mainland New Guinea and immediately adjacent island groups. Politically, this area comprises the Indonesian provinces of Papua and Papua Barat (collectively termed here Indonesian New Guinea) and the independent country of Papua New Guinea (excluding Bougainville, which is regarded as part of the Solomon Islands biogeographically [Marsh *et al.* 2009]). The Aru Islands fall outside the scope of this book as they are part of Maluku province, although other authors include them in the New Guinea region (Cámara-Leret *et al.* 2020). Nevertheless, this account covers all but one of Aru's recorded native palm species. In the text, we have attempted to limit the use of political subdivisions of New Guinea, such as provincial names. Where these names are used, the reader is referred to standard geographical information sources for further details. The provinces of Papua New Guinea have remained unchanged for many years. By contrast, the provinces of Indonesian New Guinea will be subdivided from two into six from 2024.

Arrangement

The book comprises three main sections:

- Introductory chapters that place the New Guinea palms in a broader context and set the scene.
- The taxonomic account, the core of the book, comprising complete treatments of all species in each of the 34 genera, including identification keys to genera and species. The sequence of genera follows the current systematic classification of palms (Dransfield *et al.* 2008, Baker & Dransfield 2016).
- Concluding sections, consisting of a comprehensive bibliography, three appendices containing summary checklists, a compendium of useful species, and an index.

Each genus account is structured as follows:

- Accepted genus name and synonyms, and placement within the current classification of the palm family (Dransfield *et al.* 2008, Baker & Dransfield 2016).
- Genus description, distribution and diversity both globally and in New Guinea, and general notes.
- Key characters for each genus based on Baker & Dransfield (2006a).

- Key to species (if two or more species occur in New Guinea).
- One or more species accounts.

Each species account is structured as follows:

- Accepted species name and synonyms. Summary nomenclature only is provided. Full nomenclatural details, including taxonomic literature and type citations, will be made available in a separately published checklist (Baker *et al.* in prep.)
- Species description, based on extensive examination of herbarium specimens. Within species descriptions, features of particular diagnostic value are highlighted in *italics*. For definitions of botanical terms, the reader is referred to Beentje (2016) and Dransfield *et al.* (2008).
- Distribution based on occurrence data derived from herbarium specimens (https://doi.org/10.5281/zenodo.8056512).
- Habitat, including elevation, based on herbarium specimen label data and field observations.
- Local Names, obtained mainly from herbarium specimen label data.
- Uses, obtained from herbarium specimen label data and use reports in published literature (Cámara-Leret & Dennehy 2019a).
- Conservation status, including IUCN extinction risk assessments compiled from the IUCN Red List (IUCN 2023) and preliminary assessments prepared for this book according to the IUCN Red List Categories and Criteria (IUCN 2012).
- General notes.

Maps, illustrations and photographs

Every species account is furnished with a map, a botanical illustration and, where available, photographs. The maps (generated by spatial analyst Tim Wilkinson) visualise all available occurrence records for each species individually. These occurrence records were derived from our dataset of almost 3,000 verified herbarium specimens compiled for this project. Geolocations, as latitude and longitude, were present on 43% of the herbarium specimens and were retrospectively added to a further 45% using gazetteers, maps, online tools and expert knowledge. This georeferenced specimen dataset can be freely downloaded here – https://doi.org/10.5281/zenodo.8056512.

The botanical illustrations were prepared over a period of 23 years by botanical illustrator Lucy T. Smith, commencing in 2000. Each illustration is based on meticulous examination of herbarium specimens, occasionally supplemented by photographs and observations of cultivated plants. The illustrations were created in close collaboration with many of the book's authors. Some of the illustrations have been exhibited and a selection were awarded a gold medal by the Royal Horticultural Society. They are published here as a complete collection for the first time.

Photographic credits

The photographs that accompany this book were taken by 35 individuals. Photo credits are given as initials as follows – AB: Anders Barfod, AD: Allan Damborg, AK: Anders Kjaer, AM: Andrew McRobb, AP: Axel Dalberg Poulsen, APD: Aaron Davis, AS: Andre Schuiteman, CDH: Charlie Heatubun, CH: Charles Humfrey, LJ: Laura Jennings, CL: Carl Lewis, CN: Christopher Nash, FE: Frederick Essig, GB: George Beccaloni, GH: Garrick Hitchcock, GP: Gilles Pierson, IC: Ian Cowan, JD: John Dransfield, JLD: John Dowe, JSM: Jeff and Suchin Marcus, JP: Jonah Philip, JW: Jeff Wood, LG: Lauren Gardiner, MC: Mark Coode, MF: Michael Ferrero, MS: Martin Sands, OG: Osia Gideon, PL: Penniel Lamei, SB: Sasha Barrow, SV: Stephanus Venter, SZ: Scott Zona, TU: Timothy Utteridge, WB: William Baker, ZE: Zacky Ezedin, ZJ: Zaki Jamil.

Author contributions

Multiple authors contributed to each taxonomic account, as follows – WJB: *Actinorhytis, Calamus, Dransfieldia, Heterospathe, Hydriastele, Korthalsia, Manjekia, Metroxylon, Pinanga, Rhopaloblaste, Wallaceodoxa,* genus descriptions for all accounts. ASB: *Licuala.* JLD: *Calyptrocalyx, Livistona, Saribus.* JD: *Areca, Arenga, Borassus, Calamus, Calyptrocalyx, Caryota, Corypha, Cyrtostachys, Korthalsia, Linospadix, Orania, Pigafetta, Sommieria,* genus descriptions for all accounts. CDH: *Areca, Cyrtostachys, Dransfieldia, Manjekia, Sommieria, Wallaceodoxa.* PP: *Heterospathe, Hydriastele.* SZ: *Brassiophoenix, Clinostigma, Cocos, Drymophloeus, Manjekia, Nypa, Physokentia, Ponapea, Ptychococcus, Ptychosperma, Wallaceodoxa.* RCL: uses for all accounts. JHT: conservation status and botanical illustration legends for all accounts. WJB and JD reviewed and edited all taxonomic accounts.

The introductory chapters were written by WJB, except chapter 6 (RCL) and the first draft of chapter 7 (JHT). All authors reviewed and contributed to these chapters.

Saribus papuanus, near Timika. Photo: AM.

1. INTRODUCTION TO NEW GUINEA AND ITS FLORA

New Guinea is the world's largest tropical island and a globally exceptional biodiversity hotspot (Marshall & Beehler 2007). It supports the most extensive tropical forest wilderness area in the Indo-Pacific and intact ecological gradients unsurpassed in the region, from mangroves to glacier-capped mountains (Mittermeier *et al.* 2003, Cámara-Leret *et al.* 2020). Much of this wilderness is inaccessible and yet New Guinea's documented plant diversity already exceeds that of any other island (Cámara-Leret *et al.* 2020). The fauna, in turn, has near mythic status with some of Earth's most unusual animals, such as birds of paradise, cassowaries, tree kangaroos and birdwing butterflies, and the diversity of the island's coral reef ecosystems is unrivalled globally (Veron *et al.* 2009). Within this melting pot of diversity, human culture has flourished too and over 1,100 language groups have evolved since humans colonised the island 30,000–50,000 years ago (Beehler 2007, Hope 2007, Simons & Fennig 2018).

Landscape, geology and climate

The biological and human diversity of New Guinea is driven by the diversity of the landscape (Fig. 1.1). The island's outline is often likened to a bird, some 2,500 km long from the head in the west to the tail in the east, and just over 700 km wide, north to south, at its widest point. At 785,753 km^2 in extent, New Guinea accounts for just 0.53% of the Earth's land surface area (Gaveau *et al.* 2021). Along the spine of the island runs a central cordillera from the neck of the Bird's Head Peninsula in the west to the far eastern tip, with rugged mountain ranges reaching an elevation of 4,884 m on Mount Jaya where the last equatorial glaciers in the Asia-Pacific persist. South of the central cordillera lies a vast lowland plain thick with forest and swamps. Northern New Guinea, including the Bird's Head Peninsula, is a more complex patchwork of smaller mountain ranges, basins and lowlands. Immediately surrounding mainland New Guinea are more than one thousand islands that form part of the broader New Guinea region, including the Aru Islands, the Raja Ampat Islands, the Biak Islands, Yapen, the Bismarck Archipelago, the D'Entrecasteaux Islands, the Trobriand Islands, Woodlark Island and the Louisiade Archipelago. These islands sit on and around the Sahul Shelf, the shallow continental platform that connects New Guinea and Australia (Hall 2012).

The geological origins of New Guinea are complex and a consistent interpretation of the available evidence is yet to emerge (Polhemus 2007, Hall 2012, 2017, Gold *et al.* 2020). As Australia moved northwards, it collided with South-East Asia from the early Miocene (ca. 20 million years ago [Mya]) onwards. In parallel, numerous terranes accreted to the northern margin, forming the so-called "mobile belt". Considerable uncertainty surrounds the origins and boundaries of these terranes,

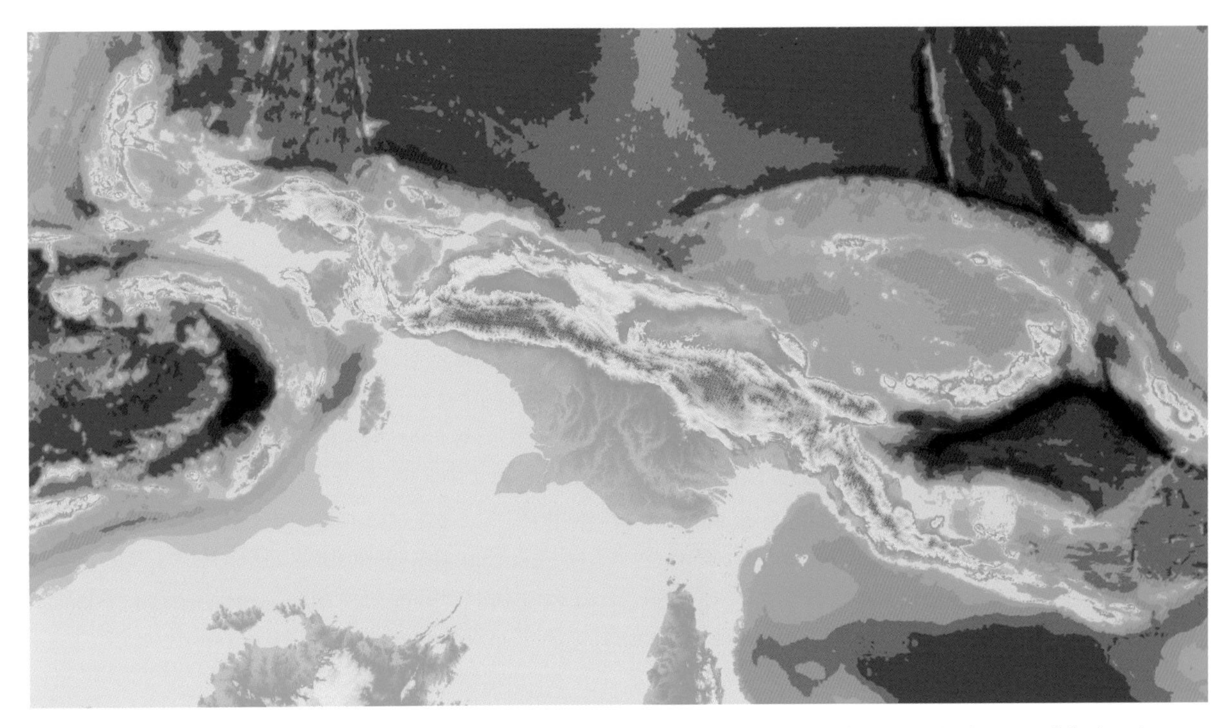

Figure 1.1. New Guinea in geological context. Relief map (GEBCO Compilation Group 2022) showing bathymetry (blue) and topography (green to brown). The shallow seas (pale blue) of the Sahul Shelf, which connects New Guinea to Australia, are clearly visible. The Torres Strait, which separates the two land masses, is mostly 10–20 m deep. The central cordillera of New Guinea reaches a maximum elevation of 4,884 m on Mount Jaya.

which were generally thought to be of Pacific origin but may in fact be derived from eastern Australia (Hall 2017, Gold *et al.* 2020). The impact of the Australian plate with the Pacific Plate and intervening Philippine Sea and Caroline Plates led to the uplift of New Guinea from around 10 Mya, with the mountains rising within the last 5 million years (Myr) (Hall 2009, 2012). Prior to this, New Guinea was largely submerged except for perhaps scattered islands in the north. In the recent past, New Guinea and Australia formed a contiguous land mass. The Torres Strait, the shallow seaway that currently separates Australia from New Guinea, formed ca. 6,000–9,700 years ago (Polhemus 2007, Dowe 2010).

Located between 0 and 11°S, New Guinea is entirely tropical today. Annual rainfall of 2,500–4,500 mm occurs across much of the island with some areas receiving up to 7,000 mm (Prentice & Hope 2007). Rainfall seasonality patterns are varied, ranging from no meaningful seasonality to strong seasonality in parts of the south. The lowlands are hot and humid (25–35°C), but there is a steep temperature gradient, with frosts becoming frequent above 2,800 m and snowfall above 3,800 m. The complexity and mountainous nature of the landscape leads to many local climate patterns. These factors also drive considerable variation in vegetation across four major zones – lowland, montane, subalpine and alpine (Utteridge & Jennings 2021). For detailed analyses of New Guinea vegetation, the reader is referred to Paijmans (1976) and Marshall and Beehler (2007), or to Utteridge and Jennings (2021) for a simplified classification.

New Guinea's extraordinary plant diversity

New Guinea's vascular plant diversity is estimated to comprise 13,634 species across 1,742 genera and 264 families (Cámara-Leret *et al.* 2020). Species diversity exceeds that of the second and third largest tropical islands, Borneo (11,165 species) and Madagascar (11,488 species), by 19% and 22%, respectively. Since 1970, 2,812 species have been described as new to science in New Guinea and the rate of species discovery shows little sign of abating (Cámara-Leret *et al.* 2020). Species endemism reaches 68% in New Guinea, making it the only Malesian island with more endemics than non-endemics. Species diversity is higher in Papua New Guinea (10,973 species) than in Indonesian New Guinea (7,616 species), and twice as many endemics can be found there. Although Papua New Guinea may be more species rich purely because it is the larger of the two entities, the figures are likely to be biased by the highly uneven collecting patterns across the island (Middleton *et al.* 2019, Cámara-Leret *et al.* 2020), with the collecting effort in Papua New Guinea reported to exceed that in Indonesian New Guinea by as much as 300% (Takeuchi 2007). In time, the species diversity gap between Papua New Guinea and Indonesian New Guinea may be reduced with investment in taxonomic capacity and strategic collecting effort. However, plant collecting has been in overall decline in New Guinea since its peak in the 1960s and 1970s (Cámara-Leret *et al.* 2020), a trend which must be reversed if a comprehensive inventory of the island's plants is to be achieved.

Biogeographic origins of the New Guinea flora

New Guinea falls within the Indo-Pacific floristic region (Slik *et al.* 2018) and sits at the eastern end of Malesia (Fig. 1.2), a coherent biogeographic unit that also includes the Malay Peninsula, Borneo, Sumatra, Java, the Philippines, Sulawesi, Maluku and the Lesser Sunda Islands (Van Steenis 1950, Raes & Van Welzen 2009). Quantitative floristic analyses (Warburg 1891, Van Steenis 1950, Van Balgooy 1976, Marsh *et al.* 2009, van Welzen & Slik 2009, Joyce *et al.* 2020) support the placement of New Guinea within Malesia, rather than with its nearest neighbour Australia. Traditionally, the Bismarck Archipelago has been excluded from Malesia, but this has been a matter of debate (Van Steenis 1950, Marsh *et al.* 2009, Raes & Van Welzen 2009).

Malesia has long been recognised as a contact zone for distinct biotas that is best understood through the lens of its complex tectonic history. Wallace's Line (Wallace 1860), a biogeographic boundary based on animal distributions that separates Borneo, Sumatra, Java and the Philippines from all islands to the east, is the most well-known formal expression of this. Botanists have struggled to demonstrate that plants adhere to these boundaries, despite adopting multiple modifications and additional biogeographic lines (Richardson *et al.* 2012). Almost half of all vascular plant genera transgress either Wallace's or Lydekker's line, which separates New Guinea from the rest of Malesia, indicating that neither represent a strong biogeographic barrier for plants (Joyce *et al.* 2020).

Recent research on floristic assembly across Malesia has focused less on biogeographic lines and more on the relative roles of the Sunda and Sahul regions, the distinct continental shelves at the western and eastern ends of Malesia and the land masses that sit on them, and intervening Wallacea, the transition

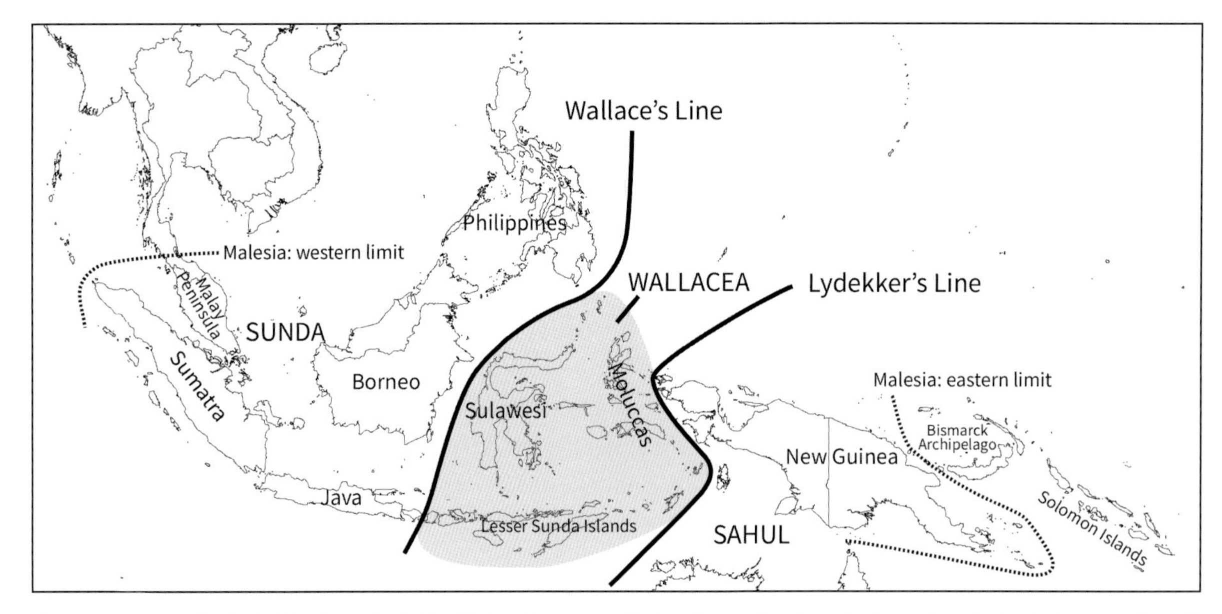

Figure 1.2. Map of Malesia (Van Steenis 1950) with key biogeographic barriers and regions (redrawn from Baker & Couvreur [2012]).

zone between the two (van Welzen & Slik 2009, Hall 2012, Richardson *et al.* 2012, Crayn *et al.* 2015, Kooyman *et al.* 2019, Joyce *et al.* 2020). The Sunda Shelf connects Borneo, Sumatra, Java and adjacent islands to the Malay Peninsula and the Asian continent, whereas the Sahul Shelf connects New Guinea to Australia. Wallacea comprises Sulawesi, Maluku and the Lesser Sunda Islands. It is well established that eastwards migration of plant groups from Sunda to Sahul is more common than the reverse, with one study indicating that 64% of New Guinea species belong to groups of Sunda Shelf origin (Richardson *et al.* 2012). Further support from dated molecular phylogenies suggests that eastwards migration from Sunda to Sahul is 2.4 times more frequent than the reverse, began between 33–1 Myr and accelerated during the past 12 Myr (Crayn *et al.* 2015), consistent with the Miocene collision of the Australian plate with South-East Asia.

Sunda Shelf immigrants dominate the lowlands of New Guinea and the northern Australian rainforests. However, the New Guinea uplands contain many austral components, which may be relicts of temperate rainforest floras of Australia and other parts of the southern hemisphere (Richardson *et al.* 2012, Kooyman *et al.* 2019). The extensive mesic flora that once existed in Australia declined as the continent dried during its northward migration. By the time New Guinea emerged, this flora had largely been replaced by a diversifying dry-adapted flora (Richardson *et al.* 2012, Crayn *et al.* 2015). Humid tropical New Guinea was colonised by immigrants from the climatically similar west, indicating that climate and niche conservatism are stronger drivers of plant distribution than geographic proximity (Richardson *et al.* 2012, Joyce *et al.* 2020). Rapid recent radiation over the past 10 Myr must also have accompanied these processes for New Guinea to have developed the extraordinarily rich flora that exists today (Shee *et al.* 2020).

2.
NEW GUINEA PALMS IN A GLOBAL CONTEXT

The New Guinea palm flora, as documented in this book, consists of 250 species (Appendix 1), comprising 247 native species and three species of unknown geographical origin (*Areca catechu*, *Arenga pinnata*, *Cocos nucifera*). This represents around 10% of the world's ca. 2,500 species and one quarter of the Malesian palm flora (Dransfield *et al.* 2008, Govaerts *et al.* 2021, POWO 2023). There are only three times more palm species in the Americas as a whole, and New Guinea exceeds the depauperate palm flora of Africa almost four-fold. Only two other tropical islands rival New Guinea, Borneo with around 300 recorded species and Madagascar with 208.

Over 85% (212 species) of New Guinea's native palm species are endemic to the New Guinea region (Fig. 2.1; Appendix 1). The Arecaceae is among just eight families that contribute 50% of New Guinea's 9,301 endemic species, along with Orchidaceae, Rubiaceae, Ericaceae, Myrtaceae, Gesneriaceae, Apocynaceae and Lauraceae (Cámara-Leret *et al.* 2020). By contrast, palm endemism on Borneo is lower (68%), with many species shared with adjacent land masses on the Sunda shelf (Govaerts *et al.* 2021, POWO 2023), reflecting their long history of connectivity. In Madagascar, which has been isolated for 84–91 Myr, an extraordinary 99% of the more than 210 native species are endemic (Dransfield & Beentje 1995, POWO 2023).

At the time of writing, 182 palm genera are recognised globally (Baker & Dransfield 2016, Hodel *et al.* 2021, Eiserhardt *et al.* 2022, Sâm *et al.* 2023). Thirty-four of these occur in New Guinea, 18% of the global total. Five genera are endemic to New Guinea (*Brassiophoenix*, *Dransfieldia*, *Manjekia*, *Sommieria* (Fig. 2.2), *Wallaceodoxa*) and a further two are nearly so (*Calyptrocalyx*, *Ptychococcus*). Only two other vascular plant families, Asteraceae and Rubiaceae, have more endemic genera in New Guinea (six and seven, respectively; Cámara-Leret *et al.* 2020). These are much larger families than the palms, emphasising that the number of endemic palm genera, which have all been verified phylogenetically, is exceptional. In contrast, none of Borneo's 25 palm genera is endemic, whereas Madagascar has 10 endemic genera and a further two that are nearly so. The varied diversity and endemism patterns among these three, the world's largest tropical islands, reflect the contrasting histories and processes that underlie the assembly of their biotas.

Systematics

New Guinea's palm genera are widely scattered across the palm tree of life (Baker *et al.* 2009) and the corresponding phylogenetic classification (Table 2.1; Dransfield *et al.* 2008, Baker & Dransfield 2016). Of the five subfamilies, only the Ceroxyloideae is not represented in New Guinea, although one genus, *Oraniopsis*, is endemic to nearby north Queensland. New Guinea palm genera are distributed among the remaining four subfamilies as follows:

Figure 2.1. Vascular plant families with >200 endemic species in New Guinea. **A.** Orchidaceae (2,464 endemic species), e.g. *Dendrobium cuthbertsonii*. **B.** Rubiaceae (669 endemic species), e.g. *Myrmecodia brassii*. **C.** Ericaceae (431 endemic species), e.g. *Rhododendron versteegii*. **D.** Myrtaceae (255 endemic species), e.g. *Decaspermum* cf. *forbesii*. **E.** Gesneriaceae (218 endemic species), e.g. *Cyrtandra vittata*. **F.** Arecaceae (212 endemic species), e.g. *Hydriastele costata*. Photos: WB.

Calamoideae – Four genera (*Calamus, Korthalsia, Metroxylon, Pigafetta*), all from just one of its three tribes, Calameae, occur in New Guinea. The rattan genus *Calamus* alone contributes one quarter of New Guinea's palm species.

Nypoideae – The sole genus and species *Nypa fruticans*, the mangrove palm, is present.

Coryphoideae – Four tribes of the fan palm subfamily are represented, namely Borasseae (*Borassus*), Caryoteae (*Arenga, Caryota*), Corypheae (*Corypha*) and Trachycarpeae (*Licuala, Livistona, Saribus*). *Licuala* and *Saribus* are notable radiations.

Arecoideae – New Guinea's diversity of the largest subfamily is outstanding in Malesia, comprising, for example, >70% more species and almost three times as many genera as are found in the Bornean arecoids (Baker & Couvreur 2012). Three arecoid tribes provide just a single genus each, namely Cocoseae (*Cocos*), Pelagodoxeae (*Sommieria*) and Oranieae (*Orania*), the last being especially species rich in New Guinea. Tribe Areceae is exceptional with 19 of its 61 genera occurring in New Guinea (*Actinorhytis, Areca, Brassiophoenix, Calyptrocalyx, Clinostigma, Cyrtostachys, Dransfieldia, Drymophloeus, Heterospathe, Hydriastele, Linospadix, Manjekia, Physokentia, Pinanga, Ponapea, Ptychococcus, Ptychosperma, Rhopaloblaste, Wallaceodoxa*). Six out of seven of the New Guinea endemic or near-endemic genera fall in tribe Areceae, as do several notable New Guinea radiations (*Calyptrocalyx, Heterospathe, Hydriastele, Ptychosperma*). This immensely widespread tribe ranges from Madagascar (and adjacent Africa) to Samoa and has a complex history of long-distance dispersal across the islands and continents of the Indo-Pacific region and numerous local radiations.

Figure 2.2. *Sommieria*, the first of New Guinea's five endemic palm genera to be described (Beccari 1877b). Only one variable species, *S. leucophylla*, is currently accepted (Heatubun 2002). Molecular evidence places the genus in an isolated lineage (tribe Pelagodoxeae) with the Pacific genus *Pelagodoxa* (Dransfield *et al.* 2008). Photo: JD.

TABLE 2.1 New Guinea palm genera in the current classification of the palm family (Dransfield *et al.* 2008, Baker & Dransfield 2016). See Dransfield *et al.* (2008) for detailed distribution maps of genera.

SUBFAMILY / Tribe / Subtribe	Genus	Distribution
CALAMOIDEAE		
Calameae		
Korthalsiinae	*Korthalsia*	India to New Guinea
Metroxylinae	*Metroxylon*	Maluku to Caroline Islands, Vanuatu & Samoa
Pigafettinae	*Pigafetta*	Sulawesi, Maluku & New Guinea
Calaminae	*Calamus*	India to S China, through Malay Archipelago to Australia & Fiji. Also Africa.
NYPOIDEAE	*Nypa*	India to S China, Australia & Solomon Islands
CORYPHOIDEAE		
Trachycarpeae		
Livistoninae	*Livistona*	Africa, Arabia, S Asia, through Malay Archipelago to Australia
	Licuala	India to S China, through Malay Archipelago to Australia & Vanuatu
	Saribus	Philippines to New Caledonia
Caryoteae	*Caryota*	India to S China, through Malay Archipelago to Australia & Vanuatu
	Arenga	India to S China, through Malay Archipelago to Australia
Corypheae	*Corypha*	India to Philippines & Australia
Borasseae		
Lataniinae	*Borassus*	Africa to Madagascar, India & New Guinea
ARECOIDEAE		
Oranieae	*Orania*	Madagascar, Thailand to New Guinea
Cocoseae		
Attaleinae	*Cocos*	Pantropical (modified by humans)
Pelagodoxeae	*Sommieria*	New Guinea endemic (mainland)
Areceae		
Archontophoenicinae	*Actinorhytis*	New Guinea & Bougainville
Arecinae	*Areca*	India to South China, through Malay Archipelago to Solomon Islands
	Pinanga	India to South China, through Malay Archipelago to New Guinea
Basseliniinae	*Physokentia*	New Britain to Fiji & Vanuatu

TABLE 2.1 (contd.)

SUBFAMILY Tribe Subtribe	Genus	Distribution
ARECOIDEAE		
Areceae (contd.)		
Laccospadicinae	*Calyptrocalyx*	Maluku to New Guinea
	Linospadix	New Guinea & Australia
Ptychospermatinae	*Ptychosperma*	New Guinea to Australia & Solomon Islands
	Ponapea	New Britain & Caroline Islands
	Drymophloeus	Maluku & New Guinea
	Brassiophoenix	New Guinea endemic (mainland)
	Ptychococcus	New Guinea to Bougainville
	Manjekia	New Guinea endemic (Biak)
	Wallaceodoxa	New Guinea endemic (Waigeo, Gag)
Unplaced Areceae	*Clinostigma*	Bonin to New Ireland, Solomon Islands, Samoa, Fiji & Vanuatu
	Cyrtostachys	Malay Peninsula, Sumatra & Borneo; New Guinea to Solomon Islands
	Dransfieldia	New Guinea endemic (mainland, Waigeo)
	Heterospathe	Philippines to New Guinea, Solomon Islands & Fiji
	Hydriastele	Sulawesi to Palau, Australia & Fiji
	Rhopaloblaste	Nicobar Islands & Malay Peninsula; Maluku to Solomon Islands

Biogeography

Floristically, New Guinea is more closely related to Malesia than to other adjacent regions, such as Australia and the South-West Pacific (Van Balgooy 1976, Marsh *et al.* 2009, van Welzen & Slik 2009, Joyce *et al.* 2020). However, the biogeography of the New Guinea palms is not so readily explained in such simple terms. Quantitative analyses of palm floras worldwide based on comprehensive global checklists and dated phylogenies (Hansen *et al.* 2021) tell a nuanced story, highlighting floristic relationships with Melanesian-Pacific areas, northern Queensland and also with the Malesian islands to the west (including Maluku, Sulawesi, Borneo and the Philippines) that diminish with increasing distance. These mixed signals reflect the complex biogeographic origins of New Guinea's palm flora, consistent with known biogeographic ambiguities at the eastern edge of Malesia (Marsh *et al.* 2009).

New Guinea's palm flora is highly phylogenetically clustered, indicating a central role for *in situ* diversification in its evolution. In contrast, adjacent areas of Wallacea are dominated by a strong signal of biotic interchange (Kissling *et al.* 2012a). Current phylogenetic evidence suggests that almost

every palm genus has colonised the island independently one or more times (Baker *et al.* 2009, Baker & Couvreur 2012, 2013a, b), and many have subsequently diversified there.

Phylogenetic evidence indicates that 13 of New Guinea's 34 palm genera are immigrants from the west (*Areca, Arenga, Borassus, Calamus, Caryota, Corypha, Cyrtostachys, Korthalsia, Licuala, Orania, Pinanga, Rhopaloblaste, Saribus*). Almost all of these genera approach their eastern distributional limits in or near to New Guinea. Six of these genera contain only one or two species, including some widespread species (e.g. *Arenga microcarpa, Caryota rumphiana, Corypha utan, Pinanga rumphiana*) as well as a few notable endemics (e.g. *Borassus heineanus, Caryota zebrina, Korthalsia zippelii*).

Seven of the western immigrant genera have diversified in New Guinea, four on a small scale (*Areca, Cyrtostachys, Rhopaloblaste, Saribus*), the others to a much greater extent (*Calamus, Licuala, Orania*). *Cyrtostachys* and *Rhopaloblaste* (Fig. 2.3) are disjunctly distributed across Wallace's Line, perhaps due to very long distance dispersals or extinction (Baker & Couvreur 2012). *Calamus* and *Licuala* display so-called bimodal distributions across Malesia (Dransfield 1987), meaning that they have two peaks of diversity in West and East Malesia with a dip through Wallacea in the middle. Bimodal diversity may simply be explained by species-area relationships (Baker & Couvreur 2012, Kissling *et al.* 2012b); however, phylogenetic evidence indicates that major radiations arising primarily from single colonisation events from the west have taken place in both genera (Bacon *et al.* 2013, Kuhnhäuser *et al.* in prep.) *Orania* is differently bimodal (or arguably trimodal); the greatest species richness is in New Guinea, with a second peak in the Philippines and a third disjunct peak in Madagascar. The New Guinea species appear to form a single lineage that reached New Guinea from the west (Baker & Couvreur 2013a, Schrödl 2020), but broader analyses are needed to clarify this unusual biogeographic pattern (Baker & Couvreur 2012).

Nineteen palm genera have reached New Guinea from areas east of Wallace's Line, primarily Australasia and the West Pacific. Thirteen of these (*Actinorhytis, Brassiophoenix, Calyptrocalyx, Dransfieldia, Drymophloeus, Heterospathe, Linospadix, Manjekia, Physokentia, Ponapea, Ptychococcus, Ptychosperma, Wallaceodoxa*) belong to a Pacific clade nested within tribe Areceae. This lineage dispersed into the Pacific from west of Wallace's Line in the Oligocene and now ranges from New Guinea and Australia to Samoa, with some outliers in Wallacea and the Philippines. All but one of New Guinea's seven endemic/near-endemic genera fall in this clade (*Brassiophoenix, Calyptrocalyx, Dransfieldia, Manjekia, Ptychococcus, Wallaceodoxa*), as do three key New Guinea radiations (*Calyptrocalyx, Heterospathe, Ptychosperma*). *Heterospathe* resembles *Orania* in its secondary peak of diversity in the Philippines, as well as in the West Pacific, but its biogeographic origins are distinct.

Two further genera of New Guinea Areceae, *Clinostigma* and *Hydriastele*, originated in the Pacific, their ancestors having dispersed there from the west independently of the main Pacific clade. A similar pattern is observed for *Metroxylon, Pigafetta* (Calameae) and *Sommieria* (Pelagodoxeae; Baker & Couvreur [2013a], Kuhnhäuser *et al.* [2021]). *Hydriastele* has diversified extensively in New Guinea. *Livistona* originated in Asia, but dispersed into Australia in the Miocene, where it radiated significantly (Crisp *et al.* 2010, Bacon *et al.* 2013). The two species of *Livistona* in New Guinea are range extensions of Australian species and are therefore treated as Australian in origin (Dowe 2010).

Figure 2.3. *Rhopaloblaste singaporensis*, endemic to Singapore and Peninsular Malaysia, represents the disjunct distribution of the genus *Rhopaloblaste*, which occurs elsewhere in Papuasia and the Nicobar Islands. *Cyrtostachys* displays a somewhat similar disjunct distribution (Baker & Couvreur 2012). Photo: WB.

The two remaining genera, *Cocos* and *Nypa*, both famous for their floating seeds, have widespread coastal distributions. *Nypa* is naturally distributed from India to the Solomon Islands, but its extensive fossil record shows that it occurred in the Americas, Africa and even Europe as far back as the Cretaceous (Dransfield *et al.* 2008). The origin of *Cocos* (Gunn *et al.* 2011) has been disputed for many years. Although regions including New Guinea have been proposed as a potential place of origin, the distribution has been extensively modified by humans.

As expected (Chapter 1; Richardson *et al.* 2012, Crayn *et al.* 2015), a significant part of New Guinea's palm flora is derived from the west; 38% of New Guinea's palm genera encompassing 141 species (56%) stem from immigrant lineages from the west of Wallace's Line. New Guinea has contributed no westward migrants across Wallace's Line. Even the two New Guinea-centred genera *Cyrtostachys* and *Rhopaloblaste*, which are present on the Sunda Shelf, appear to be eastward migrants into New Guinea. However, the New Guinea palm flora has equivalent affinities to the east, with 56% of genera containing 107 species (43%) originating from Australasian-Pacific regions, although the ancestors of these lineages also have deeper origins as immigrants from the west.

Timescale

Because New Guinea's uplift and emergence above sea level occurred relatively recently (Hall 2009, 2012, Joyce *et al.* 2020), the divergence and diversification of New Guinea lineages are expected to have taken place within the past 10 Myr. However, evidence from available dated phylogenies indicates that New Guinea's palm genera most often diverged from their nearest relatives before this time (Baker & Couvreur 2013a), suggesting a history for these lineages in areas other than New Guinea. This also applies to some endemics, such as *Sommieria* or *Dransfieldia*, although others fit well within the time window of New Guinea's uplift (the clade comprising *Brassiophoenix*, *Drymophloeus* and *Ptychococcus*). Recent diversification within the 10 Myr window has been reported in *Licuala*, *Livistona*, *Saribus* and *Orania* (Bacon *et al.* 2013, Schrödl 2020), whereas other groups pre-date this (e.g. *Calamus*, Kuhnhäuser [2021]), sometimes significantly so (e.g. *Borassus*, Bellot *et al.* [2020b]). The paradoxical mismatch of geological history and diversification timing in many New Guinea palms merits further investigation, drawing on the increasingly complete body of species-level palm phylogenies that are now emerging (Bellot *et al.* 2020a).

3.
EXPLORING THE NEW GUINEA PALM FLORA

The palms of New Guinea have been explored for millennia, ever since humans colonised New Guinea at least 30,000 years ago. The rich lexicon of local names and the variety of documented uses attest to this long relationship between people and palms. By contrast, exploration of New Guinea's palms within the western scientific worldview began less than 200 years ago, but is now essential for the future preservation and sustainable management of these remarkable and valuable plants. Botanical research, especially the identification and description of species, is dependent on access to herbarium specimens. In general, palms tend to be under-represented in herbaria because they take much more effort to collect than other groups of plants. They are thus widely avoided by general collectors, for whom one palm might take the effort of ten other plants. We are indebted to the collectors who expended time and energy to make the palm specimens that form the scientific foundations of this book.

Herbarium collections of New Guinea palms

An early task for the palms of New Guinea project was to assemble a specimen database, recording all available palm specimens from herbaria around the world. We visited key herbaria known for their New Guinea holdings and captured data from all of their palm specimens, including: A, AAU, B, BH, BM, BO, BRI, CANB, FI, K, L, LAE, MAN, MEL and NY (herbarium acronyms follow Thiers [2023]). As databasing progressed, the number of new records began to decline because many collections were duplicated across multiple herbaria. For example, the Papua New Guinea National Herbarium (LAE) traditionally shared specimen duplicates with herbaria such as BRI, K and L, among others. Our database of 3,800 specimens contains a significant majority of the New Guinea palm specimens currently available in the world's herbaria. Almost 3,000 of these specimens have been identified confidently to species level during the project, forming the basis of our understanding of variation and distribution in New Guinea palms (dataset available online at https://doi.org/10.5281/zenodo.8056512).

Over 150,000 herbarium specimens from New Guinea have been digitised in the world's herbaria (Cámara-Leret *et al.* 2020). The total dataset of 3,800 specimens (2.5%) that we report here closely corresponds to the proportion of palm species (2%) in the New Guinea flora. Thus, contrary to expectation, palms appear to be no less well-collected in New Guinea than expected on average. Around 65% of the identified palm specimens were collected in Papua New Guinea, 1.9 times more than in Indonesian New Guinea (Fig. 3.1). This collecting bias, though significant, is far lower than other authors have reported (e.g. Takeuchi 2007), perhaps due to concerted efforts during the Palms of New Guinea project to collect in Indonesian New Guinea. However, collecting density remains

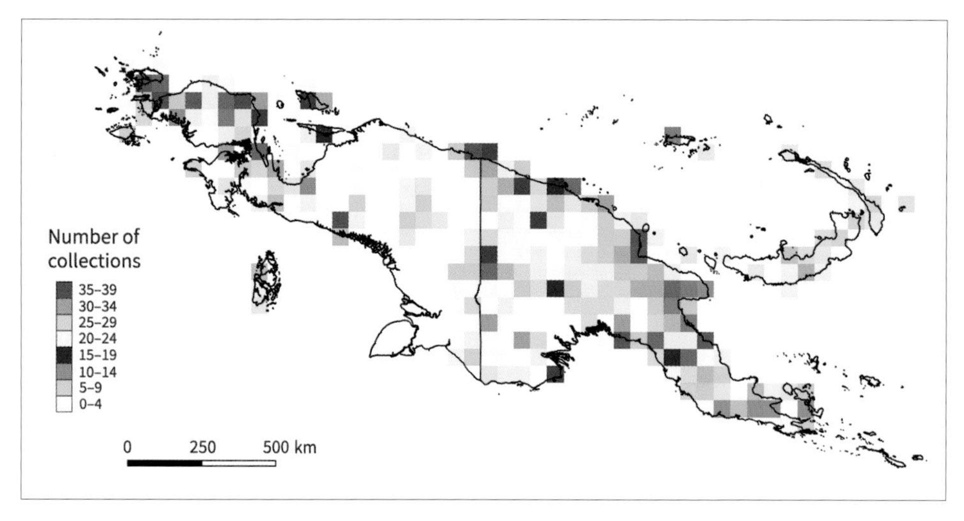

Figure 3.1. Map of the New Guinea region showing the number of digitised, georeferenced collections of palms in each grid cell of 50 × 50 km.

highly uneven across the island (Fig 3.1; Bachman *et al.* 2004) with many areas remaining severely under-explored, corresponding closely to the collecting gaps illustrated by Cámara-Leret *et al.* (2020) for the New Guinea flora as a whole.

The earliest collectors

To our knowledge, the earliest scientific botanical collection of a palm from New Guinea is the type specimen of the genus *Ptychosperma*, collected in 1792 in New Ireland by Labillardière, who described it as *P. gracile*. It was 40 years later that the first recorded mainland specimens of palms were collected by Alexander Zippel (also known as Zippelius) who visited southern New Guinea in 1828 as part of an expedition for the Commission for Natural Sciences of the Dutch East Indies. Among his collections are the type collections of important species such as *Areca macrocalyx*, *Hydriastele procera*, *Korthalsia zippelii*, *Licuala penduliflora* and *Orania regalis* (Baker *et al.* 2000a, Dowe & Latifah 2020). It was more than 40 years after Zippel that the next specimens were made by Johannes Teijsmann, who collected type material of *Ptychococcus paradoxus* and *Ptychosperma schefferi* in 1871. Then, in 1872 and 1875, Odoardo Beccari (Fig. 3.2), the prolific Italian palm botanist, collected in numerous localities in far western New Guinea, including the Arfak Mountains, Andai, Onin Peninsula, Ramoi (in present day Sorong),

Figure 3.2. Odoardo Beccari (1843–1920, photo taken in 1868), celebrated Italian palm botanist and the most significant 19th-century palm collector in New Guinea (1872, 1875). Photo: The Museum System of the University of Florence, Natural History Museum.

and Num and Yapen Islands (Van Steenis-Kruseman 1950). Beccari was the most significant collector of palm specimens from New Guinea in the 19th century, the majority of his ca. 50 collections later typifying species that he described. Other celebrated plant collectors of the 19th century include Armit, d'Albertis, Lauterbach, Loria and Sayer (see Van Steenis-Kruseman [1950]). General scientific expeditions also resulted in the collection of palm specimens. For example, Macleay's Chevert Expedition of 1875 (Dowe 2007) and William Macgregor's explorations in 1888 (Thomson 1889) returned with type specimens. The challenges that these early collectors faced in accessing plant specimens from New Guinea at that time cannot be over-stated. Overall, fewer than 100 palm specimens are known from the 19th century.

Twentieth-century collectors

In the early 20th century, few palm collections were made until the 1930s, which saw a 10-fold increase in specimens to over 200 within the decade. This decade features prominent New Guinea collectors such as Cedric Carr, and Joseph and Mary Clemens, but the leading figure was Australian botanist, Leonard Brass (Fig. 3.3), who was responsible for the collection of vascular plants on six Archbold Expeditions to New Guinea between 1933 and 1959 (Forster 1997, Cookson 2000). As a collector, Brass's contribution to New Guinea botany is unrivalled. Prior to the Palms of New Guinea project, no other botanist collected palms so prolifically. By any standards, his 236 palm specimens are exemplary and are always accompanied by excellent notes. They were described by Max Burret at Berlin over several publications, and as a result, dozens of New Guinea palm names are typified on Brass specimens. Regrettably, all eight of the New Guinea palm species names commemorating Brass are now treated as synonyms, but the endemic genus, *Brassiophoenix*, survives.

Figure 3.3. Leonard Brass (1900–1971), prolific Australian botanist, who collected plants in New Guinea during six Archbold Expeditions (1933–1959), including more than 230 palm specimens. Photo: Archbold Expeditions.

Following a hiatus in the 1940s due to the Second World War, post-war palm collecting in New Guinea increased significantly to a peak in the 1970s, tailing off somewhat in the 1980s after Papua New Guinea independence from Australia in 1975. Brass remained active through the 1950s. Under the NGF (New Guinea Forests) collecting series of the Department of Forests and the LAE series of the Papua New Guinea National Herbarium that replaced it, many botanists made important additions to the growing body of palm specimens, notably Croft, Essig, Foreman, Gideon, Henty, Katik, Kerenga, Millar, Streimann, Wiakabu, Womersley and Zieck. Collectors operating outside these series, many of them visitors to Papua New Guinea, also made notable collections, especially Darbyshire, Hoogland, Jacobs, Johns, Mente, Moore, Poudyal, Pullen, Schodde and Taurereko. Between the 1950s and 1980s, twelve times more palm specimens were made in Papua New Guinea (ca. 1,500) than in Indonesian New Guinea (ca. 130). The Indonesian collections were mostly made under the BW (Boswezen, or Forestry Service) series between 1957 and 1962 by collectors such as Dijkstra, Rappard, Schram and Sijde. Of the collectors operating outside the BW series, only van Royen stands out as being especially productive. The contrast in effort between the two halves of the island could not be more stark, possibly the result of the more limited access and greater political complexities of Indonesian New Guinea.

Of the collectors active in the 1950–1980s period, three deserve special mention – Harold Moore, Jack Zieck and Fred Essig:

The celebrated US-based palm expert Harold E. Moore Jr. achieved a long-held ambition to collect palms in New Guinea in March 1964, spending two weeks exploring in Morobe Province, Papua New Guinea in areas accessible from Lae (Moore 1966). Although Moore's visit was fleeting, resulting in just 36 specimens, he was the first palm specialist to visit New Guinea since Beccari and made important, comprehensive specimens of several species we now take for granted as widespread. He also collected two new species that he later described, *Cyrtostachys glauca* and *Ptychococcus lepidotus*.

Between 1967 and 1979, Jack Zieck made 165 collections of rattans (mainly *Calamus*, but also *Korthalsia*) in Papua New Guinea. Working for the Forest Products Research Centre in Port Moresby, his focus on rattans was driven by their economic potential, for example in the cane furniture industry. He made some of the most diligent collections of rattans available in any herbarium, each accompanied by finely typed notes and measurements that often ran to a full page or more, which also included information on cane quality. Zieck specimens typify six accepted *Calamus* species and he is commemorated in *Calamus zieckii*. His collections have been invaluable for all students of New Guinea rattans (e.g. Johns & Zibe 1989, Baker & Dransfield 2014, 2017, Henderson 2020) and are a cornerstone of the *Calamus* treatment in this book. Sadly, to our knowledge Zieck did not publish any results based on his collections, other than internal reports (e.g. Zieck 1972). He did, however, correspond with leading palm experts of the time, Harold Moore (BH) and John Dransfield (K), and deposited duplicate specimens at their respective herbaria, as well as at LAE.

Figure 3.4. Frederick B. Essig (1947–), American botanist (University of South Florida) who collected more than 240 palm specimens during field campaigns between 1971 and 1989. He also compiled the first synthetic checklist of New Guinea palms (Essig 1977).

American botanist Frederick B. Essig (Fig. 3.4) was the most productive collector of New Guinea palms between the Archbold Expeditions and the inception of the Palms of New Guinea project. Essig was a student of Moore's and attempted to make a "systematic effort" to document the palms of New

Guinea (Essig 1977, 1995, Essig & Young 1980, 1981a, b. He undertook two major expeditions in Papua New Guinea, first, in September 1971 to July 1972 and, second, with Bradford E. Young, in April-May 1978, as well as a short trip to New Britain in February 1989. Of his 241 specimens, more than a dozen are designated as types and his efforts are commemorated in *Calamus essigii* and *Licuala essigii*.

Collecting for the Palms of New Guinea project

From the 1990s until the present day, nearly 1,400 palm collections have been made in New Guinea; in the most prolific decade, the 1990s, over 800 specimens were made. Almost 1,100 of these collections fall under the umbrella of the Palms of New Guinea project. Key contributions have been made by William Baker (243 collections), Anders Barfod (146 collections), John Dransfield (51 collections), Charlie Heatubun (221 collections), Anders Kjaer (21 collections), Rudi Maturbongs (238 collections) and Scott Zona (22 collections). Roy Banka (formerly of Papua New Guinea Forest Research Institute) also played a central role in the majority of project collecting in Papua New Guinea, though he is not often listed as a primary collector. Students and collaborators of Charlie Heatubun and Rudi Maturbongs at Universitas Papua, Manokwari, such as Barnabas Desianto, Piter Gusbager, Samsoni Mehen, Elisa Wally and Jack Wanggai, have made excellent contributions in Indonesian New Guinea. Other important collectors in this period include Wanda Avé, Ary Keim, Johanis Mogea, Robert Johns and Wayne Takeuchi. For the first time, collections from Indonesian New Guinea outnumbered those from Papua New Guinea by around two-to-one.

Priorities for future exploration

Although palm collecting effort varies widely across New Guinea, with some areas known much better than others (Fig. 3.1), collecting density does not exceed 40 palm specimens per 50 × 50 km grid cell in any part of the region. Further botanical collection in almost any location would potentially yield new insights into species distributions and perhaps even species new to science. The most intensively collected grid cells are those that include Lae, Vanimo, Jayapura, Timika, Wasior, Manokwari, Sorong and parts of the Tamrau Mountains. Almost all other grid cells contain fewer than 20 palm specimen records, and many have none. The most sparsely collected areas are concentrated in the eastern half of Indonesian New Guinea, although south-western Papua New Guinea is almost as incompletely explored. These areas include vast tracts of lowland vegetation types likely to be rich in palms. Grid cells in which palms are effectively unrecorded are scattered throughout the region.

The exploration of New Guinea's palm diversity is far from complete. It is hoped that this book will provide the springboard for a new wave of targeted fieldwork, which will significantly improve our understanding of the distribution of species and the risks that they face in the wild, and will potentially lead to a new phase of discovery of species new to science.

An as-yet unidentified species of *Licuala* discovered near to Sorong. Photo: WB.

4.

DESCRIPTION AND DOCUMENTATION OF THE PALMS OF NEW GUINEA

The collection of a specimen is only the first part of the story. Identifying a specimen to species or determining that it represents a species unknown to science requires painstaking work in the herbarium and wide experience. Despite the involvement of numerous specialists in palm and Malesian botany, the description of New Guinea's palm diversity has largely unfolded in an unsystematic way. It was not until the 1970s that more synthetic attempts to describe the island's palms began to be made.

The first species new to science

Many of New Guinea's widespread, non-endemic palms have long, complex taxonomic histories dating back to the 18th century (e.g. *Areca catechu, Arenga pinnata, Cocos nucifera, Corypha utan, Drymophloeus litigiosus, D. oliviformis, Heterospathe elata, Metroxylon sagu, Nypa fruticans, Pigafetta filaris, Saribus rotundifolius*). However, the first of New Guinea's native, endemic (or near endemic) palms, *Ptychosperma gracile,* was described in 1809 (Labillardière 1809), lighting the fuse on the charting of the island's extraordinary palm riches. Renowned 19th century players in the botany of Malesia and the palm family played a small part in the incremental illumination of the palm flora (e.g. Blume, von Martius), but it was Odoardo Beccari, based at the Florence herbarium, who made the largest contribution, surpassing any taxonomist of New Guinea palms. Beccari described 72 species and an endemic genus (*Sommieria;* Fig. 4.1) that are still accepted today. His publications ranged from "new species"

Figure 4.1. Holotype specimen of *Sommieria leucophylla* at the Florence herbarium, collected in Andai near Manokwari by Odoardo Beccari in 1872 (*Beccari PP607*). Photo: WB.

papers to monumental monographs, commencing in 1877 and continuing posthumously beyond his death in 1920 until 1934. Beccari had the advantage of precedent, being the first serious student of New Guinea palms, and consequently many of his names have survived the test of time.

Twentieth-century discovery until World War 2

The early 20th century saw small numbers of species described by workers such as Lauterbach, Ridley and Schumann, but the most prolific taxonomist was Max Burret (Fig. 4.2) of the Berlin Herbarium. Burret published numerous new species between 1927 and 1943. Most were described in "new species" papers, including six papers in the series *Neue Palmen aus Neuguinea,* although he also undertook some more systematic work. He worked extensively, but not exclusively, on the collections of Leonard Brass from the Archbold Expeditions. Unfortunately, Burret appeared to work in a vacuum and often failed to make connections between the material that he studied and species that had already been described. For example, in *Calamus,* 10 out of 17 Burret names (58%) have been placed in synonymy whereas only eight of 23 Beccari names (35%) met the same fate. Of the many New Guinea species that Burret described, only 18 are still accepted today. The endemic genus *Brassiophoenix* described by Burret is also accepted.

New Guinea palm taxonomy was further complicated by the Allied bombing of the Berlin Herbarium in 1943, which destroyed a substantial proportion of the institution's collections, including many type specimens by Brass, Clemens, Hollrung, Kraemer, Lauterbach, Ledermann (Fig. 4.3), Mayr and Versteeg, among others. This affected Burret's and Beccari's names in particular. Duplicates of some, but not all of the lost specimens exist in other herbaria; the collections of Lauterbach, Ledermann and some late Clemens material were particularly badly affected. Where type material was lost, the interpretation of the affected names must be secured. Either a new type from among the surviving duplicates (lectotypification) must be designated or a completely new type must be selected if none of the author's original material survives (neotypification [Fig. 4.3]; e.g. Henderson [2020], Baker *et al.* [in prep.]). In some lucky cases, photographs of lost specimens taken by Beccari are available to inform this process. Where no material survives and the

Figure 4.2. Max Burret (1883–1964), German palm taxonomist who published many New Guinea palm species between 1927 and 1943. Unfortunately, many of his species are regarded as synonyms today. Photo: Botanic Garden and Botanical Museum Berlin, Science History Collection.

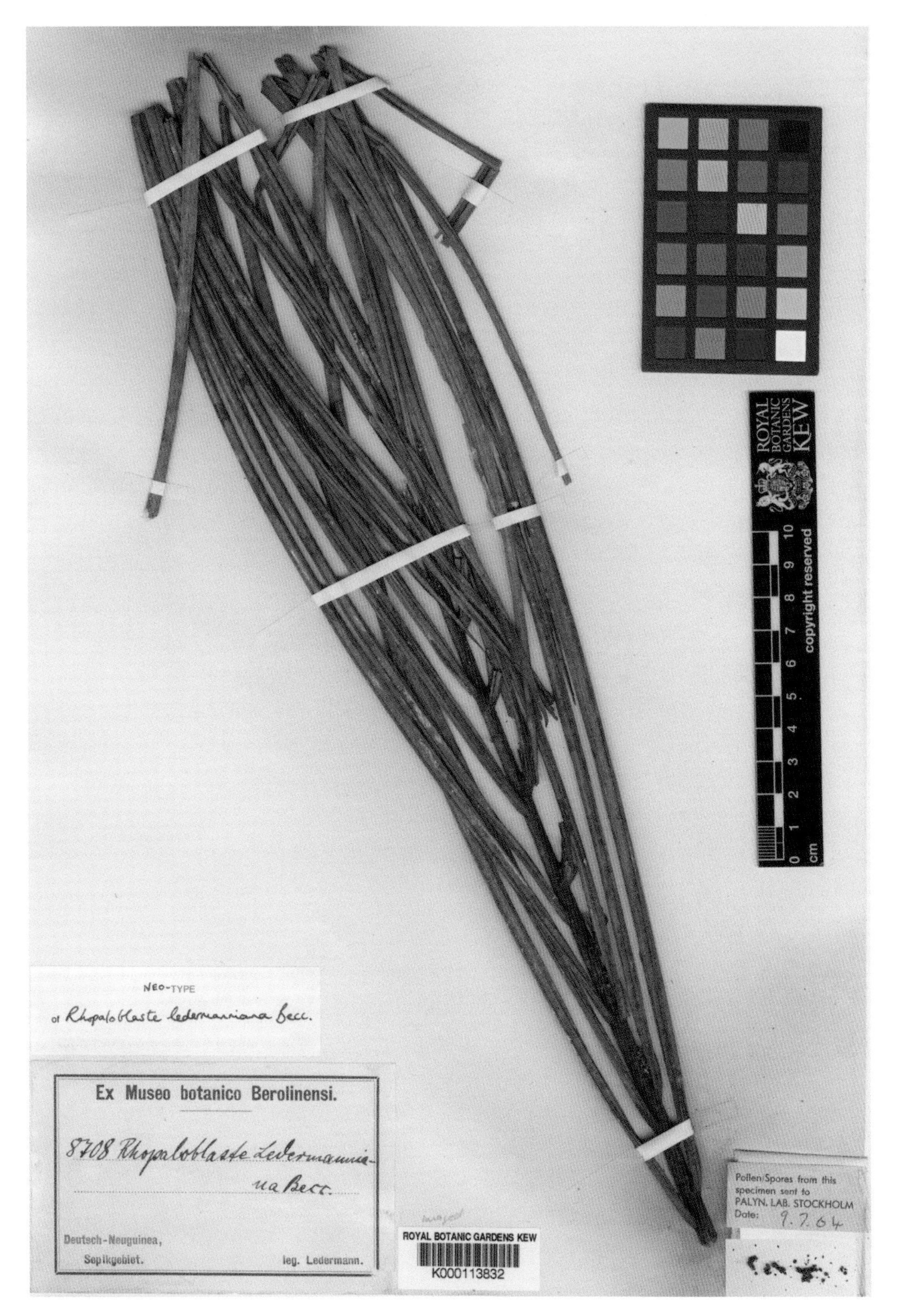

Figure 4.3. Neotype specimen of *Rhopaloblaste ledermanniana* at the Kew herbarium. The original material collected by Ledermann (*Ledermann 8648* and *9718*) and cited by Beccari in his description of this species was destroyed in the Allied bombing of the Berlin Herbarium in 1943. A third collection (*Ledermann 8708*), annotated but not cited by Beccari, was also destroyed in Berlin, but this duplicate survives at Kew and was designated as neotype by Banka and Baker (2004). Photo: Royal Botanic Gardens, Kew.

available descriptions cannot be interpreted, they are disregarded as "names of uncertain application", although they cause instability because they can be resurrected if relevant materials are discovered or the literature reinterpreted. Unsurprisingly, Burret was disheartened by the loss of the Berlin herbarium and published little more on palms after this disastrous loss.

Post-war progress

In the post-war decades, two palm taxonomists set the scene for the Palms of New Guinea project. Working from his base at the L.H. Bailey Hortorium at Cornell University, Harold Moore's contributions to New Guinea palm botany appear limited in the small number of specimens and just six accepted species. However, we know that he was especially interested in understanding palm genera in the region and that he attempted to make sense of the New Guinea rattans, inspired by the efforts of Jack Zieck, whose specimens he received. He also supervised Fred Essig in his doctoral studies on *Ptychosperma*. In practice, Moore was over-extended by the scale of his global palm research efforts and simply did not have the capacity to immerse himself in the complexities of the New Guinea palm problem.

It was Fred Essig who took on the New Guinea palm challenge from Moore, undertaking extended field visits, describing 15 species still accepted today and producing a checklist for the whole island (Essig 1977), in which he listed 270 species in 30 genera. Notwithstanding some early regional accounts (e.g. Schumann & Hollrung 1889, Schumann 1898, Schumann & Lauterbach 1900), Essig's checklist was the first synthesis of the island's palm flora. He emphasised the preliminary nature of his analysis, but it provided an invaluable baseline for subsequent work. Later, he added a checklist of the Bismarck Archipelago (Essig 1995). Essig also addressed the taxonomy of poorly understood arecoid genera in monographs of *Orania* and *Ptychosperma*, as well as in synoptic accounts of *Gronophyllum*, *Gulubia* and *Nengella*, now subsumed within *Hydriastele* (Essig 1978, 1980, 1982, Essig & Young 1985).

Essig's work was followed by a guide to the palms of Papua New Guinea by Alistair Hay (1984). This account contains informative botanical and horticultural commentary alongside species checklists, and is peppered with wit (e.g. "*Pigafetta* has chic"). Hay did not contribute formal taxonomic accounts for New Guinea palms, but his guide was an accessible and inspiring entry point for any new student of palms, including the authors of this book. A few years later, this guide was complemented by a set of preliminary reports on rattan diversity in Papuasia (Johns & Taurereko 1989a, 1989b, Johns & Zibe 1989).

The Palms of New Guinea project

The Palms of New Guinea project started formally in 1998 (see Baker [2000a, 2001a, b, 2002a] for summaries of the early history of the project). At the start of the Palms of New Guinea project, the palm flora was estimated to number approximately 250 species in 33 genera (Baker 2001a), drawing in part on an updated outline checklist of New Guinea palms (Ferrero 1997) that was illustrated with

Figure 4.4. *Saribus pendulinus*, the most recently described palm species in New Guinea, at the time of writing (Dowe & Venter 2023). Type locality near the Purari River. Photo: SV.

exciting photographs from the field. Fieldwork was the highest priority in the early stages of the project. This led to the publication of a number of synoptic accounts (Baker & Dransfield 2007, Baker *et al.* 2021) and accessible field guides (Barfod *et al.* 2001, Baker & Dransfield 2006a, b) as interim products in preparation for this book.

The early stages of the project also emphasised the testing and resolution of generic limits, usually with evidence from phylogenetic analysis of DNA sequence data. The following changes were made (in systematic order):

1. The expansion of *Calamus* to include *Daemonorops* and other small genera of Calaminae (Baker 2015).
2. The removal of the genus *Lanonia* (not in New Guinea) from *Licuala* (Henderson & Bacon 2011, Bacon *et al.* 2012).
3. The resurrection and expansion of the genus *Saribus* by inclusion of *Pritchardiopsis* and several species previously placed in *Livistona* (Bacon & Baker 2011, Bacon *et al.* 2012).
4. The sinking of *Paralinospadix* into *Calyptrocalyx* (Dowe & Ferrero 2001).
5. The removal of species from *Drymophloeus* to *Veitchia* (not in New Guinea) and *Ponapea* (Zona *et al.* 2011).
6. The description of *Manjekia* and *Wallaceodoxa* as new genera (Heatubun *et al.* 2014c).
7. The expansion of *Heterospathe* to include *Alsmithia* (not in New Guinea) and the confirmation of the delimitation of *Rhopaloblaste* (Norup 2005, Norup *et al.* 2006).
8. The description of *Dransfieldia*, which had formerly been placed in *Heterospathe, Ptychosperma* and *Rhopaloblaste* (Baker *et al.* 2006).
9. The expansion of *Hydriastele* to include *Gronophyllum*, *Gulubia* and *Siphokentia* (Baker & Loo 2004, Loo *et al.* 2006).

As a result of this work, 34 robustly defined genera are now recognised for New Guinea, including five endemic genera, *Sommieria* and *Brassiophoenix*, both previously described, and *Dransfieldia*, *Manjekia* and *Wallaceodoxa*, described during the project. Since the work of Essig (1977), *Clinostigma*, *Ponapea* and *Saribus* have also been added to the New Guinea palm flora. Complete or partial monographs of over half of New Guinea's genera have been published in preparation for this book: *Areca* (Heatubun *et al.* 2012a), *Borassus* (Bayton 2007), *Brassiophoenix* (Zona & Essig 1999), *Calyptrocalyx* (Dowe & Ferrero 2001), *Cyrtostachys* (Heatubun *et al.* 2009), *Dransfieldia* (Baker *et al.* 2006), *Drymophloeus* (Zona 1999), *Heterospathe* (Trudgen & Baker 2008, Petoe & Baker 2019), *Hydriastele* (Heatubun *et al.* 2018, Petoe *et al.* 2018a, b), *Linospadix* (Dowe & Ferrero 2001), *Livistona* including species now referred to *Saribus* (Dowe 2009), *Manjekia* (Heatubun *et al.* 2014c), *Orania* (Keim & Dransfield 2012), *Ptychococcus* (Zona 2005), *Rhopaloblaste* (Banka & Baker 2004), *Sommieria* (Heatubun 2002) and *Wallaceodoxa* (Heatubun *et al.* 2014c). A detailed checklist, including complete taxonomic literature and type citations, will be made available as a companion to this book (Baker *et al.* in prep.)

Through monographs and other papers, the project authors have described 91 species as new to science (all endemic to New Guinea), all accepted in this account and representing 37% of the native palm flora and 43% of its endemics (Baker, 47 species; Barfod, 14 species; Dowe, 11 species; Dransfield, 45 species; Heatubun, 24 species; Petoe, 7 species). These new species include 32 *Calamus* species, 12 *Licuala* species, 11 *Orania* species, 10 *Hydriastele* species, seven *Calyptrocalyx* species, five *Saribus* species and three each of *Areca*, *Cyrtostachys* and *Heterospathe*. It is striking, then, that the number of New Guinea palm species remains unchanged from the start of the project (Baker 2001a), due to the placement of many redundant names in synonymy.

In Madagascar, the palm flora has grown by over 20% since the publication of *The Palms of Madagascar* (Dransfield & Beentje 1995) because the existence of a sound taxonomic account promoted new exploration, and facilitated taxonomic identification and the discovery of species new to science. We hope that the same will happen in New Guinea as a result of this book. Indeed, it has already started: the first new species to be added to the book, *Saribus pendulinus* (Fig. 4.4; Dowe & Venter 2023), was inserted after the manuscript was submitted to the publishers. Exciting times lie ahead for New Guinea palm exploration. It is urgent that this opportunity is seized.

THE STORY OF NEW GUINEA PALM EXPLORATION — The role of *Palms*, the journal of the International Palm Society

In addition to taxonomic papers in general botanical journals, the exploration of the New Guinea palm flora over the past five decades has been especially well-documented in the journal of the International Palm Society, *Palms* (formerly *Principes* until 1999), via expedition accounts, descriptions of new taxa and profiles of species and genus discoveries (Moore 1966, Essig 1972, 1982, 1995, Essig & Young 1980, 1981a, 1981b, 1985, Sneed 1985, Baker 1997, Forster 1997, Baker *et al.* 2000a, 2000b, 2[illegible], Dowe & Ferrero 2000, Dransfield *et al.* 2000a, 2000b, Heatubun 2000, 2008, Zona 2000, Baker & Zona 2006, Bacon & Baker 2011, Baker & Heatubun 2012, Heatubun *et al.* 2012b, 2014a, 2014d, Baker & Venter 2019, Petoe *et al.* 2019, Barfod & Baker 2022, Dowe & Venter 2023). Many new species from New Guinea have been described in the journal, including several that have come to light through horticulture (e.g. Baker *et al.* 2000b, Dransfield *et al.* 2000a, Gardiner *et al.* 2012, Dransfield & Marcus 2020). A New Guinea special issue of *Palms* was published in 2000 (volume 44, part 4). Other specialist palm journals, such as *Palms and Cycads* and *The Palm Journal* have also been important in disseminating information about New Guinea palms to a wide audience.

Hydriastele procera on karst limestone in Misool, Raja Ampat Islands. Photo: GB.

5.
DISTRIBUTION AND ECOLOGY OF PALMS IN NEW GUINEA

So far we have discussed New Guinea as a single geographical unit. However, there is considerable differentiation in palm diversity, distribution and ecology across the region, which is the focus of this chapter.

Palm diversity and distributions within the New Guinea region

Of the 250 species of palms in New Guinea (Appendices 1 and 2), 150 occur in Indonesian New Guinea and 180 in Papua New Guinea (ratio of 1:1.2). This is lower than the ratio across the political divide for the flora as a whole (1:1.44, Cámara-Leret *et al.* [2020]), which may reflect the special efforts that have been made to study palms in Indonesian New Guinea since the 1990s. Endemism is more imbalanced; 95 palm species are endemic to Papua New Guinea, whereas only 58 are endemic to Indonesian New Guinea. Thirty genera (out of 34 total) occur in both regions, but there is variation between the two countries: *Brassiophoenix*, *Clinostigma*, *Ponapea* and *Physokentia* have not been found in Indonesian New Guinea, whereas *Dransfieldia*, *Drymophloeus*, *Manjekia* and *Wallaceodoxa* are not recorded in Papua New Guinea.

The distribution of species across political borders is important for national, policy and conservation reasons, but is otherwise rather problematic from a biogeographic perspective. Here we dig deeper into more natural subdivisions within the region according to the major New Guinea island groups (Appendix 2).

Mainland New Guinea

As expected, the vast majority of New Guinea palm diversity is found in mainland New Guinea. Almost 90% of species (223; Appendix 2) occur there and 156 of these are endemic to the mainland. All but five genera (*Clinostigma*, *Ponapea*, *Physokentia*, *Manjekia* and *Wallaceodoxa*) occur on the mainland. Patterns of palm distribution and diversity in New Guinea would repay careful quantitative analysis. However, informal study reveals a variety of distinctive, repeating patterns among well-sampled species.

Widespread distributions – A set of widespread and often common species can be found throughout mainland New Guinea: *Actinorhytis calapparia*, *Areca catechu*, *A. macrocalyx*, *Arenga microcarpa*, *Calamus aruensis*, *C. fertilis*, *C. lauterbachii*, *C. vitiensis*, *C. warburgii*, *C. zebrinus*, *C. zieckii*, *Caryota rumphiana*, *Cocos nucifera*, *Cyrtostachys loriae*, *Hydriastele costata*, *H. flabellata*, *H. pinangoides*, *H. wendlandiana*, *Korthalsia zippelii*, *Licuala telifera*, *Linospadix albertisianus*, *Metroxylon sagu*, *Nypa* *fruticans, Orania lauterbachiana*, and *Ptychococcus paradoxus*. Only two of these are strictly endemic to the New Guinea mainland (*C. lauterbachii, H. flabellata*). The majority of these species are robust palms and may be good dispersers, though many robust species are narrowly distributed and rare. Some are useful (e.g. *Metroxylon sagu, Ptychococcus paradoxus*) and their distribution is probably modified by humans (Essig 1977). It is striking that no species of species-rich genera such as *Calyptrocalyx, Ptychosperma, Heterospathe* or *Saribus* are included in this list.

North-south distributions – Reaching almost 5,000 m in elevation, the central range forms a hard barrier as so much of it rises above elevations tolerable to palms. Numerous distribution patterns indicate species that have not been able to cross either southwards or northwards across the central range. For example, *Borassus heineanus, Calamus vestitus, Calyptrocalyx elegans, C. hollrungii* and *Rhopaloblaste ceramica* are all found only north of the central cordillera, whereas *Calamus badius, C. bulubabi, C. distentus, Heterospathe macgregorii, Hydriastele apetiolata, Licuala brevicalyx* and *Orania archboldiana* are restricted to the south.

East-west distributions – Although comparisons across the political divide seem generally artificial, a number of quite widespread palm species appear to stop at the border, or nearly so. For example, *Hydriastele lurida, Pigafetta filaris, Sommieria leucophylla* and *Rhopaloblaste ledermanniana* are found almost exclusively in Indonesian New Guinea, whereas *Brassiophoenix* (both species), *Calamus longipinna* and *Calyptrocalyx pachystachys* are primarily Papua New Guinean. Artefacts of collecting patterns may play a role here. Western distributions that range outside New Guinea may reflect species that are still in the process of range expansion (e.g. *Calamus calapparius, Pinanga rumphiana, Rhopaloblaste ceramica*).

Species restricted to the far east or west of New Guinea are numerous. Examples of palms endemic (or nearly so) to the Bird's Head Peninsula and neck include *Areca mandacanii, Calamus sashae, Calyptrocalyx multifidus, Cyrtostachys excelsa, Dransfieldia micrantha, Hydriastele gibbsiana, H. variabilis* and *Licuala bifida*. Even more numerous are species restricted to the eastern tip of Papua New Guinea, such as *Brassiophoenix drymophloeoides, Calamus anomalus* group (except for *C. maturbongsii*), *C. cuthbertsonii, Calyptrocalyx forbesii, Cyrtostachys glauca, Heterospathe barfodii, H. obriensis, Licuala multibracteata*, at least seven *Ptychosperma* species such as *P. caryotoides*, and *Saribus chocolatinus*.

Disjunctions – A number of striking disjunct distributions have been identified between the far east and west (*Calamus barbatus, C. heteracanthus* and *C. erythrocarpus/C. maturbongsii*, a pair of species that are very likely sister taxa [Kuhnhäuser 2021] located at the east and west ends, respectively). *Calyptrocalyx albertisianus* and *Orania macropetala* are similar but are more widely distributed in Papua New Guinea. These patterns may be masking multiple cryptic species, but at least in the *Calamus* species, this seems not to be the case (Henderson 2020, Kuhnhäuser 2021).

Islands around mainland New Guinea

Raja Ampat – The Raja Ampat islands are the most species rich of the major island groups around New Guinea. Comprising over 600 islands around the western tip of New Guinea of varied geology and vegetation, Raja Ampat supports 36 palm species (Appendix 2), including three endemic to the island group, *Licuala urciflora*, *Saribus brevifolius* and *Wallaceodoxa raja-ampat*. All three endemics are restricted to Waigeo and Gag Islands (*L. urciflora* on Waigeo only).

Cenderawasih Bay Islands – Thirty palm species can be found here in total (Appendix 2), 20 on the Biak Islands (Biak, Supiori, Numfor) and 20 on Yapen. Four are endemic to the islands, recorded from Biak (*Manjekia maturbongsii*), Biak and Numfor (*Hydriastele biakensis, H. dransfieldii*) and Supiori (*Heterospathe porcata*).

Bismarck Archipelago – The Bismarck Archipelago includes the largest of New Guinea's offshore islands, notably New Britain (18 species), New Ireland (18 species) and Manus (17 species). We recognise 31 species in total for the Archipelago (Appendix 2), eight of which are endemic: on New Ireland, *Clinostigma collegarum*, *Hydriastele kasesa* and *Rhopaloblaste gideonii*; on New Britain, *Heterospathe parviflora*, *Hydriastele kasesa*, *Physokentia avia* and *Ponapea hentyi*; and on Manus, *Hydriastele manusii* and *Licuala sandsiana*. In the New Guinea region, the widespread genera *Clinostigma*, *Physokentia* and *Ponapea* are found only in the Bismarck Archipelago.

Milne Bay Islands – Encompassing the D'Entrecasteaux Islands, Trobriand Islands, Woodlark Island and the Louisiade Archipelago, the documented palm flora of the Milne Bay islands comprises 24 species (Appendix 2) including five endemics, all in the genera *Heterospathe* and *Ptychosperma*. Two are known only from Rossel Island (*Heterospathe annectens, Ptychosperma ramosissimum*), one from Sudest Island (*Ptychosperma tagulense*), while the remaining two are more widespread in the D'Entrecasteaux and Louisiade Archipelagoes (*Heterospathe pulchra, Ptychosperma rosselense*).

Aru Islands – Aru is not formally included in this book, but for completeness, we note here that 15 of its 16 recorded palm species (Appendix 2) are shared with mainland New Guinea. The additional species is *Calamus melanochaetes*, a very widespread and variable taxon (Henderson 2020).

Palm habitats

Although the typical perception of New Guinea is of endless, towering rainforest, the island has a wide diversity of habitats driven by the complex landscape and the many physical and environmental parameters that vary across it. Elevation, climate (including temperature, precipitation and seasonality), geology, soil, topography and drainage all play a role. Disturbance is also important in New Guinea, for example where geological dynamism promotes high frequencies of earthquakes and landslides. However, the relationships between these variables and palm diversity, distribution and abundance have scarcely been explored, with the exception of some analyses of palm diversity and elevation.

Palms and the elevational gradient

Elevation is a critical determinant of vegetation composition and of the distribution of species. Palm species occurrence across the elevational gradient in New Guinea varies widely, but is inherently limited by the vulnerability of palms to freezing due to their structural biology (Tomlinson 2006, Kissling *et al.* 2012b). Our herbarium specimen database contains palm records from sea level up to 2,800 m, which corresponds precisely with the elevation at which frosts become prevalent (Prentice & Hope 2007). Only ten palm species are known with certainty to occur above 2,000 m: *Areca macrocalyx, Calamus cuthbertsonii, C. depauperatus, C. klossii, C. oresbius, C. pintaudii, Heterospathe elegans, H. muelleriana, H. obriensis* and *Hydriastele gibbsiana*. The five highest reported palm specimen records are all from *Calamus klossii* at 2,600–2,800 m (Fig. 5.1). *Ptychococcus lepidotus* has been reported to occur up to 3,000 m (Ferrero 1996), but this requires confirmation (Zona 2005). If palm specimen numbers are an indicator of palm abundance in the landscape, the number of available palm specimens approximately doubles in 500 m elevation bands from 120 specimens at 1,500–2,000 m, to 216 specimens at 1,000–1,500 m and 400 specimens at 500–1,000 m. Numbers then jump dramatically to 1638 specimens at 0–500 m. Note that these crude figures have not been adjusted for the much smaller area of land at higher elevations.

Figure 5.1. *Calamus klossii*, which reaches 2,800 m, the highest elevation record in New Guinea palms. Here pictured on the slopes below Mount Jaya. Photo: WB.

Species richness declines monotonically with elevation (Bachman *et al.* 2004, Jimbo *et al.* 2023). However, species-area relationships confound these conclusions. For example, the land surface area of New Guinea between 0–500 m is more than five times greater than that between 500–1,000 m, which will have a knock-on effect on species richness. To address this, Bachman *et al.* (2004) repeated their analyses by quantifying palm species richness in equal area bands rather than equal elevation bands. This revealed a mid-elevation peak of species richness, which the authors explained primarily as a product of the geometric constraints of the range size frequency distribution (the so-called mid-domain effect), although this conclusion has been criticised subsequently (Currie & Kerr 2008). With our much-improved dataset of identified and georeferenced occurrences, it would be timely to revisit this analysis, drawing in environmental factors that might also help to explain the nuances of the elevational gradient in species richness in New Guinea palms.

Vegetation types and palms

Highly detailed classifications of New Guinea vegetation have been devised to account for the immense variation across the region (Paijmans 1976, Marshall & Beehler 2007). However, we lack a clear understanding of how palm species map on to these classifications. Here, we use a simplified classification (Utteridge & Jennings 2021) within four major zones – lowland, montane, subalpine and alpine – as a framework for an outline of palm variation across vegetation types. Only vegetation types within the lowland and montane zones are relevant to palms.

Coastal habitats of the lowlands are generally palm poor but do include some characteristic species. The mangrove palm, *Nypa fruticans,* forms vast, monodominant stands around mangrove forest, especially in parts of southern New Guinea (Fig. 5.2). These palms are of immense ecological importance, stabilising coastlines and providing breeding grounds for fish and crustaceans. Other species that are characteristic of the back of the mangrove zone include *Calamus lucysmithiae, Ptychosperma lauterbachii, P. mambare, P. lineare* and *Livistona benthamii.* Beach or coastal forest, often on limestone, also features some distinctive species, such as *Hydriastele biakensis, H. procera* and *Saribus woodfordii.*

Lowland swamp vegetation occurs where water tables sit near the soil surface and where seasonal flooding is common. Swamp vegetation ranges from open aquatic habitats through to tall forest (Fig. 5.3). The most important swamp palm is the sago palm (*Metroxylon sagu*), which dominates freshwater swamplands and is often heavily managed by humans. In practice though, many palms thrive in swampy conditions, including *Calamus fertilis, C. zebrinus* (and other rattans), *Calyptrocalyx albertisianus, Hydriastele costata, Ptychococcus paradoxus, Ptychosperma lineare* and *Saribus surru.*

Lowland rainforest is by far the most abundant vegetation type in New Guinea (Fig. 5.4). It occurs up to 700 m elevation and has a complex, multi-layered structure with a canopy to ca. 35 m and emergent

Figure 5.2. Swamp forest almost entirely dominated by *Nypa fruticans*, Kikori River. Photo: JLD.

Figure 5.3. Swamp forest on the Purari River, with tall, mixed swamp forest in the foreground (including *Hydriastele costata*), and sago-dominated (*Metroxylon sagu*) forest in the distance. Photo: PL.

Figure 5.4. Lowland forest near Sorong. *Hydriastele costata* in the foreground. Photo: WB.

trees to ca. 50 m. It supports by far the greatest number of palm species of any vegetation type, at all levels within the forest structure from understorey to canopy, including climbers. More than 90% of New Guinea's palm species are found in lowland rainforest habitats, and many of them do not exceed the 700 m boundary; examples include *Calamus fertilis, C. heteracanthus, C. warburgii, Calyptrocalyx hollrungii, Dransfieldia micrantha, Drymophloeus oliviformis, Manjekia maturbongsii, Metroxylon sagu, Pigafetta filaris, Sommieria leucophylla* and *Wallaceodoxa raja-ampat*. Very little is known about the geography of palm distributions within this vast vegetation block.

Seasonal vegetation including lowland savannah is found in the far south of central New Guinea and also in south-eastern Papua New Guinea (in the vicinity of Port Moresby). Typical marker species for this vegetation are the two species of *Livistona, L. benthamii* and *L. muelleri*. The massive hapaxanthic *Corypha utan*, also a widespread species, may also be found. Vegetation in these areas tends to be a mosaic with forest patches or gallery forest containing more resilient palm species such as *Ptychosperma propinquum* or *Hydriastele wendlandiana*. All of these species are shared with Australia, probably reflecting the recent land connection, the Torres Strait becoming inundated only 6,000–9,700 years ago (Polhemus 2007, Dowe 2010).

Lowland heath forest comparable to the renowned Bornean white sand vegetation (*kerangas*) is a rare vegetation type in New Guinea, but is prominent on the lower slopes of the central range south of Mount Jaya at around 500 m elevation (Fig. 5.5). Fine glacial outwash sands form terraces from which

Figure 5.5. Heath forest terraces at ca. 500 m in the foothills below Mount Jaya, with montane forests rising in the distance. Photo: WB.

Figure 5.6. Lowland heath forest on ultramafic soils on Gag Island, with *Saribus brevifolius* and an undescribed species of *Hydriastele*. Photo: CDH.

tannin-rich rivers drain and support dense, medium-stature, generally small-leaved forest that is rich in palms. We have surveyed the heath forest north of Timika relatively thoroughly, discovering heath forest endemics such as *Hydriastele divaricata, H. splendida* and *Orania timikae,* as well as other notable species that also occur on other vegetation types (e.g. *Calyptrocalyx micholitzii, Hydriastele longispatha, H. lurida, Licuala bakeri, L. grandiflora, Saribus papuanus*). Heath forests also occur on ultramafic soils, such as those on Gag Island (Fig. 5.6) and Waigeo. *Saribus brevifolius* and two undescribed species of *Hydriastele* are endemic to this vegetation in the Raja Ampat islands (Heatubun *et al.* 2014a).

Montane forests range from ca. 700–3,000 m elevation and are subdivided into lower montane forest (700–1,500 m), mid-montane forest (1,500 m to ca. 2,500–2,700 m; Fig. 5.7) and upper montane forest (above ca. 2,500 m). The exact boundaries of these subdivisions of montane are a matter of debate (Utteridge & Jennings 2021). Montane forest composition differs from that of lowland forests, prominently featuring gymnosperms, Nothofagaceae and Lauraceae, for example. Palms are scarcely known from upper montane forest (*Calamus klossii*). Around 120 (48%) species of palm occur in lower montane forest; genera such as *Calamus, Calyptrocalyx, Heterospathe* and *Orania* are species-rich at this elevation. Only ca. 25 palm species are known from mid-montane forest, all but three (*Calamus katikii*, *Calyptrocalyx arfakianus* and *Heterospathe obriensis*) being also reported from lower montane forest. *Brassiophoenix*, *Cyrtostachys*, *Korthalsia*, *Ptychosperma*, *Rhopaloblaste* and *Saribus*, all of which occur in lower montane forest, are absent from mid-montane forest.

Palm-dominated communities are sometimes encountered, typically where there are extremes of geology, drainage or disturbance. The colonial palms *Metroxylon sagu* (Fig. 5.8) and *Nypa fruticans* (Fig. 5.2) are especially important and form large monodominant stands in lowland freshwater and mangrove swamps. We have observed palm hyperabundance of a single or small number of species on limestone (e.g. *Hydriastele calcicola* in the Kikori River basin; *H. procera* in the Raja Ampat Islands and Bird's Head Peninsula; *Heterospathe ledermanniana* in the Southern Highlands of Papua New Guinea), on ultramafic soils as mentioned above (e.g. community including *Saribus brevifolius* and *Hydriastele* species on Gag Island) and in swamp forest (e.g. *Hydriastele costata* in peat swamp near Timika; community including *Actinorhytis calapparia*, *Orania bakeri* and *Saribus surru* in lowland swamp forest at the mouth of the Ramu River). In secondary habitats, rattans (both *Calamus* species and *Korthalsia*) can become very abundant. The circumstances that cause palms to dominate remain to be explored.

Figure 5.7. Mid-montane forest at 1600 m at Hidden Valley, near Mount Kaindi, Wau. *Hydriastele ledermanniana* dominates the ridgeline. Photo: WB.

Figure 5.8 A monodominant sago (*Metroxylon sagu*) swamp forest in the Purari River basin. Flowering stems are clearly visible, including many that are dying after flowering. Photo: PL.

Natural history of New Guinea palms

Studies of the natural history of New Guinea palms are scarce. Here we briefly summarise a selection of aspects in the hope that they may provoke new research.

Growth forms

Life form diversity across the New Guinea palm flora is rather evenly divided among canopy tree palms, mid-storey tree palms, understorey tree or shrub palms, and rattans. The existence of more than 60 species of canopy tree palms in New Guinea is exceptional in Malesia, where rattans and small palms tend to dominate in general (Muscarella *et al.* 2020). Many of the large tree palms belong to tribe Areceae, but tribe Oranieae, several coryphoid tribes and even calamoids contribute tree species to the New Guinea flora.

Rattan diversity in New Guinea is significant (64 species of *Calamus* and *Korthalsia*), although rather lower than the diversity found on comparable Sunda Shelf land masses, Borneo, Sumatra and the Malay Peninsula (Baker & Couvreur 2012). The Sunda Shelf rattan groups include many species that have evolved from climbers into small tree, shrub or stemless forms (Couvreur *et al.* 2014), but no similar phenomenon is observed in New Guinea, where all rattans retain the climbing habit.

Four taxa in New Guinea are "stemless" (*Calyptrocalyx laxiflorus, Heterospathe elegans* subsp. *humilis, H. sphaerocarpa, Licuala graminifolia*). Five of New Guinea's native palms are hapaxanthic (stems dying after flowering): *Arenga microcarpa, Caryota rumphiana, C. zebrina, Corypha utan* and *Metroxylon sagu*. These are all large tree palms, and all but *Caryota zebrina* are widespread and common.

Rheophytes

Four species of New Guinea palm are rheophytes, one rattan (*Calamus reticulatus*) and three medium-sized tree palms (*Heterospathe macgregorii, Hydriastele rheophytica, H. simbiakii* [Fig. 5.9]). These species are restricted to the flood-zones of large rivers (Baker 1997, Dowe & Ferrero 2000). These rheophyte species are strongly multi-stemmed and are thus able to regenerate easily. They have flexible stems and narrowly pinnate leaves that offer little resistance to fast-flowing water.

Plant-insect interactions

The most conspicuous plant-insect interaction that we have observed is the ant-rattan relationship that occurs in many species of *Calamus*. It is not unusual for ants to build nests among the stems, leaves and inflorescences of rattans. The spines in particular lend themselves to supporting the construction of nests enclosed within layers of carton, a paper-like material formed of chewed plant

Figure 5.9. The rheophyte *Hydriastele simbiakii* growing in the flood zone on the banks of the Sujak River in the Tamrau Mountains. Photo: WB.

tissue and ant secretions. Ants also often build nests at the junction of the leaf and leaf sheath, especially where the lowest leaflets sweep back across the stem (e.g. *Calamus retroflexus*). However, the most striking phenomenon occurs in the ocreate *Calamus* species (see under *Calamus*). These 27 species bear prominent extensions of the leaf sheath above the insertion of the petiole. The ocrea varies widely in morphology, but is often inflated and persistent (e.g. *Calamus altiscandens, C. bankae, C. heatubunii, C. kostermansii, C. lauterbachii, C. longipinna, C. macrochlamys, C. vestitus, C. wanggaii*), favouring occupation by ants (Fig. 5.10). These ocreas are remarkably similar to the inflated ocreas found in some species of *Korthalsia* (Shahimi *et al.* 2019) and *Laccosperma* (Sunderland 2012), as well as some *Calamus* species from outside New Guinea (Henderson 2020). The development of the ocrea in *Calamus longipinna* has been studied by Merklinger *et al.* (Merklinger *et al.* 2014).

Figure 5.10. Ants inhabiting the ocrea of *Calamus macrochlamys*. Photo: WB.

Reproductive ecology

Very little is known about the reproductive ecology of New Guinea palms. A review of pollination mechanisms in palms (Barfod *et al.* 2011) revealed that beetles and bees are most often responsible for transferring pollen between palm flowers. However, a strong geographical and taxonomic bias exists in palm pollination studies towards South America and the tribes Cocoseae and Phytelepheae. Only one pollination study has been published on New Guinea palms (Essig 1973), focused on *Ptychosperma propinquum, Hydriastele wendlandiana* and *Nypa fruticans*. Essig reported syrphid flies and *Nomia* bees to be significant pollen transporters in *Ptychosperma* and likely pollinators. *Hydriastele wendlandiana*, being protogynous, undergoes a very rapid transition from pistillate to staminate anthesis shortly after the inflorescence bud opens. Essig concluded that weevils (Curculionidae) were the most likely pollinators. He also observed drosophilid flies transporting pollen in *Nypa*, concluding that they may be pollinators. Recent studies conducted in Thailand and the Philippines have pointed to beetles as the most likely pollinators of *Nypa* (Mantiquilla *et al.* 2016, Straarup *et al.* 2018).

Seed dispersal research, like pollination biology, is biased towards South America, where a great diversity of seed dispersing agents has been recorded (Zona & Henderson 1989). New Guinea palms, which have scarcely been studied with respect to seed dispersal, usually produce fleshy fruit that are brightly coloured, especially red, orange and yellow, and are likely to be attractive to frugivores. A high diversity of frugivorous birds occurs in New Guinea (Terborgh & Diamond 1970, Beehler *et al.* 1986), among them cassowaries, pigeons, megapodes, parrots, starlings, cuckoos, cuckoo-shrikes, crows and birds of paradise. Palm fruits and seeds probably feature in the diets of many of these groups. Pigeons may be especially important for the dispersal of palm seeds as they are able to consume larger seeds (Terborgh & Diamond 1970). Fruit doves (*Ducula*) and Imperial pigeons (*Ptilonopus*) feed only on fruit with soft pericarp and do not digest seeds (Beehler *et al.* 1986). Ground-feeding forest birds like cassowaries, ground pigeons and megapodes may be particularly important for palms, which often drop their fruit or sometimes present them near the ground.

The importance of cassowaries as palm seed dispersers is well-established (Beccari 1877a, Pratt 1983, Zona & Henderson 1989). Being flightless, cassowaries can only consume fruit that are presented at or below head height, or that have fallen. In a study of the northern cassowary (*Casuarius unappendiculatus*), an obligate frugivore and keystone disperser in New Guinea, seeds of ten palm species were found in cassowary droppings, *Actinorhytis calapparia, Areca macrocalyx, Borassus heineanus, Caryota rumphiana, Calamus aruensis, Licuala lauterbachii, Licuala* sp., *Metroxylon sagu, Orania* sp. and *Ptychococcus paradoxus* (Pangau-Adam & Mühlenberg 2014). Although the smaller seeds of *Areca macrocalyx* and *Caryota rumphiana* exceeded those of all other species 10-fold or more in abundance, the ability of the cassowary to disperse very large seeds (such as those of *Borassus heineanus* and *Orania*) is noteworthy. Secondary dispersal and seed predation (e.g. by rodents) from cassowary droppings may be an important additional dimension to cassowary dispersal (Bradford & Westcott 2011).

Seed predation is a significant risk to palm seeds (e.g. by palm cockatoo), but has not, to our knowledge, been investigated. Several New Guinea palms encase their seeds in elaborate, thickened, bony endocarps with wings and grooves (*Ptychococcus* species, *Brassiophoenix* species, some *Licuala species – L. grandiflora, L. insignis* (Fig. 5.11), *L. longispadix* and *L. simplex*). This unusual, apparently convergent feature has been suggested to be an adaptation that protects against damage in the gut of a frugivore or the jaws of a seed predator (Barfod 2000, Baker & Dransfield 2007). More reproductive ecological studies are needed to gain a better understanding of palm-animal network interactions.

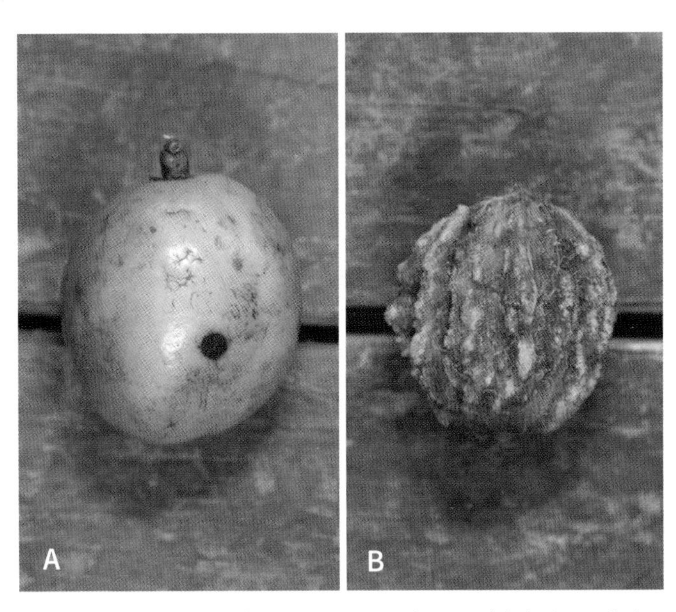

Figure 5.11. The fruit of *Licuala insignis*, showing (A) the intact fruit and (B) the elaborate, bony endocarp with wings and grooves that encases the seed. This may be an adaptation to frugivory or seed predation (Barfod 2000, Baker & Dransfield 2007). Photo: WB.

Calamus papuanus in the Tamrau Mountains. Rattans are widely utilised by the people of New Guinea. Photo: WB.

6.
ETHNOBOTANY OF PALMS IN NEW GUINEA

Recorded palm use in New Guinea dates to ca. 10,000–7,300 years before the present (Fullagar *et al.* 2006). Based on this project's compilation of herbarium specimens and a previous review of 187 references from 1885–2018 (Cámara-Leret & Dennehy 2019a), we know that at least 46% of New Guinea's palms (114 species) are used by local people (Appendix 3). Palm uses can be classified into ten different use categories; 'Utensils and tools' (84 species; Fig. 6.1), 'Construction' (69; Fig. 6.2), and 'Human Food' (43) are the most important. There are at least 70 multifunctional palm species that provide services in at least two use categories: 20 species are cited in at least five use categories and one is cited in all (*Caryota rumphiana*). By contrast, 45 species are only cited in one category, usually in 'Utensils and tools'.

The importance of multifunctional palms can be appreciated by considering their number of uses. Here, a 'palm use' is defined as "the use associated to a use category (e.g., Human Food) and subcategory (e.g., Oil) for a specific palm part for a given species" (Cámara-Leret & Dennehy 2019a). *Cocos nucifera* has the highest number of uses in New Guinea (109), followed by the betel palm (*Areca catechu* [Box 6.1]) and sago (*Metroxylon sagu* [Box 6.2], both 67), *Nypa fruticans* (60), and *Caryota rumphiana* (48).

Ecologically, the top five multifunctional palms cited above are widespread in New Guinea, and the wild origins of two of them (*Areca catechu* and *Cocos nucifera*) are unknown. This raises important, unresolved questions such as: to what degree is the distribution of multifunctional palms a product of historical management practices? Are palms generally as abundant (and therefore available to humans) in New Guinea as in other tropical rainforests of Asia or the Americas? Many more field ecological studies are needed across New Guinea to better understand how palm distribution patterns relate to human utilisation. Given the floristic uniqueness of New Guinea and the high diversity and local abundance of other multifunctional taxa critical to local livelihoods (e.g., Pandanaceae), these questions should ideally be addressed using multitaxon and trait-based approaches.

Let us now highlight certain preferences for useful palms in New Guinea. Of the 35 palm species used for **Thatch**, sago (*Metroxylon sagu*) is often preferred (Fig. 6.2). Thatch with sago leaves is geographically widespread, in communities ranging from the Kamoro people in the Arafura Sea of southwest New Guinea to the Titan people of Manus Island in eastern New Guinea. Undoubtedly, the size of sago leaves, its durability, and its local abundance play a key role in driving such preferences. For **Human Food**, sago is also one of the most cited species as its stem yields a starch that is crucial for food security in New Guinea (see Box 6.2). Six palm species are reported to have **Toxic uses** (e.g., skin irritants, fish poisons). These include *Areca catechu*, *Arenga microcarpa*, *Caryota rumphiana*, *Corypha utan*, *Hydriastele pinangoides* and *Orania palindan*. Fourteen species are reported to have **Medicinal and veterinary uses**, with just five species accounting for 89% of all medicinal reports: *Areca catechu* (70 use reports), *Cocos nucifera* (67), *Metroxylon sagu* (19), *Nypa fruticans* (15), and *Corypha utan* (14). The most frequently cited palm remedies in New Guinea are used to treat the **Digestive system** and **Skin and subcutaneous tissue** (41 use reports in each).

A
B
C
D

Figure 6.1. (OPPOSITE) Palm uses – utensils and tools.
A. Spear made from *Saribus woodfordii*, Sudest Island. **B.** Rucksack made from rattan (*Calamus*), Wandamen Peninsula.
C. Ceremonial adornment made from *Arenga microcarpa* fibres, near Finschhafen.
D. Cooking sago with *Linospadix albertisianus* sticks, Bewani.
Photos: AB (A), WB (B), JLD (C, D).

Figure 6.2. Palm uses – construction.
A. House built with sago leaf (*Metroxylon sagu*) thatch and palm stems, Torricelli Mountains.
B. Roof construction showing sago leaf (*Metroxylon sagu*) thatching panels and rattan (*Calamus*) binding, Ramu River.
C. Shelter thatched with *Saribus surru* leaves, Torricelli Mountains.
D. *Hydriastele* stems used for flooring, Morobe.
Photos: AB (A, C), WB (B), JLD (D).

BOX 6.1 Betel nut palm (*Areca catechu*)

Culturally, few palms stand out in Papuan social life as much as the betel palm. As already noted by Miklouho-Maclay in 1885, the betel palm was "cultivated in every village" (Miklouho-Maclay 1885). This remains the case to this day. The seeds of the betel palm (i.e. the betel "nut", which is in fact a seed rather than a nut) are chewed with inflorescences of *Piper betle* and lime powder (typically made from burnt coral) throughout the island as a stimulant. But widespread consumption has health consequences: Papua New Guinea has the world's highest incidence of lip/oral cancer (Johnson *et al.* 2011) because betel nut contains carcinogenic nitrosamines (Secretan *et al.* 2009). Globally, betel nut is the fourth most widely "abused" substance after caffeine, nicotine and alcohol (Norton 1998, Gupta & Warnakulasuriya 2002). See taxonomic account for further information on uses.

Betel nut (*Areca catechu*). A. Betel nut stall in Sorong market, showing fresh and dried, sliced betel nut for sale. **B.** Betel nut fruits, one fruit opened to show seed ("nut"). **C.** Betel pepper (*Piper betle*) and powdered lime (in background). **D.** *Areca macrocalyx* fruit for sale as an alternative to *A. catechu*, Finschhafen. Photos: LG (A–C), WB (D).

Much more research is needed to understand palm use patterns across New Guinea. Although palm ethnobotanical observations date back to the 19th century, the growth in ethnobotanical field studies only took off after 1975 (Fig. 6.3). Since 2001, ethnobotanical research in Indonesian New Guinea has nearly tripled, while in Papua New Guinea, it decreased considerably. Overall, most administrative areas in both countries still have a very low number of associated references and herbarium specimens with ethnobotanical information (Fig. 6.4). Culturally, just 60 or 5% of New Guinea's 1,100 indigenous groups (Simons & Fennig 2018) are represented in the literature, and most have been superficially studied: just 16 groups report over 10 palm uses, 24 groups report 2–5 uses, and 18 groups report only one use. Most studies deal with cultures whose languages are not endangered (Cámara-Leret & Dennehy 2019b), so it is urgent to prioritise future documentation efforts on threatened languages before they vanish. Also, many of New Guinea's large tribal groups are surprisingly absent from the literature or severely under-documented. Therefore, even research with large non-endangered cultures will yield much novel information. For example, it is puzzling that not a single palm ethnobotanical study exists for large groups like the Asmat of Indonesia's Papua province, who number ca. 70,000 and occupy an accessible coastal area of 20,000 km^2.

In summary, New Guinea remains the least known tropical region in palm ethnobotanical research in the world. This guide to the taxonomy, distribution, vernacular names, and uses of its palms represents a critical synthesis to facilitate more research in a region of unparalleled biocultural diversity.

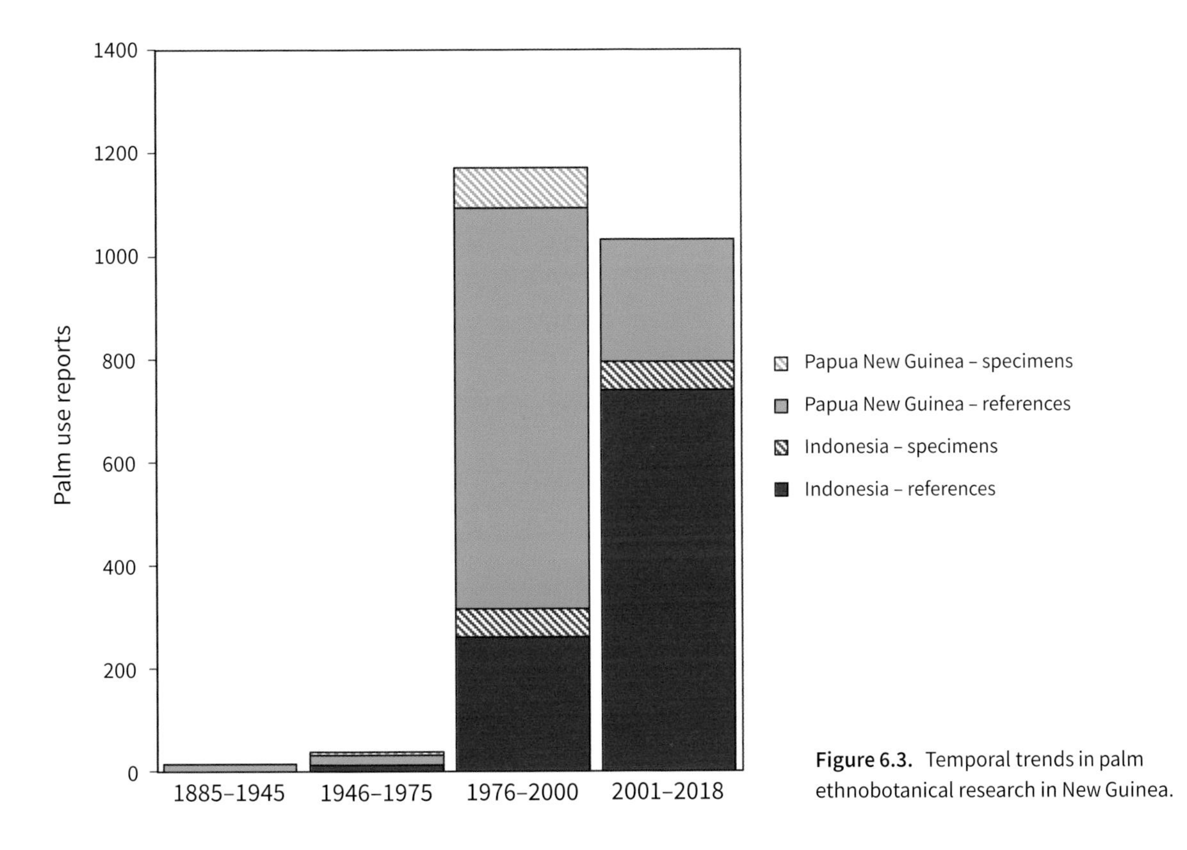

Figure 6.3. Temporal trends in palm ethnobotanical research in New Guinea.

BOX 6.2 Sago palm (*Metroxylon sagu*)

New Guinea is the centre of genetic diversity of sago, where it covers over 6 million hectares in mostly swampy lowland regions (Flach 1997, Kjaer *et al.* 2004). Sago has a life span of 11 to 20 years, and the stem is harvested for its starch before flowering commences. New Guinea's inhabitants recognize up to 20 varieties based on the pattern and shape of spines, leaflet and fruit morphology, and starch colour (Jong 1995, Kjaer *et al.* 2004). They selectively fell the largest trunks, split them lengthwise and then chip the soft pith from within using a special axe. Water is used to wash the starch from the pith. The starch settles to the bottom of the water, which is then poured away, and the mass of wet starch is packaged into parcels made of sago leaves. For consumption, boiled water is poured over the starch and this mix is stirred, yielding a glue-like porridge (see also Fig. 6.1). The starch can also be baked as cakes, cooked inside a

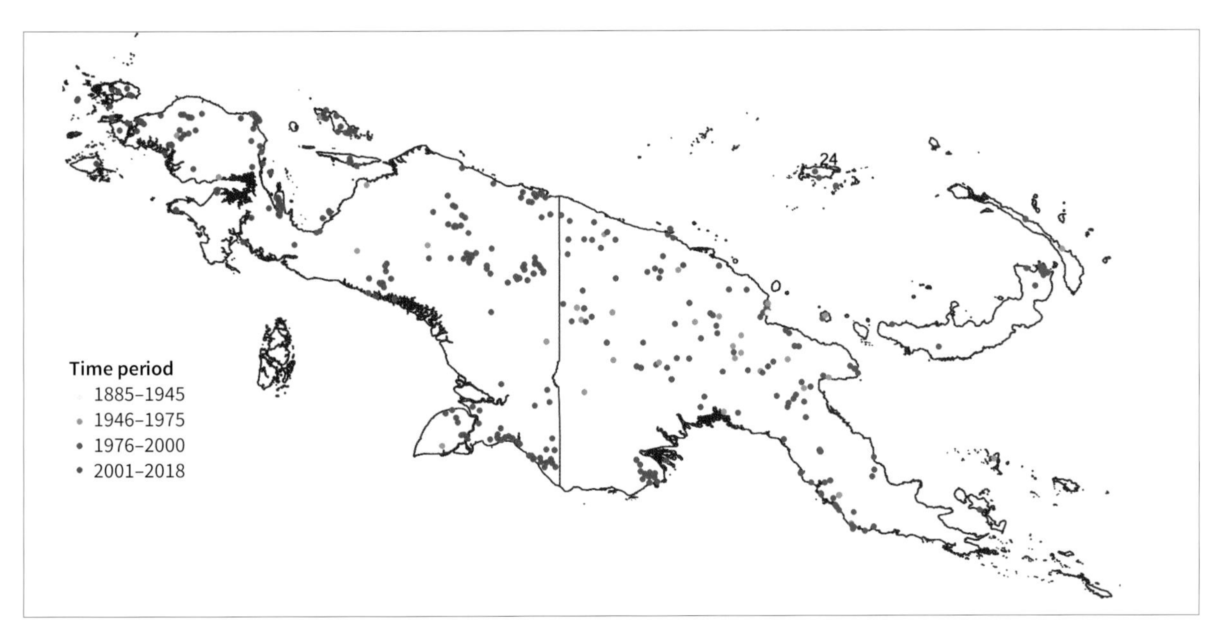

Figure 6.4. Spatiotemporal distribution of palm ethnobotanical research in New Guinea, as indicated by the geolocations of use reports on specimens and in literature.

bamboo culm, or rolled in hot ashes and consumed after peeling the black burnt layer. If dried, sago can be stored, but storage for longer than five weeks can lead to significant fungal contamination (Greenhill *et al.* 2007). A wild individual contains on average 150 kg of dry starch, and roughly 150–160 hours of work are needed to produce enough starch for one person for a whole year (Ohtsuka 1983). Such is the importance of sago as a carbohydrate source in New Guinea, that all known hunter-gatherer societies rely on it (Roscoe 2002). Sago, therefore, offers a unique window for cross-cultural research in New Guinea on palm use and management. See taxonomic account for further information on uses.

Sago extraction. A. Sago pith being chipped from a cut stem of *Metroxylon sagu* using a special axe. **B, C.** Typical method of extracting sago starch. Chipped sago pith is placed in an inclined sago leaf sheath and water poured over it (B). The starch-laden water is then squeezed from the pith (C) through a woven sieve and runs down the leaf sheath into a second sago leaf sheath, where the sago starch settles. **D–F.** Alternative method of extracting sago starch. Chipped starch is placed in a fibre bag on a frame made of sago petioles above a basin made from a moulded leaf sheath from a large tree palm (probably *Hydriastele costata*). Water is poured into the bag (D) and squeezed out using feet (E). The sago starch settles to the bottom of the basin (F).

Sago use examples. G. Sago porridge (*papeda*) served with yellow fish curry (*kuah ikan kuning*), Manokwari. **H.** Sago cakes (*sagu lempeng* or *sagu porna*; pink) with cassava cakes (white) in Sorong Market. Photos: WB (A–G), LG (H).

Road and town development at Fef in the Tamrau Mountains. Photo: WB.

7.
EXTINCTION RISK AND NEW GUINEA PALMS

New Guinea is the most extensive wilderness area remaining in the Asia-Pacific region and, alongside Amazonia and the Congo Basin, supports one of the three most significant remaining rainforests in the World (Mittermeier *et al.* 2003, Brooks *et al.* 2006, Shearman & Bryan 2011). Though not highlighted among the well-known "biodiversity hotspots for conservation priorities" of Myers *et al.* (2000) because of its high intactness, New Guinea's wilderness is an urgent priority for preservation due to its irreplaceable biodiversity and the ecosystem services that its vast forests provide globally.

Remaining vegetation and forest loss

It is estimated that 70–75% of New Guinea's vegetation remains intact (Mittermeier *et al.* 2003, Cámara-Leret *et al.* 2019). Papua New Guinea is more disturbed than Indonesian New Guinea. The Papua New Guinea National Forest Inventory determined that 78% of the country was forested in 2015, but almost 24% of the total forest area had already been degraded by human activity (Gamoga *et al.* 2021). Earlier estimates (Shearman *et al.* 2008, Bryan & Shearman 2015) are lower (71%) but this primarily reflects differences in methodology. In contrast, 83% of Indonesian New Guinea supports old-growth forest, amounting to 42% of Indonesia's remaining forests (Gaveau *et al.* 2021). Indonesian New Guinea's forests are larger and less disturbed overall than the forests of Papua New Guinea. Among the diverse forest types found in New Guinea, which are mostly lowland, the extensive remaining peat swamp and mangrove forest are of particular global importance (Miettinen *et al.* 2011, Gaveau *et al.* 2021).

New Guinea's wilderness is undergoing rapid change (Shearman *et al.* 2009). Between 2000 and 2015, almost 7% of Papua New Guinea was deforested or degraded (Gamoga *et al.* 2021). Forest transformation is uneven across Papua New Guinea, with significant hotspots in the lowlands of the Bismarck Archipelago, the Papuan Islands and the south-east of the mainland (Shearman & Bryan 2011). The lowlands are the most heavily affected due to their vast area and easier accessibility, but some upland areas are also heavily impacted, notably the Huon Peninsula and the Adelbert Range. In Indonesian New Guinea, forest loss amounted to 2% of land area between 2000 and 2019 (Gaveau *et al.* 2021). Unlike the study of Gamoga *et al*, this study does not differentiate between degradation and deforestation, but heavy degradation is included in forest loss.

Drivers of forest loss and degradation

The leading drivers of deforestation and forest degradation are commercial logging, industrial plantation (primarily oil palm) and human subsistence (Shearman *et al.* 2009, Gamoga *et al.* 2021, Gaveau *et al.* 2021). Much of New Guinea is marginal for oil palm cultivation and yet expansion of oil palm plantation is widely planned. In Papua New Guinea, extensive logging has occurred on the pretext of proposed oil palm estates that are then not planted (Nelson *et al.* 2014, Gamoga *et al.* 2021). In Indonesian New Guinea, more than 2.6 million hectares have been slated for industrial plantation, mostly for oil palm, but also for wood pulp; three quarters of this land is currently forest (Gaveau *et al.* 2021). Road construction and growth of population centres are also key drivers (Fig. 7.1). The construction of the Trans-Papua Highway has now opened up large parts of Indonesian New Guinea and has intruded into protected areas. Several new roads are planned in Papua New Guinea, such as the Epo-Kikori link, the Madang-Baiyer link and the Wau-Malalaua link, posing threats to rainforests and peatlands (Alamgir *et al.* 2019). The link between roads and future deforestation is well-established and a grave concern in New Guinea.

Mining is also notorious in the diversity and scale of environmental impacts (e.g. PT Freeport Indonesia and Ok Tedi gold and copper mines; Porgera, Lihir, Hidden Valley, Simberi and Tolukuma gold mines; Ramu nickel mine), with mine workings, tailings and pollutants devastating landscapes, especially watersheds, through poisoning, flooding and tailings (Fig. 7.1, Frazier 2007). Though generally less impactful than mining, oil and gas extraction is also a significant industrial driver in some of the island's most fragile environments (e.g. Tangguh gas field in Bintuni Bay; Frazier [2007]). Fire, which has long been used by humans for hunting and subsistence agriculture, is a growing

Figure 7.1. Examples of drivers of forest loss in New Guinea. **A.** Tailings from the PT Freeport Indonesia gold and copper mine drown lowland swamp forest on the Aikwa River. **B.** Road construction in the Tamrau Mountains opens up pristine forest to human impacts. Photos: WB.

concern as climate change progresses, especially in El Niño years, promoting both natural and anthropogenic burns (Frazier 2007, Shearman & Bryan 2011). Climate change may also intensify otherwise natural processes such as landslides and tornadoes.

Beyond large-scale habitat destruction, human activity can have more insidious impacts that reshape and degrade forest. Of particular concern is large-scale defaunation, resulting from habitat loss and targeted hunting, which can have devastating consequences for dependent organisms (Dirzo *et al.* 2014, Lim *et al.* 2020). Defaunation has significant consequences for plants, for example through the disruption of plant-frugivore networks, impacting seed dispersal and recruitment, or intensified seed predation due to uncontrolled population growth of seed predators when carnivores are lost (Galetti *et al.* 2015, Heinen *et al.* 2023). The lack of basic data on frugivory and seed dispersal in New Guinea is a major obstacle to understanding these threats. Introduced species are also a grave concern; for example the invasive neotropical shrub *Piper aduncum* has taken hold even in primary forest locations in New Guinea (Lepš *et al.* 2002), and feral pigs, which are embedded in New Guinea culture, play a key role in forest disturbance.

Protected areas

In Indonesian New Guinea, 21% of the land area is under formal protection (Fig. 7.2), although these reserves are under-resourced (Gaveau *et al.* 2021) and at risk as a result. As protected lands become more accessible, more active management of protected areas will become essential. In Papua New Guinea, just 2.8% of the land is formally protected (Shearman & Bryan 2011). Customary land ownership presents practical challenges to the establishment of protected areas in Papua New Guinea, and administrative delays have held up new wildlife management areas (Frazier 2007).

Despite their global importance, the outlook is bleak for New Guinea forests. However, there is cause for optimism in Indonesian New Guinea, cemented in the Manokwari Declaration (Cámara-Leret *et al.* 2019) in which the governors of both provinces (Papua Barat and Papua) pledged to protect 70% of forest cover. The westernmost province, Papua Barat, has already established a robust conservation and sustainability agenda (Heatubun 2022) in which 70% of forest cover and 50% of coastal and coral reef systems have been ratified as protected in the 2022–2041 provincial spatial plan. Exemplar breakthroughs include a provincial government review of oil palm concessions, resulting in the revoking of permits for 12 oil palm concessions (660,000 ha in total), which could save nearly 400,000 ha of virgin forest (Pemerintah Provinsi Papua Barat 2021). Leadership at this scale is yet to be seen in Papua New Guinea, where “there have been no coherent efforts to protect or sustainably manage the country’s forest estate” (Shearman & Bryan 2011).

Extinction risk of New Guinea palms

Efforts to protect New Guinea’s biodiversity are primarily focused at the ecosystem scale. However, forest degradation and loss directly affect species and populations, and understanding the distribution of threats to species can in turn inform the prioritisation of landscape protection (Darbyshire *et al.*

Figure 7.2. The Biak Utara (North Biak) reserve protects around 60 km² of pristine Biak-Numfor forest. Photo: WB.

2017). A recent assessment of the extinction risk faced by all palm species highlights New Guinea as a research and conservation priority. Many species lack extinction risk information and many species that are evolutionarily or functionally distinct, or are useful, are predicted to be threatened (Bellot *et al.* 2022).

To address the knowledge gap highlighted by Bellot *et al.* (2022), we have undertaken a complete assessment of the extinction risk faced by New Guinea's palms using the IUCN Red List Categories and Criteria (IUCN 2012). At the time of writing, 107 species of New Guinea palm were formally included in the IUCN Red List (IUCN 2023), many of them recent assessments arising from taxonomic work undertaken for the Palms of New Guinea project (e.g. Petoe *et al.* 2018a, Petoe & Baker 2019) or from the Global Tree Assessment for Papua New Guinea (Barstow *et al.* 2023). We completed preliminary assessments for all remaining species and updated some existing assessments where new data had become available (Table 7.1).

All preliminary assessments were conducted using criterion B (geographic range size *and* evidence of declining or fragmented population or habitat) due to the limitations of available data, as is common for Red List assessments for plants. Each species was assigned one of the nine IUCN categories by analysing our database of georeferenced herbarium specimens in GeoCAT (Bachman *et al.* 2011).

This online tool generates the spatial metrics Extent of Occurrence (EOO) and Area of Occupancy (AOO) that are essential for assessment under criterion B. For a species to be classed as Vulnerable (VU), Endangered (EN) or Critically Endangered (CR), their EOO and/or AOO must be below 20,000 km^2/2,000 km^2, 5,000 km^2/500 km^2 and 100 km^2/10 km^2, respectively. Because AOO is typically hugely underestimated due to low collecting effort, we primarily used EOO. To infer trends in continuing decline or fragmentation of habitat or population, we sought evidence of threats within the EOO polygon for each species using Google Earth and Global Forest Watch (2023), an online resource that provides data and tools for monitoring forests, including information about threats and protected areas. This information was essential to confirm a putative assessment based on the EOO alone.

TABLE 7.1. IUCN extinction risk assessments for New Guinea palms. These figures combine both published assessments accessible through the IUCN Red List (IUCN 2023) and preliminary assessments prepared for this book.

EXTINCTION RISK ASSESSMENT	SPECIES	%
Extinct in the Wild (EW)	1	0.4
Critically Endangered (CR)	50	20
Endangered (EN)	30	12
Vulnerable (VU)	10	4
Near Threatened (NT)	15	6
Least Concern (LC)	109	43
Data Deficient (DD)	37	15

We estimated that 90 (36%) of New Guinea palms are threatened with extinction (i.e. fall into categories VU, EN or CR), 20% of which are CR (Table 7.1). All but one of the threatened species are endemic. This percentage matches that of Papua New Guinea's endemic trees (Barstow *et al.* 2023) and is broadly consistent with global expectations for plants as a whole (Nic Lughadha *et al.* 2020). If all 37 (15%) Data Deficient (DD) palm species were found to be threatened, the total number of threatened species would reach 127, or just over 50%. All DD species are endemic. One species, *Ptychosperma sanderianum*, is thought to be Extinct in the Wild (EW), while 109 species (43%) are Least Concern (LC) and 15 (6%) Near Threatened (NT). Of the 35 native non-endemic species, all but one are LC (*Areca novohibernica* is Endangered), while 71 endemic species are rated as LC. The only comparable island study of palms found that 83% of 192 endemic species in Madagascar are threatened (Rakotoarinivo *et al.* 2014), reflecting the advanced state of habitat destruction in Madagascar and offering a sobering warning for New Guinea if deforestation continues unabated.

Threats and solutions

The processes that place New Guinea palm species at risk of extinction are the same as those driving habitat loss throughout the island. For example, *Calamus maturbongsii* (CR) is known mainly from logging concessions and is directly threatened by commercial timber extraction. *Wallaceodoxa raja-ampat* (CR) is recorded in few, small populations in the Raja Ampat Islands that are at risk from forest clearance for subsistence agriculture and urban expansion. *Heterospathe parviflora* (EN; Turner *et al.* 2021) is endemic to New Britain where forest conversion to oil palm plantations is likely to result in the clearing of all forest below 200m by 2060; fortunately this species also occurs at higher elevations. *Cyrtostachys bakeri* (CR; Jimbo & Kipiro 2021) is known from very few individuals in one site below the Ok Tedi gold, silver and copper mine. The only known locality is at risk of burial by mine tailings. *Cyrtostachys excelsa* (EN) grows alongside major road developments in the Tamrau Mountains that are likely to lead to roadside habitat degradation (Fig. 7.3).

Figure 7.3. *Cyrtostachys excelsa* (Endangered) is threatened by habitat degradation due to road building in the Tamrau Mountains. Photo: WB.

With a flora as rich as New Guinea's facing a wide diversity of threats, conservation action is a daunting prospect. Large-scale habitat protection opportunities still remain and will be the most effective mechanism for species conservation. The commitments made by both Papua New Guinea and Indonesia to take action under United Nations biodiversity and climate agreements and to drive initiatives such as the Manokwari Declaration suggest that the aspiration to value and protect forests exists. Only time will tell whether this can be reconciled with the conflicting demands of communities, development and the global hunger for New Guinea's natural resources (Cámara-Leret *et al.* 2019, Barstow *et al.* 2023). Papua New Guinea faces particular challenges here, given its poor track record in placing its forests under protection.

Alongside the urgent need to prioritise large-scale forest protection, species-level actions remain a crucial part of the conservation arsenal. The horticultural appeal of palms creates prospects for *ex situ* conservation opportunities. Rich palm collections exist in many botanic gardens around the world, although meaningful conservation collections are rare. Botanic gardens, especially those within the region, could play a vital role by establishing conservation populations of the most threatened species. The extensive community of amateur palm horticulturists could also materially contribute. Although collections of single specimens are of limited conservation value, a distributed *ex situ* palm conservation network could be established that capitalises on amateur horticultural capacity. For example, the Critically Endangered Malagasy palm *Tahina spectabilis* has, in a matter of a few years, been secured in cultivation through a legally implemented wild harvest and distribution of seed. If this approach were to be more actively managed through centralised standards, record keeping and reporting, it could become a highly effective conservation measure for some of the most threatened palms in the world. Seed banking as an alternative approach to *ex situ* conservation has limited potential as so many palm seeds are desiccation intolerant and therefore difficult to store, especially those from humid environments (Dickie *et al.* 1993).

Impactful, *in situ* conservation actions for individual species are harder to envisage. Barstow *et al.* (2023) outline a roadmap for the conservation of Papua New Guinea endemic tree species that is applicable to palms throughout New Guinea, which calls for the empowerment of society and local communities in their conservation, and the restoration and reinforcement of threatened populations. The first step of their roadmap, however, is the generation of information on endemics. This book addresses this requirement directly by providing the essential underpinning taxonomy, distribution data and natural history information for all New Guinea palms. We hope that our work will lay the foundations for the long-term preservation of this immeasurably valuable group of plants.

OVERLEAF: *Pigafetta filaris*, near Sorong. Photo: WB.

TAXONOMIC ACCOUNTS

Sommieria leucophylla, near Timika (WB).

KEY TO GENERA

1. Leaves palmate 2
1. Leaves pinnate or bipinnate 6

2. Stems typically <11 cm diam., leaf blade usually split to base forming wedge-shaped segments (widespread) . . . . *Licuala* (p. 234)
2. Stems typically >12 cm diam., leaf blade not split to base and lacking wedge-shaped segments . . . 3

3. Petiole base lacking large triangular cleft 4
3. Petiole base with large triangular cleft 5

4. Inflorescences comprising a single main axis, fruits purplish or bluish black (seasonal southern New Guinea) . . . . *Livistona* (p. 228)
4. Inflorescences branched at the base into 2–3 equal primary axes enclosed within a single prophyllar bract, fruit red to orange or brown, rarely blackish (widespread) . . . . . . *Saribus* (p. 288)

5. Petiole not spiny, inflorescences between leaves, rachillae thick, >2 cm diam., fruit ≥10 cm long, stems not dying after flowering (northern, lowland New Guinea) . . . . *Borassus* (p. 328)
5. Petiole spiny, inflorescences above leaves, rachillae thin, <0.5 cm diam., fruit ≤2 cm long, stems dying after flowering (seasonal southern New Guinea) . . . . *Corypha* (p. 324)

6. Leaves bipinnate (widespread) . . . . *Caryota* (p. 310)
6. Leaves pinnate 7

7. Climbing palms 8
7. Short to tall palms, not climbing 9

8. Leaflets diamond-shaped, tips jagged, stems dying after flowering (widespread) . . . . *Korthalsia* (p. 62)
8. Leaflet not diamond-shaped, tips pointed, stems not dying after flowering . . . . *Calamus* (p. 80)

9. Crownshaft present 10
9. Crownshaft absent 29

10. Leaflet tips jagged 11
10. Leaflet tips pointed, lobed or toothed, but not jagged 17

11. Inflorescence branches swept forward (or rarely unbranched), resembling a brush or horse's tail (widespread) . . . . *Hydriastele* (in part; p. 614)
11. Inflorescence branches widely spreading 12

12. Fruit >3 cm long, endocarp with conspicuous grooves and ridges 13
12. Fruit ≤3 cm long, endocarp smooth, grooved or ridged. 14

13. Stem diam. ≤7 cm, leaflets wedge-shaped, ripe fruit orange, endocarp pale (central to far eastern New Guinea) . . . . *Brassiophoenix* (p. 536)
13. Stem diam. ≥9 cm, leaflets linear, ripe fruit red, endocarp black (widespread) . . . . *Ptychococcus* (p. 542)

14. Stilt roots present, prophyll not dropping off as inflorescence expands (far western New Guinea) . . . . . . *Drymophloeus* (p. 528)

14. Stilt roots absent, prophyll dropping off as inflorescence expands 15

15. Moderately robust palms, inflorescences white (Biak) . . . . . . *Manjekia* (p. 550)

15. Slender to moderate palms, inflorescences variously coloured, but not white 16

16. Leaves not recurved, inflorescences variously coloured (green, yellow, pink, purple), endocarp usually grooved, at least obscurely so (widespread) . . . . . . *Ptychosperma* (p. 480)

16. Leaves recurved, inflorescence axes green, endocarp smooth, weakly 5-grooved (New Britain) . . . . . . *Ponapea* (p. 524)

17. Mid-storey to canopy palms, stem usually >12 cm diam. 18

17. Undergrowth to mid-storey palms, stem usually ≤12 cm diam. 23

18. Inflorescence branches swept forward, resembling a brush or horse's tail 19

18. Inflorescence branches widely spreading 20

19. Fruit symmetrical, stigma at apex (widespread) . . . . . . *Hydriastele* (in part; p. 614)

19. Fruit asymmetrical, stigma to one side of apex (New Ireland) . . . . . . *Clinostigma* (p. 558)

20. Leaf strongly arching, flowers not in pits, fruit >5 cm long, red when ripe (widespread, often cultivated) . . . . . . *Actinorhytis* (p. 390)

20. Leaf not strongly arching, fruit ≤3.5 cm long, red, yellow or black when ripe 21

21. Flowers usually in pits, fruit ≤1.2 cm long, black when ripe, endosperm homogeneous (widespread). . . . . . . *Cyrtostachys* (in part; p. 562)

21. Flowers not in pits, fruit 1.5–3.5 cm long, yellow or red when ripe, endosperm ruminate 22

22. Leaf rachis with short, dark hairs, inflorescence peduncle to ≤10 cm (widespread) . . . . . . *Rhopaloblaste* (in part; p. 672)

22. Leaf rachis (and petiole and upper leaf sheath) with thick, white, woolly indumentum, interspersed with dark, twisted fibres, inflorescence peduncle to ≥15 cm (Raja Ampat Islands) . . . . . . *Wallaceodoxa* (p. 554)

23. Some leaflets with lobed tips, mainly at leaf apex 24

23. All leaflets pointed, not lobed 25

24. Female flowers and fruits at base of inflorescence branches only, fruiting inflorescence often club-like (widespread) . . . . . . *Areca* (p. 396)

24. Female flowers and fruits from base to tip of inflorescence branches, fruiting inflorescence not club-like (western and central New Guinea) . . . . . . *Pinanga* (p. 412)

25. Single-stemmed palms, fruit yellow or red (widespread, except eastern New Guinea) 26

25. Single- or multi-stemmed palms, fruit black 27

26. Crownshaft with brown to grey indumentum, leaflets ≥59 each side of rachis, inflorescence peduncle ≤10 cm long, fruit ellipsoid or ovoid (widespread, except eastern New Guinea) . . . . . . *Rhopaloblaste* (in part; p. 672)

26. Crownshaft with powdery white indumentum, leaflets ≤56 each side of rachis, inflorescence peduncle ≥27 cm long, fruit globose or subglobose (south-eastern New Guinea, D'Entrecasteaux Islands) . . . . **Heterospathe** (in part, *H. barfodii*; p. 582)

27. Stilt roots present (New Britain) . . . . **Physokentia** (p. 414)
27. Stilt roots absent. 28

28. Single- or multi-stemmed palms, stem diam. ≥5 cm, flowers in pits, endosperm homogeneous (widespread) . . . . **Cyrtostachys** (in part; p. 562)
28. Usually multi-stemmed, stem diam. ≤5 cm, flowers not in pits, endosperm ruminate (far western New Guinea) . . . . **Dransfieldia** (p. 578)

29. Leaflet tips jagged. 30
29. Leaflet tips pointed, lobed or toothed, but not jagged 31

30. Multi-stemmed, stems dying after flowering, leaflets V-shaped in section (widespread) . . . . **Arenga** (p. 318)
30. Single-stemmed, stems not dying after flowering, leaflets Λ-shaped in section (widespread) . . . . **Orania** (p. 334)

31. Leaflets, rachis, petiole or sheath spiny, fruit scaly 32
31. Spines absent, fruit smooth or with corky warts 33

32. Single- or multi-stemmed, stems dying after flowering, inflorescences above leaves, fruit ≥2 cm long (widespread, often cultivated) . . . . **Metroxylon** (p. 68)
32. Single-stemmed, stems not dying after flowering, inflorescences between or below the leaves, fruit ≤1 cm long (western New Guinea). . . . .**Pigafetta** (p. 74)

33. Stems horizontal, branching by forking, abundant in mangrove forest (widespread). . . . .**Nypa** (p. 224)
33. Single- or multi-stemmed, stems not horizontal, rarely lacking, various habitats, but not mangrove 34

34. Single-stemmed, stem erect, to 30 cm diam., coastal, cultivated for coconuts (widely cultivated) . . . . **Cocos** (p. 382)
34. Single- or multi-stemmed, stem erect or rarely stemless, to 10 cm diam 35

35. Inflorescence unbranched (unless branching at very base) 36
35. Inflorescences branched 37

36. Peduncular bract borne at top of peduncle (widespread) . . . . **Linospadix** (p. 476)
36. Peduncular bract borne at towards base of peduncle, not at top (widespread) . . . . **Calyptrocalyx** (p. 418)

37. Leaf pinnate, rarely entire bifid, lower surface green, peduncular bract borne near base or tip of peduncle, fruit smooth (widespread) . . . . **Heterospathe** (in part; p. 582)
37. Leaf entire bifid, or divided into 2–4 leaflet pairs, lower surface white, peduncular bract at top of peduncle, fruit with corky warts (western to northern central New Guinea). . . . **Sommieria** (p. 386)

Korthalsia zippelii. Habit, near Sorong (WB).

CALAMOIDEAE | KORTHALSIINAE

Korthalsia Blume

Synonym: *Calamosagus* Griff.

Climber – leaf pinnate – spiny – leaflets diamond-shaped – leaflets jagged

Moderately robust, *multi-stemmed climbing palm, individual stems dying after flowering*, hermaphrodite. **Leaf** pinnate, armed with spines and bristles, arching; sheath tubular, usually very spiny; petiole present; *leaflets few, diamond-shaped*, with jagged tips, arranged regularly, pendulous; *spiny climbing whips arising from leaf tips (cirrus)*. **Inflorescence** above the leaves, branched to 2 orders, erect to arching; primary bracts similar, not enclosing inflorescence in late bud, not dropping off as inflorescence expands, remaining tubular, usually not spiny; peduncle shorter than inflorescence rachis; *rachillae robust, sausage-like*, straight or curved, with conspicuous bracts. **Flowers** solitary throughout the length of the rachilla, developing in pits formed by rachilla bracts. **Fruit** orange-brown, ovoid, with vertical rows of scales, stigmatic remains apical, flesh thick and pulpy. **Seed** 1, not enclosed in a fleshy coat (sarcotesta), ellipsoid, endosperm homogeneous.

DISTRIBUTION. Twenty-seven species from India to Vietnam, through South-East Asia to New Guinea, where a single species occurs.

NOTES. *Korthalsia* is the smaller of the two rattan genera present in New Guinea. It is immediately distinguished from the larger rattan genus *Calamus* by its diamond-shaped leaflets with jagged apical margins. The leaf tip of all species bears a climbing whip (cirrus). All species are hapaxanthic, meaning stems die after flowering, although being multi-stemmed this does not usually result in the death of the entire plant. In New Guinea, young plants of *Korthalsia* that are yet to climb and that display juvenile bifid leaf morphology might be confused with *Sommieria*, as both have chalky indumentum on the leaf undersurface, but *Sommieria* is easily distinguishable because it is unarmed.

A synopsis of the genus was published by Dransfield (1981) and a partial monograph has also been completed (Shahimi *et al.* 2019). A full account of the genus throughout its range is needed.

Korthalsia zippelii Blume

Synonyms: *Ceratolobus plicatus* Zipp. ex Blume, *Ceratolobus zippelii* Blume, *Korthalsia brassii* Burret, *Korthalsia zippelii* var. *aruensis* Becc.

Moderate to robust, multi-stemmed rattan climbing to 25 m. **Stem** with sheaths (12–)20–40 mm diam., without (7–)15–26 mm diam.; internodes 16–26 cm. **Leaf** cirrate, to 2.6 m long including cirrus and petiole; sheath pale green with white indumentum, spines few to numerous, 2–11 × 0.5–2 mm, triangular, brown, solitary or in small groups; *knee absent*; *ocrea 8–30 cm long, varying continuously from strictly tubular and disintegrating into a fibrous mesh to leathery and splitting longitudinally on side away from petiole*, brown, usually armed with fine, hair-like spines to 25 mm, disintegrating or persistent; petiole 16–40 cm; *leaflets 7–12 each side of rachis*, regularly arranged, diamond-shaped, mid-leaf leaflets 20–45 × 7–15 cm, unarmed, *with chalky white indumentum on undersurface*; cirrus 1–1.5 m. **Inflorescence** 24–62 cm long including 2–7 cm peduncle, 4–7 inflorescences produced simultaneously above the leaves, branched

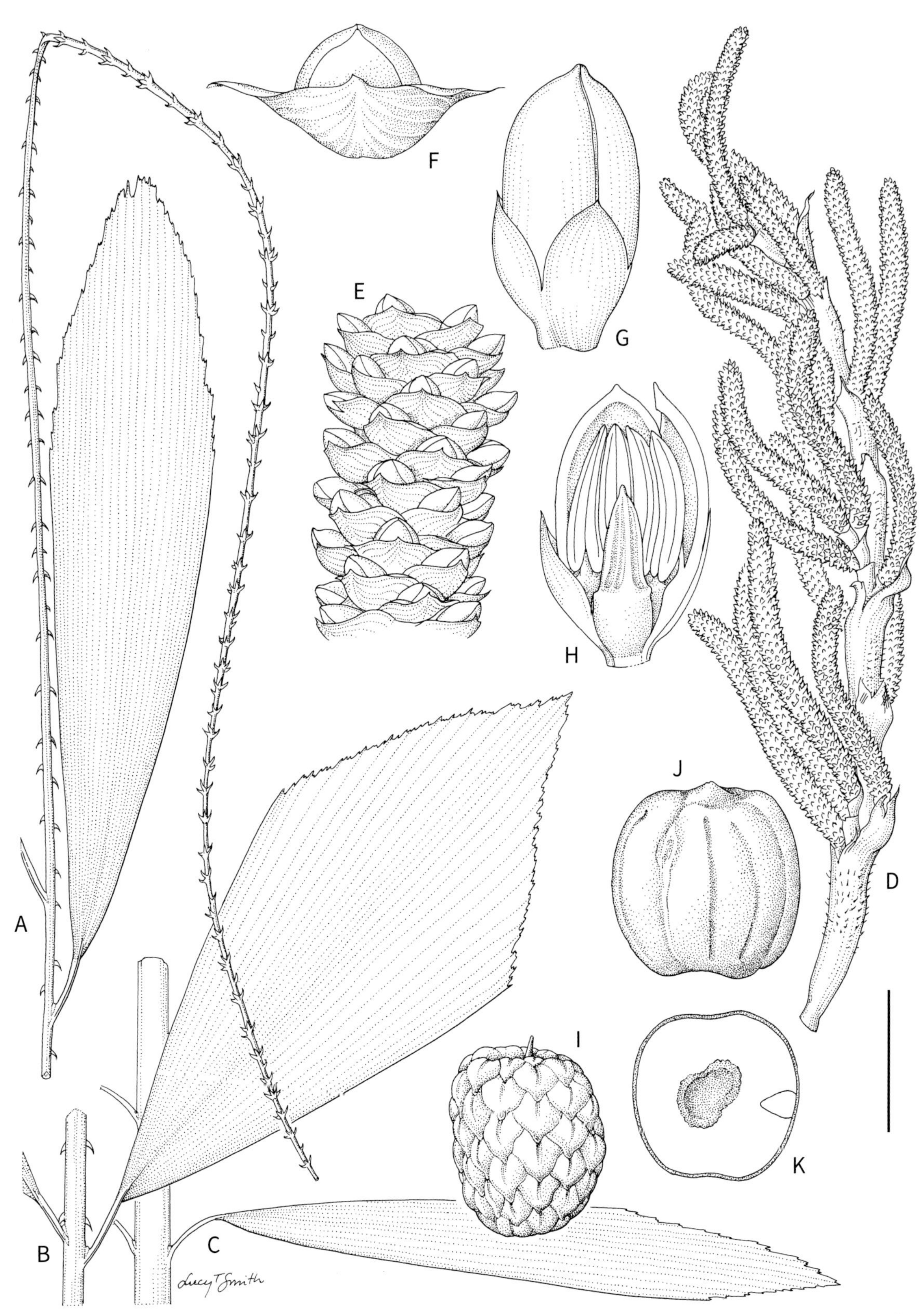

Korthalsia zippelii (Plate 1). **A.** Leaf apex with cirrus. **B.** Leaf mid-portion. **C.** Leaf base. **D.** Inflorescence. **E.** Portion of rachilla with flowers. **F.** Hermaphroditic flower in bract. **G, H.** Hermaphroditic flower whole and in longitudinal section. **I.** Fruit. **J, K.** Seed whole and in longitudinal section. Scale bar: A–D = 6 cm; E, I = 1 cm; F = 3.3 mm; G, H = 3 mm; J, K = 7 mm. A–C, I–K from *Maturbongs 262*; D–H from *Maturbongs et al. 667*. Drawn by Lucy T. Smith.

to 2 orders (very rarely 3); primary bracts funnel-shaped, somewhat inflated, unarmed or lightly armed as sheath; primary branches 7–11, to 25 cm long, rachillae 4–20 × 0.5–1.6 cm. **Fruit** obovoid or irregularly globose, 11–14 × 9–12 mm, scales pale yellow-brown. **Seed** 6–8 × 6.5–8 mm, globose.

DISTRIBUTION. Widespread throughout New Guinea and adjacent islands, including Raja Ampat and Aru Islands. Not yet recorded from the Bismarck Archipelago (except Manus Island) or east of mainland Milne Bay Province, which is the most easterly record for the genus.

HABITAT. Primary and secondary forest, including disturbed areas, 0–900 m

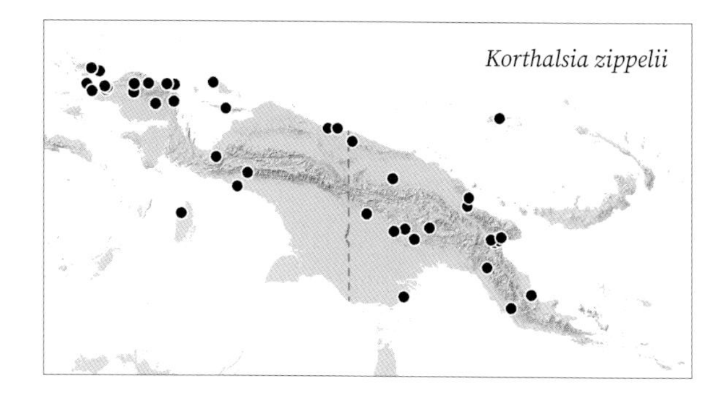

Korthalsia zippelii. LEFT TO RIGHT: habit, near Sorong (WB); stem apex with inflorescences, near Lae (WB).

LOCAL NAMES. *Age* (Ambel), *Bukaua Aalu* (Bukawa), *Fisianei* (Tanahmerah), *Gwata* (Marap), *Hajabf Kwa Nyi* (Cyclops), *Kagi* (Kutubu), *Pepenehe* (Yamur), *Pu dengoro* (Motu), *Sel-piyndaekndaek* (Wola), *To Hayuoh Nau* (Ayawasi), *Umbu Spang* (Berap), *Wakrus* (Biak), *Wil Deh* (Batanta, Dey, Tepin).

USES. Cane used as cordage in house, fence and roof construction, for matting, as bowstring and as a source of water in the forest. Cirrus used in eel and fish traps.

CONSERVATION STATUS. Least Concern.

NOTES. *Korthalsia zippelii* is one of the commonest lowland rattans in New Guinea. The numerous specimens now available reveal that the species is highly variable, especially in ocrea and inflorescence morphology. The type of *K. brassii* falls within this general range of variation and is treated as a synonym here.

Korthalsia zippelii. RIGHT: leaf sheath obscured by fibrous ocrea, near Lae (WB); BELOW: rachillae, Madang (WB).

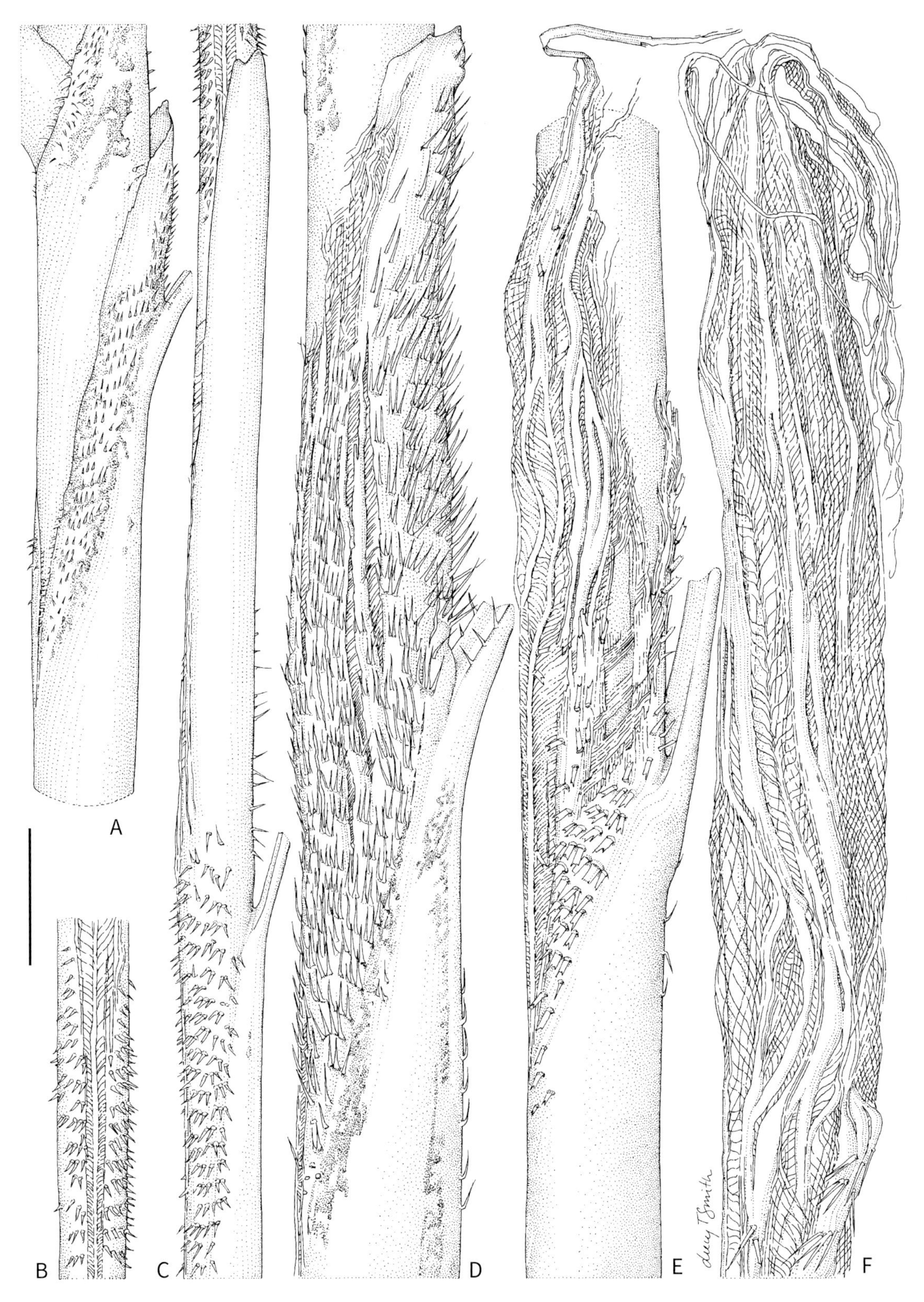

Korthalsia zippelii (Plate 2). **A–F.** Stem with leaf sheath and ocrea from five different specimens. Scale bar: A–F = 4 cm. A from *Utteridge 20*; B, C from *Upessy 5*; D from *Moore & Womersley 9267*; E from *Maturbongs 262*; F from *Dransfield et al. JD 7534*. Drawn by Lucy T. Smith.

Metroxylon sagu. Habit, flowering palms, near Lae (WB).

Metroxylon Rottb.

Synonyms: *Coelococcus* H.Wendl., *Sagus* Steck

Medium to tall – leaf pinnate – crownshaft – no spines – leaflets jagged

Robust, single- or multi-stemmed tree palm, crownshaft absent, *individual stems dying after flowering,* andromonoecious. **Leaf** pinnate, *usually armed with spines and bristles,* erect; sheath splitting to the base opposite the petiole, massive; petiole elongate; leaflets numerous, with pointed tips, arranged regularly, ± horizontal. **Inflorescence** *above the leaves, many inflorescences produced simultaneously to form a compound, terminal inflorescence,* individual inflorescences branched to 2 orders, branches widely spreading; primary bracts similar, not enclosing inflorescence in late bud, not dropping off as inflorescence expands; peduncle shorter than inflorescence rachis; *rachillae robust, sausage-like,* straight or curved, with conspicuous, densely arranged bracts. **Flowers** in pairs throughout the length of the rachilla, developing in pits formed by rachilla bracts. **Fruit** brown, globose, *with vertical rows of shiny scales,* stigmatic remains apical, flesh rather spongy and dry. **Seed** 1, enclosed in a fleshy coat (sarcotesta), globose, with sarcotesta intruding into a deep apical pit, endosperm homogeneous.

DISTRIBUTION. Seven species distributed from Maluku to the western Pacific, of which one species occurs in New Guinea. A second species (*M. salomonense*) is recorded for Papua New Guinea on Bougainville Island, but this falls outside the geographical scope of this account.

NOTES. *Metroxylon* is a robust, single- or multi-stemmed tree palm of swamp forest, usually with spiny leaves. The inflorescences are borne above the leaves and individual stems die after flowering. In New Guinea, *Metroxylon* is primarily found in lowland swamp forest at sea level, and is sometimes cultivated on swampy ground at higher elevations. It is extremely important to the people of New Guinea as the source of sago, a starch that is extracted from the stem. It is most easily confused with *Pigafetta,* which is a single stemmed canopy tree, with chalky white leaf sheaths, very small fruit (0.8–1 cm) and stems that do not die after flowering. It might also be confused with *Nypa,* which never forms an erect stem and is a palm of mangrove swamp.

The taxonomy of *Metroxylon* has been most recently revised by Rauwerdink (1985) and, in part, by McClatchey (1996).

Metroxylon sagu Rottb.

Synonyms: *Metroxylon hermaphroditum* Hassk., *Metroxylon inerme* (Roxb.) Mart., *Metroxylon laeve* (Giseke) Mart., *Metroxylon longispinum* (Giseke) Mart., *Metroxylon micracanthum* Mart., *Metroxylon oxybracteatum* Warb. ex K.Schum. & Lauterb., *Metroxylon rumphii* (Willd.) Mart., *Metroxylon sago* K.D.Koenig, *Metroxylon sagu* f. *longispinum* (Giseke) Rauwerd., *Metroxylon sagu* f. *micracanthum* (Mart.) Rauwerd., *Metroxylon sagu* f. *tuberatum* Rauwerd., *Metroxylon squarrosum* Becc., *Metroxylon sylvestre* (Giseke) Mart., *Sagus americana* Poir., *Sagus genuina* Giseke, *Sagus genuina laevis* Giseke, *Sagus genuina longispina* Giseke, *Sagus genuina sylvestris* Giseke, *Sagus inermis* Roxb., *Sagus koenigii* Griff., *Sagus laevis* Jack, *Sagus longispina* (Giseke) Blume, *Sagus micracantha* (Mart.) Blume, *Sagus rumphii* Willd., *Sagus sagu* (Rottb.) H.Karst., *Sagus spinosa* Roxb., *Sagus sylvestris* (Giseke) Blume

Robust, multi-stemmed, palm to ca. 20 m, bearing 11–16 leaves per crown. **Stem** 35–45 cm diam., internodes 8–14 cm, stem surface with short, spine-like adventitious roots. **Leaf** to 8 m long including petiole; sheath at least 1 m long, often with white indumentum, *unarmed or armed with spines up to 5 cm long, spines arranged in collars,* often detaching and leaving persistent ridges of united spine bases; petiole to 2.7 m long, armed as sheath; leaflets 60–77 each side of rachis, usually armed with marginal spines, linear elliptic; mid-leaf leaflets 122–146 × 5–11 cm wide; apical leaflets ca. 45–47 × 1–1.5 cm wide, linear, not united. **Inflorescence** up to 4 m long including peduncle to 1.5 m, widely spreading and arching upwards, *up to 27 inflorescences produced simultaneously along apical section of stem that extends up to 5 m above leaves;* primary branches up to 25 per inflorescence, to 80 cm long, with up to 16 rachillae each; rachillae 8–14 × 1.1–1.6 cm long. **Flower** 4–5 × ca. 2.5 mm in bud, green; stamens 6, orange-red. **Fruit** 30–40 × 31–46 mm diam. **Seed** 20–25 × 23–33 mm.

DISTRIBUTION. Thought to be native to Maluku and New Guinea, where it is very widespread, growing in wild and semi-cultivated or managed stands. Available specimen records (as presented on the map here) do not adequately represent the extent of the species distribution. Widely cultivated elsewhere in South-East Asia, sometimes on a plantation scale.

HABITAT. Swamp forest, including disturbed habitats, where it often occurs in vast stands, typically at very low elevations. It is reported to grow at up to 1,200 m (Ehara *et al.* 2018).

LOCAL NAMES. *Sagu, Rumbia* (Bahasa Indonesia), *Saksak* (Papua New Guinea Tok Pisin); *Metroxylon sagu* has numerous local names in New Guinea, some of which are summarised by Bintoro *et al.* (2018). For brevity, we list only a fraction here: *Aburi* (Foi), *Aiyo wo* (Kubo), *Ambe* (Kosarek Yale), *Ampehi* (Ambai), *Anane anapi* (Yawa), *Anannga* (Wandamen), *Bia* (Patpatar), *Bom* (Nobonob), *Hiywa* (Angal Heneng), *Kaniw* (Kwerba), *Keker* (Marind), *Naep* (Amanab), *Nggi* (Morori), *Õkõma* (Yamben), *Om* (Telefol), *U'tieh* (Irarutu), *Wi* (Kamoro), *Yof* (Gebe).

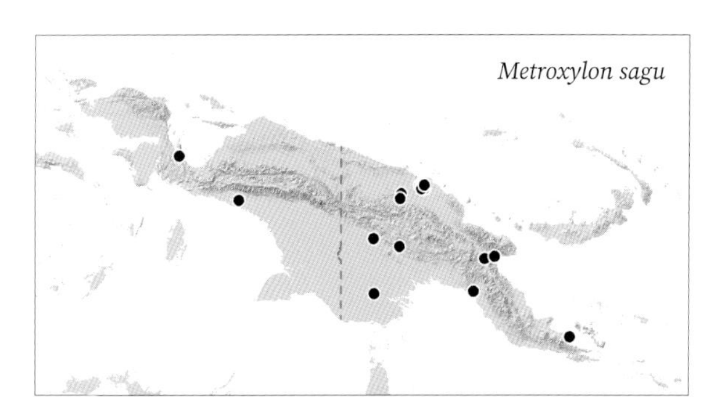

Metroxylon sagu. Habit, stems dying after flowering, Wosimi River (WB).

USES. Exceptionally important as a source of edible starch obtained from the stem, making it the main carbohydrate staple for many indigenous groups in New Guinea. The leaves are used for thatch, house walls, other construction materials, brooms, baskets, and skirts for dances. Young shoots are consumed as vegetables. The stems are used to raise edible sago grubs. Young trunks, pith and pith refuse are fed to animals. The bark of the trunk is used as timber. Sago flour is burnt and eaten to reduce stomach pain. The starch paste is also applied on skin burns. The inflorescence bracts are folded and sown to form a watertight bucket. Fruits are eaten by the northern cassowary (Pangau-Adam & Mühlenberg 2014). The roots are used to treat measles.

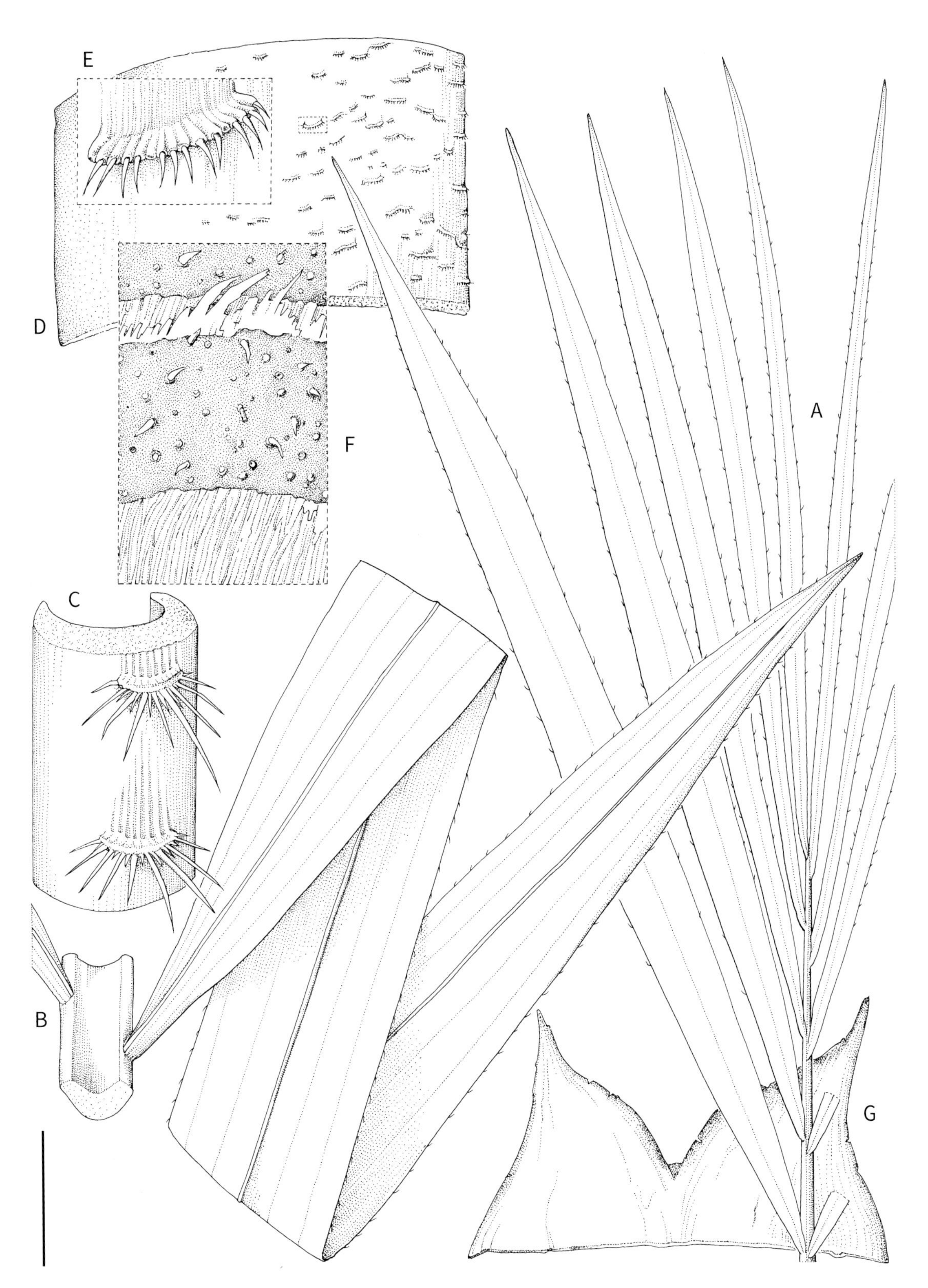

Metroxylon sagu (Plate 1). **A.** Leaf apex. **B.** Leaf mid-portion. **C.** Portion of petiole with spines. **D.** Portion of leaf sheath. **E.** Detail of leaf sheath spines. **F.** Surface of stem showing root spines. **G.** Scale leaf (reduced leaf that subtends inflorescence). Scale bar: A, B, D = 8 cm; C = 4 cm; E, F = 1.5 cm; G = 6 cm. All from *Baker et al. 881*. Drawn by Lucy T. Smith.

Metroxylon sagu. CLOCKWISE FROM TOP LEFT: leaf sheaths, Wosimi River (WB); cut stem; rachilla with flowers; fruit; inflorescence. Near Timika (WB).

CONSERVATION STATUS. Least Concern (IUCN 2018).

NOTES. The true sago palm, *M. sagu*, is fundamentally important to livelihoods in New Guinea (Ehara *et al.* 2018). The species is morphologically variable (e.g. in armature) and this, in conjunction with its economic importance, has resulted in a proliferation of names and infraspecific taxa (the synonymy provided above is only a summary of the most important names). Rauwerdink (1985) reduced all names into a single variable species, including *M. rumphii*, then distinguished as the spiny sago palm as opposed to the spineless sago palm, as *M. sagu* was interpreted at the time. He justified this mainly on the basis of well-established observations that the offspring of a spiny sago palm can be either spiny or spineless. Nevertheless, Rauwerdink formally recognised four forms with *M. sagu* corresponding to different degrees of armature, which he regarded as "only slightly different genotypes of a single species". Subsequently, however, Kjaer *et al.* (2004) demonstrated that there was no genetic correlation with armature in *M. sagu* across Papua New Guinea. Based on this, we regard *M. sagu* as a single variable species and treat all infraspecific taxa as synonyms.

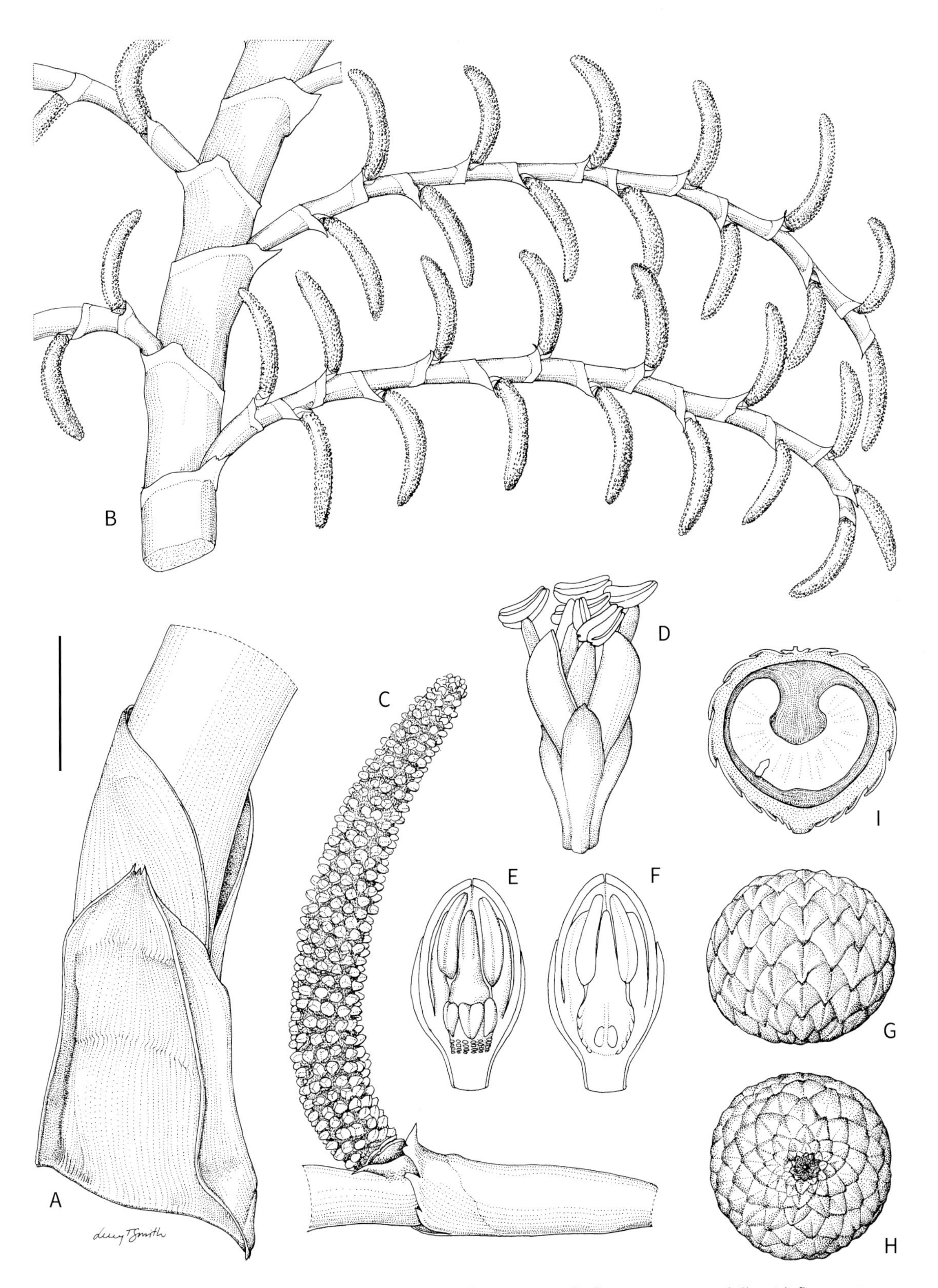

Metroxylon sagu (Plate 2). **A.** Inflorescence base with prophyll. **B.** Portion of inflorescence. **C.** Rachilla with flowers. **D.** Male flower. **E**, **F.** Hermaphroditic flower, one petal removed and in longitudinal section. **G**, **H.** Fruit in two views. **I.** Fruit in longitudinal section. Scale bar: A = 6 cm; B = 12 cm; C, I–H = 3 cm; D–F = 4 mm. All from *Baker et al. 881*. Drawn by Lucy T. Smith.

Pigafetta filaris. Habit, Biak Island (WB).

Pigafetta (Blume.) Becc.

Synonyms: *Metroxylon* section *Pigafetta* (Blume) Martius, *Sagus* section *Pigafetta* Blume

Tall – leaf pinnate – no crownshaft – spiny – leaflets pointed

Robust, single-stemmed tree palm, stem shiny green, especially in upper parts, crownshaft absent, dioecious. **Leaf** pinnate, *armed with spines and bristles, strongly arching;* sheath splitting to the base opposite the petiole, *chalky white;* petiole absent, though narrowing upper portion of sheath resembles a petiole; leaflets numerous, with pointed tips, arranged regularly, ascending. **Inflorescence** *between and below the leaves,* branched to 2 orders, *branches pendulous;* primary bracts similar, not dropping off as inflorescence expands; peduncle shorter than inflorescence rachis; rachillae moderately slender, pendulous. **Flowers** in pairs or solitary throughout the length of the rachilla, not developing in pits. **Fruit** *small, creamy white, subglobose, with vertical rows of shiny scales,* stigmatic remains apical, flesh thin. **Seed** 1, enclosed in a fleshy coat (sarcotesta), laterally flattened, endosperm homogeneous.

DISTRIBUTION. Two species, one in Sulawesi, the other in Maluku and western New Guinea.

NOTES. Until relatively recently, this was a rather poorly known genus, not well represented in herbaria or in cultivation. In the mid 1970s, new accessions of seeds were introduced into cultivation from Sulawesi and from New Guinea and new herbarium specimens made. It had been considered that there was just one species, *P. filaris,* but the distinctions between the two species that had been described long ago became obvious when cultivated plants were compared. *Pigafetta filaris* is a palm of Maluku and New Guinea whereas *P. elata* is restricted to Sulawesi (Dransfield 1998).

Pigafetta filaris (Giseke) Becc.

Synonyms: *Metroxylon microcarpum* Kunth, *Metroxylon microspermum* Kunth, *Metroxylon filare* (Giseke) Martius, *Pigafettia filifera* (Giseke) Merr., *Pigafetta papuana* Becc., *Sagus filaris* Giseke, *Sagus microcarpa* Zipp. ex Hall, *Sagus microsperma* Zipp. ex Hall

Massive, single-stemmed tree palm to 25 m or more. **Stem** 30–45 cm diam., *shiny green, especially in upper parts,* crownshaft absent, dioecious. **Leaf** pinnate, 3–6 m long, 14–20 in crown, *armed with spines and bristles, strongly arching;* sheath splitting to the base opposite the petiole, *distally with two ear-like lobes, chalky white;* petiole absent, though narrow, upper portion of sheath resembles a petiole; leaflets ca. 60 each side of leaf rachis, to 140 cm long, with pointed tips, *arranged regularly,* ascending. **Inflorescence** *between and below the leaves, branched to 2 orders,* 150–260 cm long, branches pendulous; primary bracts similar, not dropping off as inflorescence expands; *peduncle shorter than inflorescence rachis,* ca. 60 cm; rachillae moderately slender, *pendulous,* 15–30 cm long, 4–5 mm diam., *covered in brown tomentum.* **Flowers** crowded throughout the length of the rachilla. **Fruit** *creamy white at maturity,* subglobose, 0.8–1 cm × 0.6–0.7 cm, with 13–15 *vertical rows of shiny scales,* stigmatic remains apical, flesh thin. **Seed** 1, 7–8 × 5 mm, laterally flattened, somewhat angled and shallowly pitted, endosperm homogeneous.

DISTRIBUTION. North-western New Guinea, from the Bird's Head Peninsula and adjacent islands (Waigeo, Misool, Biak) to the vicinity of Vanimo. Elsewhere in Maluku.

HABITAT. Lowland forest, often in regenerating vegetation, such as along braided river banks or unstable slopes, from sea level to 250 m.

LOCAL NAMES. *Momboa* (Manokwari), *Mansinyas* (Waigeo), *Pang* (Matbat), *Rakwa* (Kaimana), *Tnang Nyi* (North Cyclops).

USES. Trunks are used in house construction, including flooring. Leaves are used for thatching and for house walls. Stems are felled to raise edible grubs.

CONSERVATION STATUS. Least Concern (IUCN 2021).

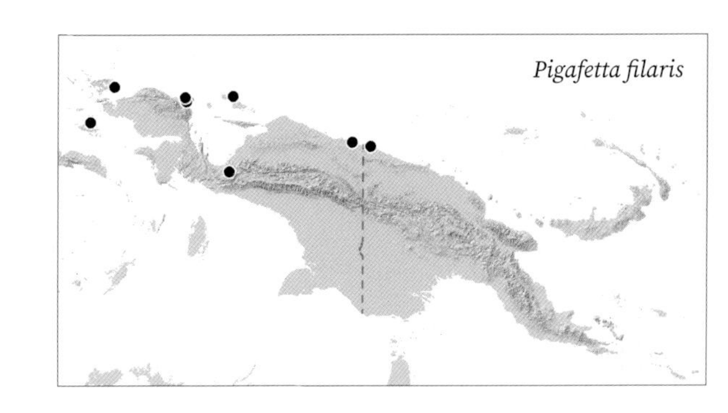

Pigafetta filaris. LEFT TO RIGHT: habit, Biak Island (WB); habit, near Sorong (WB).

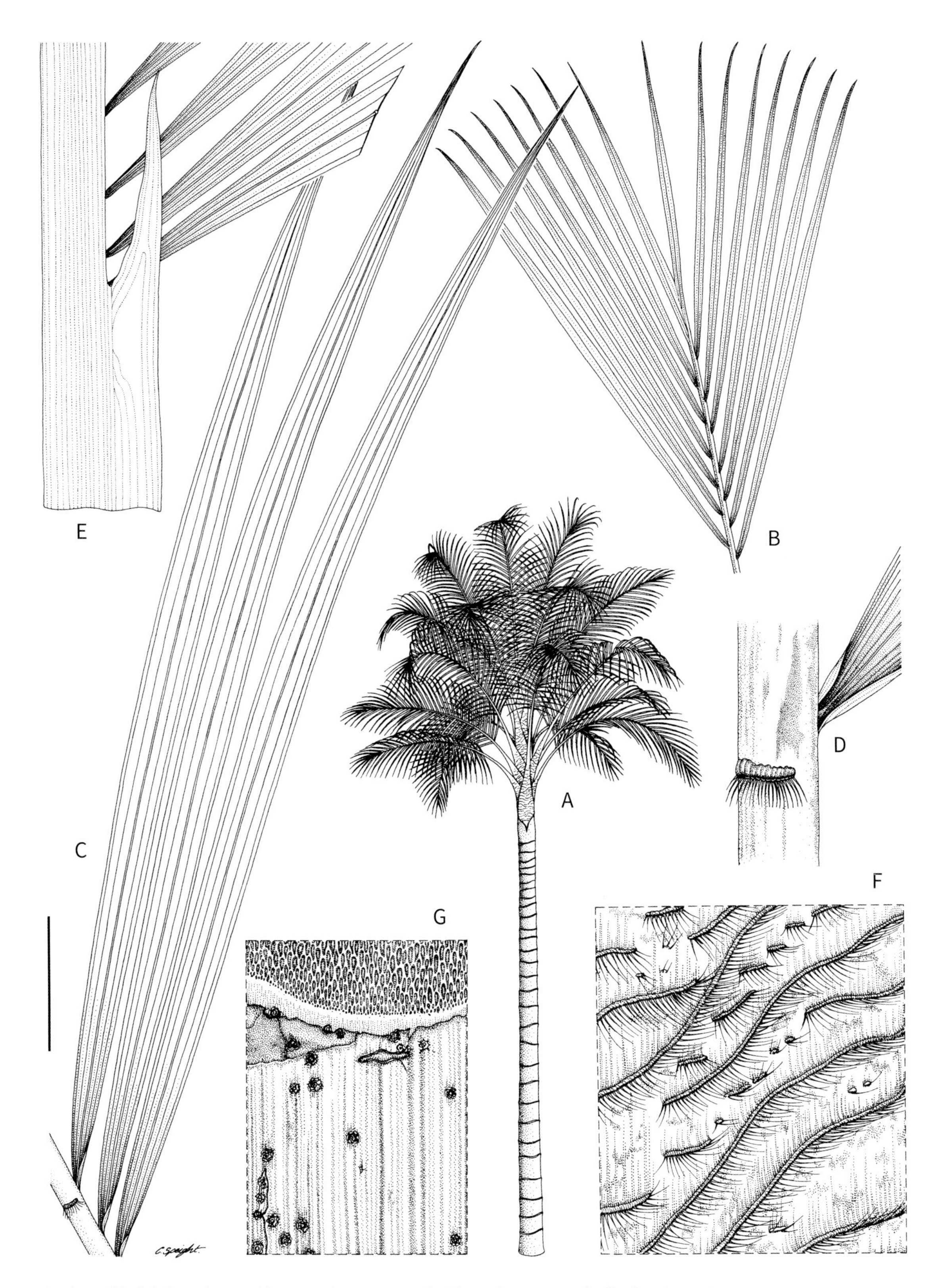

Pigafetta filaris (Plate 1). **A.** Habit. **B.** Leaf apex. **C.** Leaf mid-portion. **D.** Detail of leaf rachis with spines. **E.** Leaf base with auricle. **F.** Detail of leaf sheath spines. **G.** Stem surface. Scale bar: A = 2 m; B, C = 18 cm; D, F, G = 4 cm; E = 8 cm. All from *Dransfield JD 7610.* Drawn by Camilla Speight.

NOTES: Unmistakable, *Pigafetta filaris* is an immensely robust, spectacular, single-stemmed canopy tree with gleaming green trunks, arching, spiny leaves and chalky white spiny leaf sheaths. The inflorescences are borne between and below the leaves, reaching 2.5 m long, with long pendulous branches. The fruits are very small (0.8–1 cm) with creamy white scales when ripe. Within its range, *Pigafetta* is a common palm of lowland rainforest, including disturbed areas. It can be confused with *Metroxylon*, which is usually multi-stemmed, with leaf sheaths that are not chalky white, stems that die after flowering, and large fruit (2.5–3.5 cm).

Calamus kunzeanus Becc. is often cited as a synonym of *Pigafetta filaris*; however, it seems much more likely that the fragmentary fruit material on which the name is based indeed belongs to a species of *Calamus*.

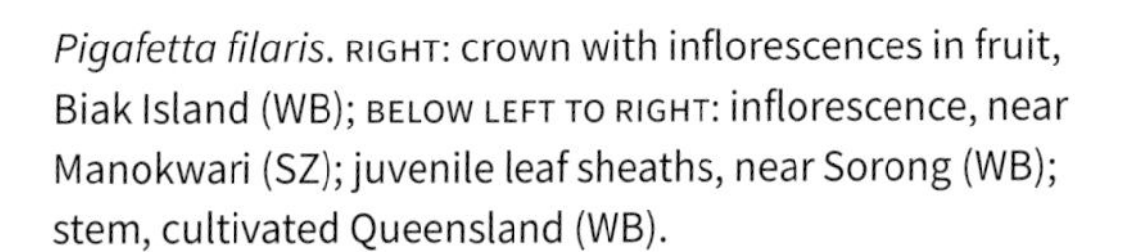

Pigafetta filaris. RIGHT: crown with inflorescences in fruit, Biak Island (WB); BELOW LEFT TO RIGHT: inflorescence, near Manokwari (SZ); juvenile leaf sheaths, near Sorong (WB); stem, cultivated Queensland (WB).

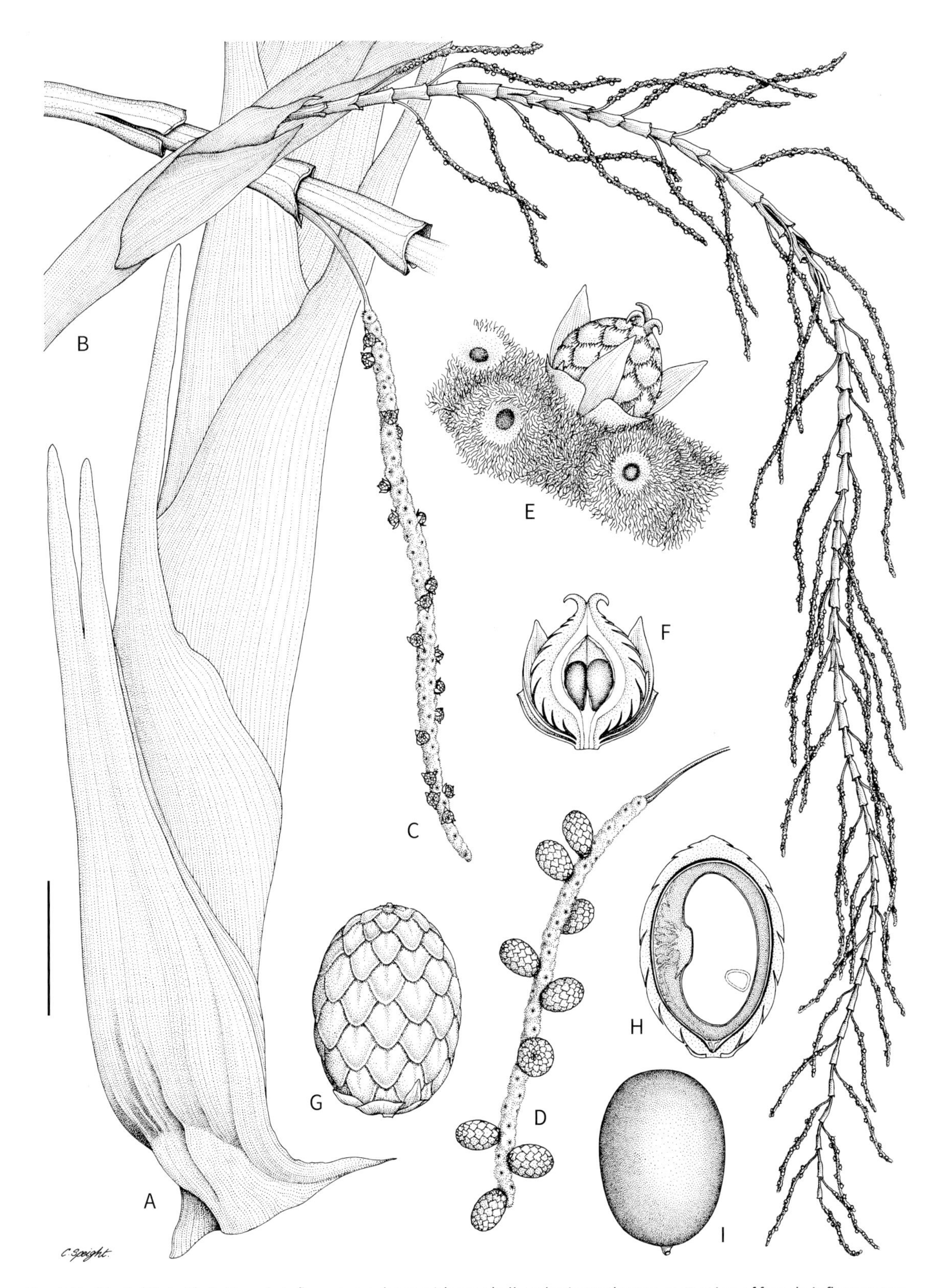

Pigafetta filaris (Plate 2). **A.** Female inflorescence base with prophyll and primary bracts. **B.** Portion of female inflorescence. **C.** Female rachilla with flowers. **D.** Female rachilla with fruit. **E.** Portion of female rachilla with flower. **F.** Female in flower in longitudinal section. **G**, **H.** Fruit whole and in longitudinal section. **I.** Seed. Scale bar: A = 6 cm; B = 12 cm; C, D = 3 cm; E, F = 4 mm; G–I = 7 mm. All from *Dransfield JD 7610.* Drawn by Camilla Speight.

Calamus zebrinus. Habit, Biak Island (WB).

CALAMOIDEAE | CALAMINAE

Calamus L.

Synonyms: *Calospatha* Becc., *Ceratolobus* Blume ex Schult. & Schult.f., *Cornera* Furtado, *Daemonorops* Blume, *Palmijuncus* Rumph. ex Kuntze, *Pogonotium* J.Dransf., *Retispatha* J.Dransf., *Rotang* Adans., *Rotanga* Boehm., *Schizospatha* Furtado, *Zalaccella* Becc.

Climber – leaf pinnate – spiny – leaflets not diamond-shaped – leaflets pointed

Slender to robust, *single- or multi-stemmed climbing palms, dioecious.* **Leaf** pinnate, *armed with spines and bristles,* straight to arching; *sheath tubular, usually very spiny;* petiole present or absent; leaflets few to numerous each side of leaf rachis, *linear, lanceolate or elliptic, with pointed tips,* arranged regularly, subregularly, irregularly or in groups, horizontal to pendulous; *spiny climbing whips arising from leaf tips (cirrus) or leaf sheaths (flagellum).* **Inflorescence** between the leaves, branched 1–3 (rarely 4–5) orders, erect, pendulous or whip-like; prophyll and peduncular bracts similar, sometimes enclosing inflorescence in late bud, sometimes splitting and dropping off as inflorescence expands, *usually remaining tubular, at least at the base,* often spiny; peduncle usually shorter than inflorescence rachis; rachillae slender, straight or curved. **Flowers** in pairs, more rarely 3s or 4s (female inflorescence) or solitary (male inflorescence) throughout the length of the rachilla, not developing in pits, usually sessile, more rarely with pedicel-like stalk. **Fruit** solitary or paired, various colours, globose to ellipsoid, *with vertical rows of scales,* stigmatic remains apical, flesh thin. **Seed** 1, *enclosed in a fleshy coat (sarcotesta),* various shapes, endosperm homogeneous (rarely ruminate).

DISTRIBUTION. Over 410 species distributed from India to South China, through South-East Asia to Australia and Fiji, with one species disjunct in Africa. Of these, 63 occur in New Guinea, 57 of which are endemic to our region.

NOTES. *Calamus* is the largest genus of palms globally. It accounts for one quarter of the total palm flora of New Guinea, more than twice as many species as the next most species-rich genus on the island (*Calyptrocalyx* with 28 species). In New Guinea, all species of *Calamus* are spiny, climbing palms (rattans) with specialised climbing whips, developed either from the tip of the leaf (cirrus) or from sterile inflorescences attached to the leaf sheath (flagellum). Non-climbing growth forms (e.g. stemless palms or small trees), which occur in *Calamus* in other parts of its range, are not found in any New Guinea species. All are dioecious, producing inflorescences that are widely varied in architecture ranging from short, erect structures around 10 cm in length to pendulous inflorescences extending to 6 m. Their fruits are scaly and the seed within is enclosed in a fleshy coat (sarcotesta). Only one other rattan genus occurs in New Guinea, *Korthalsia*, which is immediately distinguished by its diamond-shaped leaflets with jagged tips, in contrast to the linear to elliptic, pointed leaflets of *Calamus*.

During the preparation of this account, *Calamus* has undergone a number of global taxonomic treatments (Vorontsova *et al.* 2016, Henderson 2020), including significant expansion of the generic limits (Baker 2015), notably to include the genus *Daemonorops* among others. This delimitation was used in a full monograph of *Calamus* (Henderson 2020), which largely follows the species taxonomy that we present here. In preparation for this account, 33 of the 63 New Guinea species were described as new to science (Baker 2002b, Baker & Dransfield 2002a, Baker & Dransfield 2002b, Baker *et al.* 2003, Dransfield & Baker 2003, Baker & Dransfield 2014, Fernando 2014, Maturbongs *et al.* 2014, Baker & Dransfield 2017, Baker & Venter 2019). This dramatic growth in the number of New Guinea *Calamus* species is offset by the placement of 25 species in synonymy.

To aid the navigation of diversity and identification, we provide here a brief overview of New Guinea Calamus species.

Cirrate species

Calamus aruensis, C. calapparius, C. dasyacanthus, C. pachypus, C. vitiensis, C. warburgii, C. zieckii, C. zollingeri

Eight of the 63 species of New Guinea *Calamus* bear a cirrus. Within these, three unrelated groups can be recognised:

***C. aruensis* complex** – *Calamus aruensis, C. dasyacanthus, C. pachypus, C. vitiensis* (Baker *et al.* 2003).

***C. warburgii* group** – *C. warburgii, C. zieckii, C. zollingeri* (Fernando 2014).

C. calapparius – the sole New Guinea representative of the recently synonymised genus *Daemonorops*.

Flagellate species with well-developed ocreas

Calamus altiscandens, C. badius, C. baiyerensis, C. bankae, C. barfodii, C. cheirophyllus, C. distentus, C. fertilis, C. heatubunii, C. klossii, C. kostermansii, C. lauterbachii, C. longipinna, C. lucysmithiae, C. macrochlamys, C. moszkowskianus, C. nanduensis, C. nudus, C. pholidostachys, C. pintaudii, C. pseudozebrinus, C. reticulatus, C. scabrispathus, C. schlechterianus, C. vestitus, C. wanggaii, C. zebrinus

The majority of New Guinea *Calamus* bear a flagellum. Twenty-seven of these also display very well-developed ocreas, far more than in the *Calamus* species of any other region. The ocrea is an extension of the leaf sheath above the junction with the petiole (Merklinger *et al.* 2014). Well-developed ocreas in New Guinea *Calamus* range from 2.5–100 cm in length and can be completely tubular or may develop with a longitudinal opening, forming a ligule-like structure. They may clasp the stem tightly or be inflated, erect or divergent at the sheath mouth. Ocreas can consist of tough, persistent tissue, but can also be papery, net-like or fibrous, splitting or distintegrating to dry tatters or being lost entirely. Indeed, some rattans specimens may appear to lack a well-developed ocrea but the ocreas have simply been lost. Accurate observations are best made from young material. In some cases, elaborate ocreas can be occupied by ants, a remarkable convergence with some species of the unrelated rattan genera *Korthalsia* and *Laccosperma*.

Here we place some of the 27 species with well-developed ocreas into informal groups based on morphological similarities. Detailed discussion of individual species groups is included in notes under the relevant species accounts:

***C. altiscandens* group** – *C. altiscandens, C. fertilis, C. macrochlamys, C. pholidostachys, C. zebrinus* (Dransfield & Baker 2003).

***C. baiyerensis* group** – *C. baiyerensis, C. klossii, C. nanduensis, C. pintaudii* and *C. pseudozebrinus*.

***C. lauterbachii* group** – *C. heatubunii, C. lauterbachii, C. nudus* (Baker & Venter 2019).

***C. longipinna* group** – *C. bankae, C. kostermansii, C. longipinna, C. reticulatus, C. vestitus, C. wanggaii* (Baker & Dransfield 2002a).

Flagellate species lacking well-developed ocreas

C. anomalus, C. barbatus, C. bulubabi, C. capillosus, C. croftii, C. cuthbertsonii, C. depauperatus, C. erythrocarpus, C. essigii, C. heteracanthus, C. interruptus, C. jacobsii, C. johnsii, C. katikii, C. kebariensis, C. maturbongsii, C. nannostachys, C. oresbius, C. papuanus, C. pilosissimus, C. polycladus, C. retroflexus. C. sashae, C. serrulatus, C. spanostachys, C. spiculifer, C. superciliatus, C. womersleyi

In the remaining 28 flagellate New Guinea *Calamus*, the ocrea is not well-developed. It may be entirely absent, or form a low inconspicuous crest or, more rarely, a membranous collar up to 2 cm in height. Two species included here (*C. kebariensis* and *C. spanostachys*) appear to lack both cirrus and flagellum. The majority of New Guinea's most slender rattans (stems with sheaths that can be ≤10 mm) occur among these 27 species (*C. anomalus, C. croftii, C. cuthbertsonii, C. depauperatus, C. essigii, C. johnsii, C. katikii, C. kebariensis, C. nannostachys, C. oresbius, C. papuanus, C. pilosissimus, C. spanostachys, C. superciliatus*). All but three of these (*C. croftii, C. johnsii, C. nannostachys*) occur at higher elevations (>700 m). Three of these have been reported at more than 2,000 m elevation (*C. cuthbertsonii, C. depauperatus, C. oresbius*).

Only one informal group is recognised here:

***C. anomalus* group** – *C. anomalus, C. erythrocarpus, C. essigii, C. maturbongsii, C. nannostachys.*

Key to the species of *Calamus* in New Guinea

1. Spiny, whip-like structure extending from leaf apex (cirrus) 2
1. Spiny, whip-like structure attached to the leaf sheath (flagellum) or not present 10

2. Leaflets broadly elliptic, 10–25 each side of the leaf rachis 3
2. Leaflets linear or linear-lanceolate, 15–105 each side of the leaf rachis 7

3. Grapnel spines of cirrus arranged irregularly .... **3. *C. aruensis***
3. Grapnel spines of cirrus arranged regularly in whorls, resembling cat's claws 4

4. Leaf sheath entirely unarmed .... **57. *C. vitiensis* (in part)**
4. Leaf sheath densely to very lightly armed 5

5. Spines parallel-sided, somewhat papery, apices and margins distinctly jagged . **15. *C. dasyacanthus***
5. Spines triangular, not papery, apices narrow acute, margins entire 6

6. Spines red-brown to black, 1–60 mm long, in irregular partial whorls of up to 14 interspersed with solitary spines, longer spines often curving and tapering distinctly from base to a narrow apex, spine bases distinctly swollen .... **40. *C. pachypus***
6. Spines yellow-green to brown, 1–28 mm long, usually solitary, rarely also with very few, irregular, partial whorls of up to 6, longer spines not curving or tapering, spine bases not distinctly swollen. .... **57. *C. vitiensis* (in part)**

7. Leaflets >60 each side of rachis, regularly arranged, lacking thin indumentum on lower surface . . 8
7. Leaflets <40 each side of rachis, arranged in divaricate pairs or groups, rarely regular, with thin, brown or chalky indumentum on lower surface . . . . **62. *C. zieckii***

8. Sheaths densely armed with fine, easily detached, needle- or hair-like spines; fruit not blue-black . . 9
8. Sheaths armed with narrow, triangular spines of various sizes, not easily detached; fruit blue-black; Raja Ampat Islands only . . . . **63. *C. zollingeri* subsp. *zollingeri***

9. Spines brown, primary inflorescence bracts tightly sheathing, not expanding, not splitting or falling; fruit 8–11 mm diam.; endosperm homogeneous; widespread . . . . **59. *C. warburgii***
9. Spines brown to black, primary inflorescence bracts inflated, splitting along their length and often falling; fruit 22–28 mm diam.; endosperm ruminate; far western New Guinea only . . . . **10. *C. calapparius***

10. Leaf sheath mouth (in newly emerged sheaths) with well-developed ocrea 2–100 cm long, clasping, inflated, erect divergent, leathery, papery or fibrous, persistent or disintegrating 11
10. Leaf sheath mouth (in newly emerged sheaths) lacking well-developed ocrea, or with low, inconspicuous crest or membranous collar to 2 cm long 37

11. Ocrea persistent, only disintegrating with age, papery, leathery or fibrous, clasping, erect or divergent, sometimes tightly sheathing, inflated or expanded into a net-like funnel 12
11. Ocrea disintegrating into papery shreds or fibres shortly after emergence, sometimes lost entirely 28

12. Leaflets 2–19 each side of rachis, usually grouped or arranged irregularly or subregularly, (rarely regularly arranged and then leaflets <10 each side of rachis) 13
12. Leaflets 20–64 each side of rachis, usually regularly arranged (rarely grouped or subregularly arranged and then leaflets ca. 28 or more each side of rachis) 22

13. Ocrea 2–16 cm long, erect or divergent, not clasping the stem 14
13. Ocrea 15–90 cm long, usually inflated and clasping the stem 16

14. Leaflets 6–9 each side of rachis 15
14. Leaflets 14–18 each side of rachis . . . . **17. *C. distentus***

15. Leaflet with pale brown indumentum on lower surface . . . . **4. *C. badius***
15. Leaflet lacking pale brown indumentum on lower surface . . . . **32. *C. lucysmithiae***

16. Mid-leaf leaflets 2 cm wide or less, linear or linear-lanceolate; ocrea 24–29 cm long 17
16. Mid-leaf leaflets 2–10 cm wide, elliptic or linear- to broadly lanceolate; ocrea 15–90 cm long 18

17. Leaflets regularly arranged; inflorescence erect, 49–65 cm long . . . . **50. *C. scabrispathus***
17. Leaflets grouped; inflorescence whip-like and pendulous, ca. 150 cm long. . . . . **6. *C. bankae***

18. Mid-leaf leaflets 2–4 cm wide; female rachilla 5–20 cm long, bearing flowers in threes or fours (2–3 female flowers, 0–2 sterile male flowers); fruits paired 19
18. Mid-leaf leaflets 3.5–10 cm wide; female rachilla 0.6–4 cm long, bearing flowers in pairs (1 female flower, 1 sterile male flower); fruit solitary. 20

19. Leaflets 2–15 each side of rachis; ocrea unarmed or sparsely armed with scattered spines or spine whorls; inflorescence primary branching systems relatively lax .... **33. *C. macrochlamys***
19. Leaflets ca. 17–19 each side of rachis; ocrea densely armed with short spines; inflorescence primary branching systems somewhat stiff and congested .... **1. *C. altiscandens***

20. Leaflets 5–9 each side of rachis; ocrea 21–90 cm long, armed with numerous fine spines arranged in collars resembling eyelashes... **30. *C. lauterbachii***
20. Leaflets 2–5 each side of rachis; ocrea 40–55 cm long, armed with scattered to numerous solitary spines, not arranged in collars...21

21. Inflorescence 22–35 cm long, erect, with ca. 8 primary branches; seed ridged and angular .... **21. *C. heatubunii***
21. Inflorescence 160–220 cm long, whip-like, with 1–3 primary branches; seed smooth . . **38. *C. nudus***

22. Ocrea strictly tubular, encircling the stem and not open on the side opposite the petiole; leaf sheath armed with flattened, parallel-sided spines, with irregular, jagged apices..............23
22. Ocrea open on the side opposite the petiole, either congenitally or disintegrating; leaf sheath armed with stout triangular spines or fine needle-like spines, rarely unarmed 24

23. Leaflets 20–47 each side of rachis; ocrea papery, clasping the stem, unarmed or with very few minute spines; lowland forest habitats....................................... **56. *C. vestitus***
23. Leaflets 56–64 each side of rachis; ocrea fibrous, expanding distally to form a net-like funnel, armed with numerous needle-like spines; river bank habitats................. **47. *C. reticulatus***

24. Leaf sheath armed with fine, needle-like spines up to 70 mm long; ocrea < ca. 5 cm long, divergent .... **35. *C. moszkowskianus***
24. Leaf sheath armed with stout triangular spines 0.5–18 mm long; ocrea 9–45 cm long, inflated, clasping...25

25. Leaflets 23–26 each side of rachis; inflorescence erect to arching, 0.4–0.6 m long, lacking a whip-like tip .... **8. *C. barfodii***
25. Leaflets 28–50 each side of rachis; inflorescence pendulous, >1.8 m long, bearing a long, whip-like tip...26

26. Leaflets oblanceolate, arranged in three distinct groups, with conspicuous bristles 7–11 mm long on upper surface... **58. *C. wanggaii***
26. Leaflets linear or narrow lanceolate, regularly arranged (very rarely subregular or arranged in groups), with bristles up to 6 mm long on upper surface 27

27. Ocrea cleanly split longitudinally to base on side opposite the petiole; female rachillae 40–290 .... **31. *C. longipinna***
27. Ocrea disintegrating into long fibres on side opposite the petiole; female rachillae 12–35 cm...... .... **29. *C. kostermansii***

28. Spines hair- or needle-like, brittle, deciduous 29
28. Spines hair- or needle-like, or triangular, not brittle, not deciduous 31

29. Spines not united at base . . . . **36. *C. nanduensis***
29. Spines united at base to form a low ridge, ridge persistent after spines fall. 30

30. Inflorescence primary branches elongate, to 90 cm long or more; female inflorescence branched to 2 orders, female rachillae 4.5–19 cm long, bearing flowers in pairs (2 female flowers) or threes (2 female flowers, 1 sterile male flower); fruits paired. . . . . **61. *C. zebrinus***
30. Inflorescence primary branches compact, to 23 cm long; female inflorescence branched to 3 orders, female rachillae 5–8.5 cm long, bearing flowers in pairs (1 female flower, 1 sterile male flower); fruits solitary. . . . . **46. *C. pseudozebrinus***

31. Leaflets 2–30 each side of rachis 32
31. Leaflets 32–65 each side of rachis 33

32. Stems 6.5–13 mm diam.; spines 1–12 mm long; ocrea 7–13 cm long before disintegrating into fibres; leaflets 7–9 in total, arranged in a fan at petiole apex; inflorescence 14–34 cm long, slender and erect . . . . **12. *C. cheirophyllus***
32. Stems 8–32 mm diam.; spines 2–100 mm long; ocrea 12–70 cm long before disintegrating into fibres; leaflets 2–30 each side of rachis, arranged regularly, subregularly or in groups; inflorescence 80–600 cm long, whip-like and pendulous . . . . **28. *C. klossii***

33. Leaf sheath sparsely armed with few flattened, triangular spines 5–25 mm long, 2–4 mm wide . . 34
33. Leaf sheath densely armed, spines flattened, triangular, hair- or needle-like, 2–110 mm long, 1–8 mm wide (narrower if hair-like) 35

34. Stem 45–55 mm wide; leaflets 32–35 each side of rachis, mid-leaf leaflets ca. 55 × 3.5 cm, armed with scattered black bristles on upper surface and margins . . . . **5. *C. baiyerensis***
34. Stem 20–40 mm wide; leaflets ca. 60 on each side of rachis, mid-leaf leaflets 38–46 × 2.3–2.4 cm, armed with dark bristles on upper and lower surface, margins minutely bristly . . . . **42. *C. pholidostachys***

35. Spines hair-like, brown or dark brown; inflorescences branched to 3 orders in the female and 4 orders (rarely 5 orders) in the male. . . . . **51. *C. schlechterianus***
35. Spines narrow triangular to needle-like, brown, yellow-green or orange; inflorescences branched to 2 orders in the female and 3 orders in the male 36

36. Leaf sheath very densely armed with narrow triangular, yellow-green to orange spines; knee grossly swollen; leaflets 53–65 each side of rachis; inflorescence primary branches to 75 cm long, female rachilla bearing flowers in threes (2 female flowers, 1 sterile male flower); fruit paired, ovoid . . . . **20. *C. fertilis***
36. Leaf sheath densely armed with needle-like, brown spines; knee not unusually swollen; leaflets 32–35 each side of rachis; inflorescence primary branches to 35 cm long, female rachilla bearing flowers in pairs (1 female flower, 1 sterile male flower); fruit solitary, spherical . . . . **44. *C. pintaudii***

37. Leaflets arranged regularly or subregularly 38
37. Leaflets arranged irregularly or in groups 51

38. Stem 0.3–2.8 cm diam.; leaf 60 cm long or less. 39

38. Stem 1.2–4.8 cm diam.; leaf 65 cm long or more 43

39. Spiny whip-like structure attached to the leaf sheath (flagellum) absent, female inflorescence branched to 1 order 40

39. Spiny whip-like structure attached to the leaf sheath (flagellum) present, female inflorescence branched to 2–3 orders. 41

40. Leaf sheath unarmed or almost entirely so; leaflets 10–16 each side of rachis, leaflets 4–10 mm wide . . . . **27. *C. kebariensis***

40. Leaf sheath sparsely armed with narrow triangular spines to 12 mm long; leaflets 2–3 each side of rachis, leaflets 10–22 mm wide . . . . **53. *C. spanostachys***

41. Leaflets elliptic, mid-leaf leaflets 1.3–4 cm wide, apical leaflets joined from one third to one half of their length; ocrea 3–20 mm long, tightly sheathing, leathery and truncate . . . . **41. *C. papuanus* (in part)**

41. Leaflets lanceolate (or linear-lanceolate), mid-leaf leaflets 0.4–1.7 cm wide, apical leaflets not united or joined up to one quarter of their length; ocrea inconspicuous 42

42. Leaflets 25–41 each side of rachis, bristles on leaflet margins conspicuous and interlocking; inflorescence to 135 cm long, fine and whip-like . . . . **43. *C. pilosissimus***

42. Leaflets 7–14 each side of rachis, bristles on leaflet margins not conspicuous or interlocking; inflorescence 14–45 cm long, erect or trailing . . . . **14. *C. cuthbertsonii***

43. Leaflets >45 each side of rachis 44

43. Leaflets <35 each side of rachis 46

44. Stem 12–15 mm diam.; leaf sheath spines hair-like, soft; leaf 0.6–0.9 m long; leaflets 48–53 each side of rachis. . . . . **11. *C. capillosus***

44. Stem 13–45 mm diam.; leaf sheath spines flattened, triangular, stiff; leaf 1–2.5 cm long; leaflets 62–99 each side of rachis 45

45. Leaflets >70 each side of rachis, mid-leaf leaflets 16–31 cm long, 1–1.8 cm wide, armed with abundant minute spines on margins and both surfaces; leaf sheath spine tips breaking off in mature sheaths . . . . **52. *C. serrulatus***

45. Leaflets <65 each side of rachis, mid-leaf leaflets 38–40 cm long, 2.3–2.7 cm wide, sparsely armed with bristles on upper surface and margins near tip; leaf sheath spine tips not breaking off in mature sheaths . . . . **49. *C. sashae***

46. Leaflets 12 or fewer each side of rachis; leaf sheaths spines absent or few, minute spines present 47

46. Leaflets 15 or more each side of rachis; leaf sheaths spines numerous, robust (very rarely absent) 49

47. Leaflets 4–5 each side of rachis, 35–37 cm long, 5–8 cm wide; inflorescence at least 87 cm long, male inflorescence branching to 4 orders . . . . **24. *C. jacobsii***

47. Leaflets 7–12 each side of rachis, 22–34 long, 3–5.5 cm wide (apical leaflets sometimes much reduced); inflorescence 37–81 cm long, male inflorescence branching to 3 orders 48

48. Apical leaflets highly reduced, <4 cm long, the rachis forming a subcirrus at leaf apex; inflorescence 37–55 cm long .... **18. *C. erythrocarpus***
48. Apical leaflets not highly reduced, 5–12 cm long, the rachis not forming subcirrus at leaf apex; inflorescence 62–81 cm long .... **34. *C. maturbongsii***

49. Leaf 1–2.3 m long, ecirrate or subcirrate; leaflets broadly elliptic, 2–8 cm wide .... **22. *C. heteracanthus* (in part)**
49. Leaf 0.6–1.5 m long, ecirrate; leaflets narrow lanceolate, 1–2.3 cm wide 50

50. Leaf sheath spines flexible, solitary and in whorls; inflorescence 0.9–1.5 m long, branched to 2 orders in the female, primary bracts armed and often with spine tuft at mouth, rachillae with pointed, spine-like tips .... **7. *C. barbatus***
50. Leaf sheath spines stiff, solitary; inflorescence 1.2–2.2 m long, branched to 4 orders in the female, primary bracts unarmed, rachillae not with pointed tips .... **45. *C. polycladus***

51. Leaf >80 cm long 52
51. Leaf <80 cm long 56

52. Leaf sheaths with spines 1–7 mm wide, rarely unarmed 53
52. Leaf sheaths with spines <1 mm wide, not unarmed 55

53. Petiole absent; leaflets linear to narrow lanceolate, basal leaflets reflexed across the stem .... **48. *C. retroflexus* (in part)**
53. Petiole 1.5 cm long or more; leaflets elliptic, basal leaflets not reflexed across the stem 54

54. Apical leaflets 0.7–13 cm long, the rachis often forming a subcirrus at the leaf apex .... **22. *C. heteracanthus* (in part)**
54. Apical leaflets 18–30 cm long, subcirrus absent .... **23. *C. interruptus***

55. Leaf sheath armed with fine hair-like spines to 25 mm long; leaflets 10–16 each side of rachis, mid-leaf leaflets 3.5–4.2 cm wide .... **9. *C. bulubabi***
55. Leaf sheath armed with minute, detachable spicules to 1.5 mm long; leaflets 9–11 each side of rachis, mid-leaf leaflets 5–6 cm wide .... **54. *C. spiculifer***

56. Leaflets 16 or more each side of rachis 57
56. Leaflets 12 or fewer each side of rachis 58

57. Petiole absent; basal leaflets reflexed across the stem; ocrea scarcely developed; inflorescence to 4 m long .... **48. *C. retroflexus* (in part)**
57. Petiole 2–7 cm long; basal leaflets not reflexed across the stem; ocrea to 14 mm long, strictly tubular and tightly sheathing, truncate; inflorescence to 1 m long .... **60. *C. womersleyi***

58. Leaflets 3–8 each side of rachis; leaf sheath spines (if present) 1–3 mm long; inflorescence 10–60 cm long, erect to arching, unarmed (except prophyll in some instances), primary bracts deeply split or punctured by emerging primary branches 59
58. Leaflets 4–12 each side of rachis; leaf sheath spines 1–32 mm long; inflorescences 40–220 cm long, arching, pendulous or whip-like, armed, primary bracts strictly tubular, not split by emerging primary branches 61

59. Stem 9–17 mm diam.; leaflets 5–8 each side of rachis, arranged irregularly (rarely in well-defined groups); petiole 3–18 cm long. . . . . **37. *C. nannostachys***

59. Stem 4–12 mm diam.; leaflets 3–5 each side of rachis, arranged in two widely spaced groups; petiole 0–3 cm long. 60

60. Stem 5–12 mm diam.; leaflets narrow lanceolate, 11 mm wide or more, apical leaflet united from one tenth to one third of their length; inflorescence branched to 3 orders in the male and 2 orders in the female . . . . **29. *C. anomalus***

60. Stem 4–5 mm diam.; leaflets narrow linear, 8 mm wide or less, apical leaflet pair free or united to one tenth of their length; inflorescence branched to 1–2 orders in the male and 1 order in the female . . . . **19. *C. essigii***

61. Longest leaf sheath spines >15 mm long 62

61. Longest leaf sheath spines not exceeding 15 mm long, usually >6 mm long, or sheath unarmed 64

62. Leaflets 9 or more each side of rachis; leaf sheath densely armed with flattened rusty spines to 32 mm long, with a dense tuft of spines at the sheath mouth; inflorescence to 2.2 m, fine, whip-like . . . . **55. *C. superciliatus***

62. Leaflets 8 or fewer each side of rachis; leaf sheath moderately armed with spines up to 27 mm long, lacking dense tuft of spines at sheath mouth, inflorescence to 170 cm or less 63

63. Stem 10–15 mm diam.; inflorescence to ca. 1 m long, lacking a flagelliform tip; fruit pear-shaped, pointed, ca. 23 mm long. . . . . **13. *C. croftii***

63. Stem 4.5–20 mm diam.; inflorescence to 1.7 m long including including flagelliform tip; fruit broadly ellipsoid, 12–20 mm long . . . . **25. *C. johnsii***

64. Stem 3–11 mm diam.; leaflets narrow lanceolate, 0.3–2 cm wide 65

64. Stem 4–28 mm diam.; leaflets elliptic to oblanceolate, 1–4.5 cm wide 66

65. Inflorescence to 120 cm long, primary branches ca. 2–5, to 20 cm long; fruit 12.5–17 mm long . . . . **16. *C. depauperatus***

65. Inflorescence ca. 40 cm long, primary branches ca. 2, to 3.7 cm long; fruit ca. 20 mm long . . . . **26. *C. katikii***

66. Stem 4–18 mm diam.; ocrea 2–5 mm long; leaflets 3–6 each side of rachis, arranged in two groups or in a single apical group, basal leaflets not reduced or reflexed; inflorescence 1–2 m long. . . . . **39. *C. oresbius***

66. Stem 7–28 mm diam.; ocrea 3–20 mm long, leaflets 4–11 each side of rachis, variously grouped, basal leaflets reduced and reflexed across stem; inflorescence 0.4–1 m long . . . . **41. *C. papuanus* (in part)**

1. *Calamus altiscandens* Burret

Moderately robust rattan climbing to the forest canopy, stem suckering not recorded. **Stem** with sheaths 28–35 mm diam., without sheaths to 16 mm diam.; internodes ca. 15 cm long. **Leaf** ecirrate, to 1.6 m long including petiole; sheath drying pale greenish brown, with dark brown indumentum; spines very dense, 5–18 × 1–4 mm, golden, soft and flattened, the base swollen and persistent, the rest of the spine frequently breaking off just above the swollen base mostly evenly distributed, occasionally joined in short horizontal groups; knee ca. 50 mm long, ca. 30 mm wide, armed as sheath; *ocrea 18 × 4.5 cm, coriaceous, tubular, inflated, somewhat bilobed and open opposite the petiole, very densely armed with short, slender spines, persistent*; flagellum to at least 3 m long; petiole to 23 cm long; *leaflets 17–19 on each side of rachis, lanceolate, arranged in distant groups of 1s, 2s or 3s*, longest leaflet in mid-leaf 50 × 3.5 cm, apical leaflets 16–18 × 1.2–1.5 cm, united for about one quarter of their length, leaflets armed with very small spines along the margins, unarmed on surfaces. **Inflorescence** length unknown, probably at least 1 m long, probably lacking a flagelliform tip; branched to 2 orders in the female, male not seen; primary bracts tubular, split distally to form an expanded limb to 10 × 4 cm with a triangular tip, sparsely armed; primary branches at least 4, to ca. 21 cm long, male rachillae not seen, female rachillae 55–120 × ca. 3.5 mm. **Fruit** *not seen, but very likely paired*. **Seed** not seen.

DISTRIBUTION. Only one confirmed record from central New Guinea.

HABITAT. Lowland forest, particularly abundant in gullies between ridges at ca. 100 m.

LOCAL NAMES. None recorded.

USES. None recorded.

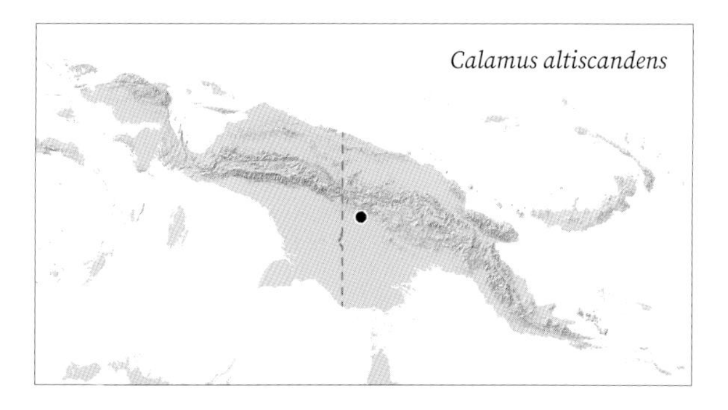

Calamus aff. *altiscandens*. Leaf sheaths with ocreas, near Kikori (WB).

CONSERVATION STATUS. Data Deficient. This species is known only from the type locality, where there is limited information on ongoing threats. Due to insufficient evidence, a single extinction risk category cannot be selected.

NOTES. *Calamus altiscandens* is one of five species that make up the *C. altiscandens* group (*C. altiscandens, C. fertilis, C. macrochlamys, C. pholidostachys, C. zebrinus*; Henderson 2020), previously termed the paired fruit group (Dransfield & Baker 2003). The female floral cluster in *Calamus* is a sympodial pair of flowers (Dransfield *et al.* 2008) comprising a sterile male flower in the terminal position and a functional female flower in the lateral position. (The male floral cluster consists of a solitary male flower.) Each cluster in the female inflorescence can thus produce a single fruit.

Calamus altiscandens. **A.** Stem with leaf sheath and ocrea. **B.** Whole leaf diagram. **C.** Leaf apex. **D.** Portion of female inflorescence. **E.** Portion of female rachilla showing scars of two female flowers and one sterile male flower. **F.** Portion of female rachilla showing scars of three female flowers. Scale bar: A = 4 cm; B = 0.5 m; C = 8 cm; D = 3 cm; E, F = 4 mm. All from *Brass 7327*. Drawn by Lucy T. Smith.

Calamus aff. *altiscandens*. Habit, near Kikori (WB).

Eleven species, including the five New Guinea species listed above, break this highly consistent structural pattern by bearing female clusters comprising various combinations of two (rarely three) female flowers with zero, one or two sterile male flowers. This results in floral clusters of usually three (but also two or four) flowers and fruit in pairs (or even threes) within a single rachilla bract (Dransfield & Baker 2003). Although the unusual floral structure suggests that the species may be related, they are very diverse in vegetative morphology. All possess elaborate ocreas that are varied in structure.

Calamus altiscandens is a moderately robust flagellate rattan that is easily distinguished by its leaves with strongly grouped leaflets, densely armed sheaths and well-developed, densely armed, persistent ocrea. The female rachillae are crowded in each primary branching system and are held erect. Flowers are borne in threes of two females and a sterile male, in quadrads of two females and two sterile males, and very rarely, in a group of three females. This species is most easily confused with *C. macrochlamys*, but differs in its distinctive inflorescence architecture and ocrea armature.

2. *Calamus anomalus* Burret

Synonyms: *Calamus setiger* Burret, *Schizospatha setigera* (Burret) Furtado

Very slender, multi-stemmed rattan to 20 m. **Stem** with sheaths 5–12 mm diam., without sheaths 3–6 mm diam.; internodes 8–27 cm. **Leaf** ecirrate, 25–43 cm long, petiole absent or very short; sheath with thin, grey-brown indumentum, with few to numerous, very fine, brittle, dark brown spines 1–3 mm long, spines more numerous and erect at sheath mouth, to 5 mm long; knee to 12 mm long, inconspicuous, unarmed or armed as sheath; ocrea absent or minute; flagellum to 1.3 m, very fine; *petiole 0–1 cm long; leaflets 4 on each side of rachis, arranged in 2 widely-spaced groups of 2 opposite, divaricate pairs, narrow lanceolate,* longest leaflet near base 14.5–29 × 1.1–4.5 cm, apical leaflets 12.5–16 × 1.1–1.8 cm, *apical leaflet pair united to one third of their length,* few bristles on margins; cirrus absent. **Inflorescence** *erect to arching, 26–53 cm long* including 5–22 cm peduncle, lacking flagelliform or sterile tip, branched to 2 orders in the female and 3 orders in the male; primary bracts initially tubular, *split to base by emerging primary branches,* sometimes remaining tubular distally, unarmed or lightly armed on lowermost bract; primary branches 3–7, to 10 cm long, male rachillae 2–20 × 1 mm, female rachillae 4–17 × 1 mm. **Fruit** solitary, ellipsoid, ca. 12 × 9 mm, orange-brown. **Seed** (sarcotesta removed) ca. 10 × 8 × 6 mm, subglobose with deep lateral pit; endosperm homogeneous.

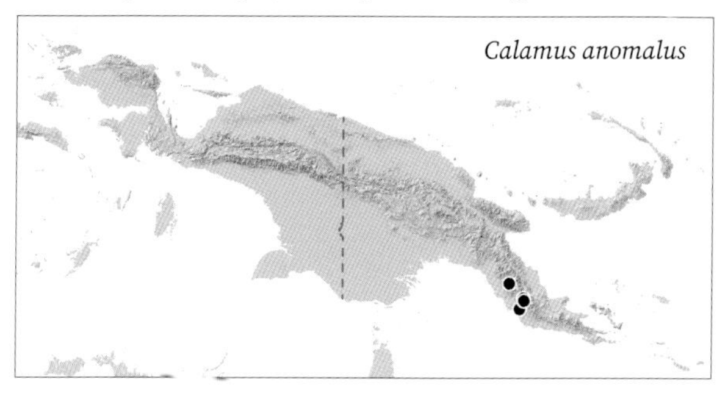

DISTRIBUTION. Owen Stanley Range, south-eastern New Guinea.

HABITAT. Montane forest at 910–1830 m

LOCAL NAMES. *Seseni* (Koiari-Mauari).

USES. None recorded.

CONSERVATION STATUS. Endangered. *Calamus anomalus* has a restricted distribution. Deforestation due to logging concessions is a major threat in its

Calamus anomalus. **A.** Stem with leaf sheath, leaf and female inflorescence. **B.** Male inflorescence. **C.** Female inflorescence. **D.** Detail of male inflorescence. **E.** Male flower. **F.** Female flower. **G.** Fruit. **H**, **I.** Seed in two views. **J.** Seed in longitudinal section. Scale bar: A = 4 cm; B, C = 2 cm; D, H–J = 7 mm; E, F = 2.5 mm; G = 1 cm. A, C, F from *Carr 14422*; B, D, E from *Carr 14421*; G–J from *Carr 13123*. Drawn by Lucy T. Smith.

range. Its close proximity to Port Moresby may also constitute a potential risk.

NOTES. The morphological features that define the *Calamus anomalus* group are so distinct that Furtado (1955) considered them sufficient to recognise a new genus, *Schizospatha,* now a synonym of *Calamus* (Baker 2002b). The five species in the group (*C. anomalus, C. erythrocarpus, C. essigii, C. nannostachys, C. maturbongsii,* all flagellate and lacking well-developed ocreas) produce short inflorescences that are unarmed (except for the prophyll in some instances), lack a flagelliform tip, and bear fragile papery bracts that overlap with the bract above and that are punctured or split deeply by the emerging inflorescence branches. Furtado included only *Calamus setiger* (as *Schizospatha setigera* (Burret) Furtado) in the new genus, overlooking Burret's (1936) comparison of the species with *Calamus anomalus,* whereas we consider the two to be conspecific (Baker 2002b). Burret (1935), in describing *C. anomalus,* noted a link to *C. nannostachys,* among other species, but this was not picked up by Furtado (1955). Henderson (2020) also connects *C. barfodii* to this group, but this species has a well-developed ocrea, which the five species listed above do not. He also hints at possible relationships to *C. croftii, C. jacobsii, C. johnsii* and *C. lucysmithiae.*

The *C. anomalus* group has a narrow distribution restricted to the Owen Stanley Range in Central Province and adjacent Northern and Morobe Provinces in south-eastern Papua New Guinea, with the exception of the remarkably disjunct *C. erythrocarpus,* which is endemic to the Bird's Head Peninsula.

Calamus anomalus, C. essigii and *C. nannostachys* are very closely related within the *C. anomalus* group. They are slender to moderate rattans with eight leaflets or fewer each side of the leaf rachis, with orange-brown fruit with smooth-margined scales. *Calamus anomalus* can be distinguished from the other two species by the leaf with very short or no petiole and typically four pairs of narrow lanceolate leaflets arranged in two divaricate groups, the terminal pair united from one tenth to one third of their length. The three species may have arisen along an elevational gradient, *C. anomalus* growing at the highest elevations of the three. Somewhat intermediate specimens have been noted in between *C. anomalus* and *C. nannostachys* (e.g. *Zieck & Kumal NGF 36510, Foreman & Vinas LAE 60146*). Nevertheless, the boundaries between the species are generally clear and easily determined.

3. *Calamus aruensis* Becc.

Synonyms: *Calamus hollrungii* Becc., *Calamus latisectus* Burret, *Palmijuncus aruensis* (Becc.) Kuntze

Robust, single-stemmed or rarely multi-stemmed rattan climbing to 50 m. **Stem** with sheaths 20–50 mm diam., without sheaths 10–30 mm diam.; internodes 16–39 cm. **Leaf** *cirrate,* to 5 m long including cirrus and petiole; sheath greyish green to dark green, with thin grey-brown indumentum, *spines absent to numerous, 4–26 × 1–7 mm, flattened, triangular, stiff, rarely with multiple points, dark green to black,* sometimes arranged in irregular partial whorls of up to 5; knee 44–125 mm long, rarely armed; ocrea 3–9 mm, a low, woody, brown, rarely armed, persistent collar; flagellum absent; petiole 7–90 mm; leaflets 13–23 each side of rachis, arranged regularly or subregularly, broadly elliptic, hooded, longest leaflets near middle of leaf, 26–66 × 6–12.5 cm, apical leaflets 10–22 × 0.3–0.4 cm, basal leaflets small and often reflexing, unarmed or

Calamus aruensis. Habit, Kikori River (WB).

Calamus aruensis. **A.** Stem with leaf sheath. **B.** Leaf apex with cirrus. **C.** Leaf mid-portion. **D.** Portion of male inflorescence. **E.** Male rachilla with flowers. **F**, **G.** Fruit whole and in longitudinal section. **H**, **I.** Seed in two views. Scale bar: A = 4 cm; B, C = 8 cm; D = 6 cm; E–I = 1 cm. A–E from *Baker & Utteridge 584*; F–I from *Baker et al. 610*. Drawn by Lucy T. Smith.

Calamus aruensis. LEFT TO RIGHT: female inflorescence, near Timika (WB); base of male inflorescence, Ramu River (WB).

with few bristles on margins and upper surface; cirrus 1–2.5 m, *cirrus grapnel spines arranged irregularly*. **Inflorescence** arching, to 3 m long including 18–47 cm peduncle and 14–120 cm sterile tip, branched to 2 orders in the female and 3 orders in the male; primary bracts strictly tubular, with acute, triangular limb to one side, unarmed or lightly armed; primary branches 6–10, to 55 cm long, male rachillae 5–33 × 0.6–1.2 mm, female rachillae 23–140 × 1–2 mm. **Fruit** solitary, shortly stalked, globose, 10.5–14 × 10–11.5 mm, scales cream-white. **Seed** (sarcotesta removed) 7.5–9.5 × 7–9 × 6–7 mm, globose, with a deep, narrow pit on one side, the surface covered with numerous deep pits and irregular channels; endosperm homogeneous.

DISTRIBUTION. Widespread in New Guinea from the Raja Ampat Islands to the Bismarck Archipelago. Also recorded from Aru Islands, the Solomon Islands and from the tip of the Cape York Peninsula, Australia.

HABITAT. Primary and secondary forest from sea level to 1,200 m, most frequently recorded in lowland forest below 500 m.

LOCAL NAMES· *Akalane* (Namasalang), *Apo gui* (Wamesa), *Bu* (Madang), *Busep* (Jal), *Dou* (Waigeo), *Futepa* (Arandai dialect, Dombano language), *Hekek punam* (Yamben), *eke kuru* (Yamben), *Kalaua* (Pala

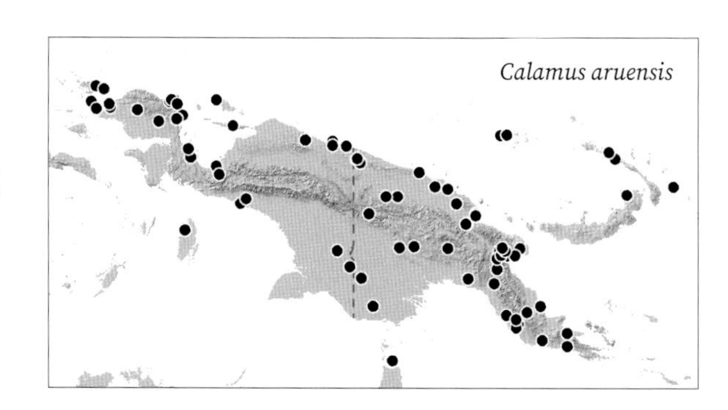

dialect, Patpatar language), *Kanda* (Papua New Guinea Tok Pisin), *Kaunor* (Waigeo), *Kerowa* (Karimui), *Kou* (Senderawoi), *Kuminang* (Kote), *Magu* (Kuanua), *Mambile* (Angguruk Yali), *Mbunom* (Marori), *Meya senga* (Nuni), *Minge* (Bembi), *Mumuni* (Orokaiva), *Ni* (Kara), *Proway* (Berap), *Sate* (Rawa), *Sel* (Angal Heneng), *Sauwe* (Mata), *Sehp* (Usino), *Sough* (Taka), *Tek niali* (Mianmin), *To mur* (Ayawasi), *To puot* (Ayawasi), *Tub* (Yei), *Wap* (Wampar), *Wampwang* (Kaigorin), *War* (Fas), *Wil dow* (Batanta, Salawati), *Wub* (Awyi), *Yapl* (Bewani).

USES. General cordage for pulling heavy objects and building bridges, split cane for tying house timbers, cane for furniture, cirrus for catching eels, leaves as wrapping for sago and in walls of houses, young shoot edible. Grass skirts from the young shoots. An infusion of an unspecified part is used to treat eye and ear infections in the Papua New Guinea southern highlands. Fruits eaten by the northern cassowary (Pangau-Adam & Mühlenberg 2014).

CONSERVATION STATUS. Least Concern.

NOTES. The *Calamus aruensis* complex comprises four New Guinea species, *Calamus aruensis*, *C. dasyacanthus*, *C. pachypus* and *C. vitiensis* (Baker *et al.* 2003), and falls within Henderson's (2020) *C. adspersus* Blume group. All species of the complex are typically single-stemmed and bear rather broad, hooded leaflets that are only lightly armed on their surfaces and margins. Inflorescences lack a flagelliform tip and the bracts on the primary axis are long, robust, strictly tubular and possess an acute, triangular limb to one side of the mouth. The bracts of the first-order branches (and second-order branches in male inflorescences) are very different from the primary bracts, being short and funnel-shaped. Female flowers and fruit are borne on a short, pedicel-like stalk. The seed is characteristically globose with numerous pits and irregular channels throughout the surface.

Calamus aruensis. LEFT TO RIGHT: leaf sheath, cultivated National Botanic Garden, Lae (WB); juvenile in forest understorey, Lake Kutubu (WB).

Calamus aruensis is the most common and widespread of all rattans in New Guinea. It bears cirrate leaves with typically regularly arranged leaflets and is armed with numerous stiff, dark, triangular spines throughout the sheath. However, armature density can vary widely and unarmed forms are well known, sometimes occurring together with armed forms. It is distinguished from all other members of the *C. aruensis* complex by its cirrus, which is armed with recurved, grapnel-spines that are distributed irregularly, rather than being regularly arranged in neat whorls, as is typical of cirrate *Calamus* species (Baker *et al.* 2003). This feature is not seen in any other *Calamus* species in New Guinea or the West Pacific, though it does occur in a few species from elsewhere, such as *C. pogonacanthus* Becc. ex H.Winkl. from Borneo.

4. *Calamus badius* J.Dransf. & W.J.Baker

Medium-sized, multi-stemmed rattan climbing to 25 m. **Stem** with sheaths 20–28 mm diam., without sheaths to 14–15 mm diam.; internodes 22–28 cm. **Leaf** ecirrate, to 50–100 cm long including petiole; sheath mid-green, with thin pale brown indumentum, sheath spines usually abundant, 2.5–6 × 1–1.5 mm, usually rather uniform, with swollen bases, black, densely covered in indumentum when newly emerged, horizontal, solitary or very rarely paired; knee 41–47 × 7–9.5 mm, armed as sheath; *ocrea 33–80 × 14–26 mm, erect or slightly diverging from sheath, open on the far side of the sheath to the petiole, coriaceous, persistent, same colour as sheath, armed with scattered spines as the sheath, usually more densely so*; flagellum present, to 3–4 m long; petiole 12–20 cm long; *leaflets 6–8 on each side of rachis, regularly arranged or in distant groups of 2, broadly elliptic*, longest leaflet in mid-leaf 33–41 × 7.5–9 cm, apical leaflets 21–26 × 3.5–6 cm, usually grouped in a fan of 4, apical leaflet pair united for half their length, *lower surface with dense pale brown-coloured indumentum*, leaflet unarmed except for short bristles on margins. **Inflorescence** pendulous to 2.5 m long including peduncle 20–37 cm long and a short flagelliform tip; branched to 2 orders in the female and 3 orders in the male; primary bracts closely sheathing, splitting briefly at tip to form a short, triangular, apical lobe, armed with scattered short triangular spines; primary branches 13, to at least 36 cm long, male rachillae 10–40 × ca. 1.5 mm, female rachillae 2–6.5 × 1.5 mm. **Fruit** not seen. **Seed** not seen.

Calamus badius. Leaf sheath with ocrea, near Timika (JD).

DISTRIBUTION. South-western New Guinea.

HABITAT. Lowland forest at 30–100 m.

LOCAL NAMES. *Bobnong* (Kati).

USES. None recorded.

CONSERVATION STATUS. Vulnerable. *Calamus badius* has a restricted distribution. Deforestation due to human activity in a mining concession within its distribution is a major threat in its range.

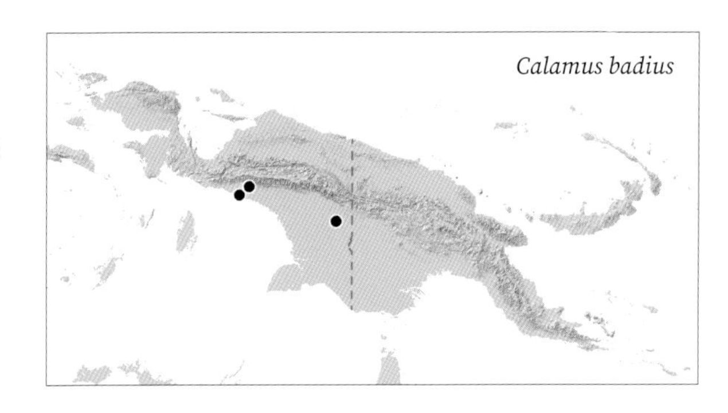

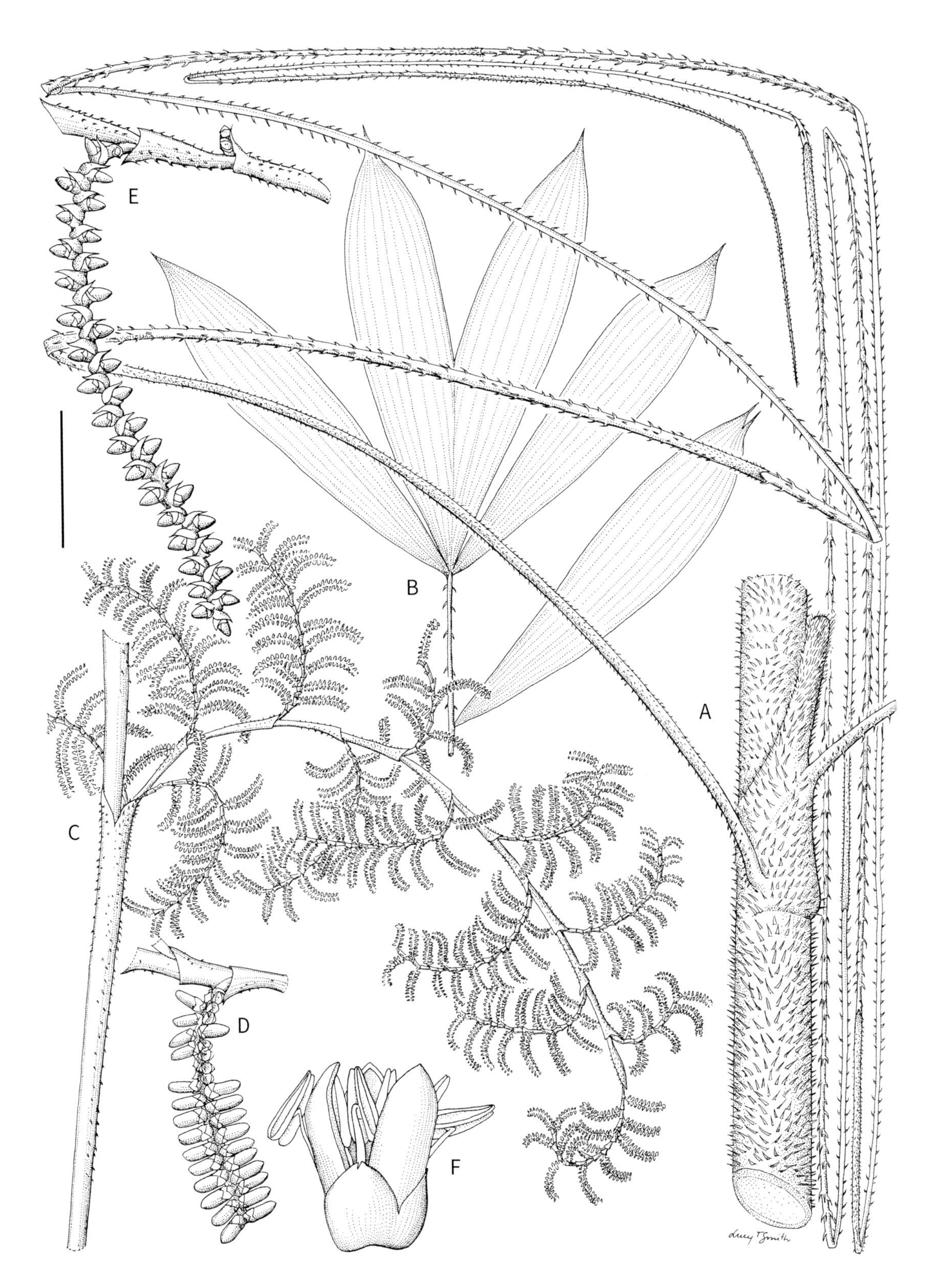

Calamus badius. **A.** Stem with leaf sheath and flagellum. **B.** Leaf apex. **C.** Male inflorescence. **D.** Male rachilla with flowers. **E.** Female rachilla with flowers. **F.** Male flower. Scale bar: A = 4 cm; B = 8 cm; C = 6 cm; D, E = 1.5 cm; F = 2 mm. A–D, F from *Dransfield et al. JD 7661*; F from *Dransfield et al. JD 7650*. Drawn by Lucy T. Smith.

NOTES. This flagellate species is distinguished by its few, broadly elliptic leaflets that are strongly discolorous (mid to dark green on the upper surface and bearing beige-brown indumentum on the lower surface), a conspicuous, leathery, slightly divergent ocrea up to 8 cm long, and dense stout spines throughout the sheath. It is one of only two species of *Calamus* in New Guinea with discolorous leaflets, the other being the unrelated *C. zieckii*, which differs in its cirrate leaves with grouped, linear leaflets and non-flagellate inflorescences.

5. *Calamus baiyerensis* W.J.Baker & J.Dransf.

Very robust, multi-stemmed rattan climbing to ca. 20 m. Stem with sheaths 45–55 mm diam., without sheaths to 15–18 mm diam.; internodes to at least 18 cm. Leaf ecirrate, to 1.7 m long including petiole; sheath green, with patchy brown-black indumentum, *spines sparse, to 10 × 2 mm, flattened, triangular, solitary,* sheath unarmed in parts; knee ca. 80 mm long, unarmed; *ocrea to ca. 18 cm long, encircling stem, disintegrating into dry, brown fibres, unarmed*; flagellum 5–8 m long, very robust; petiole length unknown; leaflets 32–35 each side of rachis, regularly to subregularly arranged, linear-lanceolate, longest leaflets near base ca. 60 × 3.5 cm, mid-leaf leaflets ca. 55 × 3.5 cm, apical leaflets ca. 26 × 1.4 cm, apical leaflet pair united to one quarter of their length, armed with scattered black bristles on upper surface and margins. Inflorescence very robust, length not known, female not seen, branched to 3 orders in the male; primary bracts tubular and tattering at mouth, armed with grapnel spines; primary branch number not known, to at least 58 cm long, male rachillae 15–43 × 5–6 mm, female rachillae not seen. Fruit not seen. Seed not seen.

DISTRIBUTION. Baiyer River valley, Western Highlands Province, Papua New Guinea.

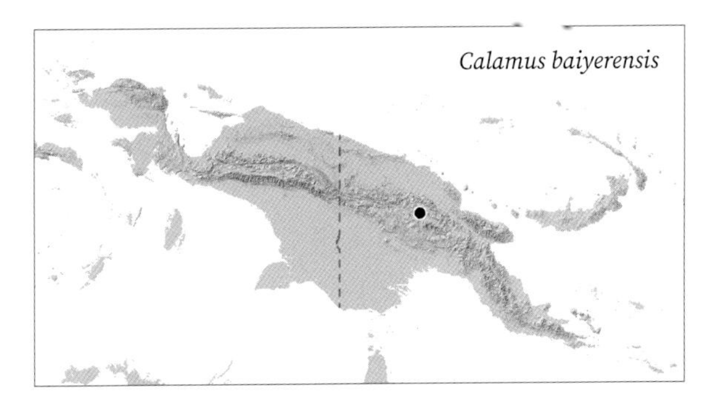

HABITAT. Disturbed mid-montane forest, 1,200 m.

LOCAL NAMES. *Kela, Sintsch* (Baiyer River valley).

USES. None recorded.

CONSERVATION STATUS. Critically Endangered. The only known site for *C. baiyerensis* is threatened by its close proximity to a gold mine.

NOTES. Several higher elevation, flagellate New Guinea *Calamus* species (the *C. baiyerensis* group) are united by a set of features found in *C. baiyerensis*, namely well-developed, disintegrating ocreas, stout rachillae with prominent funnel-shaped rachilla bracts and somewhat similar leaves and lanceolate leaflets, although leaflet arrangement varies (Baker & Dransfield 2014, 2017). It is not however clear whether these species (*C. baiyerensis, C. klossii, C. nanduensis, C. pintaudii, C. pseudozebrinus*) are actually related to each other. *Calamus womersleyi* also resembles this group, although its ocrea is not well-developed, being short, tubular, truncate and persistent. *Calamus sashae* also shares similar rachilla morphology, but this lower elevation species apparently lacks a well-developed ocrea.

Calamus baiyerensis is a massive rattan, perhaps mostly easily confused with *C. pintaudii* (see Baker & Dransfield (2017) for detailed comparison). In addition to the shared characters described above, it is distinguished by its very few, triangular spines, the encircling, straw-coloured ocrea that tatters to fibres in *C. baiyerensis*, and the robust leaflets.

6. *Calamus bankae* W.J.Baker & J.Dransf.

Moderately robust, multi-stemmed rattan climbing to 15 m. **Stem** with sheaths ca. 11 mm diam., without sheaths to ca. 9 mm diam.; internodes ca. 25 cm. **Leaf** ecirrate to ca. 74 cm long including petiole; sheath pale yellowish green, with patches of very thin, orange-brown indumentum, spines numerous, 1–2.5 × 1–1.5 mm, triangular, colour as sheath, but with black tips, solitary, some deflexed, very few erect; knee 20–22.5 mm long, armed as sheath, *ocrea 24–28.5 × 1.6 cm, inflated, tubular, splitting longitudinally with age, clasping and usually obscuring sheath, papery, brown, with numerous needle-like spines 2–4 mm long, tattering to fibres, eventually disintegrating completely;* flagellum to ca. 150 cm long; petiole ca. 20 cm; *leaflets ca. 19 each side of rachis, arranged in three widely-spaced groups of 5–8 leaflets,* regularly spaced within each

Calamus baiyerensis. **A.** Stem with leaf sheath and petiole. **B.** Leaf apex. **C.** Leaf mid-portion. **D.** Portion of male inflorescence. Scale bar: A = 3 cm; B, C = 6 cm; D = 4 cm. All from *Zieck NGF 36252*. Drawn by Lucy T. Smith.

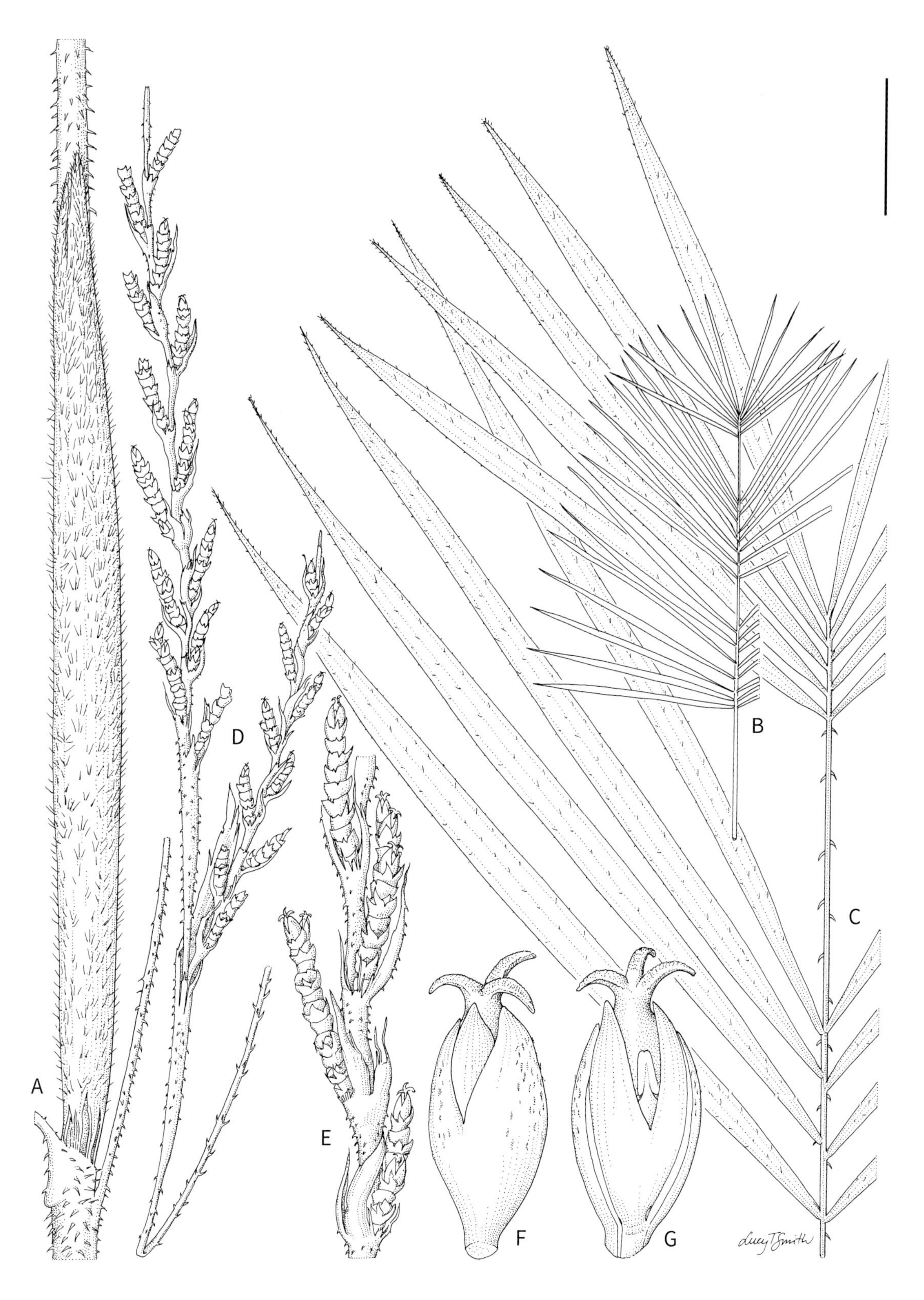

Calamus bankae. **A.** Stem with leaf sheath and ocrea. **B.** Whole leaf diagram. **C.** Leaf apex. **D.** Female inflorescence. **E.** Portion of female inflorescence. **F.** Female flower. **G.** Female flower, one sepal removed. Scale bar: A, C = 4 cm; B = 18 cm; D = 3 cm; E = 1.5 cm; F, G = 2.2 mm. All from *Baker et al. 1097*. Drawn by Lucy T. Smith.

group, but slightly divaricate, linear, longest leaflet towards base 29.5 × 1.5 cm, mid-leaf leaflets 29 × 1.2 cm, apical leaflets 18 × 0.9 cm, apical leaflet pair not united, margins and both surfaces armed with bristles to 3 mm. **Inflorescence** very slender, 1.5 m long including 1.4 m peduncle, lacking flagelliform tip, branched to 2 orders in the female, male not seen; primary bracts tubular, not splitting, tattering at mouth, with minute spines and scattered grapnel spines; primary branches ca. 2 (incomplete material seen), *to 16.5 cm long, rather congested,* male rachillae not seen, female rachillae 8.5–20 mm × ca. 4 mm. **Fruit** not seen.

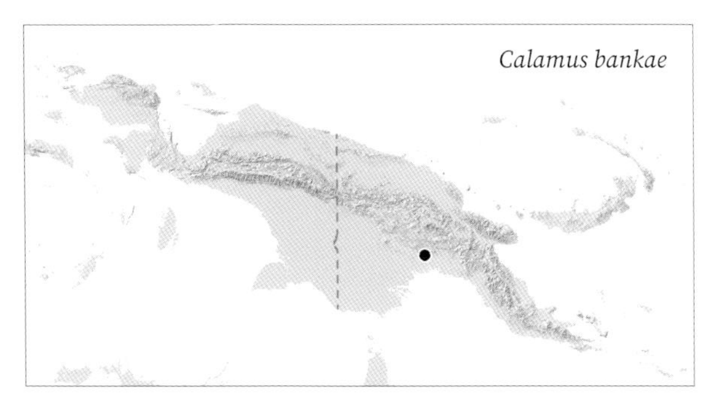

DISTRIBUTION. Kikori River catchment, Papua New Guinea.

HABITAT. Lowland forest on limestone, at 50 m.

LOCAL NAMES. None recorded.

USES. None recorded.

CONSERVATION STATUS. Critically Endangered. *Calamus bankae* is known from only one site where it is threatened by logging concessions.

NOTES. This flagellate species belongs to the *C. longipinna* group (see under *C. longipinna*). Its tubular, inflated ocrea is armed with fine, needle-like spines and readily tatters. It has rather few leaflets (ca. 19 each side of the rachis), which are arranged in widely spaced groups, and inflorescences with relatively congested primary branching systems. It is most easily confused with *C. longipinna*, *C. vestitus* and *C. wanggaii*.

7. *Calamus barbatus* Zipp. ex Blume

Synonyms: *Daemonorops barbata* (Zipp. ex Blume) Mart., *Palmijuncus barbatus* (Zipp. ex Blume) Kuntze, *Rotang barbatus* (Zipp. ex Blume) Baill.

Slender to moderately robust, multi-stemmed rattan climbing to 15 m. Stem with sheaths 15–30 mm diam., without 8–10 mm diam.; internodes 10–15 cm. Leaf ecirrate, 0.6–1.1 m long including petiole; sheath green to brown, with dense, thin, dark indumentum, *spines numerous, 1–36 × 1–3 mm, various sizes, flexible,* sometimes sinuous, dark, solitary and connected in whorls, sheath mouth densely armed with longer, more erect spines; knee 2–8 mm long, armed or unarmed; ocrea insonspicuous or lacking; flagellum to 2.3 m; petiole 1–15 cm; *leaflets 23–43 each side of rachis, arranged regularly,* or sometimes subregularly with some sections of rachis lacking leaflets, narrow lanceolate, longest leaflet at mid-leaf position 13–33 × 1–2 cm, apical leaflets 7–14 × 0.4–1 cm, apical leaflet pair scarcely united or joined for up to half of their length, leaflets armed with dark bristles on margins and upper surface, sometimes also on lower surface. Inflorescence erect to arching, 90–150 cm long including 11–35 cm peduncle and 35–70 cm flagelliform tip, branched to 2 orders in the female and

Calamus barbatus. Habit, near Sorong (WB).

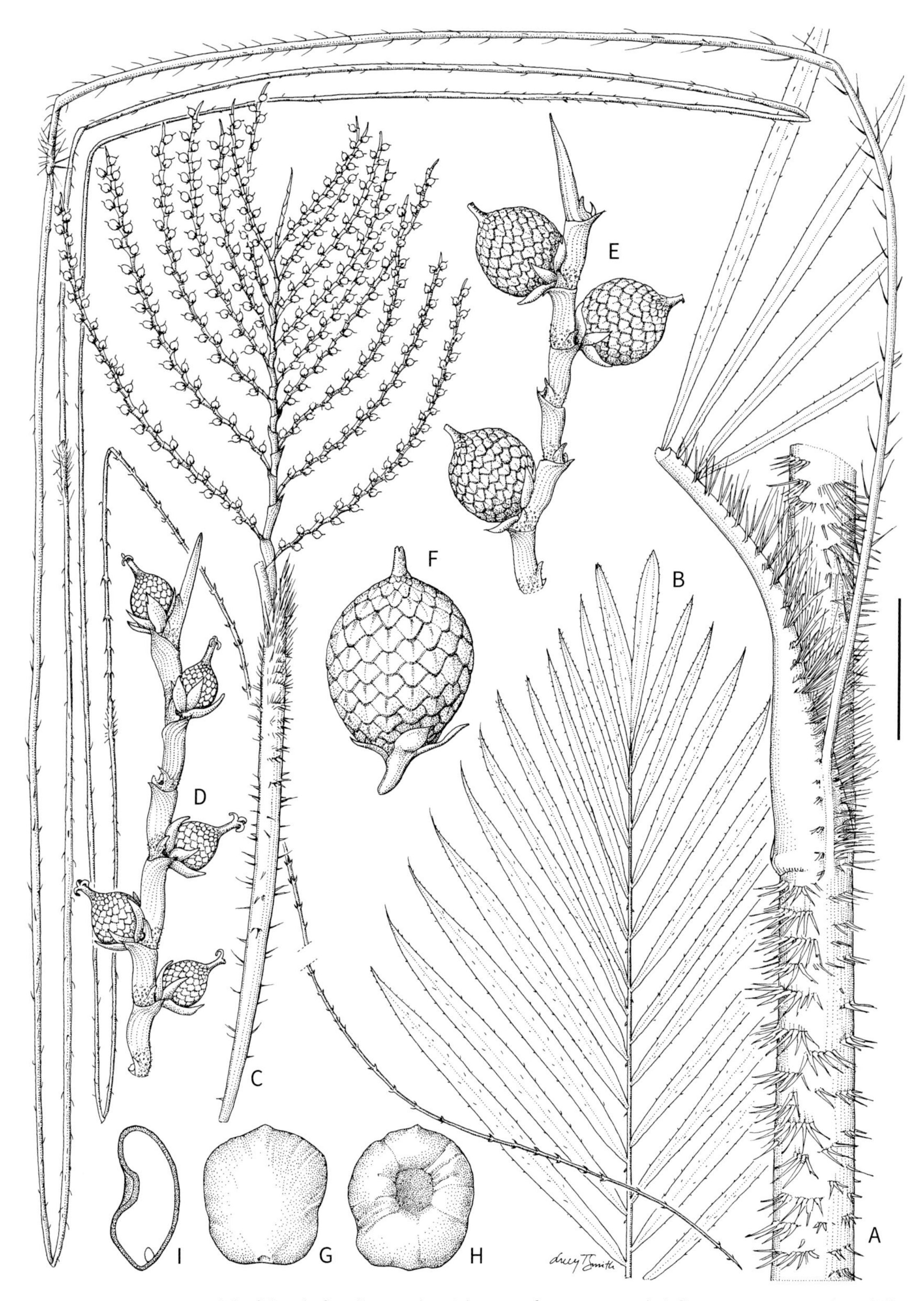

Calamus barbatus. **A.** Stem with leaf sheath, flagellum and petiole. **B.** Leaf apex. **C.** Female inflorescence. **D.** Female rachilla with immature fruit. **E.** Portion of female rachilla with fruit. **F.** Fruit. **G**, **H.** Seed in two views. **I.** Seed in longitudinal section. Scale bar: A = 4 cm; B, C = 6 cm; D, E = 1.5 cm; F = 1 cm; G–I = 7 mm. A from *Dransfield et al. JD 7543*; B–I from *Dransfield et al. JD 7550*. Drawn by Lucy T. Smith.

Calamus barbatus. Leaf sheaths and base of Inflorescence, near Sorong (WB).

3 orders in the male; primary bracts tubular, splitting asymmetrically at the mouth, armed, *typically with tuft of longer denser spines at bract mouth*; primary branches 2–6, to 35 cm long, male rachillae 3–10 × 0.4–0.5 cm, female rachillae 7–25 × 0.4–0.6 cm, *rachillae with pointed tips formed by empty rachilla bracts*. Fruit solitary, globose with prominent beak, 14–17 × 11–12 mm, scales yellow with dark margin. Seed (sarcotesta removed) 7–9.5 × 7–8.5 × 4–5.5 mm, globose, with deep lateral pit and shallow furrows; endosperm homogeneous.

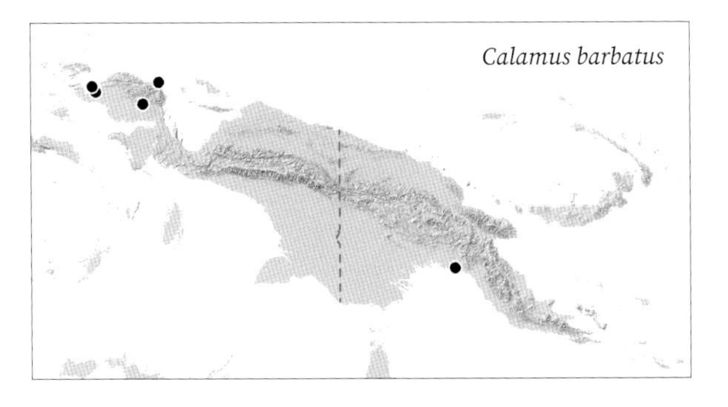

DISTRIBUTION. Bird's Head Peninsula with one outlier record in south-eastern New Guinea.

HABITAT. Mostly lowland forest from sea level to 630 m.

LOCAL NAMES. *Bomeika* (Unspecified language, Ransiki area), *Meyah mocosu* (Meyah).

USES. Used for binding.

CONSERVATION STATUS. Least Concern. However, the species may be impacted by oil palm plantations and logging concessions in its range.

NOTES. This distinctive, flagellate species was one of the first New Guinean species of rattan to be described scientifically. Its leaves bear regularly or slightly irregularly arranged leaflets, the leaf sheaths are densely armed with fine spines and the inflorescence primary bracts bear a tuft of spines at their tips. The rachillae are somewhat elongate due to the relatively long, tubular rachilla bracts that separate the flowers. The distalmost rachilla bracts are empty and form a pointed tip to the rachilla.

8. *Calamus barfodii* W.J.Baker & J.Dransf.

Moderately slender, multi-stemmed rattan climbing to 30 m. **Stem** with sheaths 8–16 mm diam., without sheaths 5.5–10 mm diam. **Leaf** ecirrate, to 72–78 cm long including petiole; sheath pale green, glaucous, with very thin pale indumentum, spines numerous, 0.5–4 × 0.5–1.2 mm, triangular, tips sometimes dark, solitary; knee 17–27 mm long, armed as sheath; *ocrea 9–17 × 1–2.7 cm, inflated, boat-shaped, open longitudinally to base on side opposite petiole insertion, clasping and usually obscuring sheath, tough, brown, dark purple when young, armed as sheath, persistent, apparently inhabited by ants*; flagellum 90–140 cm long; petiole ca. 2–4 cm long; leaflets 23–26 each side of rachis, regularly arranged, linear lanceolate, longest leaflet in mid-leaf 19–26 × 1.5–2 cm, apical leaflets 3.5–10 × 0.3–1.3 cm, apical leaflet pair united up to one fifth of their length or not united, very sparsely armed with dark bristles on margins and upper surface near tip. **Inflorescence** *erect to arching, 40–59 cm long including 21–40 cm peduncle, lacking flagelliform tip,* branched to 2–3 orders in the female and 3 orders in the male; primary bracts strictly tubular and somewhat inflated, slightly asymmetric at apex and sometimes splitting slightly, sparsely to densely armed as sheath; primary branches ca. 4, to 7 cm long, 5–8 cm apart, compact; male rachillae not seen, *female rachillae 4–13 mm ×*

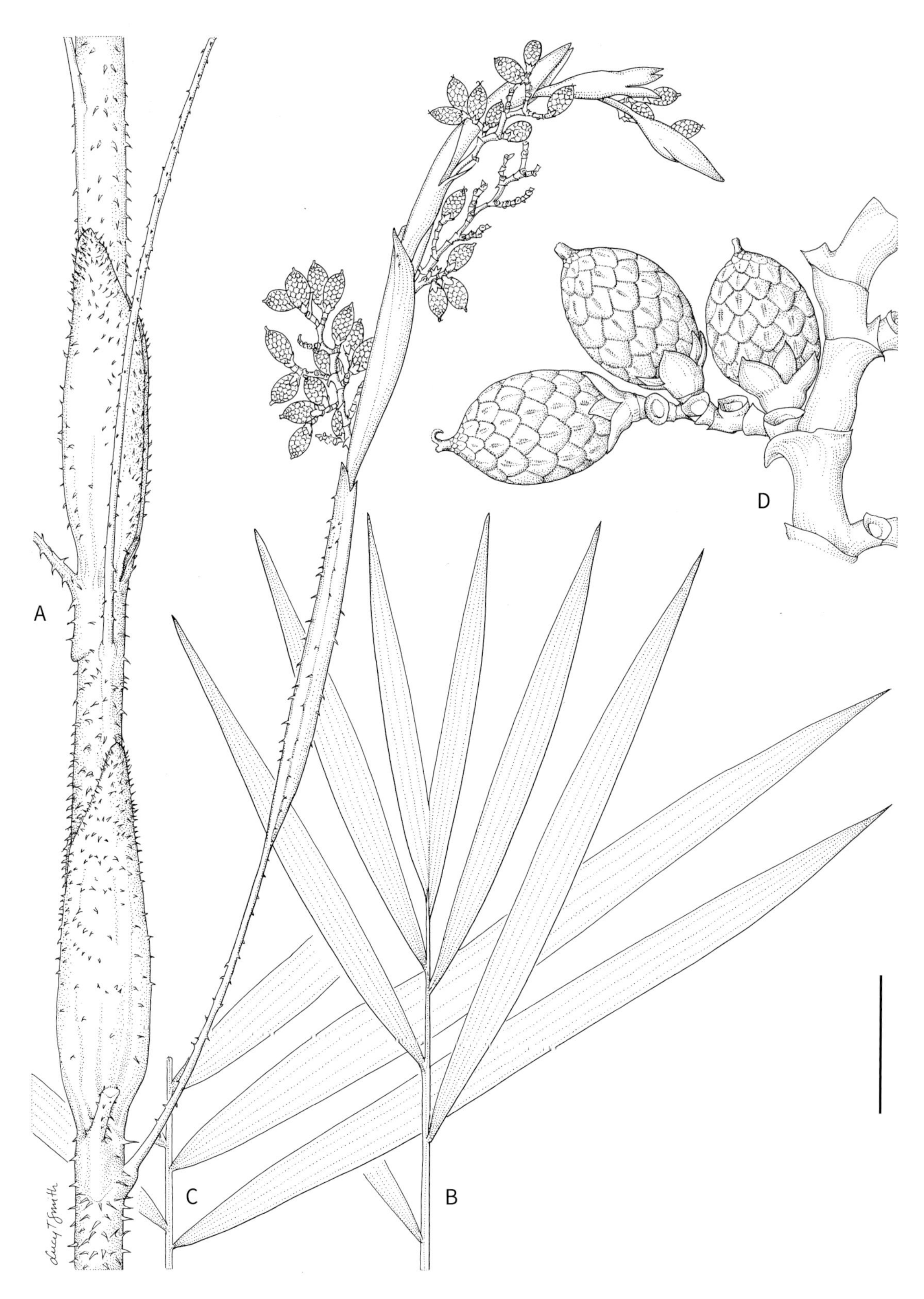

Calamus barfodii. **A.** Stem with leaf sheath, ocreas and female inflorescence. **B.** Leaf apex. **C.** Leaf mid-portion. **D.** Portion of female rachilla with fruit. Scale bar: A–C = 4 cm; D = 7 mm. A, D from *Zieck NGF 36596*; B, C from *Barfod 464*. Drawn by Lucy T. Smith.

1–1.3 mm. **Fruit** solitary (only immature specimen seen), ellipsoid, ca. 10 × 6 mm, scales light brown to yellow. **Seed** not seen.

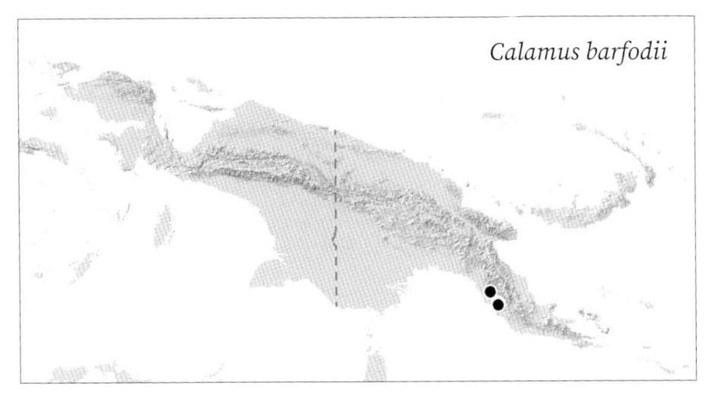

DISTRIBUTION. South side of the Owen Stanley Range, south-eastern New Guinea.

HABITAT. Lowland forest, 50–460 m.

LOCAL NAMES. None recorded.

USES. None recorded.

CONSERVATION STATUS. Endangered. *Calamus barfodii* has a restricted distribution. Deforestation due to logging concessions is a major threat in its range. Its close proximity to Port Moresby may also constitute a potential risk.

NOTES. *Calamus barfodii* is a flagellate species that displays an unusual combination of characters, consisting of a well-developed, inflated, persistent ocrea that clasps the stem (resembling that of *C. altiscandens*, *C. macrochlamys* or *C. longipinna* and similar species), a rather short, erect to arching inflorescence bearing stout, inflated bracts that split to the base by the emergence of very compact primary branching systems (similar to the *C. anomalus* group, see under *C. anomalus*) and very short rachillae.

9. *Calamus bulubabi* W.J.Baker & J.Dransf.

Slender to moderately robust, multi-stemmed rattan climbing to 30 m. **Stem** with sheaths 11–25 mm diam., without sheaths to 9–14 mm diam.; internodes 10–36 cm. **Leaf** ecirrate, to 0.8–1.3 m long including petiole; *sheath dark green, sometimes with thick, white, woolly indumentum, spines dense, fine, flexible, hair-like, to 25 mm long, brown to black*; knee 26–48 mm long, armed as sheath; ocrea not well-developed, a low crest to 5 mm high; flagellum to 2.5 m long; petiole 1–22 cm; *leaflets 10–16 each side of rachis, mainly arranged in divaricate pairs, elliptic, hooded*, longest leaflets at base or mid-leaf position, mid-leaf leaflets 14–33 × 3.5–4.2

Calamus bulubabi. LEFT TO RIGHT: habit; leaf sheaths. Near Tabubil (WB).

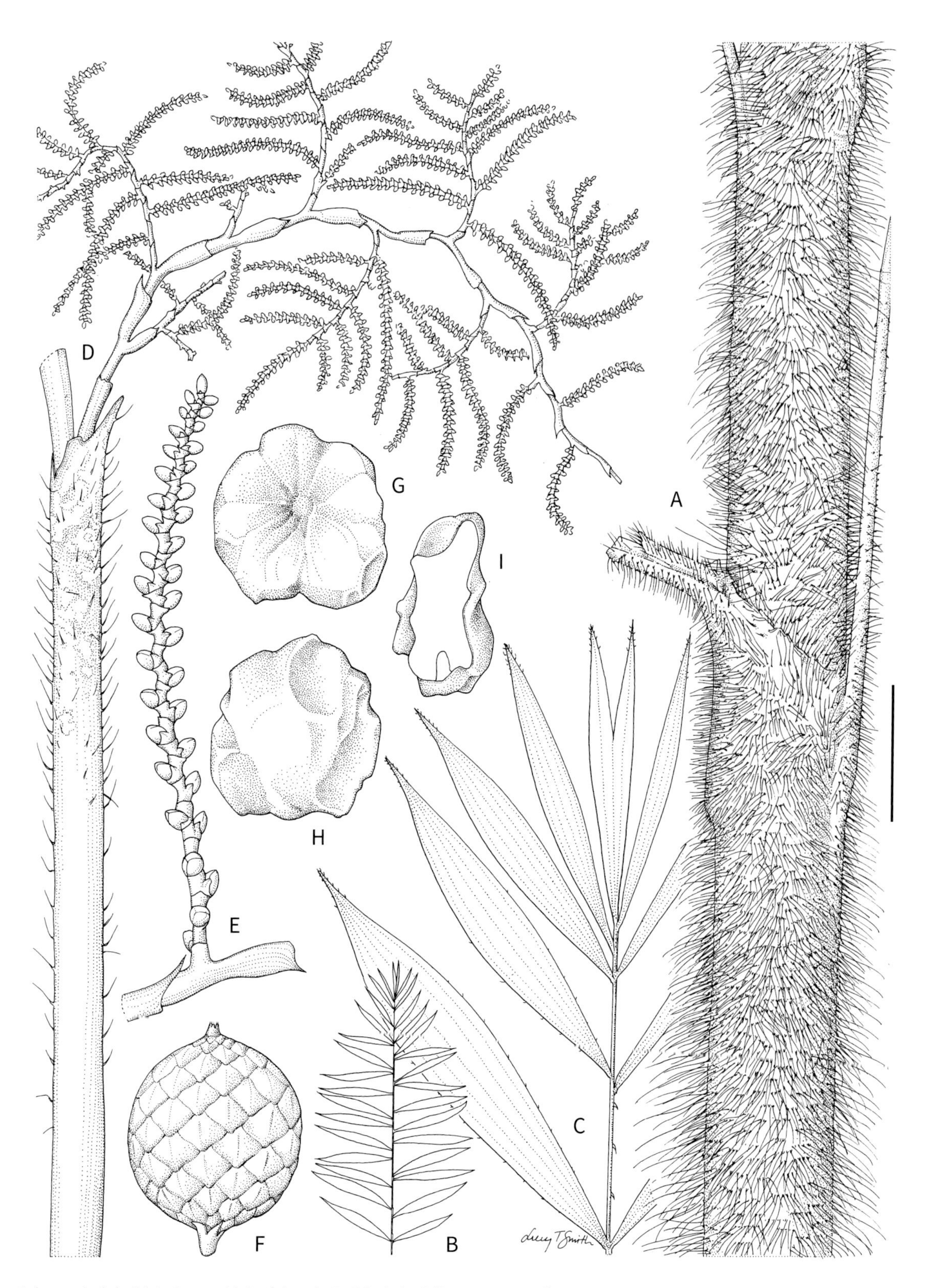

Calamus bulubabi. **A.** Stem with leaf sheath. **B.** Whole leaf diagram. **C.** Leaf apex. **D.** Portion of male inflorescence. **E.** Male rachilla with flowers. **F.** Fruit. **G, H.** Seed in two views. **I.** Seed in longitudinal section. Scale bar: A = 3 cm; B = 0.6 m; C = 6 cm; D = 4 cm; E, F = 1 cm; G–I = 7 mm. A–E from *Baker et al. 1128*; F–I from *Baker et al. 1094*. Drawn by Lucy T. Smith.

cm, apical leaflets 12–15 × 1–3 cm, apical leaflet pair united from half to two thirds of their length, leaflets armed with very few bristles on margin and upper surface, leaflets sometimes bearing woolly indumentum. **Inflorescence** arching, 0.6–2.2 m long including 15–33 cm peduncle, lacking a flagelliform tip, branched to 2 orders in the female (sometimes 3 orders at base) and 3 orders in the male (sometimes 4 orders at base); primary bracts tubular, somewhat inflated, opening asymmetrically at apex, unarmed or sparsely armed and indumentose as sheath; primary branches 6–13, to 30 cm long, male rachillae 30–50 × ca. 1 mm, female rachillae 25–90 × 1.3–2 mm. **Fruit** solitary, spherical, ca. 13 × 13 mm scales brown. **Seed** (sarcotesta removed) 9 × 9 × 4.5 mm, *discoid, concave and smoother on one side, convex and with angular sculpturing on the other side*; endosperm homogeneous.

DISTRIBUTION. Widespread in southern New Guinea.

HABITAT. Lowland primary and secondary forest habitats up to 300 m.

LOCAL NAMES. *Donggieb* (Kati), *Kurni* (Biaru), *Tipa* (Awin).

USES. For tying timbers in construction.

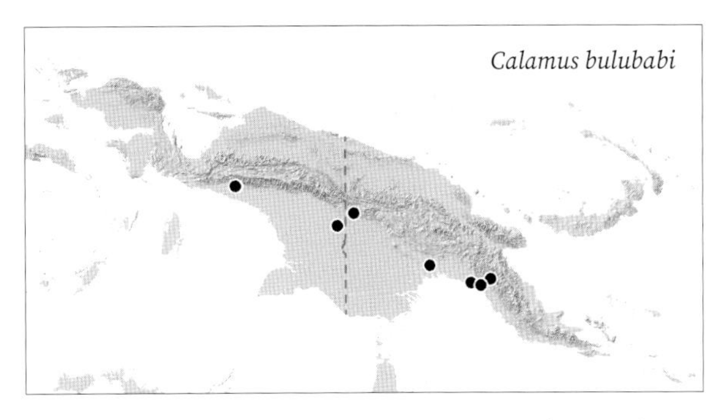

CONSERVATION STATUS. Least Concern. The species may be affected by logging concessions within its distribution.

NOTES. *Calamus bulubabi* is a flagellate species recognised by the combination of broadly elliptic leaflets grouped in divaricate pairs and "hairy" leaf sheaths that are densely covered in fine, flexible, hair-like spines, often with dense, matted, white indumentum between the spines and on other parts. In leaf, leaflet and inflorescence morphology it resembles *C. papuanus* and *C. spiculifer*, but it can be immediately distinguished by the hair-like spines and distinctive angular seed. It also bears a resemblance vegetatively to *C. distentus* though this species has a well-developed ocrea, unlike *C. bulubabi*.

Calamus bulubabi. LEFT TO RIGHT: male inflorescence, near Tabubil (WB); male flowers, near Timika (WB).

10. *Calamus calapparius* Mart.

Synonyms: *Calamus amboinensis* Miq., *Calamus beguinii* (Burret) W.J.Baker, *Calamus komsaryi* (Maturb., J.Dransf. & Mogea) W.J.Baker, *Daemonorops beguinii* Burret, *Daemonorops calapparia* (Mart.) Blume, *Daemonorops komsaryi* Maturb., J.Dransf. & Mogea, *Palmijuncus amboinensis* (Miq.) Kuntze, *Palmijuncus calapparius* (Mart.) Kuntze

Moderately robust to very robust, multi-stemmed rattan climbing to 25 m high. **Stem** with sheaths 40–80 mm diam., without sheaths to 23–30 mm diam.; internodes 16–25 cm long. **Leaf** *cirrate,* to 5.6 m long including petiole and cirrus; sheath yellowish brown, with cream to brown indumentum, *densely armed with rather irregular partial whorls of needle-like, brown to black spines,* the longest to 50 mm long, usually much less, caducous, irregular spine bases persisting, black spinules also present; knee 90–60 mm long, armed; ocrea inconspicuous; flagellum absent; petiole 20–55 cm; leaflets 68–91 on each side of rachis, regularly arranged, linear, acuminate, the longest leaflet 64 × 3.5 cm, mid-leaf leaflets 26–49 × 1.2–3.4 cm, with bristles on the lower surface and margins; cirrus 1–2 m long, cirrus grapnel spines arranged regularly. **Inflorescence** 60–135 cm long including 3–15 cm peduncle, curving away from the stem, branched to 2 orders in the female and 3 orders in the male; *primary bracts boat-shaped, sometimes tubular, splitting down one side and often falling at anthesis,* leathery, densely armed with fine brown to black spines; primary branches 5–10, to 35 cm long, male rachillae not seen, female rachillae 30–70 × 1–3 mm. **Fruit** solitary, with short, stout stalk, spherical, 22–28 × 18–21 mm with yellowish to purple-red scales. **Seed** (sarcotesta removed) 13–16 × 15–16 × 10–11 mm, spherical, shallowly channeled from the top to the base, with a deep pit at base; *endosperm ruminate.*

DISTRIBUTION. North-western Bird's Head Peninsula and the Raja Ampat Islands, Maluku (Ambon, Bacan, Buru, Halmahera, Obi) and Talaud Islands.

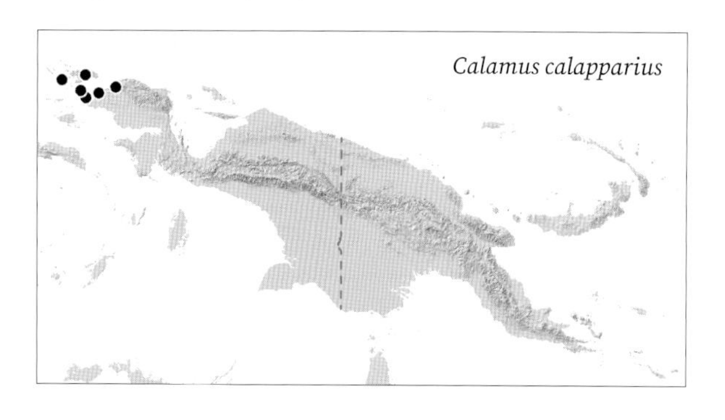

Calamus calapparius. FROM TOP: habit; fruit. Tamrau Mountains (WB).

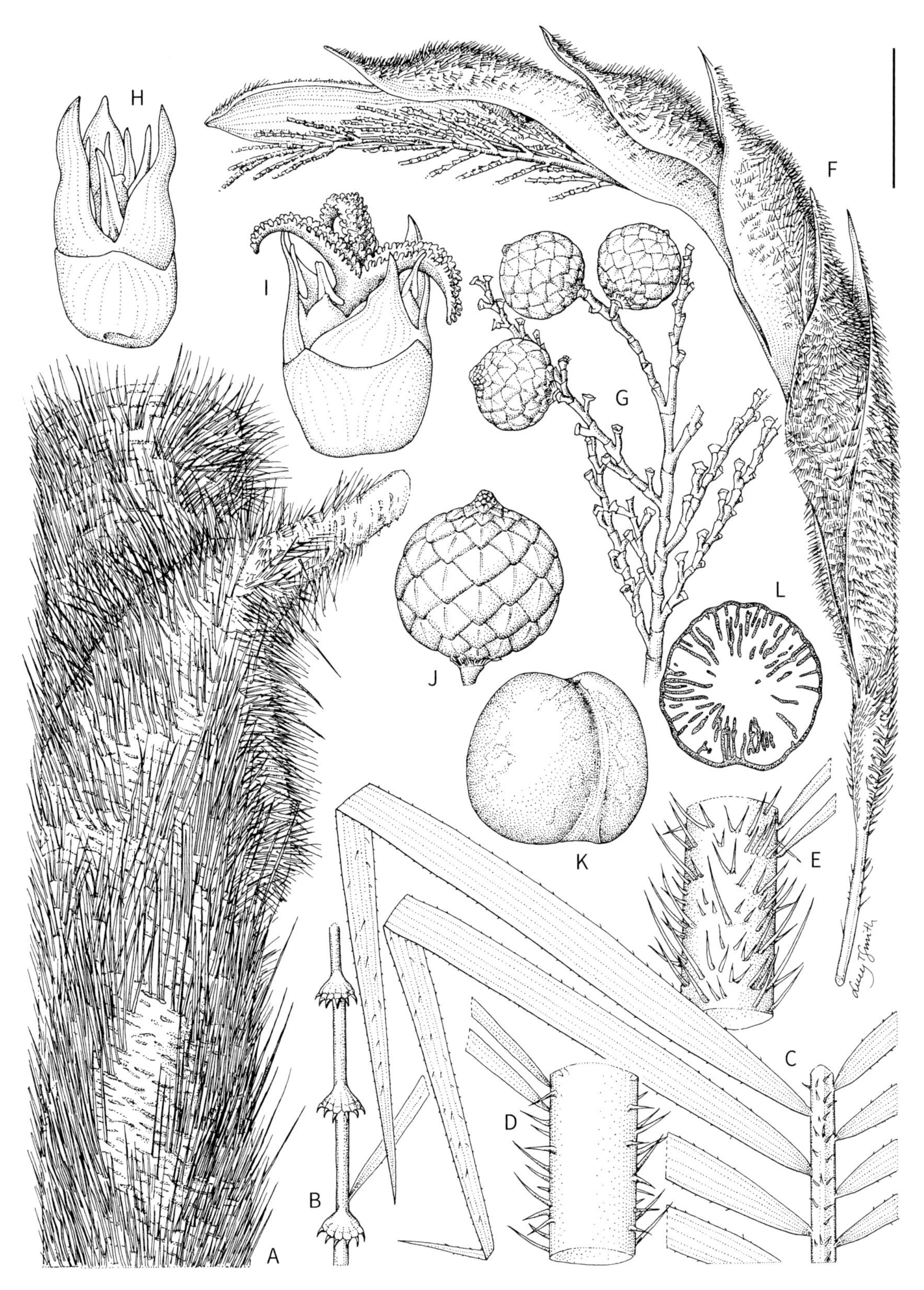

Calamus calapparius. **A.** Stem with leaf sheath and petiole. **B.** Leaf apex with cirrus. **C.** Leaf mid-portion. **D, E.** Upper and lower surface of petiole. **F.** Male inflorescence. **G.** Portion of female inflorescence with fruit. **H.** Male flower. **I.** Female flower. **J.** Fruit. **K, L.** Seed whole and in transverse section. Scale bar: A, D, E = 4 cm; B, F = 6 cm; C = 8 cm; G = 2.5 cm; H = 3.3 mm; I = 5 mm; J–L = 1.5 cm. A–F, H, I from *Maturbongs 513*; G, J–L from *Maturbongs 282*. Drawn by Lucy T. Smith.

Calamus calapparius. LEFT TO RIGHT: female and sterile male flowers; female inflorescence. Tamrau Mountains (WB)

HABITAT. Primary and secondary forest at 30–120 m.

LOCAL NAMES. *Dou-aise* (Waigeo), *Wil-he* (Batanta), *Wil-hne* (Salawati), Sane (Gebe).

USES. The cane is used for furniture. In Batanta Island, the leaf is used for roofing traditional houses.

CONSERVATION STATUS. Least Concern.

NOTES. *Calamus calapparius* is the only species of the previously accepted genus *Daemonorops* in New Guinea. *Daemonorops* was placed in synonymy under *Calamus* based on abundant molecular evidence (Baker 2015). Nevertheless, the morphological features of the *Daemonorops* group render *C. calapparius* highly distinctive among other New Guinea palms, especially the splitting, deciduous primary primary inflorescence bracts and ruminate endosperm. When sterile, *C. calapparius* might be confused with *C. warburgii*, as a robust, cirrate rattan with fine needle-like leaf sheath spines that are readily detached, although the spines of *C. calapparius* tend to be blacker in colour.

Maturbongs *et al.* (2015) erected an endemic species, *C. komsaryi*, for the New Guinea records of this species. However, subsequent research on a broader geographical scale (Henderson 2020) has led to the placement of *C. komsaryi* in synonymy with other species under a broadly defined *C. calapparius*.

11. *Calamus capillosus* W.J.Baker & J.Dransf.

Slender, single-stemmed rattan climbing to ca. 10 m. **Stem** with sheaths 12–15 mm diam., without sheaths 5–6 mm diam.; internodes 14–17 cm. **Leaf** ecirrate to 65–90 cm long including petiole; sheath pale green, with scattered brown indumentum, *spines dense, to ca. 35 mm long, hair-like, reddish-brown*; knee 20–30 mm long, inconspicuous, obscured by spines; ocrea absent; flagellum to at least 1.9 m long; petiole 8–11 cm long; *leaflets 48–53 each side of rachis, arranged regularly, linear,* longest leaflets at mid-leaf position 16.5–18 × 0.9–1.1 cm, apical leaflets ca. 10 × 0.4–0.6 cm, apical leaflets not united, leaflets armed with numerous brown-black bristles on margins and both surfaces. **Inflorescence** *flagelliform, ca. 3.2 m long,* flagelliform tip only weakly developed, branched to 2 orders in the male (limited material seen), female not seen; primary bracts strictly tubular, splitting slightly at apex, armed with fine spines and grapnel spines; primary branches 7, to 42 cm long; male rachillae 30–55 × 0.8–1 mm, female rachillae not seen. **Fruit** not seen. **Seed** not seen.

DISTRIBUTION. Known only from one record in the central Bird's Head Peninsula.

HABITAT. Disturbed forest at ca. 450 m.

LOCAL NAMES. None recorded.

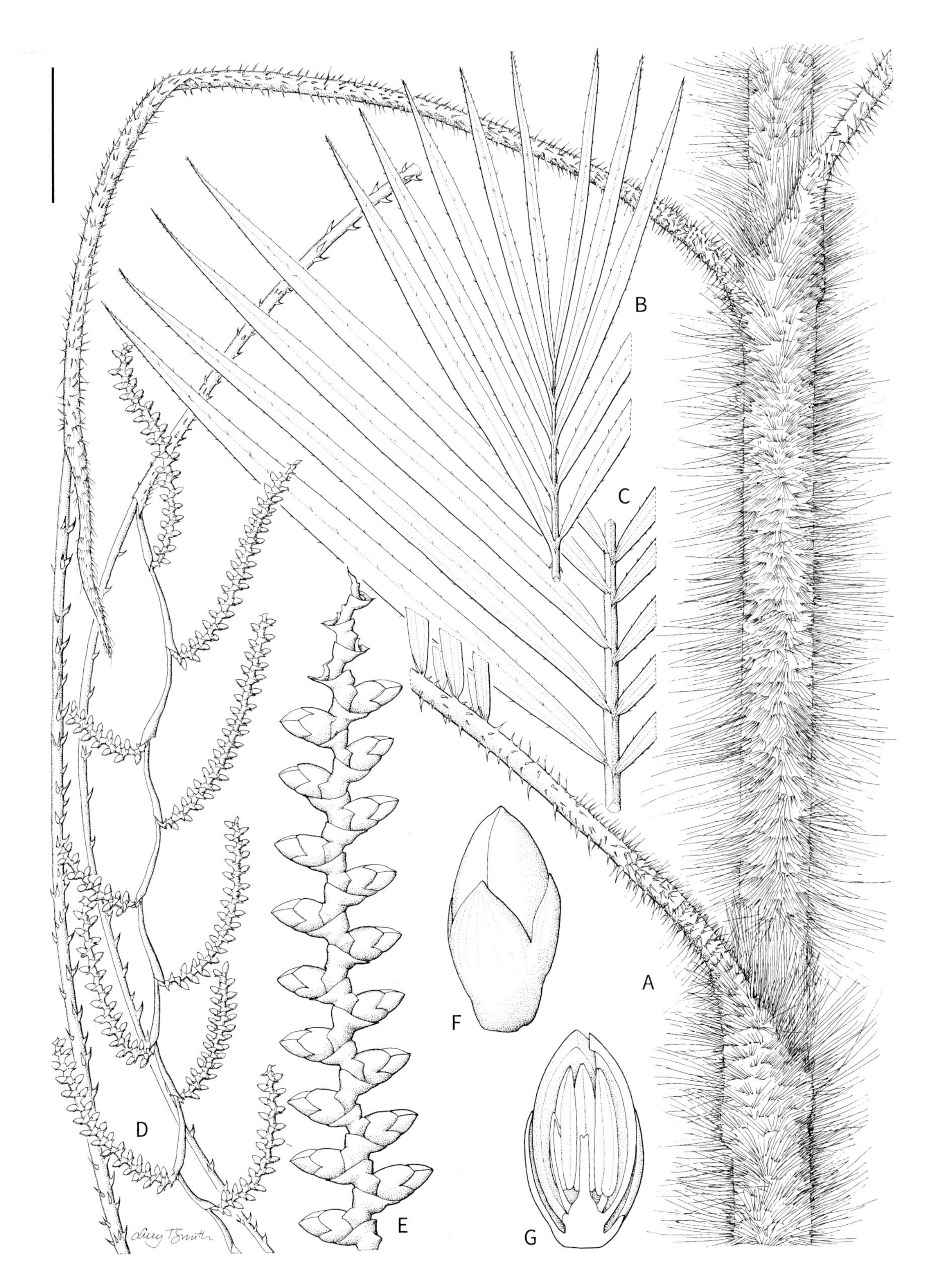

Calamus capillosus. **A.** Stem with leaf sheath, petiole and male inflorescence. **B.** Leaf apex. **C.** Leaf mid-portion. **D.** Portion of male inflorescence. **E.** Male rachilla with flowers. **F, G.** Male flower whole and in longitudinal section. Scale bar: A, D = 2.5 cm; B, C = 4 cm; E = 7 mm; F, G = 1.6 mm. All from *Avé 4048*. Drawn by Lucy T. Smith.

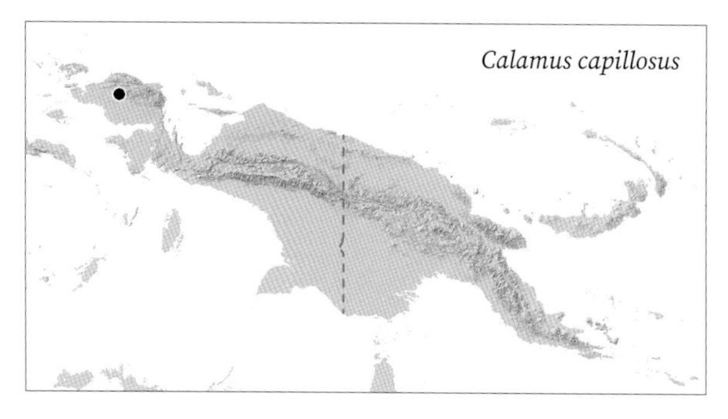

USES. None recorded.

CONSERVATION STATUS. Critically Endangered. Deforestation due to oil palm plantations and logging concessions is a major threat in the vicinity of the type locality.

NOTES. *Calamus capillosus* is a flagellate species that is likely to be confused only with *C. schlechterianus* on account of its sheaths densely armed with very fine, hair-like spines and regularly pinnate leaves with linear leaflets. The species is much more slender than *C. schlechterianus*, however, and is entirely different in reproductive form, producing very long flagelliform inflorescences with a long peduncle, and primary branches that are elongate, lax and branched only to two orders in the male material seen by us. In contrast, *Calamus schlechterianus* has much more robust, non-flagelliform inflorescences, with relatively compact and highly branched primary branching systems. It is also superficially similar to the Australian *C. australis* Mart.

12. *Calamus cheirophyllus* J.Dransf. & W.J.Baker

Slender rattan climbing to 4.5 m only, multi-stemmed (up to 50 stems per clump). **Stem** with sheaths 6.5–13 mm diam., without sheaths to 6 mm diam.; internodes to 11cm. **Leaf** ecirrate, 30–50 cm long; sheath pale green, with mid-brown indumentum, sheath spines abundant, slender, persistent, very varied in length, 1–12 × 0.5–2 mm, narrow triangular, flattened, straw-coloured, scattered or sometimes in horizontal groups, horizontal or slightly reflexed, spines around the leaf sheath mouth crowded, sometimes much larger than elsewhere and erect, to 15 mm; knee inconspicuous, to 18 mm, armed as sheath; *ocrea 7–13 × 0.5–1 cm, papery, unarmed, tubular, soon splitting and disintegrating into a fibrous network*; flagellum present, to 1.8 m long; petiole 9–12 cm long; *leaflets 7–9 in total, arranged in a fan at the tip of the petiole*, narrow elliptic, 24–35 × 1–3 cm, unarmed apart from minute black bristles on margins and upper surfaces; cirrus absent. **Inflorescence** *erect, 14–34 cm long including peduncle 7–23 cm*, branched to 2 orders in the female, male not seen; primary bracts tubular and loosely sheathing, papery, with a truncate tip, becoming lacerate, unarmed or lightly armed; primary branches ca. 4, 2–6 cm long, crowded together, male rachillae not seen, female rachillae to ca. 20 × 1 mm. **Fruit** solitary, globose, ca. 10 × 10 mm, scales pale straw-coloured. **Seed** (sarcotesta removed) 7.5 × 7.5 × 6.5 mm, globose, with a longitudinal groove on one side, seed surface very shallowly grooved and dimpled, endosperm homogeneous.

DISTRIBUTION. Two localities in the Eastern Highlands and Southern Highlands of Papua New Guinea.

HABITAT. Montane forest at ca. 1,000–1,400 m.

LOCAL NAMES. *Tiwi* (Koijari).

USES. None recorded.

CONSERVATION STATUS. Data Deficient. Owing to insufficient information about ongoing threats, a single extinction risk category cannot be selected.

NOTES. This slender, montane, flagellate rattan is recognised by the superficially palmate appearance of its leaves, in which the 7–9 pinnately arranged leaflets are crowded at the apex of a relatively long petiole. A conspicuous, unarmed, inflated ocrea encircles the stem and rapidly disintegrates, and the inflorescence is short and erect with slightly inflated, papery primary bracts that tend to tatter. The species resembles *C. lauterbachii*, but differs in its smaller stature, the narrow leaflets, with less conspicuous cross veins, the less robust, unarmed ocrea, and the more laxly arranged (rather than densely congested) rachillae.

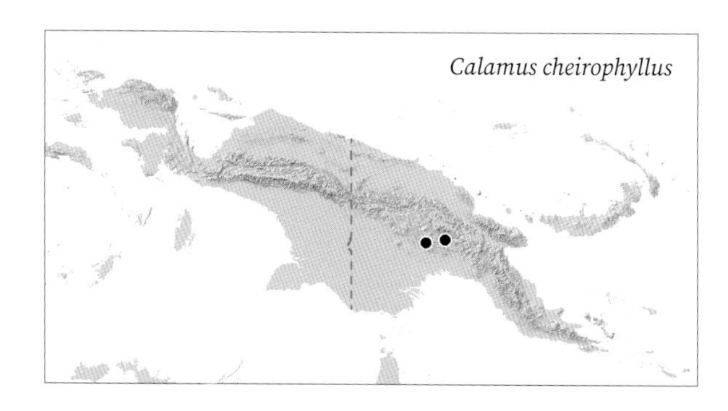

Calamus cheirophyllus. **A.** Stem with leaf sheath, leaves, male inflorescence and ocrea. **B.** Male inflorescence attached to stem. **C.** Female inflorescence with fruit attached to stem. **D.** Portion of male rachilla with flowers. **E.** Fruit. **F**, **G.** Seed in two views. **H.** Seed in longitudinal section. Scale bar: A = 6 cm; B = 2.5 cm; C = 3 cm; D = 2.5 mm; E = 1 cm; F–H = 7 mm. All from *Zieck NGF 36515*. Drawn by Lucy T. Smith.

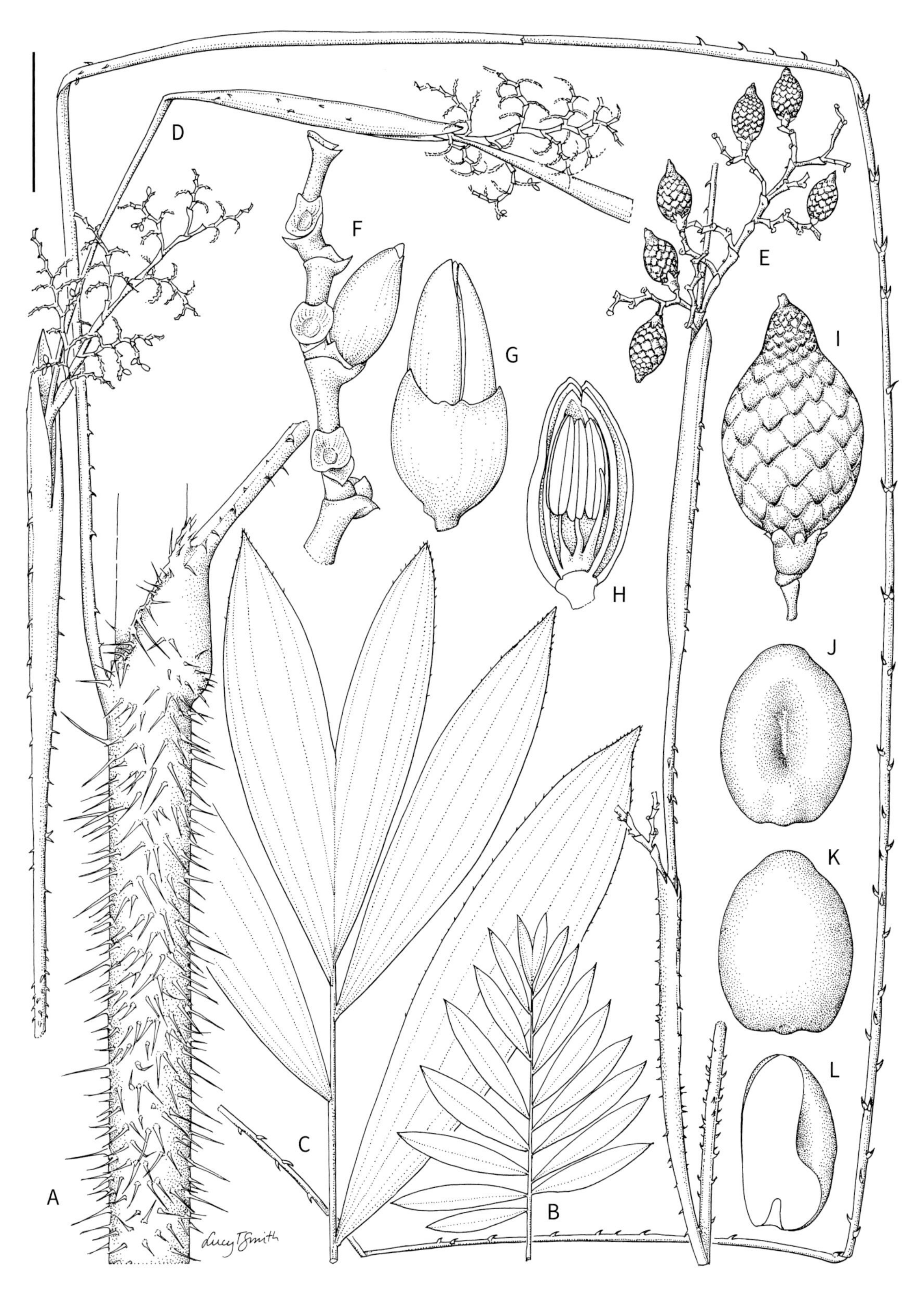

Calamus croftii. **A.** Stem with leaf sheath and flagellum. **B.** Whole leaf diagram. **C.** Leaf apex. **D.** Portion of male inflorescence. **E.** Female inflorescence with fruit. **F.** Male rachilla with flower. **G, H.** Male flower whole and in longitudinal section. **I.** Fruit. **J, K.** Seed in two views. **L.** Seed in longitudinal section. Scale bar: A = 2.5 cm; B = 24 cm; C = 4 cm; D, E = 6 cm; F = 3.3 mm; G = 2.2 mm; H = 2 mm; I = 1.5 cm; J–L = 1 cm. All from *Croft & Lelean LAE 68504*. Drawn by Lucy T. Smith.

13. *Calamus croftii* J.Dransf. & W.J.Baker

Slender rattan, climbing to 30 m or more, stem suckering not recorded. **Stem** with sheaths 10–15 mm diam., without sheaths to 6–8 mm diam.; internodes 15–31cm. **Leaf** ecirrate, to 75 cm long including petiole; sheath drying dull greenish brown, with abundant, pale grey indumentum, *spines abundant, of varying length, 3–27 × 1–2 mm, shiny straw-coloured with black tips, often edged with woolly indumentum*, evenly distributed, crowded around sheath mouth; knee to ca. 40 mm long, armed as sheath; ocrea to 10 mm long, armed as the sheath; flagellum ca. 1.25 m; petiole 10–15 cm long; *leaflets 4–8 on each side of rachis, irregularly arranged, usually in 2–4 groups, groups not always evident, broadly lanceolate with acute tips*, longest leaflet in mid-leaf 23–30 × 4.5–5.5 cm, basalmost leaflets 17–28 × 3–4.5 cm, apical leaflets 9–14 × 2–2.5 cm, apical leaflet pair united for one third of their length, leaflets armed with sinuous black bristles on margins near the tip. **Inflorescence** to ca. 1 m long, including ca. 27 cm peduncle, apparently lacking a flagelliform tip, branched to 2 orders in the female and 3 orders in the male; primary bracts tubular basally, somewhat inflated distally with an expanded triangular limb at mouth, very sparsely armed; primary branches at least 3–4, to ca. 15 cm long; male rachillae 10–30 × ca. 0.8 mm, female to ca. 30 × 1.5 mm, *female rachillae zigzag*. **Fruit** solitary, *pyriform, pointed, ca. 23 × 13 mm*, scales mid-brown with darker margins. **Seed** (sarcotesta removed) ca. 13 × 8 × 7 mm, smooth with a very deep groove on one side; endosperm homogeneous.

DISTRIBUTION. North coast of south-eastern New Guinea.

HABITAT. Lowland forest up to 150 m.

LOCAL NAMES. None recorded.

USES. None recorded.

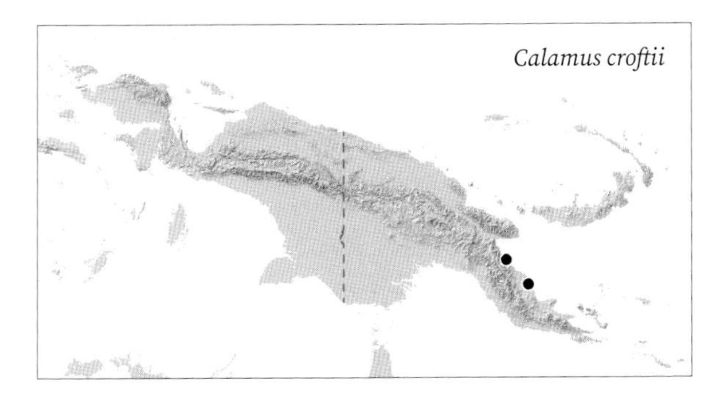

CONSERVATION STATUS. Critically Endangered. *Calamus croftii* is known from only two sites. Deforestation due to logging is a major threat to this species.

NOTES. *Calamus croftii* is a flagellate species that is recognised by its slender stems, leaf sheaths that are rather densely covered with horizontal, triangular spines, leaves with irregularly arranged lanceolate leaflets and, most conspicuously, by the large, pointed fruit widely spaced on short, zigzag rachillae. *Calamus lucysmithiae*, *C. jacobsii* and *C. johnsii* are similar in leaf and sheath morphology (though *C. jacobsii* has unarmed sheaths). It most closely resembles *C. lucysmithiae*, but this species bears a well-developed, divergent ocrea and much smaller fruit that are densely crowded on strongly recurving rachillae.

14. *Calamus cuthbertsonii* Becc.

Synonym: *Calamus brassii* Burret

Slender to very slender, multi-stemmed rattan climbing to 50 m. **Stem** with sheaths 3.5–11 mm diam., without 2.5–7 mm diam.; internodes 10–30 cm. **Leaf** ecirrate, to 26–50 cm long including petiole; sheath green, with thin, caducous, brown indumentum, spines usually numerous, 0.5–15 × 0.3–0.5 mm, triangular or needle-like, green or yellow-brown, solitary; knee 10–20 mm long, inconspicuous, colour and armature as sheath; ocrea inconspicuous, a low leathery crest, most pronounced either side of petiole base, persistent; flagellum to ca. 1.3 m; petiole 1–11 cm; *leaflets 7–14 each side of rachis, arranged regularly or subregularly, lanceolate*, longest leaflet at mid-leaf position 11–26 × 0.6–1.7 cm, apical leaflets 4.5–18 × 0.5–1.7 cm, apical leaflet pair free or united to one quarter of their length, leaflet armed with brown bristles on margins and upper surface, lower surface unarmed. **Inflorescence** *erect or trailing, 14–45 cm long including 5.5–23 cm peduncle, flagelliform tip lacking* (rarely extremely short), branched to 2–3 orders in the female and 3 orders in the male; primary bracts tubular and closely sheathing, inflating somewhat distally, lightly armed or unarmed; primary branches 2–7, to 2–12 cm long, male rachillae 6–26 mm × 0.3–0.5 mm, female rachillae 5–40 mm × 0.6–1.5 mm. **Fruit** solitary, broadly ellipsoid, 13–15 × 9–11 mm, yellow-orange. **Seed** (sarcotesta removed) 8–9.5 × 7–8 × 5.5–7 mm, globose, smooth, with deep lateral pit; endosperm homogeneous.

***Calamus cuthbertsonii*. A.** Stem with leaf sheath, leaf base and female inflorescence. **B.** Leaf. **C.** Male rachilla. **D.** Female rachilla. **E.** Fruit. **F, G.** Seed in two views. **H.** Seed in longitudinal section. Scale bar: A, B = 3 cm; C–E = 7 mm; F–H = 5 mm. A, B, D from *Henty NGF 29057*; C from *Kairo & Streimann NGF 35637*; E–H from *Womersley NGF 24887*. Drawn by Lucy T. Smith.

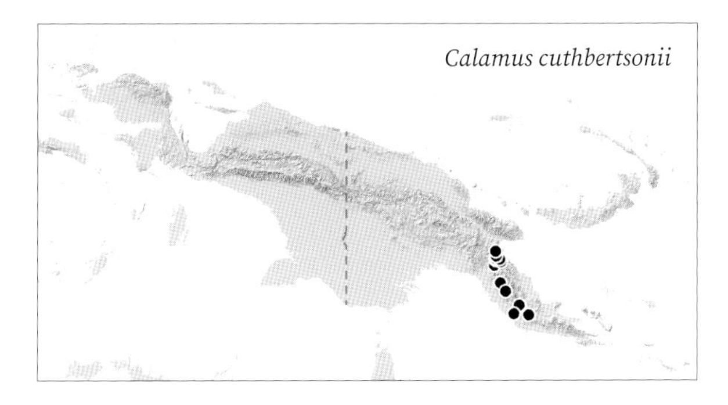

DISTRIBUTION. Owen Stanley Range in south-eastern New Guinea from the vicinity of Wau in the north to Mount Obree in the south.

HABITAT. Montane forest from 1,300–2,500 m.

LOCAL NAMES. *Kaingan* (Biaru), *Rurivi* (Goilala-Tapini), *Sesena* (Efogi), *Sinsin* (Biaru), *Siri* (Koiari-Mauari).

USES. None recorded.

CONSERVATION STATUS. Vulnerable. *Calamus cuthbertsonii* has a restricted distribution. Deforestation due to logging and gold mining is a major threat within its range.

NOTES. This flagellate, montane rattan is distinguished by its very slender stem and flagella, leaves to ca. 40 cm long with 5–13 pairs of regularly to subregularly arranged narrow lanceolate leaflets, and stout, non-flagellate inflorescences that are generally shorter than the leaves. In this area, *C. cuthbertsonii* most closely resembles *C. katikii* (see under *C. katikii*), but it could also be confused with *C. anomalus* or *C. essigii*, which are readily distinguished by their grouped leaflets and highly unusual inflorescence morphology.

15. *Calamus dasyacanthus* W.J.Baker, Bayton, J.Dransf. & Maturb.

Robust, single-stemmed rattan climbing to 15 m. **Stem** with sheaths 36–45 mm diam., without sheaths 20–23 mm diam.; internodes 10–25 cm. **Leaf** *cirrate*, to 5 m long including cirrus and petiole; sheath green, with thin, brown indumentum, *spines very numerous, 2.5–47 × 1–10 mm, orange-brown, flattened, parallel-sided, flexible, somewhat papery, apices truncate, apices and margins distinctly jagged*, forming partial whorls of few to many spines; knee 50–70 mm long, moderately to densely armed; ocrea 8–10 mm, forming a low, woody, brown, armed, persistent collar; flagellum absent; petiole 0–30 mm; leaflets 14–25 each side of rachis, *arranged irregularly or in widely spaced pairs*, pairs sometimes slightly divergent, broadly elliptic, hooded, longest leaflets near middle of leaf, 28.5–40 × 4–6.8 cm, apical leaflets 20.5–29 × 2–3 cm, leaflets lightly armed on upper surface and margins; cirrus 1.5–2 m, cirrus grapnel spines arranged regularly. **Inflorescence** arching, to 2 m long including 26–44 cm peduncle and 31–48 cm sterile tip, branched to 2 orders in the female and 3 orders in the male; primary bracts strictly tubular, with narrow, acute, triangular limb to one side, moderately armed; primary branches 4–8, to 36 cm long, male rachillae ca. 4.5–21 × 1 mm, female rachillae 35–155 × 1–2 mm. **Fruit** solitary, shortly stalked, globose, 9–12.8 × 8.5–10 mm, scales white to pale yellow. **Seed** (sarcotesta removed) 7–7.8 × 7–7.8 × 5.6–6 mm, globose, with a deep, narrow pit on one side, the surface covered with numerous deep pits and irregular channels; endosperm homogeneous.

Calamus dasyacanthus. Stem apex, near Timika (WB).

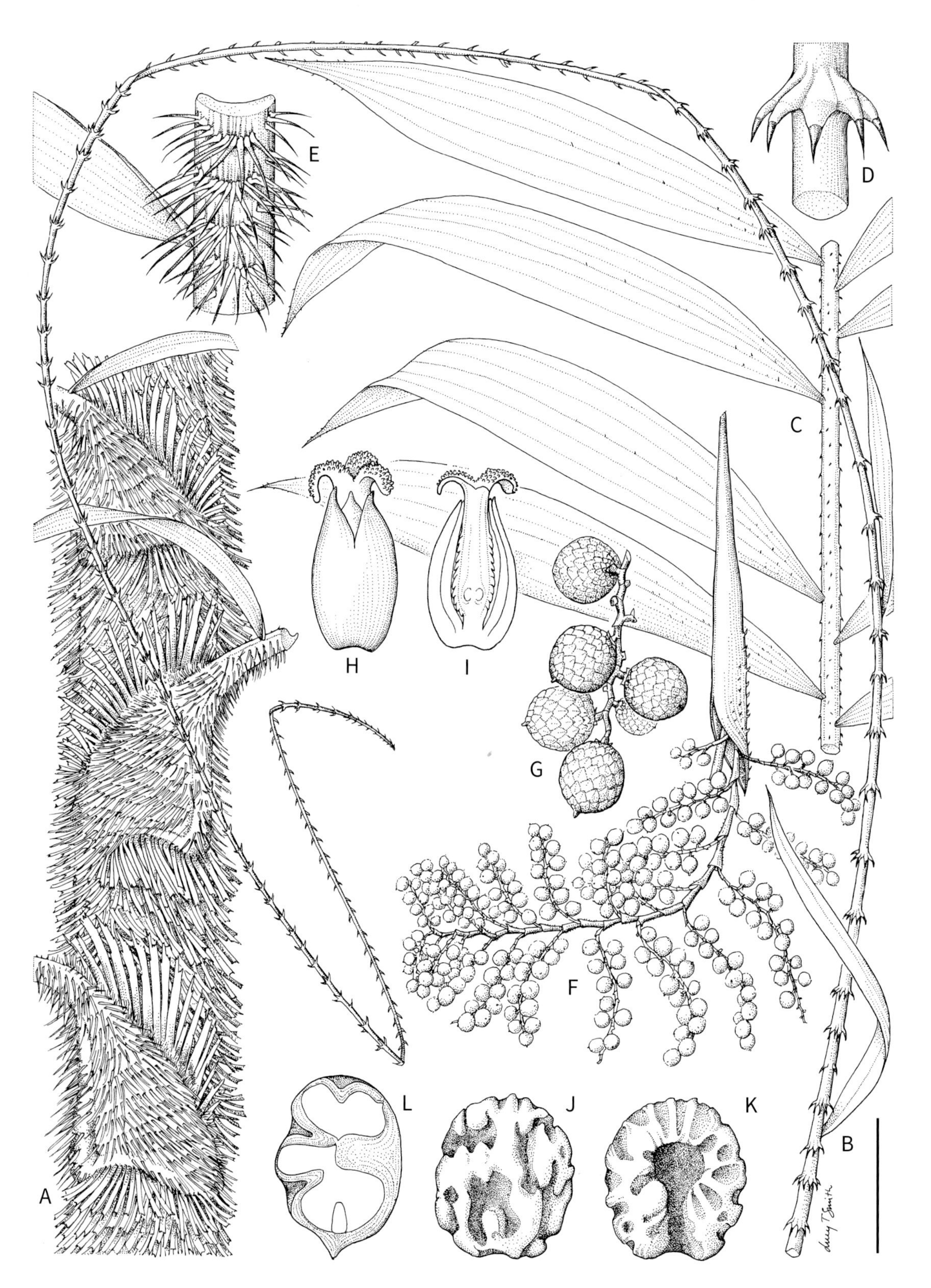

Calamus dasyacanthus. **A.** Stem with leaf sheath. **B.** Leaf apex with cirrus. **C.** Leaf mid-portion. **D.** Portion of cirrus with detail showing spines. **E.** Petiole. **F.** Portion of female inflorescence with fruit. **G.** Female rachilla with fruit. **H, I.** Female flower whole and in longitudinal section. **J, K.** Seed in two views. **L.** Seed in longitudinal section. Scale bar: A = 6 cm; B, C, F = 8 cm; D = 1 cm; E = 3 cm; G = 2 cm; H, I = 3 mm; J–L = 7 mm. A–G, J–L from *Baker et al. 827*; H, I from *Baker et al. 983*. Drawn by Lucy T. Smith.

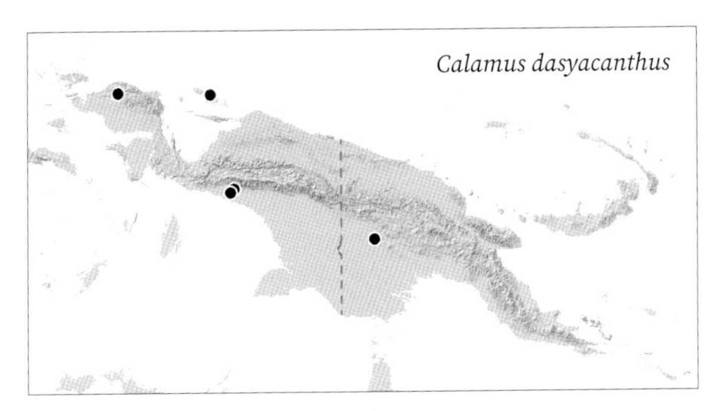

DISTRIBUTION. Widely scattered localities in southern New Guinea, the Bird's Head Peninsula and Biak Island.

HABITAT. Various types of primary and secondary forest vegetation at 30–650 m.

LOCAL NAMES. *Warar* (Biak).

USES. None recorded.

CONSERVATION STATUS. Least Concern.

NOTES. This cirrate species is distinguished from other members of the *Calamus aruensis* complex (see under *C. aruensis*) by its leaflets being arranged irregularly or in divaricate pairs, and the unusual, jagged leaf sheath spines.

Calamus dasyacanthus. Habit, near Timika (WB).

Calamus dasyacanthus. LEFT TO RIGHT: leaf sheaths; female inflorescence. Near Timika (WB).

16. *Calamus depauperatus* Ridl.

Synonyms: *Calamus arfakianus* Becc., *Calamus arfakianus* var. *imberbis* Becc., *Calamus mayrii* Burret, *Calamus prattianus* Becc.

Slender to very slender, single-stemmed or multi-stemmed rattan climbing to 6 m. **Stem** with sheaths 3–11 mm diam., without 2–8 mm diam.; internodes 9–32 cm. **Leaf** ecirrate, 13–43 cm long including petiole; sheath pale green, sometimes with grey to dark, caducous indumentum, unarmed to spines numerous, spines to 6 mm long, fine, hair-like, solitary or in small groups, often forming a dense tuft of longer, erect spines at sheath mouth to 40 mm long, sometimes matted; knee 5–12 mm long, unarmed or armed as sheath; ocrea inconspicuous, a low, encircling crest to 5 mm high, persistent; flagellum to 120 cm; petiole 0–20 mm; *leaflets 4–9 each side of rachis, arranged in 2–4 divaricate groups, lanceolate to narrow lanceolate,* longest leaflet in lower or mid-leaf position, 6–26 × 0.3–2 cm, apical leaflets 4–17 × 0.2–1.7 cm, apical leaflet pair united for one tenth to one quarter of their length, leaflet almost unarmed, very few bristles present on margins towards tip and very occasionally on upper surface. **Inflorescence** *flagelliform, to 120 cm long including peduncle to 60 cm and flagelliform tip to 30 cm,* branched to 2 (rarely 1) orders in the female and 2 orders in the male; primary bracts tightly sheath, narrow, armed (as flagellum); primary branches 2–5, to 20 cm long, male rachillae 4–27 × ca. 1 mm, female rachillae 6–45 × 1.5–2 mm. **Fruit** solitary, spheroidal, 12.5–17 × 11–15 mm, scales yellow. **Seed** (sarcotesta removed) 8.5–11 × 8–9.5 × 7–7.5 mm, ellipsoid, smooth, with a shallow lateral pit; endosperm homogeneous.

DISTRIBUTION. Scattered localities across north central and north-western New Guinea.

HABITAT. Montane forest at 800–2,500 m. Two lowland records at 120–150 m near Jayapura are also included here.

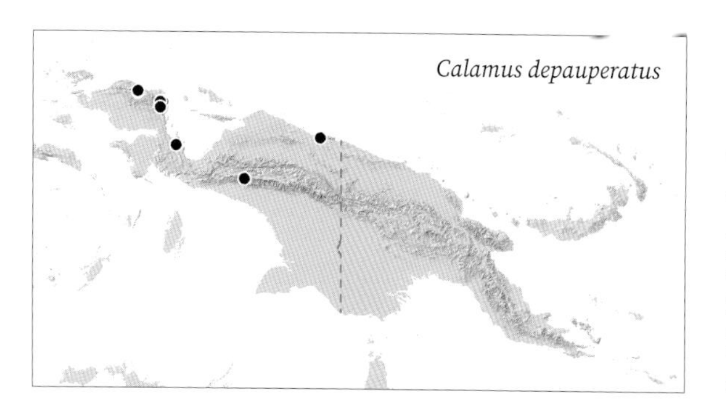

Calamus depauperatus. Habit, Arfak Mountains (AS).

LOCAL NAMES. *Makagam* (Damal), *Mul* (Dani).

USES. A valuable rattan used for rope, binding, and building by the Amungme people (Papua province). Canes are used to make bowstrings, male bracelets, baskets, agricultural tools, and weapons by the Dani people.

CONSERVATION STATUS. Least Concern.

NOTES. *Calamus depauperatus* is a confusingly variable, flagellate, montane rattan recorded from widespread localities in western New Guinea. Vegetatively, it is superficially similar to other slender, montane rattans with leaflets that are grouped or irregularly arranged and that lack conspicuous ocreas (e.g. *C. anomalus, C. cuthbertsonii, C. essigii, C. katikii, C. superciliatus*). Among these, it resembles juvenile *C. superciliatus* most closely in its long, fine, flagellum-like inflorescences (the other species have short inflorescences), but it is more slender, bears fewer leaflets (4–9 each side of the rachis), and is distinct in spine and seed morphology.

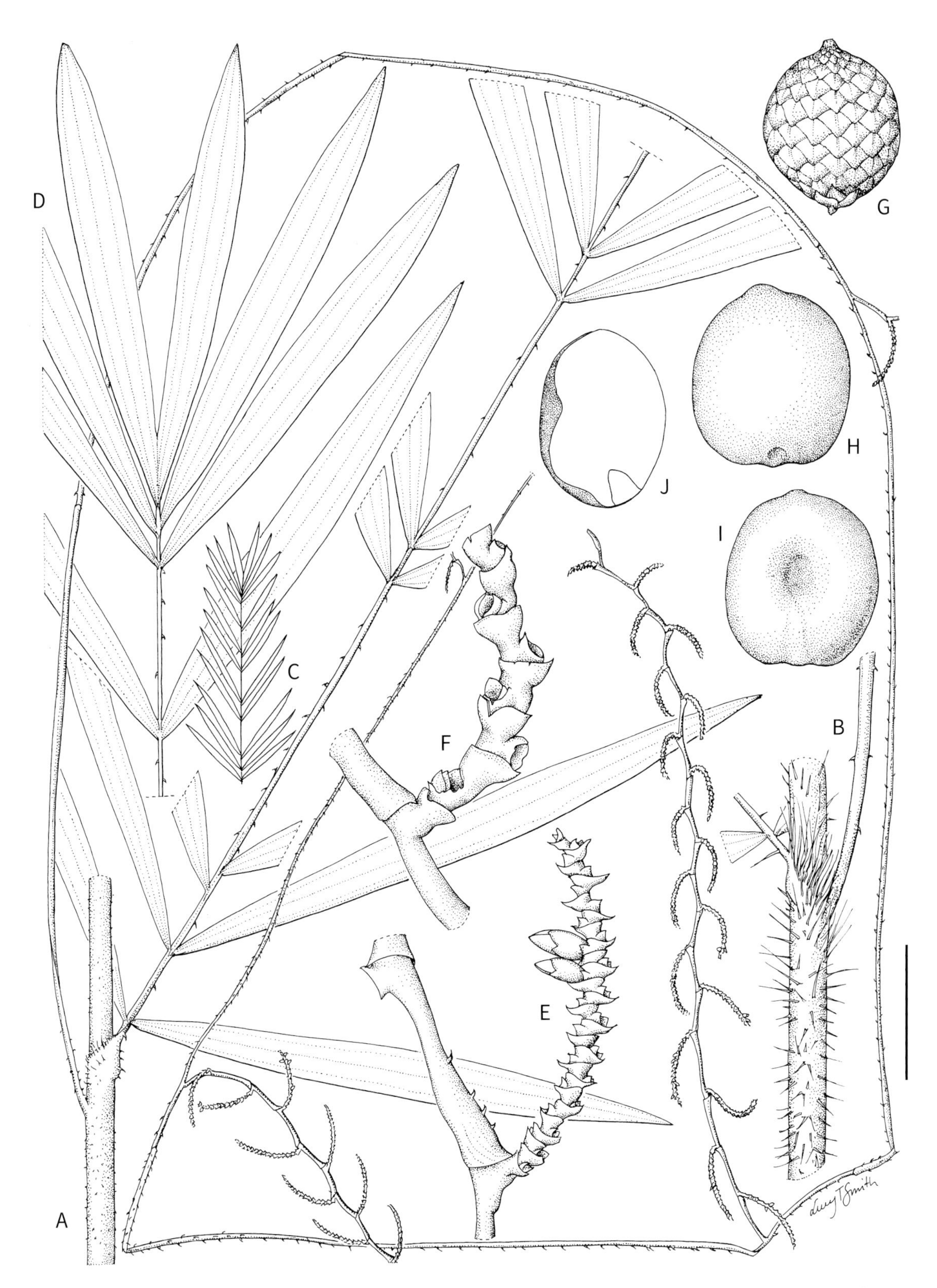

***Calamus depauperatus*. A.** Stem with leaf sheath, leaf base and male inflorescence. **B.** Stem with leaf sheath and petiole. **C.** Whole leaf diagram. **D.** Leaf apex. **E.** Male rachilla with flowers. **F.** Female rachilla. **G.** Fruit. **H**, **I.** Seed in two views. **J.** Seed in longitudinal section. Scale bar: A, B, D = 3 cm; C = 24 cm; E = 5 mm; F, H–J = 7 mm; G = 1 cm. A, C–E from *Gibbs 6144*; B, F from *Gibbs 5977*; G–J from *Stefels BW 2043*. Drawn by Lucy T. Smith.

The type of *Calamus depauperatus* is very incomplete but is sufficient to be matched to the types of *C. arfakianus* and *C. prattianus*, especially now that our understanding of variation has been improved by increased specimen availability. The type of *C. mayrii* could not be located and is presumed destroyed, but the protologue and new collections from the type locality in the Wondiwoi Mountains match *C. depauperatus* well. Thus, *C. arfakianus* (including var. *imberbis*), *C. mayrii*, *C. prattianus* are treated as synonyms here. Henderson (2020) follows this approach.

17. *Calamus distentus* Burret

Slender, multi-stemmed rattan climbing to 10 m. **Stem** with sheaths 9–15 mm diam., without 4–8 mm diam.; internodes 27–33 cm. **Leaf** ecirrate, 44–84 cm long including petiole; sheath green, with caducous brown indumentum, *spines numerous, 2–45 mm, hair-like, brown, solitary*; knee14–25 mm long, colour and armature as sheath; *ocrea 7–16 × 1–1.5 cm, ligule-like, adjacent to petiole, erect, standing proud of the sheath, leathery, green, armed as sheath, persistent*; flagellum to 250 cm; petiole 0.5–10; *leaflets 14–18 each side of rachis, arranged subregularly or in divaricate groups of 4 leaflets*, upper leaflets sometimes regularly arranged, lanceolate, longest leaflet towards base 19–29 × 1.9–2.7 cm, mid-leaf leaflets slightly smaller, apical leaflets 11–13 × 0.5–2.5 cm, apical leaflet pair free or united up to one quarter of their length, leaflet margins sparsely armed with bristles towards tip, upper and lower surface unarmed or sometimes sparsely armed. **Inflorescence** flagelliform, to 3.5 m long including 0.7–1.5 m peduncle and flagelliform tip to ca. 0.5 m, branched to 2 orders in the female and 2 orders in the male; primary bracts narrow and tightly sheathing as flagellum; primary branches 2–6, to 32 cm long, male rachillae 14–38 mm × ca. 1 mm, female rachillae 12–40 mm × ca. 1.5 mm. **Fruit** solitary, spherical, ca. 11.5 × 9 mm, scales pale brown. **Seed** (sarcotesta removed) not seen at maturity, spheroidal with a shallow lateral pit; endosperm homogeneous.

DISTRIBUTION. Few localities in south-eastern central New Guinea.

HABITAT. Lowland forest at 100–520 m, sometimes reported from limestone areas.

LOCAL NAMES. None recorded.

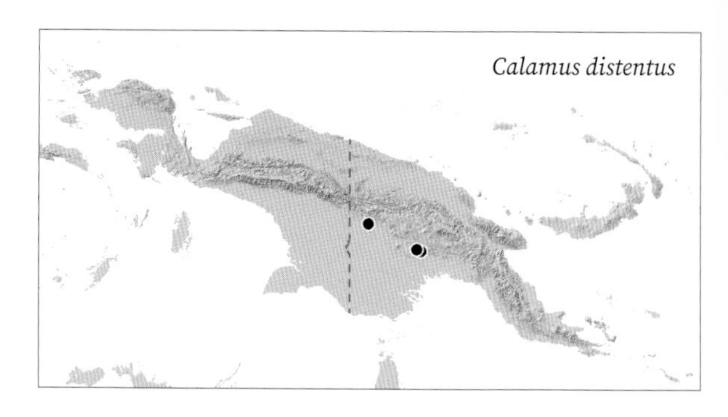

USES. None recorded.

CONSERVATION STATUS. Data Deficient. Due to insufficient evidence of ongoing threats, a single extinction risk category cannot be selected.

NOTES. This is a distinctive, relatively slender, flagellate rattan that is readily recognised by its well-developed ocrea, which stands slightly proud of or divergent from the leaf sheath and is densely armed,

Calamus distentus. Leaf sheath with ocrea, Southern Highlands (WB).

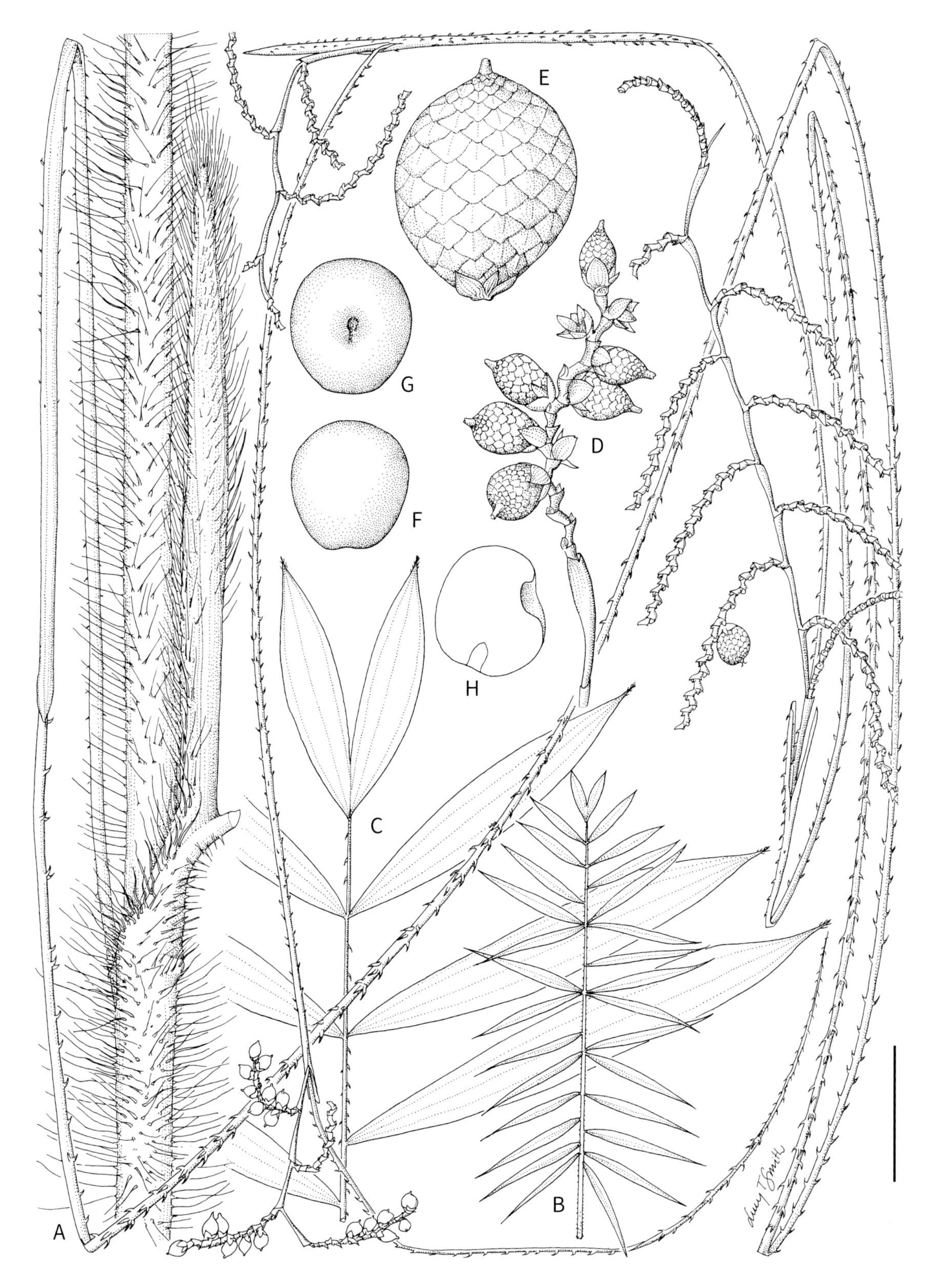

Calamus distentus. **A.** Stem with leaf sheath, ocrea and female inflorescence with fruit. **B.** Whole leaf diagram. **C.** Leaf apex. **D.** Female rachilla with immature fruit. **E.** Fruit. **F**, **G.** Seed in two views. **H.** Seed in longitudinal section. Scale bar: A = 3 cm; B = 30 cm; C = 6 cm; D = 1 cm; E–H = 7 mm. A–D from *Baker et al. 641*; E–H from *Brass 7151*. Drawn by Lucy T. Smith.

Calamus distentus. Habit, Southern Highlands (WB).

like the sheath, with long, fine spines. The leaflets are arranged subregularly to irregularly, are widely spaced and typically include some divaricate groups of two pairs. The inflorescence is fine, flagelliform and sparsely branched. *Calamus distentus* most closely resembles the little-known *C. moszkowskianus*, which also bears a similar ocrea that is armed with bristly spines like the sheaths and a similar inflorescence. However, the latter is more robust, with leaves bearing numerous fine, regularly arranged leaflets and a shorter ocrea (to ca. 8 cm). Vegetatively, there is also some similarity with *C. bulubabi*, although this species lacks a well-developed ocrea.

18. *Calamus erythrocarpus* W.J.Baker & J.Dransf.

Moderately robust, multi-stemmed rattan climbing to ca. 40 m. **Stem** with sheaths 12–23 mm diam., without sheaths to 9–13 mm diam.; internodes 24–40 cm. **Leaf** subcirrate, 70–97 cm long including subcirrus and petiole; sheath mid-green when dry, with thin, pale grey indumentum, almost unarmed, with very few, scattered, minute, upward-pointing spines to ca. 0.5 mm long; knee 24–28 mm long, unarmed; ocrea 7–13 mm, a low, tightly sheathing, unarmed, leathery crest, persistent; flagellum to 2.2 m; petiole 2–7 cm long; *leaflets 8–11 each side of rachis, arranged subregularly, rather widely spaced, broadly lanceolate, usually hooded, longest at base* 23–37 × 2–5 cm, mid-leaf leaflets 22–34 × 3–5 cm, *apical leaflets highly reduced to a remnant 2.2–3.7 × 1–2 cm, or even reduced to a fibre-like structure ca. 1.2 × 0.1 cm,* leaflets unarmed, with the exception of a very few, marginal bristles near apex of some leaflets; cirrus absent. **Inflorescence** *erect, possibly arching when mature, 37–55 cm long* including ca. 9–20 cm peduncle, lacking flagelliform tip, branched to 2 orders in the female and 3 orders in the male; *primary bracts tubular, split to base by emerging primary branches,* usually remaining tubular at tip, unarmed; primary branches 4–6, to 6 cm long, male rachillae 3–10 × ca. 0.5 mm, female rachillae 3–27 mm × 1–2 mm. **Fruit** solitary, ellipsoid, ca. 16 × 11 mm, *scales red, with eroded margins.* **Seed** (sarcotesta removed) ca. 9 × 8 × 5.5 mm, ellipsoid with deep lateral pit; endosperm homogeneous.

DISTRIBUTION. Foothills of the Owen Stanley Range, south-eastern New Guinea.

HABITAT. Lowland forest on lower slopes and valley bottoms, ca. 460 m.

LOCAL NAMES. *Ohana* (Goari).

USES. None recorded.

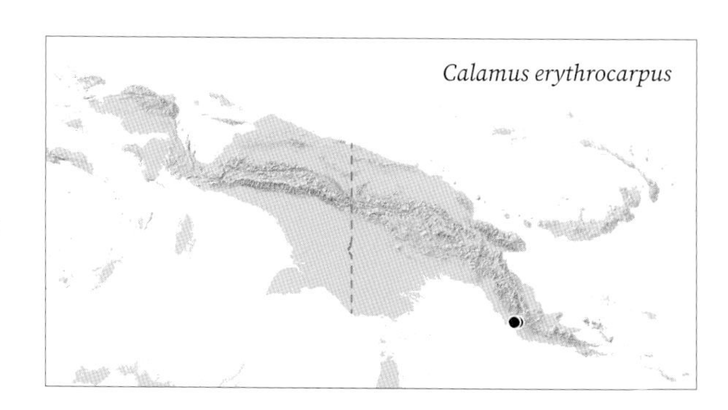

Calamus erythrocarpus. **A.** Stem with leaf sheath, petiole and flagellum. **B.** Stem with leaf sheath, leaf base and male inflorescence. **C.** Whole leaf diagram. **D.** Leaf apex. **E.** Portion of female inflorescence with fruit. **F.** Fruit. **G, H.** Seed in two views. **I.** Seed in longitudinal section. Scale bar: A, B, D, E = 4 cm; C = 24 cm; F–I = 7.5 mm. A, B, D from *Zieck NGF 36176*; C, E–I from *Zieck NGF 36181*. Drawn by Lucy T. Smith.

CONSERVATION STATUS. Critically Endangered. *Calamus erythrocarpus* is known from only two sites. Deforestation due to logging concessions is a major threat to this species. Its close proximity to Port Moresby may also pose additional risks.

NOTES. *Calamus erythrocarpus* is a member of the *C. anomalus* group (see under *C. anomalus*). *Calamus erythrocarpus* and *C. maturbongsii* appear to be very close relatives within the group, although both share the unusual inflorescence morphology with all other species in the group (Baker & Dransfield 2002b, Baker & Dransfield 2017). Similarities that unite *C. erythrocarpus* and *C. maturbongsii* include the moderately robust, multi-stemmed habit, the leaf structure (few broadly lanceolate, hooded leaflets, longest leaflets at the base of the leaf), the almost unarmed leaf sheaths with only few, minute spines, and the robust inflorescence relative to other members of the group. Their fruit are also similar in shape, in orange-red colour and in the structure of their rather flat scales with erose margins.

Calamus erythrocarpus can be immediately distinguished by its apical leaflets that are highly reduced to a remnant no more than 4 cm long, which in one specimen is reduced to a fibre-like structure, the leaf apex then resembling a short cirrus. The apical leaflets of *C. maturbongsii* are more typical and may be free or united up to one quarter of their length. The inflorescences of *C. erythrocarpus* are recorded to be shorter than those of *C. maturbongsii* (37–55 cm vs. 62–81 cm). The splitting rachis bracts of *C. erythrocarpus* do not tatter as they do in *C. maturbongsii*. The primary branches of *C. erythrocarpus* female inflorescences bear fewer rachillae (up to nine in *C. erythocarpus*, compared to up to 17 in *C. maturbongsii*). It is remarkable that these two very similar species are disjunctly distributed at opposite ends of New Guinea. *Calamus erythrocarpus*, on the other hand, is reported to occur within a few kilometres of other species in the *C. anomalus* group.

19. *Calamus essigii* W.J.Baker

Very slender, multi-stemmed, short-stemmed rattan to 8 m. **Stem** with sheaths 4–5 mm diam., without sheaths to 2–2.5 mm diam.; internodes 4.5–10 cm. **Leaf** ecirrate, 12–23 cm long including petiole; sheath with thin, brown indumentum, unarmed or armed with numerous, very fine spines to 3 mm long, spines most numerous near to sheath mouth, to 5 mm long, erect; knee to 7 mm long, unarmed or armed as sheath; ocrea absent or minute; flagellum to 36 cm, readily detached; *petiole 1–3 cm long; leaflets 3–5 on each side of rachis, arranged subregularly or in two distinct subregular groups, narrow linear*, longest leaflet near base 12–21.5 × 0.5–0.8 cm, apical leaflets 11.5–21 × 0.4–0.6 cm, apical leaflet pair united to one tenth of their length or not at all, few minute spines on leaflet margins, very few on upper surface of midrib; cirrus absent. **Inflorescence** *erect, 10.5–17.5 cm long* including 15–45 mm peduncle, lacking sterile tip, branched to 1 order in the female and 2 orders (rarely 1 order) in the male; primary bracts initially tubular, *split to base by emerging primary branches*, sometimes remaining tubular distally, sometimes armed with few reflexed grapnel spines; primary branches up to 5, to 45 mm long, male rachillae 2–15 × 0.3–1 mm, female rachillae 4–20 × 1–2 mm. **Fruit** solitary, globose, 9.5 × 9 mm, scale orange-brown. **Seed** (sarcotesta removed) 7 × 5.5 × 5 mm, ellipsoid with shallow lateral depression; endosperm homogeneous.

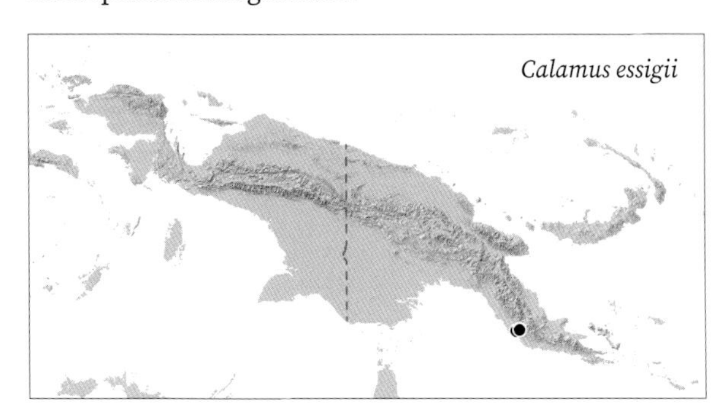

DISTRIBUTION. Owen Stanley Range, south-eastern New Guinea.

HABITAT. Lower montane oak forest at 650–900 m.

LOCAL NAMES. *Hulawarra* (Goari).

USES. None recorded.

CONSERVATION STATUS. Critically Endangered. *Calamus essigii* is known from only three sites. Deforestation due to logging concessions is a major threat to this species. Its close proximity to Port Moresby may also intensify risks.

NOTES. *Calamus essigii* belongs to the *C. anomalus* group (see under *C. anomalus*). It is most closely related to *C. anomalus* and *C. nannostachys*, but can be distinguished by its very slender habit and its 3–5 pairs of extremely narrow linear leaflets, the terminal leaflet

Calamus essigii. **A.** Habit with female inflorescences. **B.** Male inflorescence. **C**, **D.** Male flower whole and in longitudinal section. **E.** Fruit. Scale bar: A, B = 3 cm; C, D = 1.3 mm; E = 7 mm. A, B from *Essig & Womersley LAE 55180*; B–D from *Larivita & Maru LAE 70623*; E from *Zieck NGF 36158*. Drawn by Lucy T. Smith.

pair being free or united to only one tenth of their length. In addition, its inflorescences are branched to one order in the female and two orders (rarely one) in the male.

20. *Calamus fertilis* Becc.

Very robust, single-stemmed or multi-stemmed rattan climbing to 30 m. **Stem** with sheaths 31–45mm diam., without sheaths to 18–32 mm diam.; internodes 13–32 cm. **Leaf** ecirrate, to 1.8–2.6 m long including petiole; sheath pale green to yellowish green, usually with sparse to abundant indumentum, *spines very dense and of various sizes, 2–110 × 1–8 mm, narrow triangular, flattened, sometimes papery, yellow-green to orange, mostly horizontal, with very long downward pointing spines directly below the knee; knee 53–78 cm long, 28–45 mm, grossly swollen, usually unarmed or very sparsely armed; ocrea 2.5–13 × 4–6 cm, entire with a shallow cleft next to the petiole, soon tattering and disintegrating into fibres,* armed with slender spines to 5 mm, or unarmed; flagellum to 6–8 m long; petiole 20–40 cm long; *leaflets 53–65 each side of rachis, regularly arranged, lanceolate,* longest leaflets in mid-leaf 28–56 × 1.3–3.2 cm, apical leaflets 9–15 × 0.5–1.0 cm, apical leaflet pair united from one tenth to one half of their length, leaflets armed with pale bristles on upper and lower surfaces. **Inflorescence** pendulous, to 4.4 m long including peduncle to 58 cm and flagelliform tip to 90 cm, branched to 2 orders in the female and 3 orders in the male; primary bracts closely sheathing, with a triangular tip, armed along margins or unarmed; primary branches up to 9, to 75 cm long, male rachillae 50–110 × 2 mm, female rachillae 30–260 × 2–5 mm. **Fruit** *paired,* ovoid, 12–18 × 8–10 mm scales pale yellowish to brown. **Seed** (sarcotesta removed) 8–10 × 6–8 × 5–7 mm, ellipsoid, rather deeply scalloped and pitted, with a shallow to deep pit; endosperm homogeneous.

Calamus fertilis. LEFT: habit (*C. fertilis* centre, *C. reticulatus* to left and near water), Kikori River (OG); RIGHT: leaf sheath with swollen knee, Mubi River (WB).

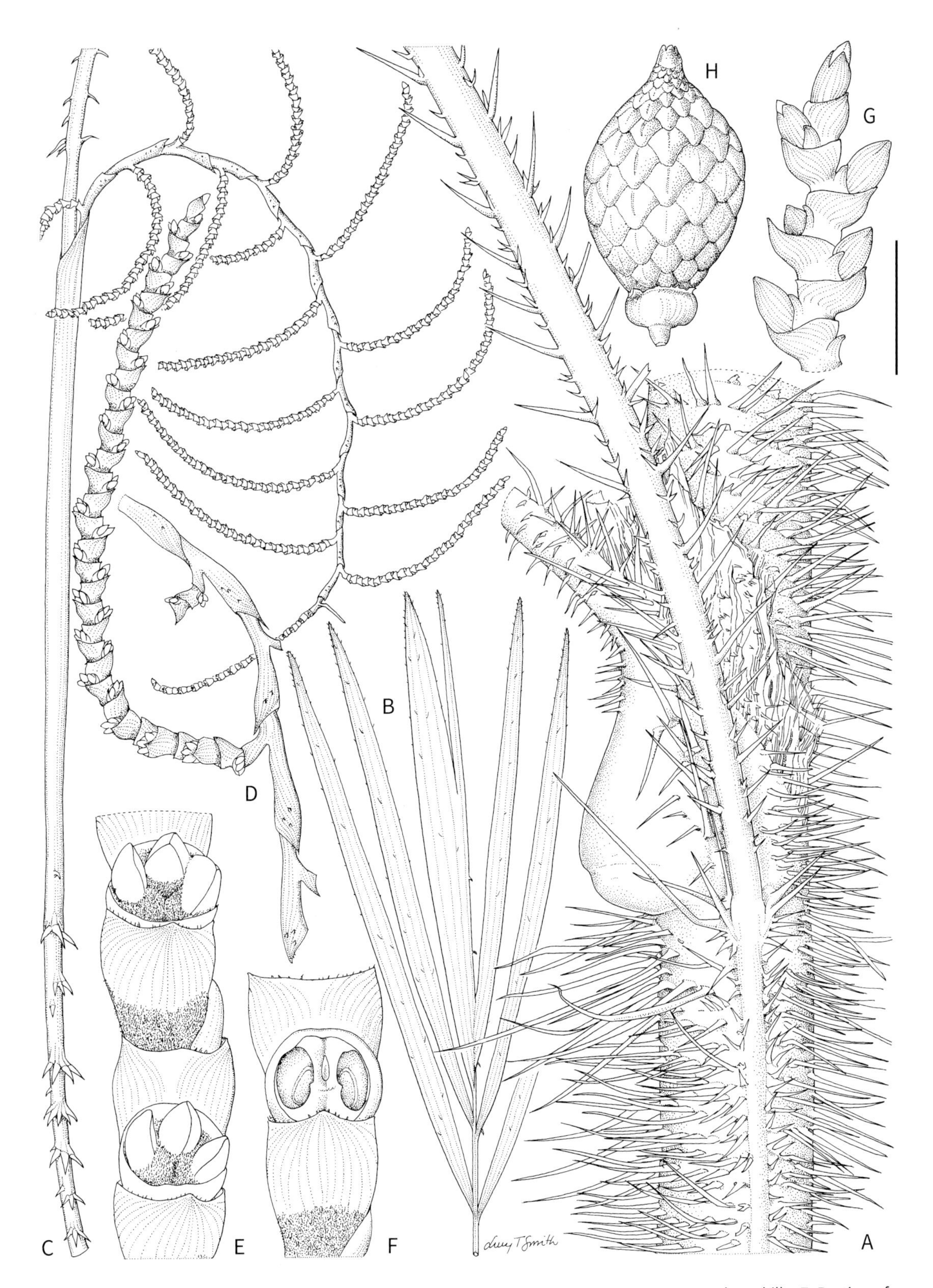

Calamus fertilis. **A.** Stem with leaf sheath. **B.** Leaf apex. **C.** Portion of female inflorescence. **D.** Female rachilla. **E.** Portion of female rachilla with buds of two female flowers and one sterile male flower. **F.** Portion of female rachilla with flowers removed, showing scars. **G.** Portion of male rachilla. **H.** Fruit. Scale bar: A, B = 4 cm; C = 6 cm; D = 2 cm; E–G = 3.3 mm; H = 7 mm. A–F from *Baker et al. 1125*; G from *Dransfield et al. JD 7669*; H from *Poudyal et al. 83*. Drawn by Lucy T. Smith.

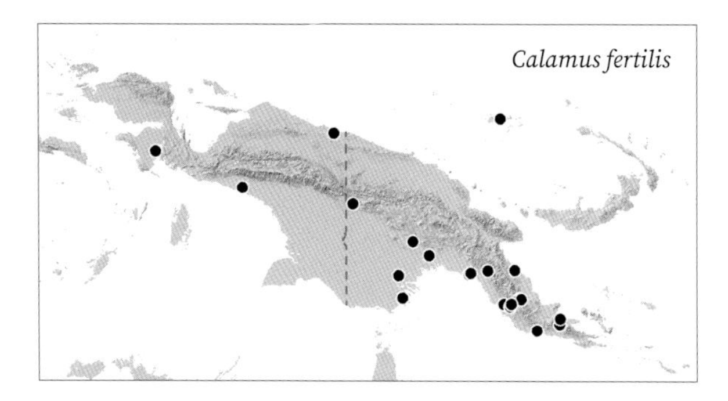

DISTRIBUTION. Widespread in mainland New Guinea and Manus Island.

HABITAT. In lowland forest, usually on riverbanks or in swamp forest at altitudes up to 450 m.

LOCAL NAMES. *Okou* (Gulf), *Hurumau* (Kerema), *Kema* (Yamur).

USES. None recorded.

CONSERVATION STATUS. Least Concern.

NOTES. This flagellate species belongs to the *C. altiscandens* group (see notes under *C. altiscandens*). It is one of the largest of all rattans in New Guinea and is easily distinguished by its regularly arranged leaflets, long flagella, fiercely armed leaf sheaths, and grossly swollen knees that are usually unarmed, but are fringed with large orange or yellow, downward-pointing spines just below the knee. The ocrea is relatively well-developed, but disintegrates. In the female inflorescence, the flowers are borne in threes, comprising two female flowers and a sterile male flower.

21. *Calamus heatubunii* W.J.Baker & J.Dransf.

Moderately robust, multi-stemmed rattan climbing to 20 m. **Stem** with sheaths 11–18 mm diam., without sheaths 7–11 mm diam.; internodes 13–28 cm. **Leaf** ecirrate, 70–90 cm long including petiole; sheath dark green, with grey indumentum, densely and evenly armed with solitary spines 2–4 × 0.5–1.5 mm, triangular, orange-brown tipped; knee 20–33 mm long, armed; *ocrea 18–50 × 2–2.5 cm, boat-shaped, somewhat inflated and clasping, papery, dark purple-green, drying brown, armed with numerous, scattered solitary spines, open to base opposite the petiole, persistent, but eventually distintegrating*; flagellum to 2.5 m long; petiole 20–26 cm; *leaflets 2–5 each side of rachis, clustered in a single, fan-like group (typically of four leaflets) at the petiole apex, less commonly an*

Calamus heatubunii. FROM TOP: habit; female inflorescence. Near Sorong (WB).

Calamus heatubunii. **A.** Stem with leaf sheath, flagellum and female inflorescence with fruit. **B.** Whole leaf diagram. **C.** Leaf. **D.** Male rachilla. **E.** Female rachillae. **F**, **G.** Female flower whole and in longitudinal section. **H.** Fruit. **I**, **J.** Seed in two views. **K.** Seed in longitudinal section. Scale bar: A = 4 cm; B = 30 cm; C = 8 cm; D, E = 8.5 mm; F, G = 3 mm; H–K = 7.5 mm. A, C, H–K from *Baker et al. 1392*; B, D–G from *Baker et al. 1394*. Drawn by Lucy T. Smith.

Calamus heatubunii. LEFT TO RIGHT: male inflorescence; male flower buds. Waigeo Island (CDH).

additional divaricate group of leaflets also present, broadly lanceolate, hooded, apical leaflets 33–52 × 6–10 cm, apical leaflet pair united from one half to two thirds of their length, remaining leaflets 32–50 × 5–9 cm, armed with fine, stiff spines on leaflet margins; cirrus absent. **Inflorescence** *erect, compact, held close to stem, 22–35 cm long including ca. 6–17 cm peduncle,* lacking sterile tip, branched to 2 orders in the female and 3 orders in the male; primary bracts narrow tubular, with asymmetric, acute, distal limb; primary branches ca. 8, 6–14 cm long, male rachillae 20–40 mm × 2–3 mm, female rachillae 21–37 mm × 3–5 mm. **Fruit** solitary, spherical, 13 × 10–11 mm, scales brown. **Seed** (sarcotesta removed) ca. 8.5 × 7.5 × 4 mm, rounded, but bilaterally compressed and concave on one side, with irregular channels and ridges; endosperm homogeneous.

DISTRIBUTION. Vicinity of Sorong and the Raja Ampat Islands (Waigeo).

HABITAT. Lowland forest, including secondary, hill and swamp forest at 45–180 m.

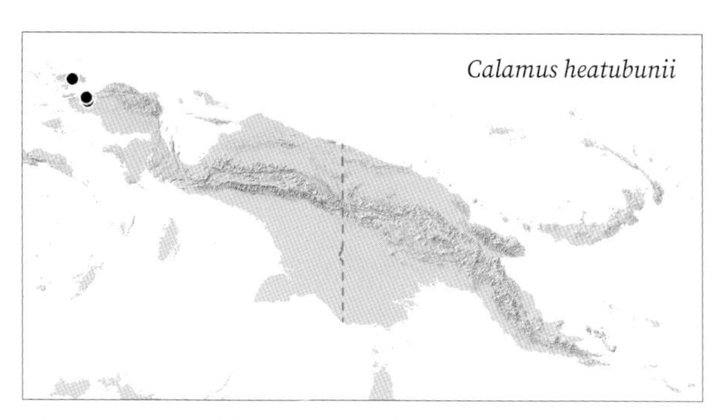

LOCAL NAMES. None recorded.

USES. None recorded.

CONSERVATION STATUS. Endangered. *Calamus heatubunii* has a restricted distribution. Deforestation is a major threat within its range.

NOTES. *Calamus heatubunii* is a member of the *Calamus lauterbachii* group (see under *C. lauterbachii*). It is a moderately robust, flagellate species with very few leaflets, which are most often arranged in a single, fan-like group at the leaf apex. Its well-developed, purple-brown ocrea can reach ca. 50 cm in length and is armed with numerous, solitary triangular spines, similar to those found on the leaf sheath (but not eyelash-like, as in *C. lauterbachii*). The inflorescences are short, lacking a flagelliform tip, and erect, with compact, though not congested branching.

22. *Calamus heteracanthus* Zipp. ex Blume

Synonyms: *Calamus hartmannii* Becc., *Calamus keyensis* Becc., *Daemonorops heteracantha* Blume, *Palmijuncus heteracanthus* (Zipp. ex Blume) Kuntze

Moderately slender to robust, single- or multi-stemmed rattan climbing to 25 m. **Stem** with sheaths 12–48 mm diam., without 7–28 mm diam.; internodes 20–40 cm. **Leaf** *subcirrate or ecirrate,* 1–2.3 m long including petiole; sheath green, with very thin, grey, white or brown indumentum, forming spine shadows, *spines sparse to moderately numerous (rarely unarmed), 9–25 × 2–7 mm, triangular, yellow-orange to black-brown, usually grouped in partial collars or otherwise solitary*; knee 25–60 mm long, 12–16 mm wide, colour as sheath, unarmed; ocrea a low, encircling crest to 5 mm high, leathery, brown, persistent; flagellum present, to 5.3 m; petiole 1.5–45 cm; leaflets 15–32 each side of rachis, *subregularly (rarely regular) or irregularly arranged in groups,* broadly elliptic, longest leaflet at mid-leaf position 11–34 × 2–8 cm, apical leaflets 0.7–13 × 0.2–1.3 cm, apical leaflet pair united up to three quarters of their length, leaflets margins scarcely to sparsely armed with black bristles, surfaces unarmed. **Inflorescence** erect to arching, 0.8–3 m long including 25–30 cm peduncle and flagelliform tip to 60 cm or lacking, *branched to 3 orders in both the female and the male*; primary bracts tubular, robust, armed with scattered, fine, triangular spines, especially near tip, more rarely unarmed; primary branches 2–7, to 20 cm long, male rachillae 2–25 × 0.3–0.5 mm, female rachillae 3–50 × 0.5–1.5 mm, *female flowers pedicellate.* **Fruit** solitary, stalked, spheroidal, 9–13 × 7–12 mm, scales white with dark margins. **Seed** (sarcotesta removed) 6.5–9 × 6–8.5 × 5–7 mm, globose, surface smooth with a lateral pit; endosperm homogeneous.

Calamus heteracanthus. LEFT TO RIGHT: habit; leaf sheath. Biak Island (WB).

Calamus heteracanthus. **A.** Stem with leaf sheath and inflorescence base. **B.** Leaf apex. **C.** Portion of female inflorescence with fruit. **D.** Portion of male inflorescence. **E.** Female rachilla with dyads. **F.** Floral dyad. **G.** Female flower in longitudinal section. **H.** Fruit. **I**, **J.** Seed in two views. **K.** Seed in longitudinal section. Scale bar: A = 3 cm; B = 6 cm; C = 4 cm; D = 2 cm; E = 1.5 cm; F = 2.5 mm; G = 2 mm; H = 1 cm; I–K = 7 mm. A–C from *Jebb CRI 721*; D–G from *Maturbongs 287*; H–K from *Zieck NGF 36106*. Drawn by Lucy T. Smith.

Calamus heteracanthus. LEFT TO RIGHT: leaf apex with subcirrus, near Kikori (WB); male inflorescence, Biak Island (WB).

DISTRIBUTION. Divided into two disjunct subdistributions in the south-eastern and north-western extremes of the New Guinea (including Yapen, Biak and the Raja Ampat islands). Also known from the Kai Islands. Records from Maluku (Bacan, Halmahera, Morotai, Obi) have been reclassified as *C. heteracanthopsis* A.J.Hend. (Henderson 2020).

HABITAT. Lowland forest from sea-level to 650 m.

LOCAL NAMES. *Arotan tibi* (Arandai dialect, Dombano language), *Dou bako* (Ambel), *Karpiapa manawayen* (Woi), *Kou* (Wondama), *Mufun mayahgah* (Nuni), *To pam* (Ayawasi), *Wapar* (Biak), *Wirot jaman* (Irarutu).

USES. Used for weaving, cane split for bow strings, spiny leaf sheath used as a grater. Stem used for floors and beds. Young shoot apex edible.

CONSERVATION STATUS. Least Concern (IUCN 2013).

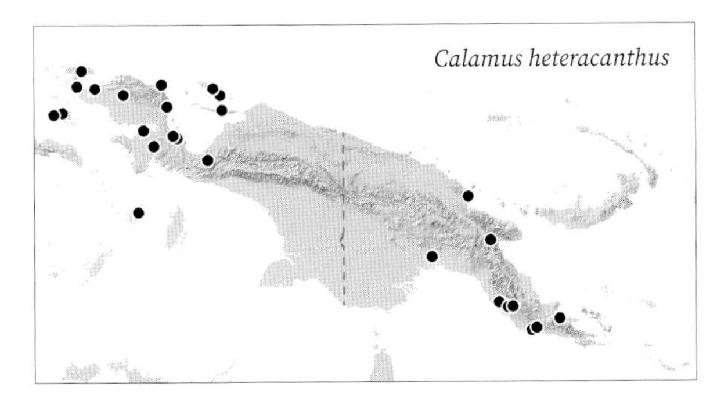

NOTES. This widespread flagellate species has a striking, disjunct distribution across New Guinea. This may be an artifact of collecting patterns, although records of the species are rather numerous. Alternatively, this could point to the species comprising more than one taxon. However, Henderson (2020) identified very limited morphological differentiation between the two ends of the distribution.

Calamus heteracanthus is characterised by leaves that are often subcirrate and bear broadly elliptic leaflets. Its leaf sheaths are armed with stout triangular spines, which are usually grouped in partial collars. The female inflorescences are branched to three orders (rather than two; see under *C. polycladus*) and the female flowers are borne on pedicel-like stalks. In vegetative and inflorescence form, it bears a superficial resemblance to the cirrate *C. aruensis*, to which it is unrelated; it is immediately distinguished from this species by the presence of a flagellum.

Henderson (2020) reassigned specimens of *C. heteracanthus* from Maluku (Bacan, Halmahera, Morotai, Obi) to the rather similar *C. heteracanthopsis* A.J.Hend. A single New Guinea specimen (*Dransfield et al. 7575*) from the Bird's Head Peninsula was also identified as *C. heteracanthopsis*. According to Henderson, *C. heteracanthopsis* is distinguished by its inflorescence with a flagelliform tip (rather than a tubular tip comprising empty rachis bracts), with female inflorescences branched to two orders (rather than three) bearing longer rachillae than *C. heteracanthus* (up to 10 cm rather than 5 cm). However, we note that the isotype of *C. heteracanthopsis* illustrated by Henderson (2020) does not have a flagelliform tip, and we have observed well-developed flagelliform tips in specimens on some New Guinean *C. heteracanthus*. The branching order of female inflorescences does seem to differ consistently in the specimens of the two species cited by Henderson, although the rachilla length difference is a correlate of this. Flowers aside, the structure of the female second-order branches of *C. heteracanthus* corresponds closely to that of the female second-order branches (i.e. rachillae) of *C. heteracanthopsis*. Other differentiating characters raised by Henderson (spine arrangement, leaflet arrangement, leaflet pulvinus) do not appear to be robust in the specimens examined by us. As for *Dransfield et al. 7575*, except for its female inflorescences branched to two orders and the concomitant longer rachillae, the specimen is a close match for *C. heteracanthus* in all other respects. The circumscription of *Calamus heteracanthopsis* and its presence in New Guinea, which is not accepted here, requires further examination.

23. *Calamus interruptus* Becc.

Synonyms: *Calamus docilis* (Becc.) Becc., *Calamus interruptus* var. *docilis* Becc., *Palmijuncus interruptus* (Becc.) Kuntze

Moderate, multi-stemmed rattan climbing to 15 m. **Stem** with sheaths ca. 15–20 mm diam., without ca. 10 mm diam.; internodes 20–30 cm. **Leaf** ecirrate, to 0.9–1.5 m long including petiole; sheath pale green, *drying dark brown*, with abundant thin, dark, caducous indumentum, *spines very few or absent, 15–20 mm long, narrow triangular, solitary, arranged in a single vertical line*; knee 34–37 mm long, ca. 13 mm wide, colour as sheath, unarmed; *ocrea a low, papery, encircling crest, ca. 4 mm high, extending along petiole for 5 cm, persistent*; flagellum ca. 2 m; petiole 20–37 cm; leaflets ca. 14–19 each side of rachis, arranged in irregular, divaricate groups of 2–3, narrow elliptic, longest leaflet at mid-leaf position ca. 23–36 × 2.5–3 cm, apical leaflets ca. 18–30 × 2.5–3 cm, *apical leaflet pair united for half to two thirds of their length*, leaflet unarmed except for very few marginal bristles at tip, leaflets drying dark brown. **Inflorescence** not seen. **Fruit** not seen. **Seed** not seen.

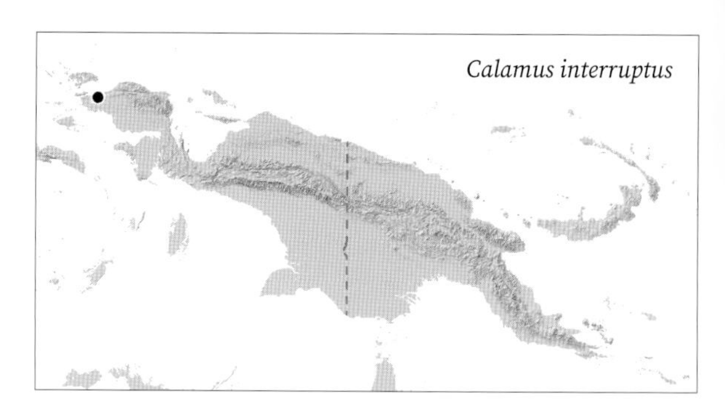

DISTRIBUTION. Vicinity of Sorong, Bird's Head Peninsula.

HABITAT. Recorded from lowland forest up to 80 m.

LOCAL NAMES. None recorded.

USES. None recorded.

CONSERVATION STATUS. Endangered. *Calamus interruptus* has a restricted distribution. Deforestation is a major threat in its range.

Calamus interruptus. Habit, near Sorong (WB).

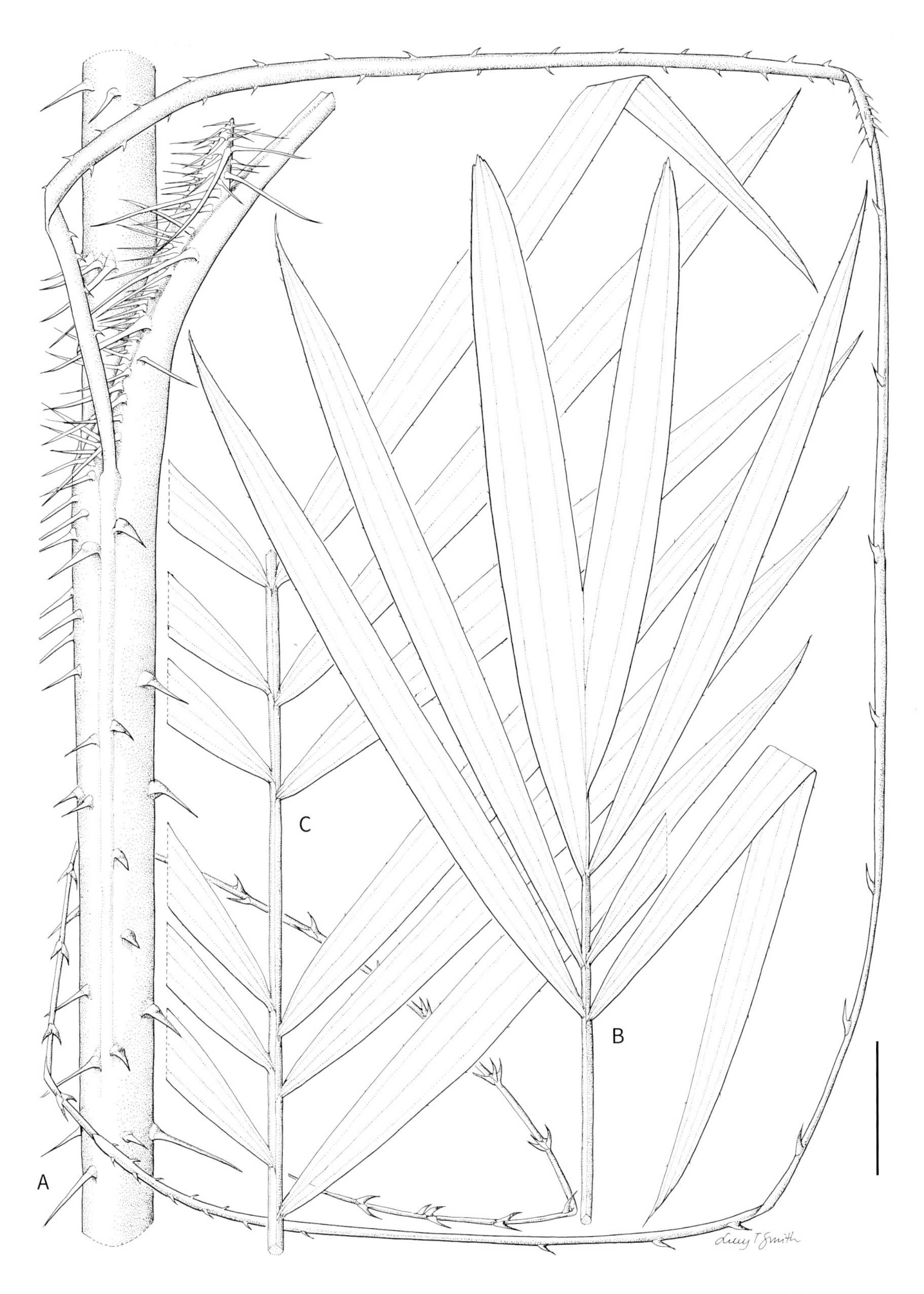

Calamus interruptus. **A.** Stem with leaf sheath, petiole and flagellum. **B.** Leaf apex. **C.** Leaf mid-portion. Scale bar: A = 2 cm; B, C = 4 cm. All from *Maturbongs 515*. Drawn by Lucy T. Smith.

Calamus interruptus. Leaf sheaths, near Sorong (WB).

NOTES. This poorly known species has not been observed in a fertile state. Although morphologically unremarkable, it is rather distinct from other New Guinea species.

Calamus interruptus is a moderate, flagellate rattan that bears leaves up to 1.5 m long with up to 16 rather narrow elliptic leaflets arranged in groups, the apical leaflet pair being joined for much of their length. The sheaths are unarmed or sparsely armed with slender, stiff spines to ca. 2 cm long (the sole distinguishing feature of the synonymous *Calamus interruptus* var. *docilis* is the lack of sheath spines). The sheath mouth is oblique and bears a low papery ocrea that runs along the upper surface of the petiole for a short distance. The leaf sheath dries a distinctive brown colour. The lack of inflorescences and certain vegetative features suggests that the available specimens may have been collected from juvenile plants of a known species. It perhaps resembles most closely *C. heteracanthus*, but is distinguished by its larger apical leaflets, sheath drying brown rather than green, and its dull, matt leaflets when dry (as opposed to glossy green with conspicuous veinlets). In the absence of strong evidence to the contrary, we provisionally accept the species here.

24. *Calamus jacobsii* W.J.Baker & J.Dransf.

Moderately robust rattan climbing to 15 m, stem suckering not recorded. **Stem** with sheaths 18–24 mm diam., stem without sheaths not seen; internodes ca. 35 cm. **Leaf** ecirrate, to 76 cm long including petiole; *sheath brown when dried, with thin indumentum, unarmed*; knee 25 mm long, unarmed; ocrea 4–6 mm high, a bony crest, persistent; flagellum to ca. 220 cm long; *petiole ca. 5 mm long; leaflets ca. 4–5 each side of rachis, subregularly arranged, broadly elliptic*, longest leaflet at mid-leaf position ca. 37 × 8 cm, apical leaflets 35 × 5 cm, apical leaflet pair united to ca. two thirds of their length, leaflets unarmed except for very rare minute bristles at apex margins. **Inflorescence** arching, at least 87 cm long (complete inflorescence not seen), *apparently lacking flagelliform tip, branched to 4 orders in the male*, female inflorescence not seen; Primary bracts strictly tubular with asymmetric pointed tips, not splitting, unarmed; primary branches at least 3, male rachillae 5–16 × 0.5–1 mm, female rachillae not seen. **Fruit** not seen. **Seed** not seen.

DISTRIBUTION. Mountains south of Lae, south-eastern New Guinea.

HABITAT. Primary forest at 500–600 m.

LOCAL NAMES. None recorded.

USES. None recorded.

CONSERVATION STATUS. Critically Endangered. *Calamus jacobsii* is known from only two sites which are threatened by gold mining and logging concessions.

NOTES. *Calamus jacobsii* is a flagellate species known only from relatively incomplete material. The leaf sheath is unarmed and covered with thin, but dense, caducous indumentum, which also occurs on other organs. The petiole is very short and the leaf bears very few, large, elliptic leaflets, which are almost

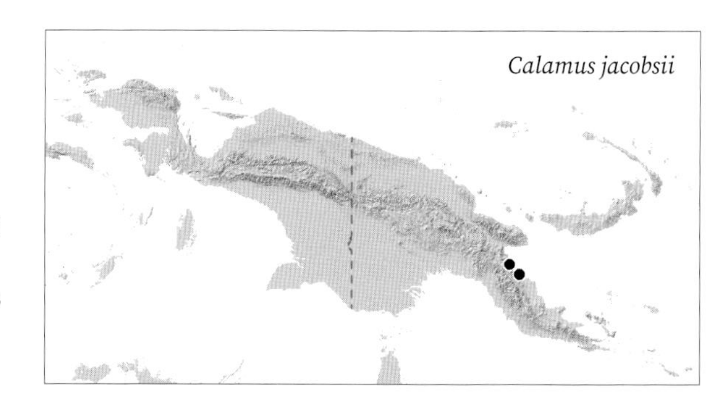

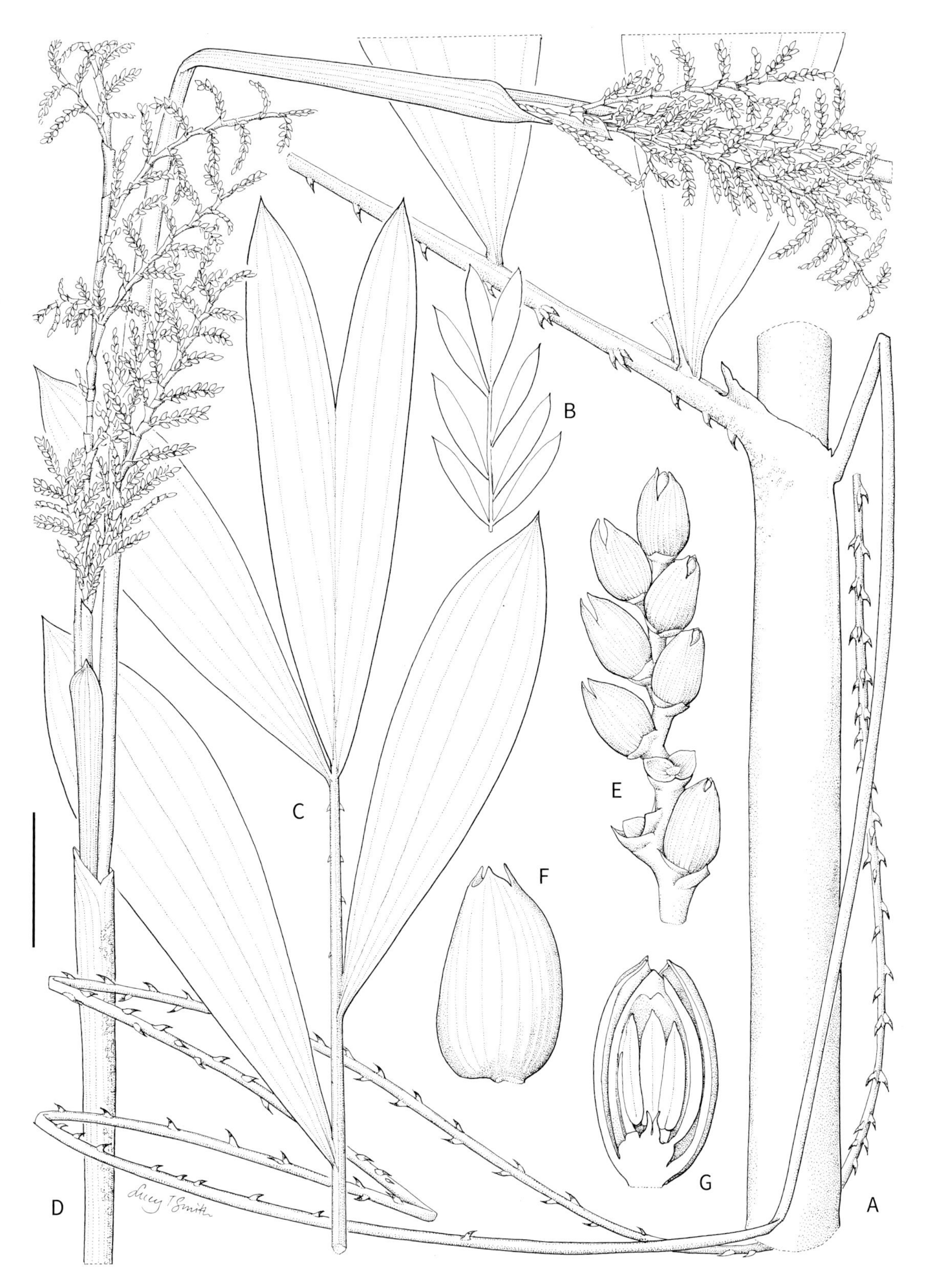

Calamus jacobsii. **A.** Stem with leaf base and flagellum. **B.** Whole leaf diagram. **C.** Leaf apex. **D.** Portion of male inflorescence. **E.** Male rachilla with flowers. **F**, **G.** Male flower whole and in longitudinal section. Scale bar: A, D = 3 cm; B = 35 cm; C = 8 cm; E = 16 mm; F, G = 1.5 mm. All from *Jacobs 9561*. Drawn by Lucy T. Smith.

entirely unarmed. The male inflorescence seen by us is not flagelliform and is branched to four orders (see discussion under *C. polycladus*). Specimens dry a distinctive brown colour. It somewhat resembles *C. croftii*, *C. johnsii* and *C. lucysmithiae* vegetatively, although these species are spinier. It is perhaps most easily confused with *C. johnsii*, a robust form of which also occurs in the vicinity of the type locality of *C. jacobsii* (*Jacobs 9698*). The two species are similar in leaflet size, shape and number and in the lack of a petiole. *Calamus johnsii*, however, is more slender, dries pale green, is armed on its leaf sheath with flattened, black spines, and bears slender, whip-like inflorescences that have a flagelliform tip and that are branched to three orders in the male. *Calamus johnsii* is also not known from above 350 m elevation.

25. *Calamus johnsii* W.J.Baker & J.Dransf.

Slender, multi-stemmed rattan climbing to 20 m. **Stem** with sheaths 4.5–20 mm diam., without sheaths 3.5–7 mm diam.; internodes to 23cm. **Leaf** ecirrate, to ca. 75 cm long including petiole; sheath pale green, with thin grey indumentum, *sheath spines sparse to abundant (rarely absent), 2–25 × 1–2 mm, very narrow triangular, flattened, scattered or sometimes in horizontal groups of 2–3, shiny black with conspicuously swollen pale green bases, spines sometimes crowded around the leaf sheath mouth*; knee to 36 mm long, unarmed or sparsely armed as sheath; ocrea scarcely developed; flagellum to at least 1.1m long; petiole very short or absent (rarely up to 4.2 cm long); *leaflets 5–7 on each side of rachis, arranged irregularly or grouped, the basalmost pair reflexed across the stem, broadly elliptic (narrow elliptic in Sudest Island form)*, with conspicuous cross veins, longest leaflet in mid-leaf 15–41 × 1.5–8 cm, apical leaflets 11–38 × 1–6.5 cm, apical leaflets united for one fifth to half their length, leaflets armed with very short sparse black bristles on margins near tips. **Inflorescence** pendulous, to 1.7 m long including peduncle to 88 cm and ca. 45 cm flagelliform tip, branched to 2 orders in the female and 3 orders in the male; primary bracts tubular and closely sheathing, with entire, triangular apex and bearing scattered reflexed spines; primary branches 2–5, to 21 cm long, male rachillae 3–20 × ca. 1 mm, female rachillae to 50 × 2 mm. **Fruit** solitary, broadly ellipsoid, *12–20 × 10–14 mm, scales pale brown with darker margins*. **Seed** (sarcotesta removed) 10–12 × 7.5–9.5 × 6–7 mm, with a longitudinal groove on one side, seed surface mostly smooth; endosperm homogeneous.

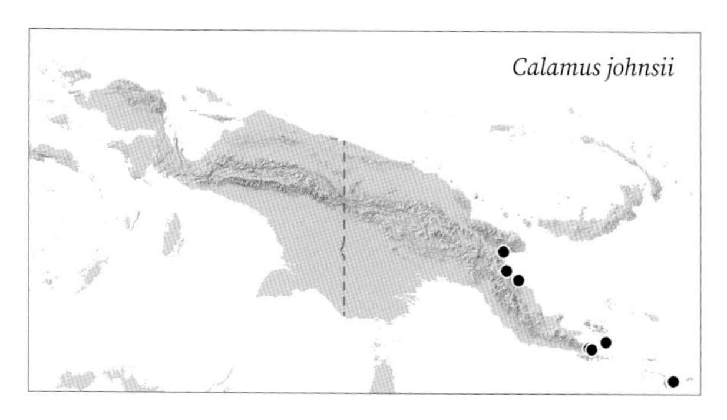

DISTRIBUTION. South-eastern New Guinea to the D'Entrecasteux Islands and Louisiade Archipelago.

HABITAT. Lowland primary and secondary forest from sea level to 350 m.

LOCAL NAMES. *Jeje* (Dzjedrje).

USES. None recorded.

CONSERVATION STATUS. Least Concern.

NOTES. This relatively widespread, lowland, flagellate rattan is distinguished by its slender stem, sheaths with fine, scattered black spines with somewhat swollen bases, by its leaves with the petiole very short or absent and few, irregularly arranged, broadly elliptic leaflets with conspicuous cross veins, and by the inflorescences with few, small first-order branches and relatively robust primary bracts. The fruits are also distinctive in being among the largest in New Guinea *Calamus* species and being covered in scales that are usually pale with darker margins. *Calamus johnsii* displays superficial vegetative similarities to *C. anomalus*, *C. croftii*, *C. jacobsii*, *C. katikii*, *C. lucysmithiae*, *C. nannostachys* and *C. oresbius* (see under those species and Baker & Dransfield 2014 for further discussion)

26. *Calamus katikii* W.J.Baker & J.Dransf.

Very slender rattan climbing to 5 m. **Stem** with sheaths 3.5–5 mm diam., without sheaths ca. 3 mm diam.; internodes 9.5–12 cm. **Leaf** ecirrate, to 24 cm long including petiole; sheath orange-green, with sparse, dark indumentum, spines sparse, 2–4 mm long, needle-like, brown-black; knee ca. 8 mm long, ca. 4 mm wide, with spines as sheath; ocrea ca. 1.5 mm high, inconspicuous, tightly sheathing, armature as sheath; flagellum at least 35 cm long (incomplete

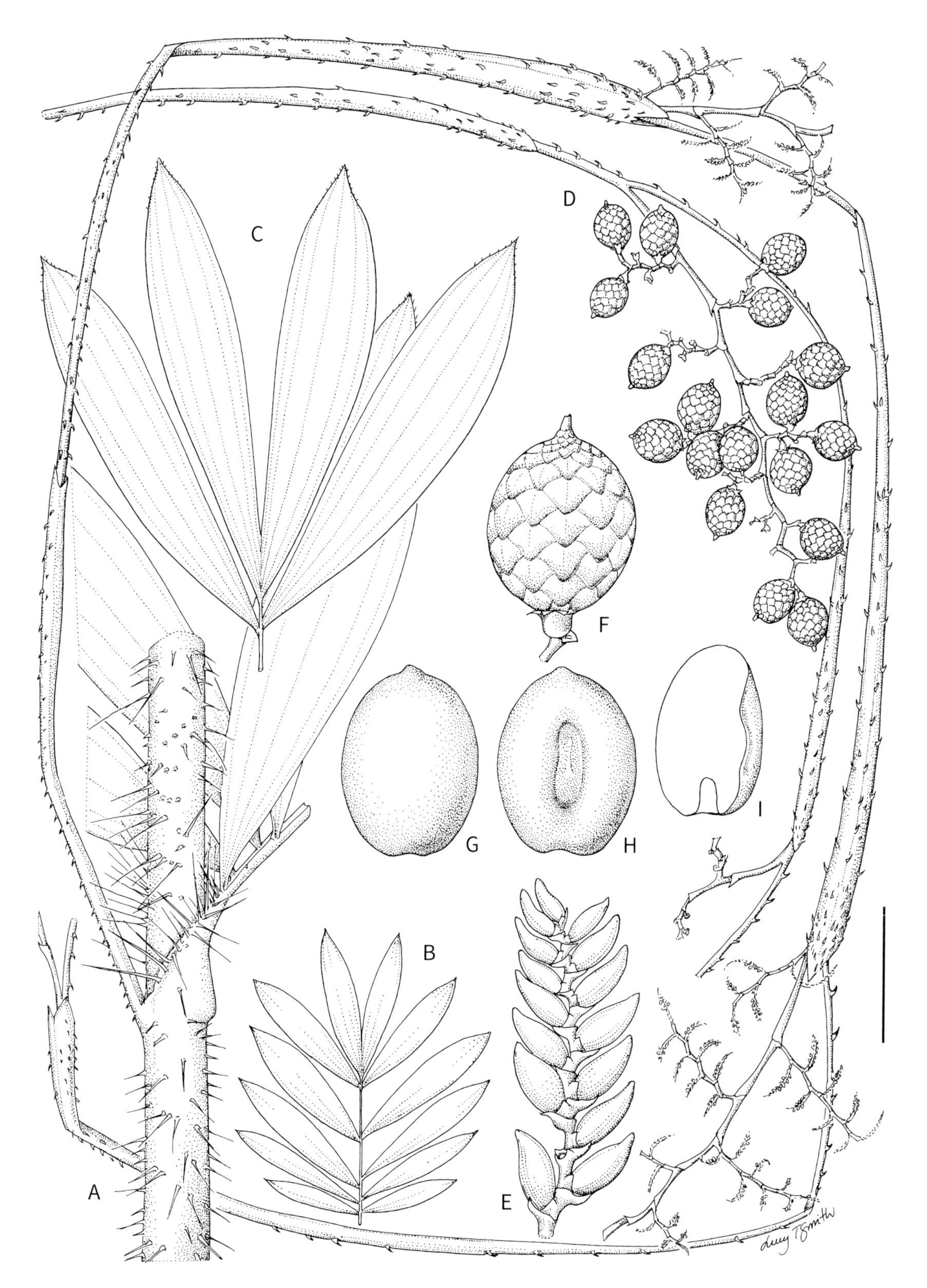

Calamus johnsii. **A.** Stem with leaf base and male inflorescence. **B.** Whole leaf diagram. **C.** Leaf apex. **D.** Portion of female inflorescence with fruit. **E.** Male rachilla with flowers. **F.** Fruit. **G**, **H.** Seed in two views. **I.** Seed in longitudinal section. Scale bar: A = 3 cm; B = 18 cm; C = 6 cm; D = 4 cm; E = 3.3 cm; F = 1 cm; G–I = 7 mm. A–C, E from *Essig LAE 55061*; D, F–I from *Brass 25397*. Drawn by Lucy T. Smith.

Calamus katikii. **A.** Stem with leaf sheath, leaf and female inflorescence with fruit. **B.** Fruit. **C, D.** Seed in two views. **E.** Seed in longitudinal section. Scale bar: A = 3 cm; B–E = 1 cm. All from *Katik LAE 74954*. Drawn by Lucy T. Smith.

material seen); petiole 16–18 mm long; *leaflets 4 each side of rachis, arranged in two widely spaced groups, narrow elliptic*, longest leaflets in lower group ca. 13 × 1.2–1.5 cm, apical leaflets 11.5–13 × 1.2–1.4 cm, apical leaflet pair united to one fifth of their length, leaflets almost unarmed except for minute dark bristles on margins at apex. **Inflorescence** *arching, ca. 40 cm long including 11 cm peduncle and 19 cm flagelliform tip*, branched to 2 orders in the female, male not seen; primary bracts strictly tubular, armed as sheath; primary branches 2, to 3.7 cm long; male rachillae not seen, female rachillae 5–12 × ca. 2 mm. **Fruit** solitary, *globose, ca. 20 × 15 mm scales cream-orange.* **Seed** (sarcotesta removed) ca. 12 × 6 × 9 mm, ellipsoid with broad, shallow pit; endosperm homogeneous.

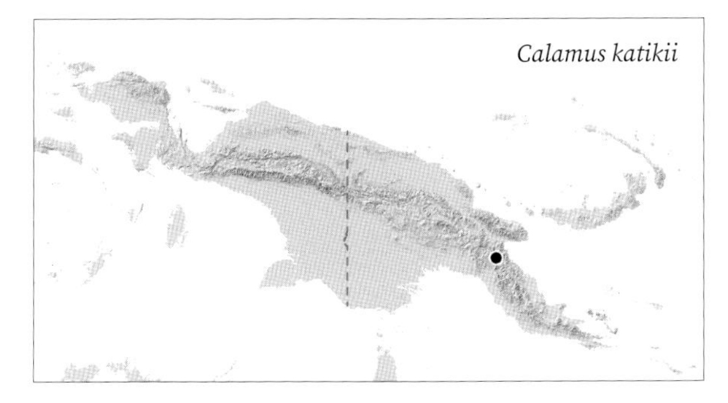

DISTRIBUTION. Kodama Range, south-eastern New Guinea.

HABITAT. Montane, mossy forest at ca. 1,800 m.

LOCAL NAMES. None recorded.

USES. None recorded.

CONSERVATION STATUS. Critically Endangered. The only known site for *C. katikii* is disturbed by gold mining.

NOTES. *Calamus katikii* is a poorly known, high-elevation, montane, flagellate rattan, very slender in habit, with few, grouped, narrow leaflets, short inflorescence and disproportionately robust fruit up to 20 mm long. It most closely resembles *C. cuthbertsonii*, but this species has more leaflets, regularly or subregularly arranged (not grouped), inflorescences lacking a pronounced flagelliform tip and smaller fruit. Vegetatively, *C. katikii* could also be confused with *C. anomalus*. *C. depauperatus* and *C. essigii*, although they differ strongly in inflorescence morphology. It is also similar to some forms of the low elevation species *C. johnsii* (see Baker & Dransfield 2017 for discussion).

27. *Calamus kebariensis* Maturb., J.Dransf. & W.J.Baker

Very slender, multi-stemmed rattan, erect, to 1.5 m. **Stem** with sheaths 3–5 mm diam., without sheaths 1.5–3.5 mm diam.; internodes 4–10.5 cm. **Leaf** ecirrate, to 36 cm long including petiole; sheath green, with thin, patchy brown indumentum, unarmed or very sparsely armed with minute spines between inflorescence and sheath; knee 6.5–9 mm long, inconspicuous, unarmed; ocrea to 5 mm high, scarcely developed, tightly sheathing, densely armed with fine brown bristles; *flagellum absent (in material seen)*; petiole 4–10 cm; *leaflets 10–16 each side of rachis, regularly arranged*, narrow elliptic to linear, longest leaflets at mid-leaf position 6.5–9 × 0.4–1 cm, apical leaflets 7–7.5 × 0.4–0.5 cm, apical leaflet pair not or scarcely united at base, leaflets with fine bristles on margin and upper surface. **Inflorescence** *erect, 12–16 cm long* including 2.5–9 cm peduncle and 2–3.5 cm flagelliform tip, *branched to 1 order in the female*, male not seen; primary bracts, tightly sheathing, very sparsely armed with minute recurved spines; primary branches 3–5; male rachillae not seen, female rachillae 12–42 × 1–3 mm. **Fruit** solitary (or rarely paired), ellipsoid, 16 × 11 mm, scales yellowish. **Seed** (sarcotesta removed) 10 × 8 × 6 mm, ellipsoid, with a shallow pit on one side, seed surface smooth; endosperm homogeneous.

DISTRIBUTION. Mount Nutoti in the Kebar Valley, Bird's Head Peninsula.

HABITAT. Lower montane forest at 1,240–1,500 m.

LOCAL NAMES. *Ibuam* (Mpur).

USES. None recorded.

CONSERVATION STATUS. Critically Endangered. *Calamus kebariensis* is known from only one site where it is threatened by deforestation.

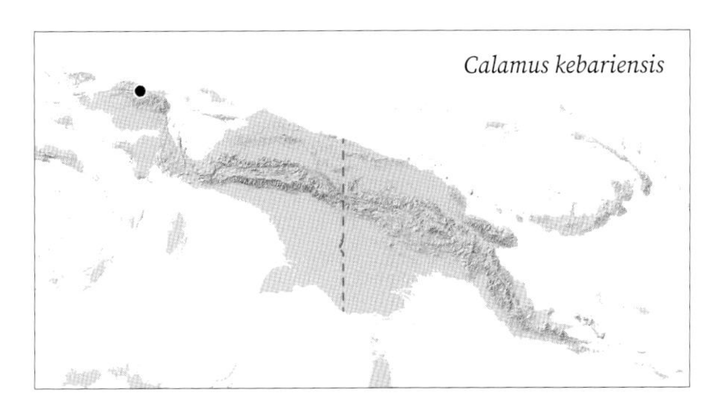

Calamus kebariensis. **A.** Habit with female inflorescences. **B.** Female inflorescence with fruit attached to stem. **C.** Female rachilla with flowers. **D.** Fruit. **E**, **F.** Seed in two views. **G.** Seed in longitudinal section. Scale bar: A = 4 cm; B = 2 cm; C = 7 mm; D–G = 1 cm. A–C from *Maturbongs 73*; D–G from *Maturbongs 75*. Drawn by Lucy T. Smith.

Calamus kebariensis. Female inflorescence with fruit, Kebar Valley (APD).

NOTES. Like *C. spanostachys* from the Sudirman Mountains, which it resembles most closely, *C. kebariensis* is extremely slender, short-stemmed, lacks flagella or cirri, and bears a short, erect inflorescence, that is branched to only one order in female material. *Calamus kebariensis* differs from this species in its leaves, which are finely pinnate with up to 16 regularly arranged leaflets per side with scattered bristles on margins and upper surface (compared to the 2 or 3, largely glabrous leaflet pairs of *C. spanostachys*). The sheaths of *C. kebariensis* are also almost entirely unarmed except for the short, bristly ocrea, which is lacking in *C. spanostachys*. There is also some resemblance between *C. kebariensis* and *C. cuthbertsonii*, though the latter species is more robust with densely spiny leaf sheaths and larger inflorescences that are branched to two orders in female material.

28. *Calamus klossii* Ridl.

Slender to robust, multi-stemmed rattan climbing to 15 m. **Stem** with sheaths 8–32 mm diam., without 4–17 mm diam.; internodes 12–25 cm. **Leaf** ecirrate, to 30–180 cm long including petiole; sheath often covered in dense dark brown, red-brown or blackish indumentum, spines abundant (more rarely moderately numerous), 2–100 × 0.2–2 mm, needle-like or narrow triangular, straw-coloured to golden, arranged in partial collars, pointing in different directions; knee 10–65 mm long, usually obsured by spines; *ocrea 12–70 × 1.5–2 cm, tubular, papery, unarmed or densely armed towards base with crowded, fine spines, soon disintegrating to fibres*; flagellum to 5 m, often shorter; petiole 5–28 cm; leaflets 2–30 each side of rachis, arranged regularly, subregularly or in groups, lanceolate or elliptic, longest leaflet position near base 10–56 × 1–5 cm, mid-leaf leaflets 10–46 × 0.8–4 cm, apical leaflets 10–28 × 0.4–3 cm, apical leaflet pair free or united to one quarter of their length, with long, dark bristles on leaflet margin and (usually) upper surface, lower surface unarmed. **Inflorescence** flagelliform, 0.8–6 m long including peduncle to ca. 0.6 m or more and flagelliform tip to ca. 0.9 m or more, branched to 2 orders in the female and 3 orders in the male; *primary bracts tubular and tightly sheathing, bracts tattering conspicuosly at apex throughout inflorescence*; primary branches 1–5, to 35 cm long, *male rachillae 11–90 × 3–7 mm, female rachillae 5–50 × 3–6 mm*. **Fruit** solitary, spherical, 11–13 × 12–13 mm, scales yellow. **Seed** (sarcotesta removed) ca. 8–8.5 × 7–8.5 × 5.5–6.5 mm, globose, smooth, with deep lateral pit; endosperm homogeneous.

DISTRIBUTION. Central mountain ranges of west and central New Guinea.

HABITAT. Montane forest including secondary vegetation at 435–2,800 m.

LOCAL NAMES. *Mul* (Airgaram), *Nangan* (Weng, Busilmin), *Sebente* (Yali), *Singpa* (Yali), *Towar* (Eipomek).

USES. The stem is used for tying timbers in houses, bindings, arrow construction and making bracelets.

CONSERVATION STATUS. Least Concern.

NOTES. *Calamus klossii* is a member of the *C. baiyerensis* group (see under *Calamus baiyerensis*). It is a highly variable, flagellate species and the taxonomic solution presented here is preliminary. When further specimens become available, it may be possible to subdivide it into more uniform units within a *C. klossii* complex, but a pragmatic delimitation is adopted here.

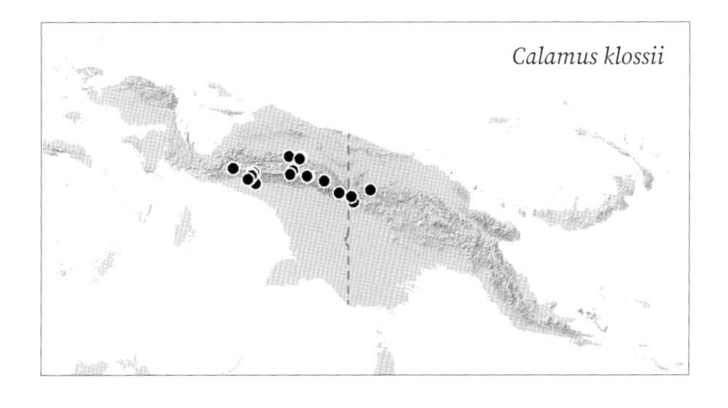

Calamus klossii. **A.** Stem with leaf sheath and petiole. **B.** Leaf apex. **C.** Portion of male inflorescence. **D.** Male rachilla with flowers. **E**, **F.** Male flower whole and in longitudinal section. **G.** Fruit. **H**, **I.** Seed in two views. **J.** Seed in longitudinal section. Scale bar: A = 3 cm; B = 8 cm; C = 4 cm; D = 2 cm; E = 4 mm; F = 3.3 mm; G = 1 cm; H–J = 7 mm. A–F from *Johns et al. 10597*; G–J from *Baker et al. 840*. Drawn by Lucy T. Smith.

Calamus klossii. LEFT TO RIGHT: leaf, robust form; leaf sheath with papery ocrea, robust form. Near Tabubil (WB).

The defining features of *C. klossii* are the slender to moderate habit, the well-developed, disintegrating, papery ocrea and the flagelliform inflorescence, typically bearing tattering primary and secondary bracts. Brown to blackish-brown indumentum is often present on the sheath, leaf rachis and inflorescence. The rachillae are stout, especially the female rachillae, which bear broad, funnel-shaped rachilla bracts. It varies widely in stature (from slender to robust), in leaf length (from 30 to 180 cm), in leaflet number (from 2 to 30 each side) and arrangement (regular, in groups or in fascicles in more than one plane).

Calamus fuscus Becc. (1923) was described as a montane rattan with dense indumentum on the leaf rachis. However, the type of this species has been lost and we are unable to relate the protologue confidently to the species that we describe here. Here we regard *Calamus fuscus* as a name of uncertain application.

29. *Calamus kostermansii* W.J.Baker & J.Dransf.

Moderately robust, multi-stemmed rattan climbing to ca. 10 m. **Stem** with sheaths 18–27 mm diam., without sheaths to 15–18 mm diam.; internodes 16–31 cm. **Leaf** ecirrate, to ca. 1 m long including petiole; sheath yellowish green, *covered with dense, caducous, brown indumentum*, spines scattered, 2–7 mm, triangular with slightly swollen bases, pale green with brown tips, spines solitary or rarely in groups of up to 3; knee ca. 35 mm long, unarmed or with scattered spines; *ocrea ca. 16 × 2.5 cm, elongate, tubular and enclosing the stem, thick papery in texture next to the petiole and leaf rachis, splitting and disintegrating into long fibres on the opposite side, straw-coloured, unarmed or with few scattered spines*; flagellum 2–3 m; petiole very short or absent; *leaflets ca. 50 on each side of rachis, regularly arranged, narrow lanceolate*, longest leaflet in mid leaf ca. 30 × 1.4 cm, basal leaflets ca. 14 × 0.3 cm, apical leaflets ca. 18 ×

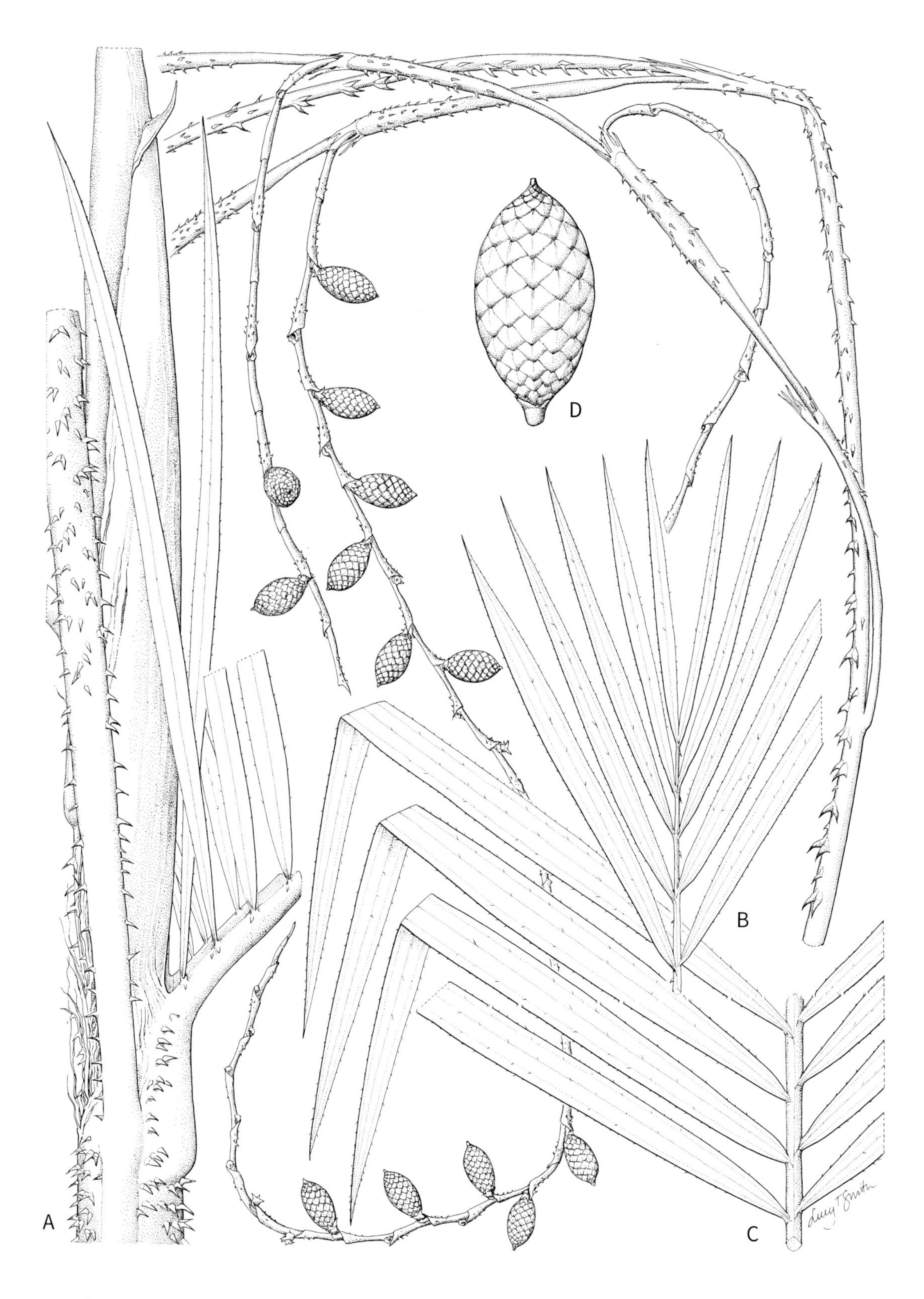

***Calamus kostermansii*. A.** Stem with leaf sheath, leaf base, ocrea and female inflorescence with fruit. **B.** Leaf apex. **C.** Leaf mid-portion. **D.** Fruit. Scale bar: A = 3 cm; B, C = 4 cm; D = 1 cm. All from *Kostermans K10*. Drawn by Lucy T. Smith.

0.7 cm, apical leaflet pair not united, leaflets armed with conspicuous black bristles on both surfaces and margins. **Inflorescence** pendulous, flagelliform, ca. 4 m long including ca. 1 m peduncle and ca. 2 m flagelliform tip, branched to 2 orders in the female, male not seen; primary bracts tubular and tightly sheathing, splitting and eroding into fibres distally, sparsely armed; primary branches ca. 4, to ca. 105 cm long, male rachillae not seen, *female rachillae 12–35 × 0.2–0.3 cm*. **Fruit** solitary, ellipsoid, ca. 15 × 7 mm, scales pale brown with dark margins. **Seed** (sarcotesta removed) ca. 8 × 3 × 2 mm, ellipsoid; endosperm probably homogeneous.

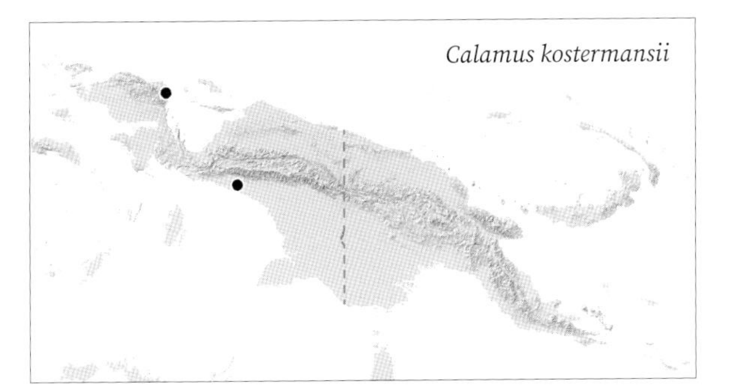

DISTRIBUTION. Two known localities in the Bird's Head Peninsula and south-western New Guinea.

HABITAT. Riverine and swamp habitats, ca. 10 m.

LOCAL NAMES. None recorded.

USES. None recorded.

CONSERVATION STATUS. Critically Endangered. *Calamus kostermansii* is threatened by mining activities.

NOTES. This poorly known flagellate species is a member of the *C. longipinna* group (see under *C. longipinna*). It resembles *Calamus longipinna*, but is greatly disjunct from that species in its distribution. It is distinctive in its well-developed ocrea that partially disintegrates into fibres, abundant indumentum on the sheath and very elongate female rachilla. Henderson (2020) reports that *C. kostermansii* has the longest female rachillae of any *Calamus* species.

30. *Calamus lauterbachii* Becc.

Synonym: *Calamus humboldtianus* Becc.

Moderately robust, multi-stemmed rattan climbing to 25 m. **Stem** with sheaths 14–28 mm diam., without 9–17 mm diam.; internodes 8–32 cm. **Leaf** ecirrate, 40–130 cm long including petiole; sheath green to bluish green, with scattered dark indumentum, spines numerous, 0.5–5 × 0.5–3 mm, triangular, stiff, green to brown, solitary or in collars; knee 18–27 mm long, unarmed or armed as sheath; *ocrea 21–90 × 3–5 cm, boat-shaped, somewhat inflated and clasping, erect, leathery, purple-brown, drying brown, armed with numerous fine spines arranged in collars (resembling eyelashes), persistent*, but eventually tattering; flagellum 1–4 m; petiole 14–52 cm; *leaflets 5–9 each side of rachis, arranged in 2–3 divaricate groups, elliptic to linear-lanceolate*, often strongly hooded, leathery, longest leaflet at mid-leaf position 26–44 × 2.8–8 cm, apical leaflets 16–36 × 2.5–8 cm, apical leaflet pair united for half to three quarters of their length, leaflet surfaces almost always unarmed, margins armed with stout bristles. **Inflorescence** erect or flagelliform, 0.5–2.6 m long including 0.2–0.7 m peduncle, flagelliform tip either lacking or well-developed to 180 cm, branched to 2 orders in the female and 2–3 orders in the male; primary bracts papery, inflated, splitting, tattering and disintegrating (especially where subtending a primary branch); *primary branches 1–2, to 32 cm long,*

Calamus lauterbachii. Habit, Jivewaneng (WB).

Calamus lauterbachii. LEFT TO RIGHT: base of male inflorescence, Jivewaneng (WB); female inflorescence with fruit, near Timika (JD); fruit, near Timika (JD).

compact and/or congested, male rachillae 5–44 mm × 1–2.5 mm, female rachillae 6–25 mm × 3–4 mm. **Fruit** solitary, ellipsoid or obovoid, 20–30 × 13–14 mm, scales yellow to reddish brown. **Seed** (sarcotesta removed) 11–15 × 9–11 × 6–8 mm, ellipsoid or obovoid, with deep lateral pit and shallow to deeper rounded furrows; endosperm homogeneous.

DISTRIBUTION. Widespread in mainland New Guinea and Yapen Island.

HABITAT. Lowland forest, from sea level to 700 m.

LOCAL NAMES. *Ajabf kwa nyi* (Ormu), *Angulim* (Bogia), *Kobing* (Kotte), *Kolaben* (Maia), *Nongkoek* (Kati), *Onda* (Kabori), *Woy* (Arso).

USES. Cane used for tying timbers in buildings and as bowstring. Chewing leaves yields blue dye. Stem water can be drunk.

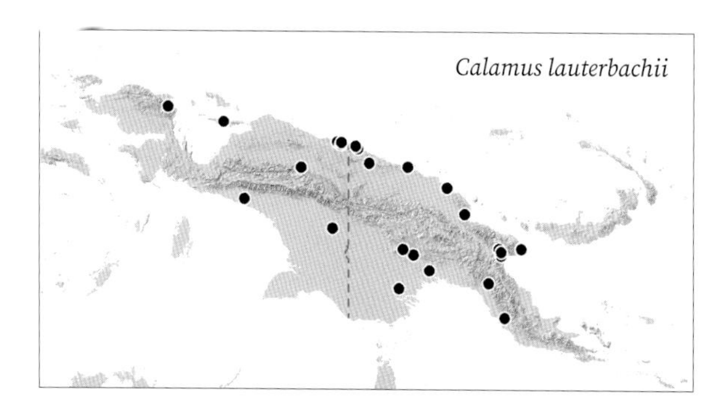

CONSERVATION STATUS. Least Concern.

NOTES. This species is the commonest member of the *Calamus lauterbachii* group, which comprises three flagellate species (including also *C. heatubunii* and *C. nudus*) that share strongly grouped, rather broad leaflets, conspicuous, persistent, papery to leathery ocreas and inflorescences sometimes with congested primary branching systems (Baker & Dransfield 2017). *Calamus lauterbachii* itself is a highly distinctive rattan, characterised by its broad, hooded leaflets, which are few in number and arranged in irregular, divaricate groups, and its very long, leathery ocrea, which is often purple-brown in colour and armed with low, finely spiny crests that resemble eyelashes. Its inflorescence can be either flagelliform or erect and not flagelliform. Typically, the inflorescence bears only one or two primary branches, which are very condensed and subtended by papery primary bracts that split and tatter. Because the rachillae are so congested, the inflorescence can appear club-like in fruit.

The holotype of *C. lauterbachii* appears to have been destroyed in Berlin. However, an image of the specimen is available (Beccari 1908) and an isotype, comprising fragments of male rachilla and flowers, is held in Florence. These closely match the type of *Calamus humboldtianus*, the later name under which this rattan was previously widely known.

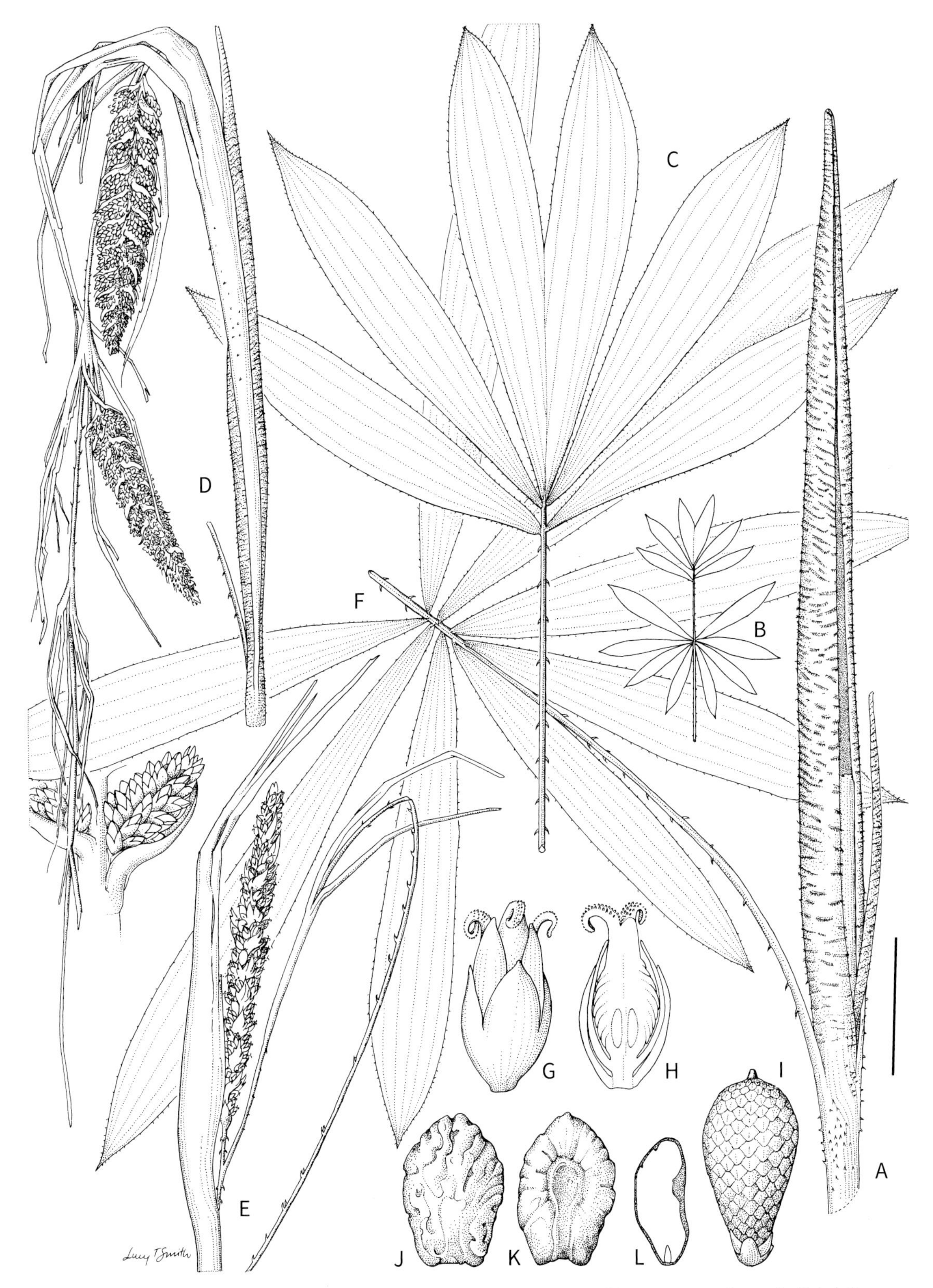

Calamus lauterbachii. **A.** Stem with leaf sheath, leaf base and ocrea. **B.** Whole leaf diagram. **C.** Leaf apex. **D.** Male inflorescence attached to stem. **E.** Portion of female inflorescence. **F.** Detail of male inflorescence. **G**, **H.** Female flower whole and in longitudinal section. **I.** Fruit. **J**, **K.** Seed in two views. **L.** Seed in longitudinal section. Scale bar: A, C = 6 cm; B = 36 cm; D = 8 cm; E = 4 cm; F, I, L = 2 cm; G, H = 4 mm; J, K = 1.5 cm. A–D, F from *Zieck NGF 36185*; E, G, H from *Brass 8964*; I–L from *Dransfield JD 7663*. Drawn by Lucy T. Smith.

31. *Calamus longipinna* K.Schum. & Lauterb.

Synonym: *Calamus ralumensis* Warb. ex K.Schum. (invalid name)

Moderately robust, multi-stemmed rattan climbing to 20 m. **Stem** with sheaths 14–30 mm diam., without sheaths 11–21 mm diam.; internodes 18–30 cm. **Leaf** ecirrate to 1.2 m long including petiole; sheath green, sometimes rather pale, with variable, purple-brown or white indumentum, *spines very few to numerous, 2.5–18 × 1–2.5 mm, triangular, pale green with yellow-orange bases and brown tips, solitary, not arranged in groups*; knee 24–58 mm long, unarmed or sparsely armed; *ocrea 10.5–45 × 1.5–5 cm, inflated, boat-shaped, open longitudinally to base on side opposite petiole, clasping and usually obscuring sheath, papery, brown, unarmed or armed with needle-like spines to 5 mm long,* often inhabited by ants, persistent, rarely separating into fibres; flagellum to 4 m; petiole 1–13 cm; leaflets 33–50 each side of rachis, *arranged regularly,* very rarely subregularly or in three regular groups, linear, longest leaflet towards base of leaf, 20–46 × 1–2.5 cm, mid-leaf leaflets 20–40 × 1.5–2.5 cm, apical leaflets 5.6–15 × 0.3–0.8 cm, apical leaflet pair not or only briefly united, margins and both surfaces armed with short bristles to 6 mm. **Inflorescence** pendulous, up to 4 m long including 17–104 cm peduncle and 6–104 cm flagelliform tip, branched to 2 orders in the female and 3 orders (very rarely 4 orders) in the male; *primary bracts tubular, not splitting, but often tattering at mouth, unarmed or with few short grapnel spines*; primary branches up to 10, to 110 cm long, male rachillae 16–73 × 2.5–4.5 mm, female rachillae 40–290 × 2–4 mm. **Fruit** solitary, oblong to subspherical, 10–12 × 8–9 mm scales blue-green. **Seed** (sarcotesta removed) 7 × 5–6 × 4.5–5 mm, ellipsoid to subspherical, with a narrow pit on one side, with irregular, rounded ridges and pits throughout surface; endosperm homogeneous.

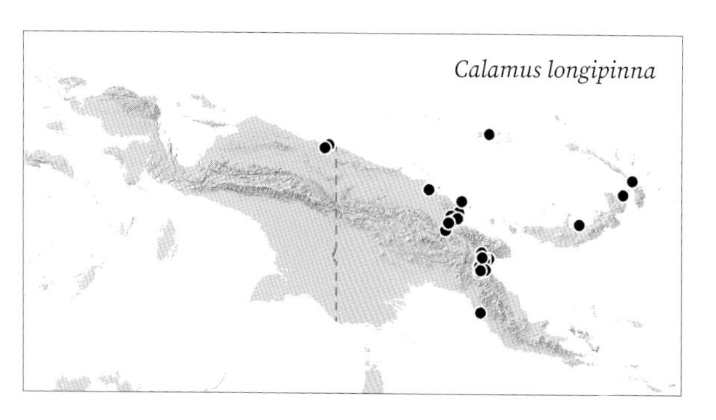

Calamus longipinna. Habit, Wau (WB).

DISTRIBUTION. Widespread in north-eastern and and south-eastern New Guinea, the Bismarck Archipelago and the Solomon Islands.

HABITAT. Lowland forest up to 800 m, often in degraded habitats.

LOCAL NAMES. *Buso* (Pala dialect, Patpatar language), *Iogel* (Kara), *Kada* (Kuanua), *Kanda* (Papua New Guinea Tok Pisin), *Kikis* (Kaigorin), *Mazzas* (dialect from vicinity of Markham River bridge near Lae), *Muli* (Bembi), *Sating* (Rawa), *Sep* (Jal), *Seribi* (Bo Village, Namatanai), *Sirei* (Madang), *Sos* (Usino).

USES. General use as cordage.

CONSERVATION STATUS. Least Concern.

NOTES. The *Calamus longipinna* group comprises six flagellate species (*C. bankae, C. kostermansii, C. longipinna, C. reticulatus, C. vestitus, C. wanggaii*) characterised by the presence of well-developed, inflated papery or fibrous ocreas up to ca. 45 cm long that clasp the stem (with the exception of *C. reticulatus*), flagelliform inflorescences and female flowers borne

Calamus longipinna. **A.** Stem with leaf sheath and ocrea. **B.** Leaf apex. **C.** Portion of male inflorescence. **D.** Portion of female rachilla with fruit. **E.** Male flower. **F.** Female flower. **G.** Female flower one petal removed. **H, I.** Seed in two views. **J.** Seed in longitudinal section. Scale bar: A, C = 4 cm; B = 6 cm; D, H–J = 1 cm; E–G = 1.8 mm. A–C from *Baker et al. 593*; D, H–J from *Baker & Utteridge 585*; E–G from *Lauterbach 242*. Drawn by Lucy T. Smith.

Calamus longipinna. FROM TOP: leaf sheath with ocrea, Lae (JD); fruit, Ramu River (WB).

in pairs (sterile male together with fertile female), resulting in solitary fruit.

Calamus longipinna is well known as it is common in heavily collected areas of Papua New Guinea (Madang and Morobe provinces), although its occurrence outside these areas seems to be patchier. It is readily distinguished by its usually unarmed, papery ocrea that is open longitudinally on the side opposite the petiole, and leaflets that are typically regularly arranged. The sheath is armed with triangular spines or sometimes unarmed, and the primary inflorescence bracts are unarmed or only sparsely armed. It is most easily confused with *C. vestitus*, though this species is immediately distinguished by its completely tubular ocrea and spines with jagged, rather than pointed apices.

32. *Calamus lucysmithiae* W.J.Baker & J.Dransf.

Slender rattan climbing to 30 m. **Stem** with sheaths 6–11 mm diam., without sheaths to 4–8 mm diam. **Leaf** ecirrate, to 45–80 m long including petiole; sheath glaucous green to reddish brown, with scattered to dense, thin brown indumentum, spines moderately numerous, 1–13 mm long, needle-like, stiff, dark, scattered; knee 15–24 mm long, unarmed or lightly armed as sheath; *ocrea 20–48 × 4–10 mm, divergent, triangular with edges inrolled, stiff, brown, unarmed or densely armed as sheath, especially at apex, persistent*; flagellum 120–200 cm long; petiole 3–50 mm long; *leaflets 6–8 each side of rachis, regularly to subregularly arranged, rather widely spaced especially at tip, broadly elliptic*, longest leaflets in mid-leaf 14–24 × 3.5–4.5 cm, apical leaflets 7–13 × 0.6–2 cm, apical leaflet pair not united or united up to half their length, leaflets armed with very few minute bristles on margins. **Inflorescence** *trailing, 170–230 cm long including 24–36 cm peduncle and 35–120 cm flagelliform tip*, branched to 2–3 orders in the female, male not seen; primary bracts strictly tubular, opening symmetrically or asymmetrically at apex, unarmed or with scattered grapnel spines; primary branches 3–7, to 27 cm long, male rachillae not seen, female rachillae 5–70 mm × 0.5–1.2 mm, *female flowers sometimes pedicellate*. **Fruit** solitary, ellipsoid, ca. 14.5 × 10 mm, scales pale with brown margins. **Seed** (sarcotesta removed) ca. 8 × 6.5 × 5.5 mm, ellipsoid with with a shallow pit on one side, seed surface smooth; endosperm homogeneous.

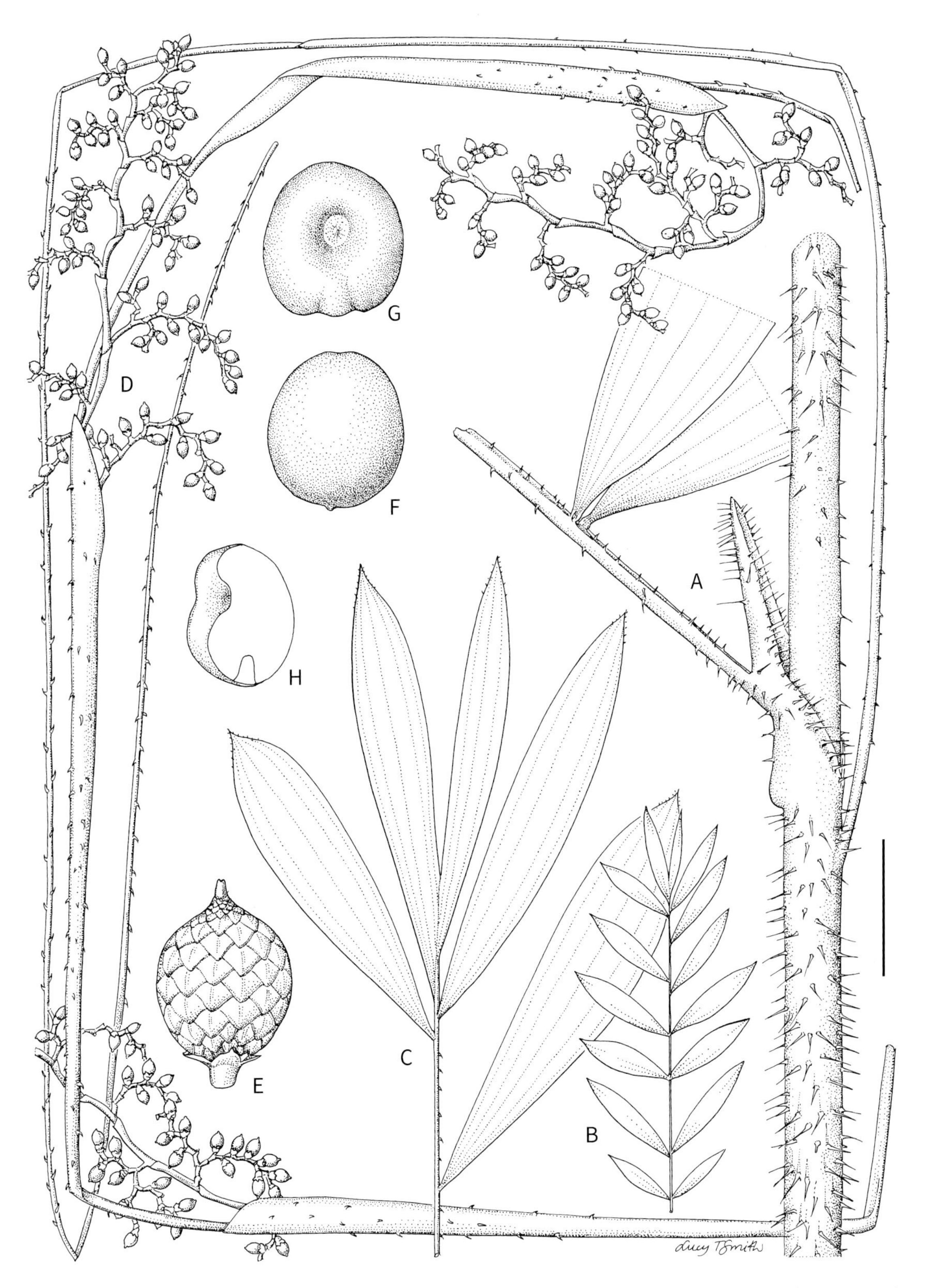

Calamus lucysmithiae. **A.** Stem with leaf sheath, leaf base, ocrea and flagellum. **B.** Whole leaf diagram. **C.** Leaf apex. **D.** Portion of female inflorescence with fruit. **E.** Fruit. **F**, **G.** Seed in two views. **H.** Seed in longitudinal section. Scale bar: A, D = 3 cm; B = 18 cm; C = 6 cm; E = 1 cm; F–H = 7 mm. A, E–H from *Rahiria & Zieck NGF 36577*; B, C from *Rahiria & Zieck NGF 36576*; D from *Zieck NGF 36555*. Drawn by Lucy T. Smith.

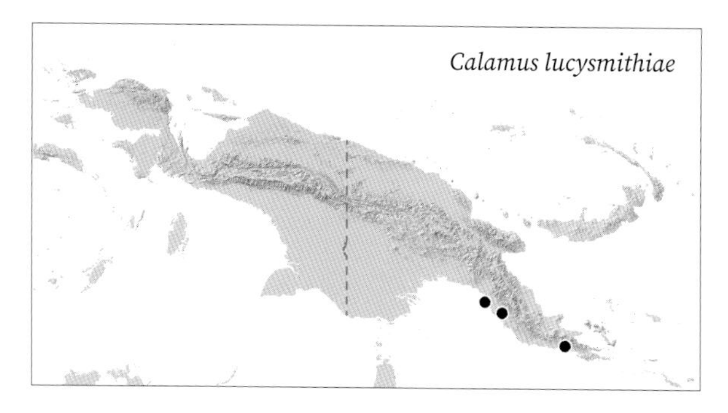

DISTRIBUTION. South coast of south-eastern New Guinea.

HABITAT. Lowland forest from sea level to 90 m, on hills, in valleys and at the edge of mangroves.

LOCAL NAMES. *Eraharo* (Toaripi).

USES. None recorded.

CONSERVATION STATUS. Endangered. *Calamus lucysmithiae* has a restricted distribution. Deforestation due to logging concessions is a major threat in its range.

NOTES. *Calamus lucysmithiae* is a flagellate species that is easily distinguished by its short, but well-developed, divergent ocrea, its leaf with widely spaced, broadly elliptic leaflets and its long, trailing inflorescence with strongly recurving rachillae with female flowers sometimes borne on pedicel-like stalks (comprising bracteole and floral axis). It is similar to *C. croftii*, *C. jacobsii* and *C. johnsii* in leaf and sheath morphology (though *C. jacobsii* has unarmed sheaths), but none of these bears a conspicuous ocrea or stalked female flowers.

33. *Calamus macrochlamys* Becc.

Synonym: *Calamus macrospadix* Burret

Moderate, multi-stemmed rattan climbing to 50 m. **Stem** with sheaths 9–26 mm diam., without sheaths to 5–16 mm diam., internodes 6–38 cm. **Leaf** ecirrate, to 95 cm long including petiole; sheath green, with sparse to abundant brown indumentum, spines usually abundant, rarely unarmed, 5–10 × 2–3.5 mm, rather uniform, with swollen bases, pale brown with black tips, solitary or very rarely in horizontal groups; knee 25–90 mm long, 5–90 mm wide, unarmed or sparsely armed as the rest of the sheath; *ocrea to 15–48 × 2.5–4.5 cm, erect, swollen, open on the side away from the petiole, papery or leathery, pale to mid-brown, unarmed or armed with scattered spines or combs of bristles, persistent*; flagellum to 3.7 m long; petiole 8–35 cm long; *leaflets 2–15 on each side of rachis, arranged in divaricate groups (2–4 leaflets per side), apical leaflets in a fan of 2–6 on each side,* broadly lanceolate (rarely narrower), longest leaflet often near the base 35–38 × 2.5–4.5 cm, mid-leaf leaflets 20–37 × 2–4 cm, apical leaflets 9–38 × 0.7–6 cm, apical leaflet pair united for one third to three quarters their length, surfaces unarmed except for very short sparse bristles along margins. **Inflorescence** 0.6–3 m long including 23–35 cm peduncle and 6–72 cm flagelliform tip, branched to 2 orders in the female and 3 orders in the male; primary bracts closely sheathing, splitting apically; primary branches 7–10, 6–30 cm long, male rachillae 8–45 × 1–1.5 mm, female rachillae 30–200 × 2–3 mm. **Fruit** *usually paired*, 10–11.5 × 10–11.5 mm, including beak to 1.5 × 1.5 mm, scales pale yellowish brown. **Seed** (sarcotesta removed) 6–8 × 5–7 × 4–5 mm, irregularly ridged and with a deep pit on one side; endosperm homogeneous.

DISTRIBUTION. Widespread in mainland New Guinea.

HABITAT. In lowland and upland forest, including swamp forest, from sea level to 1,200 m.

LOCAL NAMES. *Baea* (Karimui), *Arompotto* (Kati), *Kotowo* (Goilala-Tapini).

USES. None recorded.

CONSERVATION STATUS. Least Concern.

NOTES. This flagellate rattan is a member of the *C. altiscandens* group (see notes under *C. altiscandens*). It is easily distinguished by its strongly grouped generally broad lanceolate leaflets, and distinctive rather leathery, persistent, inflated, clasping ocrea that does not usually tatter. The flowers are borne in threes comprising a pair of female flowers and a sterile male flower. It is similar to the much rarer *C. altiscandens*, but has more lax inflorescence architecture and less densely armed (or unarmed) ocreas than this species.

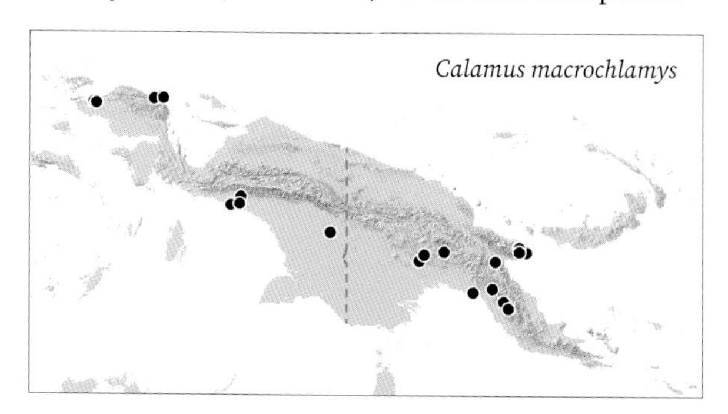

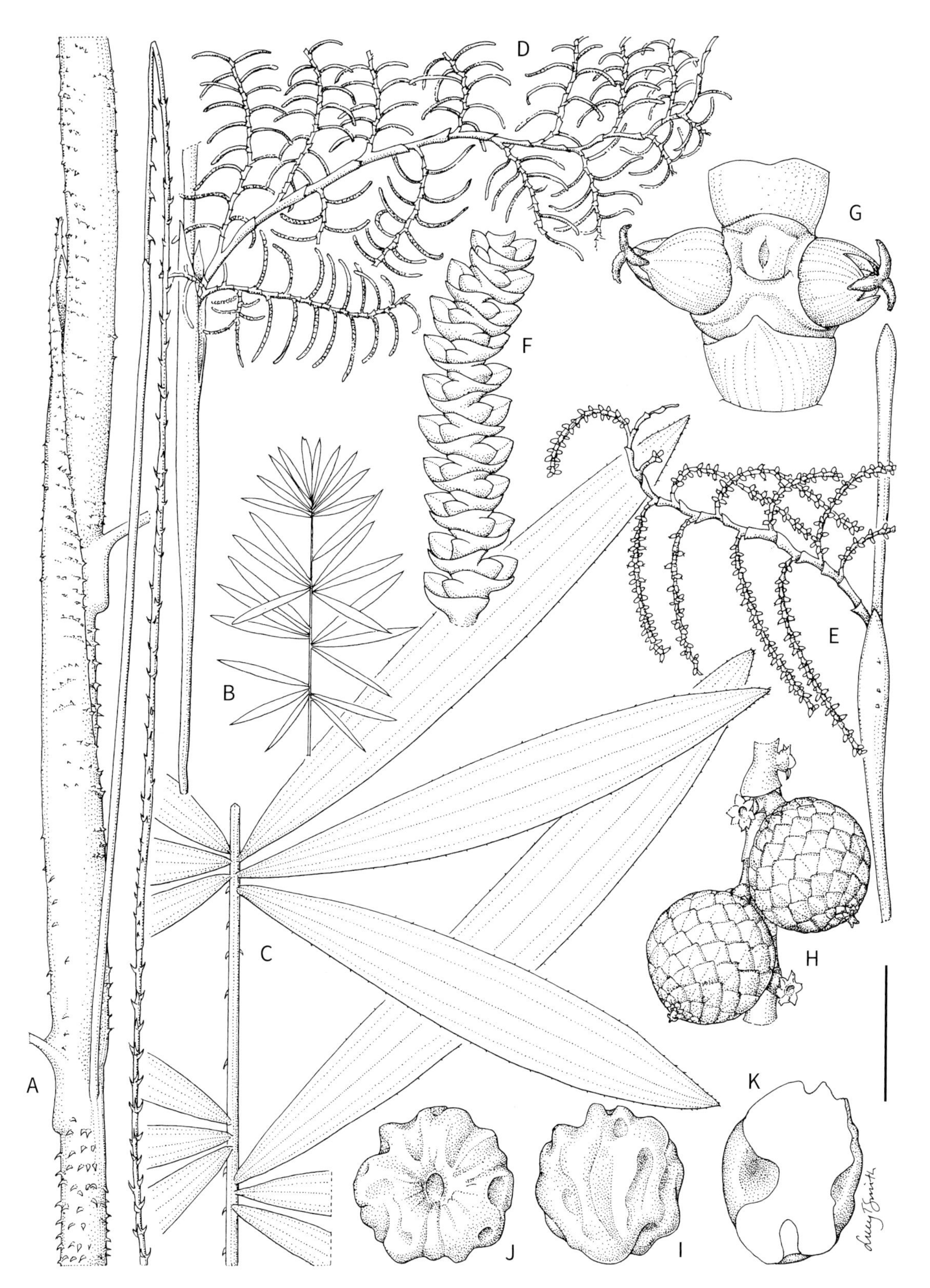

Calamus macrochlamys. **A.** Stem with leaf sheath, ocrea and flagellum. **B.** Whole leaf diagram. **C.** Leaf mid-portion. **D.** Portion of male inflorescence. **E.** Portion of female inflorescence. **F.** Portion of male rachilla with flowers. **G.** Detail of female rachilla with two female flowers and sterile male flower scar. **H.** Fruit. **I, J.** Seed in two views. **K.** Seed in longitudinal section. Scale bar: A, D, E = 6 cm; B = 18 cm; C = 8 cm; F, K = 5 mm; G = 3.3 mm; H = 1.5 cm; I, J = 7 mm. A–C from *Baker et al. 627*; D, F from *Baker et al. 1124*; E, G from *Zieck NGF 36513*; H–J from *Maturbongs 283*. Drawn by Lucy T. Smith.

34. *Calamus maturbongsii* W.J.Baker & J.Dransf.

Moderately robust, multi-stemmed rattan climbing to 30 m. **Stem** with sheaths 16–22 mm diam., without sheaths to 10–11 mm diam.; internodes ca. 48 cm. **Leaf** ecirrate, to 1 m long including petiole; sheath with very thin, white, indumentum, unarmed or armed with very few, scattered, minute, easily detached, triangular spines to 1mm; knee 13–22 mm long, unarmed, with conspicuous ridge at the base; ocrea 3–5 mm, forming a low collar, persistent, drying woody, unarmed; flagellum to 1.7 m; petiole 3–5 mm long; *leaflets 7–12 each side of rachis, arranged regularly, lanceolate with hooded apices, longest leaflet near leaf base* 26–36 × 4.5–5 cm, mid-leaf leaflets 24.5–26 × 4.3–5.5 cm, apical leaflets 5.5–12 × 0.7–2 cm, apical *leaflet pair not united or united to one quarter of their length,* leaflets unarmed or with very few bristles on margins or tip; cirrus absent. **Inflorescence** *62–81 cm long* including 33–46 cm peduncle and 5–14 cm sterile tip, branched to 2 orders in the female, male not seen; primary bracts tubular, split to base and tattering by emerging primary branches, sometimes remaining tubular distally, unarmed; primary branches 3–7, to 7 cm long, male rachillae not seen, female rachillae 3–20 × ca. 1 mm. **Fruit** solitary, subspherical to broadly ellipsoid, ca. 15 × 12 mm, *scales orange-red with erose margins.* **Seed** (sarcotesta removed) ca. 9 × 8 × 5 mm, globose with deep lateral pit, with a small, rounded apical appendage; endosperm homogeneous.

DISTRIBUTION. Vicinity of Sorong in the Bird's Head Peninsula.

HABITAT. Lowland forest at 100–200 m.

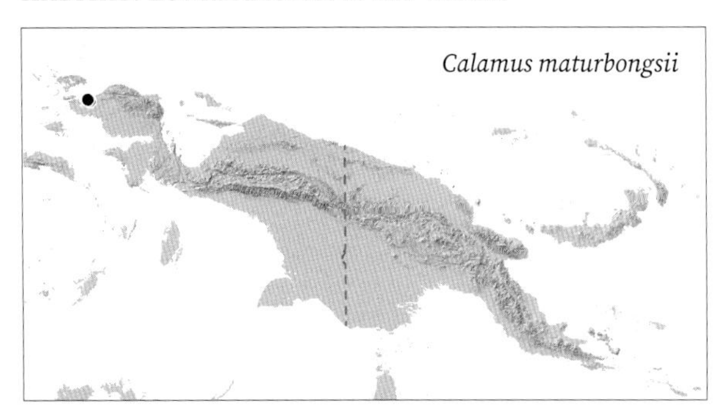

LOCAL NAMES. None recorded.

USES. None recorded.

CONSERVATION STATUS. Critically Endangered. Deforestation due to oil palm plantations and logging concessions is a major threat in its limited range.

NOTES. *Calamus maturbongsii* is a member of the *C. anomalus* groups (see under *C. anomalus*). See notes under *C. erythrocarpus* for further discussion.

35. *Calamus moszkowskianus* Becc.

Moderately robust rattan, climbing height not recorded. **Stem** with sheaths 9–18 mm diam., without ca. 7.5 mm diam.; internodes ca. 15 cm. **Leaf** ecirrate, 50–70 cm long including petiole; sheath brown when dry, with thin, caducous, brown indumentum, *spines numerous, various lengths, to 70 mm, needle-like, brown, solitary, sheath mouth armed with long, ascending spines*; knee 30–50 mm long, unarmed except for a central line of long spines and marginal spines, knee margins strongly oblique; *ocrea 4–5 × ca. 1.5 cm, ligule-like, adjacent to petiole, erect, hard-leathery, brown, armed as sheath, persistent*; flagellum present, length not known; petiole 0–5 cm; *leaflets 32–40 (at least) each side of rachis, regularly arranged, lanceolate,* longest leaflet in mid-leaf position 15–33 × 1–1.2 cm, apical leaflets 3.6–13 × 0.4–1 cm, apical leaflet pair united to ca. one seventh of their length, leaflets armed with bristles on upper and lower surface, and on margins at tip. **Inflorescence** pendulous or flagelliform, up to 2.3 m long including up to ca. 1.5 m peduncle and short flagelliform tip, branched to 2 orders in the female and 3 orders in the male; primary bracts tightly sheathing, armed; primary branches 1–6, to ca. 15 cm long, male rachillae 10–30 mm × ca. 1 mm, female rachillae ca. 50 mm × ca. 1.5 mm. **Fruit** solitary, spherical, 12–13 × 10 mm, scales brown. **Seed** (sarcotesta removed) ca. 6.5 × 6.5 × 5 mm, spheroidal with a shallow lateral pit; endosperm ruminate.

DISTRIBUTION. Van Rees Mountains in western New Guinea.

HABITAT. Lowland forest between 70 and 300 m.

LOCAL NAMES. None recorded.

USES. None recorded.

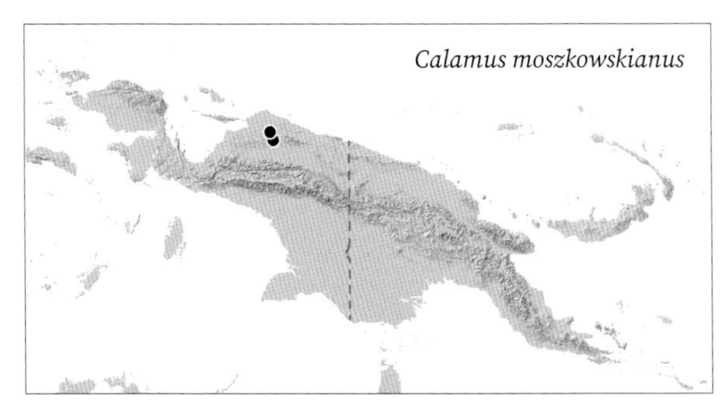

***Calamus maturbongsii*. A.** Stem with leaf sheath, leaf base and flagellum. **B.** Leaf apex. **C, D.** Female inflorescence with fruit and peduncular bract, two views. **E.** Female inflorescence base showing prophyll. **F.** Portion of female inflorescence with fruit. **G.** Portion of female rachilla with fruit. **H.** Fruit. **I, J.** Seed in two views. **K.** Seed in longitudinal section. Scale bar: A = 3 cm; B = 6 cm; C–E = 4 cm; F = 2 cm; G, H = 1 cm; I–K = 7 mm. All from *Maturbongs 286*. Drawn by Lucy T. Smith.

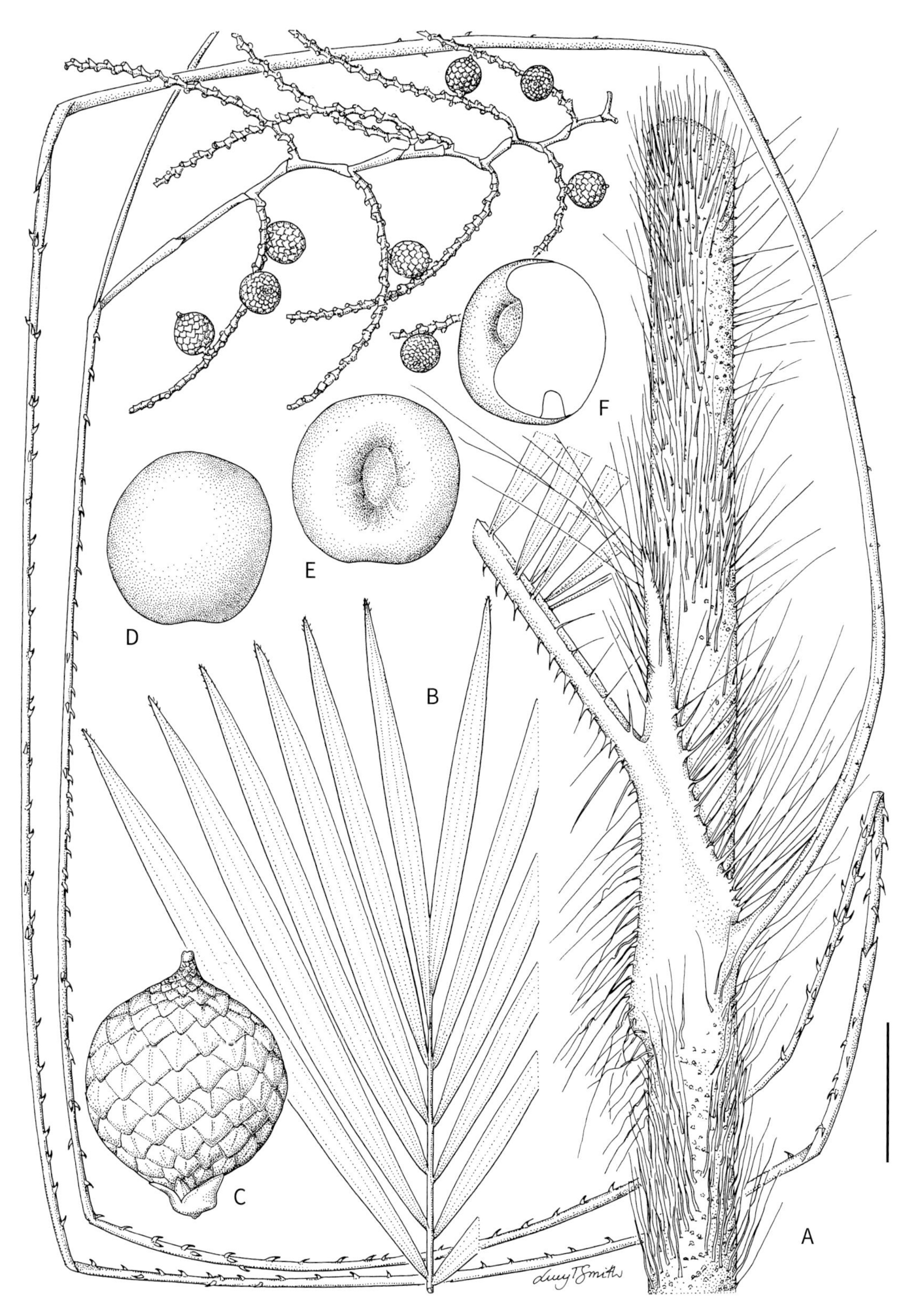

Calamus moszkowskianus. **A.** Stem with leaf sheath, ocrea and female inflorescence with fruit. **B.** Leaf apex. **C.** Fruit. **D**, **E.** Seed in two views. **F.** Seed in longitudinal section. Scale bar: A = 3 cm; B = 4 cm; C = 7 mm; D–F = 5 mm. All from *Lam 672*. Drawn by Lucy T. Smith.

CONSERVATION STATUS. Critically Endangered. *Calamus moszkowskianus* is known from only two adjacent sites that are threatened by deforestation.

NOTES. *Calamus moszkowskianus* is a little-known, flagellate species recognised by its moderately slender habit, numerous fine, regularly arranged leaflets and its stiff, conspicuous ocrea to ca. 5 cm, which is armed with long, fine spines like the leaf sheath. It is most similar to *C. distentus* (see under *C. distentus*).

36. *Calamus nanduensis* W.J.Baker & J.Dransf.

Robust, multi-stemmed rattan climbing to 20 m. **Stem** with sheaths 14–35 mm diam., without sheaths to 9–20 mm diam., internodes 6–17 cm. **Leaf** ecirrate, to 0.7–1.2 m long including petiole; sheaths dull to dark green, with *dense, grey-brown indumentum, spines numerous, very fine, to 15 mm long, needle-like, sinuous, grey-brown, solitary, readily detached, some spines erect and adpressed to sheath*; knee 30–65 mm long, more or less unarmed; *ocrea 10–20 cm, ligule-like or splitting into auricles, well-developed adjacent to petiole, not developed opposite the petiole, papery, brown, unarmed, fragile, disintegrating*; flagellum 1.8–3 m long; petiole 12–16 cm; leaflets 21–29 each side of rachis, regularly arranged, linear-lanceolate, longest leaflet near to base 31–54 × 2–2.5 cm, mid-leaf leaflets 29–45 × 2–2.8 cm, apical leaflets 8.5–12 × 0.5–0.7 cm, apical leaflet pair united to one fifth of their length or not united, leaflets armed with few dark bristles on margins and upper surface. **Inflorescence** pendulous, ca. 5.2 m long including ca. 0.6 m peduncle and ca. 2.3 m flagelliform tip, branched to 2 orders in the female, male not seen; primary bracts tubular, opening asymmetrically at apex to form acute triangular limb, armed with scattered, robust grapnel spines; primary branches ca. 7, to ca. 40 cm long, male rachillae not seen, female rachillae 60–120 × ca. 4 mm. **Fruit** not seen. **Seed** not seen.

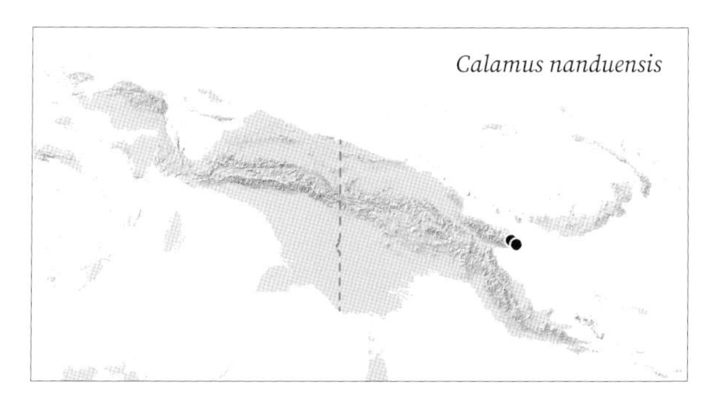

Calamus nanduensis. Leaf sheath with papery ocrea, Nanduo (WB).

DISTRIBUTION. Eastern Huon Peninsula.

HABITAT. Primary and secondary montane forest, 1,100–1,200 m.

LOCAL NAMES. *Kobing* (Kotte).

USES. None recorded.

CONSERVATION STATUS. Data Deficient. This species is known from only two sites. Due to insufficient evidence of ongoing threats, a single extinction risk category cannot be selected.

NOTES. *Calamus nanduensis* is a member of the *C. baiyerensis* group (see under *C. baiyerensis*). It is a flagellate species that is distinguished by its robust, multi-stemmed habit, leaf sheaths with numerous very fine, easily dislodged spines and dense, grey or brown indumentum (also present on other organs)

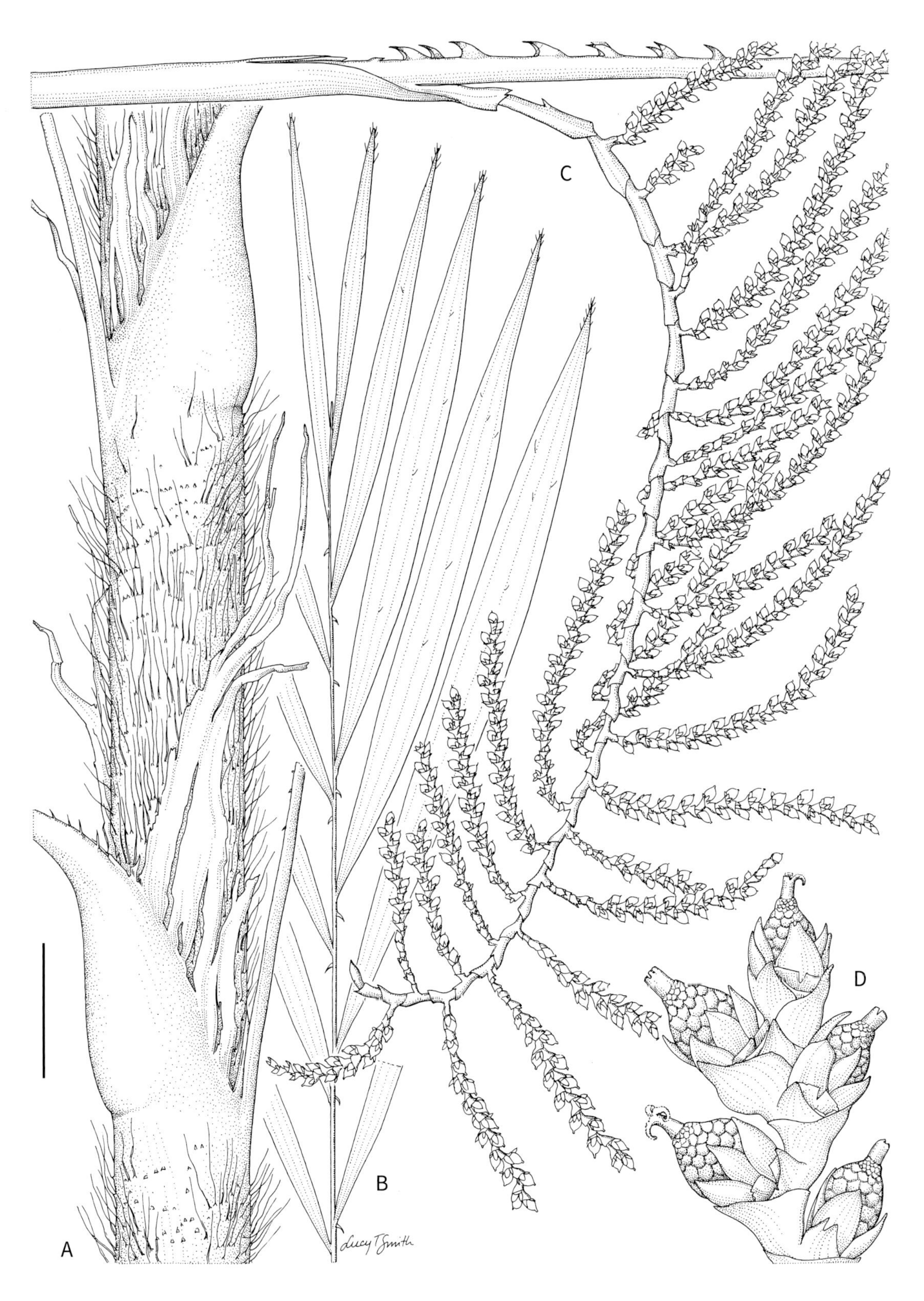

***Calamus nanduensis*. A.** Stem with leaf sheath and ocrea. **B.** Leaf apex. **C.** Portion of female inflorescence. **D.** Portion of female rachilla with fruit. Scale bar: A = 3 cm; B, C = 4 cm; D = 7 mm. All from *Banka et al. 2010*. Drawn by Lucy T. Smith.

Calamus nanduensis. LEFT TO RIGHT: female inflorescence; base of female inflorescence. Nanduo (WB).

and a well-developed, but fragile ocrea. *Calamus pintaudii* was included as a synonym of *C. nanduensis* by Henderson (2020), but we consider the available material to be insufficient to support this decision.

37. *Calamus nannostachys* Burret

Moderately slender, multi-stemmed rattan climbing to 45 m. **Stem** with sheaths 9–17 mm diam., without sheaths 6–9 mm diam.; internodes 14–23 cm. **Leaf** ecirrate, 30–75 cm long including petiole; sheath with thin grey-brown indumentum, spines moderately numerous to almost absent, to 2 mm, stout, dark-tipped, solitary, evenly distributed; knee 20–33 mm long, unarmed or armed as sheath; ocrea a low encircling collar to 5 mm high, brown, unarmed or lightly armed, persistent; flagellum 65–160 cm; *petiole 3–18 cm; leaflets 5–8 each side of rachis, irregularly arranged, rarely in well-defined groups, broadly elliptic or oblanceolate,* hooded, longest leaflet at mid-leaf position 8–38 × 2–5.5 cm, apical leaflets 7–25 × 1.5–6 cm, apical leaflet pair united from one quarter to one half of their length, leaflets with short bristles in margins and very few bristles on upper surface, lower surface glabrous or rarely with orange brown indumentum. **Inflorescence** *erect to arching, 30–60 cm long* including 8–54 cm peduncle, sterile tip lacking (or almost so), branched to 2 orders in the female and 3 orders in the male; *primary bracts relatively short and splitting to base by expanding primary branches*; primary branches 1–9, to 8 cm long, male rachillae 1–20 × 0.5–1 mm, female rachillae 7–14 × 1–2 mm. **Fruit** solitary, globose, 17–20 × 12–15 mm, scales yellow-orange. **Seed** (sarcotesta removed) ca. 9.5 × 9.5 × 6.5 mm, ellipsoid, smooth, with deep lateral pit to one side; endosperm homogeneous.

DISTRIBUTION. Lower slopes of the Owen Stanley Range, south-eastern New Guinea.

HABITAT. Lowland forest at 45–350 m.

LOCAL NAMES. *Durodo* (Kurandi).

USES. None recorded.

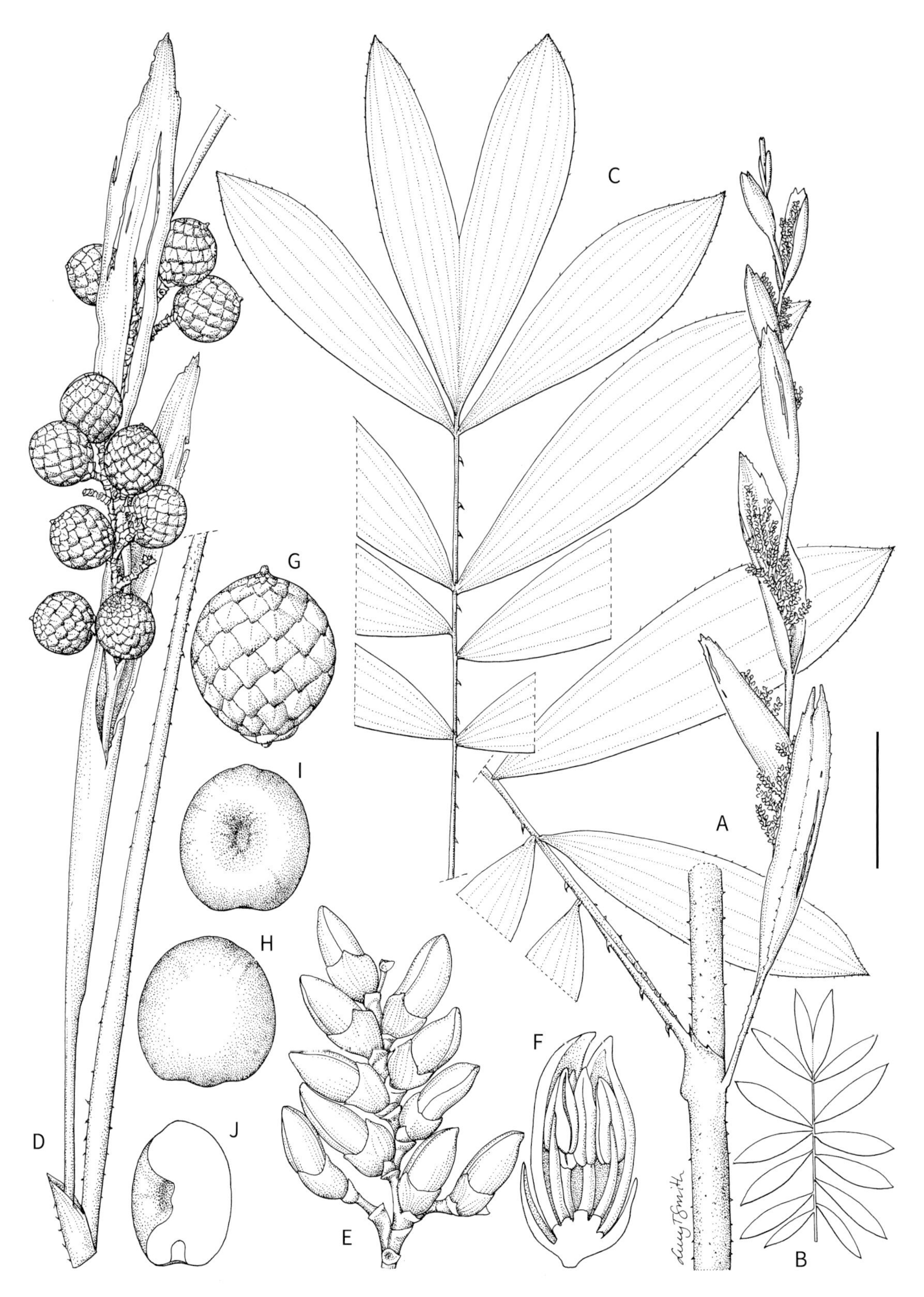

***Calamus nannostachys*. A.** Stem with leaf sheath, leaf base and male inflorescence. **B.** Whole leaf diagram. **C.** Leaf apex. **D.** Portion of female inflorescence with fruit. **E.** Male rachilla with flowers. **F.** Male flower in longitudinal section. **G.** Fruit. **H**, **I.** Seed in two views. **J.** Seed in longitudinal section. Scale bar: A, C = 6 cm; B = 30 cm; D = 3 cm; E = 4 mm; F = 2 mm; G–J = 1 cm. All from *Zieck NGF 36544*. Drawn by Lucy T. Smith.

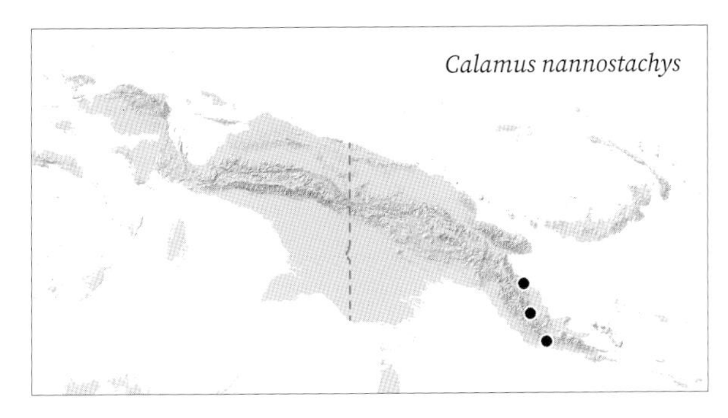

CONSERVATION STATUS. Endangered. *Calamus nannostachys* has a restricted distribution. Deforestation due to gold mining and logging concessions is a major threat in its range.

NOTES. *Calamus nannostachys* belongs to the *C. anomalus* group (see under *C. anomalus*). It is most closely related to *C. anomalus* and *C. essigii*, but can be distinguished by its leaf with 5–8 pairs of irregularly arranged broadly elliptic or oblanceolate leaflets, the terminal pair united for one quarter to half of their length or more, and by the more robust inflorescence up to 60 cm in length. It occurs at low elevation relative to the other two species. A specimen of a robust, unarmed form (*Essig LAE 55164*) from low elevation on the Mo River near to Lae is attributed to *C. nannostachys* here but may represent a further undescribed species. It bears very large leaflets in two widely spaced groups, the undersurface of which are covered with thin orange-brown indumentum. Additional material is required to confirm this.

38. *Calamus nudus* W.J.Baker & S.Venter

Moderately slender, multi-stemmed rattan climbing to 30 m. **Stem** with sheaths 9–15 mm diam., without sheaths to 7–9 mm diam.; internodes 10–28 cm. **Leaf** ecirrate, to 64 cm long including petiole; sheath green, with scattered dark scales, spines numerous, 1–4 × ca. 1 mm, narrow triangular, stiff, green, solitary; knee 18–20 mm long, lightly armed as sheath; *ocrea 40–55 × 1–2.5 cm, erect, inflated, boat-shaped, open longitudinally to base, papery, brown, with numerous solitary spines, similar to sheath spines, but finer, ocrea tattering to dry fibres, sometimes disintegrating*; flagellum 100–200 cm; petiole to ca. 13 cm; *leaflets 3–5 each side of rachis, arranged in a single fan-like group, lanceolate, leathery, longest leaflets at apex*, apical leaflets 44–51 × 3.5–4 cm, lower leaflets slightly smaller, apical leaflet pair united from one third to one half of their length, leaflets armed only on margins with stout bristles. **Inflorescence** flagelliform, 1.6–2.2 m long including 0–4 cm peduncle, the remainder being flagellum-like, branched to 2 orders in the female, male not seen; primary bracts very narrow and tightly sheathing as in flagellum, tattering at mouth; primary branches 1 (sometimes up to 3), to 21 cm long, *first primary branch emerging directly from inflorescence base and not subtended by any primary bract*, male rachillae not seen, female rachillae 14–25 mm × 3–4 mm. **Fruit** solitary, ellipsoid, ca. 20 × 12.5 mm, scales orange-brown. **Seed** (sarcotesta removed) ca. 10 × 8 × 6 mm, ellipsoid, smooth, with deep lateral pit; endosperm homogeneous.

DISTRIBUTION. Scattered sites in southern central New Guinea.

HABITAT. Lowland forest at 30–146 m.

LOCAL NAMES. *Aliopoi* (Bamu), *Arompotto* (Kati).

USES. None recorded.

Calamus nudus. Leaf, Purari River (SV).

Calamus nudus. **A.** Stem with leaf sheath, leaf base and female inflorescence with fruit. **B.** Stem with leaf sheath and female inflorescence. **C.** Stem with leaf sheath, ocrea and emerging inflorescence. **D.** Whole leaf diagram. **E.** Portion of female rachilla with flowers. **F.** Sterile male flower in longitudinal section. **G.** Female flower in longitudinal section. **H.** Fruit. **I, J.** Seed whole and in longitudinal section. Scale bar: A–C = 4 cm; D = 30 cm; E = 5 mm; F, G = 2.5 mm; H = 1.5 cm; I, J = 7.5 mm. A, D–J from *Dijkstra BW 6635*; B–C from *Akivi & Zieck NGF 36324*. Drawn by Lucy T. Smith.

Calamus nudus. LEFT TO RIGHT: female and sterile male flowers; female inflorescence branch; base of female inflorescence with fruit. Purari River (SV).

CONSERVATION STATUS. Near Threatened. *Calamus nudus* is threatened at some of its sites by logging concessions.

NOTES. *Calamus nudus* belongs to the flagellate *C. lauterbachii* group (see under *C. lauterbachii*). It is unique among *Calamus* species (and indeed palms) in that its first primary inflorescence branch emerges directly from the inflorescence base and is not subtended by any primary bract (including a prophyll). Primary bracts are evident in the distal, flagellum-like part of the inflorescence (which is long and flagellum-like). Vegetatively, *C. nudus* is similar to *C. heatubunii* in the almost palmate leaves comprising very few leaflets and the pronounced ocrea armed with numerous solitary spines.

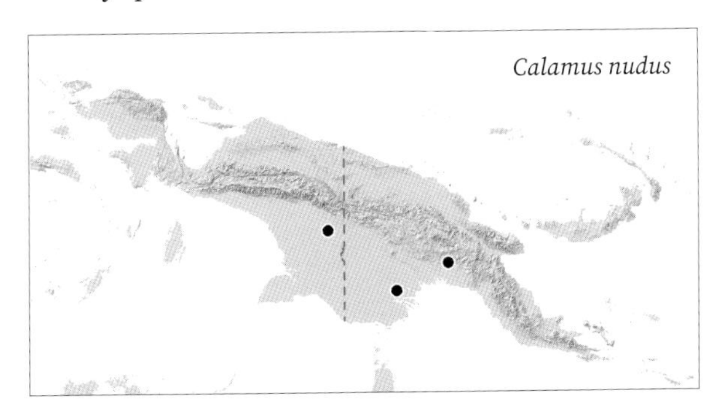

39. *Calamus oresbius* W.J.Baker & J.Dransf.

Synonym: *Calamus oresbiopsis* A.J.Hend.

Slender, multi-stemmed rattan climbing to ca. 10 m, sometimes to 25 m. **Stem** with sheaths 4–18 mm diam., without sheaths to 3–8 mm diam.; internodes 8–22 cm. **Leaf** ecirrate, 16–54 cm long including petiole; sheath dull green to yellow-green, with thin, grey-brown indumentum, spines sparse to dense, 0.5–15 mm long, fine, needle-like, grey to brown, sometimes longer and more dense near sheath mouth (to 48 mm); knee ca. 7–20 mm long, ca. 4–9 mm wide, unarmed or armed as sheath; ocrea 2–5 mm high, forming a persistent, closely sheathing, dry crest, brown, unarmed or armed as sheath; flagellum 0.3–2 m long (ca. 10 cm in very slender forms); petiole 0.5–12 cm long; *leaflets 3–5 (rarely 6) leaflets each side of rachis, typically clustered at leaf apex with 1–2 pairs at base*, sometimes with additional leaflets at mid-leaf postion or more rarely clustered in a single group at the leaf apex, *elliptic to oblanceolate, sometimes hooded, leathery*, basal leaflets 9–32 × 1–4.5 cm, longest leaflets at leaf apex 11–36 × 2.5–4.5 cm, apical leaflet pair united from one fifth to half their length, leaflet with scattered bristles on margin, especially near tip. **Inflorescence** *flagelliform, 1–2 m long* including ca. 20–100 cm peduncle and ca.

5–20 cm flagelliform tip, branched to 2 orders in the female, 2 or sometimes 3 orders in the male; primary bracts strictly tubular and tightly sheathing, armed as flagellum; primary branches 1–8, to 54 cm long, male rachillae 5.5–45 × ca. 1 mm, female rachillae 10–48 × 1–1.5 mm. **Fruit** solitary, globose, ca. 10–18 × 8–14 mm scales yellow to orange. **Seed** (sarcotesta removed) 6–12 × 6–11 × 4.5–8 mm, globose, seed surface smooth, with a deep, lateral depression; endosperm homogeneous.

DISTRIBUTION. Widespread in the eastern half of the Central Range.

HABITAT. Montane forest from 700 to 2,200 m, including secondary forest. Also cultivated in some areas.

LOCAL NAMES. *Ep* (Mendi), *Karrikaribu* (Koijari), *Kral* (Bolin), *Kurnin* (Biaru), *Waiang* (Bolin, Ganja, Narak, Nondugl, Noltubi), *Waiangl* (Kuman), *Waiink* (Bolin), *Wajam* (Yoowi; *Wajamumum* for fruits), *Wi-kiral* (Sinasina).

USES. Weaving baskets, armbands, waistbands and finer rattan work. Used as a long-lasting binding in fence- and house-building.

CONSERVATION STATUS. Least Concern.

NOTES. *Calamus oresbius* is a variable, yet distinctive, slender, flagellate species that is characterised by its short leaves (up to ca. 54 cm long) with 3–6 pairs of leaflets, which, in most forms, are arranged in two widely spaced groups, one clustered at the apex and one or two pairs grouped at the base. The leaflets are broadly elliptic to oblanceolate and leathery, the apical pair being fused for one fifth to half their length. Its inflorescence is narrow and flagelliform, with fine rachillae, and is branched to two orders (or sometimes to three orders in some male material). It is potentially confused with *Calamus johnsii*, which occurs at much lower elevation and is easily distinguished by its relatively robust inflorescence, or *Calamus anomalus*, which is similar in stature and leaflet arrangement, but has very different inflorescence morphology.

Calamus oresbius. LEFT TO RIGHT: habit; fruit. Southern Highlands (WB).

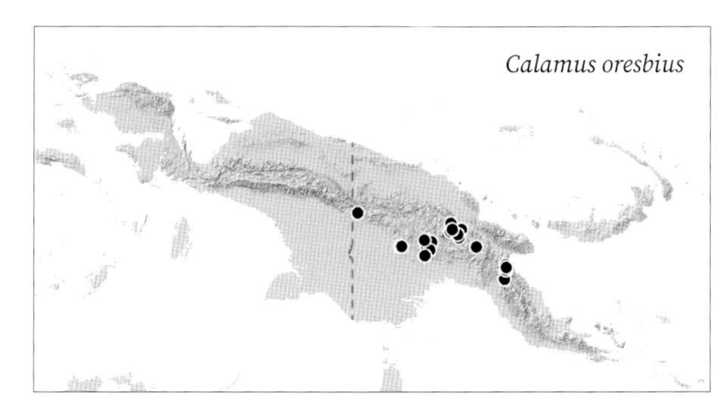

Henderson (2020) split *C. oresbius* into two, describing a new species *C. oresbiopsis*, which he differentiated on its needle-like spines, membranous ocrea, thinner stems and higher mean elevation. In doing so, Henderson effectively narrowed *C. oresbius* into a rather uniform species, while *C. oresbiopsis* remained highly variable. Spines in some forms of *C. oresbiopsis* are indeed needle-like, though some are more similar to the very short spines of *C. oresbius sensu* Henderson, and some forms are unarmed. We could not readily distinguish the ocreas of *C. oresbiopis*, although some forms were armed with needle spines. The difference in mean elevational range may be an artifact of collecting bias, and in any case, the elevation ranges overlap. In light of these observations, we treat *C. oresbiopsis* as a synonym of *C. oresbius* here, adhering to the species concept as defined in our original publication of the species (Baker & Dransfield 2014).

40. *Calamus pachypus* W.J.Baker, Bayton, J.Dransf. & Maturb.

Robust, single-stemmed rattan climbing to 26 m. **Stem** with sheaths 25–60 mm diam., without sheaths 13–30 mm diam.; internodes 18–33 cm. **Leaf** *cirrate*, to 4 m long including cirrus and petiole; sheath dark green, with brown indumentum, *spines few to numerous, 1–60 × 0.3–5 mm, red-brown to black, flattened, triangular, flexible, often curving, tapering to a narrow acute apex, spine bases swollen, spines arranged in irregular partial whorls interspersed with solitary spines*; knee 60–100 mm long, unarmed or lightly armed; ocrea 2–11 mm, forming a low, woody, brown, lightly armed, persistent collar; flagellum absent; petiole 5–90 mm; leaflets 10–17 each side of rachis, *arranged in widely spaced, sometimes divergent pairs*, rarely regular or subregular, broadly elliptic, hooded, longest leaflets near middle of leaf, 27–46 × 4.4–6.5 cm, apical leaflets 13–30 × 0.6–4.8 cm, leaflets lightly armed with very few bristles on upper surface and margins; cirrus 80–160 cm, cirrus grapnel spines arranged regularly. **Inflorescence** arching, to 4 m long including 27–72.5 cm peduncle and 25–90 cm sterile tip, branched to 2 orders in the female and 3 orders in the male; primary bracts strictly tubular, with acute, triangular limb to one side, moderately armed; primary branches 6–9, to 70 cm long, male rachillae 3–44 × 0.5–2 mm, female rachillae 40–200 × 2 mm. **Fruit** solitary, shortly stalked, globose, 10–15 × 8.5–13.5 mm, scales light green to white. **Seed** (sarcotesta removed) 7.3–8 × 7–9.5 × 6–8 mm, globose, with a deep, narrow pit on one side, the surface covered with numerous deep pits and irregular channels; endosperm homogeneous.

DISTRIBUTION. Widespread throughout mainland New Guinea, one record from New Ireland.

HABITAT. Primary and secondary forest, 100–1,500 m with more than half of the records above 600m.

LOCAL NAMES. *Hele bu* (Yali), *Kour* (Biaru), *Kur* (Karkar), *Mambile* (Yali), *Meya* (Arfak Plains), *Tendu mundu* (Berap).

USES. Cane used for making bridges and waist hoops, split cane for general cordage, for making arrows and bow strings, and for fire-making. Sometimes planted.

CONSERVATION STATUS. Least Concern.

NOTES. *Calamus pachypus* is distinguished from other cirrate species in the *C. aruensis* complex (see under *C. aruensis*) by its leaf sheath spines, which are flexible, triangular and distinctly swollen at the base. While both large and small spines occur on the sheath, very long spines (up to 60 mm) are almost always present. Leaflet arrangement is variable, but the most frequent form bears leaflets grouped in pairs. Surprisingly, Henderson (2020) treats this distinctive species as a synonym of *C. vitiensis*, overlooking basic features that differentiate the two described by Baker *et al.* (2003). Though recorded from low elevations, *C. pachypus* is more frequently found in submontane and montane vegetations.

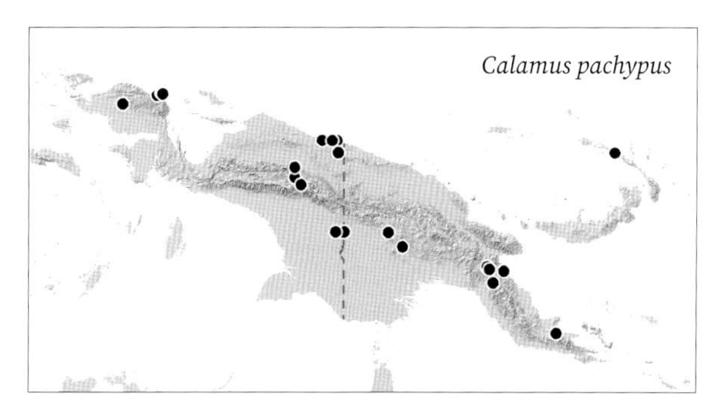

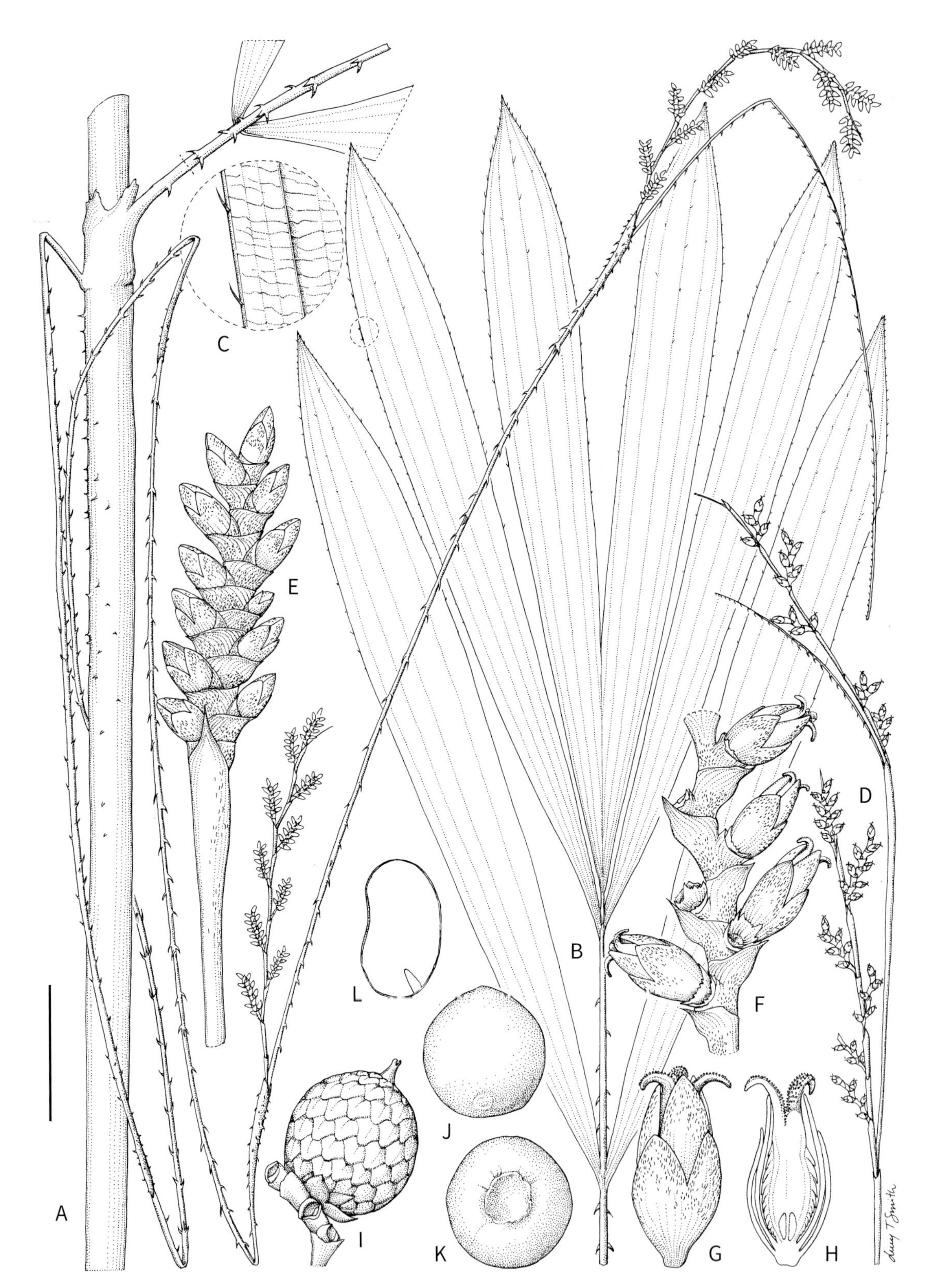

Calamus oresbius. **A.** Stem with leaf sheath, leaf base and male inflorescence. **B.** Leaf apex. **C.** Leaflet detail showing transverse veinlets. **D.** Female inflorescence. **E.** Male rachilla with flowers. **F.** Portion of female rachilla with flowers. **G, H.** Female flower whole and in longitudinal section. **I.** Fruit. **J, K.** Seed in two views. **L.** Seed in longitudinal section. Scale bar: A, D = 3 cm; B = 6 cm; C, I = 1 cm; E, F = 5 mm; G, H = 3 mm; J–L = 7 mm. A–C, E from *Baker et al. 627*; D, F–H from *Baker et al. 609*; I–L from *Baker et al. 624*. Drawn by Lucy T. Smith.

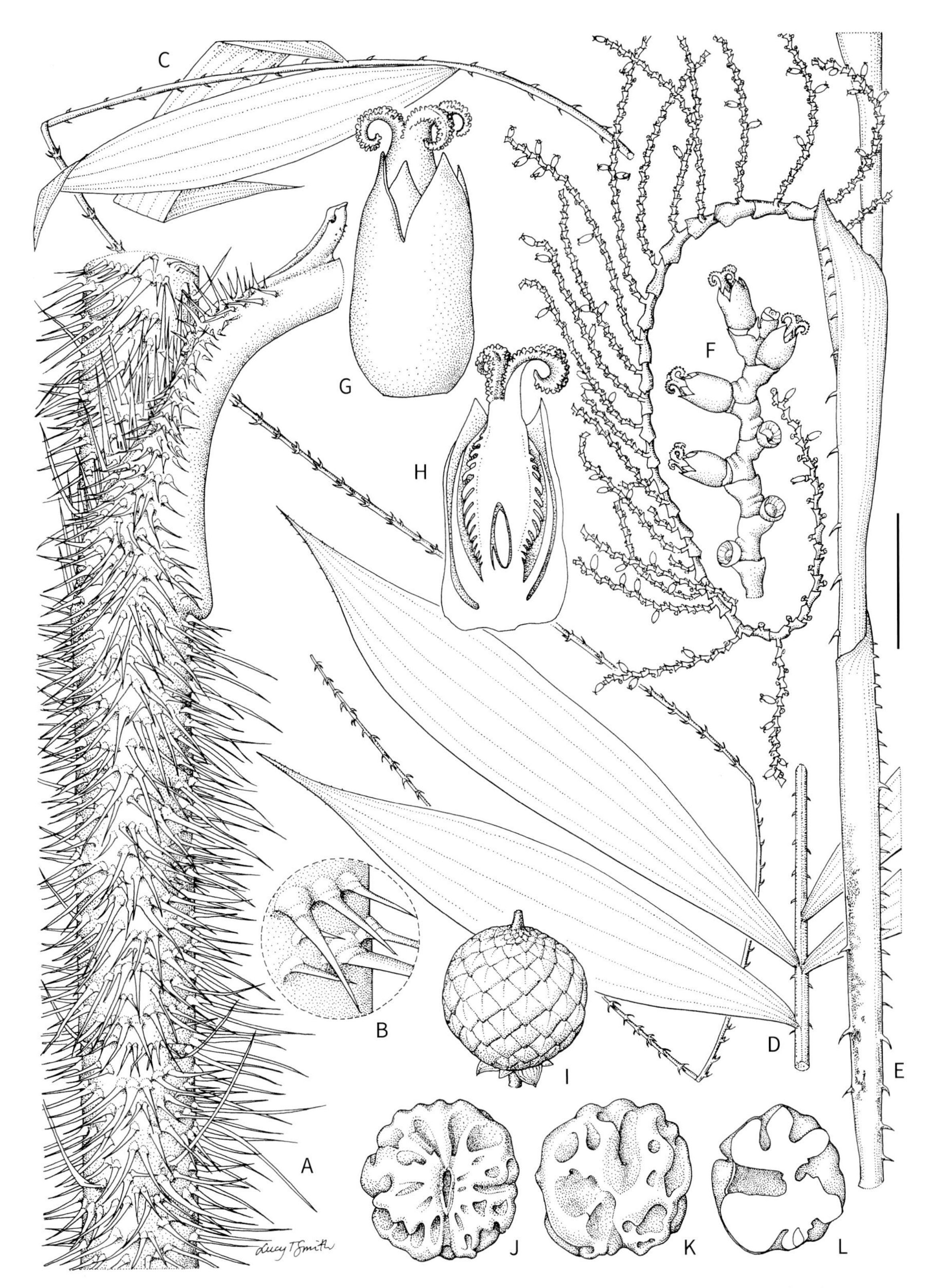

Calamus pachypus. **A.** Stem with leaf sheath and petiole. **B.** Detail spines. **C.** Leaf apex with cirrus. **D.** Leaf mid-portion. **E.** Portion of female inflorescence. **F.** Female rachilla with flowers. **G, H.** Female flower whole and in longitudinal section. **I.** Fruit. **J, K.** Seed in two views. **L.** Seed in longitudinal section. Scale bar: A = 3 cm; B = 1.5 cm; C, D = 8 cm; E = 4 cm; F, I = 1 cm; G, H = 2 mm; J–L = 7 mm. A–H from *Dransfield et al. JD 7600*; I–L from *Maturbongs 47*. Drawn by Lucy T. Smith.

41. *Calamus papuanus* Becc.

Synonym: *Palmijuncus papuanus* (Becc.) Kuntze

Slender to very slender, single-stemmed or multi-stemmed rattan climbing to 30 m. **Stem** with sheaths 7–28 mm diam., without 5.5–12 mm diam.; internodes 10–35 cm. **Leaf** ecirrate, to 23–60 cm long including petiole; sheath green, with thin grey-brown indumentum, *spines few to numerous, 1–3 × ca. 1 mm (rarely longer to ca. 7 × 3 mm), stout, triangular, often somewhat upward-pointing,* solitary; knee 12–25 mm long, colour and armature as sheath; ocrea 3–20 mm high, tightly sheathing, leathery, obliquely truncate, unarmed or sometimes armed as sheath, sometimes disintegrating into fibres; flagellum 0.8–2.5 m; petiole 0–7 cm; *leaflets 4–11 each side of rachis, grouped in 2s and 4s (rarely subregularly arranged), elliptic, often hooded,* longest leaflet in mid-leaf position 8–24 × 1.3–4 cm, apical leaflets 9–22 × 1.2–3.5 cm, *apical leaflet pair united from one third to one half of their length,* basal leaflets reduced and usually reflexed across stem, *leaflet unarmed except for very few marginal bristles.* **Inflorescence** arching, 42–100 cm long including 20–75 cm peduncle, flagelliform tip brief or lacking, branched to 2 orders in the female and 3 orders in the male; primary bracts strictly tubular, somewhat inflated, lightly armed; primary branches 2–9, 7–20 cm long, male rachillae 18–50 × 0.5–1 mm, female rachillae 15–90 × ca. 1.5 mm. **Fruit** solitary, spherical, 11–13 × 10–12 mm, scales cream-yellow with dark margins. **Seed** (sarcotesta removed) 8–9 × 8–9 × 5.5–6.5 mm, *globose with a deep lateral pit and uneven surface;* endosperm homogeneous.

DISTRIBUTION. Bird's Head Peninsula, Misool and vicinity of Timika.

HABITAT. Lowland and montane forest from sea level to 1050 m.

LOCAL NAMES. *Lamtei* (Matbat), *Movendja* (Sidai), *Otonoma* (Sumuri), *Su* (Madik), *To sao, To sau* (Ayawasi).

Calamus papuanus. LEFT TO RIGHT: habit; female and sterile male flowers. Tamrau Mountains (WB).

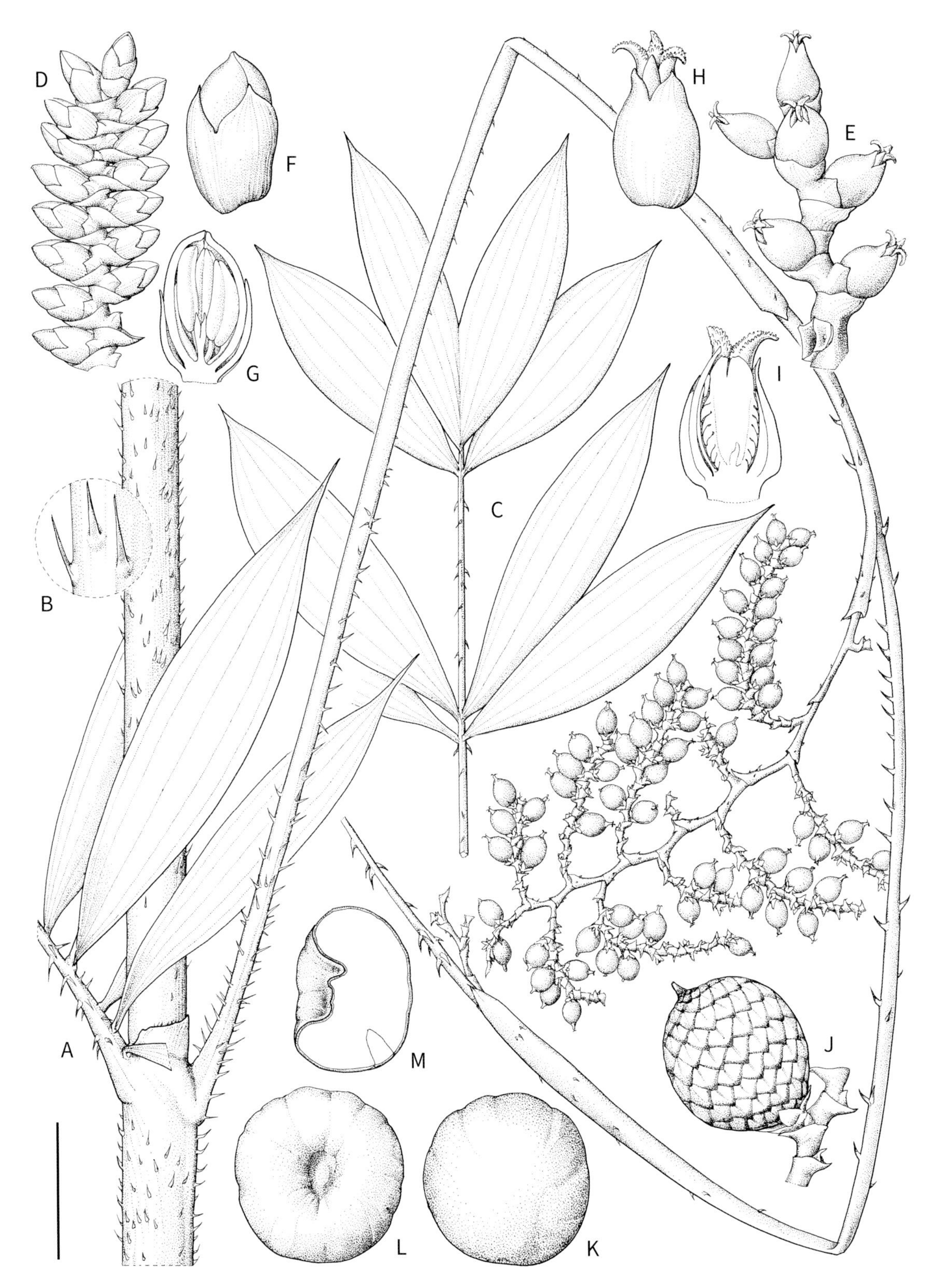

Calamus papuanus. **A.** Stem with leaf sheath, leaf base and female inflorescence with fruit. **B.** Detail spines. **C.** Leaf apex. **D.** Portion of male rachilla with flowers. **E.** Portion of female rachilla with flowers. **F**, **G.** Male flower whole and in longitudinal section. **H**, **I.** Female flower whole and in longitudinal section. **J.** Fruit. **K**, **L.** Seed in two views. **M.** Seed in longitudinal section. Scale bar: A = 3 cm; B = 7 mm; C = 4 cm; D, E, J–M = 7.5 mm; F, G = 3.3 mm; H, I = 4 mm. A–C, E, H–M from *Avé 4157*; D, F, G from *Avé 4077*. Drawn by Lucy T. Smith.

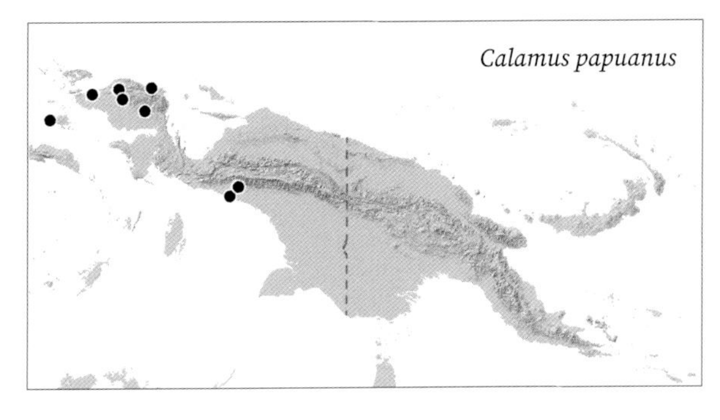

USES. Cane for tying houses.

CONSERVATION STATUS. Least Concern. However, monitoring is required due to oil palm plantations and logging concessions in the range of this species.

NOTES. *Calamus papuanus* is a rather variable, flagellate species. Its leaves bear 4–11 leaflets each side of the rachis, which are broadly elliptic with acuminate tips, scarcely armed and are arranged in groups of two to four, or rarely subregularly. The tip of the leaf always bears a group of four leaflets, the apical pair being fused by at least one third of their length. In some forms, the lowermost leaflets are borne at the very base of the leaf and reflexed across the stem. The leaf sheaths are armed with small, upward pointing spines (occasionally unarmed) and bear a short, tightly sheathing, truncate ocrea. The inflorescences are robust for such a relatively slender rattan, with conspicuous primary bracts and stout branches. Beccari (1908) likened this species to the common west Malesian *C. javensis* Blume to which it bears a superficial similarity in vegetative form, especially the leaves. In New Guinea, it is most similar to the poorly known *C. spiculifer*, but this is a much bigger palm, with an inconspicuous ocrea and deeply scalloped seeds.

42. *Calamus pholidostachys* J.Dransf. & W.J.Baker

Moderately robust, single-stemmed rattan climbing to 30 m. **Stem** with sheaths 20–40 mm diam., without sheaths to 9–15 mm diam.; internodes 12–28 cm. **Leaf** ecirrate, to 2 m long including petiole; sheath green, with sparse to very dense, brown indumentum; *spines rather sparse, 5–25 × 2–4 mm, sometimes much longer around sheath mouth, rigid, flattened, black, drying grey, rather uniform, scattered or arranged in horizontal groups*; knee ca. 80 mm long, 10 mm wide, green, *unarmed*; *ocrea 15 × 2 cm, tubular at first, splitting to form two auricles, disintegating to fibres later*, unarmed or with a few spines near sheath mouth; flagellum to 2.5–3 m long; petiole 7–12 cm long; leaflets ca. 60 on each side of rachis, *regularly arranged*, lanceolate, longest leaflet in mid-leaf 38–46 × 2.3–2.4 cm, apical leaflets 8–12 × 0.4–0.9 cm, apical leaflet pair united to one tenth of their length, leaflets armed with dark bristles on upper and lower surface, margins minutely bristly. **Inflorescence** pendulous, 4.5–8.2 m long including peduncle to 3.9 m long, flagelliform tip not seen, branching to 2 orders in the female, male not seen; primary bracts closely sheathing, armed with robust grapnel-like hooks; primary branches 5–7, to 50 cm long, male rachillae not seen, female rachillae 15–30 × 0.3–0.5 cm. **Fruit** *usually paired*, 14–16.5 × 9.5–12.5 mm, dull pale brown. **Seed** (sarcotesta removed) ca. 11 × 9 × 7 mm, ellipsoid, with a deep lateral pit, the surface covered in shallow scalloped depressions; endosperm homogeneous.

DISTRIBUTION. South-eastern New Guinea.

HABITAT. In lowland forest, usually on riverbanks at 50–750 m.

LOCAL NAMES. None recorded.

USES. None recorded.

CONSERVATION STATUS. Vulnerable. *Calamus pholidostachys* has a restricted distribution. Deforestation due to logging concessions is a major threat in its range. Its close proximity to Port Moresby may also present additional risks.

NOTES. This robust, flagellate rattan belongs to the *C. altiscandens* group (see notes under *C. altiscandens*). It is recognised by its regularly arranged leaflets, somewhat sparsely armed sheaths with unarmed knee, well-developed, fibrous ocrea and distinctive female rachillae with their almost overlapping expanded rachilla bracts that lend an almost scaly appearance to the rachilla.

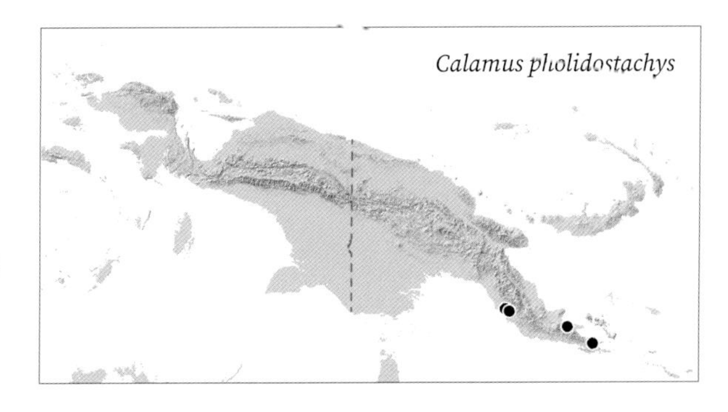

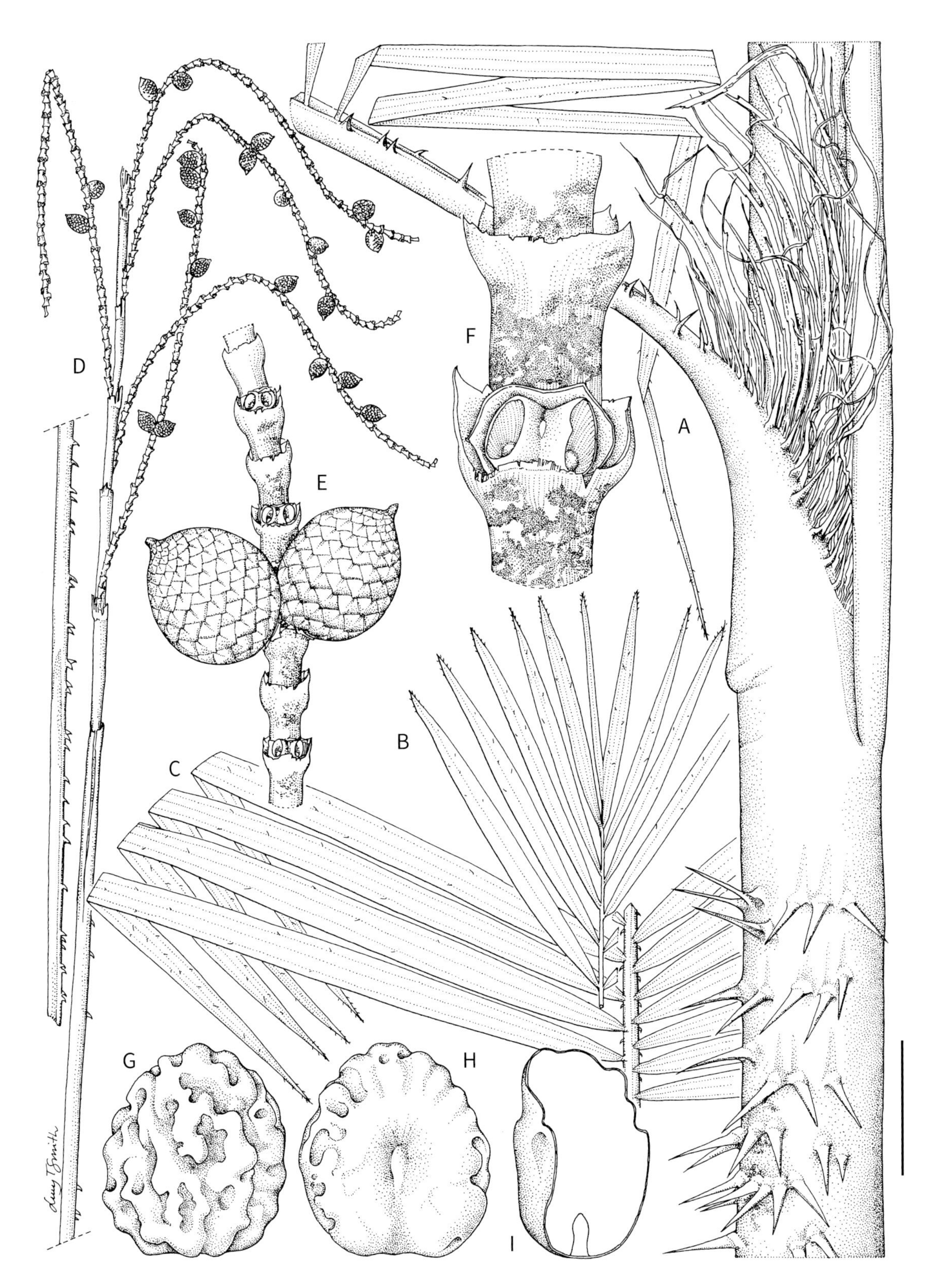

Calamus pholidostachys. **A.** Stem with leaf sheath and petiole. **B.** Leaf apex. **C.** Leaf mid-portion. **D.** Female inflorescence with fruit. **E.** Portion of female rachilla with fruit. **F.** Female rachilla detail showing two female flower scars and a sterile male flower scar. **G**, **H.** Seed in two views. **I.** Seed in longitudinal section. Scale bar: A = 3 cm; B–D = 8 cm; E = 1.5 cm; F = 5 mm; G–I = 7 mm. All from *Brass 24209*. Drawn by Lucy T. Smith.

43. *Calamus pilosissimus* Becc.

Slender to very slender, multi-stemmed rattan climbing to 8 m. **Stem** with sheaths 4–8 mm diam., without 3–6.5 mm diam.; internodes 11–20 cm. **Leaf** ecirrate, *to 20–55 cm long including petiole (where present);* sheath green with scattered grey-white indumentum, spines numerous, various lengths to 25 mm, fine to hair-like, brown, solitary or grouped in fine collars, sometimes with long, erect, sometimes matted hairs at sheath mouth to 70 mm; knee 6–21 mm long, armed as sheath; ocrea inconspicuous; flagellum to ca. 1 m; petiole absent (rarely to 10 cm); *leaflets 25–41 each side of rachis, arranged regularly, lanceolate,* longest leaflet in mid-leaf position 7.5–21 × 0.4–0.8 cm, apical leaflets 2.5–10 × 0.1–0.4 cm, apical leaflet pair not united, *leaflets armed with conspicuous interlocking bristles throughout margins ca. 4mm long,* bristles also present on upper surface, lower surface unarmed. **Inflorescence** flagelliform, to 135 cm long including 42–60 cm peduncle and ca. 20 cm flagelliform tip, branched to 2 orders in the female and 2 orders in the male; primary bracts narrow, tightly sheathing and armed (as flagellum); primary branches 2–5, to 17 cm long, male rachillae 18–32 × ca. 0.5 mm, female rachillae 24–41 × ca. 1 mm. **Fruit** solitary, mature fruit not seen. **Seed** not seen.

Calamus pilosissimus. Leaflets with marginal hairs, Lake Kutubu (WB).

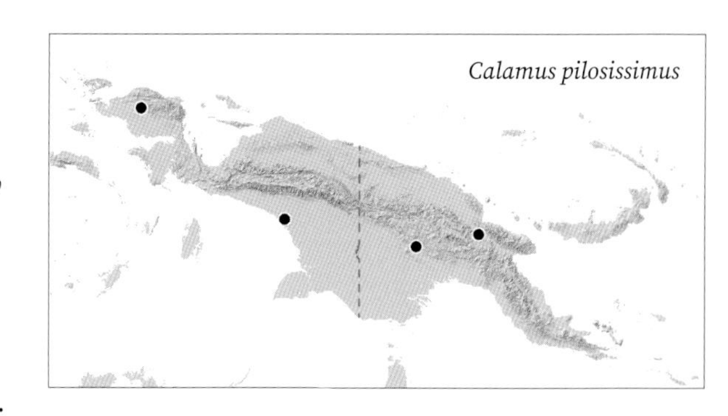

DISTRIBUTION. Scattered records across New Guinea.

HABITAT. Submontane and montane forest, 450–1350 m.

LOCAL NAMES. *To fro mawian* (Ayawasi).

USES. None recorded.

CONSERVATION STATUS. Least Concern.

NOTES. This slender, montane, flagellate rattan is distinguished by its small leaves, with numerous, fine leaflets, regularly arranged and with conspicuous interlocking bristles on the margins. The inflorescence is fine and flagelliform, and the leaves usually lack petioles. No other rattan like *C. pilosissimus* is known in New Guinea, but it does superficially resemble unrelated flagellate species from the Sunda region, such as *C. ciliaris* Blume from Java and Sumatra or *C. pilosellus* Becc. from Borneo in its slender form and leaf morphology with bristles.

44. *Calamus pintaudii* W.J.Baker & J.Dransf.

Robust, multi-stemmed rattan climbing to 20 m. **Stem** with sheaths 20–55 mm diam., without sheaths 10–20 mm diam.; internodes 35–45 cm. **Leaf** ecirrate, to ca. 2 m long including petiole; sheath orange-brown when dry, with dense white to brown indumentum, *spines numerous, to 50 mm long, needle-like, brown, varying in length;* knee 55 mm long, unarmed or armed as sheath; *ocrea 23–30 cm long, ligule-like, almost encircling the sheath, but open on the side opposite the petiole, papery, fragile, rusty brown, unarmed, tattering and disintegrating;* flagellum 3–5 m long, robust; petiole 8–28 cm; leaflets 32–35 each side of rachis, arranged regularly, linear lanceolate, longest leaflet at mid-leaf position 36–40 × 1.8–2.2 cm, apical leaflets 6–14 × 0.2–0.6 cm, apical leaflet pair free to united by one quarter of their length, *leaflets armed on upper surface and margins with dark bristles to 7 mm long.* **Inflorescence** pendulous ca. 4.5 m long including

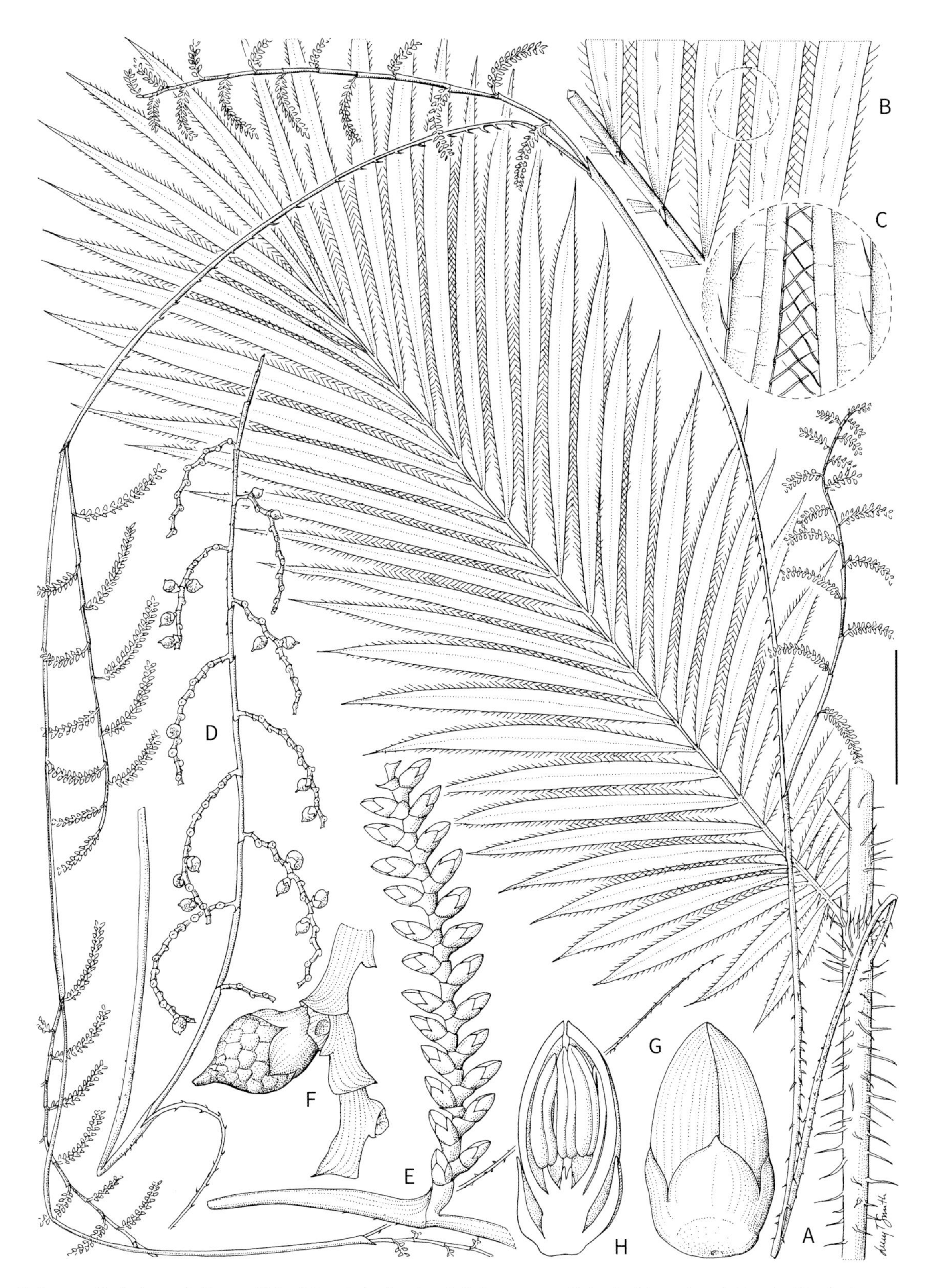

Calamus pilosissimus. **A.** Stem with leaf sheath, leaf and male inflorescence. **B.** Leaf mid-portion. **C.** Marginal leaflet hairs detail. **D.** Portion of female inflorescence. **E.** Male rachilla with flowers. **F.** Portion of female rachilla with immature fruit. **G, H.** Male flower whole and in longitudinal section. Scale bar: A = 4 cm; B = 2 cm; C, E = 7 mm; D = 3 cm; F = 4 mm; G, H = 1.3 mm. A–C, E, G, H from *Baker & Kage 653*; D, F from *Baker & Kage 657*. Drawn by Lucy T. Smith.

***Calamus pintaudii*. A.** Stem with leaf sheath and petiole. **B.** Leaf apex. **C.** Leaf mid-portion. **D.** Portion of female inflorescence. **E.** Male rachilla. **F.** Female rachilla with fruit. **G.** Fruit. **H, I.** Seed in two views. **J.** Seed in longitudinal section. Scale bar: A, F = 3 cm; B–D = 4 cm; E, G–J = 1.5 cm. A, E from *Zieck NGF 36189*; B–D, F–J from *Pintaud et al. 671*. Drawn by Lucy T. Smith.

peduncle and ca. 2 m flagelliform tip, branched to 2 orders in the female and 3 orders in the male; primary bracts tubular, asymmetric at tip and sometimes deeply split to ca. 15 cm, sparsely to moderately armed with stout spines and grapnels; primary branches 4–8, to 35 cm long, erect, male rachillae 30–60 × ca. 8 mm, female rachillae 90–160 × 7–8 mm. **Fruit** solitary, spherical, 15.5–17.5 × 12.5–13 mm, scales brown with dark-margins. **Seed** (sarcotesta removed) 8.5–10 × 8–8.5 × 8–8.5 mm, spheroidal, sculptured with deep pits and grooves; endosperm homogeneous.

DISTRIBUTION. Scattered records in south-eastern end of the Central Range between Mount Wilhelm and Mount Suckling.

HABITAT. Primary montane forest at 600–1,400 m.

LOCAL NAMES. *Kapurna* (Goilala-Tapini).

USES. Locally used in the construction of suspension bridges, binding fences and village houses.

CONSERVATION STATUS. Vulnerable. Deforestation due to logging concessions is a major threat within the restricted range of *C. pintaudii*.

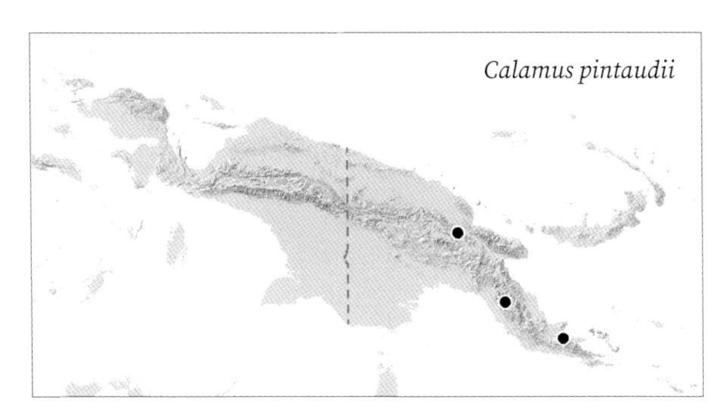

NOTES. *Calamus pintaudii* is a member of the *C. baiyerensis* group (see under *C. baiyerensis*). It is distinguished from similar flagellate species by its crowded, numerous, needle-like spines and the well-developed, papery, fragile, dark brown ocrea that is cleft opposite the petiole. Henderson (2020) placed *C. pintaudii* in synonymy with *C. nanduensis*. While further work is required to understand the taxonomy of this and similar species, the data are insufficient at this time to support Henderson's taxonomic conclusion.

45. *Calamus polycladus* Burret

Moderately robust, multi-stemmed rattan climbing to ca. 10 m. **Stem** with sheaths 15–19 mm diam., without 6–8 mm diam.; internodes ca. 26 cm. **Leaf** ecirrate, 0.9–1.5 m long including petiole; *sheath with abundant grey-brown indumentum, spines numerous, 3–6 × 2–2.5 mm, triangular, stiff, brown, solitary*; knee 25–33 mm long, largely unarmed; ocrea low, encircling, to 4 mm high leathery, unarmed, persistent; flagellum 2.5–3 m; petiole 18 cm; *leaflets ca. 24 each side of rachis, regularly arranged, lanceolate*, longest leaflet near base 31–36 × 1.5–2 cm, mid-leaf leaflets 28–30 × 1.5–2.3 cm, apical leaflets 9–11 × 0.5–0.8 cm, apical leaflet pair united from one tenth to two thirds of their length, leaflets armed with dark bristles on margins and upper surface, lower surface unarmed. **Inflorescence** arching, 1.2–2.2 m long including 17–21 cm peduncle and ca. 80 cm flagelliform tip, *branched to 4 orders in the female,* male not seen; primary bracts tubular, closely sheathing, with an asymmetrc tip; primary branches 5–6, to 30 cm long, male rachillae not seen, female rachillae 6–41 × ca. 1 mm. **Fruit** solitary, spherical, ca. 10 × 10 mm, scales orange-brown. **Seed** (sarcotesta removed) 8 × 8 × 7 mm, globose, with irregular channels and a lateral pit; endosperm homogeneous.

DISTRIBUTION. Huon Peninsula.

HABITAT. Montane forest, including secondary forest, at 1,300–1,500 m.

LOCAL NAMES. None recorded.

USES. None recorded.

CONSERVATION STATUS. Endangered. *Calamus polycladus* is known from only three sites. Deforestation due to logging concessions is a major threat in its range.

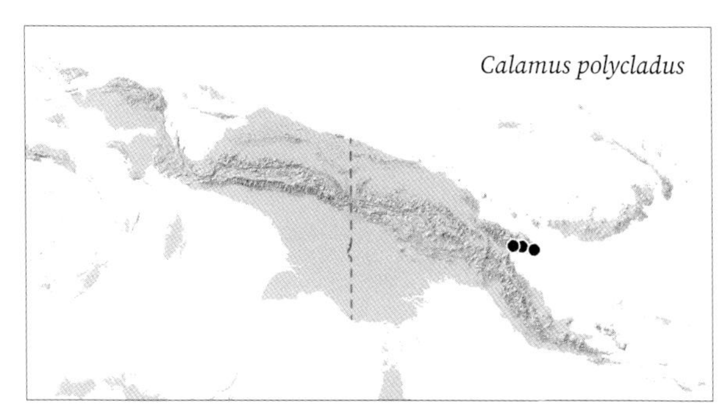

Calamus polycladus. **A.** Stem with leaf sheath, petiole and flagellum. **B.** Leaf apex. **C.** Leaf mid-portion. **D.** Portion of female inflorescence. **E.** Female rachilla with fruit. **F.** Fruit. **G, H.** Seed in two views. **I.** Seed in longitudinal section. Scale bar: A = 3 cm; B, C = 6 cm; D = 4 cm; E, G–I = 7 mm; F = 1 cm. All from *Baker et al. 680*. Drawn by Lucy T. Smith.

NOTES. *Calamus polycladus* is a moderately robust, flagellate rattan with grey-brown leaf sheaths, irregularly arranged spreading spines up to 1 cm long and numerous regularly arranged leaflets. Its most striking feature, however, is the female inflorescence branched to four orders. Typically, *Calamus* inflorescences are branched to two orders in the female and three orders in the male. In New Guinea *Calamus*, female inflorescences branched up to three orders are reported in *C. heteracanthus*, *C. pseudozebrinus* and *C. schlechterianus*. Several other species are known to have female inflorescences branched to either two or three orders (*C. barfodii*, *C. bulubabi*, *C. cuthbertsonii*, *C. lucysmithiae*). Male inflorescences branched to four orders are known in *C. jacobsii* and *C. schlechterianus*, the latter in rare cases branching to an exceptional five orders, and either three or four orders in *C. longipinna*. The picture is far from complete, however, as female inflorescences of *C. jacobsii* and male inflorescences of *C. polycladus*, *C. pseudozebrinus* and *C. lucysmithiae* have not been seen. Moreover, accurate observation of orders of branching is difficult as the maximum number of orders tends only to be represented in the lowermost primary branching systems.

The type of *C. polycladus* was destroyed in Berlin. Henderson (2020) neotypified the name on a specimen (*Baker et al. 680*) collected relatively near to the type locality that matched the original description.

46. *Calamus pseudozebrinus* Burret

Robust rattan, high-climbing. **Stem** with sheaths ca. 30 mm diam., without sheaths not seen; internodes at least 20 cm. **Leaf** ecirrate, more than 113 cm long including petiole; sheath green, with caducous brown indumentum, *spines numerous, 4–20 × 0.5 mm, needle-like, readily detached, brown, arranged in horizontal groups with bases united in low ridges, the ridges persisting after spines falling*; knee ca. 50 mm long, armed as sheath; *well-developed ocrea apparently present*, only a low, bony, eroded crest-like remnant present in available material; flagellum present; petiole ca. 15 cm; leaflets ca. 25 each side of rachis, arranged regularly, lanceolate, longest leaflet at base ca. 52 × 2.5, mid-leaf leaflets ca. 49 × 2.5 cm, apical leaflets 11–13 × 0.5–1 cm, apical leaflet pair united to ca. one third of their length, leaflet margins sparsely armed with bristles 0.5–1 mm, leaflet surfaces unarmed. **Inflorescence** arching, to at least 2.4 m long including at least ca. 17 cm peduncle, flagelliform tip apparently present, *branched to 3 orders in the female*, male not seen; primary bracts robust, tubular, with asymmetric apical limb, lightly armed; primary branches several, to 23 cm long, male rachillae not seen, female rachillae 50–85 × 3–5 mm. **Fruit** *solitary*, ellipsoid, immature. **Seed** not seen.

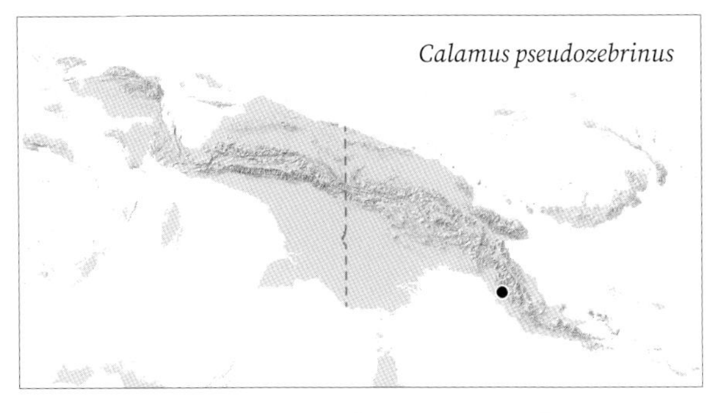

DISTRIBUTION. Owen Stanley Range, south-eastern New Guinea.

HABITAT. Forested hills at 700 m.

LOCAL NAMES. None recorded.

USES. None recorded.

CONSERVATION STATUS. Critically Endangered. The only known site of *C. pseudozebrinus* is threatened by logging concessions.

NOTES. *Calamus pseudozebrinus* is a member of the *C. baiyerensis* group (see under *C. baiyerensis*). This scarcely known flagellate species is aptly named, due to its striking similarity in vegetative morphology to the widespread *C. zebrinus*. The two species are similar as moderately robust, flagellate high climbers, bearing regularly pinnate leaves and leaf sheaths with collars of fine, brittle, caducous spines. The striking papery ocrea of *C. zebrinus*, which can reach 1 m in length, does not appear to be present in *C. pseudozebrinus* although the remnants of an ocrea are visible on the leaf sheath mouth in the type duplicates we have seen. Burret (1935) also refers to an ocrea in the protologue. In reproductive morphology, *C. pseudozebrinus* departs from *C. zebrinus* substantially. It bears solitary female flowers (and therefore fruit) rather than paired female flowers, the rachillae are shorter and stouter with much smaller, unarmed rachilla bracts, and the inflorescence, though robust, appears less elongate and is only very sparsely armed on the primary bracts. The female inflorescences are branched up to three

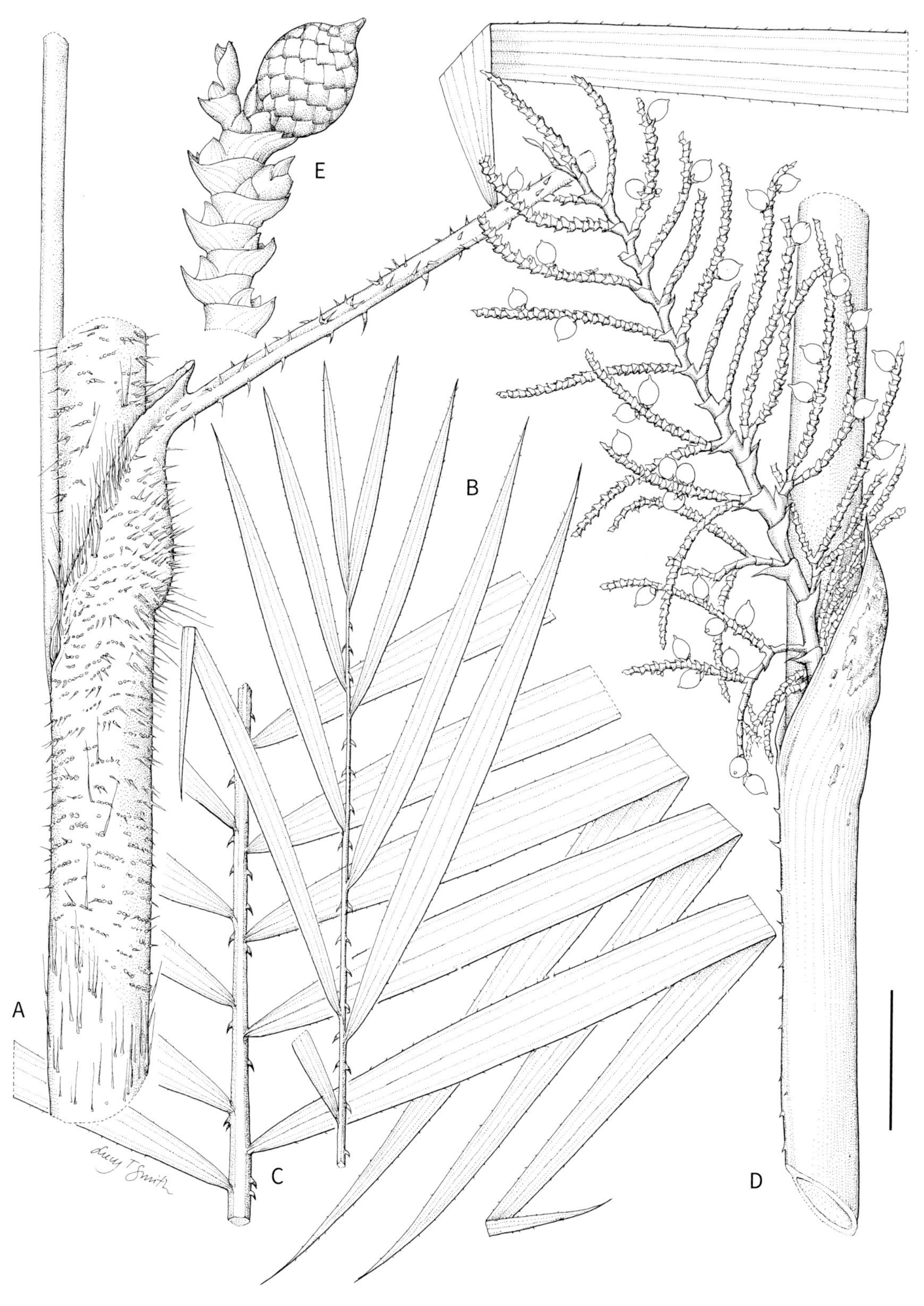

Calamus pseudozebrinus. **A.** Stem with leaf sheath and petiole. **B.** Leaf apex. **C.** Leaf mid-portion. **D.** Female inflorescence with fruit. **E.** Portion of female rachilla with fruit. Scale bar: A, D = 4 cm; B, C = 6 cm; E = 7 mm. All from *Brass 3923*. Drawn by Lucy T. Smith.

orders, instead of the usual two (see discussion under *C. polycladus*). Brass's collection notes also indicate that the leaflets are glossy, which is evident in the type material and is not the case in *C. zebrinus*.

47. *Calamus reticulatus* Burret

Moderately robust, multi-stemmed rattan climbing to c.10 m. **Stem** with sheaths 12.5–23 mm diam., without sheaths 8.5–18 mm diam.; internodes 18–30 cm. **Leaf** ecirrate to 1.4 m long including petiole; sheath pale green, with variable purple-brown indumentum, spines numerous, *spines 0.5–8.5 × 0.5–5 mm, flattened, with parallel-sides and irregular, jagged apices, of various sizes,* pale green with darker apices, not arranged in groups; knee 21–30 mm long, armed; *ocrea 25–31 × 3.5–4 cm, strictly tubular, encircling and obscuring sheath, stretching around the petiole and flagellum above to form a funnel, composed of a fine, network of fibres, brown, with numerous, needle-like spines,* 2–15.5 mm long, unarmed distally, persistent in upper parts of stem, disintegrating; flagellum 3–4 m; petiole 5–18 cm; leaflets 56–64 each side of rachis, arranged regularly, linear, longest leaflet near middle of leaf, 17–31 × 1–1.4 cm, apical leaflets 7–12.5 × 0.5–0.6 cm, apical leaflets not united or united to one quarter of their length, margins and both surfaces armed with short bristles to 7.5 mm. **Inflorescence** pendulous, 1.7–4 m long including 50–60 cm peduncle and up to ca. 1.7 m flagelliform tip, branched to 2 orders in the female, male not seen; primary bracts tubular, but tattering to fibres at mouth, armed with numerous reflexed spines; primary branches ca. 7, to c. 60 cm long, male rachillae not seen, female rachillae 4.8–16 × 0.3–0.5 cm. **Fruit** solitary, spherical to oblong, 13–15 × 11–12 mm, scales yellow. **Seed** (sarcotesta removed) 10–11 × 8–8.5 × 5.5–6 mm, *flattened subellipsoid, with numerous angular ornamentations on one side,* an open, shallow pit on the opposite side; endosperm homogeneous.

Calamus reticulatus. LEFT TO RIGHT: habit, Kikori River (WB); net-like ocrea, Mubi River (WB).

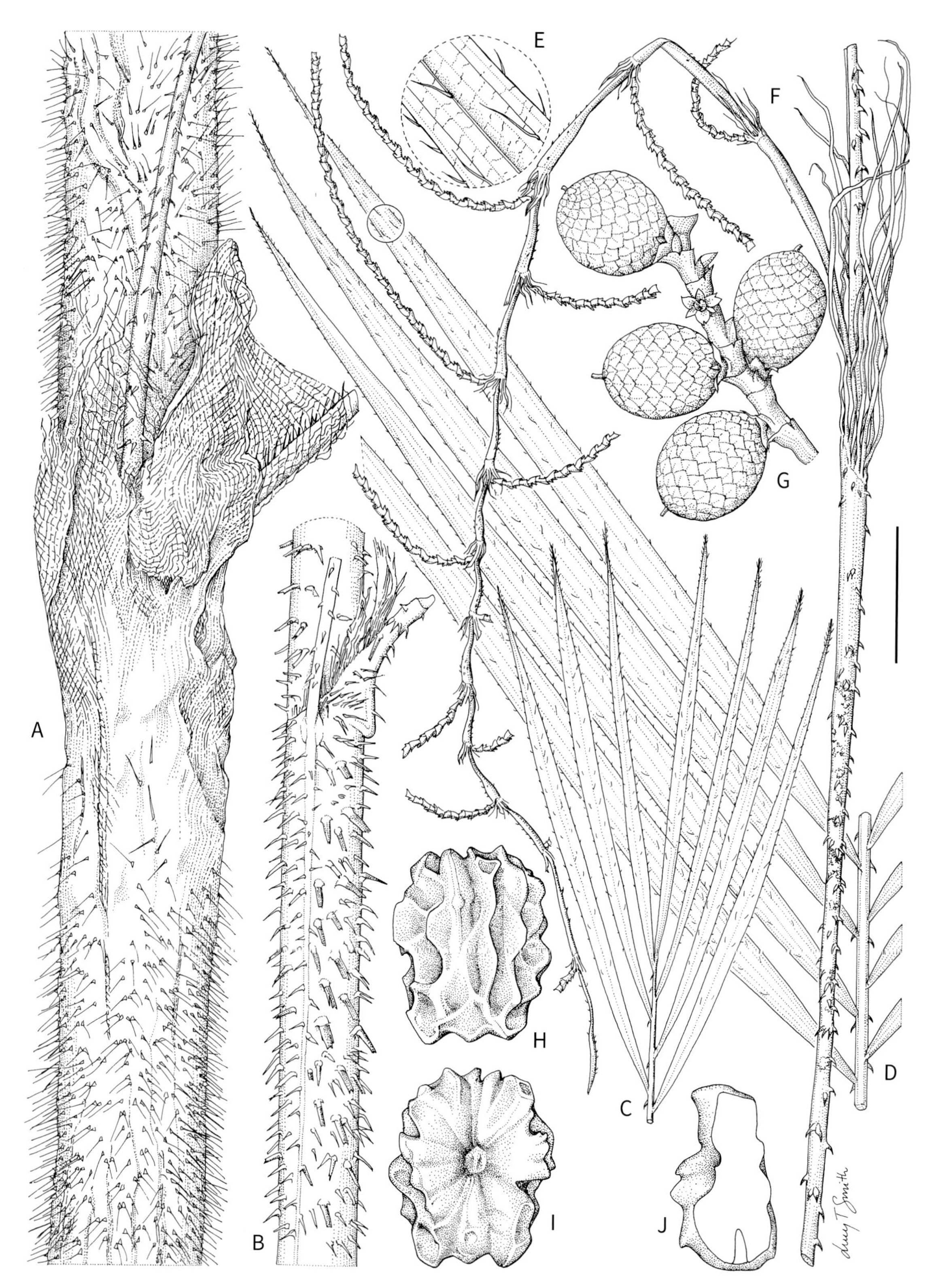

Calamus reticulatus. **A.** Stem with leaf sheath and ocrea. **B.** Stem with leaf sheath and petiole (ocrea removed). **C.** Leaf apex. **D.** Leaf mid-portion. **E.** Detail indumentum. **F.** Female inflorescence. **G.** Portion of female rachilla with fruit. **H, I.** Seed in two views. **J.** Seed in longitudinal section. Scale bar: A, B = 3 cm; C, D, F = 4 cm; E = 1 cm; G = 1.5 cm; H–J = 7 mm. A, B from *Baker et al. 644*; C–F from *Baker & Auge 650*; G–J from *Brass 6811*. Drawn by Lucy T. Smith.

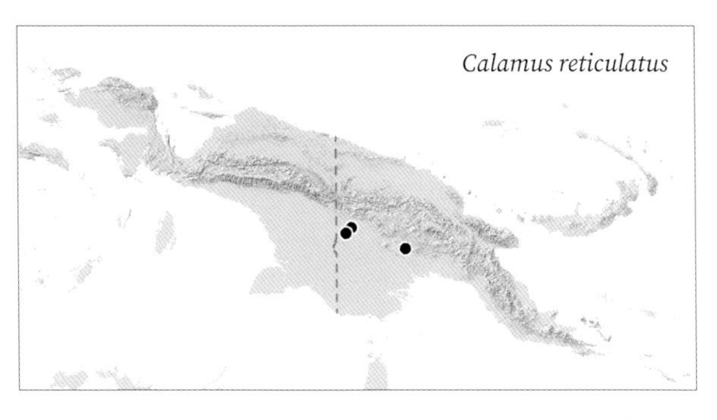

DISTRIBUTION. Eastern part of southern New Guinea.

HABITAT. Restricted to river margins in sunlit, open vegetation, up to 450 m.

LOCAL NAMES. None recorded.

USES. None recorded.

CONSERVATION STATUS. Least Concern. Despite a restricted distribution, *C. reticulatus* is known to occur in great abundance and to persist in riverine vegetation with limited threats.

NOTES. This flagellate species, a member of the *C. longipinna* group (see under *C. longipinna*), forms dense thickets on riversides and appears to be an obligate rheophyte. It is unique in *Calamus* in bearing a well-developed ocrea that is composed of a fibrous network, armed with needle-like spines, that extends to stretch around the petiole and flagellum of the leaf sheath above to form a fibrous funnel (remarkably similar to the unrelated Bornean rattan *Korthalsia jala* J.Dransf.). Other features indicate a close relationship to *C. vestitus*, on account of their similar leaf sheath armature, inflorescence morphology and seed shape (see under *C. vestitus*). The ranges of these two species do not appear to overlap.

48. *Calamus retroflexus* J.Dransf. & W.J.Baker

Moderate, multi-stemmed rattan climbing to 30 m. **Stem** with sheaths 18–21 mm diam., without sheaths to 6–15 mm diam.; internodes 23–40 cm. **Leaf** ecirrate, 0.6–1.5 m long; sheath mid-green, with thin pale brown indumentum, *spines usually abundant, 4–40 × 1–2 mm, very narrow triangular, flattened, pale green, drying yellowish, solitary, horizontal*; knee to 27 mm long, armed as sheath; ocrea scarcely developed; flagellum 1.5–2 m; petiole absent; *leaflets 23–30 on each side of rachis, arranged in 4–8 distant groups, crowded within groups, basalmost leaflets reflexed across the stem, forming a space sometimes occupied by ants, linear or very narrow lanceolate*, longest leaflet in mid-leaf 21–31 × 1.5–3 cm, basalmost leaflets to 5.5 × 0.6 cm, apical leaflets 9–10 × 1–1.2 cm, apical leaflets not united, leaflets armed with short, dark spinules on both surfaces and margins. **Inflorescence** *pendulous, to 4 m long* including peduncle to 2.5 m (flagelliform tip not seen), branched to 2 orders in the female and 3 orders in the male; primary bracts tightly sheathing, splitting neatly apically with a triangular lobe, sometimes lacerate, unarmed or with scattered reflexed spines; primary branches up to 12, to ca. 90 cm long, male rachillae 4–24 × ca. 2 mm, female rachillae 55–100 × ca. 2.5 mm. **Fruit** solitary, globose, ca. 10 × 10 mm, scales pale brown. **Seed** not seen.

Calamus retroflexus. Habit with male inflorescence, Wosimi River (SB).

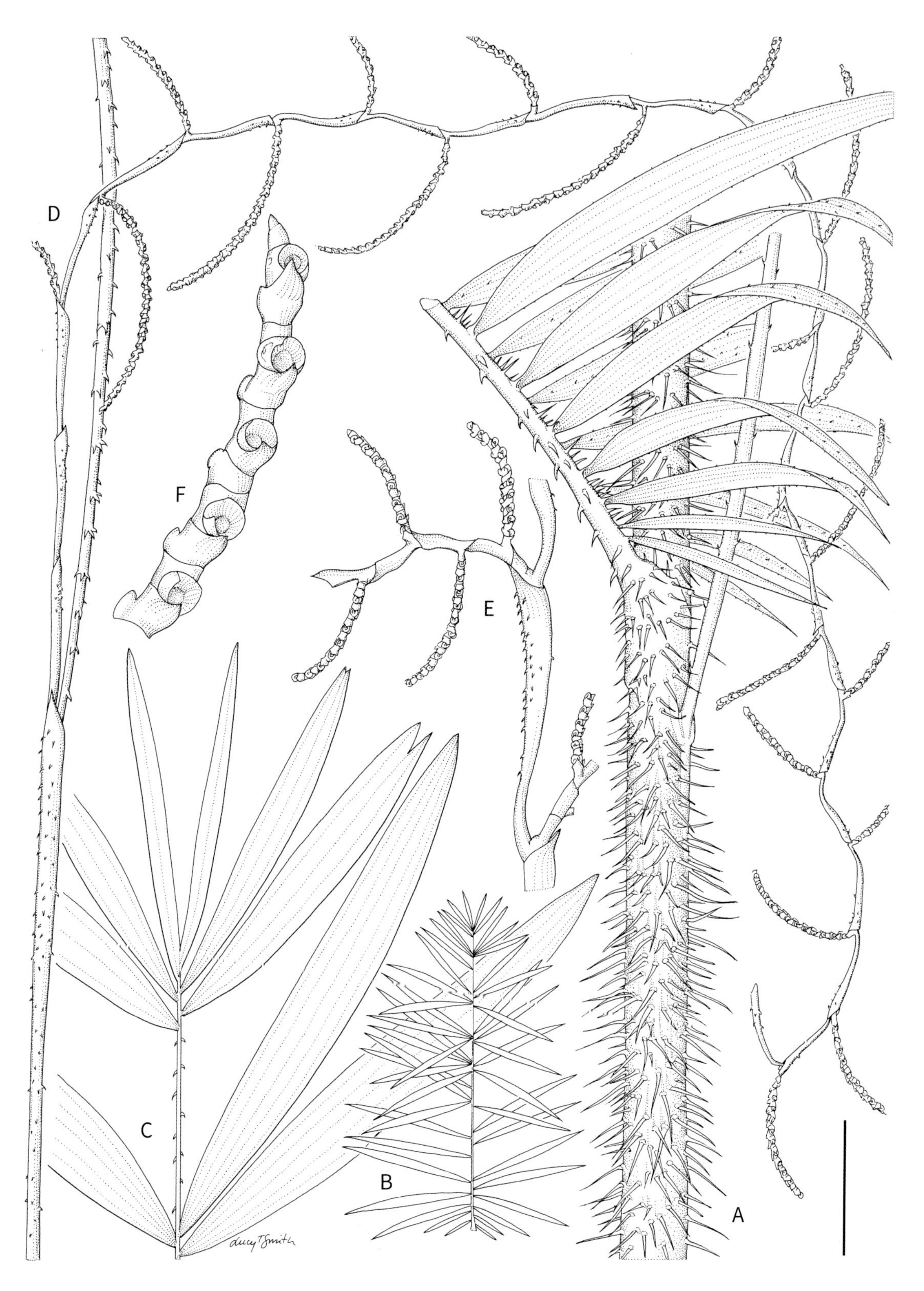

Calamus retroflexus. **A.** Stem with leaf sheath and base. **B.** Whole leaf diagram. **C.** Leaf apex. **D.** Portion of male inflorescence. **E.** Portion of female inflorescence. **F.** Portion of female rachilla. Scale bar: A, C = 4 cm; B = 30 cm; D = 1.5 cm; E = 6 cm; F = 7 mm. A, E, F from *Dransfield et al. JD 7723*; B–D from *Maturbongs et al. 651*. Drawn by Lucy T. Smith.

Calamus retroflexus. Leaf sheaths with reflexed basal leaflets, Wosimi River (WB).

DISTRIBUTION. Scattered records from across mainland New Guinea.

HABITAT. Lowland forest from sea level to 640 m.

LOCAL NAMES. *Are* (Sayal), *Sinsin* (Merigem).

USES. Used for cordage and thatching.

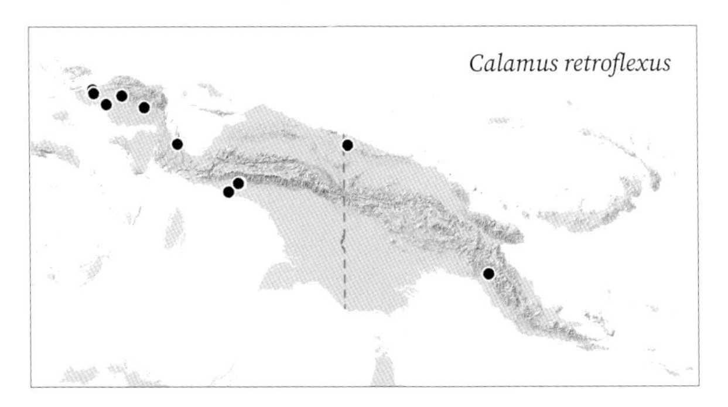

CONSERVATION STATUS. Least Concern.

NOTES. *Calamus retroflexus* is a flagellate species that is easily distinguished by its leaves with numerous, strongly grouped, narrow leaflets, the basalmost of which are swept back across the stem (hence the specific epithet), forming a chamber in which ants make nests. It is unlikely to be confused with other species in New Guinea.

49. *Calamus sashae* J.Dransf. & W.J.Baker

Robust, single-stemmed or multi-stemmed rattan climbing to 20 m. **Stem** with sheaths 27–35 mm diam., without sheaths to 15–19 mm diam.; internodes to ca. 37 cm. **Leaf** ecirrate, to 2.5 m long including petiole; sheath mid-green, with dense mid-brown indumentum, spines abundant, 20–40 × 3–5 mm, robust, narrow triangular, flattened, with swollen bases, usually rather uniform, pale brown with reddish bases, solitary or grouped in partial whorls, horizontal or reflexed; *knee to 60 mm long, unarmed; ocrea poorly developed*, membranous; flagellum to 4 m long; petiole 10–40 cm long; *leaflets to 62 on each side of rachis, regularly arranged, narrow lanceolate*, longest leaflet in mid-leaf 38–40 × 2.3–2.7 cm, apical leaflets 7–9 × 0.7–0.9 cm, apical leaflets joined in basal 2 cm, leaflets armed with bristles on upper surface and margins (at tip). **Inflorescence** *pendulous, to ca. 6.1 m long,* including peduncle to 3.7 m (flagelliform tip not seen), branched to 2 orders in the female, male not seen; primary bracts strictly tubular, armed with abundant reflexed spines; primary branches ca. 3, to ca. 65 cm long, male rachillae not seen, female rachillae 20–25 × 0.4 cm. **Fruit** solitary, ovoid, ca. 15 × 12 mm, scales pale brown. **Seed** not seen.

DISTRIBUTION. Foothills of the Arfak Mountains and the Wandamen Peninsula.

HABITAT. Lowland forest at 350–400 m.

LOCAL NAMES. None recorded.

USES. None recorded.

CONSERVATION STATUS. Critically Endangered. Deforestation due to oil palm plantations and logging concessions is a major threat in its limited range.

NOTES. This robust, flagellate rattan is recognised by its leaves with regularly arranged rather narrow leaflets, the sheaths armed with large, scattered or grouped spines, the unarmed knee, and the very robust female inflorescence with only few very

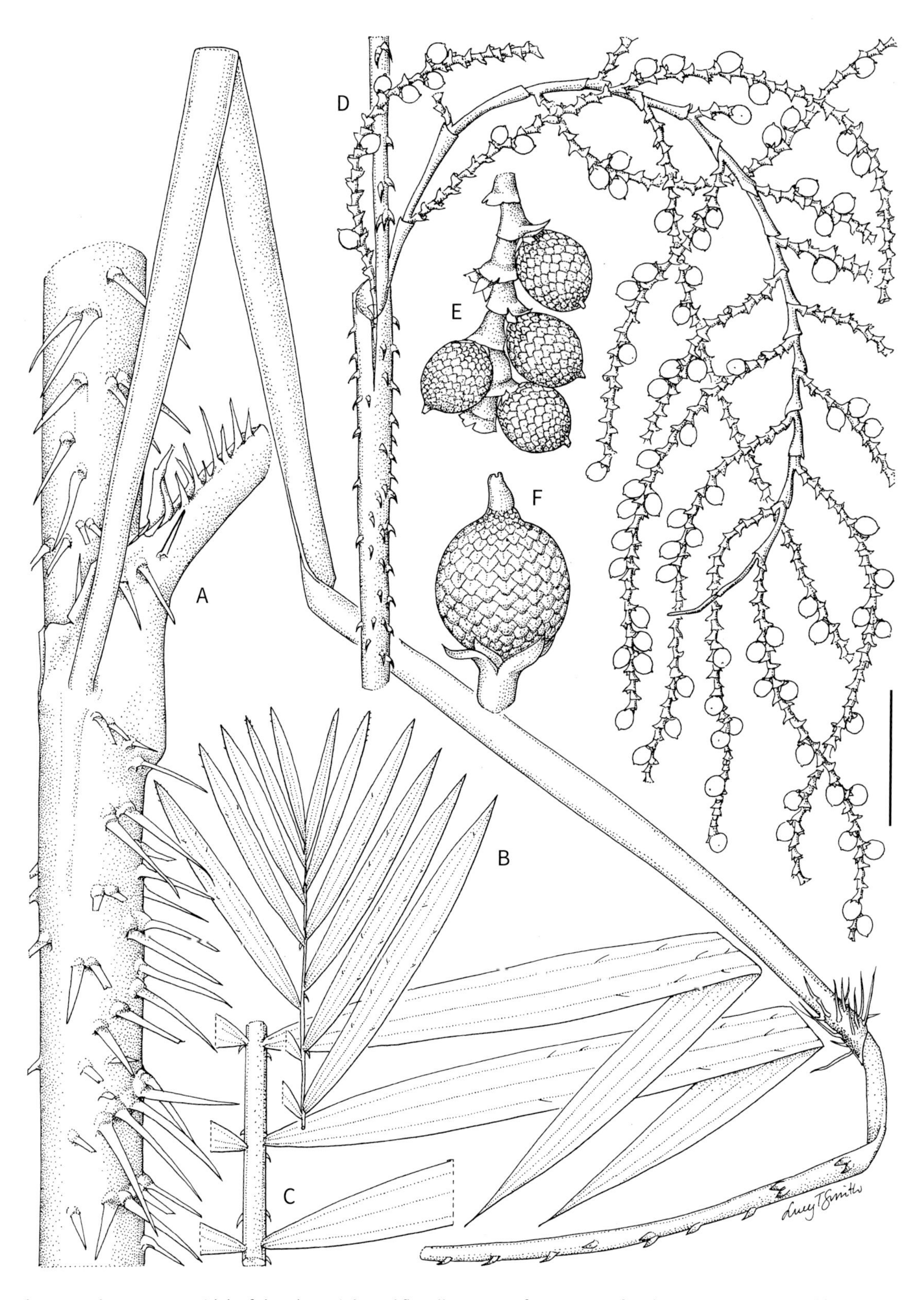

Calamus sashae. **A.** Stem with leaf sheath, petiole and flagellum. **B.** Leaf apex. **C.** Leaf mid-portion. **D.** Portion of female inflorescence with fruit. **E.** Portion of female rachilla with fruit. **F.** Fruit. Scale bar: A = 4 cm; B–D = 6 cm; E = 2 cm; F = 1 cm. All from *Dransfield et al. JD 7601*. Drawn by Lucy T. Smith.

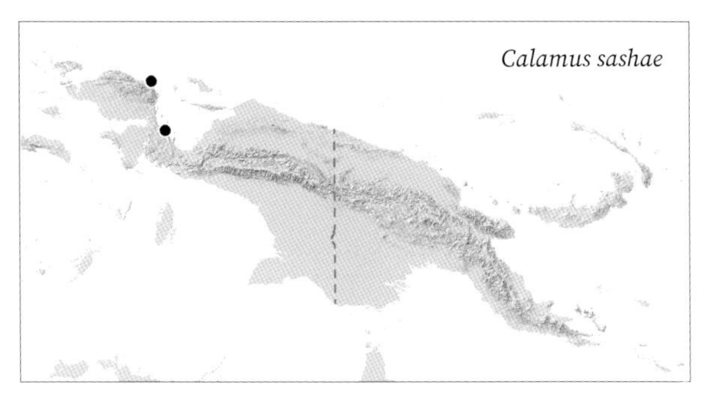

distant partial inflorescences, the rachillae with funnel-shaped, almost overlapping dark bracts. Its general inflorescence morphology and leaf sheath armature resemble those of *C. pholidostachys*, but unlike *C. sashae* that species bears paired fruit and a well-developed ocrea. Funnel-shaped rachilla bracts also occur in the *C. baiyerensis* group (see under *C. baiyerensis*, and also Baker & Dransfield 2014 for further discussion). However, unlike these species, *C. sashae* does not have a well-developed ocrea.

50. *Calamus scabrispathus* Becc.

Synonym: *Calamus papyraceus* W.J.Baker & J.Dransf.

Slender rattan climbing to 6.5 m. **Stem** with sheaths 10–12.5 mm diam., without sheaths to 6–7 mm diam.; internodes not seen. **Leaf** ecirrate, 0.7–2.5 m long including petiole; sheath pale brown when dry, with scattered brown indumentum, *densely armed with fine, brown, hair-like spines to 30 mm long, spines grouped in short whorls and fused into a ridge at the base*; knee ca. 30 mm long, ca. 10 mm wide, less densely armed than sheath; *ocrea ca. 27 × 1.2 cm, erect, split longitudinally opposite the petiole insertion, papery, brown, armature similar to sheath, but less dense, disintegrating to fibres*; flagellum ca. 1 m long; petiole 14–26 cm; leaflets 15–22 on each side of rachis, arranged regularly, linear-lanceolate, longest leaflet at mid-leaf position 20.5–46 × 1.5–2.2 cm, apical leaflets 10–26 × 0.6–1.5 cm, apical leaflet pair united to one third of their length, leaflets armed with conspicuous black bristles on margins and both surfaces. **Inflorescence** *erect, 49–65 cm long* including 13–30 cm peduncle and 8–10 cm sterile tip, branched to 2 orders in the female, male not seen; primary bracts tubular, papery and tattering deeply distally, armed with whorls of fine brown spines; primary branches 3–4, to 12 cm long, male rachillae not seen, female rachillae 17–35 × 2.5–3.5 mm. **Fruit** *solitary*, spherical, ca. 13 × 9.5 mm, scales yellow. **Seed** (sarcotesta removed) 7 × 6 × 6 mm, globose with shallow lateral pit; endosperm homogeneous.

DISTRIBUTION. Hunstein and Prince Alexander Ranges, north-eastern New Guinea.

HABITAT. Montane forest on steep, stony slope, 600–1,000 m.

LOCAL NAMES. *Ipis* (Jal, Madang), *Khaza* (Ambakanja, Maprik).

USES. None recorded.

CONSERVATION STATUS. Critically Endangered. Deforestation due to logging concessions is a major threat in its narrow range.

NOTES. This flagellate species resembles a slender form of *C. zebrinus* Becc. in its regularly pinnate leaves, leaf sheaths with collars of fine, brittle spines, and the long, erect papery ocrea that is armed with fine spines and disintegrates. It is also similar in some reproductive features (Baker & Dransfield 2017). However, it is distinct in its smaller stature, shorter ocrea, and erect, congested inflorescence that lacks a flagelliform tip and bears papery, tattering bracts throughout. *Calamus scabrispathus* also bears conventional floral clusters that comprise a sterile male flower and a functional female flower, resulting in a single fruit per floral cluster, unlike *C. zebrinus* with its anomalous floral cluster with an additional female flower resulting in paired fruits (Dransfield & Baker 2003).

The original type of *C. scabrispathus* was destroyed in Berlin. Based on his interpretation of the protologue, Henderson (2020) neotypified the name using a specimen cited in our description of *C. papyraceus* (Baker & Dransfield 2017), rendering this newer species a synonym of *C. scabrispathus*.

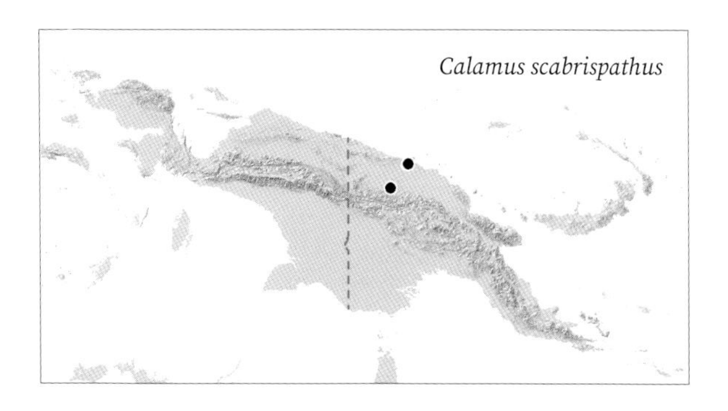

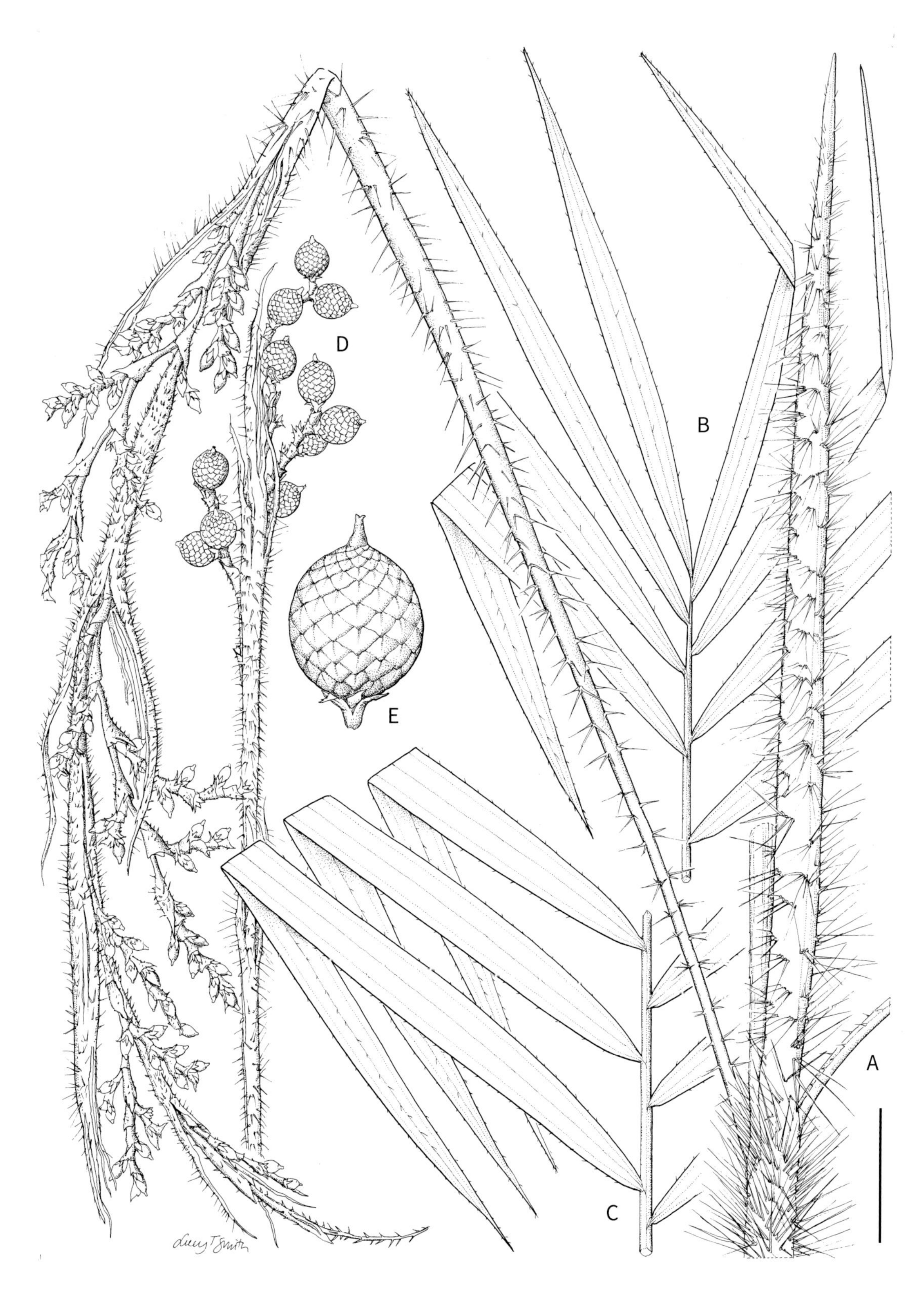

Calamus scabrispathus. **A.** Stem with leaf sheath, ocrea and female inflorescence. **B.** Leaf apex. **C.** Leaf mid-portion. **D.** Portion of female inflorescence with fruit. **E.** Fruit. Scale bar: A, D = 3 cm; B, C = 6 cm; E = 1 cm. All from *Pullen 1506*. Drawn by Lucy T. Smith.

51. *Calamus schlechterianus* Becc.

Synonyms: *Calamus multisetosus* Burret, *Calamus myriocarpus* Burret, *Calamus wariwariensis* Becc.

Moderately robust to robust, multi-stemmed rattan climbing to more than 20 m. **Stem** with sheaths 11–33 mm diam., without 8–15 mm diam.; internodes 10–24 cm. **Leaf** ecirrate, 0.7–1.4 m long including petiole; sheath dark green, sometimes with dense grey-brown indumentum, *spines very numerous, 5–60 mm, hair-like (rarely flattened to ca. 1 mm wide), brown or dark brown, arranged in partial collars united at base, sometimes solitary*; knee 37–70 mm long, colour and armature as sheath; *ocrea 4–9 cm long, splitting to form auricles or a ligule*, adjacent to or flanking petiole, fragile papery or fibrous, brown, armed as sheath, disintegrating to fibres; flagellum 110–350 cm long; petiole 10–27 cm; *leaflets 33–56 each side of rachis, arranged regularly*, lanceolate, longest leaflet at mid-leaf position 18–32 × 1.4–2 cm, apical leaflets 8–12 × 0.4–0.8 cm, apical leaflet pair not united, leaflets armed on margins and (usually) both surfaces with black bristles, conspicuously so on upper surface. **Inflorescence** pendulous to flagelliform, 0.9–4 m long including 27–50 m peduncle and flagelliform tip absent or to 2 m, *branched to 3 orders in the female and 4 orders (rarely 5 orders) in the male; primary bracts tubular and rather inflated*, unarmed or lightly armed with spines as sheath; primary branches 3–12, to 30 cm long, male rachillae 4–40 × ca. 0.5 mm, female rachillae 10–60 × 1–1.5 mm. **Fruit** solitary, spherical, 10–12 × 8–9 mm, scales yellow. **Seed** (sarcotesta removed) 7–8 × 7–8 × 6–7 mm, globose, smooth, with deep lateral pit; endosperm homogeneous.

DISTRIBUTION. Widespread in eastern New Guinea.

Calamus schlechterianus. LEFT TO RIGHT: habit; leaf sheaths with fibrous ocreas. Southern Highlands (WB).

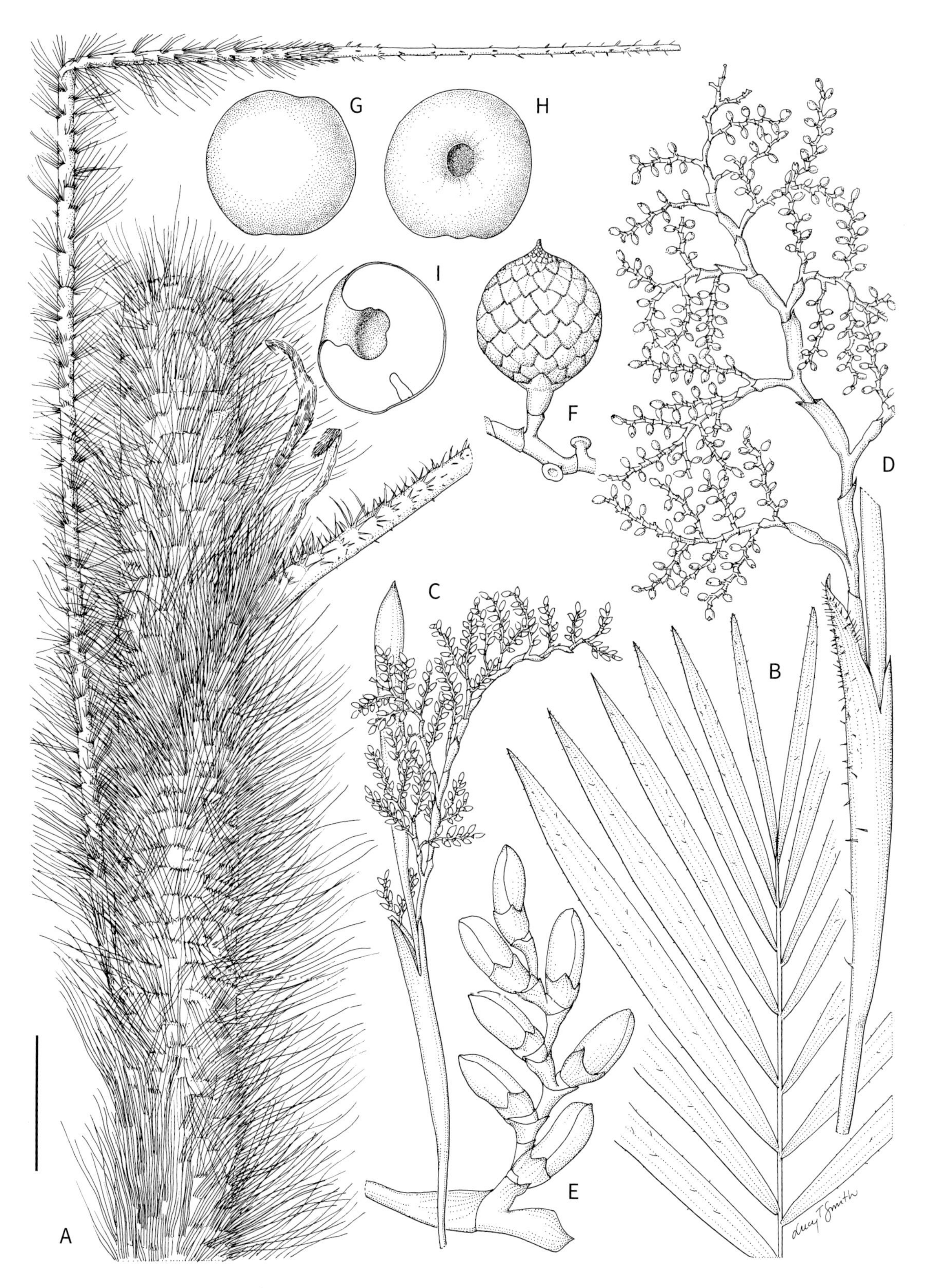

Calamus schlechterianus. **A.** Stem with leaf sheath and petiole. **B.** Leaf apex. **C.** Portion of male inflorescence. **D.** Portion of female inflorescence. **E.** Male rachilla with flowers. **F.** Fruit. **G**, **H.** Seed in two views. **I.** Seed in longitudinal section. Scale bar: A–D = 4 cm; E = 5 mm; F = 1 cm; G–I = 7 mm. A from *Zieck NGF 36175*; B, D from *Hartley 10288*; C, E from *Sayers NGF 19604*; F–I from *Zieck et al. NGF 36595*. Drawn by Lucy T. Smith.

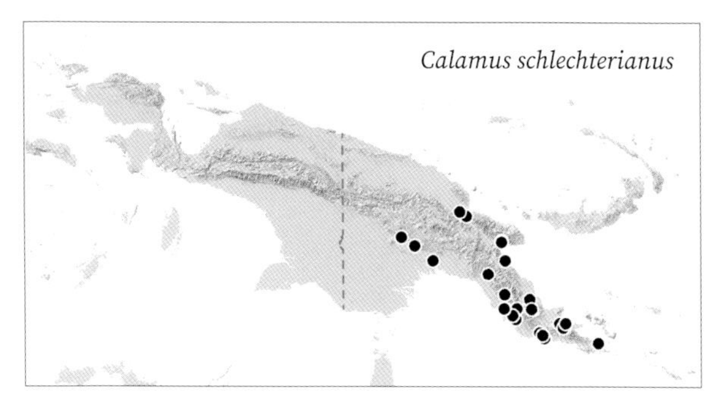

HABITAT. Lowland and montane forest from sea level to 1350 m.

LOCAL NAMES. *Kuwe* (Alotau), *Siri* (Goari).

USES. A medium quality rattan, used by the Koiari people for basketry.

CONSERVATION STATUS. Least Concern.

NOTES. *Calamus schlechterianus* is recognisable among the more robust, flagellate rattans by its dense, dark, hair-like spines, disintegrating ocrea, regularly arranged leaflets, and inflorescence with rather inflated primary bracts and branched to three orders in the female and four orders (sometimes five orders) in the male (see also under *C. polycladus*). According to Henderson (2020), *C. schlechterianus* has the most highly branched male inflorescence in the genus. The poorly known *C. capillosus* bears some resemblance vegetatively, but is more slender and bears a fine, flagelliform inflorescence.

52. *Calamus serrulatus* Becc.

Synonyms: *Calamus eximius* Burret, *Palmijuncus serrulatus* (Becc.) Kuntze

Very robust to moderately slender, single-stemmed or multi-stemmed rattan climbing to 40 m. **Stem** with sheaths 13–45 mm diam., without 9–25 mm diam.; internodes 21–41 cm. **Leaf** ecirrate, 1–2 m long including petiole, rachis with indumentum as sheath and numerous grapnels usually yellow-orange; sheath dull to bluish green, with thin, caducous, grey-white indumentum, sometimes with thin, brown indumentum beneath, *spines few to numerous, 3–40 × 1–5 mm, flattened, triangular, stiff, tips breaking off in older parts*, green to brown, solitary or grouped in short, slanting collars; knee 38–95 cm long, green to yellowish, unarmed or scarcely armed; ocrea inconspicuous, a low papery or leathery crest, persistent; flagellum 2–3 m, with indumentum as sheath; petiole 1–60 cm; *leaflets 73–99 each side of rachis, regularly arranged (sometimes in different planes), linear, abruptly constricting at base*, longest leaflet in mid-leaf position 16–31 × 1–1.8 cm, apical leaflets 4–12 × 0.3–1 cm, apical leaflet pair free or united up to one third of their length, *leaflet armed with numerous, closely spaced, minute black bristles throughout margin, surfaces armed with black bristles (in up to three rows)*, long bristles on upper surface, shorter bristles on lower surface. **Inflorescence** erect to arching, up to 220 cm long including ca. 40 cm peduncle and 17–45 cm flagelliform tip, branched to 2 orders in the female and 3 orders in the male; primary bracts robust, somewhat inflated, splitting deeply at apex and tattering somewhat, lightly armed to unarmed, indumentum as sheath; primary branches up to ca. 5, to 40 cm long, *male rachillae 55–130 × ca. 6 mm, female rachillae 60–130 × 4–7 mm.* **Fruit** solitary, spheroidal with a pronounced beak, ca. 11 × 8 mm (immature), scales rusty brown. **Seed** not seen.

Calamus serrulatus. Habit, near Tabubil (WB).

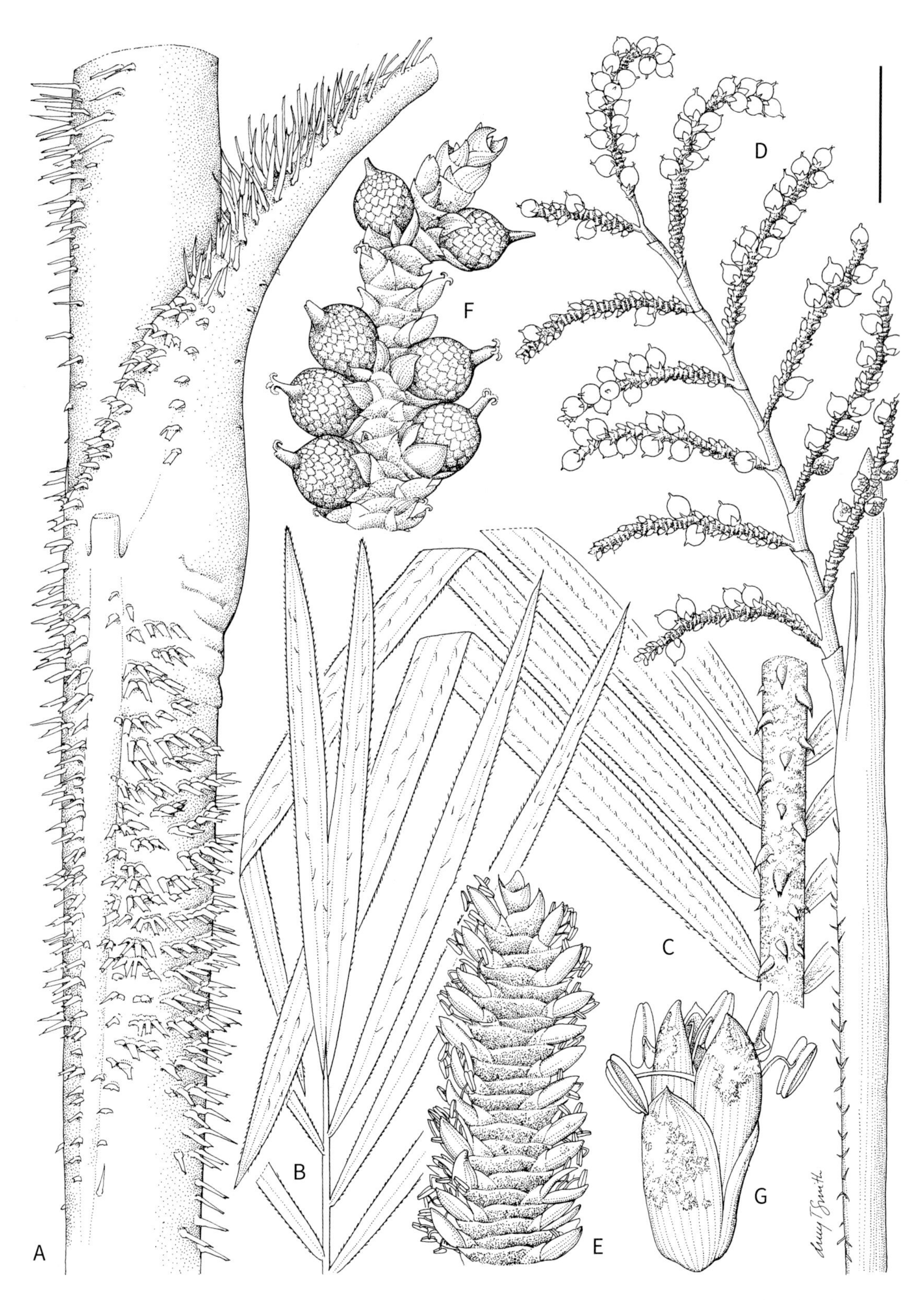

Calamus serrulatus. **A.** Stem with leaf sheath and petiole. **B.** Leaf apex. **C.** Leaf mid-portion. **D.** Portion of female inflorescence with fruit. **E.** Portion of male rachilla with flowers. **F.** Portion of female rachilla with fruit. **G.** Male flower. Scale bar: A–C = 3 cm; D = 6 cm; E = 1 cm; F = 1.5 cm; G = 3.3 mm. A–D, F from *Dransfield et al. JD 7653*; E, G from *Widjaja et al. 7057*. Drawn by Lucy T. Smith.

Calamus serrulatus. LEFT TO RIGHT: leaf sheath, juvenile stem, Tamrau Mountains (WB); female inflorescence, near Tabubil (WB); female flowers, near Tabubil (WB).

DISTRIBUTION. Widespread throughout western and central New Guinea (including Yapen Island).

HABITAT. Often recorded from montane forest, but also known in lowlands from 100–1,390 m.

LOCAL NAMES. *Arotan furama* (Arandai dialect, Dombano language), *Buga* (Yapen), *Hajabfkwa Nyi* (Cyclops), *Rotan to kawia* (Kawia), *Tek kana* (Mianmin), *To kawia* (Ayawasi).

USES. Cane used for construction and main cables of bridges.

CONSERVATION STATUS. Least Concern.

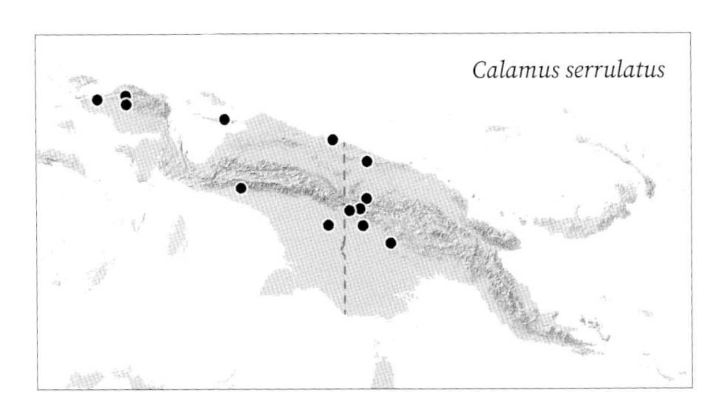

NOTES. *Calamus serrulatus* is a distinctive, robust, flagellate rattan that is widespread, but not frequently seen. It bears long, arching leaves with very numerous short, fine leaflets. The leaflets are armed with abundant minute spines along their margins. The leaf sheaths are densely spiny, although the spine tips often break off in mature sheaths. The inflorescences are robust and erect, with stout rachillae bearing conspicuous, overlapping rachilla bracts. Dense indumentum of various colours covers many of the surfaces of the leaf rachis, leaf sheath and inflorescence bracts.

53. *Calamus spanostachys* W.J.Baker & J.Dransf.

Very slender rattan to 1.5 m. **Stem** with sheaths 3–5 mm diam., without sheaths 2–3 mm diam.; internodes 7–11 cm. **Leaf** ecirrate, to 37 cm long including petiole; sheath mid-green, with scattered indumentum on young sheaths, sparsely armed in distal portion of sheath, spines to 12 mm long, 0.5 mm long at the base, narrow triangular; knee scarcely developed, unarmed or sparsely spiny; ocrea scarcely developed; *flagellum*

Calamus spanostachys. **A.** Stem with leaf sheath, leaf and female inflorescence. **B.** Male inflorescence. **C.** Portion of male rachilla with flowers. **D.** Portion of female rachilla with fruit. **E, F.** Male flower whole and in longitudinal section. **G.** Fruit. Scale bar: A = 4 cm; B = 3 cm; C = 2.5 mm; D = 4 mm; E, F = 1.1 mm; G = 7 mm. All from *van Leeuwen 10473*. Drawn by Lucy T. Smith.

absent (in material seen); petiole 7–13 mm long; *leaflets 2–3 on each side of rachis, subregular, lanceolate*, basal and mid-leaf leaflets 12–19 × 1–2.2 cm, apical leaflets 22–29 × 2.5–4 cm, apical leaflet pair united to one quarter of their length, leaflets sparsely armed with minute spines on margins and lower surface. **Inflorescence** *erect, 16–36 cm long* including 3–8 cm peduncle and 1–6 cm flagelliform tip, *branched to 1 order in the female and 2 orders in the male*; primary bracts closely sheathing, opening eccentrically at the apex, sparsely armed as rachis; primary branches 2–4, male rachillae 8–21 × ca. 1 mm, female rachillae 3.3–4 × ca. 2 mm. **Fruit** solitary globose, ca. 11 × 8 mm (immature material seen), scales brown when dried. **Seed** not seen.

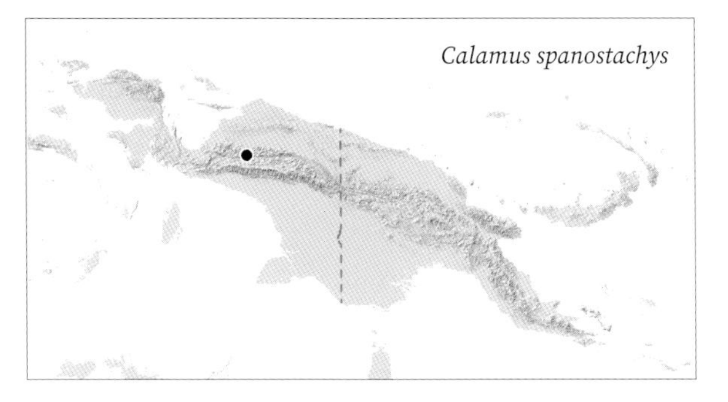

DISTRIBUTION. Known only from the upper Rouffaer River in the Sudirman Mountains.

HABITAT. Primary forest at 700 m.

LOCAL NAMES. None recorded.

USES. None recorded.

CONSERVATION STATUS. Data Deficient. Due to insufficient information about ongoing threats in the type locality, a single extinction risk category cannot be selected.

NOTES. This rattan is an extremely slender, short-stemmed species, lacking flagella (or indeed cirri) in the material seen by us, with only two or three pairs of leaflets, the apical pair being substantially larger than the basal ones. Its inflorescences are erect and short, with a poorly developed flagelliform tip, and are branched to only one (female) or two (male) orders (hence the species epithet). *Calamus spanostachys* is most similar to *C. kebariensis* from the Bird's Head Peninsula, which differs in being largely unarmed and bearing leaves with numerous, fine, regularly arranged leaflets.

54. *Calamus spiculifer* J.Dransf. & W.J.Baker

Moderately robust rattan, climbing to 35 m or more, stem suckering not recorded. **Stem** with sheaths 21–28 mm diam., without sheaths to ca. 15 mm diam.; internodes to at least 23 cm. **Leaf** ecirrate, 80–90 cm long including petiole; sheath pale greenish brown (when dry), with thin grey indumentum, *spines minute, 1.5 × 0.1 mm, easily detached, uniformly distributed, black with minute swollen pale bases, pointing upward*; knee to ca. 40 mm long, unarmed; ocrea scarcely developed; flagellum ca. 2.7 m long; petiole absent or to 2 cm long; *leaflets 9–11 on each side of rachis, irregularly arranged in pairs, the basal 1–3 leaflets on each side not grouped, very broad elliptic, hooded*, longest leaflet in mid-leaf 19–28 × 5–6 cm, apical leaflets 18–19 × 3–4 cm, apical leaflets joined for half their length, leaflets armed with black bristles on margins near tip and sometimes on lower surface. **Inflorescence** pendulous, to ca. 2.7 m long, including peduncle ca. 35 cm and flagelliform tip

Calamus aff. *spiculifer*. Habit, Tamrau Mountains (WB).

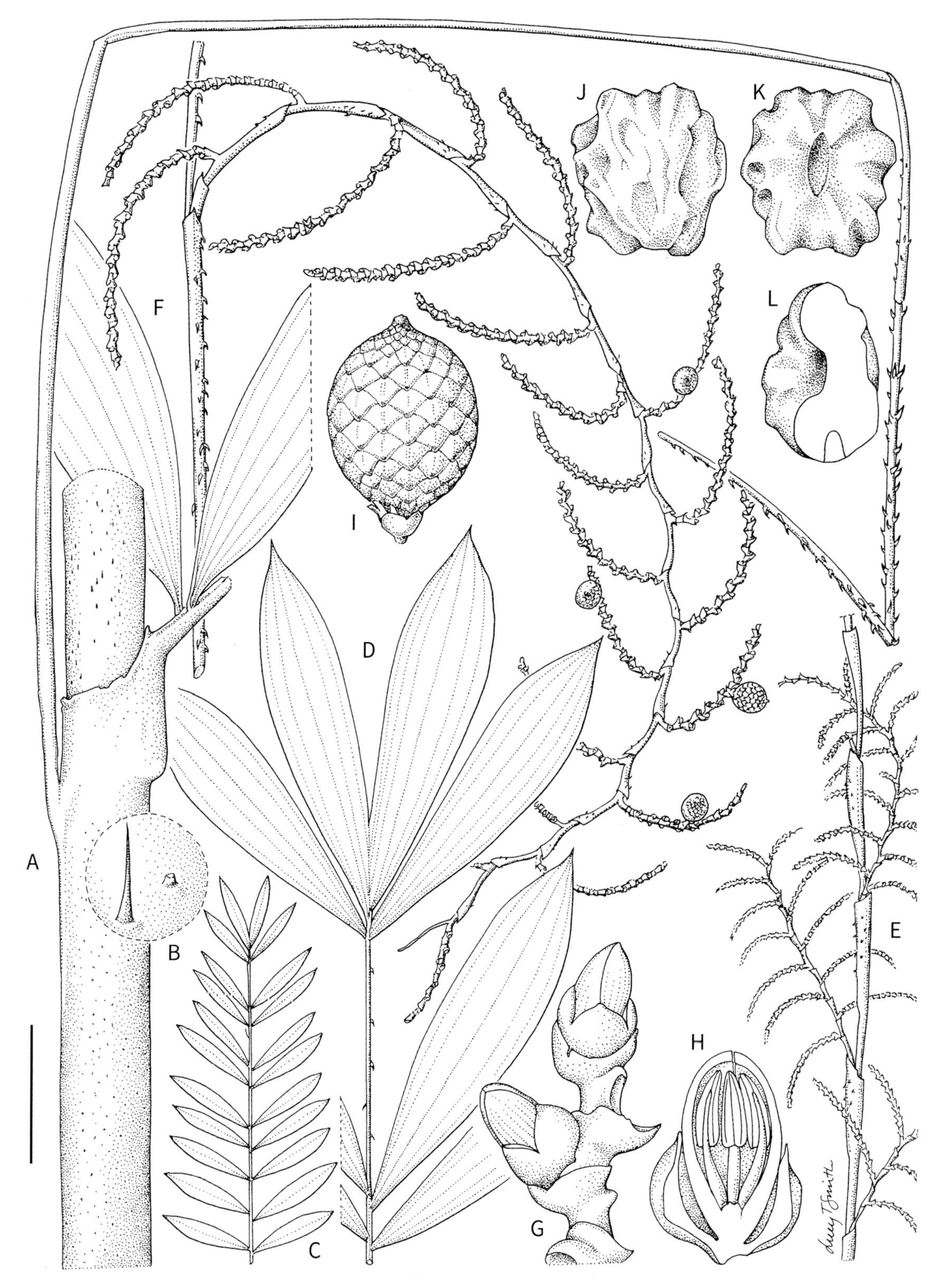

Calamus spiculifer. **A.** Stem with leaf sheath, leaf base and flagellum. **B.** Detail spicules. **C.** Whole leaf diagram. **D.** Leaf apex. **E.** Portion of male inflorescence. **F.** Portion of female inflorescence with fruit. **G.** Portion of male rachillla with flowers. **H.** Male flower in longitudinal section. **I.** Fruit. **J**, **K.** Seed in two views. **L.** Seed in longitudinal section. Scale bar: A = 4 cm; B = 1.8 mm; C = 30 cm; D–F = 6 cm; G = 3.3 mm; H = 2.2 mm; I = 1 cm; J–L = 7 mm. A, B, E, G, H from *Zieck NGF 36558*; C, D, F, I–L from *Brass 13341*. Drawn by Lucy T. Smith.

Calamus aff. *spiculifer*. LEFT TO RIGHT: leaf; ocrea. Tamrau Mountains (WB).

(length not known), branched to 2 orders in the female and 3 orders in the male; primary bracts tubular, with entire apices, armed with reflexed hooks; primary branches 5–7, to at least 75 cm long, male rachillae ca. 25–50 × 1.2 mm diam., female rachillae to ca. 120 × 3 mm. **Fruit** solitary, globose, 13 × 10 mm, scales mid-brown scales with darker margins. **Seed** (sarcotesta removed) 8 × 7 × 6 mm, with a very deep groove on one side, *seed surface deeply scalloped and pitted*; endosperm homogeneous.

DISTRIBUTION. Central northern New Guinea.

HABITAT. Submontane forest at 700–850 m.

LOCAL NAMES. *Nelmo* (Kabori).

USES. None recorded.

CONSERVATION STATUS. Data Deficient. This species is known from only two sites, where there is limited information on ongoing threats. Owing to insufficient evidence, a single extinction risk category cannot be selected.

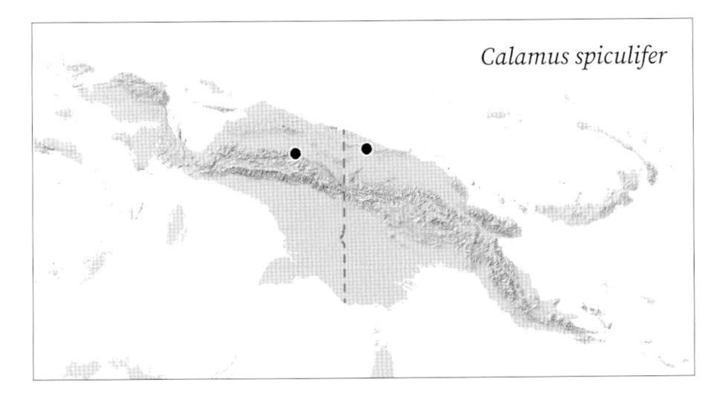

NOTES. This flagellate rattan is distinguished by its moderately robust stem, the minute, easily detached spicules on the leaf sheath (hence the epithet) and the ecirrate leaves with very broad hooded leaflets borne in pairs in the upper part of the leaf. In leaf and inflorescence morphology, it resembles *C. bulubabi* and *C. papuanus*. *Calamus bulubabi* is immediately distinguished by the abundant, hair-like spines and by the apical leaflets diminishing in size (compared to the large leaflets grouped at the leaf apex of *C. spiculifer*). *Calamus papuanus* is a more slender palm, variously armed, with a short, but well-defined, truncate, tightly sheathing ocrea and uneven, not pitted seed surface.

55. *Calamus superciliatus* W.J.Baker & J.Dransf.

Slender, multi-stemmed rattan climbing to 8 m. **Stem** with sheaths 8–13 mm diam., without sheaths to 7–8 mm diam.; internodes 17–22 cm; juvenile stems to ca. 1.5 m, bearing reduced leaves as short as ca. 9.5 cm in length. **Leaf** ecirrate, to 74–80 cm long including petiole; sheath dark green, with scattered purple-brown indumentum, *spines dense, to ca. 32 × 0.5–0.9 mm, fine, flattened, rusty-brown spines very dense, paler, longer (to 58 mm) and erect around sheath mouth*, spines falling in older sheaths; knee 28–33 mm long, armed as sheath; ocrea ca. 2–3 mm high,

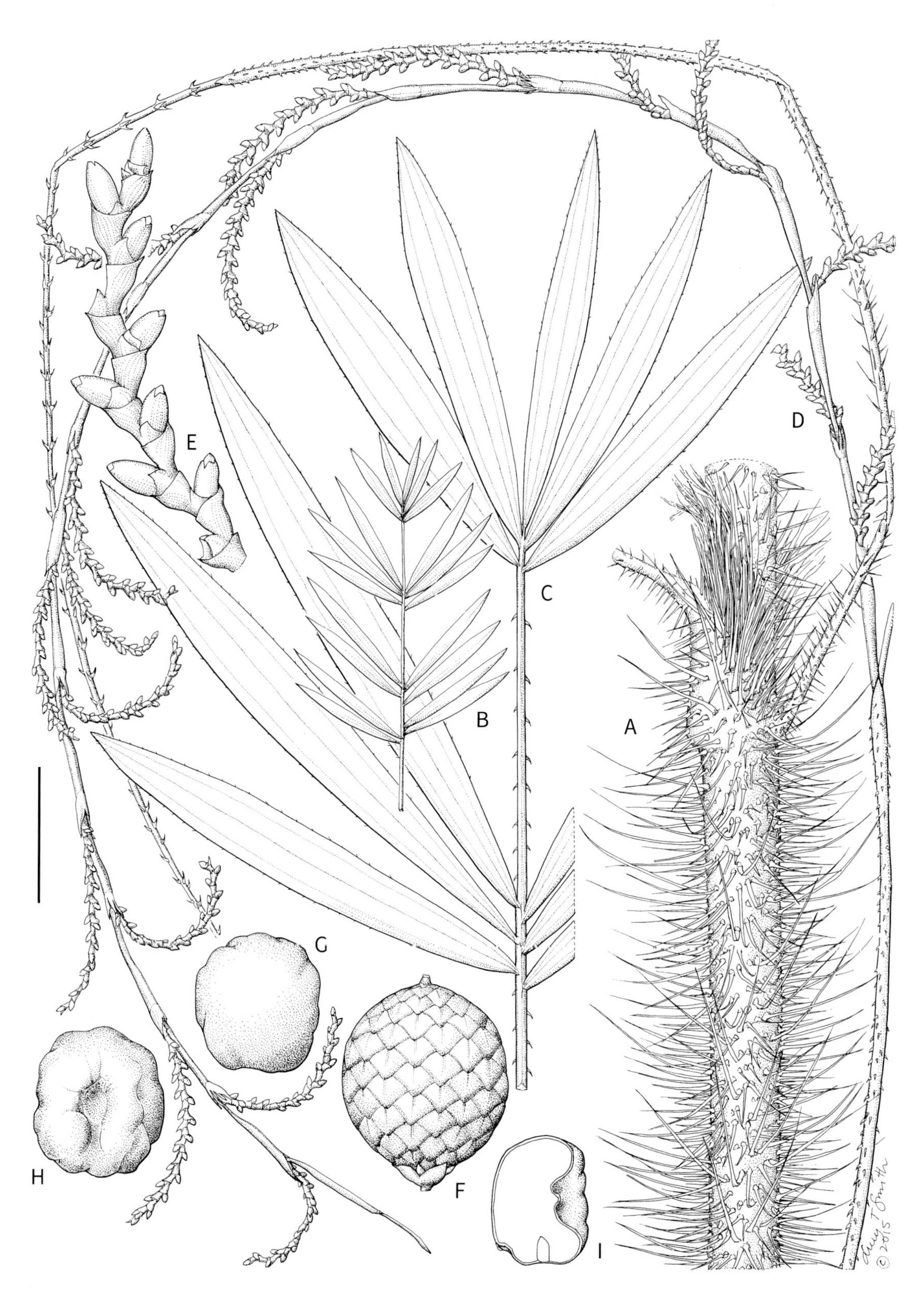

Calamus superciliatus. **A.** Stem with leaf sheath, petiole and flagellum. **B.** Whole leaf diagram. **C.** Leaf apex. **D.** Male inflorescence. **E.** Male rachilla with flowers. **F.** Fruit. **G, H.** Seed in two views. **I.** Seed in longitudinal section. Scale bar: A, D = 3 cm; B = 30 cm; C = 6 cm; E = 7 mm; F–I = 8 mm. A–E from *Baker 1370*; F–I from *Baker et al. 1385*. Drawn by Lucy T. Smith.

Calamus superciliatus. CLOCKWISE FROM LEFT: habit, juvenile stem; leaf sheaths with conspicuous sheath mouth spines; leaf. Tamrau Mountains (WB).

forming an inconspicuous, dry, papery crest extending into the petiole base, persistent, obscured by spines; flagellum to ca. 1.5 m long; petiole 11–15 cm long; *leaflets 9–12 each side of rachis, arranged in 3–5 widely spaced, divaricate groups, linear lanceolate, longest leaflet near base* 27–29 × 2.3–2.5 cm, mid-leaf leaflets 24–27 × 2.1–3 cm, apical leaflets 18–19 × 1.7–2.8 cm, apical leaflet pair not or scarcely united at the base, leaflets very sparsely armed with few bristles on margins and upper surface. **Inflorescence** *flagelliform, to ca. 2.2 m long* including ca. 1.4 m peduncle and ca. 35 cm flagelliform tip, branched to 2 orders in the female, male not seen; primary bracts strictly tubular, splitting slightly at apex, armed as sheath, though spines shorter; primary branches 1–3, to 55 cm long, male rachillae not seen, female rachillae 20–70 × ca. 1.5 mm. **Fruit** solitary, globose, ca. 13 × 8 mm, scales cream-white. **Seed** (sarcotesta removed) ca. 8.5 × 7.8 × 5.7 mm, globose with deep lateral intrusion, shallowly channeled; endosperm homogeneous.

DISTRIBUTION. Tamrau Mountains, Bird's Head Peninsula.

HABITAT. Lower montane forest at 700–900 m.

LOCAL NAMES. None recorded.

USES. None recorded.

CONSERVATION STATUS. Critically Endangered. *Calamus superciliatus* has a restricted distribution. Deforestation driven by road building is a major threat in its range.

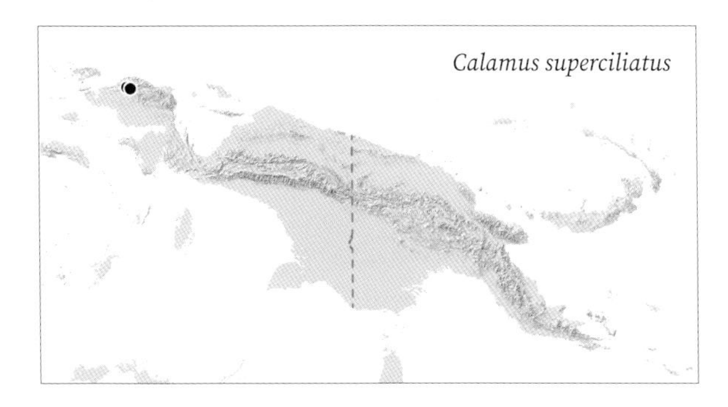

NOTES. *Calamus superciliatus* is a flagellate species that is distinguished by its dense, fine, flattened leaf sheath spines that are particularly numerous and longer at the sheath mouth, forming a tuft. The leaves bear only 9–12 pairs of leaflets, which are clustered in few, divaricate groups, the longer leaflets being located near the base, and are very sparsely armed with bristles. The inflorescence is long, fine and flagellum-like and bears few primary branches. Reduced, very slender juvenile stems might be confused with *C. depauperatus*. In adult form, however, *C. superciliatus* is likely to be confused only with *C. retroflexus* on account of leaflet shape and arrangement, and spine shape, but the latter species is more robust, has more numerous leaflets that are swept back across the stem and lacks both a petiole and the conspicuous tuft of spines at the leaf sheath mouth.

56. *Calamus vestitus* Becc.

Synonym: *Palmijuncus vestitus* (Becc.) Kuntze

Moderately robust, multi-stemmed rattan climbing to 20 m. **Stem** with sheaths 11–24 mm diam., without sheaths to 7.5–12.5 mm diam.; internodes 10–31 cm. **Leaf** ecirrate to 1.2 m long including petiole; sheath green, with variable dark indumentum, *spines numerous, 1.5–7 × 0.5–3 mm, mostly flattened, parallel-sided, with irregular, jagged apices,* simple spines also present, spines drying brown with dark tips, arranged randomly with some partial collars; knee 12–27 mm long, armed as sheath; *ocrea 15–40 × 1.7–2.5 cm, inflated, tubular, encircling the sheath, papery but often separating into a network of fibres, unarmed or with very few, dry, brown spines to 3.5 mm long, persistent;* flagellum 1.5–2 m long; petiole to 16.5 cm long or absent; leaflets 20–47 each side of rachis, arranged regularly, linear, longest leaflet near leaf base 12–36 × 0.8–2.4 cm, mid-leaf leaflets slightly smaller, apical leaflets 4.5–12 × 0.3–0.7 cm, apical leaflet pair barely united or united up to one fifth of their length, leaflets with bristles on surfaces and margins (fewer on lower surface). **Inflorescence** pendulous, up to 2.5 m long including 29–76 cm peduncle and 14–45 cm flagelliform tip, branched to 2 orders in the female and 3 orders in the male; primary bracts tubular, sometimes splitting longitudinally and tattering to fibres at mouth, with numerous, short, reflexed spines; primary branches up to 8, to 72 cm long, male rachillae 26–90 × 1–1.5 mm, female rachillae 15–130 × 1.5–3 mm. **Fruit** solitary, globose to obovoid, 14–18.5 × 8.5–11.5 mm, scales pale yellow to green. **Seed** (sarcotesta removed) 6.8–10 × 6–7.5 × 4–5.8 mm, *ellipsoid, with a deep, rather open pit on one side, with irregular, angular ornamentations;* endosperm homogeneous.

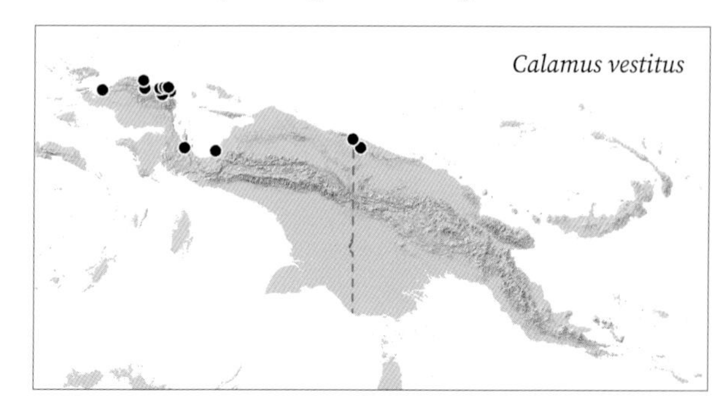

DISTRIBUTION. North-western to central northern New Guinea.

HABITAT. Lowland forest, including disturbed areas, often on alluvium near rivers, sea-level to 600 m.

LOCAL NAMES. *Balsa* (Pinai-Hagahai), *Ensyake* (Nuni), *Keboe* (Nimboran), *Kumisi* (Yamur), *Mafoni ngekobra* (Wariori and Waramoi Rivers), *P(u)ruk* (Melpa), *Samsa* (Awyi).

USES. General use as cordage, for tying roofs, cane split for use as bush nails. Sap from cut stem used to treat eye inflammation. Fruit eaten for its sweet taste.

CONSERVATION STATUS. Least Concern.

NOTES. This flagellate species belongs to the *C. longipinna* group (see under *C. longipinna*). It is characterised by the tubular, largely unarmed ocrea that is papery in texture, but separates into fibres, often disintegrating. The leaf sheaths are heavily armed with flattened spines of various sizes with parallel sides and jagged apices. The seed is also distinctive with angular grooves and ridges on one side and a shallow open pit on the other. These seed and sheath spine features are shared with *C. reticulatus,* to which *C. vestitus* is undoubtedly closely related. Another similar species, *C. longipinna,* is differentiated from *C. vestitus* by its open ocrea and pointed, rather than jagged, leaf sheath spines.

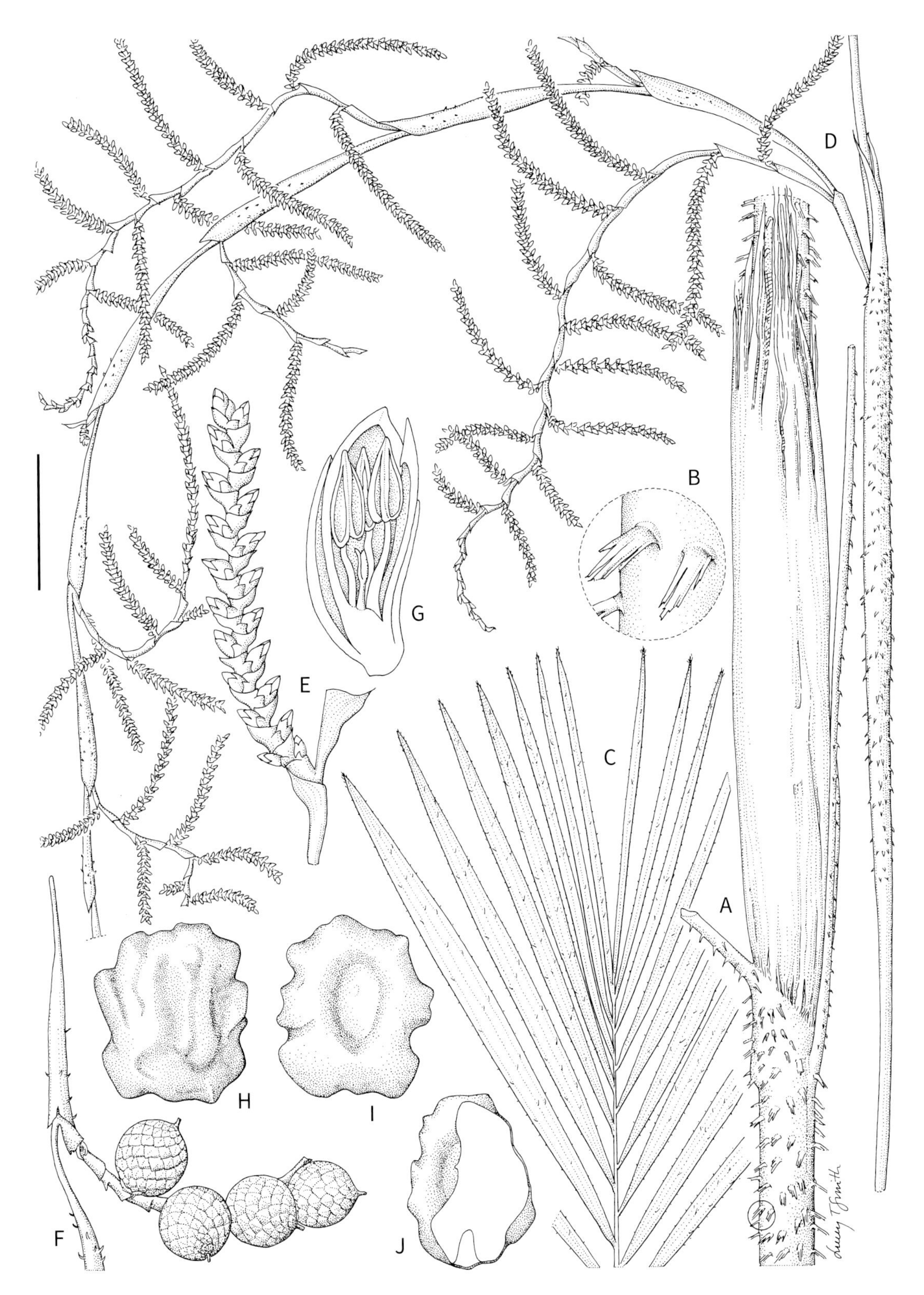

Calamus vestitus. **A.** Stem with leaf sheath, petiole and ocrea. **B.** Detail spines. **C.** Leaf apex. **D.** Male inflorescence. **E.** Male rachilla with flowers. **F.** Female rachilla with fruit. **G.** Male flower in longitudinal section. **H, I.** Seed in two views. **J.** Seed in longitudinal section. Scale bar: A, C, D = 4 cm; B, E = 1 cm; F = 2 cm; G = 1.8 mm; H–J = 7 mm. All from *Sands et al. 6616*. Drawn by Lucy T. Smith.

57. *Calamus vitiensis* Warb. ex Becc.

Synonyms: *Calamus stipitatus* Burret, *Calamus vanuatuensis* Dowe

Slender to robust, single-stemmed rattan climbing to 15 m. **Stem** with sheaths 10–50 mm diam., without sheaths 7–22 mm diam.; internodes 12.5–33 cm. **Leaf** *cirrate,* to 3 m long including cirrus and petiole; sheath dark green, with grey-brown indumentum, *spines absent to numerous, 1–28 × 0.3–2.5 mm, yellow-green to brown, flattened, triangular, longer spines flexible, spine bases sometimes slightly swollen, spines usually solitary*; knee 17–60 mm long, unarmed or lightly armed; ocrea 2–4.5 mm, forming a low, woody, brown, unarmed or lightly armed, persistent collar; flagellum absent; petiole 0–45 mm; leaflets 10–22 each side of rachis, *arranged regularly or in widely spaced, sometimes divergent pairs*, broadly elliptic, hooded, longest leaflets near middle of leaf, 18–43 × 3.5–7 cm, apical leaflets 7–23.5 × 0.9–1.5 cm, leaflet surfaces and margins unarmed or with very few bristles on upper surface; cirrus 60–125 cm, cirrus grapnel spines arranged regularly. **Inflorescence** arching, to ca. 2 m long including 26–33 cm peduncle and sterile tip to ca. 50 cm, branched to 2 orders in the female and 3 orders in the male; primary bracts strictly tubular, with acute, triangular limb to one side, unarmed or lightly armed; primary branches 6–12, to 28 cm long, male rachillae not seen, female rachillae 9–95 × 1–2 mm. **Fruit** solitary, shortly stalked, globose, 10–11 × 8.5–9.5 mm scales white. **Seed** (sarcotesta removed) 6.8–8 × 6.5–7 × 5.2–6.5 mm, globose, with a deep, narrow pit on one side, the surface covered with numerous shallow pits and irregular channels; endosperm homogeneous.

DISTRIBUTION. Scattered records throughout New Guinea, including Biak and Manus Islands. Also recorded from northern Australia (Queensland), Solomon Islands, Vanuatu and Fiji. The most easterly occurring species in the genus *Calamus*. Henderson (2020) reports the species from Maluku, but we have been unable to confirm this.

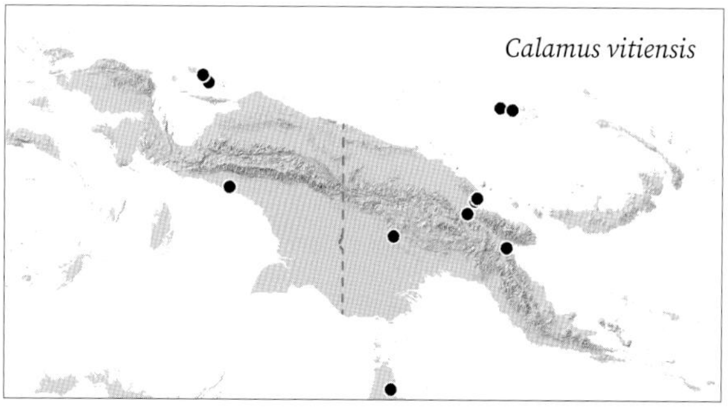

Calamus vitiensis

HABITAT. Primary and secondary forest at 60–750 m.

LOCAL NAMES. *Wusiu* (Manus).

USES. General cordage, cane for tying houses, and for making swings for children. Sap from cut stem used for curing eye ailments.

CONSERVATION STATUS. Least Concern.

NOTES. This broadly defined and variable cirrate species is recognised by its simple triangular spines, which are usually of rather uniform size and are only slightly swollen at the base, and by the regular arrangement of cirrus grapnel spines. In addition, the sheath spines in most forms are largely solitary, whereas the sheath spines of other members of the *C. aruensis* complex (see under *C. aruensis*) are usually at least partly organised into partial whorls with solitary spines interspersed among the whorls. As in *C. aruensis*, unarmed forms occur in *C. vitiensis*, but cirrus morphology can be used to distinguish the two species (see Baker *et al.* (2003) for further discussion).

Our circumscription of *C. vitiensis* differs somewhat from that of Henderson (2020), who includes two additional synonyms *C. pachypus* and *C. ledermannianus*. We regard the former as a distinct species (see under *C. pachypus*) and the latter as a name of uncertain application, the type having been destroyed in Berlin.

58. *Calamus wanggaii* W.J.Baker & J.Dransf.

Moderately robust, multi-stemmed rattan climbing to 25 m. **Stem** with sheaths ca. 15 mm diam., without sheaths to 7–9 mm diam.; internodes ca. 35 cm. **Leaf** ecirrate to 75 cm long including petiole; sheath dark green, with abundant, thin brown indumentum, *spines numerous, 2–6 × 0.5–1 mm, solitary, narrow triangular*; knee 28 mm long, armed as sheath; *ocrea 22 × 2.6 cm, inflated, boat-shaped, open longitudinally to base on side opposite petiole, clasping and usually obscuring sheath, papery, tattering, armed as sheath, persistent*; flagellum to ca. 2 m; petiole ca. 13 cm; *28 leaflets each side of rachis, arranged in three widely-spaced groups of 9–11 leaflets, leaflets regularly spaced and divaricate within groups, oblanceolate, longest leaflet in upper part of lowest group,* 33 × 2 cm, mid-leaf leaflets 26.5 × 2 cm, apical leaflets 17 × 1.2 cm, apical leaflet pair briefly united, leaflets armed with conspicuous bristles on upper surface (few

Calamus vitiensis. **A.** Stem with leaf sheath and leaf base. **B.** Detail spines. **C.** Stem with leaf sheath and petiole. **D.** Detail spine. **E.** Leaf apex with cirrus. **F.** Leaf mid-portion. **G.** Detail grapnels. **H.** Portion of female inflorescence. **I.** Female rachilla with flowers. **J**, **K.** Female flower whole and in longitudinal section. Scale bar: A, C, H = 3 cm; B = 3.3 mm; D = 1.5 mm; E, F = 4 cm; G = 7 mm; I = 5 mm; J, K = 1.8 mm. A from *Baker & Utteridge 574*; B–K from *Sands et al. 2744*. Drawn by Lucy T. Smith.

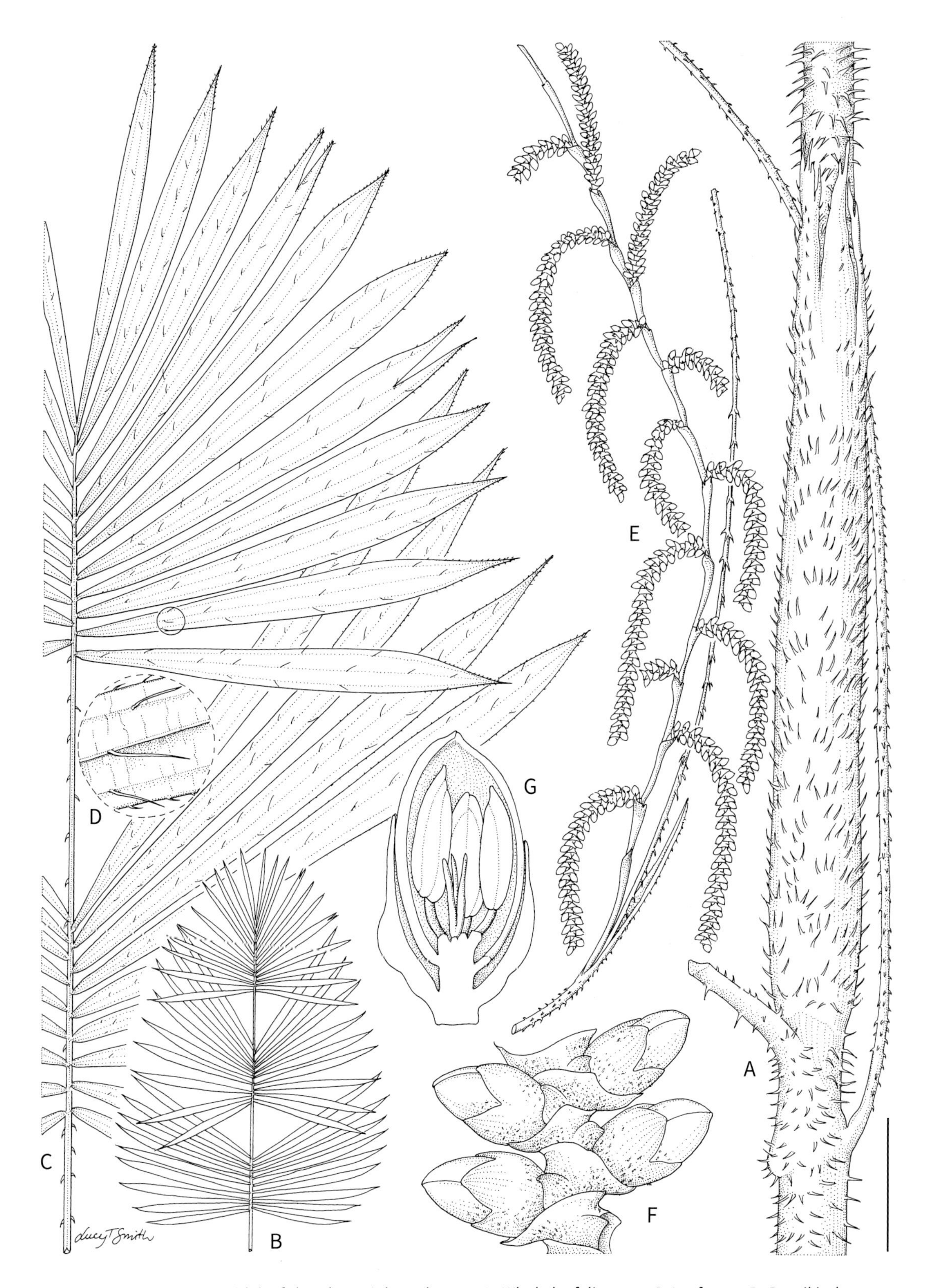

Calamus wanggaii. **A.** Stem with leaf sheath, petiole and ocrea. **B.** Whole leaf diagram. **C.** Leaf apex. **D.** Detail indumentum. **E.** Male inflorescence. **F.** Portion of male rachilla with flowers. **G.** Male flower in longitudinal section. Scale bar: A = 3 cm; B = 30 cm; C = 6 cm; D = 1.5 cm; E = 2.5 cm; F = 3 mm; G = 1.3 mm. All from *Barrow et al. 129*. Drawn by Lucy T. Smith.

or absent on lower surface), numerous short bristles on margins. **Inflorescence** pendulous, ca. 2.2 m long including ca. 1.9 m peduncle and ca. 25 cm flagelliform tip, branched to 3 orders in the male, female not seen; primary bracts, strictly tubular, splitting very briefly at apex, armed with numerous grapnel spines; primary branches ?1 (incomplete material seen), to at least 19 cm long, male rachillae 26–40 mm × 1.5 mm, female rachillae not seen. **Fruit** not seen.

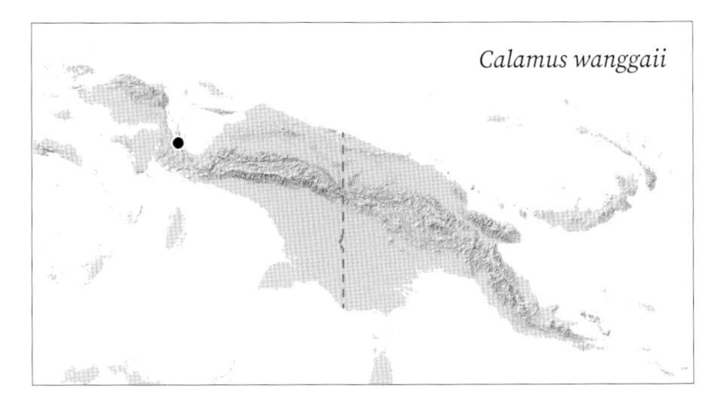

DISTRIBUTION. Known from a single collection near the Wosimi River, south of the Wandamen Peninsula.

HABITAT. Lowland forest, ca. 30 m.

LOCAL NAMES. None recorded.

USES. None recorded.

CONSERVATION STATUS. Critically Endangered. Logging is a major threat in its range and the type locality has been assessed for timber extraction.

NOTES. This poorly known flagellate species is a member of the *C. longipinna* group (see under *C. longipinna*). It is similar to *C. longipinna*, but is distinguished by its well-developed ocrea with a longitudinal opening on the side opposite the petiole that is heavily armed with short triangular spines, similar to those on the leaf sheath. It also has unusual, grouped, oblanceolate leaflets with very long bristles on the upper surface, the leaflets within each group being regularly arranged and divaricate.

59. *Calamus warburgii* K.Schum.

Synonym: *Calamus sepikensis* Becc.

Moderate to very robust, single- or multi-stemmed rattan climbing to 25 m. **Stem** with sheaths 20–52 mm diam., without 12–30 mm diam.; internodes 15–40 cm. **Leaf** *cirrate*, 2–6.8 m long including cirrus and petiole; sheath green to yellow-green, with brown, caducous, woolly indumentum between spines of younger sheaths, *spines very abundant (or rarely less numerous), 5–70 mm long, flattened needle-shaped, brown, dry, easily detached, often upward-pointing, united at base in green, persistent, closely spaced collars 1–5 mm high* (spines more rarely solitary); knee 6–9 cm long, 2–5 cm wide, colour as sheath, unarmed or sparsely armed as sheath; ocrea not well developed, comprising two, low, papery auricles flanking petiole base, sometimes lightly armed as sheath, fragile; flagellum absent; petiole 0.5–12 cm; leaflets 62–82 each side of rachis, *regularly arranged*, pendulous, linear-lanceolate, longest leaflets shortly above the basal leaflets to ca. 50 cm long, mid-leaf leaflets 22–44 × 1.5–2.5 cm, apical leaflets 20–30 × 1–2 cm, leaflets with numerous, stout, black marginal bristles and longer, black bristles on main veins of upper and lower surface; cirrus 0.7–3.9 m, sometimes with scattered vestigial leaflets, cirrus grapnel spines arranged regularly. **Inflorescence** erect and widely spreading, 100–150 cm long including 14–30 cm peduncle, terminating with a rachilla or very short empty bract, branched to 2 orders in the female and 3 orders in the male; primary bracts not significantly differentiated in shape from other bracts in the inflorescence, stout, truncate, tightly sheathing, lightly armed; primary branches 9–15, to 95 cm long, *with pronounced stalk concealed by primary bract*, male rachillae 2–40 × 1–1.5 mm, female rachillae 25–130 × 1.5–3 mm, *female rachillae stalked, stalk concealed by bract*. **Fruit** solitary, ellipsoid, 8–11 × 7.5–8 mm, scales green. **Seed** (sarcotesta removed) 5.5–7 × 4–5 × 4–4.5 mm, ellipsoid, surface with irregular channels and pits; endosperm homogeneous.

DISTRIBUTION. Widespread on mainland New Guinea. Also recorded from Manus Island and Cape York Peninsula, Australia. Henderson (2020) also records the species from Seram, but we have been unable to confirm this report.

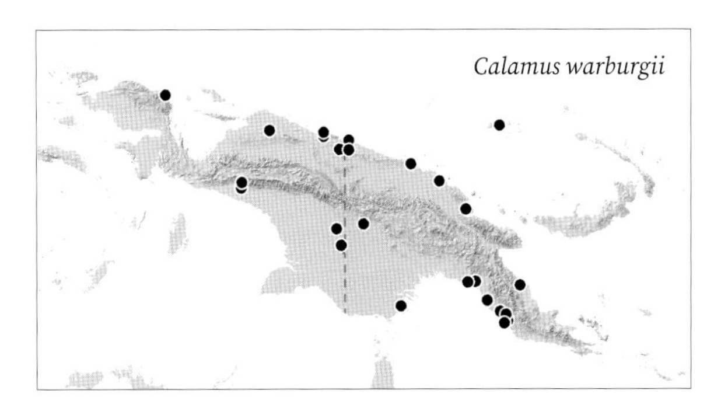

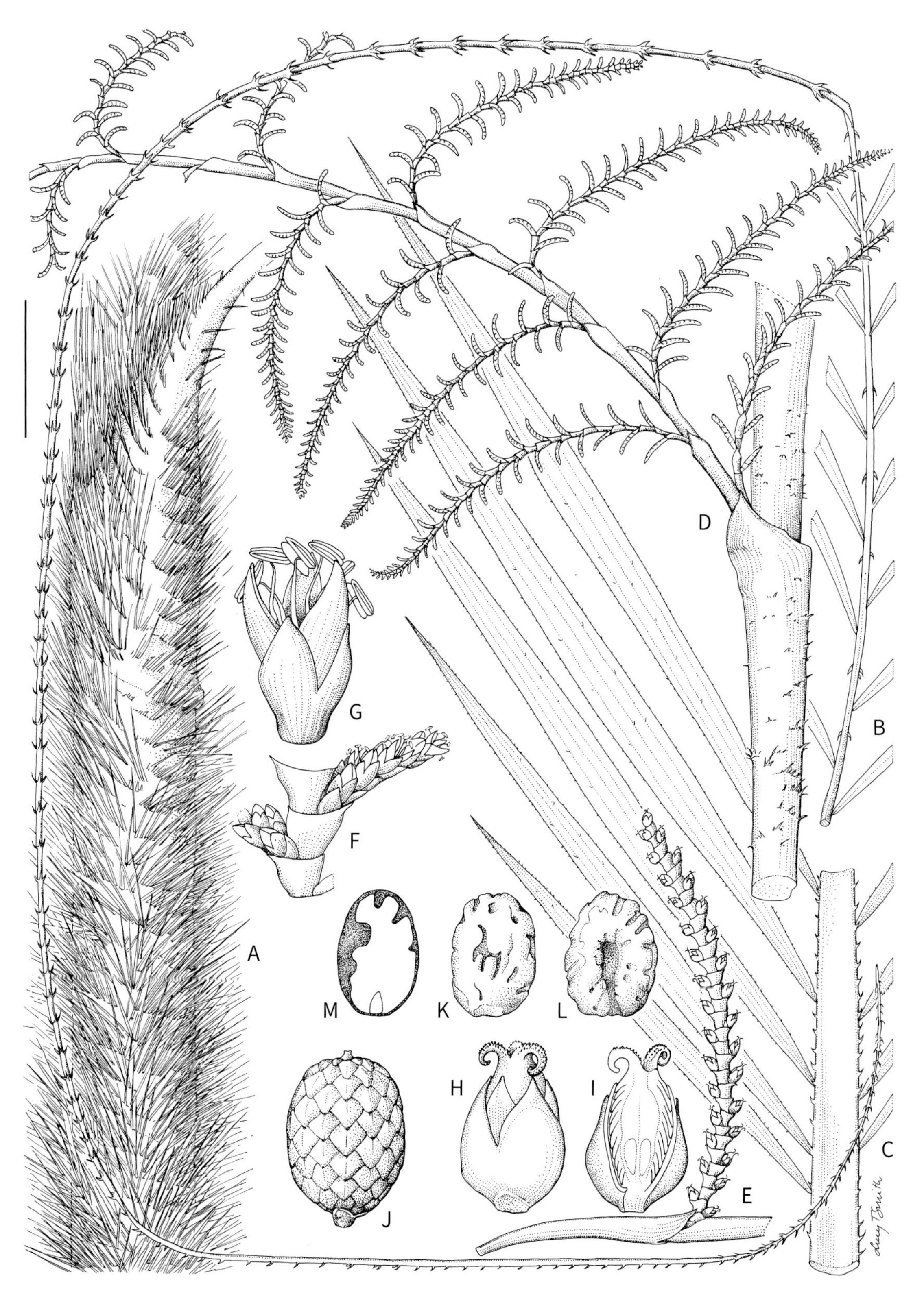

***Calamus warburgii*.** **A.** Stem with leaf sheath and petiole. **B.** Leaf apex with cirrus. **C.** Leaf mid-portion. **D.** Male inflorescence. **E.** Female rachilla with flowers. **F.** Portion of male rachilla with flowers. **G.** Male flower. **H**, **I.** Female flower whole and in longitudinal section. **J.** Fruit. **K**, **L.** Seed in two views. **M.** Seed in longitudinal section. Scale bar: A = 4 cm; B = 8 cm; C, D = 6 cm; E = 2 cm; F, J–M = 1 cm; G–I = 2.5 mm. A–D from *Baker 578*; E, H, I from *Poudyal et al. 84*; F, G from *Baker & Utteridge 587*; J–M from *Maturbongs 194*. Drawn by Lucy T. Smith.

HABITAT. Lowland forest, both primary and secondary, especially flat areas and riverine vegetation from sea level to 350 m.

LOCAL NAMES. *Hareha* (Kerema), *Hekek sirakemi* (Yamben), *Moroue* (Toaripe), *Wuryawur* (Marap).

USES. Split stems used for bowstrings, binding material and for making the framework of baskets. A section of the cane with spines is used to grate cassava. Juicy flesh around the seed is eaten raw.

CONSERVATION STATUS. Least Concern.

NOTES. *Calamus warburgii* group comprises three closely related, cirrate species (*C. warburgii*, *C. zieckii*, *C. zollingeri*). They are characterised by inflorescences that are spreading rather than flagelliform and have an undifferentiated appearance on account of the similarity in morphology of the inflorescence bracts on all axes, except for the rachillae. These bracts are tubular, tightly sheathing and relatively short compared to those of other *Calamus* species. The female rachillae bear pronounced stalks (concealed by the subtending bract), a key character of Furtado's (1956) section *Podocephalus* and Henderson's *C. andamanicus* Kurz group. The three species in New Guinea differ primarily in vegetative morphology.

Calamus warburgii is among the most widespread rattans in New Guinea and is sometimes locally abundant. It is a robust, single- or multi-stemmed species, with a cirrate leaf bearing numerous, pendulous, regularly arranged linear leaflets. The sheath is densely armed with long, fine, brown spines that form collars and are readily detached. Sometimes, spines are lost entirely, leaving ridges where the bases of the spine collars persist. We follow Henderson (2020) in including the little-known *C. sepikensis* as a synonym of *C. warburgii*.

Calamus warburgii. CLOCKWISE FROM LEFT: habit, near Timika (WB); base of male inflorescence, near Madang (WB); male flowers, Ramu River (WB).

60. *Calamus womersleyi* J.Dransf. & W.J.Baker

Slender, multi-stemmed rattan, total length not recorded. **Stem** with sheaths 13–15 mm diam., without sheaths to 6.5 mm diam.; internodes to 16–19 cm. **Leaf** ecirrate, 45–69 cm long including petiole; *sheath pale brown when dry, with dark, patchy indumentum, spines almost absent to abundant,* 5–19 × 1–1.5 mm, very narrow triangular, flattened, rather uniform, with conspicuously swollen bases, pale straw coloured, easily detached, horizontal; knee to 15 mm long, more or less unarmed or with a few short spines; *ocrea to 14 mm long, strictly tubular, truncate, membranous, drying dark brown, unarmed;* flagellum ca. 1.3 m; petiole 2–7 cm long; *leaflets 16–24 on each side of rachis, irregularly arranged in 2–4 groups, regularly arranged within the groups, narrow lanceolate,* abruptly constricted at the base, longest leaflet near the base 20–21 × 1.4–2 cm, mid-leaf leaflets 16.5–29 × 1.6–2.2 cm, apical leaflets 10–11 × 1.5 cm, apical leaflets joined for one fifth of their length, leaflets armed with black bristles on upper surface and margins near the tip. **Inflorescence** to ca. 1 m long including peduncle 18–43 cm, flagelliform tip not seen, branched to 2 orders in the female and 3 orders in the male; primary bracts tubular and closely sheathing, with a triangular tip, sparsely armed with short spines; primary branches 3–6, to 28 cm long, male rachillae to 50 × 2.5 mm; rachillae, female rachillae to ca. 50 × 3 mm. **Fruit** solitary, globose, 6 × 6 mm, scales pale brown scales with darker margins. **Seed** (sarcotesta removed) 7 × 6.5 × 5.5 mm, with a deep pit on one side, seed surface smooth, endosperm homogeneous.

DISTRIBUTION. North-western end of the Owen Stanley Range.

HABITAT. Montane forest at ca. 330–1,500 m.

LOCAL NAMES. *Ren* (Biaru).

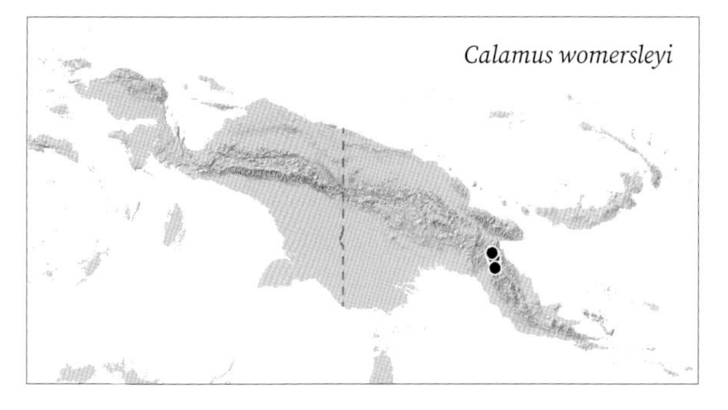

USES. None recorded.

CONSERVATION STATUS. Endangered. Deforestation due to gold mining concessions is a major threat within the restricted range of *C. womersleyi.*

NOTES. This flagellate rattan is distinguished by its relatively slender stem, the sheaths armed with rather sparse, pale spines, by the well-developed, short, tightly sheathing, truncate, membranous ocrea, and by the ecirrate leaves with narrow lanceolate leaflets that are arranged in irregular groups, but usually regularly arranged within the groups. Dark caducous indumentum is present in patches on leaf sheath, rachis and inflorescence bracts, lending the specimens a dirty appearance. *Calamus womersleyi* resembles members of the *C. baiyerensis* group (see under *C. baiyerensis*), but lacks a well-developed ocrea.

61. *Calamus zebrinus* Becc.

Synonyms: *Calamus laceratus* Burret, *Calamus steenisii* Furtado, *Palmijuncus zebrinus* (Becc.) Kuntze

Moderately robust, multi-stemmed rattan climbing to 35 m. **Stem** with sheaths 13–36 mm diam., without sheaths to 8–18 mm diam.; internodes 8–50 cm. **Leaf** ecirrate to 2.2 m long including petiole; sheath green, with abundant brown indumentum, *sheath spines usually numerous, 2–30 × 0.5–1 mm, needle-like, pale brown, easily detached, solitary or more usually in horizontal groups with bases united in low collars ca. 1 mm high, the collars persisting after spines falling;* knee 25–70 mm long, 5–10 mm wide, unarmed or armed as sheath; *ocrea to 100 × 4 cm, usually poorly preserved, erect at first, linear-lanceolate, papery, pale brown, armed with scattered bristles, the ocrea soon disintegrating to fibres or lost completely;* flagellum to 5.5 m long; petiole 3–20 cm long; *leaflets 23–64 each side of rachis, regularly arranged, linear,* longest leaflet in mid-leaf 23–47 × 1.2–2.5 cm, apical leaflets 7–15 × 0.4–1 cm, apical leaflet pair not united, armed with bristles on both surfaces and margins (more densely on lower surface). **Inflorescence** *pendulous, 1.9–6 m long* including 0.45–3.8 m peduncle and 0.5–2 m flagelliform tip, branched to 2 orders in the female and 3 orders in the male; primary bracts closely sheathing, splitting and sometimes disintegrating to fibres at apex, armed with scattered and grouped stout triangular spines to 4 mm long; primary branches to 7, to 90 cm long, male rachillae 2–5 × 2.5–3 mm, female rachillae

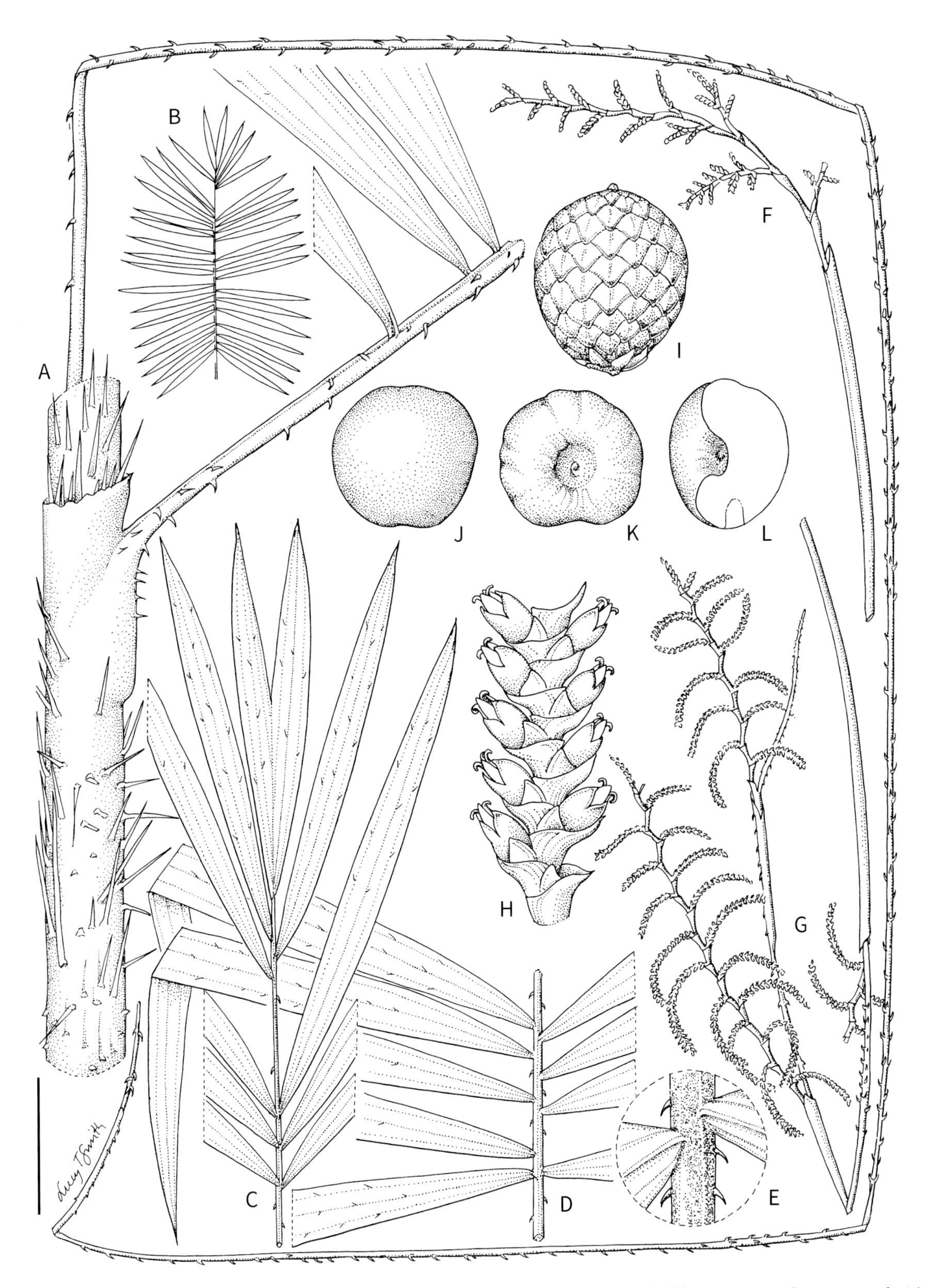

Calamus womersleyi. **A.** Stem with leaf sheath, leaf base, ocrea and flagellum. **B.** Whole leaf diagram. **C.** Leaf apex. **D.** Leaf mid-portion. **E.** Detail rachis. **F.** Portion of male inflorescence. **G.** Portion of female inflorescence. **H.** Portion of female rachilla with flowers. **I.** Fruit. **J**, **K.** Seed in two views. **L.** Seed in longitudinal section. Scale bar: A = 2.5 cm; B = 24 cm; C, D = 6 cm; E = 1.5 cm; F, G = 8 cm; H, J–L = 7 mm; I = 1 cm. A, C, D, F from *Moore & Womersley 9285*; B, E, G, H from *Wormersley & Thorne NGF 12753*; I–L from *Womersley & Thorne 12758*. Drawn by Lucy T. Smith.

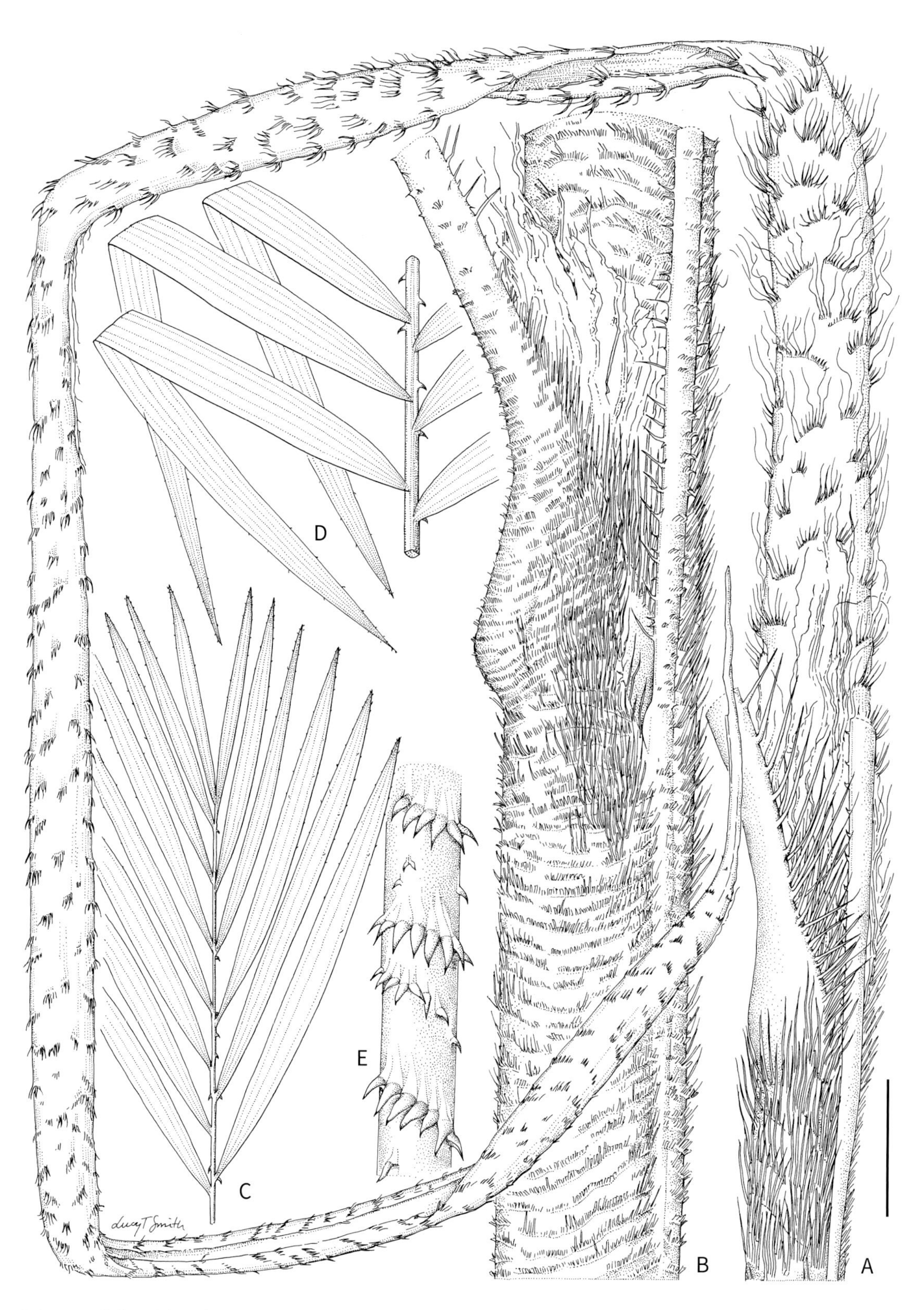

Calamus zebrinus (Plate 1). **A.** Young stem with leaf sheath and ocrea. **B.** Mature stem with leaf sheath, ocrea eroded and many spines fallen. **C.** Leaf apex. **D.** Leaf mid-portion. **E.** Prophyll. Scale bar: A, B = 3 cm; C, D = 6 cm; E = 2 cm. A, C, D from *Banka et al. 2006*; B, E from *Baker et al. 636*. Drawn by Lucy T. Smith.

45–190 × 2.5–3 mm. **Fruit** *usually paired*, spherical or sometimes oblate, 8–15 × 8–11, yellowish brown. **Seed** (sarcotesta removed) to ca. 9 × 8 × 7 mm, ellipsoid with a pronounced longitudinal pit; endosperm homogeneous.

DISTRIBUTION. Widespread in mainland New Guinea, also recorded in the Raja Ampat Islands, Biak Islands, Yapen and New Ireland.

HABITAT. Lowland forest, especially on riverbanks or in swamp forest from sea level to 1,500 m.

LOCAL NAMES. *Ahal* (Yali), *Ainan* (Yali), *Ajenkondo* (Kati), *Bugamare* (Unate), *Kil* (Kavieng), *Lamtei* (Matbat), *Marokwakwa* (meaning "kills trees"; Kotte), *Otonoma* (Sumuri), *Saeb* (Yali), *Sesawi* (Marap), *Si* (Yali), *To fanesif* (Ayawasi), *Warar* (Biak), *Wil su* (Ambel).

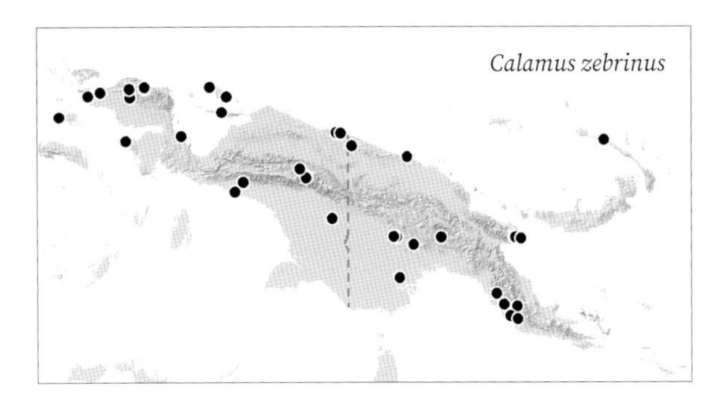

USES. Stems used for straps, armbands, fire-making, waist-hoops, tying, house construction and bow strings by the Yali People of the Snow Mountains region, and as other kinds of bindings. In the Baliem Valley area, it is reported to be planted sometimes.

Calamus zebrinus. LEFT TO RIGHT: female inflorescence; leaf sheath with tattering, papery ocrea. Mount Bosavi (WB).

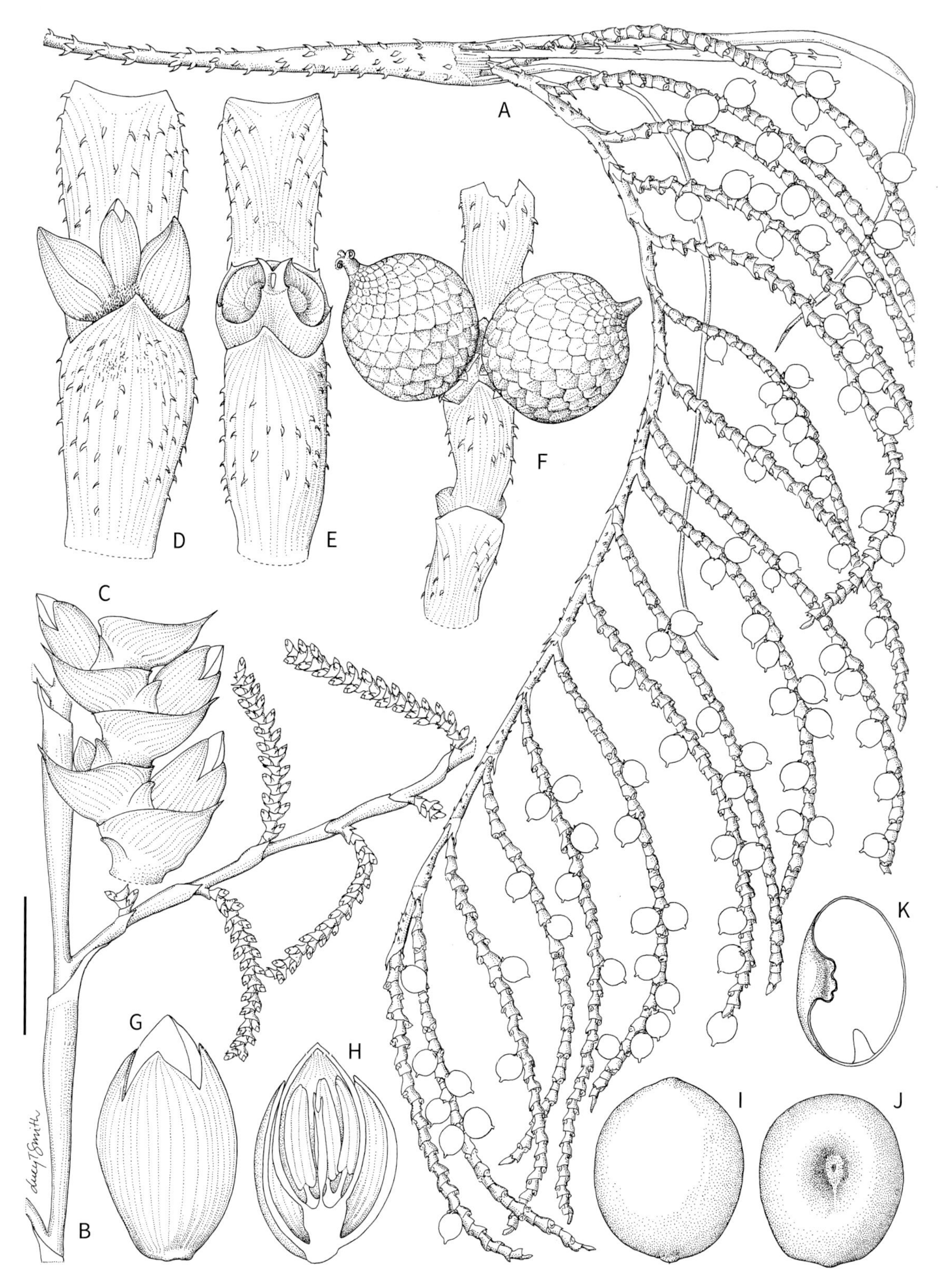

Calamus zebrinus (Plate 2). **A.** Portion of female inflorescence with fruit. **B.** Portion of male inflorescence. **C.** Portion of male rachilla with flowers. **D, E.** Portion of female rachilla with flowers and flowers removed, showing two female flowers and one sterile male flower. **E.** Portion of rachilla. **F.** Portion of rachilla with fruit. **G, H.** Male flower whole and in longitudinal section. **I, J.** Seed in two views. **K.** Seed in longitudinal section. Scale bar: A = 6 cm; B = 3 cm; C–E = 4 mm; F = 1 cm; G, H = 2 mm; I–K = 7 mm. A, F, I–K from *Banka et al. 2006*; B, C, G, H from *Avé 4080*; D, E from *Dransfield et al. JD 7657*. Drawn by Lucy T. Smith.

Calamus zebrinus. Fruit, borne in pairs, Mount Bosavi (WB).

Calamus zebrinus. LEFT TO RIGHT: inflorescence spines, Mount Bosavi (WB); female rachilla with flower buds in threes, near Timika (JD).

CONSERVATION STATUS. Least Concern.

NOTES. *Calamus zebrinus* belongs to the *C. altiscandens* group (see notes under *C. altiscandens*). It is one of the most common lowland, flagellate rattans in New Guinea. Vegetatively, it is highly distinctive because of its regularly arranged leaflets, very long papery ocrea (probably the longest among New Guinea *Calamus*) that disintegrates to fibres, and the fine, brown spines that coalesce at the base into low collars, which persist when the spines dry and fall early. Thus, specimens of this species can appear to lack spines and ocreas altogether. The inflorescence can be extremely long and the female flowers are borne in pairs or more often threes comprising two female flowers and a sterile male flower.

62. *Calamus zieckii* Fernando

Moderately slender to robust, multi-stemmed rattan climbing to 30 m. **Stem** with sheaths 15–35 mm diam., without 8–15 mm diam.; internodes 15–27 cm. **Leaf** *cirrate*, to 2–3.5 m long including cirrus and petiole; sheath light green, with caducous, woolly indumentum of matted white and brown hairs between spines on younger sheaths, *spines numerous, of various lengths to 5 mm, needle-like, pale brown to black, arranged in collars (often densely so) and united at base forming low crests, spines easily detached, crests persisting, spines often lost completely in older sheaths*; knee 30–66 mm

Calamus zieckii. **A.** Stem with leaf sheath and petiole. **B.** Leaf apex with cirrus. **C.** Leaf mid-portion. **D.** Portion of female inflorescence. **E.** Portion of rachilla with immature fruit. **F.** Immature fruit. **G.** Fruit. **H, I.** Seed in two views. **J.** Seed in longitudinal section. Scale bar: A = 3 cm; B, D = 4 cm; C = 6 cm; E, H–J = 7 mm; F = 3.3 mm; G = 1 cm. A–D from *Baker et al. 852*; E–J from *Maturbongs 541*. Drawn by Lucy T. Smith.

long, 15–43 mm wide, colour as sheath, unarmed or lightly armed as sheath; ocrea not well developed, comprising 2, low, inconspicuous auricles flanking petiole base, lightly armed as sheath; flagellum absent; petiole 3–15 cm long; leaflets 15–38 each side of rachis, *usually arranged in divaricate pairs or grouped in 3s–6s*, rarely regularly arranged, linear-lanceolate, longest leaflet at mid-leaf position 25–34 × 1.8–3.5 cm, apical leaflets 15–18 × 1.5–2 cm, leaflet margins sparsely armed with short, black bristles, surfaces unarmed or with scattered hair-like bristles on primary veins of either surface, *thin brown (rarely chalky) indumentum on lower surface*; cirrus 0.8–1.5 m, cirrus grapnel spines arranged regularly. **Inflorescence** spreading, held horizontally, 70–200 m long including 23–46 cm peduncle, terminating with a very short empty bract, branched to 2 orders in the female and 3 orders in the male; primary bracts not significantly differentiated in shape from other bracts in the inflorescence, stout, truncate, tightly sheathing, unarmed or lightly armed as sheath; primary branches 6–20, to 48 cm long, *with pronounced stalk concealed by primary bract*, male rachillae 1.5–30 × 1–1.5 mm, female rachillae 4–60 × 1.5–2 mm, *female rachillae stalked, stalk concealed by bract*. **Fruit** solitary, ellipsoid, 9–10 × 8–9 mm, scales yellow-orange. **Seed** (sarcotesta removed) 6.5–8 × 6–7 × 4.5–5.5 mm, ellipsoid with few shallow furrows and a lateral depression; endosperm homogeneous.

Calamus zieckii. LEFT TO RIGHT: habit, Kikori River (WB); leaf showing pale leaflet undersurface, near Sorong (WB).

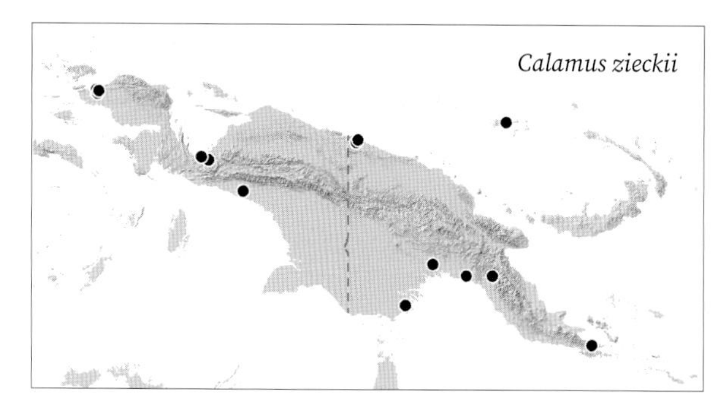

DISTRIBUTION. Widespread in mainland New Guinea and Manus Island.

HABITAT. In lowland and swamp forest, including disturbed areas, sea level to 200 m.

LOCAL NAMES. *Seato* (Kala), *Swampa soft* (Kolkamuk), *Wasam* (Krisa).

USES. None recorded.

CONSERVATION STATUS. Least Concern.

NOTES. This cirrate species is less robust than its relatives in the *C. warburgii* group (see notes under *C. warburgii*). It is immediately distinguished from *C. warburgii* and *C. zollingeri* by its leaflets that are arranged in divaricate pairs (and sometimes other groupings) with brown indumentum on the lower surface. Its sheaths bear fine, readily detached spines that are similar to those of *C. warburgii*, but are darker and less dense.

Calamus zieckii. Habit, near Timika (WB).

Calamus zieckii. Leaf sheaths and female inflorescence, near Timika (WB).

63. *Calamus zollingeri* Becc. subsp. *zollingeri*

Robust, single-stemmed or multi-stemmed rattan climbing to 20 m. **Stem** with sheaths 29–62 mm diam., without 22–51 mm diam.; internodes ca. 20 cm. **Leaf** *cirrate,* to ca. 4.7 m long including cirrus and petiole; sheath green, with thin, matted, brown, caducous indumentum, *spines rather numerous, of various sizes, 1–70 × 0.2–3.5 mm, flattened, narrow triangular, flexible, brown, solitary and in partial collars united at the base;* knee 7–13 cm long, 3–3.5 cm mm wide, unarmed or sparsely armed as sheath; ocrea not well developed, comprising two, low, papery auricles flanking petiole base, sometimes lightly armed as sheath, fragile; flagellum absent; petiole 35–45 cm; leaflets 60–105 each side of rachis, *regularly arranged,* pendulous, linear lanceolate, longest leaflet at mid-leaf position 24.5–68.5 × 1–3 cm, apical leaflets 18–28 × 0.8–1 cm, leaflets with scattered black bristles on margins, few long hair-like spines on main veins of upper and lower surfaces; cirrus ca. 120–175 cm, sometimes with scattered vestigial leaflets, cirrus grapnel spines arranged regularly. **Inflorescence** spreading, 80–200 cm long including ca. 11 cm peduncle, branched to 2 orders in the female and 3 orders in the male; primary bracts not significantly differentiated in shape from other bracts in the inflorescence, stout, truncate, tightly sheathing, lightly armed; primary branches number not recorded, to ca. 82 cm long, *with pronounced stalk concealed by primary bract,* male rachillae 4–55 mm × 4–2.5 mm, female rachillae 42–140 mm × 2.5–3.5 mm,

Calamus zollingeri. LEFT: habit, cultivated Bogor Botanic Garden, Indonesia (WB); RIGHT: leaf sheath, Gag Island (CDH).

Calamus zollingeri subsp. *zollingeri*. **A.** Stem with leaf sheath and petiole. **B.** Leaf apex with cirrus. **C.** Leaf mid-portion. **D.** Male inflorescence. **E.** Portion of male rachilla with flowers. **F, G.** Male flower whole and in longitudinal section. Scale bar: A, D = 4 cm; B, C = 6 cm; E = 7 mm; F, G = 3 mm. A–C from *Heatubun et al. 733*; D–G from *Heatubun et al. 734*. Drawn by Lucy T. Smith.

female rachillae stalked, stalk concealed by bract. **Fruit** solitary, ellipsoid, 10–13 × 7–11 mm, scales blue-black or black. **Seed** not seen.

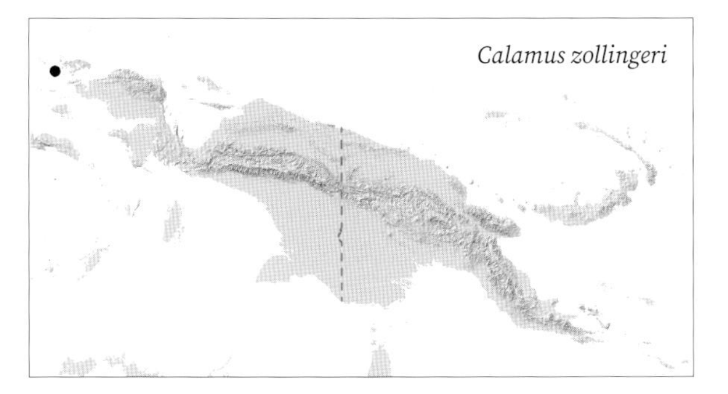

DISTRIBUTION. Widespread in Sulawesi and Maluku (Henderson 2020). In our area, known only from Gag Island in the Raja Ampat Islands.

HABITAT. In the New Guinea area, in lowland forest at 30 m.

LOCAL NAMES. *Waladou* (Gebe).

USES. Canes used for rope and as binding material for traditional houses.

CONSERVATION STATUS. Least Concern.

NOTES. This robust cirrate rattan differs from other members of the *C. warburgii* group (see notes under *C. warburgii*) in the much more robust leaf sheath spines that are solitary or partially grouped. Although the inflorescence appears to contain no differentiating characters, the blue-black fruit colour of *C. zollingeri* is unusual in the group. Henderson (2020) recognises two further subspecies, *C. zollingeri* subsp. *foxworthyi* (Becc.) A.J.Hend. and *C. zollingeri* subsp. *merrillii* (Becc.) A.J.Hend., both occurring in the Philippines. The record of *C. zollingeri* subsp. *zollingeri* on Gag Island is the easternmost limit of the species.

NAMES OF UNCERTAIN APPLICATION

Calamus brevifolius Becc.

The holotype was destroyed in Berlin and it is not possible to assign the name to any known species.

Calamus fuscus Becc.

The holotype was destroyed in Berlin and it is not possible to assign the name to any known species. See discussion under *C. klossii*.

Calamus gogolensis Becc.

The holotype of *C. gogolensis* was destroyed in Berlin. A photographic plate from Beccari's *Calamus* monograph (Beccari 1908) shows that it comprised a leaf and a leaf sheath section, and lacked fertile parts. An isotype consisting of a single leaf is extant at the Leiden herbarium. The leaflets are arranged in widely spaced divaricate groups and the sheath bears a medium-sized, bristly ocrea. It somewhat resembles *C. altiscandens* and *C. macrochlamys*. Henderson (2020) places it in synonymy with *C. macrochlamys*, but in our opinion, it is not possible to make this taxonomic judgement with certainty. Our view is consistent with Beccari's (1908), who linked it to *C. macrochlamys*, *C. ralumensis* (= *C. longipinna*) and *C. vestitus*.

Calamus ledermannianus Becc.

The holotype specimen was destroyed in Berlin and it is not possible to assign the name to any known species. Henderson (2020) places *C. ledermannianus* (without neotypification) in synonymy with *C. vitiensis*, based on the protologue and locality, but we do not consider the evidence strong enough to substantiate this attribution.

Calamus macgregorii Becc.

The holotype of *C. macgregorii* in the Florence herbarium comprises some fragments of leaves, which do not bear a cirrus, and some inflorescence branches with ripe fruit. The inflorescence material seems to correspond well with that of *C. warburgii* in fruit morphology and the apparently stalked rachillae, but the absence of a cirrus is inconsistent with that. If we are correctly attributing the fertile parts, it would appear that the specimen may be a mixture of species. See also commentary in Henderson (2020).

Calamus sessilifolius Burret

The type specimen was destroyed in Berlin and it is not possible to assign the name to any known species.

Nypa fruticans. Habit, Wosimi River (SB).

NYPOIDEAE

Nypa Steck

Synonym: *Nipa* Thunb.

Stemless – leaf pinnate – no crownshaft – no spines – leaflets pointed

Robust palm with creeping, forking stem, abundant in mangrove, monoecious. **Leaf** pinnate, erect; sheath open at base; petiole long; leaflets numerous each side of leaf rachis, with pointed tips, reduplicate, arranged regularly, horizontal, forked scales (ramenta) abundant on lower surface of leaflet midrib. **Inflorescence** *between the leaves,* branched to 5 orders; primary bracts and all subsequent *bracts similar, orange-brown and rubbery,* not dropping off as inflorescence expands; peduncle longer than inflorescence rachis. **Inflorescence** *ending in a round head of female flowers, male rachillae borne on lateral branches,* sausage-like, densely covered in flowers. **Flowers** male small, bright yellow, female larger, brown. **Fruit** dark brown, *in a large head at the end of the peduncle, individual fruits wedge-like, deeply grooved, husk fibrous,* endocarp hard and thick. **Seed** 1, endosperm homogeneous or rarely ruminate.

DISTRIBUTION. One species from South China to India, through South-East Asia to Australia and Solomon Islands, widespread in New Guinea and adjacent islands.

NOTES. *Nypa* is the only creeping, dichotomously branched palm found in mangrove habitats. Its inflorescence structure, consisting of a terminal club-like head of female flowers and lateral branches densely packed with male flowers, is unique in the palm family. The fruits float and disperse by water, often with the seedling emerging. *Nypa* might be confused with a juvenile *Metroxylon,* but unlike *Metroxylon, Nypa* is unarmed.

The taxonomic treatment of Dransfield *et al.* (2008) is followed here.

Nypa fruticans Wurmb

Synonyms: *Cocos nypa* Lour., *Nipa arborescens* Wurmb ex H.Wendl., *Nipa fruticans* (Wurmb) Thunb., *Nipa litoralis* Blanco, *Nypa fruticans* var. *neameana* F.M.Bailey

Dichotomously branching, creeping palm with 6–12 leaves in crown. **Stem** creeping at or just below ground level, forming adventitious roots on the underside. **Leaf** ca. 5–11 m long, unarmed, sheath ca. 235 cm long, petiole 30–350 cm long; rachis up to 10 m long; leaflets 71–179 × 3.3–9.5 cm, concolorous, 60–75 leaflets each side of the rachis, bearing *scattered dark brown forked scales (ramenta) along the mid-vein lower surface.* **Inflorescence** 120–235 cm long, largest peduncular bract 66–100 cm long. **Male flowers** yellow, bearing *3 connate stamens that appear as a single stamen; pistillode lacking.* **Female flowers** brown, *densely packed and forming a head* 20–22 cm diam.; *staminode lacking,* carpels 3, with *cup-like stigmas.* **Fruit** 11–14 × 8 cm at its widest, *angular (from mutual compression), fibrous, buoyant.* **Seed** ellipsoid, ca. 4.5 × 2.5 cm; often *viviparous.*

DISTRIBUTION. Widespread throughout New Guinea.

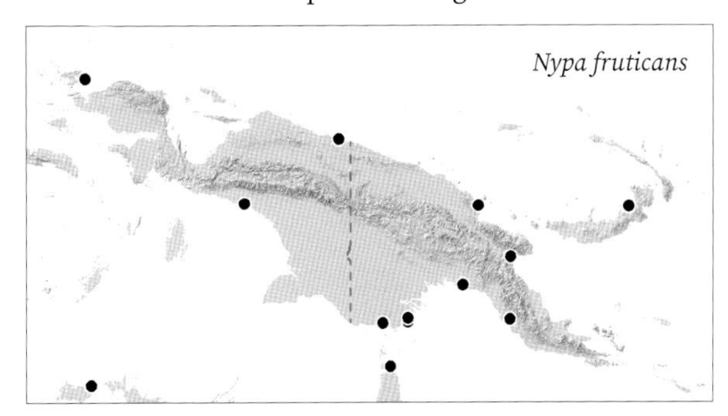

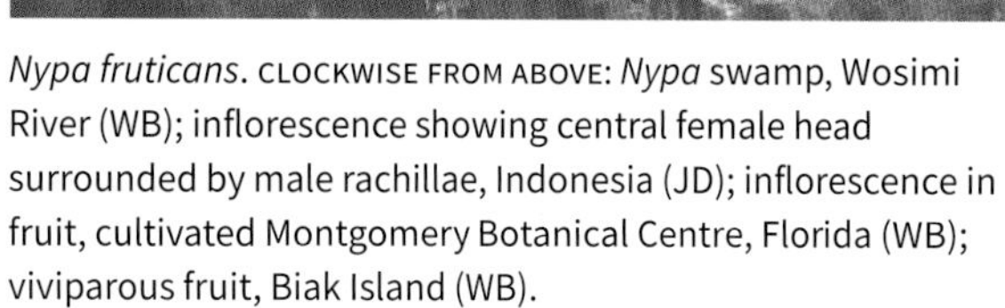

Nypa fruticans. CLOCKWISE FROM ABOVE: *Nypa* swamp, Wosimi River (WB); inflorescence showing central female head surrounded by male rachillae, Indonesia (JD); inflorescence in fruit, cultivated Montgomery Botanical Centre, Florida (WB); viviparous fruit, Biak Island (WB).

HABITAT. A major component of the mangrove fringe, forming dense stands, at or near sea level.

LOCAL NAMES. *Bigi, Korikis* (Waigeo), *Parema.*

USES. Leaves are used for thatch, umbrellas, sun hats, raincoats, coarse baskets, mats and bags. The leaves placed in sea are said to attract fish when fishing. Leaf ash is used to heal stingray wounds (Kamoro people, Mimika). Leaves and stems are used to treat hernias. Petiole is used for house walls. Leaflet epidermis is used for cigarette paper. Inflorescences are tapped for sugary sap. Immature seeds are eaten as food. Dye from fruits and leaves is used to colour hats and bags.

CONSERVATION STATUS. Least Concern (IUCN 2010).

NOTES. *Nypa* is a critical component of the coastal mangrove belt, which not only serves as a nursery for marine ecosystems but also protects terrestrial areas from storm surges. Although the *Nypa* stands in New Guinea are still extensive, destruction of the *Nypa* mangrove belt has had disastrous effects elsewhere in Southeast Asia, e.g. the Indian Ocean tsunami of December 2004, in which property destruction and environmental devastation were attributed to the absence of healthy *Nypa* stands (Hirashi 2008).

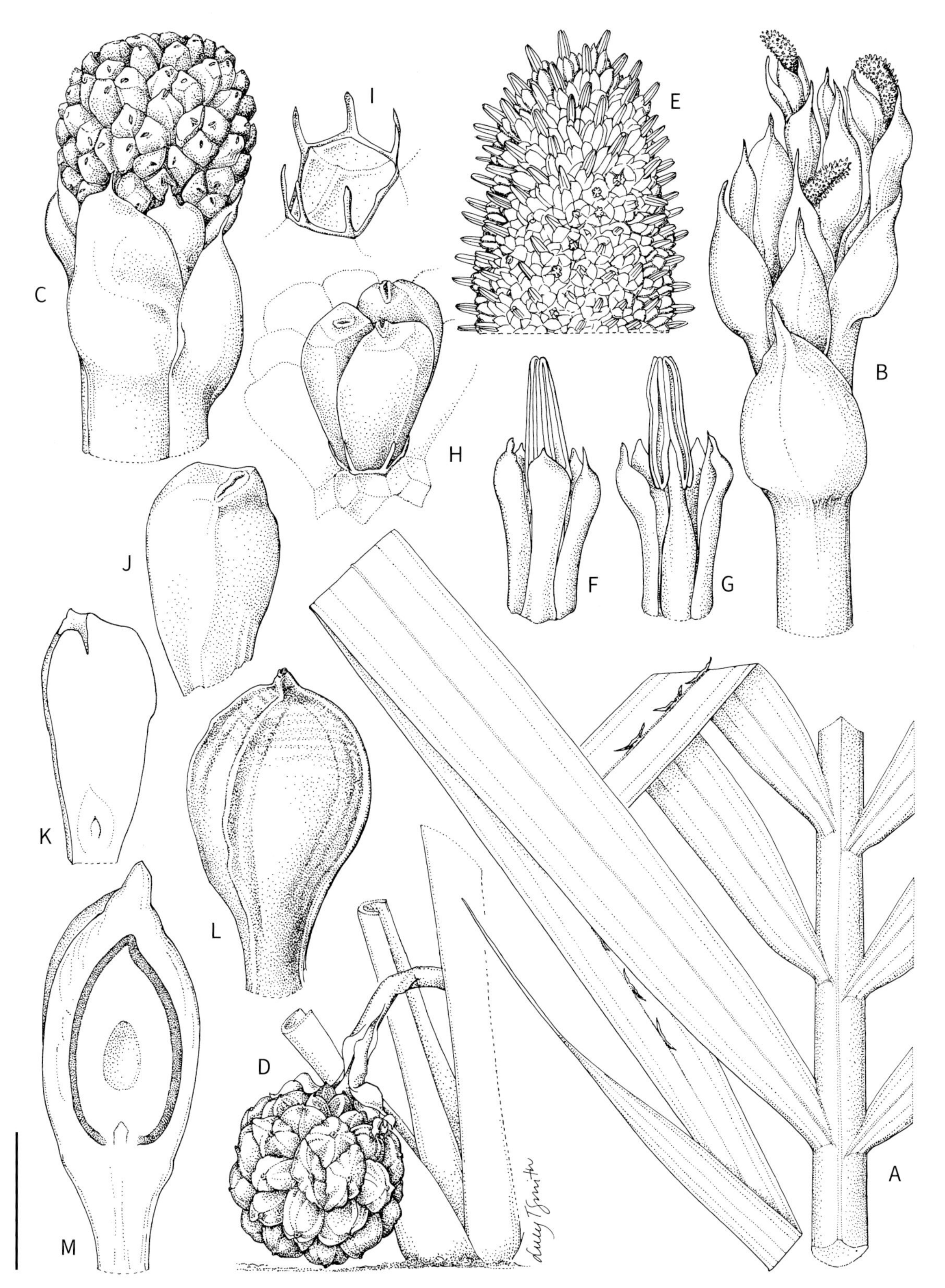

Nypa fruticans. **A.** Leaf mid-portion with ramenta. **B.** Male portion of inflorescence. **C.** Female portion of inflorescence. **D.** Inflorescence with fruit emerging from leaf bases. **E.** Male rachilla. **F**, **G.** Male flower whole and with one petal removed. **H.** Female flower. **I.** Female flower with carpels removed showing perianth. **J**, **K.** Carpel whole and in longitudinal section. **L**, **M.** Fruit whole and in longitudinal section. Scale bar: A, B = 8 cm; C = 4.5 cm; D = 30 cm; E–G = 7 mm; H = 2.5 cm; I = 1 cm; J, K = 1.5 cm; L = 6 cm; M = 4 cm. A from *Desianto 10*; B–J from *Johns et al. 9791*; L, M from *Craven & Schodde 875*. Drawn by Lucy T. Smith.

Livistona benthamii. Habit, Bensbach River (GH).

Livistona R.Br.

Synonym: *Wissmannia* Burret

Medium to tall – leaf palmate – petiole usually spiny – leaf segments not wedge-shaped

Moderate or robust, single-stemmed fan palms, hermaphrodite or dioecious. **Leaf** palmate; *sheath disintegrating into a brown fibrous network;* petiole elongate, with spiny margins; blade split to ca. one third or more of its radius, with pointed tips. **Inflorescence** *between the leaves, comprising a single axis, not divided at base,* branched 2–4 orders; primary bracts similar, tubular, few to many, not enclosing inflorescence in late bud, not dropping off as inflorescence expands; peduncle shorter than inflorescence rachis; rachillae generally slender, straight or curved. **Flowers** generally small, appearing hermaphrodite, borne in groups of up to three or solitary. **Fruit** *purplish or bluish black,* globose, ellipsoid or pyriform, stigmatic remains apical, flesh juicy or fibrous, endocarp smooth. **Seed** 1, endosperm homogeneous, with seed coat intruding into one side of the seed.

DISTRIBUTION. Twenty-eight species in north-east Africa, southern Arabia, south-east and east Asia, Malesia and Australia, with two species in New Guinea that are outliers of the Australian clade (Dowe 2010). Disjunction of the populations of species shared with Australia was most likely a result of inundation of the Australia-New Guinea land-bridge during the Holocene (Sloss *et al.* 2018).

NOTES. *Livistona,* across its entire distribution, is a variable genus of single-stemmed, short to tall, erect, hermaphrodite or dioecious fan palms. Ecological requirements are broad with species occurring in rainforest, montane forest, swamp forest, monsoon forest, coastal forest, semi-closed to open woodlands and savanna, semi-arid woodlands, and riparian and riverine forest. They grow on various soils, including limestone and peat, in permanently or seasonally wet situations, or in semi-arid to arid situations where they are usually associated with permanent ground water or at the base of cliffs where subsurface moisture is seasonally available. In New Guinea, *Livistona* species occur only in seasonal, monsoonal areas in Western and Central provinces in Papua New Guinea and the Merauke area of Indonesian New Guinea.

Until recently, *Livistona* also included species now assigned to *Saribus* (Bacon & Baker 2011). *Livistona* is distinguished by an inflorescence composed of a single axis, whilst *Saribus* usually has an inflorescence that bifurcates or trifurcates at the base, with the axes more or less similar and all sharing a common prophyll and with each axis bearing its own peduncular bract or bracts. *Livistona* has fruit that are green, blue, purple, brown or black at maturity, whereas those of *Saribus* are orange, orange-brown, red or dark violet at maturity. A monograph of *Livistona* (including *Saribus*) has been completed (Dowe 2009).

Livistona and *Saribus* may be confused with *Borassus* and *Corypha,* both of which can be distinguished by the triangular cleft at the base of the petiole. *Borassus* in New Guinea also has sharp, non-spiny petiole margins and very large fruit (>10 cm long). *Corypha* is distinguished by the inflorescences borne above the leaves and stems dying after flowering.

Key to the species of *Livistona* in New Guinea

1. Tall fan palm to 18 m; leaf segment apices pendulous; rachillae green-yellow; sepals and petals whitish–pale yellow to green; fruit globose to obovoid or pyriform 9–13 × 9–11 mm; epicarp purple-black . . . . 1. *L. benthamii*
1. Moderate fan palm to 12 m; leaf segment apices rigid; rachillae maroon to red; sepals maroon, petals yellow; fruit ellipsoid 10–12 × 8.5–10 mm.; epicarp powdery blue to bluish black . . . . 2. *L. muelleri*

1. *Livistona benthamii* F.M. Bailey

Synonyms: *Livistona holtzei* Becc., *Livistona melanocarpa* Burret

Robust, single-stemmed tree palm to 18 m. **Stem** 13–20 cm diam.; leaf scars raised; internodes grey. **Leaves** 30–50 in a globose crown; leaf blade circular, 90–160 cm diam., dark green on the upper surface, lighter green, glossy and non–waxy on the lower surface; segments 50–80, free for 60–75% of their length, depth of apical cleft 60–75% of the length of the free portion; *segment apices pendulous*, lobes acute; petiole 1.2–2 m long, 10–18 mm wide, margins with single curved black spines congested in the proximal portion. **Inflorescences** 1.2–2.1 m long; 7–9 first-order branches, branched to 2 orders; prophyll 12–22 × 4–5 × 1–2 cm, dorsiventrally compressed, with sparse silver scales; peduncular bract lacking; rachis bracts tightly sheathing, light brown, with sparse scurfy silver scales; rachillae 5–12 cm long, green-yellow, patchily pubescent. **Flowers** solitary or in clusters of 2–3; sepals triangular, 0.8–2 mm long, membranous, acute, *whitish–pale yellow to green*; petals broadly triangular, 1.0–1.3 mm long, fleshy, acute, whitish–pale yellow to green; stamens ca. 1 mm long. **Fruit** globose to obovoid or pyriform, 9–13 × 9–11 mm; *epicarp smooth, purple–black, with a powdery bloom.* **Seed** ovoid, 8–9 mm long.

DISTRIBUTION. Southern and south-eastern New Guinea. Also north Northern Territory and Cape York Peninsula, Australia.

HABITAT. Swamp forests, on alluvial flats, on the landward side of mangroves and in seasonally inundated areas, in monsoonal thickets, swamp forest and riparian forests at low elevations. Grows in large or small colonies and as scattered individuals, as a canopy emergent, sea level to 100 m.

LOCAL NAMES. None recorded.

USES. None recorded.

CONSERVATION STATUS. Least Concern (IUCN 2018).

Livistona benthamii. FROM TOP: habit, Brown River (JLD); stem with remnant leaf bases, Bensbach River (GH).

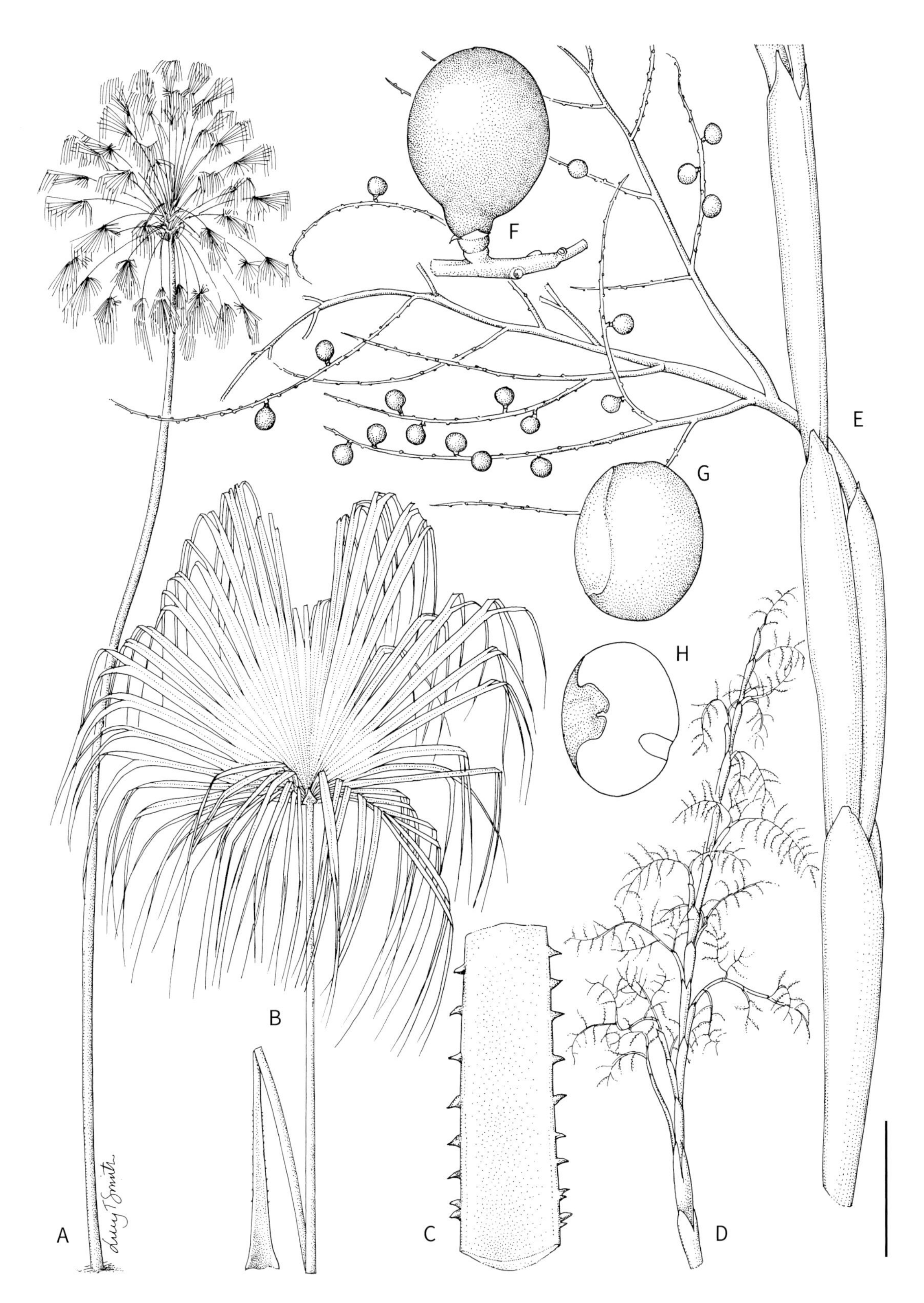

***Livistona benthamii*. A.** Habit. **B.** Leaf. **C.** Detail petiole spines. **D.** Inflorescence. **E.** Portion of inflorescence with fruit. **F.** Fruit. **G, H.** Seed whole and in longitudinal section. Scale bar: A = 2 m; B = 40 cm; C = 16 mm; D = 36 cm; E = 4 cm; F = 7 mm; G, H = 5 mm. A from photograph; B–D from *Derbyshire 685*; E–H from *Maturbongs et al. 661*. Drawn by Lucy T. Smith.

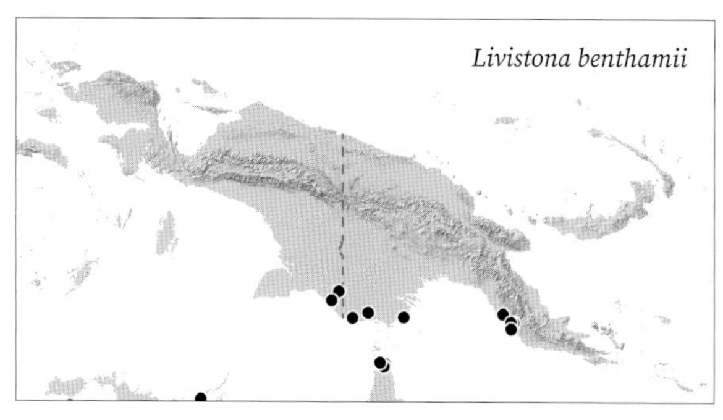

NOTES. *Livistona benthamii* is uncommon in New Guinea. It is distinguished by its relatively small leaves and the abscission of the petiole at about its midpoint. This section of erect protruding petiole may, in time, deteriorate proximally to a distinctive bulbous-rounded base. The leaf segments are pendulous or at least drooping from the base of the free segments. It bears a slender inflorescence in which the rachis bracts are tightly tubular, smooth and non-fibrous, and has small fruit.

2. *Livistona muelleri* F.M.Bailey

Synonyms: *Livistona brassii* Burret, *Livistona crustacea* Burret, *Livistona humilis* var. *novoguineensis* Becc., *Livistona humilis* var. *sclerophylla* Becc.

Moderately robust, single-stemmed tree palm to 12 m. **Stem** 15–25 cm diam.; leaf scars narrow, raised; internodes narrow, grey; petiole remnants persistent at the base, or deciduous with age or if burnt. **Leaves** 25–35 in a globose crown, held erect; leaf blade circular, 60–90 cm diam., rigid, flat, olive green to grey green on the upper surface, dull bluish green and glabrous except for a few scales on the ribs on the lower surface; *segments 48–60, rigid,* free for 50–65% of their length, depth of apical cleft 5–14% of the length of the free portion; apical lobes acute, rigid; petiole 0.7–1 m long, 1.4–2 cm wide, margins with single curved reddish to black spines 0.2–1.2 cm long throughout, larger and closer spaced in the proximal portion; both the upper and lower surfaces with rows of persistent corky scales, at first red-brown aging to grey. **Inflorescences** 0.8–1.6 m long; first-order branches 5–10, branched to 4 orders; peduncular bract lacking; rachis bracts loosely tubular, with silver scales, splitting and disintegrating with age; rachillae 2–13 cm long, papillose, *maroon to red*. **Flowers** solitary or in clusters of 2–3, 1.3–1.6 mm long in bud; *sepals broadly triangular, 0.8–1 mm long, maroon, fleshy, cuspidate; petals ovate, 1.3–1.6 mm long, subacute, yellow*; stamens ca 1.4 mm long, yellow. **Fruit** ellipsoid, 10–12 × 8.5–10 mm; *epicarp smooth, powdery blue or bluish black*. Seed globose, 8–9 mm diam.

Livistona muelleri. Habit, Bensbach River (GH).

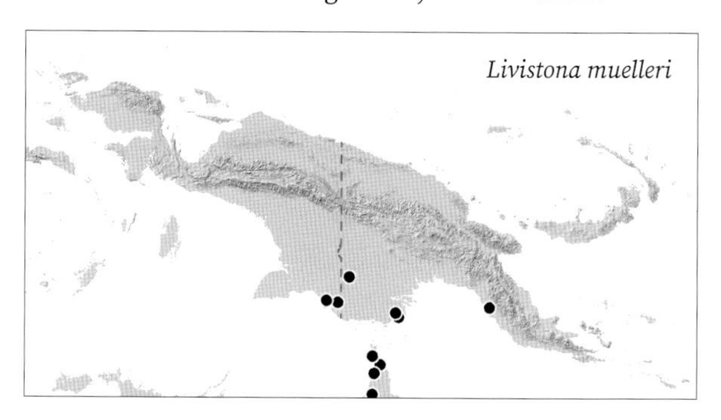

DISTRIBUTION. Southern and south-eastern New Guinea. Also Cape York Peninsula, Australia.

HABITAT. In grassy open forest, woodland, moist sclerophyll forest, and less commonly on the margins of vine thickets, sea level to 300 m, and most common in areas that have a strongly seasonal rainfall pattern.

LOCAL NAMES. None recorded.

USES. Leaves infrequently used as thatch for simple temporary huts. Young shoot (heart-of-palm) edible. Young leaves placed on the back of the dancer as a tassel while performing a traditional dance (Merauke).

CONSERVATION STATUS. Least Concern.

NOTES: *Livistona muelleri* is a moderate sub-canopy palm. The leaves are regularly segmented with the segment apices rigid. The inflorescence bracts are loosely sheathing. The colour combination of the rachillae and flowers is very distinctive. Flowers have maroon sepals, yellow petals and mauve carpels, and the rachillae are maroon to red. Fruit are ellipsoid, pyriform, to obovoid, and powdery blue, reddish black to bluish black at maturity.

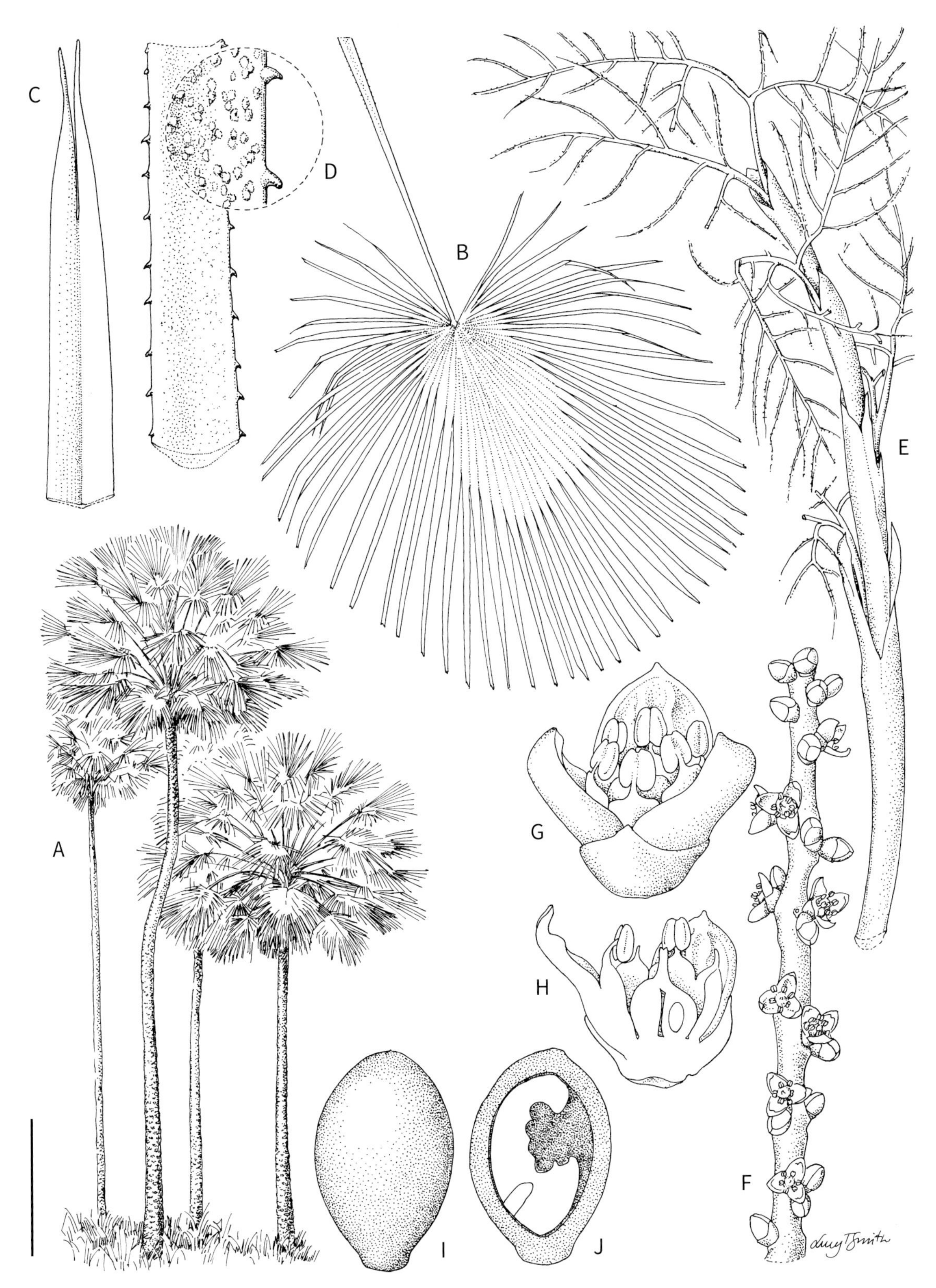

***Livistona muelleri*. A.** Habit. **B.** Leaf. **C.** Leaflet tip. **D.** Petiole with detail showing scales. **E.** Portion of inflorescence. **F.** Rachilla with flowers. **G, H.** Flower whole and in longitudinal section. **I, J.** Fruit whole and in longitudinal section. Scale bar: A = 22 cm; B = 25 cm; C = 2 cm; D, E = 8 cm; F = 7 mm; G, H = 1.3 mm; I, J = 1 cm. A, B from *Brass 7668*; C from *Maturbongs et al. 659*; D, E from *Rodd 3170*; F–J from *Dowe s.n.* Drawn by Lucy T. Smith.

Licuala lauterbachii subsp. *lauterbachii*. Habit, Biak Island (WB).

Licuala Wurmb..

Synonyms: *Dammera* K.Schum. & Lauterb., *Pericycla* Blume

Small to medium – leaf palmate – petiole usually spiny – leaf segments wedge-shaped

Under- to mid-storey, single- or multi-stemmed fan palms, some species stemless, hermaphrodite (rarely dioecious). **Leaf** palmate; sheath disintegrating into a brown fibrous network; petiole long, usually with spines along the margins; *blade bifid or divided to base into 3–25 wedge-shaped segments, with toothed or lobed tips,* sometimes of very varying width, horizontal. **Inflorescence** *between the leaves,* unbranched or branched 1–3 orders; first-order branches one to many, decreasing in size distally, usually subtended by tubular bracts thereby appearing as distinct units; rachillae usually slender. **Flowers** variable in size, solitary or in groups of 2–3; stamens adnate to corolla at the base, distally united into a staminal ring, with anthers inserted in one level (uniseriate) or two levels (biseriate). **Fruit** variable in size, red to brown, globose to ovoid or ellipsoid; stigmatic remains apical, flesh juicy or fibrous, endocarp smooth or sometimes heavily ridged or grooved. **Seed** 1, endosperm homogeneous, with seed coat often intruding into one side of the seed.

DISTRIBUTION. Around 130 species from South China to India, through South-East Asia to Vanuatu and Australia, with 25 species in New Guinea. The three main centres of species diversity are Borneo, Malay Peninsula and New Guinea.

NOTES. One of the most species-rich palm genera in New Guinea. Its palmate leaf, which splits to the base to form wedge-shaped segments, is highly distinctive and is not found in any other New Guinea genus. *Licuala* species are found in moist forest all over New Guinea. A few species (e.g. *Licuala graminifolia*) occur in savanna-like habitats. The highest diversity is found in the central part of the island especially in the Bewani and Cyclops mountain ranges. Most species of *Licuala* are sensitive to disturbance and thus good indicators of human impact.

The fruits of a number of species such as *L. grandiflora*, *L. insignis*, *L. longispadix* and *L. simplex* are unusual in having thick endocarps reinforced with longitudinal ridges. These fruits are the largest within the genus and in the last two species they are presented near or on the forest floor. These attributes are probably an adaptation to dispersal by ground-dwelling animals such as cassowaries, which are keystone species in the forests of New Guinea and are known to consume palm fruits (Pangau-Adam & Mühlenberg 2014). Whether the species with ridged endocarps constitute a single lineage or whether the character complex has arisen several times independently deserves further investigation.

Licuala lacks a recent monograph. Prior to this account, New Guinea species were mainly described in two bursts, with 14 species published by Beccari between 1877 and 1921 (summarised in Beccari 1931) and a further 13 species by Burret between 1933 and 1941 (e.g. Burret 1941). Many of these species, especially those of Burret, are regarded as synonyms here. In preparation of this account, we have described a series of species new to science (Barfod 2000, Banka & Barfod 2004, Heatubun & Barfod 2008, Barfod & Heatubun 2009, Barfod & Baker 2022, Barfod & Heatubun 2022).

Key to the species of *Licuala* in New Guinea

1. All leaves entire and bifid. . . . . **6. *L. bifida***
1. At least some leaves divided into three or more segments 2

2. Mid-segment of leaf bifid. 3
2. Mid-segment of leaf not bifid. 6

3. Leaf blade <25 cm diam . . . . **5. *L. bellatula***
3. Leaf blade >25 cm diam 4

4. Rachis bracts inflated, first-order branch partly contained within rachis bract at anthesis 5
4. Rachis bracts not inflated, first-order branch fully exposed at anthesis. . . . . **24. *L. telifera***

5. Flower pedicels <1 mm . . . . **25. *L. urciflora***
5. Flower pedicels >1 mm . . . . **1. *L. anomala***

6. Inflorescence >3 m long. . . . . **16. *L. longispadix***
6. Inflorescence <3 m long. 7

7. Inflorescence unbranched or, if branched, then only 1 rachilla fully developed. 8
7. Inflorescence branched to one or more orders, with >1 fully developed rachilla 9

8. Flowers in groups of 2–3, >4 mm long, floral bract inconspicuous. . . . . **12. *L. grandiflora***
8. Flowers mostly solitary, <4 mm long, floral bract conspicuous. . . . . **11. *L. graminifolia***

9. The bracts subtending first-order branches missing or not tubular 10
9. The bracts subtending first-order branches present and tubular 11

10. First-order branches <5; flowers >10 mm long; fruit >1.5 cm long, endocarp grooved. . **22. *L. simplex***
10. First-order branches >10; flowers <10 mm long; fruit <1.5 cm long, endocarp smooth . . . . **19. *L. parviflora***

11. Flowers >15 mm long. . . . . **13. *L. heatubunii***
11. Flowers <15 mm long. 12

12. Flower pedicel >1 mm 13
12. Flower pedicel <1 mm or lacking 20

13. Fruits >3 cm long . . . . **14. *L. insignis***
13. Fruit <3 cm long 14

14. Multi-stemmed understorey palm <2 m tall; inflorescence with <5 first-order branches. 15
14. Single-stemmed tree palms >2 m tall; inflorescence with >5 first-order branches 16

15. Leaf mid-segment not wider than the rest; rachilla with paired flowers to above the middle, pedicel not red, calyx breaking up irregularly, almost truncate, stamens uniseriate . . . . **2. *L. bacularia***
15. Leaf mid-segment wider than the rest; rachilla with mostly solitary flowers, pedicel red, calyx cleanly divided to ⅓–½, stamens biseriate. . . . . **8. *L. coccinisedes***

16. Inflorescence axis zigzag towards the apex, rachis bracts <7, loosely sheathing to inflated 17
16. Inflorescence axis straight throughout, rachis bracts >7, tightly sheathing. 18

17. Inflorescence glabrous to sub-glabrous, rachis bracts splitting neatly, calyx loosely fitting .... 7. *L. brevicalyx*

17. Inflorescence covered with rust-coloured hairs, rachis bracts splitting into fibrous mesh, calyx tightly fitting .... 4. *L. bankae*

18. Flowers yellow, longest pedicels longer than flower, peduncular bracts 0 .... 20. *L. penduliflora*

18. Flowers white, longest pedicels shorter than flower, peduncular bracts 1–2 19

19. Calyx divided to one third of its length; fruit mesocarp ca. 1 mm thick, endocarp smooth .... 18. *L. multibracteata*

19. Calyx divided halfway or more; fruit mesocarp >2 mm thick, endocarp shallowly furrowed. .... 23. *L. suprafolia*

20. Flowers globose, <2 mm long, stamens uniseriate .... 3. *L. bakeri*

20. Flowers bullet-shaped, >2 mm long, stamens biseriate 21

21. Leaf mid-segment not noticeably wider than the rest .... 17. *L. montana*

21. Leaf mid-segment noticeably wider than the rest 22

22. Calyx splitting regularly, with well-defined lobes. 23

22. Calyx splitting irregularly, with ill-defined lobes 24

23. Petiole armed basally; calyx sparsely pubescent, splitting to less than $^{1}/_{10}$ .... 9. *L. essigii*

23. Petiole armed almost throughout; calyx glabrous, splitting to more than $^{1}/_{5}$ .... 10. *L. flexuosa*

24. Aborted flowers not persistent after anthesis .... 21 *L. sandsiana*

24. Aborted flowers persistent after anthesis 25

25. Basal first-order branch of inflorescence with >15 rachillae, New Guinea mainland .... 15a. *L. lauterbachii* **subsp.** *lauterbachii*

25. Basal first-order branch of inflorescence with <15, rachillae, New Britain, New Ireland .... 15b. *L. lauterbachii* **subsp.** *peekelii*

1. *Licuala anomala* Becc.

Small, single-stemmed palm to 2 m. **Stem** ca. 3 cm diam. **Leaf** sheath distally detached from petiole in 2 strap-like extensions to 40 cm long; petiole 80–100 cm long, armed on lower third; *blade divided into (3–)5–7 segments, uneven in width,* mid-segment 40–55 cm long, truncate or rounded, *variably split apically even within the same individual,* 10–30 cm wide at the apex. **Inflorescence** 50–70 cm long, branched to 2–3 orders, erect with 4–5 first-order branches; peduncle 15–30 cm long, contained in prophyll; peduncular bract lacking; rachis zigzag*; rachis bracts inflated; first-order branches partly contained at anthesis, erect*, the proximal ones with 3–4 cm long main axis, bearing 5–7, to 9 cm long, thick rachillae. **Flowers** densely inserted in clusters of up to three, *borne on 1–1.5 mm long pedicels,* bullet-shaped to rounded, 4–4.5 mm long; calyx ca. 2 mm long, bell-shaped with stalked base, splitting ca. halfway into 3 lobes, *corolla ca. 3 mm long, yellow,* fleshy; *stamens biseriate,* filaments as long as anthers; ovary obovate ca. 1 mm long; style 1–1.2 mm long. **Fruit** globose, *12–14 mm across,* mesocarp 1–2 mm thick, endocarp thin, smooth. **Seed** 7–8 mm diam.

DISTRIBUTION. From the Bird's Head Peninsula to south-western New Guinea.

HABITAT. In hill forest and on river banks at sea level to 750 m.

LOCAL NAMES. *Beimeji* (Sougb), *Sifenete* (Sumuri).

USES. Tobacco pipes and arrow heads are carved

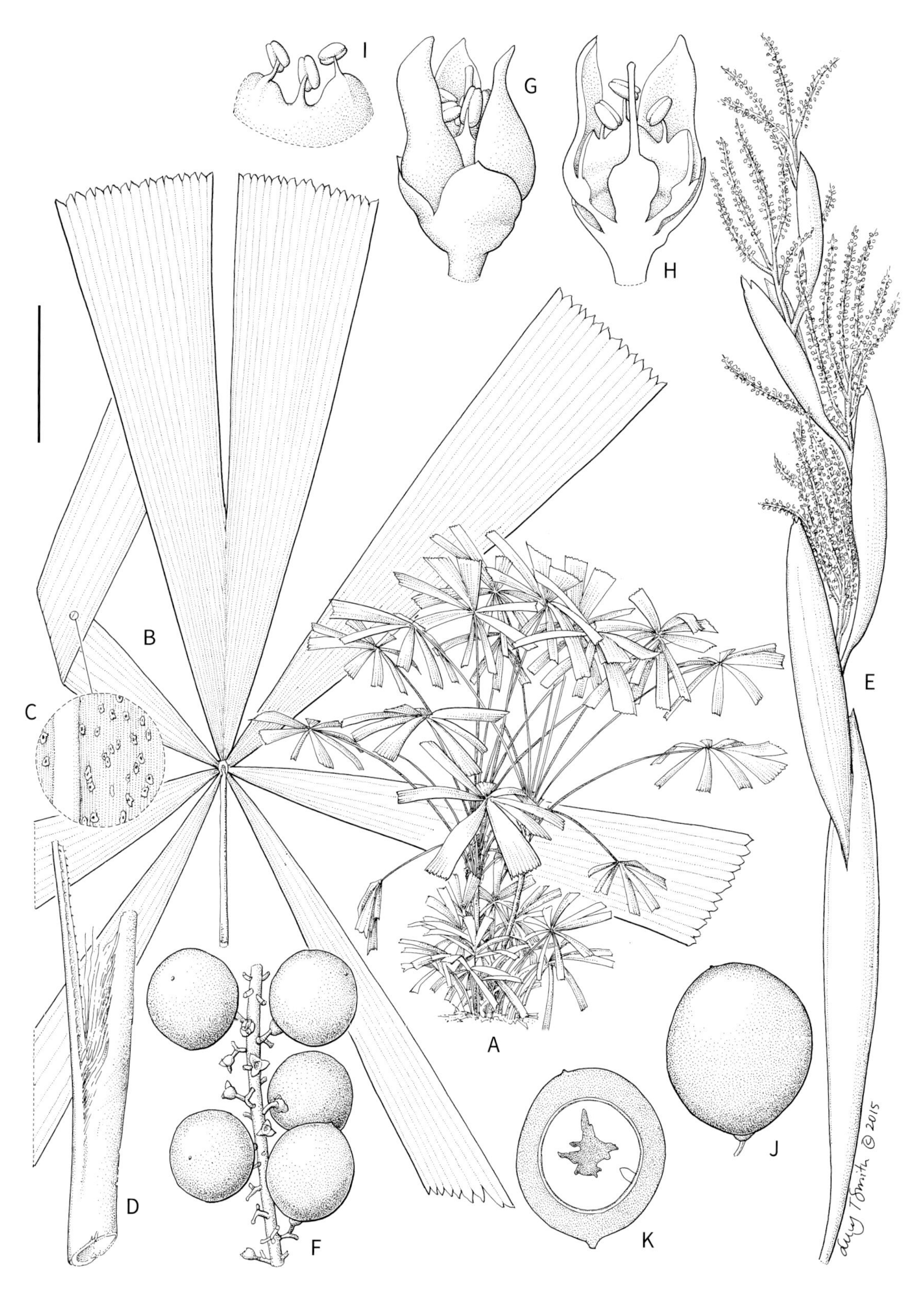

***Licuala anomala*.** **A.** Habit. **B.** Leaf. **C.** Detail scales abaxial leaf surface. **D.** Petiole and leaf sheath. **E.** Inflorescence. **F.** Portion of rachilla with fruit. **G**, **H.** Flower whole and in longitudinal section. **I.** Staminal ring. **J**, **K.** Fruit whole and in longitudinal section. Scale bar: A = 54 cm; B, D = 12 cm; C = 1.5 mm; E = 4 cm; F–I = 2 mm; J, K = 1 cm. A from photograph; B–E from *Heatubun et al. 1038*; F–K from *Maturbongs et al. 725*. Drawn by Lucy T. Smith.

from the stem. Leaves often used for thatching temporary shelters.

CONSERVATION STATUS. Near Threatened. *Licuala anomala* has a relatively restricted distribution. Deforestation due to mining, oil palm plantations and logging concessions is a major threat within its distribution.

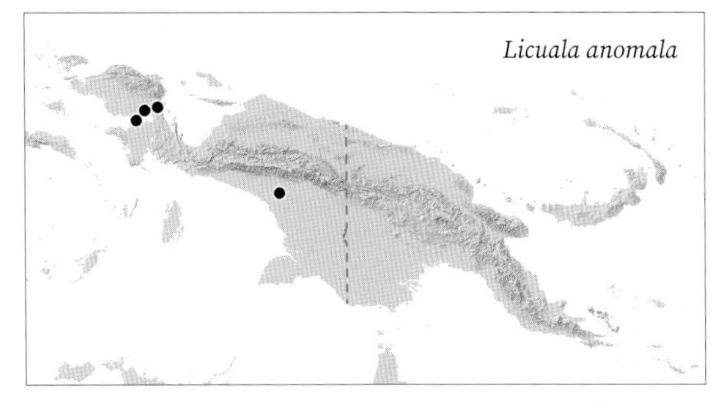

NOTES. First collected in 1907, under the Dutch expedition to the Lorentz River in Indonesian New Guinea, *Licuala anomala* has recently been recollected near Bintuni in the south-eastern part of the Bird's head Peninsula. It is superficially similar to *L. urciflora* from Waigeo Island, but characterised by having clearly pedicellate, rounded flowers, with a membranous, bell-shaped and basally stalked calyx. The split mid-segment and the pedicellate flowers reveal its affinity to the *L. telifera* complex (*L. anomala, L. bellatula, L. bifida, L. telifera, L. urciflora*). The former feature, however, is not constant and even within the same plant various types of splitting may be observed. The leaf of the type collection is aberrant and may represent a juvenile leaf.

2. *Licuala bacularia* Becc.

Synonym: *Licuala debilis* Becc.

Multi-stemmed palm, usually less than 2 m. **Stem** ca. 3 cm diam. **Leaf** sheath distally detached from petiole in 2 strap-like extensions ca. 10 cm long; petiole 90–120 cm long, armed on basal 15–20 cm; blade divisions age-dependent from 3–5 segments in young shoots to 15 in mature shoots, *splitting of mid-segment variable but mid-segment usually not conspicuously wider than other segments*; mid-segment 55–60 cm long, sides often slightly curved, 3–8 cm wide at the apex, truncate or oblique. **Inflorescence** 60–100 cm long, branched to 2 orders, erect with *3–4 first-order branches*; peduncle

Licuala bacularia. Inflorescence, near Sorong (WB).

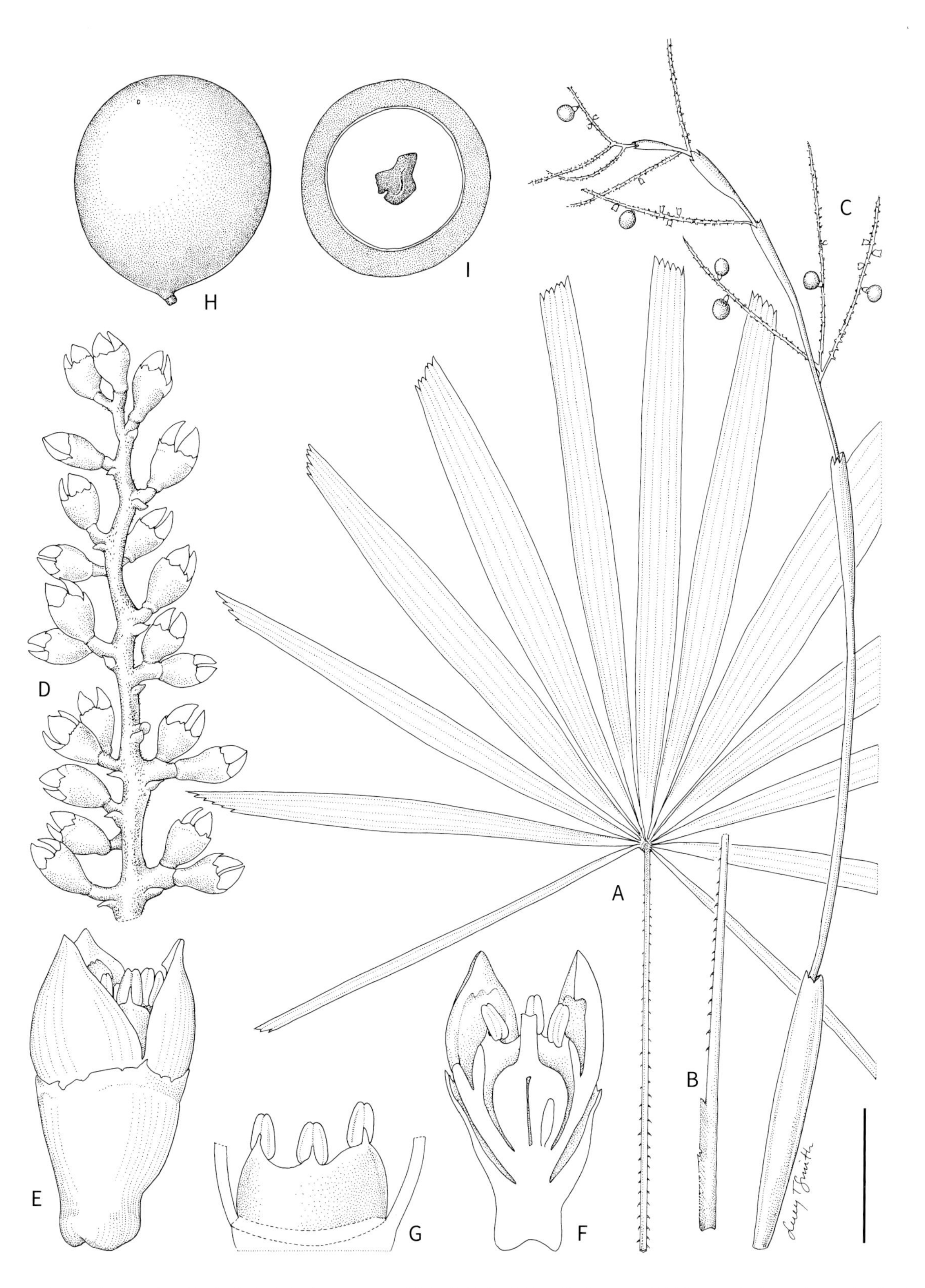

***Licuala bacularia*. A.** Leaf. **B.** Petiole. **C.** Inflorescence with fruit. **D.** Rachilla with flowers. **E, F.** Flower whole and in longitudinal section. **G.** Staminal ring. **H, I.** Fruit whole and in transverse section. Scale bar: A, B = 8 cm; C = 6 cm; D = 3.3 mm; E, F = 1.2 mm; G = 1 mm; H, I = 7 mm. All from *van Royen 3009*. Drawn by Lucy T. Smith.

Licuala bacularia. Habit, near Sorong (WB).

25–35 cm long; peduncular bract lacking; rachis more or less straight; first-order branch erect, the proximal one with 2–4 cm main axis, bearing 3–12 rachillae, these to 10 cm long, spreading and covered with scattered to dense tomentum. **Flowers** paired to solitary above the middle of the rachilla, borne on 1 mm or longer pedicel, bullet-shaped, 3–3.5 mm long; *calyx ca. 2 mm long, truncate, splitting irregularly*; stamens fused to corolla for 1.1–1.3 mm, staminal ring ca. 0.8 mm high; *stamens uniseriate*; ovary obovate, ca. 1 mm long; style ca. 0.5 mm long. **Fruit** globose, 8–10 mm across, orange at maturity, mesocarp ca. 1 mm thick, *endocarp thin, smooth*. **Seed** 6–8 mm diam.

DISTRIBUTION. From the Bird's Head Peninsula to south-western New Guinea.

HABITAT. Typically found in light gaps in lowland forest at sea level to 600 m.

LOCAL NAMES. *Kempyar* (Sayal).

USES. The stems are used as spears.

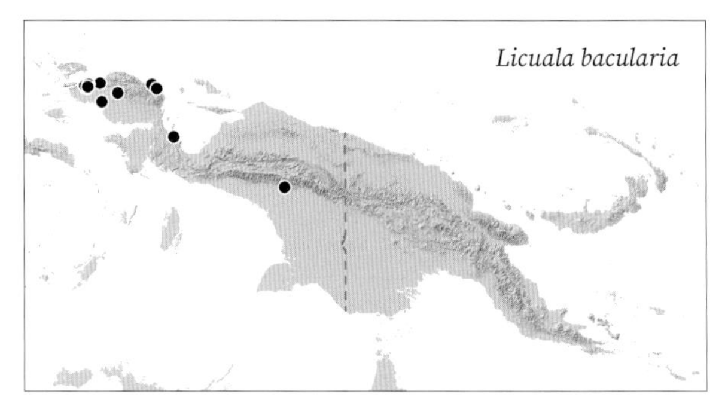

CONSERVATION STATUS. Least Concern. *Licuala bacularia* is known to be locally common but monitoring is required due to mining, oil palm plantations and logging concessions within its range.

NOTES. This species stands out in having a combination of a mid-segment that is not bifid and pedicellate flowers. It is difficult to identify this species in a sterile condition due to variation in the relative width of the leaf segments.

3. *Licuala bakeri* Barfod & Heatubun

Single-stemmed, understorey palm *to 1–2 m*. **Stem** ca. 3 cm diam. **Leaf** sheath breaking up early into a fine, brown, fibrous mesh; petiole 60–90 cm long, *unarmed or with minute spines at the very base*; *blade divided into 31–45 segments*, segments 30–35 cm long, truncate, increasing in apical width from 4 mm to 12 mm in mid-segment. **Inflorescence** *35–45 cm long*, erect to curved, branched to 2 orders, with 5–6 first-order branches; peduncle 10–12 cm long; peduncular bracts lacking; rachis erect to curved, straight to somewhat zigzag; proximal first-order branches with 2–9 cm long main axis, bearing 8–25 rachillae, these 3–9 cm long, spreading, sparsely pubescent to glabrous. **Flowers** *solitary throughout, subsessile*, elliptic to obovate, *1.5–2 mm long*; calyx ca. 1 mm long, breaking up regularly half way into 3, apically obtuse lobes; *corolla noticeably longer than calyx*, stamens fused to corolla for 0.4–0.5 mm, staminal ring ca. 0.2 mm high; *stamens uniseriate*; ovary globose, 0.5–0.6 mm diam.; style ca. 0.5 mm long. **Fruit** globose, 8–10 mm across, mesocarp ca. 1 mm thick, *endocarp brittle, smooth*. **Seed** 6–8 mm diam.

DISTRIBUTION. Western New Guinea including Yapen Island.

HABITAT. In light-open places, such as ridgetops and heath-like vegetation, at 250–950 m.

LOCAL NAMES. *Ansuni* (Yapen).

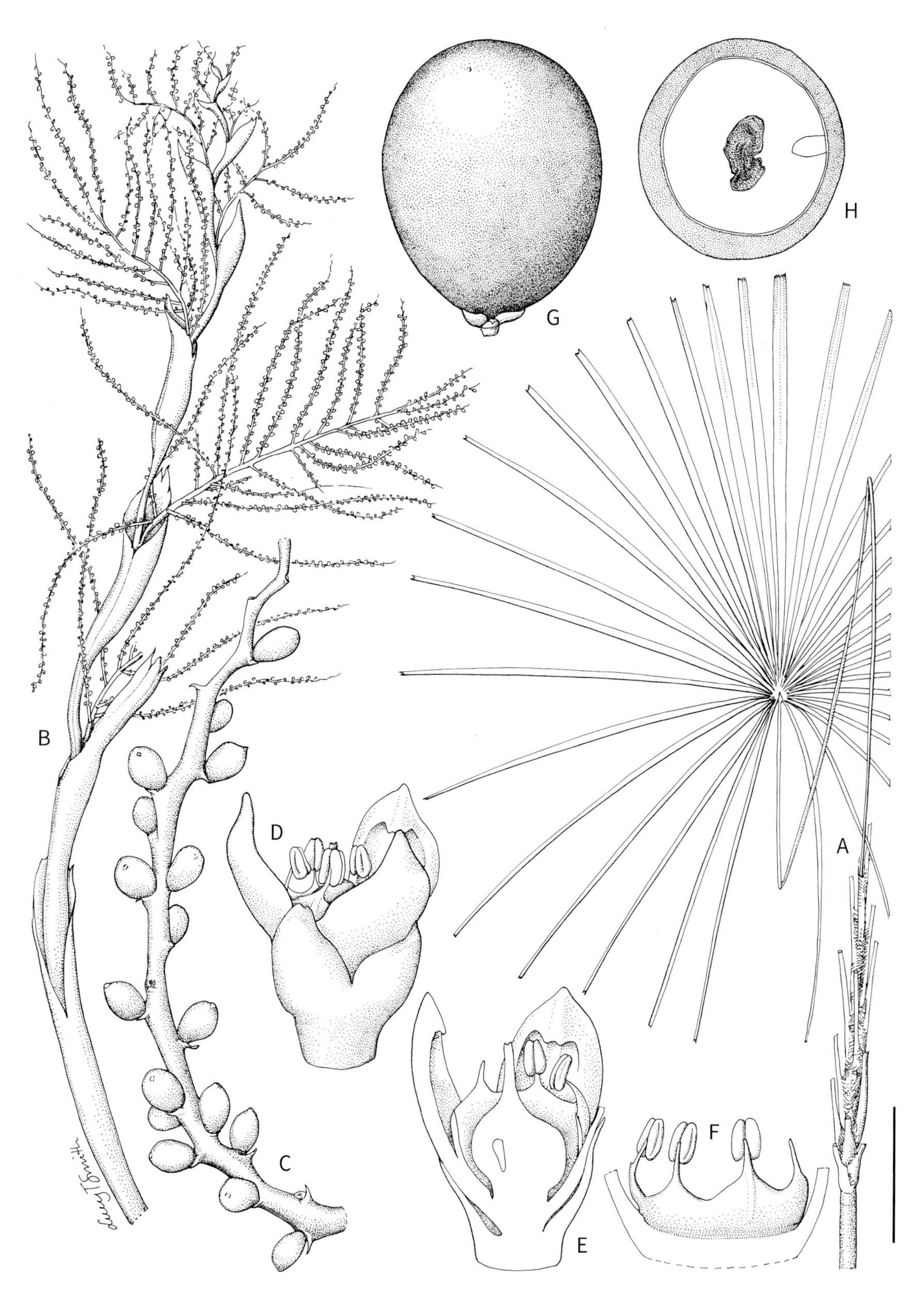

Licuala bakeri. **A.** Stem with leaf. **B.** Inflorescence. **C.** Rachilla with fruit. **D**, **E.** Flower whole and in longitudinal section. **F.** Staminal ring. **G**, **H.** Fruit whole and in transverse section. Scale bar: A = 12 cm; B = 4 cm; C = 5 mm; D, E = 1.2 mm; F = 1.1 mm; G, H = 6.3 mm. All from *Baker et al. 1059*. Drawn by Lucy T. Smith.

Licuala bakeri. LEFT TO RIGHT: habit; inflorescence with fruit. Wandamen Peninsula (WB).

USES. The stem is used for arrows.

CONSERVATION STATUS. Near Threatened. *Licuala bakeri* has a relatively restricted distribution. Deforestation due to mining concessions is a major threat in its distribution range.

NOTES. *Licuala bakeri* is highly distinctive with its unarmed petioles, finely divided leaves and delicate inflorescence. It has the smallest flowers recorded in the genus.

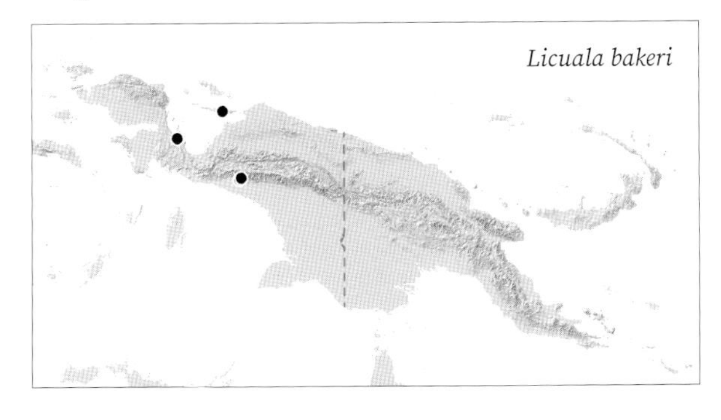

4. *Licuala bankae* Barfod & Heatubun

Single-stemmed palm to 7 m. **Stem** ca. 10 cm diam. **Leaf** sheath 30–40 cm long, disintegrating into a fibrous mesh; petiole variable in length, to 1.8 m long in fully developed leaves, *covered by patches of rust-coloured, woolly tomentum*, armed in lower third; blade divided into 17–19 segments; mid-segment 65–70 cm long, 18–20 cm wide and rounded apically. **Inflorescence** 1–1.2 m long, branched to 2 orders, *with 10–11 first-order branches, curved at anthesis and zigzag*, contained within the crown at anthesis, pendent in fruit; peduncle 40–50 cm long, contained in prophyll; *peduncular bract 1*; rachis 70–80 cm long; *rachis bracts fibrous, loosely sheathing*, proximal first-order branch with ca. 30 cm long main axis, bearing ca. 40 rachillae, *both rachis and rachillae covered in patches of rust-coloured hairs*. **Flowers** *solitary*, 4–4.5 mm long, borne on 1–2.5 mm pedicels with rust-coloured hairs; calyx bell-shaped, ca. 2.5 mm long, glabrous, greenish with

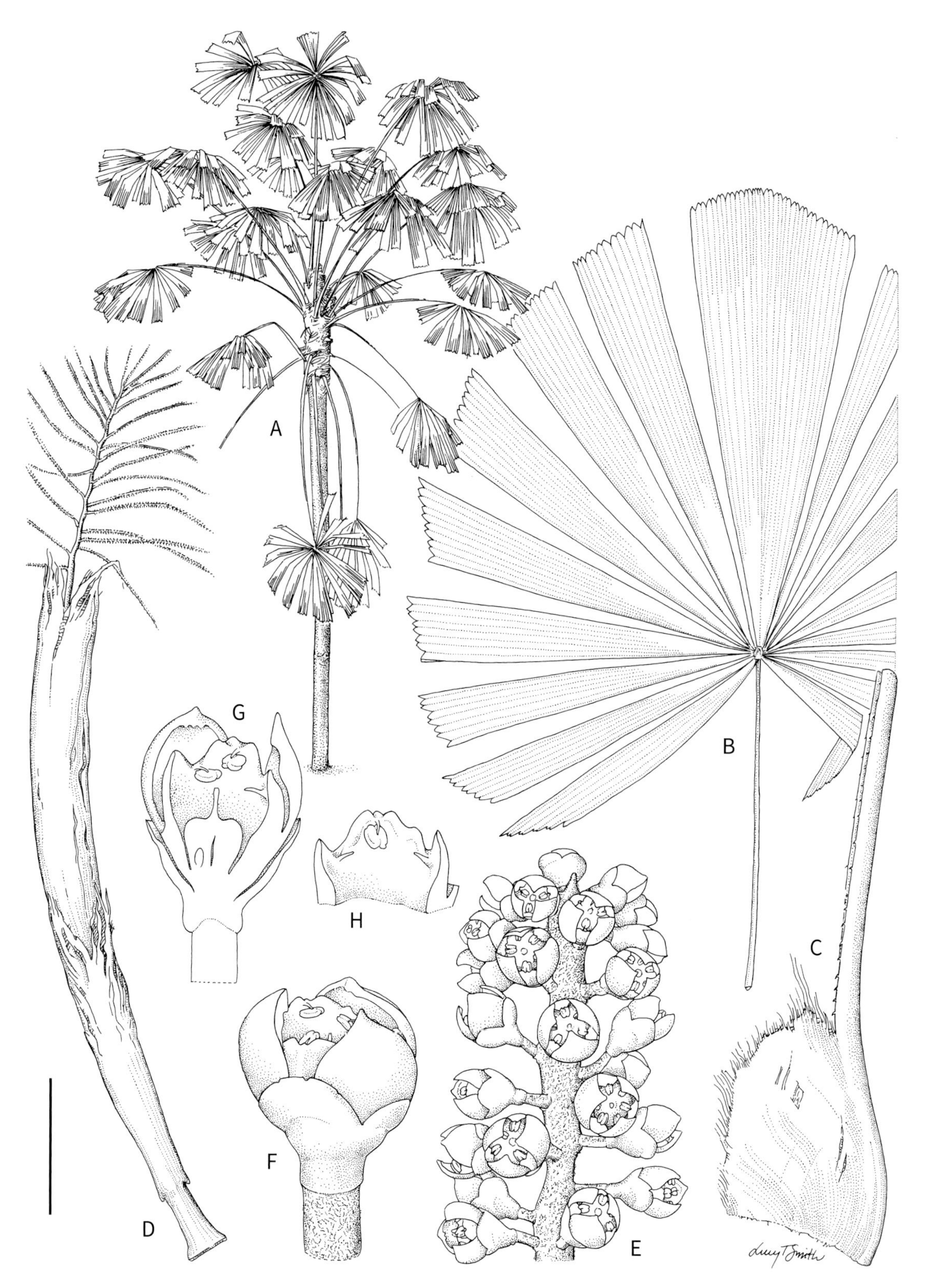

Licuala bankae. **A.** Habit. **B.** Leaf. **C.** Leaf sheath. **D.** Portion of inflorescence. **E.** Portion of rachilla with flowers. **F, G.** Flower whole and in longitudinal section. **H.** Staminal ring. Scale bar: A = 1.3 m; B = 18 cm; C, D = 12 cm; E = 7 mm; F–H = 2.5 mm. All from *Barfod et al. 449*. Drawn by Lucy T. Smith.

Licuala bankae. LEFT TO RIGHT: habit; flowers. Alotau (AB).

brown margins, breaking up regularly into 3, ca. 1 mm long, *rounded lobes; corolla yellow,*
ca. 4 mm long; *stamens biseriate*; ovary glabrous, 0.8–1 mm long, truncate to rounded apically; style 0.5–0.6 mm long. **Fruit** not seen. **Seed** not seen.

DISTRIBUTION. Far south-eastern New Guinea.

HABITAT. Lowland forest at sea level.

LOCAL NAMES. None recorded.

USES. None recorded.

CONSERVATION STATUS. Critically Endangered. *Licuala bankae* is only known from one site, where it was locally common along the fringes of land cleared for oil palm plantation.

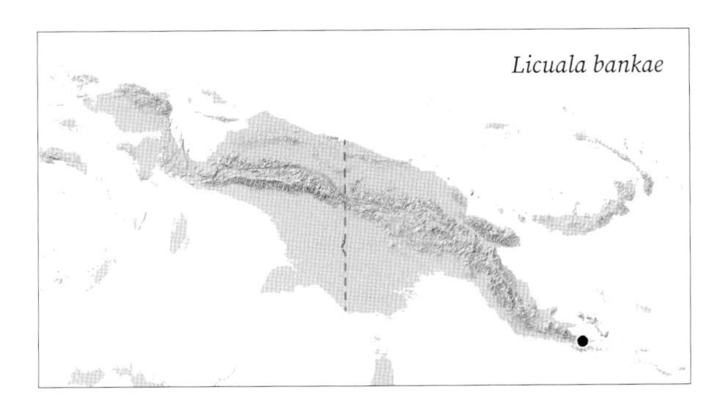

NOTES. The species belongs to the *L. penduliflora* complex (*L. bankae, L. brevicalyx, L. penduliflora, L. multibracteata, L. suprafolia*). It is distinguished from the other species in this group by having fibrous and loosely sheathing rachis bracts, rachillae and pedicels with rust-coloured hairs, and rounded calyx lobes.

5. *Licuala bellatula* Becc.

Single- or multi-stemmed understorey palm, 1–2 m. **Stem** 7–12 mm diam. **Leaf** sheath 8–15 cm long, brown, breaking up in fine fibrous mesh, sometimes detached distally from petiole in 2 strap-like extensions 20–25 cm long; petiole 15–25 cm long, unarmed to armed in lower ¼–½; blade divided into 3–5 segments, underneath with *minute rust-coloured scales and hairs* and woolly indumentum along major veins; *mid-segment 10–15 cm long, deeply split to usually ⅔ of length,* the 2 lobes equal in width and slanted to truncate. **Inflorescence** *25–35 cm long,* elongating after anthesis to 30–40 cm long, branched to 2 orders, erect with

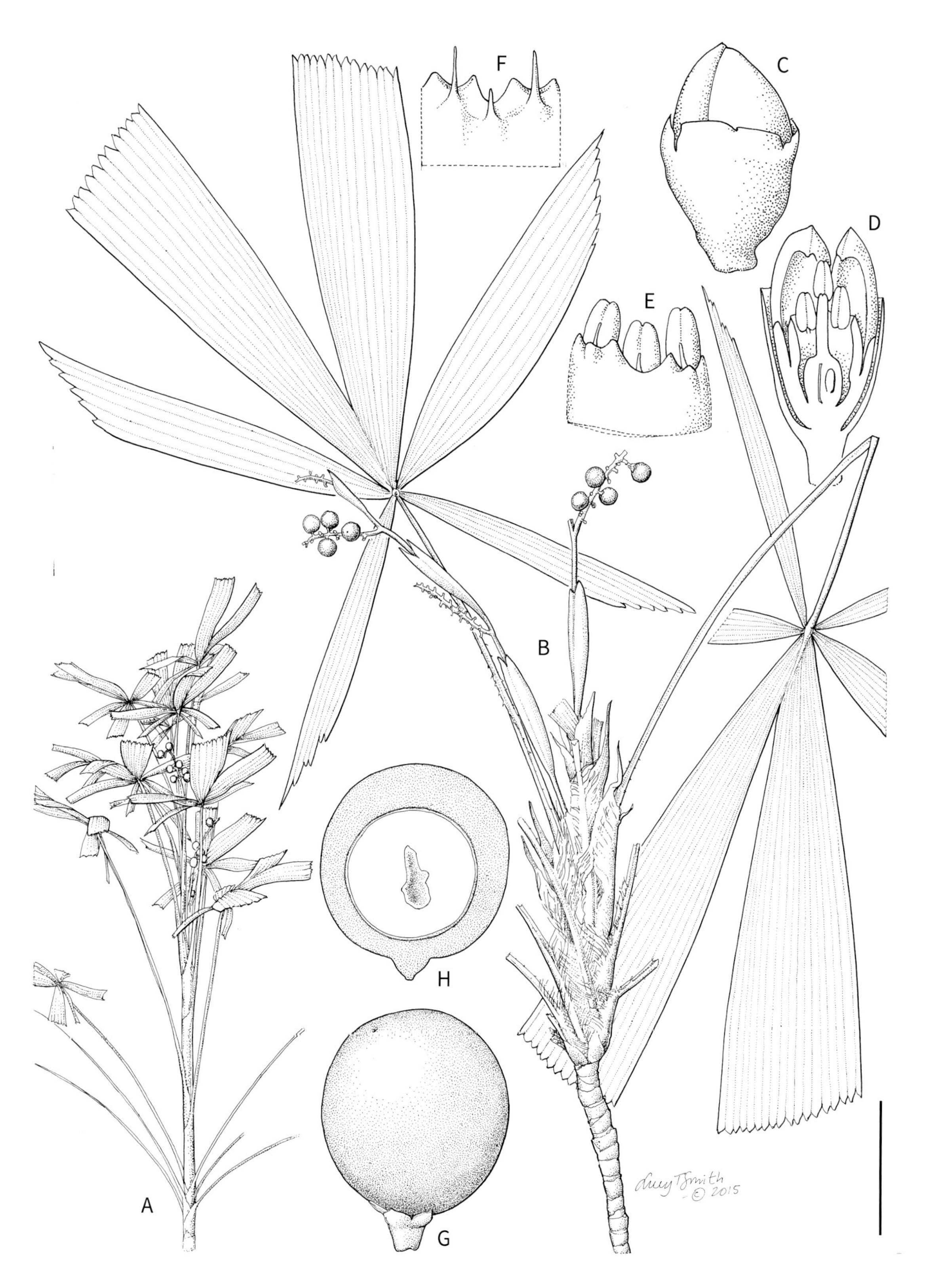

Licuala bellatula. **A.** Habit. **B.** Stem with leaf sheath, leaves and inflorescence with fruit. **C, D.** Flower whole and in longitudinal section. **E.** Staminal ring. **F.** Staminal ring inner surface and filaments. **G, H.** Fruit whole and in longitudinal section. Scale bar: A = 30 cm; B = 4 cm; C, D = 1.6 mm; E, F = 1 mm; G, H = 7.5 mm. All from *Baker et al. 1054*. Drawn by Lucy T. Smith.

Licuala bellatula. Habit, Wandamen Peninsula (WB).

3–5 first-order branches; peduncle 7–15 cm long, not exposed at anthesis; peduncular bract lacking; rachis straight to slightly curved; proximal first-order branch, with 1–2 cm peduncle, erect, spicate or bearing to 5 rachillae, these 2–4 cm long, minutely pubescent, erect. **Flowers** solitary or paired basally on the rachilla, pedicel 0.6–0.8 mm long; calyx bell-shaped 1.3–1.5 mm long, striate when dry, with *rust-coloured hairs at the base, breaking up irregularly* to almost truncate; corolla 2.4–2.6 mm long, *pale cream*, lobes 1.5–2 mm long; *stamens biseriate*; style 1.3–1.5 mm long. **Fruit** globose, *12–14 mm diam.*, red, *mesocarp 3–4 mm thick, endocarp thin, smooth*. **Seed** 6–8 mm diam.

DISTRIBUTION. North-western to central New Guinea.

HABITAT. Lowland forest on well-drained soil at sea level to 650 m.

LOCAL NAMES. *Tengir* (Wandamen).

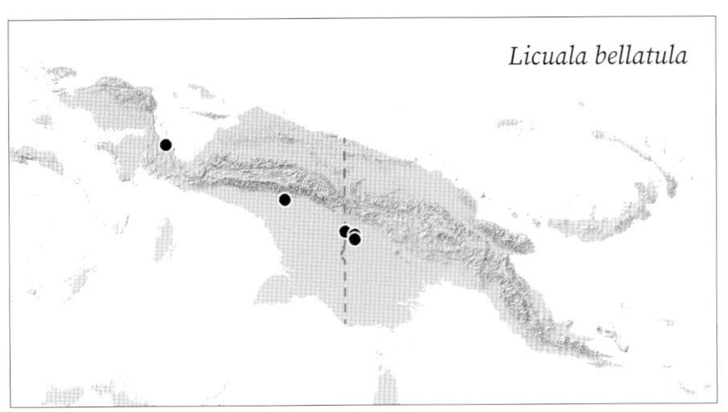

USES. None recorded.

CONSERVATION STATUS. Near Threatened. *Licuala bellatula* has a restricted distribution and parts of its distribution range are threatened by logging and mining concessions. However, much of its range is not currently at risk.

NOTES. The leaves of this species are among the smallest in the genus. The fruits were not available to Beccari when he originally described this species. They appear disproportionately large relative to the moderate size of the palm. This species belongs to the *L. telifera* complex (*L. anomala, L. bellatula, L. bifida, L. telifera, L. urciflora*). It is characterised by the combination of small leaves with a deeply bifid mid-segment and large fruits.

6. *Licuala bifida* Heatubun & Barfod

Single-stemmed palm to 2 m. **Stem** to 3 cm diam. **Leaf** petiole 13–30 cm long, armed; *blade entire, diamond-shaped, bifid*, 80–110 cm long, 17–20 cm wide, apically truncate, central split 30–50 cm deep, *light green above, whitish below*. **Inflorescence** 80–90 cm long, *with 2–3 spicate first-order branches*; peduncle 67–75 cm long; *peduncular bract missing but lower rachis bract 8–12 cm away from the first-order branch being subtended*; rachillae 8–12 cm long, 2–3 mm wide. **Flowers** not seen. **Fruit** globose, 7–10 mm diam., orange to scarlet when mature, *endocarp smooth*. **Seed** globose, 5–8 mm diam.

DISTRIBUTION. Bird's Head Peninsula, north-western New Guinea.

HABITAT. Lowland forest at sea level.

CONSERVATION STATUS. Endangered. *Licuala bifida* is known from only three sites. Deforestation due to mining, oil palm plantations and logging concessions is a major threat in its distribution range.

LOCAL NAME. None recorded.

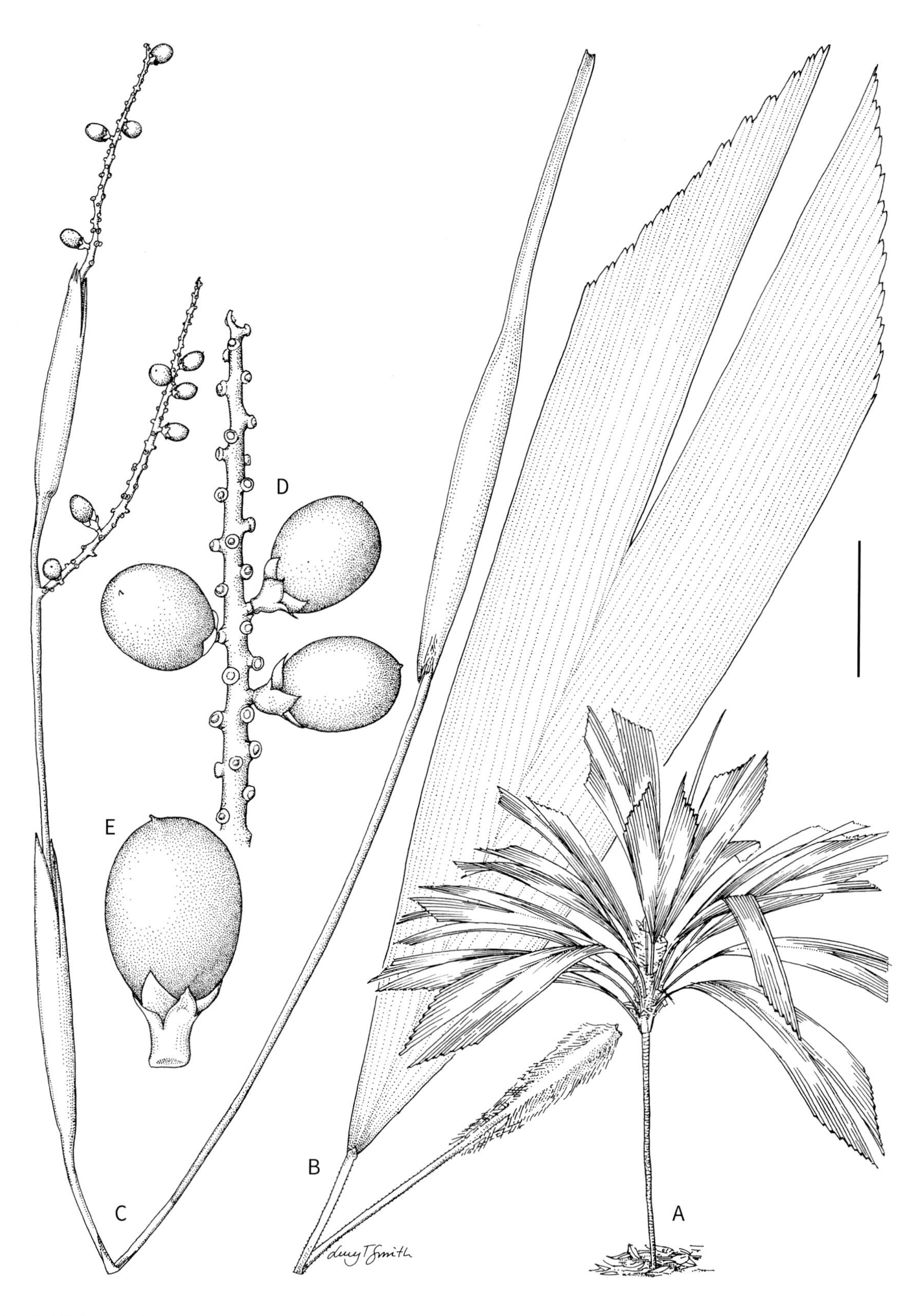

Licuala bifida. **A.** Habit. **B.** Leaf. **C.** Inflorescence with fruit. **D.** Portion of rachilla with fruit. **E.** Fruit. Scale bar: A = 72 cm; B = 8 cm; C = 4 cm; D = 1 cm; E = 7 mm. All from *Wally 839*. Drawn by Lucy T. Smith.

Licuala bifida. Habit, Bird's Head Peninsula (CDH).

USES. None recorded.

NOTES. An incompletely known member of the *L. telifera* complex (*L. anomala*, *L. bellatula*, *L. bifida*, *L. telifera*, *L. urciflora*). The entire, diamond-shaped, bifid leaf, reminiscent of *Sommieria* and some species of *Calyptrocalyx*, is a unique feature among the New Guinean *Licuala* species, except for some aberrant forms of *L. telifera*. The inflorescence is erect among the leaves and branched to 1 order, not 2 orders as in *L. telifera*. The flowers are unknown but calyx remnants on the fruits are wide basally and hairy.

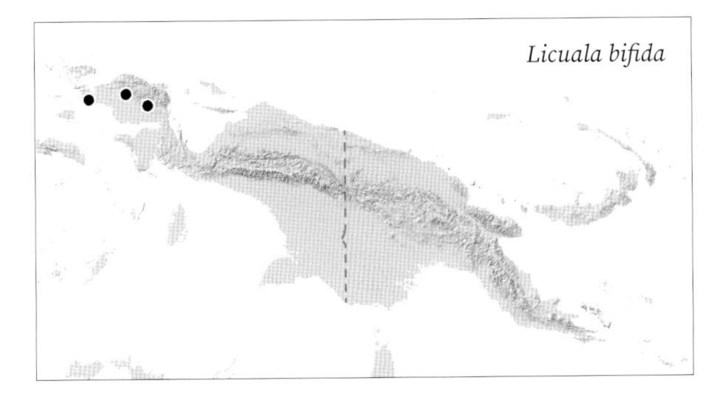

7. *Licuala brevicalyx* Becc.

Single-stemmed palm to 3–4 m. **Stem** 7–10 cm diam. **Leaf** sheath 10–15 cm long, breaking up distally in fibrous mesh; petiole 1.2–2 m long, armed in lower ¼–⅓; divided into 15–35 segments, with scattered minute, dark brown scales underneath; mid-segment 40–90 cm long, irregularly fused basally with adjacent segments, truncate apically. **Inflorescence** 1.5–1.8 m long, with *6–8 first-order branches*, curved; peduncle 30–35 cm long, not exposed at anthesis; peduncular bract lacking; *rachis zigzag, bracts loosely sheathing distally*; proximal first-order branch with straight to curved main axis, peduncle 10–13 cm long, rachis 9–12 cm long, bearing up to 30 rachillae, these to 15 cm long, straight to curved, minutely pubescent; increasing in robustness after anthesis. **Flowers** solitary or paired; *pedicel 0.6–1 mm long; calyx bell-shaped, 2.4–2.7 mm long, loosely sheathing, parchment-like, semi-transparent in dried condition,* glabrous, apically truncate or with 3, up to 1 mm long lobes; *stamens biseriate*; ovary 0.8–1 mm long; style 0.4–0.5 mm long. **Fruit** globose, 0.8–1 cm diam., *endocarp brittle, smooth*. **Seed** globose, 0.6–0.8 mm diam.

Licuala brevicalyx. LEFT TO RIGHT: habit; leaf detail; inflorescence. Kikori (WB).

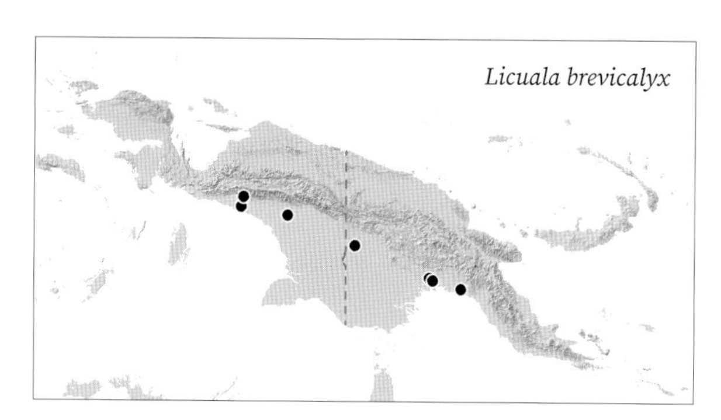

Licuala brevicalyx. Leaf, Kikori (WB).

DISTRIBUTION. Widely distributed south of the Central Range.

HABITAT. Lowland forest at sea level to 450 m.

LOCAL NAMES. *Filil* (Amele), *Meo* (Kairi).

USES. Timber for construction, axe handles and bows. Leaves for thatch.

CONSERVATION STATUS. Near Threatened. *Licuala brevicalyx* has a relatively restricted distribution. Deforestation due to logging and mining concessions is a major threat in its distribution range.

NOTES. A distinctive member of the *L. penduliflora* complex (*L. bankae, L. brevicalyx, L. multibracteata, L. penduliflora, L. suprafolia)* in its large inflorescences, with zigzag main axis and inflated tomentose rachis bracts. The flowers are delicate with a loosely sheathing, membranous calyx.

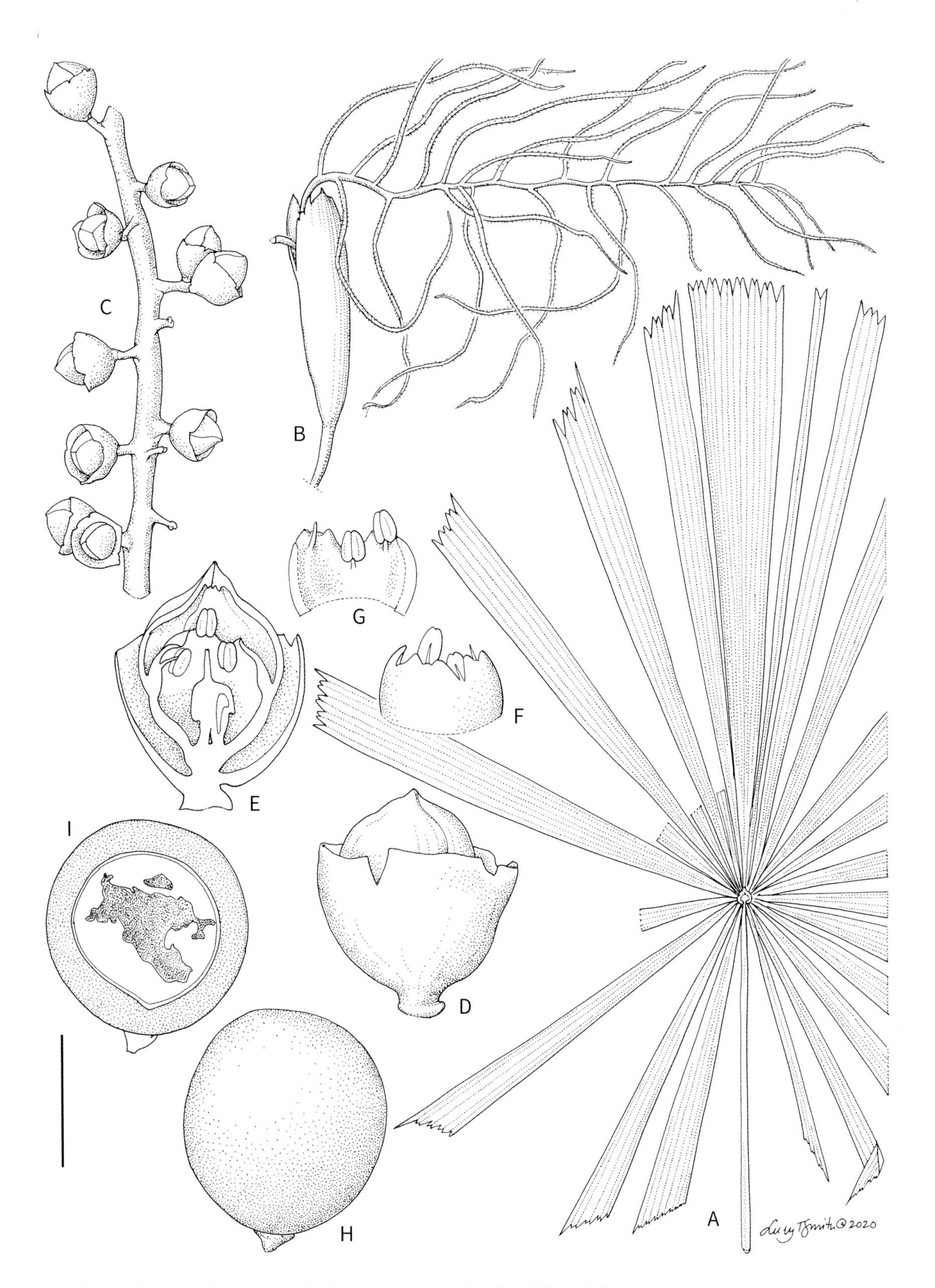

Licuala brevicalyx. **A.** Leaf. **B.** Portion of inflorescence. **C.** Portion of rachilla with flowers. **D**, **E.** Flower whole and in longitudinal section. **F**, **G.** Staminal ring in two views. **H**, **I.** Fruit whole and in longitudinal section. Scale bar: A = 21 cm; B = 8 cm; C = 5 mm; D–G = 1.5 mm; H, I = 7 mm. A, H, I from *Baker et al. 1102*; B–G from *Versteeg 1135*. Drawn by Lucy T. Smith.

8. *Licuala coccinisedes* Barfod & Heatubun

Single-stemmed, shrubby palm to 2 m. **Stem** 2–2.5 cm diam. **Leaves** 12–20 in crown; leaf sheath 15–20 cm long, brown, breaking up in fibrous mesh, sometimes detached from petiole distally in 2 strap-like extensions 20–25 long; petiole 40–100 cm long, armed on lower half; blade divided into 7–11 segments, with scattered, minute, rust-coloured, possibly glandular hairs on lower surface, otherwise glabrous; *mid-segment 35–55 cm long, not bifid, 15–20 cm across at the apex, clearly wider than remaining segments,* rounded to truncate. **Inflorescence** 60–90 cm long, erect, branched to 2 orders, with 5–6 first-order branches; peduncle 22–28 cm long, not or shortly exposed distally; peduncular bract lacking; rachis straight to curved distally; proximal first-order branch, erect with 1–2 cm long peduncle and 0.5–1 cm long rachis, bearing 3–5 rachillae, each 6–10 cm long, minutely pubescent, spreading. **Flowers** *solitary,* pedicel, 0.5–0.6 mm long, *pink to crimson red at anthesis;* calyx bell-shaped, 1.2–1.5 mm long, glabrous, *neatly splitting in 3 up to 0.5–0.7 mm, triangular lobes;* corolla 1.6–1.8 mm long, pale cream, lobes 1–1.2 mm long; *stamens biseriate;* style 0.7–0.8 mm long. **Fruit** globose, 9–11 mm diam., red mesocarp ca. 1 mm thick, *endocarp brittle, smooth.* **Seed** 7–9 mm diam.

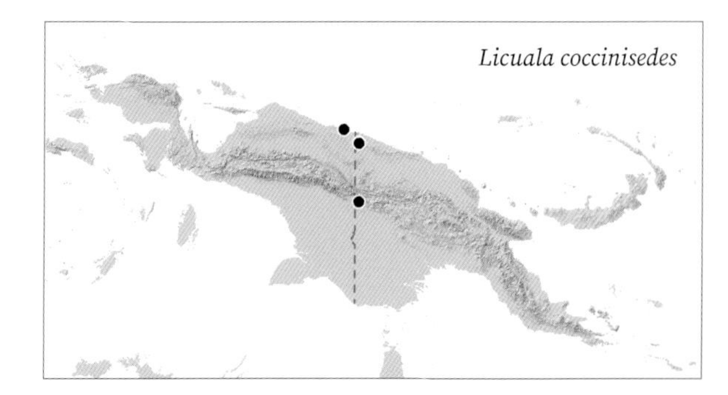

Licuala coccinisedes. LEFT TO RIGHT: habit, near Tabubil (WB); flowers, Bewani (AB).

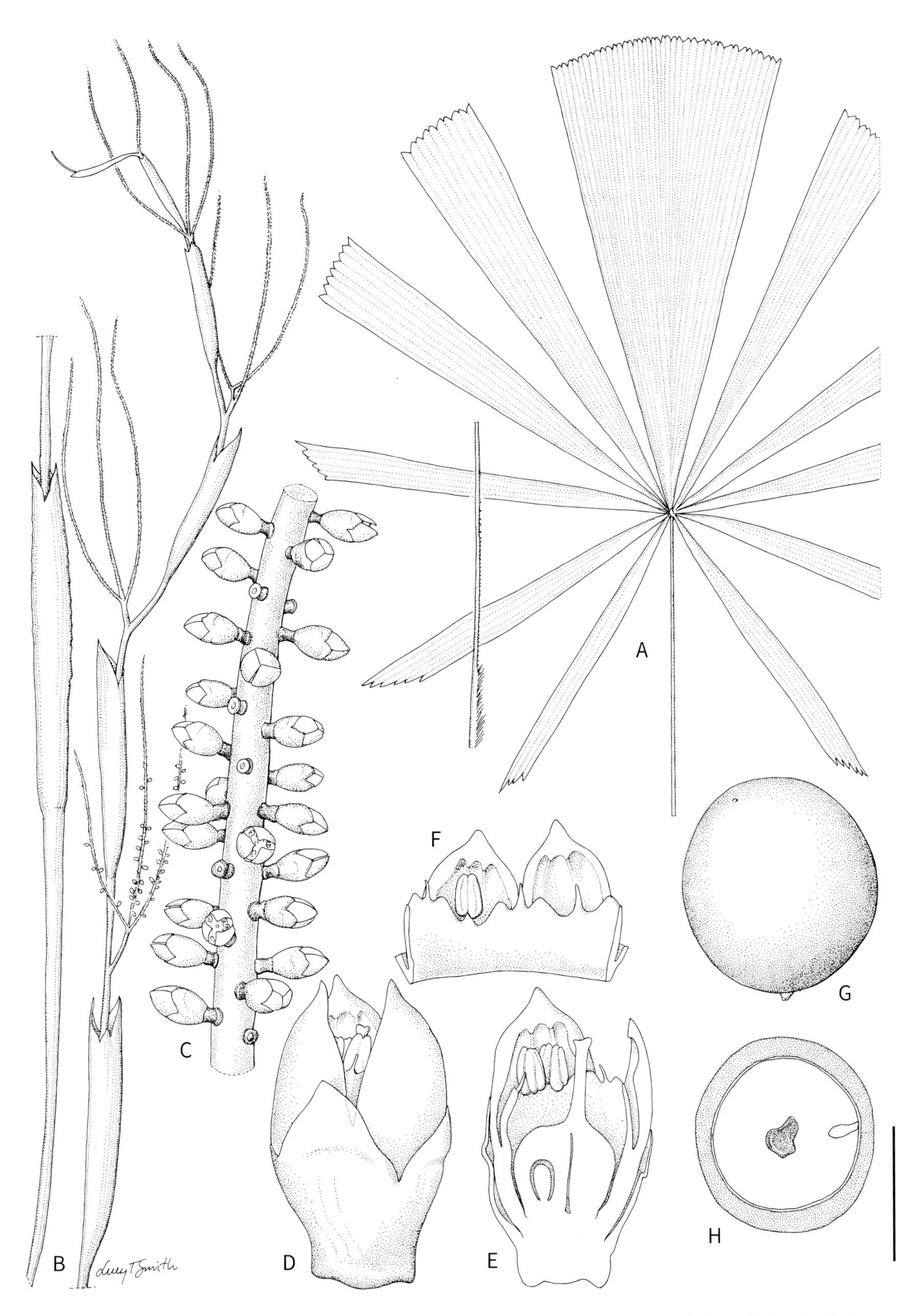

Licuala coccinisedes. **A.** Leaf. **B.** Inflorescence. **C.** Portion of rachilla with flowers. **D**, **E.** Flower whole and in longitudinal section. **F.** Staminal ring. **G**, **H.** Fruit whole and in transverse section. Scale bar: A = 12 cm; B = 4 cm; C = 7 mm; D, E = 1.5 mm; F = 1.6 mm; G, H = 5 mm. All from *Barfod et al. 484*. Drawn by Lucy T. Smith.

DISTRIBUTION. Northern central New Guinea, in Cyclops and Bewani Mountains, and Ok Tedi area.

HABITAT. Forest understorey at sea level to 750 m.

LOCAL NAMES. *Lump Bral* (Bewani), *Yal Bral* (Bewani).

USES. None recorded.

CONSERVATION STATUS. Vulnerable. *Licuala coccinisedes* has a restricted distribution. Deforestation due to mining and logging concessions is a major threat in its distribution range.

NOTES. *Licuala coccinisedes* resembles *L. bacularia* but is distinctive in having white, mostly solitary flowers borne on red pedicels. The leaf segments are characterized by having straight margins, as opposed to slightly convex margins in *L. bacularia*.

9. *Licuala essigii* Barfod & Heatubun

Single-stemmed, medium-sized palm to 3–6 m. **Stem** ca. 3–5 cm diam. **Leaf** sheath 20–30 cm long, disintegrating into fibrous mesh apically and along the margins; petiole 120–150 cm long, *armed on lower 1/10*; blade divided in 13–17 segments, below with scattered rust-coloured indumentum towards the base; *mid-segment 55–70 cm long, 25–30 cm wide at the apex,* truncate. **Inflorescence** *70–80 cm long,* erect with 4–5 first-order branches; peduncle 32–38 cm long, covered or only slightly exposed at anthesis; peduncular bract lacking; rachis straight to slightly curved distally; proximal first-order branch erect, basally hidden in the rachis bract, with *4–6 cm long peduncle bearing 10–12, to 15 cm long, slightly curved, pubescent rachillae.* **Flowers** mostly solitary, pedicels 0.3–0.5 mm long; calyx urn-shaped, *3–3.5 mm long, distally membranous and striate after drying, slightly pubescent at the base and distally, breaking up irregularly,* almost truncate; *stamens biseriate*; ovary 1.6–1.8 mm long, style ca. 1 mm long. **Fruit** not seen. **Seed** not seen.

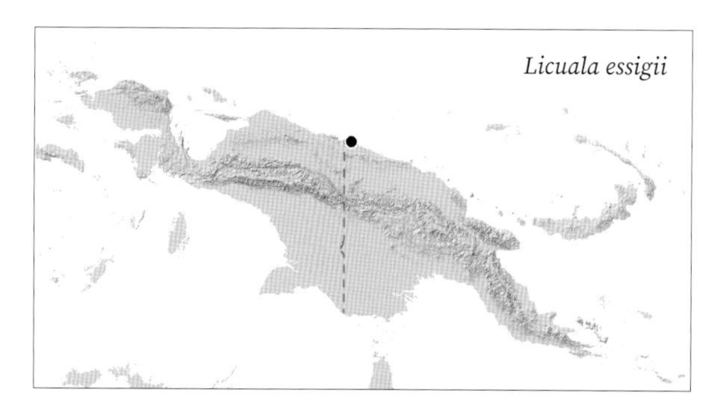

DISTRIBUTION. Near Vanimo.

HABITAT. Lowland forest at sea level to 50 m.

LOCAL NAMES. None recorded.

USES. None recorded.

CONSERVATION STATUS. Critically Endangered. *Licuala essigii* is known from only one site where it is threatened by a logging concession.

NOTES. The leaves of this understorey palm are characterized by having non-bifid mid-segments that are much wider than the remaining segments. The flowers are bullet-shaped, sub-sessile and sparsely hairy. This species is similar to *Licuala flexuosa* from Halmahera and Misool Islands but differs by having petioles armed basally only, first-order branches bearing more branches and smaller flowers with a pubescent calyx and shallow lobes.

10. *Licuala flexuosa* Burret

Single-stemmed palm, variable in height to 3.5 m. **Stem** lacking or to 2.5 m. **Leaves** sheath 25–40 cm long, disintegrating into fibrous mesh, distally detached in 2 strap-like extensions 20–25 cm long; *petiole 0.9–1.2 m long, armed throughout*; blade divided in 7–9 segments, fewer on young shoots; *mid-segment 25–37 cm long, apex 4–14 cm wide, slightly curved.* **Inflorescence** *0.8–1.2 m long,* erect with 4–6 first-order branches; peduncle 30–50 cm long, flattened in cross section; peduncular bract lacking; proximal first-order branch separated from the subtending bract, *peduncle short,* rachis to 6 cm long, *bearing 3–9 rachillae, these to 15 cm long, often zigzag distally,* covered with caducous, scattered to dense tomentum, *inflorescence increasing in robustness and size after anthesis,* reaching to 150 cm long in fruit. **Flowers** solitary, pedicel to 1 mm long; *calyx bell-shaped, 3.7–4.2 mm long, gradually narrowing and truncate at the base, glabrous, more or less regularly split apically into 3, to 1 mm long, rounded to obtuse lobes*; corolla cream-coloured, *stamens biseriate*; ovary 1.5–1.8 mm long, style 0.8–1 mm long. **Fruit** red at maturity, 1.2–1.4 cm diam., endocarp thin, brittle, with 8–10 inconspicuous, rounded ridges. **Seed** 9.5–11.5 mm diam.

DISTRIBUTION. Misool Island and Halmahera.

HABITAT. Lowland forest on alluvial, clay soils at sea level to 25 m.

LOCAL NAMES. *Lanchat* (Matbat), *Wekamanauro* (Tobaro).

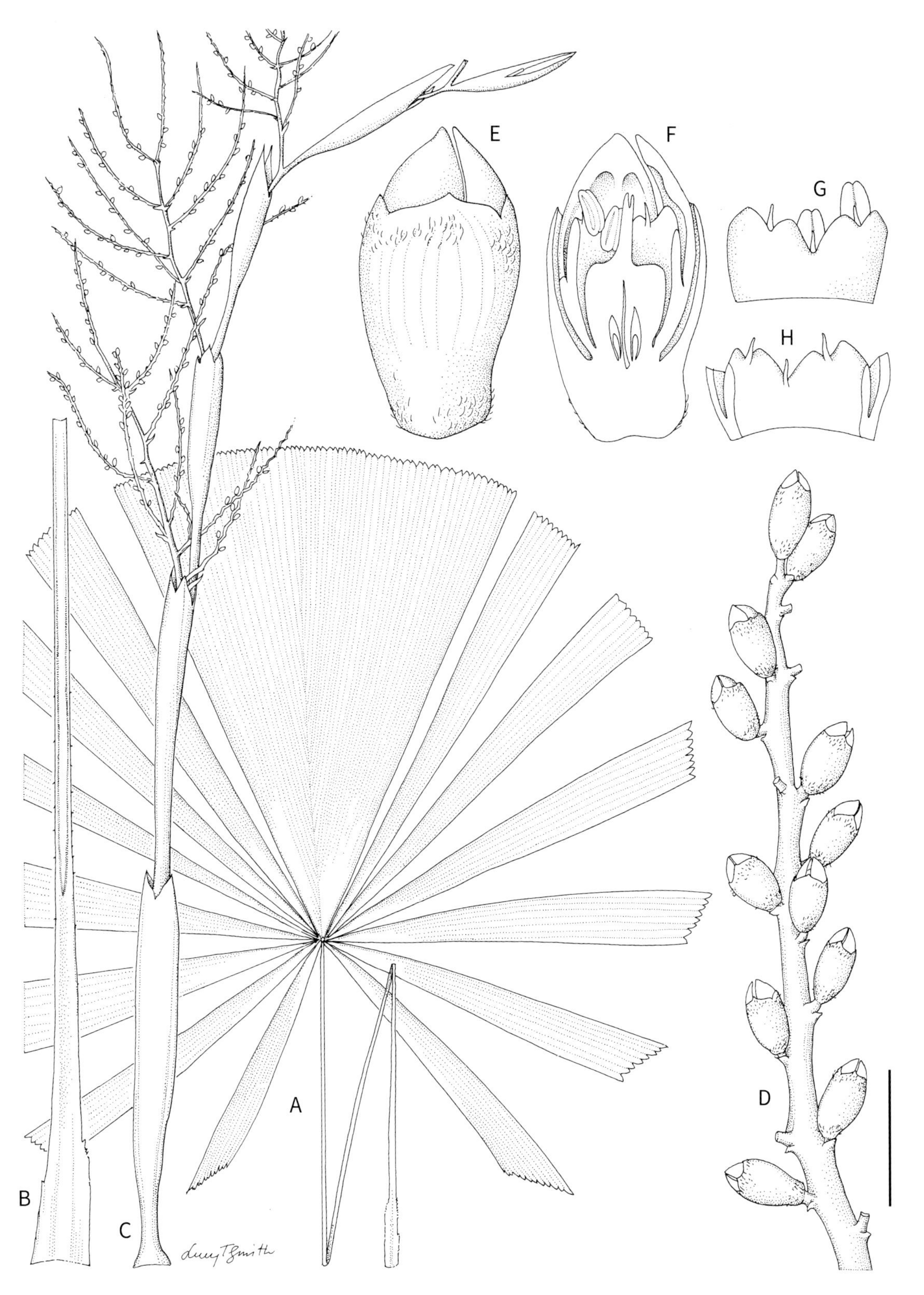

Licuala essigii. **A.** Leaf. **B.** Petiole. **C.** Inflorescence. **D.** Rachilla with flowers. **E**, **F.** Flower whole and in longitudinal section. **G**, **H.** Staminal ring in two views. Scale bar: A = 18 cm; B, C = 8 cm; D = 5 mm; E–H = 2 mm. All from *Essig LAE 55080*. Drawn by Lucy T. Smith.

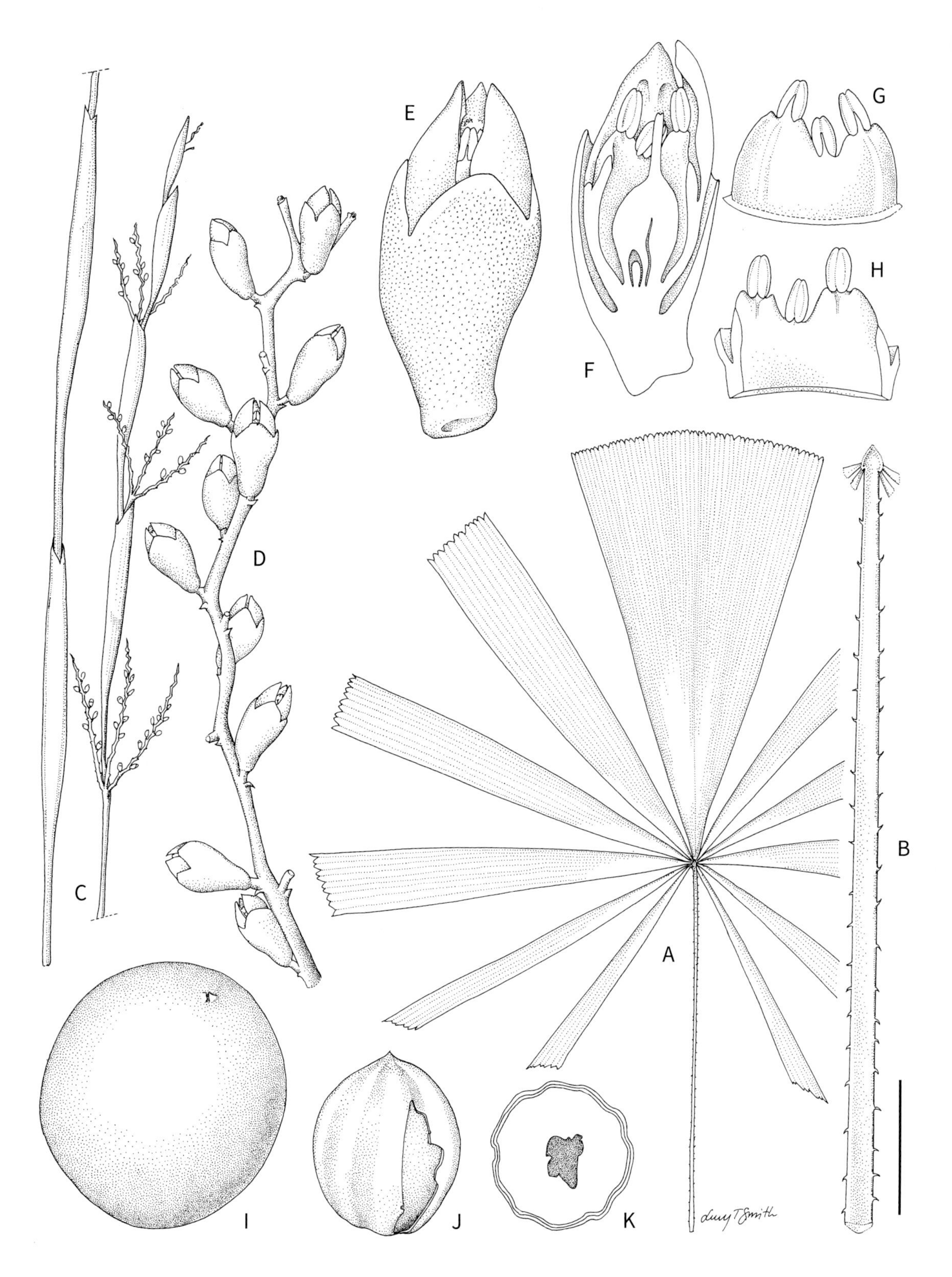

Licuala flexuosa. **A.** Leaf. **B.** Petiole. **C.** Inflorescence. **D.** Rachilla with flowers. **E, F.** Flower whole and in longitudinal section. **G, H.** Staminal ring in two views. **I.** Fruit. **J, K.** Seed whole and in transverse section. Scale bar: A = 12 cm; B = 3 cm; C = 6 cm; D, I–K = 7 mm; E–H = 2 mm. All from *de Vogel 3074*. Drawn by Lucy T. Smith.

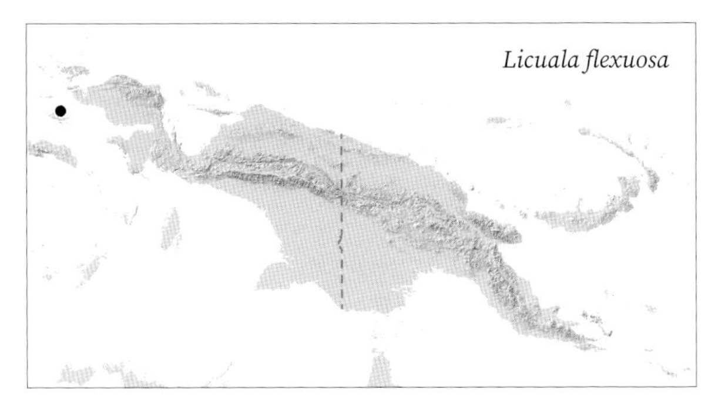

USES. None recorded.

CONSERVATION STATUS. Least Concern. Narrowly distributed within the New Guinea area.

NOTES. This species has its main distribution on the island of Halmahera. See note under *Licuala essigii.*

11. *Licuala graminifolia* Heatubun & Barfod

Single-stemmed palm, acaulescent, to 1 m. **Stem** ca. 1.5 cm diam. **Leaf** sheath disintegrating into a loose mesh of fine fibres; petiole 40–65 cm long, *unarmed; blade divided into 5–11 segments, all segments more or less equal, 20–35 cm long, 6–9 mm wide at apex, narrowly wedge-shaped or with slightly curved lateral margins.* **Inflorescence** 20–60 cm long, *spicate*, or, if branched, then only 1 rachilla fully developed; peduncle 8–20 cm long; peduncular bracts lacking; flower bearing part 6–8 cm long, *flower-subtending bracts conspicuous.* **Flowers** solitary, bullet-shaped 2.5–3 mm long, borne on tubercles, *pedicel very short*; calyx 2.2–2.8 mm long, translucent distally, with 3 0.4–0.6 mm long pointed lobes; *stamens uniseriate to inconspicuously biseriate*; ovary 0.7–0.8 mm long, attenuate apically; style 0.6–0.7 mm long. **Fruit** globose, 8–10 mm diam.; *endocarp smooth.* **Seed** 6–8 mm diam.

DISTRIBUTION. Kebar Valley, and sight records from Lake Yamur area.

HABITAT. Lowland forest at sea level to 800 m.

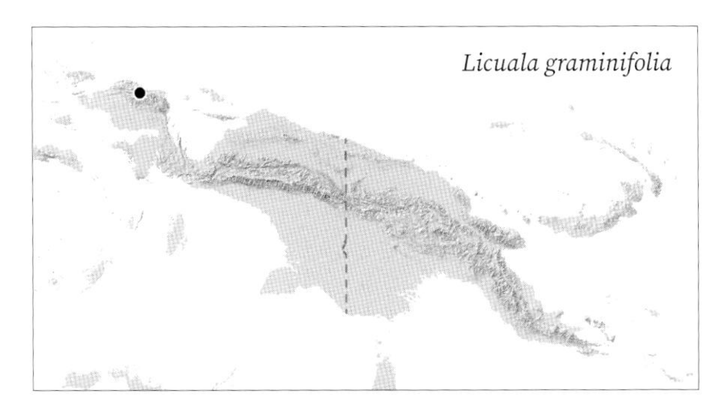

Licuala graminifolia. FROM TOP: habit; inflorescence. Bird's Head Peninsula (LJ).

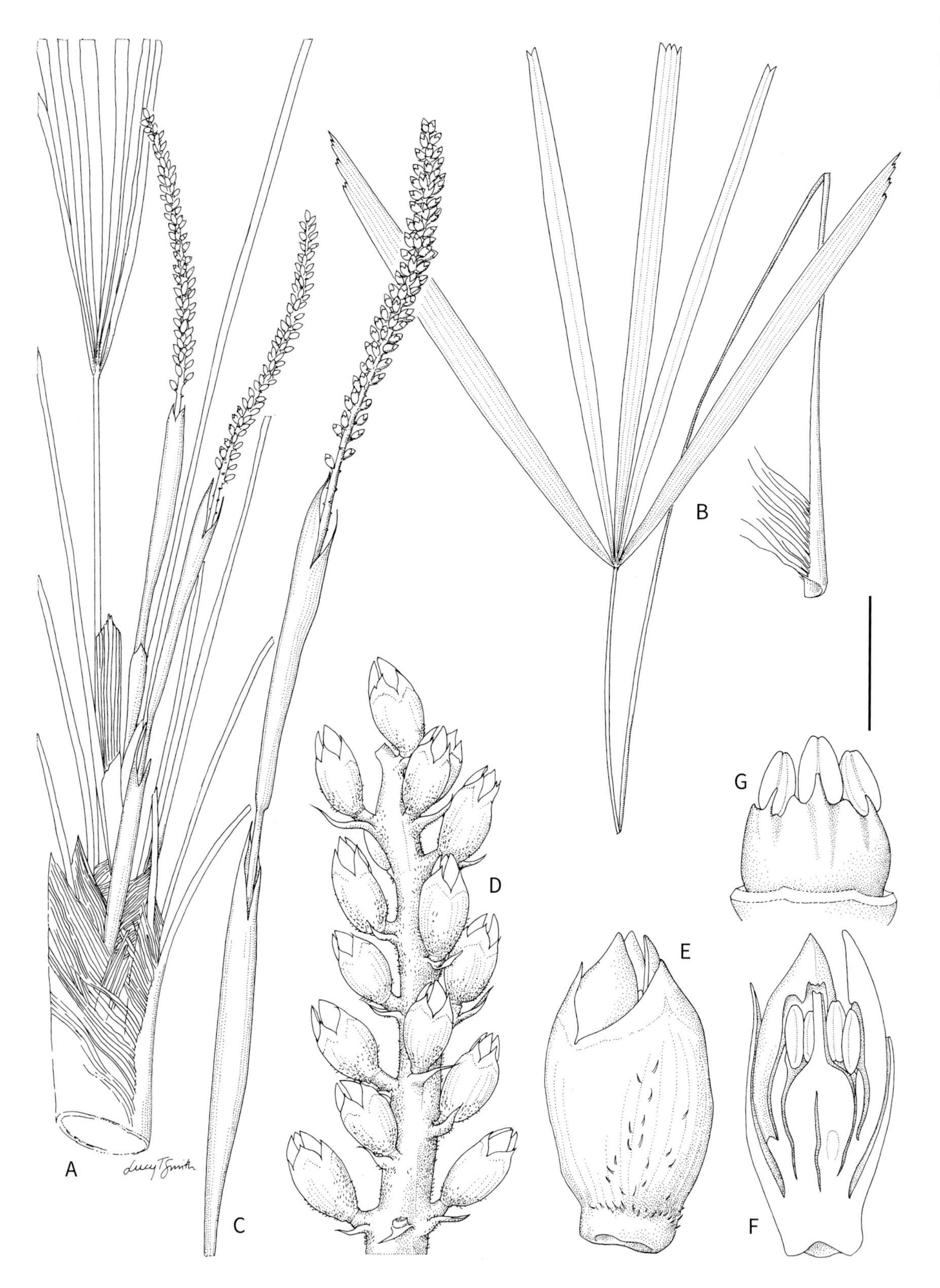

Licuala graminifolia. **A.** Stem with leaf base and inflorescence. **B.** Leaf with petiole. **C.** Inflorescence. **D.** Portion of inflorescence with flowers. **E, F.** Flower whole and in longitudinal section. **G.** Staminal ring. Scale bar: A = 2.5 cm; B = 6 cm; C = 2 cm; D = 3 mm; E, F = 1.2 mm; G = 1 mm. All from *van Royen & Sleumer 7345*. Drawn by Lucy T. Smith.

CONSERVATION STATUS. Data Deficient. The distribution and status of this species is insufficiently known. It is said to be locally common.

LOCAL NAME. *Pupuru* (Yamur dialect; a collective term for species of *Licuala* in the Lake Yamur area).

USES. None recorded.

NOTES. Remarkable for its narrow leaf segments reminiscent of grass leaves, hence the scientific name. *Licuala graminifolia* resembles *L. bifida* in the inflorescence structure, but in contrast to this species, the inflorescence is unbranched.

12. *Licuala grandiflora* Ridl.

Single-stemmed palm to 2 m. **Stem** ca. 5 cm diam., often with a cone of adventitious roots basally. **Leaf** sheath 40–50 cm long, breaking up in coarse fibrous mesh distally; petiole 50–100 cm long, armed in lower ½ to ⅔; blade divided into 7–19 segments, underneath with scattered rust-coloured, brown scales; *mid-segment 50–70 cm long, 3–10 cm across at widest part, rounded and slightly slanted apically.* **Inflorescence** *0.4–1.2 m long, branched to 1 order, erect, with 1–2 first-order branches that each comprise a single rachilla*; peduncle 30–80 cm long, exposed at anthesis; peduncular bract 1; rachis bract non-sheathing, strap-shaped, variable in length, to 10 cm long, rachilla 8–30 cm long, *flowers densely inserted in clusters of 2–3 proximally to solitary distally*, individual clusters subtended by 0.8–1.2 mm long bracts. **Flowers** *subsessile; calyx urn-shaped, 5–5.5 mm long, finely hairy on exposed faces*, split into 3 pointed lobes; stamens uniseriate; style 3–3.5 mm long. **Fruit** rounded to elliptic, *surrounded by the persistent calyx and corolla*, both more or less split to the base, 1.6–2 × 1.2–1.4 cm long, *dark brown to red at maturity, endocarp thick, with five, conspicuous longitudinal ridges.* **Seed** *1.2–1.4 × 0.7–0.9 cm.*

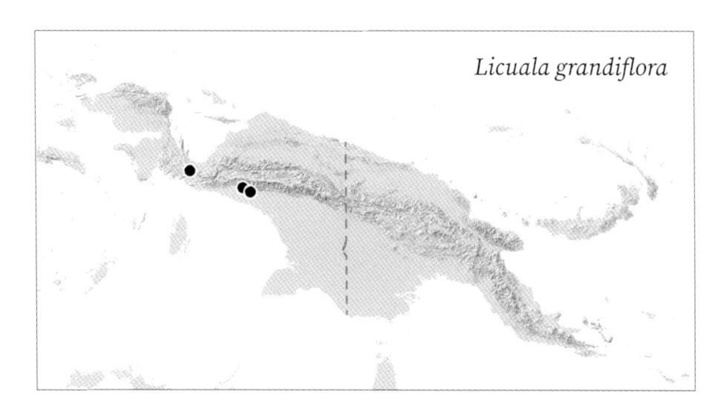

Licuala grandiflora. Habit, Aifat (CDH).

DISTRIBUTION. Lowland areas around the western end of the Central Range.

HABITAT. Lowland forest on alluvial soil, heath forest and peat swamp forest at sea level to 100 m.

LOCAL NAMES. *Pupuru* (Yamur dialect; a collective term for species of *Licuala* in the Lake Yamur area).

USES. None recorded.

CONSERVATION STATUS. Endangered. *Licuala grandiflora* is known from only three sites, one of which is in a mining concession. Specialised habitat preferences may put this species further at risk.

NOTES. A short-stemmed palm, characterised by inflorescences bearing one or two first-order branches that are themselves unbranched, with large flowers, densely inserted in groups of two to three throughout the rachilla. The stamens are inserted in one level

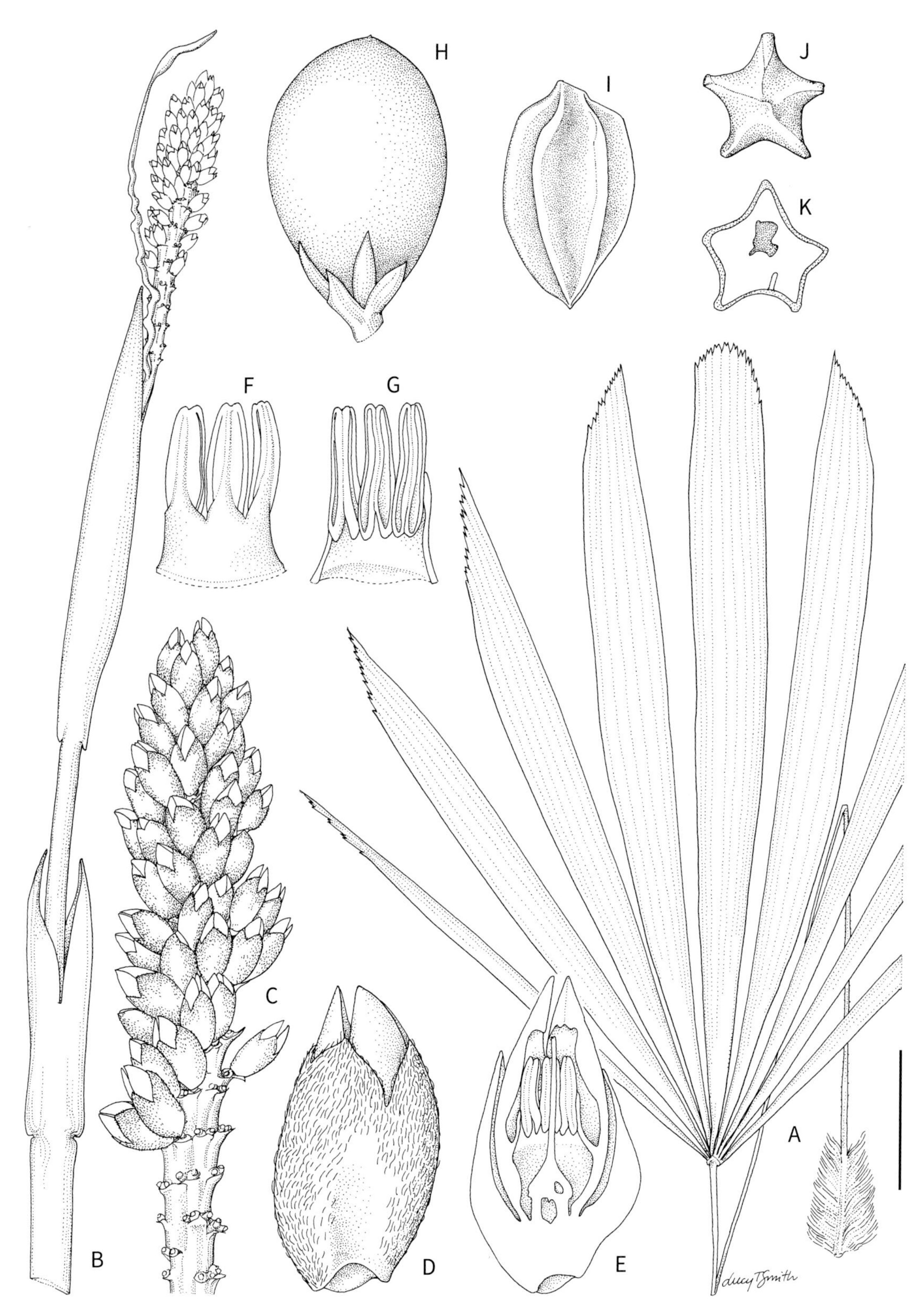

***Licuala grandiflora*. A.** Leaf with petiole. **B.** Inflorescence. **C.** Rachilla with flowers. **D, E.** Flower whole and in longitudinal section. **F, G.** Staminal ring in two views. **H.** Fruit. **I, J.** Endocarp in two views. **K.** Endocarp in transverse section. Scale bar: A = 12 cm; B = 4 cm; C, H–K = 1 cm; D, E = 3.3 mm; F, G = 2 mm. A–G from *Heatubun et al. 337*; H–K from *Heatubun et al. 339*. Drawn by Lucy T. Smith.

Licuala grandiflora. LEFT TO RIGHT: inflorescence; flowers; fruit. Aifat (CDH).

(uniseriate). The fruits are brown or red at maturity, almost 2 cm in diameter, with a reinforced and deeply furrowed endocarp surrounding the seed.

13. *Licuala heatubunii* Barfod & W.J. Baker

Single-stemmed palm, ca. 2 m. **Stem** ca. 7 cm diam. **Leaf** petiole variable in length to 170 cm long, armed in basal 40–50 cm; blade divided into 7–11 segments, glabrous above, with minute rust-coloured scales below, scales increasing in density towards major veins and towards the centre; *mid-segment 65–75 cm long, 28–30 cm wide and truncate apically*. **Inflorescence** 0.8–1 m long, with 4–5 first-order branches; peduncle 20–25 cm long, covered in short rust-coloured, felt-like indumentum; peduncular bract lacking; rachis slightly sinuous, not exposed; *proximal rachis bracts 23–27 cm long, tightly sheathing proximally for 8–10 cm to loosely sheathing distally*; main axis of proximal first-order branch 1–5 cm long, bearing 3–5 rachillae, *each 20–30 cm long, with 50–80 flowers and covered with scattered rust-coloured indumentum*. **Flowers** solitary, borne on flattened tubercles, *20–24 mm long, cylindrical; calyx 9–11 mm long, expanded and loosely sheathing, glabrous, truncate to slightly sinuous apically*; corolla white; *stamens uniseriate*; ovary glabrous, globose, 1.5–2.5 mm long; *style ca. 15 mm long*. **Fruit** not seen. **Seed** not seen.

DISTRIBUTION. Lake Sentani.

HABITAT. Periodically flooded forest on clay soil at sea level to 100 m.

LOCAL NAMES. *Jaiboh* (Anitaal).

USES. None recorded.

CONSERVATION STATUS. Critically Endangered. The area along the south bank of Lake Sentani where *Licuala heatubunii* was collected in 1957 has since been converted into vast grasslands. It is likely that this species is severely threatened or even extinct in the wild.

NOTES. A small palm, with a non-bifid mid-segment, which is much wider than other segments in the leaf. It is easily recognised by having white, 2–2.5 cm long flowers, the largest recorded in the genus so far.

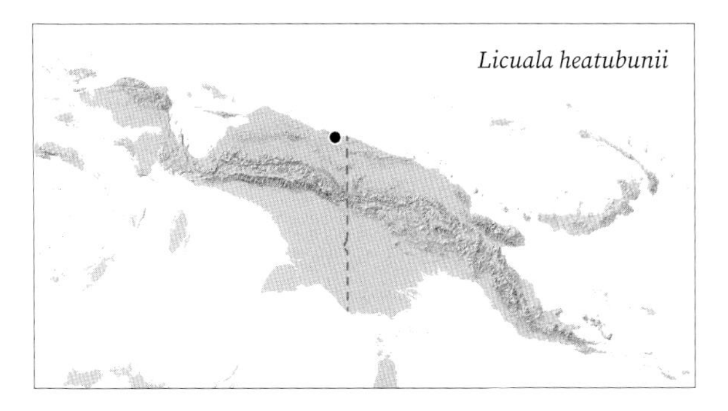

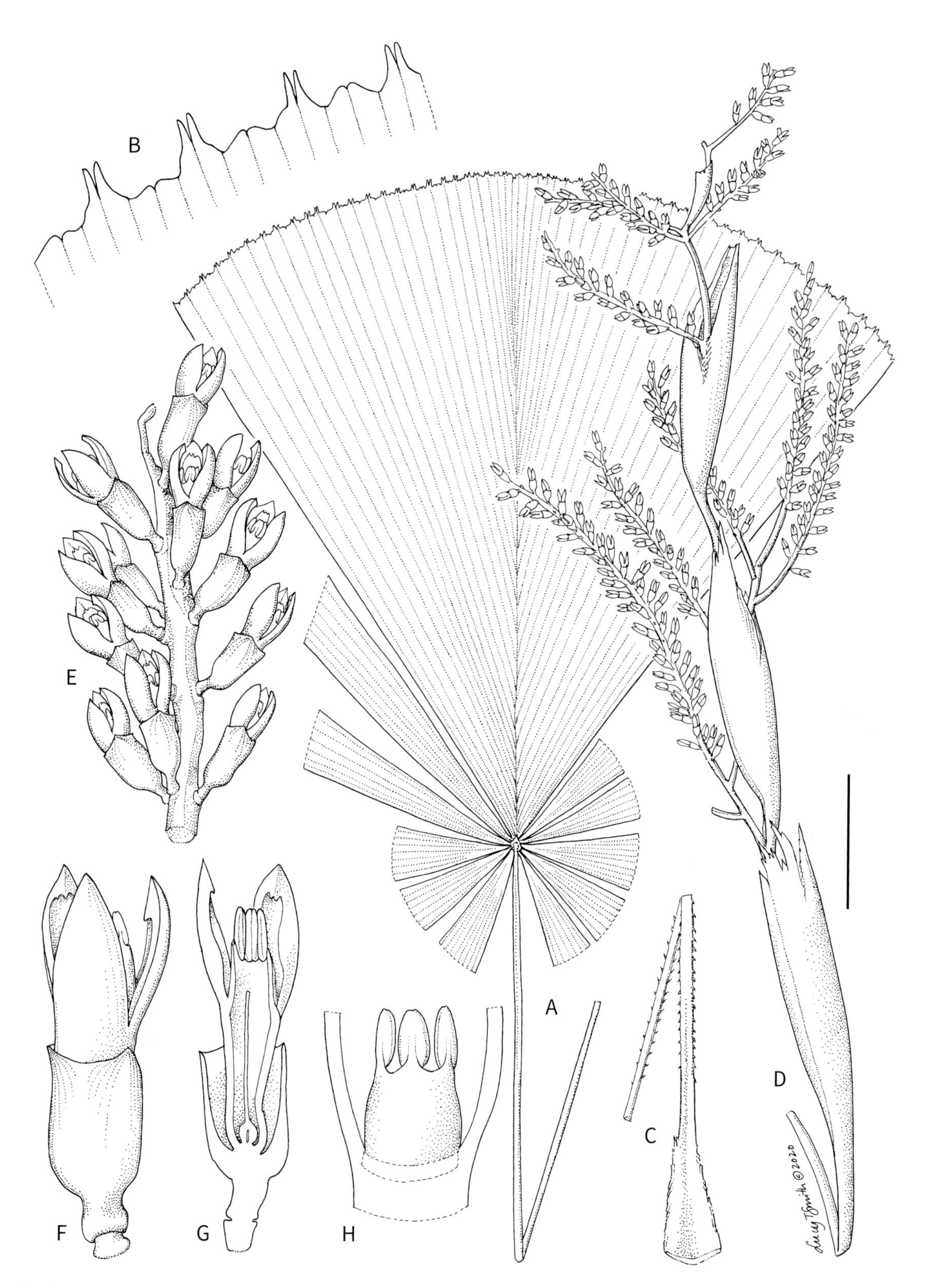

***Licuala heatubunii*. A.** Leaf. **B.** Detail leaf tips. **C.** Petiole. **D.** Inflorescence. **E.** Portion of rachilla with flowers. **F, G.** Flower whole and in longitudinal section. **H.** Staminal ring. Scale bar: A, C = 15 cm; B = 3 cm; D = 9 cm; E = 15 mm; F, G = 7.5 mm; H = 4 mm. All from *Versteegh BW 4711.* Drawn by Lucy T. Smith.

14. *Licuala insignis* Becc.

Synonyms: *Licuala crassiflora* Barfod, *Licuala moszkowskiana* Becc.

Single-stemmed palm to 7 m. **Stem** ca. 7 cm diam. **Leaf** sheath 50–60 cm long, tubular, breaking up irregularly in brown fibrous mesh; *petiole variable in length to 2.2 m long, with few minute spines at the base*; blade divided into 13–40 segments, with rust-coloured scales on lower surface of major veins; mid-segment 90–95 cm long, 28–30 cm wide and truncate apically. **Inflorescence** 1.7–1.9 m long, pendent, *with 7–9 first-order branches*; peduncle 45 cm long, with patches of long rust-coloured hairs; peduncular bracts lacking; *rachis zigzag*; proximal rachis bract 15 cm long; proximal first-order branch with 5–6 cm main axis, bearing 4–5 rachillae, these 25–26 cm long, bearing 80–90 flowers and covered with minute stellate hairs. **Flowers** solitary or inserted in pairs, borne on tubercles, bud ovoid, *12–14 mm long, fleshy,* pointed apically; calyx 6–7 mm long, shiny green, glabrous, lobes rounded ca. 2 mm long; *corolla fleshy, whitish green with reddish tinge at the base of the exposed parts; stamens biseriate*; ovary glabrous, 2.5–3 ×1.2–1.5 mm; style ca. 1.5 mm long. **Fruit** *yellow at maturity, globose, sometimes almond-shaped, 3.5–4.5 × 2–2.5 cm,* apically obtuse to rounded; endocarp ca. 3–4 cm long, *with 10–11 longitudinal ridges,* the one running along the raphe usually wider. **Seed** *1.2–1.4 × 0.7–0.9 cm.*

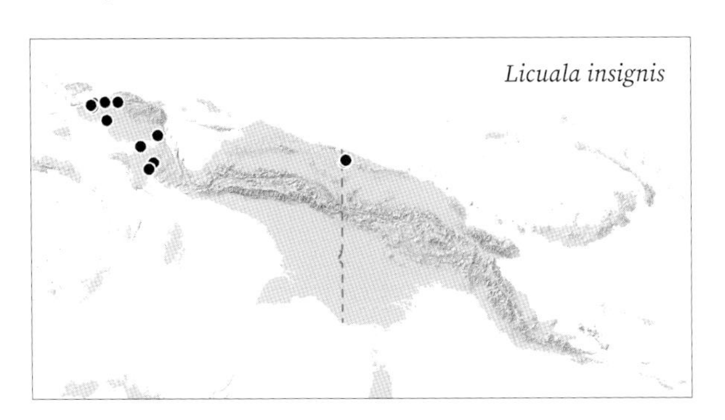

DISTRIBUTION. Widespread in the Bird's Head and Bomberai Peninsulas, with disjunct records in the Bewani region.

HABITAT. Lowland forest at sea level to 100 m.

LOCAL NAMES. *Bre* (Miyah), *Brubenei Bral* (Bewani), *Brunei bral* (Bewani), *Kar* (Sayal), *Sasin* (Moi), *Sifenate* (Sumuri, collective term for *Licuala*), *Unawa* (Irarutu).

Licuala insignis. FROM TOP: habit, Tamrau Mountains (WB); flowers, Bewani (AB).

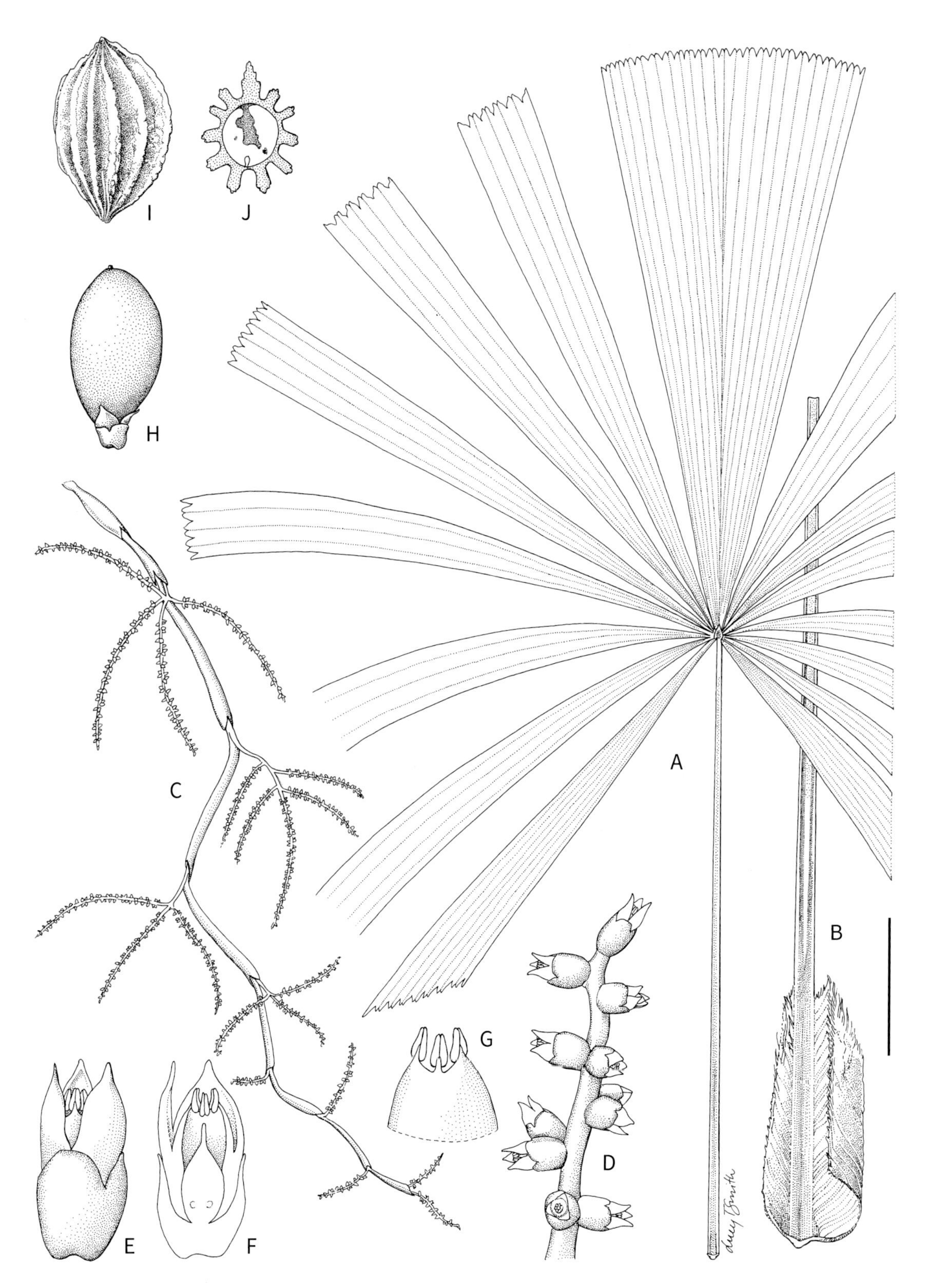

Licuala insignis. **A.** Leaf. **B.** Leaf sheath. **C.** Inflorescence. **D.** Portion of rachilla with flowers. **E**, **F.** Flower whole and in longitudinal section. **G.** Staminal ring. **H.** Immature fruit. **I**, **J**. Endocarp whole and in transverse section. Scale bar: A, B = 21 cm; C = 30 cm; D = 2 cm; E, F = 1 cm; G = 7 mm; H–J = 2.5 cm. A, B from photograph; C–H from *Barfod et al. 402*; I, J from *Ferrero* s.n. Drawn by Lucy T. Smith.

Licuala insignis. LEFT TO RIGHT: inflorescence with fruit; fruit (whole and in transverse section), and seed. Tamrau Mountains (WB).

USES. Stems are used for house posts. Occasionally, bows, spears or digging sticks are produced from the peripheral layers of the stem. Leaves are sometimes used for thatching of temporary shelters in the forest.

CONSERVATION STATUS. Least Concern. However, monitoring is required due to mining, oil palm plantations and logging concessions in the distribution range of this species.

NOTES. Medium-sized tree palm that belongs to the group of New Guinean species with ornate endocarps. As circumscribed here, it constitutes a broad species varying in the dimensions of the inflorescence. The collections from Bintuni and the Mamberamo River in Indonesian New Guinea have larger inflorescences with more rachillae. Sometimes one flower produces two, flattened and almond-shaped fruits.

15. *Licuala lauterbachii* Dammer & K.Schum.

Single-stemmed tree palm to 1.5–10(–15) m. **Stem** 7–10 cm diam., basally often with pointed leaf remnants. **Leaf** sheath 25–50 cm long; *fully developed petiole 150–200 cm long, unarmed or armed on lower 10–20 cm; blade divided into 19–35 segments; mid-segment 70–90 cm long, sometimes with a stalk to 10 cm long, truncate to curved apically.* **Inflorescence** *1.5–2.0 m long, curved, drooping, with 7–9 first-order branches*; peduncle 50–90 cm long, hardly exposed at anthesis; peduncular bract lacking; rachis zigzag; proximal rachis bract 25–50 cm long; *tightly sheathing proximally to slightly inflated distally,* splitting irregularly but cleanly, green; proximal first-order branch curved basally otherwise straight, peduncle, 1–2 cm long, main axis 20–25 cm long, *bearing up to 30 rachillae,* often subtended by strap-shaped, caducous bracts 4 cm long; *rachillae to 25 cm long, straight to curved, pubescent*; infructescence maintaining overall dimensions but increasing in robustness after anthesis. **Flowers** solitary or paired; pedicel 0.2–0.3 mm long, calyx urn-shaped to cylindrical, 2.8–3 mm long, glabrous, irregularly splitting apically or shallowly lobed; *corolla white; stamens biseriate*; ovary 1.3–1.5 mm long, style 0.4–0.5 mm long. **Fruit** orange at maturity, globose, 10–12 cm diam., *endocarp brittle, slightly grooved longitudinally*. **Seed** 0.8–1 cm diam.

NOTES. The numerous collections of *L. lauterbachii* from our region are referred to two subspecies that reflect significant variation in stature and dimensions of the inflorescence correlated with geographic distribution. Specimens from adjacent areas of the Solomon Islands may be referable to a third subspecies.

15a. *Licuala lauterbachii* subsp. *lauterbachii*

Synonyms: *Licuala gjellerupii* Becc., *L. magna* Burret., *Licuala robusta* Warb. ex K.Schum. & Lauterb.

Medium-sized tree palm *to 8–10(–15) m. Basal first-order branch of inflorescence with more than 15 rachillae*, longest rachillae rarely more than 20 cm long. Flowers with whitish-beige corolla. Infructescence often with undeveloped fruits persisting among the mature fruits and in some cases even outnumbering these.

DISTRIBUTION. Widely distributed north of the Central Range from Supiori Island to Milne Bay.

HABITAT. Lowland and montane forest at sea level to 1,200 m.

LOCAL NAMES. *Sepekir* (Yamben), *Som* (Biak).

USES. Fruit eaten by the northern cassowary (*Casuarius unappendiculatus*) in Jayapura area. Mature leaves are sometimes used to thatch houses, young leaves are tied to the upper arms of men as decorations during traditional gatherings. The peripheral part of the stem is used for making bows and arrow tips (all in the Adelbert Range).

CONSERVATION STATUS. Least Concern.

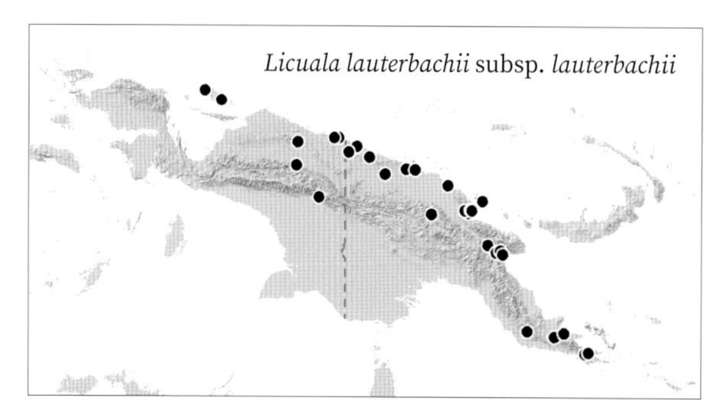

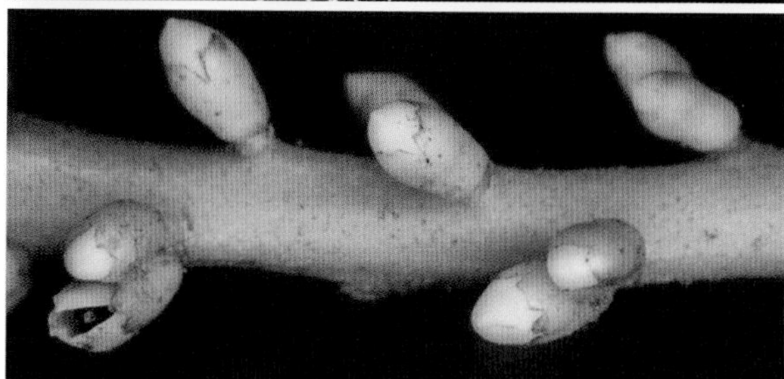

Licuala lauterbachii subsp. *lauterbachii*. FROM TOP: Leaf and inflorescence, Frieda River (FE); flowers, Vanimo (AB).

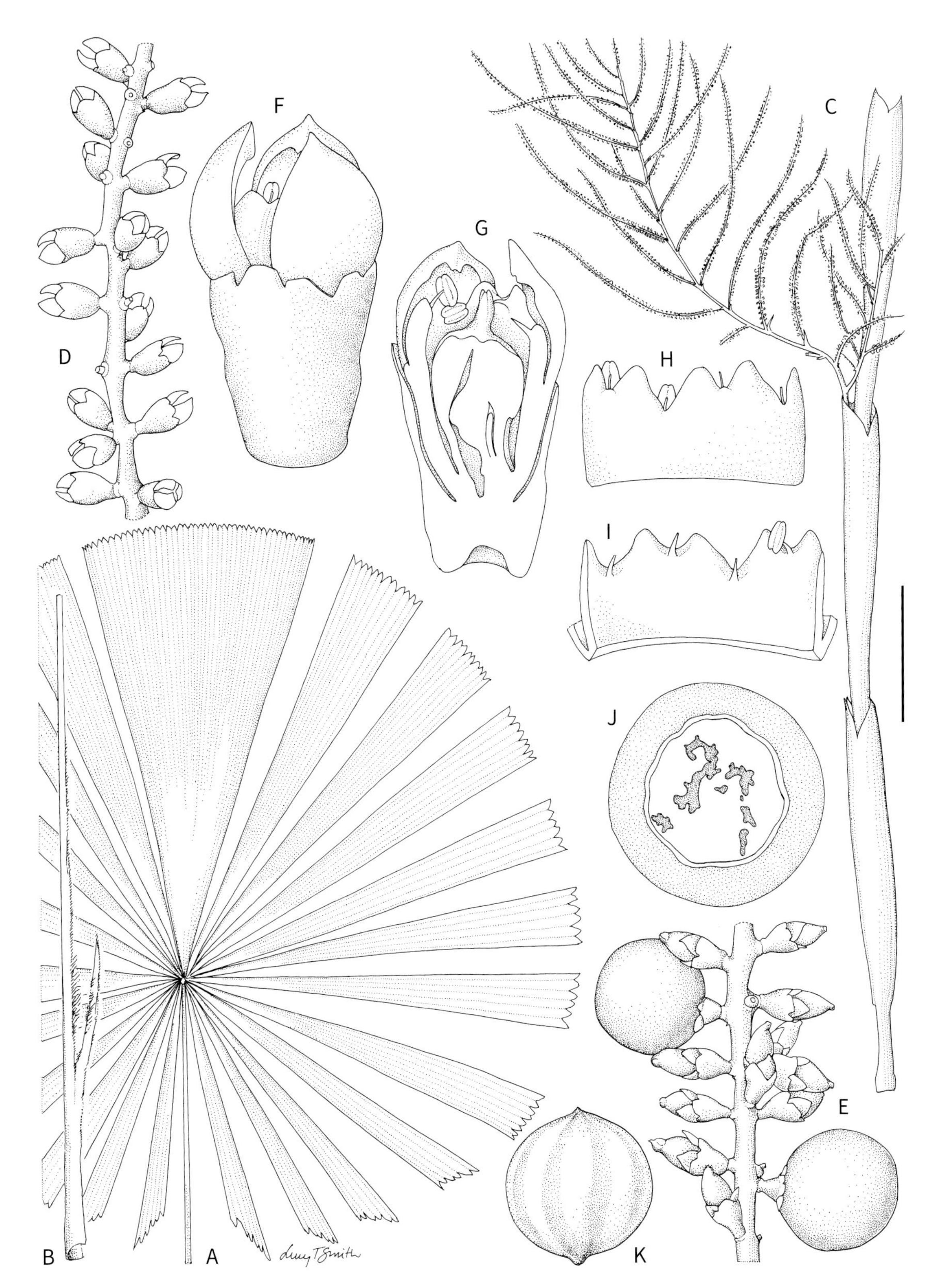

Licuala lauterbachii **subsp.** *lauterbachii.* **A.** Leaf. **B.** Leaf sheath. **C.** Portion of inflorescence. **D.** Portion of rachilla with flowers. **E.** Portion of rachilla with fruit. **F**, **G.** Flower whole and in longitudinal section. **H**, **I.** Staminal ring in two views. **J.** Fruit in transverse section. **K.** Endocarp. Scale bar: A, B = 24 cm; C = 12 cm; D, E = 1 cm; F–I = 2 mm; J, K = 7 mm. A–D, F–I from *Barfod 361*; E, J, K from *Pullen 1139*. Drawn by Lucy T. Smith.

15b. *Licuala lauterbachii* subsp. *peekelii* (Lauterb.) Barfod

Synonyms: *Licuala lauterbachii* ssp. *bougainvillensis* Becc., *Licuala peekelii* Lauterb.

Small tree palm *to (1.5–)3–5 m. Basal first-order branch of inflorescence with 5–10(–15) rachillae,* longest rachillae 15–25 cm. Flowers with yellowish corolla. Adjacent fruits all at the same developmental stage.

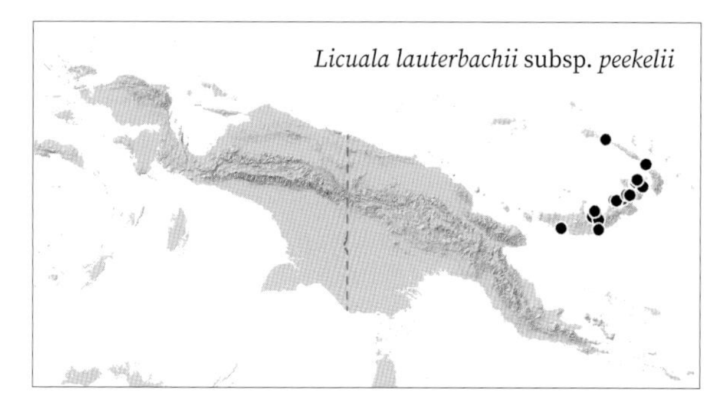

DISTRIBUTION. Widely distributed in New Britain and New Ireland, reaching Bougainville Island.

HABITAT. Lowland forest, often in light, open places at sea level to 610 m.

LOCAL NAMES. *Elaviluvilu* (West Nakanai), *Lavillavilla* (Mora Mora), *Salaho* (Pala dialect of Patpatar language), *Uban* (Kuanua).

USES. Leaves used for decoration at ceremonies.

CONSERVATION STATUS. Least Concern.

16. *Licuala longispadix* Banka & Barfod

Single-stemmed palm to 3–4 m. **Stem** ca. 14 cm diam. **Leaf** sheath ca. 20 cm long, extended on the opposite side of the petiole into a brown, papery ligule to 76 cm long; *petiole variable in length, exceeding 3 m in fully developed leaves,* the basal 60 cm armed; *blade divided into 36 segments*; mid-segment 1.2 m long, 45 cm wide at the apex. **Inflorescence** *4–4.5 m long, pendent often reaching the ground, with 12–13 first-order branches;* peduncle 82 cm long, one peduncular bract 60–80 cm long, splitting irregularly at the apex; *rachis 3.4–3.7 m long*; first-order branch with ca. 40 cm long main axis, bearing 40–45 pendent rachillae. **Flowers** solitary, borne on raised points or tubercles; calyx cylindrical, 2.5–3 mm long, breaking up regularly in 3, ca. 1 mm long lobes; stamens uniseriate; ovary glabrous 2–2.5 mm long, truncate apically; style ca. 0.5 mm long. **Fruit** *globose, 2–2.5 × 1–1.5 cm, red at maturity; endocarp reinforced with 3, to 5 mm wide, longitudinal ridges.* **Seed** *1.9–2.1 × 0.9–1.1 cm.*

DISTRIBUTION. Central northern New Guinea.

HABITAT. Lowland forest on alluvial plains at sea level to 30 m.

LOCAL NAMES. *Marim* (Amanab).

USES. None recorded.

CONSERVATION STATUS. Critically Endangered (IUCN 2021). *Licuala longispadix* is known from only two sites that have since been logged intensively. In the type locality near Pual River, a careful search within a radius of 100 m revealed only one additional individual and no regeneration.

Licuala longispadix. Leaf, Pual River (AB).

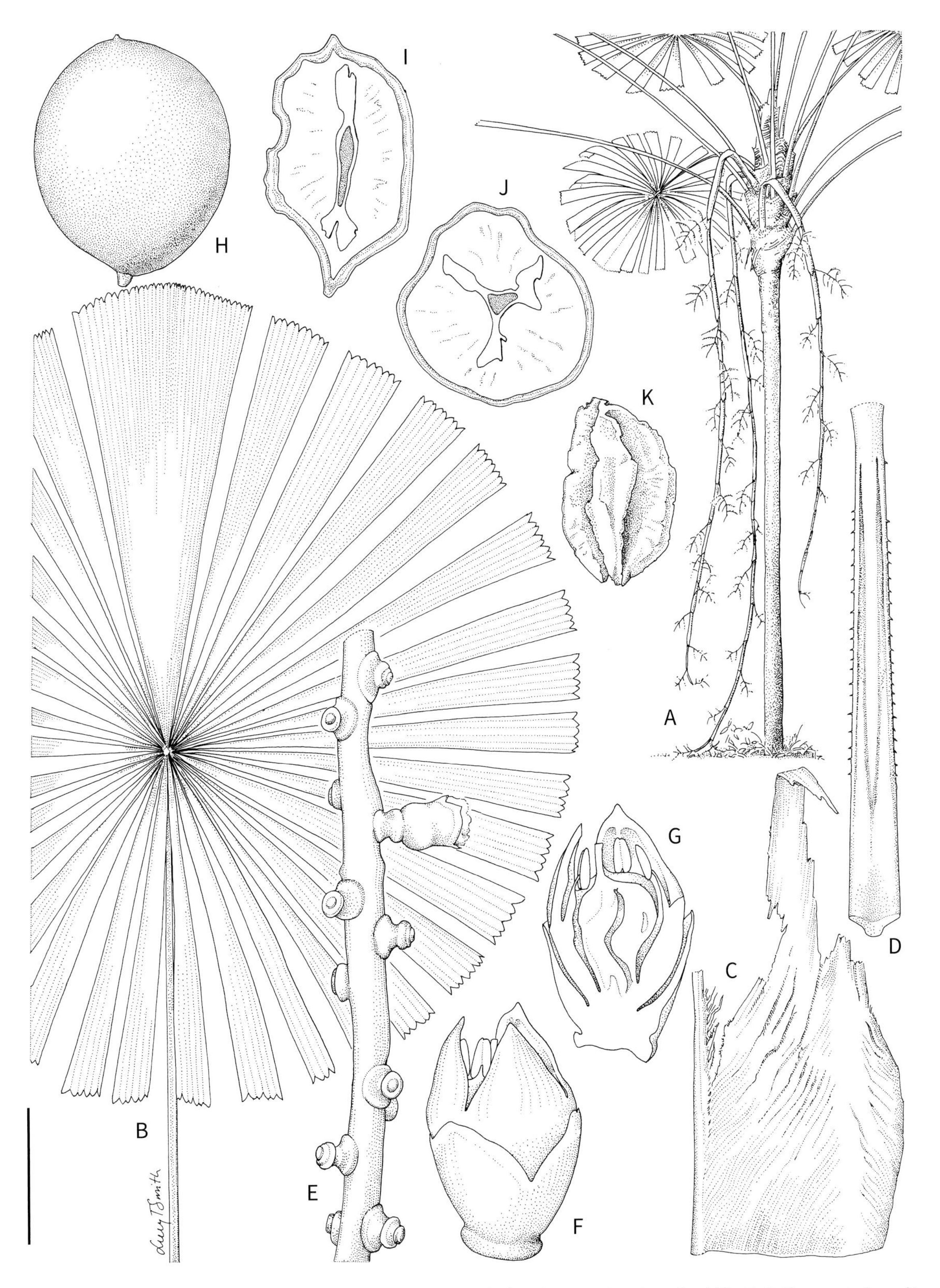

***Licuala longispadix*. A.** Habit. **B.** Leaf. **C.** Leaf sheath. **D.** Petiole with sheath scar. **E.** Portion of rachilla. **F, G.** Flower whole and in longitudinal section. H–**J.** Fruit whole, in longitudinal section (dried) and in transverse section (dried). **K.** Endocarp. Scale bar: A = 65 cm; B = 36 cm; C = 18 cm; D = 6 cm; E = 7 mm; F, G = 2.5 mm; H–K = 1.5 cm. A–G from *Barfod 508*; H–K from *Wiakabu & Warimbangu 75798*. Drawn by Lucy T. Smith.

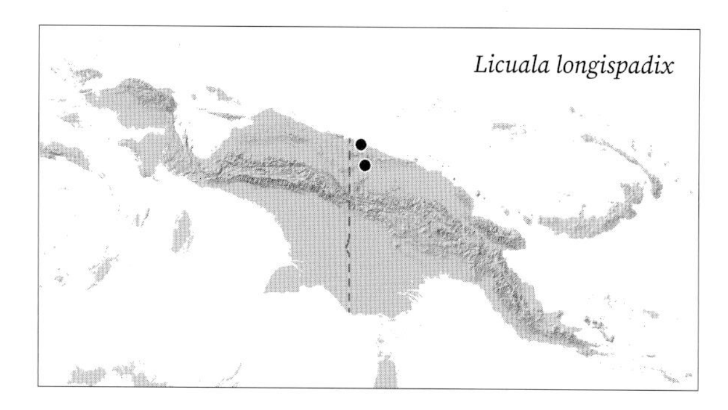

NOTES. The collections available to us comprise only old inflorescences and fruiting material. This species has the longest inflorescences of any *Licuala*. The distal parts of the infructescence reach the forest floor where the large fruits are presented.

17. *Licuala montana* Dammer & K.Schum.

Synonyms: *Licuala micrantha* Becc., *Licuala naumoniensis* Becc.

Single-stemmed palm to 1–4 m. **Stem** ca. 2 cm diam. **Leaf** sheath disintegrating into messy tufts of fibres; petiole 60–70 cm long, armed on lower ⅓; blade divided into *even or uneven number* of *12–18 segments; mid-segment 45–50 long, clearly wider in leaves with an uneven number of segments, here to 4.5 cm wide, otherwise almost identical to the adjacent leaf segments.* **Inflorescence** 0.7–1.3 m long, erect with *3–4 first-order branches*; peduncle 40–60 cm long at anthesis, elongating to 60–90 cm in fruit; peduncular bract lacking; rachis more or less straight; proximal rachis bract 10–13 cm long, inserted 30–90 cm above prophyll, grayish green, with rust-coloured indumentum; proximal first-order branch with 1–2 cm long main axis, bearing 3 spreading rachillae, to 9 cm long, covered with light brown tomentum. **Flowers** solitary throughout, *pedicel ca. 1 mm long, densely tomentose*; *calyx stalked, 30–35 mm long, with scattered hairs at anthesis,* breaking up regularly into 3, ca. 1 mm long, rounded to acute lobes; *stamens biseriate*; ovary glabrous, 1.5–2 mm long, truncate to rounded apically; style 0.5–0.6 mm long. **Fruit** globose 9–11 mm diam., orange at maturity, mesocarp 1.2–1.5 mm thick, *endocarp brittle, smooth.* **Seed** 7.5–9.5 mm diam, endosperm often with central cavity.

DISTRIBUTION. Widely distributed mainly north of the Central Range from the Bird's Head Peninsula to far south-eastern New Guinea.

Licuala montana. Habit, East Sepik (AB).

HABITAT. Lowland forest on hilly ground at sea level to 900 m.

LOCAL NAMES. *Grere* (Marap), *Ol bral* (Bewani), *Sifenete* (Sumuri, collective term for *Licuala*).

USES. None recorded.

CONSERVATION STATUS. Least Concern.

NOTES. A species that stands out by having rather stiff, narrow leaf segments with a greyish tinge. The species epithet is misleading since *L. montana* occurs largely in lowland areas.

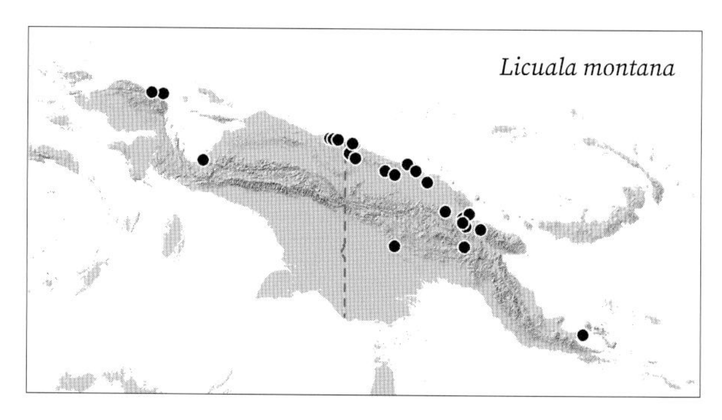

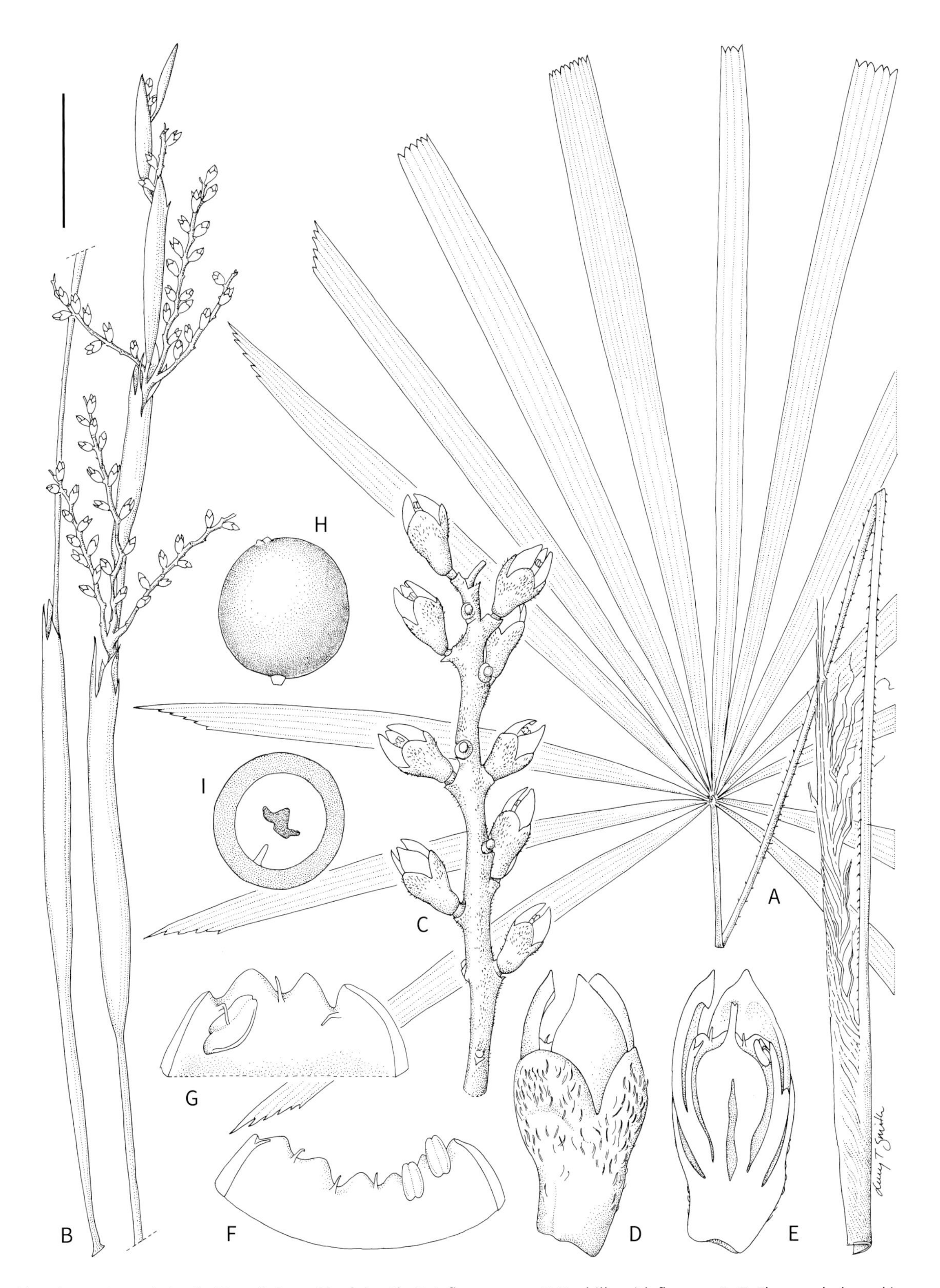

Licuala montana. **A.** Leaf with petiole and leaf sheath. **B.** Inflorescence. **C.** Rachilla with flowers. **D, E.** Flower whole and in longitudinal section. **F.** Staminal ring. **G.** Staminal ring detail. **H, I.** Fruit whole and in transverse section. Scale bar: A, B = 8 cm; C = 3 cm; D, E = 2.5 mm; F = 2 mm; G = 1.3 mm; H, I = 1 cm. A–G from *Barfod 494*; H, I from *Barfod et al. 382.* Drawn by Lucy T. Smith.

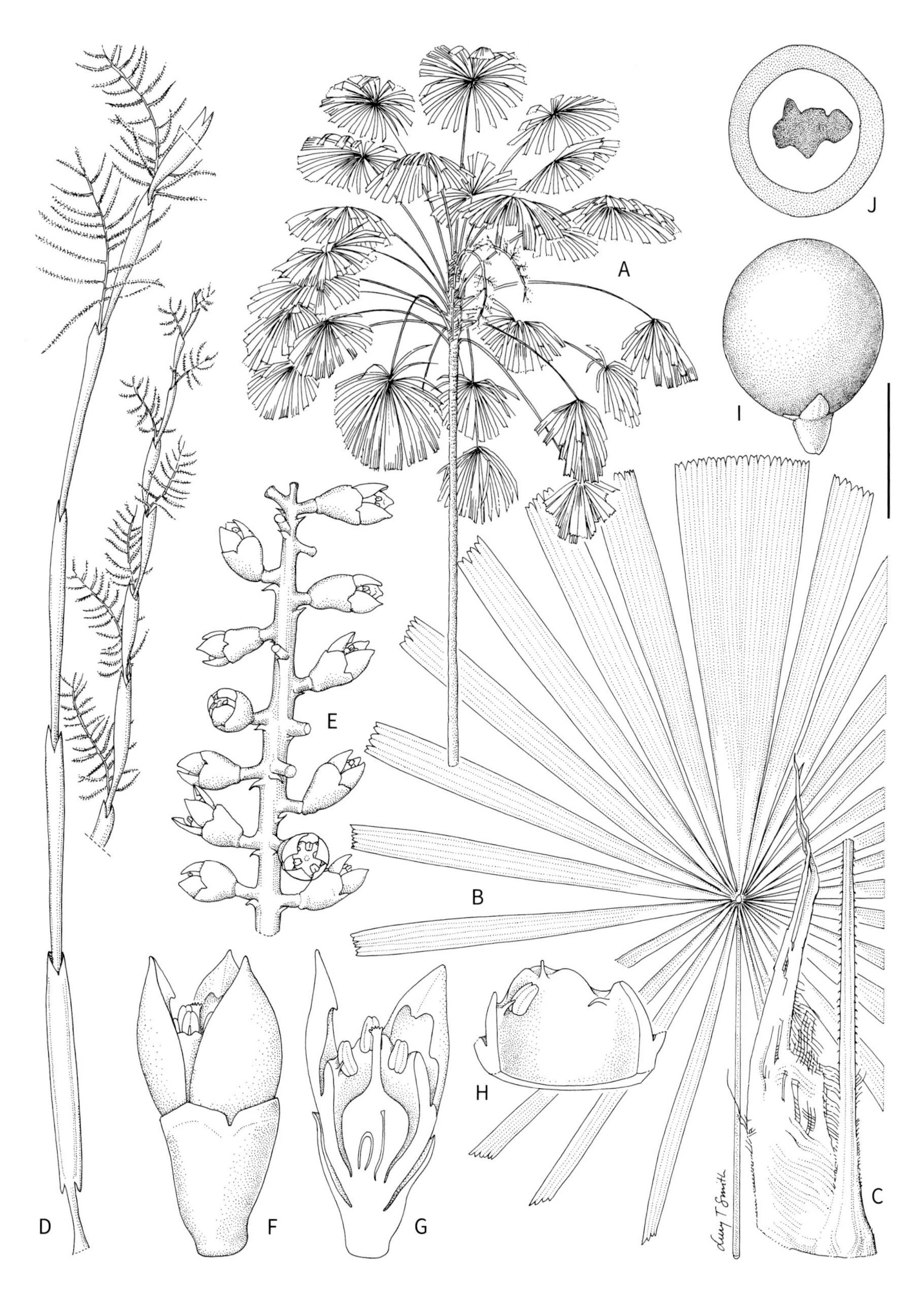

***Licuala multibracteata*.** **A.** Habit. **B.** Leaf. **C.** Petiole with leaf sheath. **D.** Inflorescence. **E.** Portion of rachilla with flowers. **F**, **G.** Flower whole and in longitudinal section. **H.** Staminal ring. **I**, **J.** Fruit whole and in longitudinal section. Scale bar: A = 74 cm; B, C = 18 cm; D = 11 cm; E = 7.5; F, G = 2 mm; H = 1.6 mm; I, J = 5 mm. All from *Craven & Schodde 932*. Drawn by Lucy T. Smith.

18. *Licuala multibracteata* Barfod & Heatubun

Single-stemmed tree palm to 4 m. **Stem** ca. 7–8 cm diam. **Leaf** sheath 25–30 cm long, disintegrating into a fibrous mesh, distally detached from petiole in 2 strap-like extensions 20–25 cm long; petiole 1.5–2 m long, flattened, covered distally by patches of rust-coloured tomentum, lower $^{1}/_{10}$ armed; blade divided into 25–35 segments; mid-segment 60–70 cm long, 13–17 cm wide and truncate apically. **Inflorescence** *1.4–1.6 m long, branched to 2 orders, with 9–11 first-order branches,* straight and curved at anthesis, peduncle 60–80 cm long, covered by prophyll and peduncular bracts; *peduncular bracts 2, 18–22 cm long*; rachis 75–90 cm long; proximal first-order branch 5–15 cm long, bearing ca. 10–20 rachillae, both rachis and rachillae covered in scattered to dense rust-coloured tomentum. **Flowers** mostly single, pedicel 1.5–1.7 mm long, glabrous; calyx funnel-shaped, 3 mm long, fused with receptacle for ca. 1 mm, glabrous, breaking up in 0.2–0.3 mm long, pointed lobes; corolla 4.7–5 mm long, white, splitting in ca. 3 mm long, pointed lobes; staminal ring 1.1- 1.3 mm high, *anthers biseriate*; *ovary rounded to slightly truncate apically, 1.4–1.6 mm long; style, 1.3–1.5 mm long.* **Fruit** orange to red at maturity, globose, 8–10 mm diam., remnant calyx conspicuous; *endocarp smooth.* **Seed** *6–8 mm diam.*

DISTRIBUTION. Vicinity of Port Moresby.

HABITAT. Lowland swamp forest on alluvial flats from sea level to 50 m.

LOCAL NAMES. None recorded

USES. None recorded.

CONSERVATION STATUS. Endangered. *Licuala multibracteata* is known from only four sites, three of which are closely adjacent. Deforestation due to logging concessions is a major threat in its distribution range.

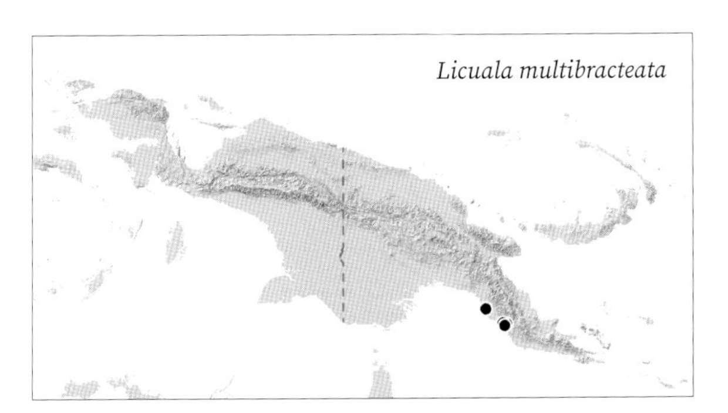

NOTES. *Licuala multibracteata* most closely resembles *L. suprafolia,* with which it shares the numerous, rather tightly sheathing peduncular and rachis bracts that split neatly. The fruit of *L. suprafolia* has a thicker mesocarp and a slightly grooved endocarp.

19. *Licuala parviflora* Dammer ex Becc.

Synonym: *Licuala tanycola* H.E.Moore

Multi-stemmed palm to 3 m. **Stem** prostrate to erect, 5–7 cm diam. **Leaf** sheath 30–40 cm long, eventually disintegrating into a fine fibrous mesh, distally detached from petiole in 2 strap-like extensions 5–10 cm long; petiole 80–120 cm long, armed in lower 3/4; blade divided into 11–19 segments, covered below in scattered, minute, rust-coloured indumentum; *mid-segment 45–60 cm long, with curved margins.* **Inflorescence** 0.8–1.1 m long, erect, *branched to 1–2 orders*; peduncle 60–90 cm long, flattened; prophyll 20–25 cm long, *covered with rust-coloured indumentum,*

Licuala parviflora. Habit, near Tabubil (WB).

Licuala parviflora. **A.** Leaf. **B.** Leaf sheath with petiole. **C.** Inflorescence. **D.** Portion of rachilla with flowers. **E, F.** Flower whole and in longitudinal section. **G.** Staminal ring. **H.** Fruit whole. **I.** Seed in transverse section. Scale bar: A, B = 9 cm; C = 4 cm; D, H, I = 7 mm; E–G = 2.5 mm. A–G from *Brass 1310A*; H, I from *Jacobs 8755*. Drawn by Lucy T. Smith.

Licuala parviflora. LEFT TO RIGHT: inflorescence, near Tabubil (WB); fruit, Bewani (AB).

peduncular bract reduced and contained in prophyll or lacking; proximal first-order branches often subtended by up to 15 mm long, narrow, strap-shaped bracts; basal first-order branches often branched to 2 orders, rachillae 10–50, spreading, 6–18 cm long. **Flowers** solitary throughout; pedicel no longer than 0.3 mm long, on tubercle; calyx stalked, 1.8–2.5 mm long, breaking up regularly into 3, ca. 1 mm long, pointed to shortly acuminate lobes; *stamens uniseriate;* ovary glabrous, 1.2–1.5 mm long, rounded to attenuate apically; style 0.6–0.8 mm long. **Fruit** globose to ellipsoid at maturity, 10–14 × 10–12 mm diam., orange to red, mesocarp 1.2–1.5 mm thick, *endocarp brittle, smooth.* **Seed** 8–12.5 mm diam.

DISTRIBUTION. Central Range of New Guinea from the Weyland Mountains in the west to Southern Highlands of Papua New Guinea in the east.

HABITAT. Montane forest at 500 m, perhaps to 1,800 m.

LOCAL NAMES. *Holia* (Yali), *Janggeam këlèlpa* (Yali).

USES. None recorded.

CONSERVATION STATUS. Least Concern.

NOTES. The identity of this species has remained uncertain because of a sketchy original description based on a single specimen from Etappenberg collected by Ledermann in 1912 in the upper Sepik area and later destroyed at Berlin. It stands out in having a racemose inflorescence with first-order branches that are subtended by strap-shaped bracts.

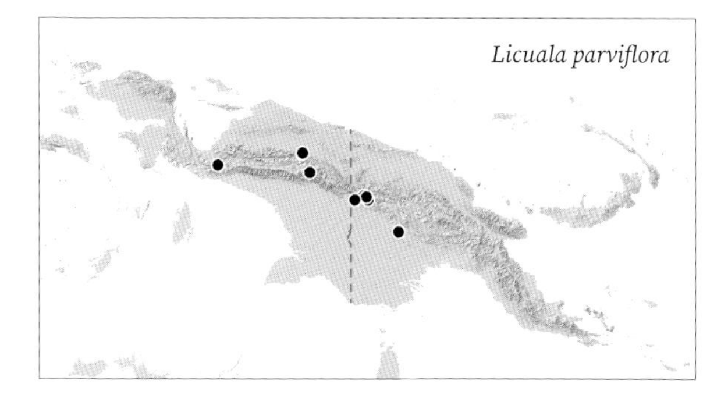

20. *Licuala penduliflora* (Blume) Zipp. ex Blume

Synonyms: *Licuala aruensis* Becc., *Licuala penduliflora* var. *aruensis* (Becc.) Becc., *Licuala penduliflora* var. *australis* Becc., *Pericycla penduliflora* Blume

Single-stemmed palm to 2–10 m. **Stem** 6–8 cm diam. **Leaf** sheath 40–60 cm long, disintegrating into bands of fibrous mesh proximally, with 2 early caducous, strap-like extensions; petiole 150–170 cm long, lower ⅓ armed; blade divided into 15–25 segments; mid-segment 50–75 cm long, 4–16 cm wide, sometimes with additional, irregular split reaching 5–10 cm above the hastula. **Inflorescence** *1.5–2.5 m long, erect to slightly curved, all parts with rust-coloured hairs,* branched to 2 orders, *with 5–9 first-order branches;* peduncle 20–25 cm long, contained in prophyll; peduncular bract lacking; *rachis 1.6–2.3 cm long, straight;* basal rachis bract 27–34 cm long; proximal first-order branch with 5–8 cm long peduncle and 15–20 cm long main axis bearing 15–20 rachillae. **Flowers** in pairs basally and solitary distally, subtending bracts to 1 mm long, *pedicels 2–3 mm long, with scattered rust-coloured hairs;* calyx 2.5–3 mm long, breaking up in 0.4–0.6 mm long, rounded lobes; *corolla yellow,* 3–4 mm long; stamens *distinctly biseriate;* ovary glabrous, 0.8–1 mm long, truncate to rounded apically; style 0.5–0.6 mm long. **Fruit** globose, 8–10 mm diam., orange at maturity, mesocarp 1 mm thick, *endocarp brittle, smooth.* **Seed** endosperm with sinuous cavities resembling ruminations.

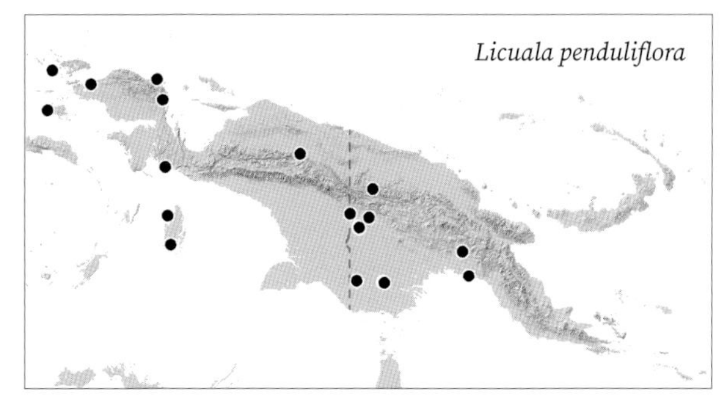

Licuala penduliflora. Habit, Prafi River (WB).

DISTRIBUTION. Widely distributed in New Guinea, the Raja Ampat Islands and Aru.

HABITAT. Lowland forest at sea level to 400 m.

LOCAL NAMES. *Erela* (Orokolo), *Kaa* (Uruaru), *Kie* (Tmindemanuk), *Nú* (Matbat), *Pesem* (Gebe), *Sinlo* (Sough).

USES. For construction. Edible palm heart.

CONSERVATION STATUS. Least Concern.

NOTES. A variable species recognised by the combination of long pedicellate flowers and a distinctly biseriate staminal ring with a narrow opening into the flower. The latter is probably closely linked to the pollination mechanism.

21. *Licuala sandsiana* Barfod & Heatubun

Single-stemmed palm to 4 m. **Stem** ca. 6–8 cm diam. **Leaf** sheath 20–25 cm long, disintegrating into a fibrous mesh, later lost; petiole 50–80(–100) cm, *lower ⅓ to ½ armed;* blade divided into 14–17 segments, mid-segment 45–50 cm long, 8–12 cm wide and truncate at the apex. **Inflorescence** 60–80 cm, branched to 2 orders, *with 6–7 first-order branches,* straight to slightly curved; *peduncle 15–20 cm long, contained within or barely exserted from the prophyll, covered in patches of dense rust-coloured tomentum;* peduncular bract missing; rachis 45–60 cm long; rachis bracts inflated distally; *proximal first-order branch with peduncle hidden in subtending bract,* 2–3 cm long rachis, bearing

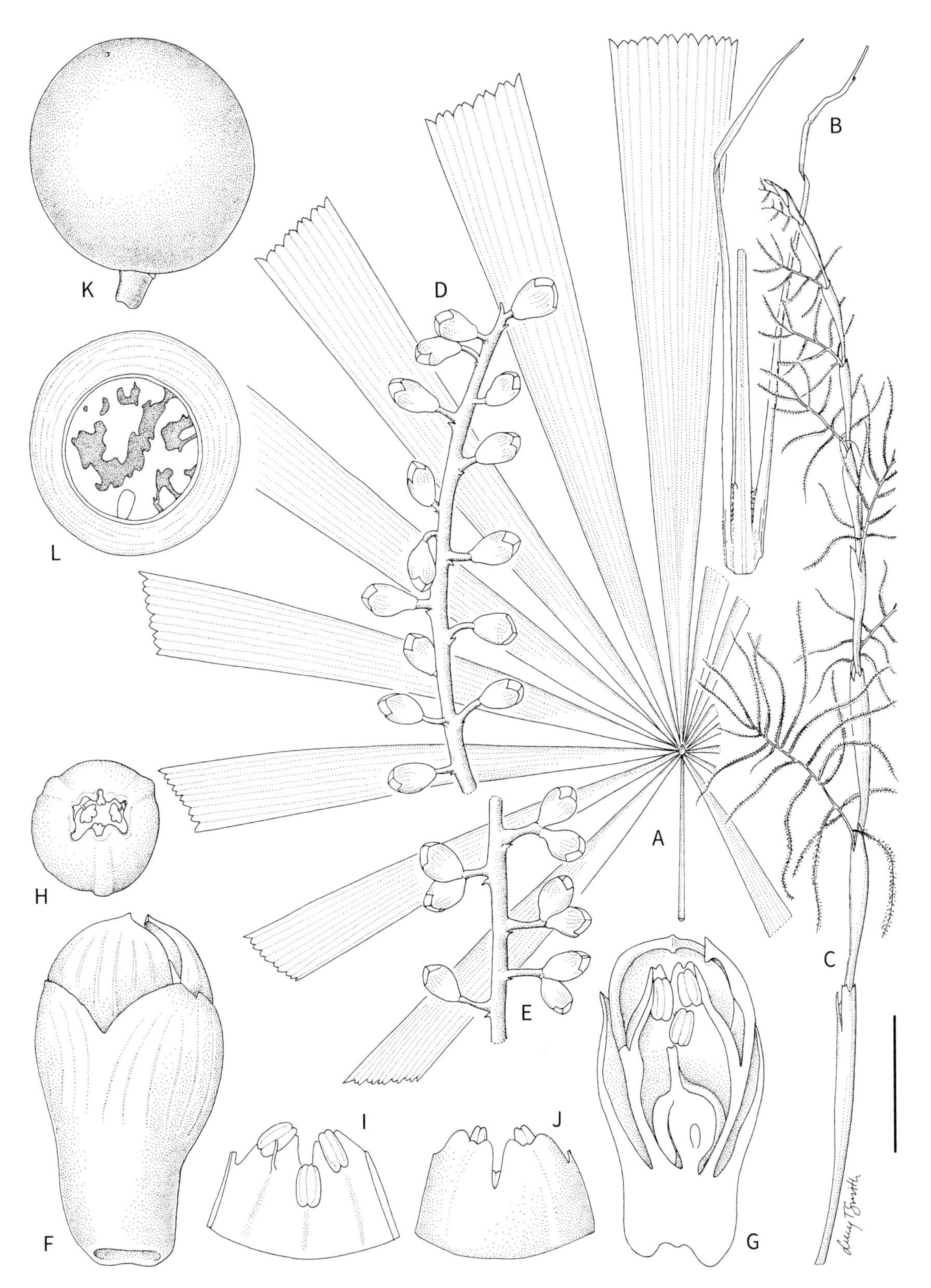

Licuala penduliflora. **A.** Leaf. **B.** Leaf sheath. **C.** Inflorescence. **D**, **E.** Upper and lower portion of rachilla with flowers. **F**, **G.** Flower whole and in longitudinal section. **H.** Staminal ring from above. **I**, **J.** Staminal ring in two views. **K**, **L.** Fruit whole and in transverse section. Scale bar: A = 18 cm; B, C = 12 cm; D, E, K, L = 7 mm; F–J = 1.2 mm. A–J from *Zippelius 265*; K, L from *Brass 13907*. Drawn by Lucy T. Smith.

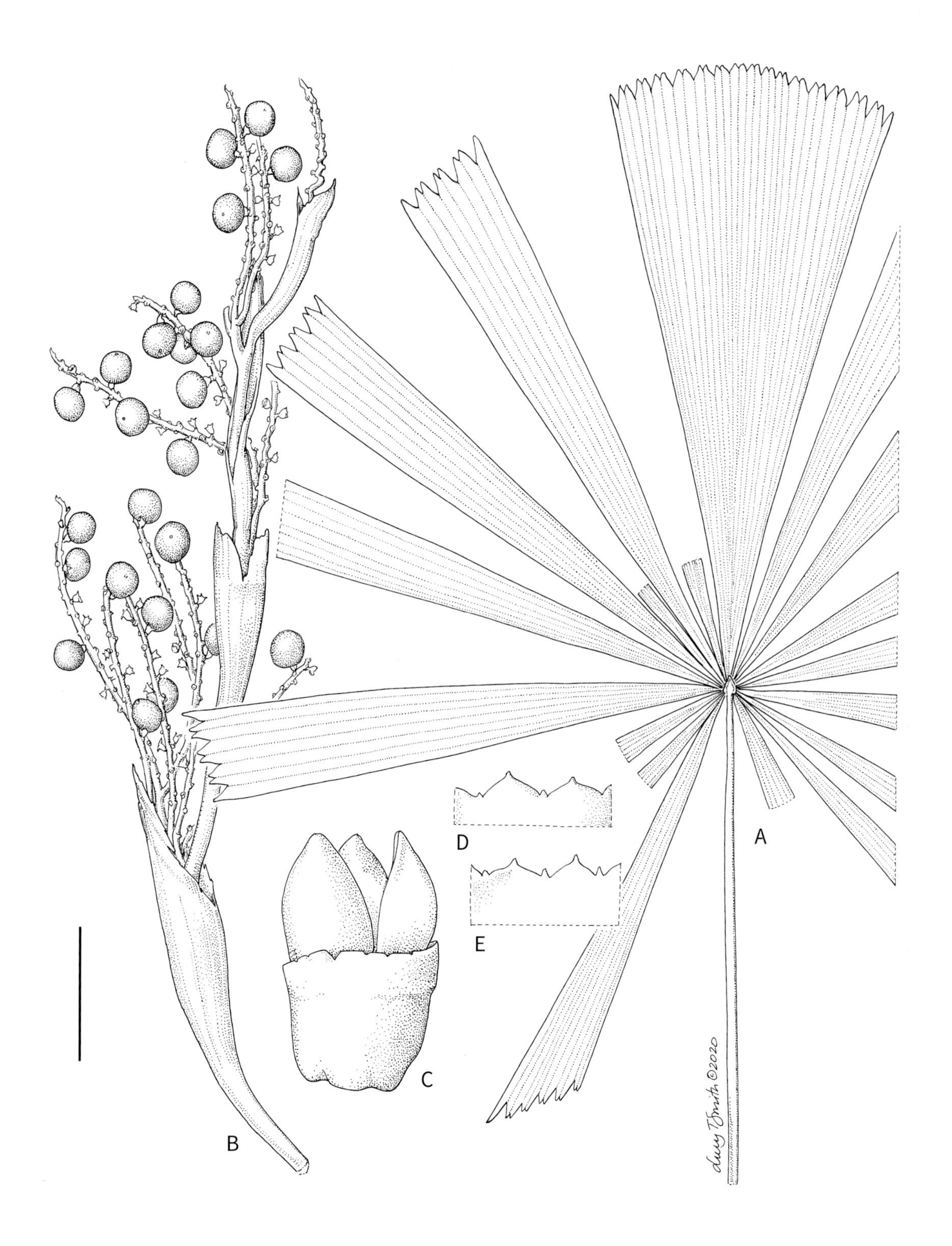

Licuala sandsiana. **A.** Leaf. **B.** Portion of inflorescence with fruit. **C.** Flower (incomplete). **D, E.** Remnants of staminal ring (reconstructed). Scale bar: A = 12 cm; B = 4 cm; C–E = 2.5 mm. A, C–E from *Sands et al. 2768*; B from *Kerenga LAE 77563*. Drawn by Lucy T. Smith.

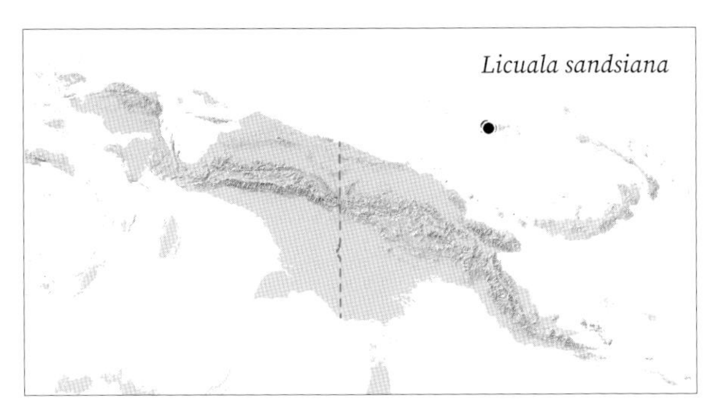

7–9 robust rachillae, 5–15 cm long, both rachis and rachillae covered in rust-coloured tomentum. **Flowers** in pairs basally, solitary distally, subtending bracts a shallow rim to 0.5 mm long, pedicels ca. 0.5 mm long; calyx ca. 3 mm long, breaking up irregularly, truncate, stamens ca. 1 mm long, *biseriate*. **Fruit** globose, *12–15 mm diam.*, yellow to red at maturity, *mesocarp 3–4 mm thick, endocarp brittle, slightly grooved longitudinally*. **Seed** 6-8 mm diam.

DISTRIBUTION. Manus Island.

HABITAT. Lowland forest at sea level to 125 m.

LOCAL NAMES. *Sanin* (Manus).

USES. None recorded.

CONSERVATION STATUS. Endangered. *Licuala sandsiana* has a restricted distribution. Deforestation due to logging concessions is a major threat in its distribution range.

NOTES. A member of the *L. lauterbachii* complex (*L. essigii, L. flexuosa, L. lauterbachii, L. sandsiana*). Distinguished by its moderate size and short petioles. The inflorescence is short, bearing 7–9, robust, first-order branches, which are partly contained in the subtending bract. The material available to us comprised only a few flowers past anthesis. The fruits are relatively large with a thick mesocarp.

22. *Licuala simplex* (Lauterb. and K.Schum.) Becc.

Synonyms: *Dammera ramosa* Lauterb. & K.Schum., *Dammera simplex* Lauterb. & K.Schum., *Licuala beccariana* (Lauterb. & K.Schum.) Furtado, *Licuala macrantha* Burret, *Licuala ramosa* (Lauterb. & K.Schum.) Becc. (not *Licuala ramosa* Blume)

Single-stemmed, tree palm, rarely more than 1.5 m. **Stem** 5–6 cm diam., erect or procumbent. **Leaf** sheath 30–40 cm long, disintegrating into a fibrous mesh, distal part often detached from petiole in 2 strap-like appendices ca. 10 cm long; petiole variable in length, to 1.8 m in fully developed leaves, armed on lower half or sometimes entirely; *blade divided into 7–9 segments; mid-segment bifid*, sometimes petiolulate, 60–95 cm long, lobes asymmetrically truncate at the apex. **Inflorescence** *0.6–1.5 m long, unbranched or bearing of 2–4-spicate first-order branches (rachillae), erect at flowering, later curved to pendent by the weight of the fruits*; peduncle 20–60 cm long, flattened, sparsely to densely covered by rust-coloured indumentum; peduncular bract lacking; rachis to 15 cm long in branched inflorescences; *rachilla 8–45 cm long, covered with patches of rust-coloured indumentum*. **Flowers** solitary, *10–15 mm long, spindle-shaped, borne on pedicels decreasing in length along the rachilla from up to 20 mm basally to 1–2 mm distally*; calyx rounded to triquetrous, truncate or with 3, 8–10 mm long lobes, covered with rust-coloured hairs except on the membranous apical rim; corolla cream; *stamens uniseriate*; ovary glabrous, ca. 1.5 mm long, rounded to ellipsoid, truncate to attenuate apically; *style 4–5 mm long*. **Fruit** *globose to ellipsoid, 1.9–2.2 × 1.7–2.0 cm, dark red to reddish brown at maturity; endocarp 1.5–1.8 × 0.9–1.6, with ca. 10 up to 1.5 mm wide, longitudinal ridges*. **Seed** 8–10 mm diam.

Licuala simplex. Habit, East Sepik (AK).

***Licuala simplex*. A.** Habit. **B.** Leaf. **C.** Inflorescence with fruit. **D.** Portion of rachilla with flowers. **E**, **F.** Flower whole and in longitudinal section. **G.** Ovary. H–**J.** Fruit whole, in longitudinal section and in transverse section. **K.** Endocarp. Scale bar: A = 30 cm; B = 12 cm; C = 8 cm; D, H, K = 2 cm; E–G = 1 cm; I, J = 3 cm. All from *Baker & Utteridge 589*. Drawn by Lucy T. Smith.

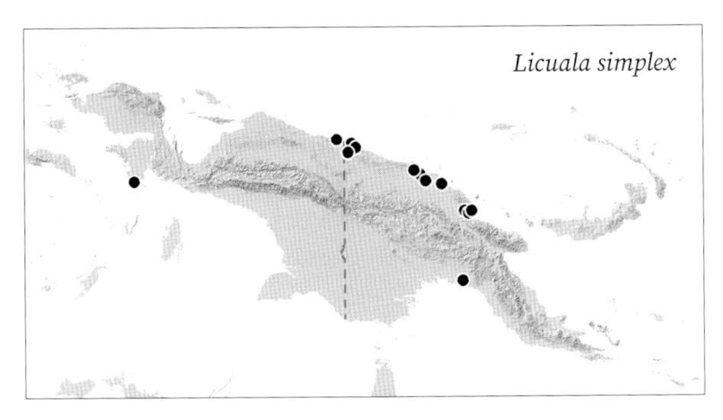

Licuala simplex. Inflorescence, Papua New Guinea (AB).

DISTRIBUTION. Widely distributed in north-eastern New Guinea, with outlying records in the south and west.

HABITAT. Lowland forest understorey, often on well-drained sandy soils in river beds at sea level to 300 m.

LOCAL NAMES. *Bruk Bral* (Bewani), *Dipwarr* (Angoram), *Sarr* (Madang), *Wiika* (Weewak).

USES. Leaves are used for thatching bush shelters and mature fruits as a betel substitute by the Maia people of Madang province.

CONSERVATION STATUS. Least Concern.

NOTES. Easily recognised due to its short, rather thick stem, bifid mid-segment of the leaf and large flowers. The inflorescence is erect at anthesis, but later bends over due to the weight of the relatively large fruits. *Licuala simplex* is one of the several species of *Licuala* endemic to New Guinea with ornate endocarps. This species has been described several times under different names. When the genus *Dammera* was merged into *Licuala* by Beccari, a homonym was created as the name *Licuala ramosa* Blume already existed. Furtado was the first to realise this and coined the name *Licuala beccariana*.

Licuala simplex. LEFT TO RIGHT: habit, East Sepik (AB); fruit, Wewak (JLD).

Licuala suprafolia. **A.** Habit. **B.** Leaf. **C.** Petiole. **D.** Inflorescence. **E.** Portion of rachilla with flowers. **F, G.** Flower whole and in longitudinal section. **H.** Staminal ring. **I, J.** Fruit whole and in transverse section. **K.** Seed. Scale bar: A = 1.1 m; B = 24 cm; C, D = 18 cm; E, K = 5 mm; F–H = 1.6 mm; I, J = 7 mm. All from *Barfod et al. 462*. Drawn by Lucy T. Smith.

23. *Licuala suprafolia* Barfod & Heatubun

Single-stemmed palm to 7 m. **Stem** ca. 8 cm diam. **Leaf** bases persistent to the ground in shorter individuals; sheath ca. 30–40 *cm long, early disintegrating into a delicate deciduous, fibrous, brown mesh; petiole 100–120 cm long, covered with patches of dense, adpressed, rust-coloured indumentum, armed in lower half*; blade, divided into 19–23 segments; mid-segment 75–90 cm long, 12–15 cm wide and truncate apically. **Inflorescence** *2.5–3.2 m long, erect and arching away from the crown centre, with 10–11 first-order branches*; peduncle 1.1–1.2 cm long, covered by the prophyll and peduncular bracts; *2 peduncular bracts*, 30–40 cm long; rachis 140–200 cm long; peduncle of basal first-order branch hidden in subtending bract, main axis 25–30 cm long, proximally with rust-coloured tomentum in patches, bearing 25–30 rachillae, with minute hairs. **Flowers** in pairs proximally to mostly single distally on the rachillae; pedicels of single flowers 0.8–1.2 mm long, those of paired flowers 0.2–0.4 mm long; *calyx bell-shaped, glabrous, breaking up into 3, ca. 1.5 mm long, pointed lobes*; corolla white; *stamens biseriate*; ovary glabrous, ca. 1 mm long, more or less truncate apically; style 0.4–0.6 mm long. **Fruit** globose, 9–12 mm diam., *mesocarp ca. 2.5 mm thick, endocarp brittle, slightly furrowed.* **Seed** 5–7 mm diam.

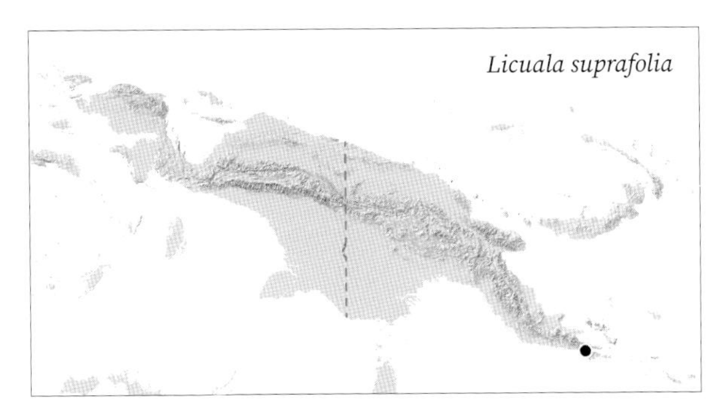

DISTRIBUTION. One site west of Alotau in far south-eastern New Guinea.

HABITAT. Periodically inundated forest patches surrounded by grassland at sea level.

LOCAL NAMES. None recorded.

USES. None recorded.

CONSERVATION STATUS. Critically Endangered. The only known site of *Licuala suprafolia* is threatened by a logging concession.

NOTES. Characterised by its long persistent leaf bases and heavily armed petioles. The inflorescence is more than 2.5 m long and exposed above the crown at anthesis. It has two peduncular bracts and numerous first-order branches. See note under *Licuala multibracteata*.

24. *Licuala telifera* Becc.

Synonyms: *Licuala angustiloba* Burret, *Licuala concinna* Burret, *Licuala flavida* Ridl., *Licuala klossii* Ridl., *Licuala leprosa* Lauterb. & K.Schum., *Licuala leptocalyx* Burret, *Licuala linearis* Burret, *Licuala oninensis* Becc., *Licuala paucisecta* Burret, *Licuala platydactyla* Becc.

Single-stemmed, *dioecious*, understorey palm, to 1.5 m. **Stem** 2–3.5 cm diam., covered by persistent leaf bases. **Leaf** sheath 25–35 cm long, brown, breaking up in fine fibrous mesh, extended apically to form, 2 early caducous, strap-like appendices, 10 cm long; petiole 50–110 cm long, armed on lower 20–25 cm; *blade characteristically stiff, divided into 5–11 segments, with minute, rust-coloured indumentum underneath*; mid-segment 25–40 cm long, *deeply split more than ⅔ of length*, truncate. **Inflorescence** 45–80 cm long, erect, peduncle 35–50 cm long, flattened in cross section, first-order branches 3–6, the proximal one with 2–4 cm long peduncle and a 1–2 cm long main axis, rachillae pubescent, spreading; male inflorescence usually without peduncular bract and 5–6 first-order branches, the proximal one bearing 3–6 rachillae, these being 8–10 cm long; female inflorescence with 1–2 peduncular bracts and 3–4 first-order branches, the proximal one bearing 1–4 rachillae, these being shorter and thicker the male rachillae. **Flowers** paired to solitary distally on

Licuala telifera. Flowers, Tamrau Mountains (WB).

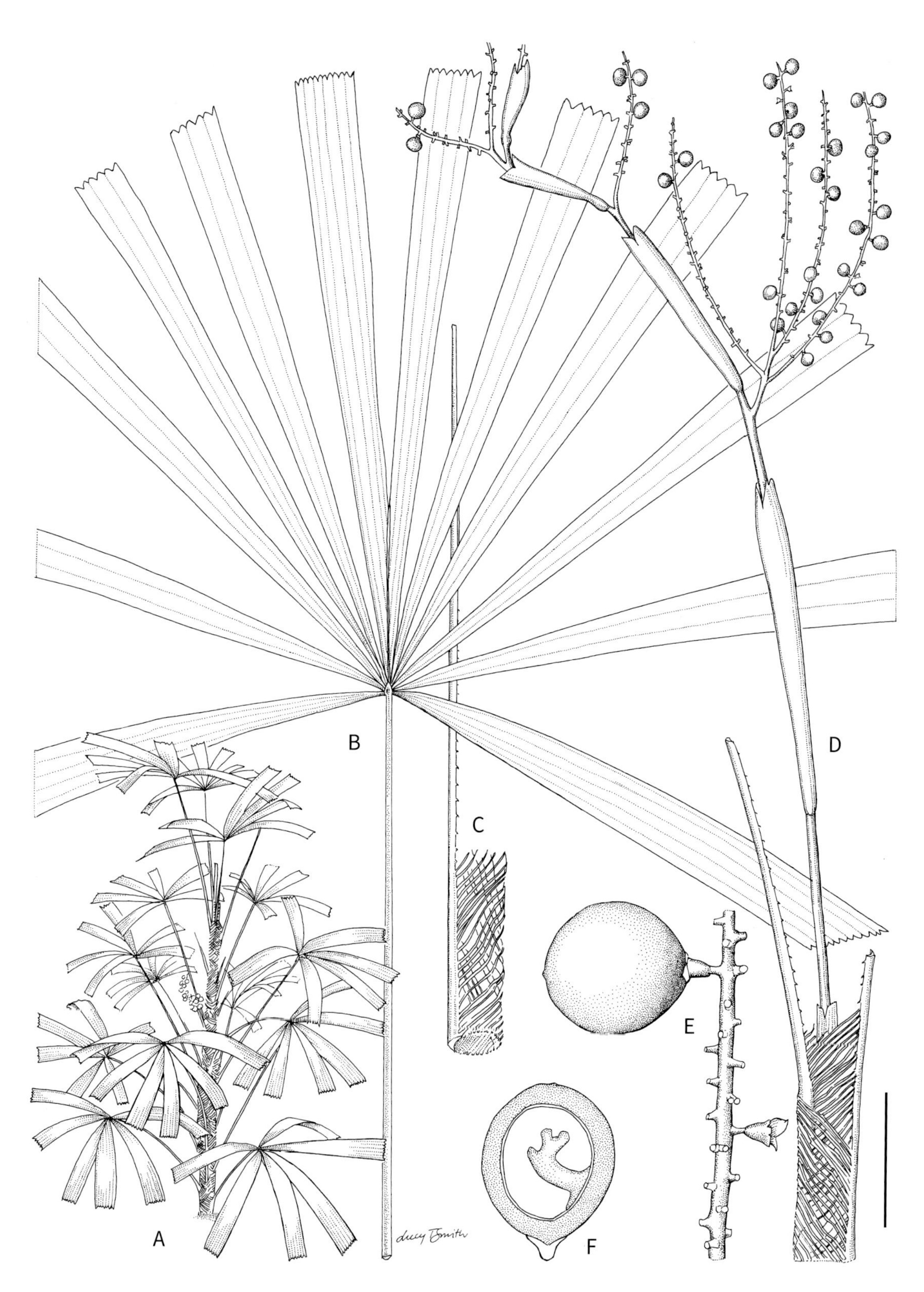

Licuala telifera. **A.** Habit. **B.** Leaf. **C.** Petiole. **D.** Inflorescence with fruit. **E.** Portion of rachilla with fruit. **F.** Fruit in longitudinal section. Scale bar: A = 42 cm; B, C = 6 cm; D = 4.5 cm; E, F = 1 cm. A from photograph; B–D from *Baker et al. 626*; E, F from *Baker et al. 608*. Drawn by Lucy T. Smith.

Licuala telifera. LEFT TO RIGHT: habit, Tamrau Mountains (WB); inflorescence, Papua New Guinea (AB).

the rachilla; pedicel 1–5 mm long; calyx bell-shaped, 2–2.2 mm long, striate after drying, with rust-coloured hairs at the base, split regularly into 3, ca. 1 mm long, pointed to shortly acuminate lobes; corolla white to cream; *stamens biseriate*; ovary 2.2–2.5 mm long, rounded, glabrous, rudimentary ovary in male flowers narrowly obclavate. **Fruit** globose to ellipsoid, 8–10 × 7–9 mm, orange to red at maturity, mesocarp ca. 1 mm thick, *endocarp brittle, smooth*. **Seed** 6–8 cm diam., endosperm with irregularly shaped cavity.

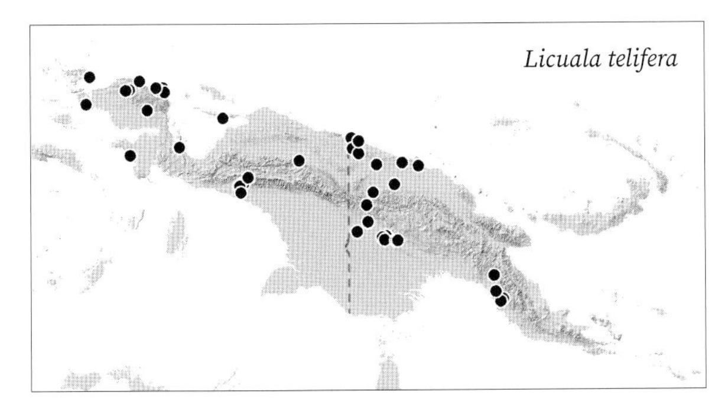

DISTRIBUTION. Widely distributed in New Guinea.

HABITAT. Lowland forest in light open places such as creeks and disturbed forest at sea level to 900 m.

LOCAL NAMES. *Amorau* (Wandama), *Murap* (Samu Kundi), *Nogusah* (Waskuk), *Pupuru* (Yamur, collective term for *Licuala*), *Pingu* (Mekeo), *Poetjem* (Kebar), *Sambul mun* (Bewani), *Touku* (Wagu), *Wah* (Bewani).

USES. Peripheral sclerified layers of the stem used for tools such as harpoons, arrow tips and sago scraping tools.

CONSERVATION STATUS. Least Concern.

NOTES. This species has been misunderstood due to variation in leaf morphology and the sexual dimorphism of the inflorescences. With more material becoming available, it is now clear that many of the characters used to delimit this species are highly variable, even within the same population. This species is clearly a candidate for studies of evolution of sexual expression in general and dioecy in particular.

25. *Licuala urciflora* Barfod & Heatubun

Single-stemmed, medium-sized tree palm to 3 m. **Stem** 4–5 cm diam. **Leaf** sheath 20–25 cm long, disintegrating into loose, fibrous mesh; petiole to 120 cm, armed on lower half; *blade divided into 30–36 segments, mid-segment 55–65 cm long, deeply split to*

Licuala urciflora. **A.** Leaf. **B.** Leaf sheath. **C.** Inflorescence. **D.** Rachilla with flowers. **E**, **F.** Flower whole and in longitudinal section. **G.** Staminal ring. **H.** Filaments on a ring. Scale bar: A, B = 12 cm; C = 6 cm; D = 7 mm; E, F = 2.5 mm; G, H = 1.6 mm. All from *van Royen 5158*. Drawn by Lucy T. Smith.

4–6 cm above the base, lobes truncate, 1.5–3 cm wide at the apex. **Inflorescence** 0.8–1.2 cm long, erect with 4–6 first-order branches; peduncle ca. 30 cm long, flattened in cross section; peduncular bracts lacking; rachis erect and more or less straight 50–70 cm long; *rachis bracts inflated, urn-shaped, covered with deciduous, rust-coloured indumentum, splitting neatly to ca. the middle along one side;* proximal first-order branch with 4 cm long main axis, bearing 5–7 rachillae, to 6 cm long, pubescent and densely covered with flowers. **Flowers** solitary, *subsessile on elevated points, 6–6.5 mm long, bullet-shaped;* calyx 4–4.5 mm long, breaking up into 3, sometimes notched lobes; *stamens biseriate;* ovary 2.0–2.4 mm long, flattened, rounded to attenuate apically; style ca. 1.3–1.5 mm long. **Fruit** ellipsoid, 8–10 mm long. **Seed** 6–8 mm diam.

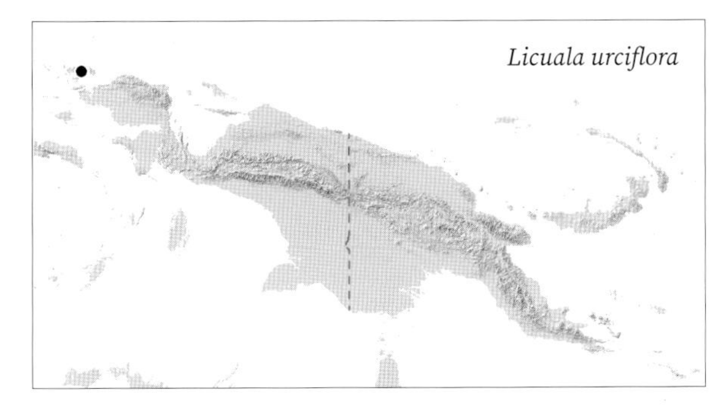

DISTRIBUTION. Waigeo Island.

HABITAT. Recorded in *Vatica-Horsfieldia* forest along river banks at sea level.

LOCAL NAMES. *Sil* (Ambel).

USES. Stem used as a traditional sago scraping tool.

CONSERVATION STATUS. Data Deficient. This species is known only from one site, where there is limited information on ongoing threats. As there is insufficient evidence, a single extinction risk category cannot be selected.

NOTES. *Licuala urciflora* shows affinities to *Licuala flexuosa* Burret from Halmahera. They both have a wide leaf mid-segment and the flowers are similar. *Licuala urciflora* is easily distinguished, however, by the mid-segment being divided almost to the base (as opposed to not split), an inflated rachis bracts split neatly to the middle (as opposed to not inflated and shortly split), rachilla straight with closely inserted flowers (as opposed to rachilla zigzag with loosely inserted flowers).

Names of uncertain application

Licuala naumannii Burret

This species was originally described by Burret (1935) from Bougainville Island. The type collection was destroyed in Berlin and we have not been able to locate duplicates elsewhere. Without the type material, an accurate interpretation is not possible. The original description suggests it may be a synonym of *Licuala lauterbachii* subsp. *peekelii.*

Licuala pachycalyx Burret

Burret (1933) described this species from 1,300 m altitude in the Wondiwoi Mountains on the Wandamen Peninsula. The type collection was destroyed in Berlin and we have not been able to locate duplicates elsewhere. Thus the identity of this name is determined entirely by the original description. Burret (1933) points to an affinity with *Licuala naumoniensis,* which is here considered a synonym of *L. montana.* This may well be a synonym of this species.

Licuala polyschista K.Schum & Lauterbach

The holotype was destroyed in Berlin. A duplicate is kept at FI, which is probably a mixed collection of *L. lauterbachii* (leaves and fruits) and *L. simplex* (old rachillae).

Licuala pulchella Burret

The type collection was destroyed in Berlin and not duplicated elsewhere. Burret (1935) points to a close affinity with *Licuala bacularia.* Without type material an accurate interpretation is not possible.

Licuala steinii Burret

This species is based on a specimen collected by Stein at 1,800 m altitude in the Weyland Mountains in Indonesian New Guinea. The type collection was destroyed in Berlin and not duplicated elsewhere. The original description of Burret (1933) and the altitude of the type locality both suggest a close affinity with *L. parviflora.* However, without type material, an accurate interpretation is not possible.

Saribus pendulinus. Habit, Purari River (SV).

Saribus Blume

Synonym: *Pritchardiopsis* Becc.

Medium to tall – leaf palmate – petiole usually spiny – leaf segments not wedge-shaped

Moderate or robust, single-stemmed fan palms, hermaphrodite, rarely dioecious. Leaf palmate; *sheath disintegrating into a brown fibrous network*; petiole long, with spiny margins; blade split to ca. one quarter or more of its radius, with pointed tips. **Inflorescence** between the leaves, *divided into 2–3 equal axes at base, all enclosed within a common prophyll*, individual axes branched 3–4 orders; primary bracts similar, tubular, few to many, not enclosing inflorescence in late bud, not dropping off as inflorescence expands; peduncle shorter than inflorescence rachis; rachillae generally slender, straight or curved. **Flowers** generally small, hermaphrodite (rarely unisexual) borne in groups of up to six, or solitary. **Fruit** *red, orange brown or rarely blackish, globose to obovoid*, stigmatic remains apical, flesh juicy or fibrous, endocarp smooth. **Seed** 1, endosperm homogeneous, with seed coat intruding into one side of the seed.

DISTRIBUTION. Ten species, in the Philippines, Borneo (Banggi Island), Sulawesi, Maluku, Solomon Islands, New Caledonia, and with eight species (six endemic) in New Guinea. Not recorded from the Bismarck Archipelago.

NOTES. *Saribus* was segregated from *Livistona* on molecular evidence (Bacon & Baker 2011) although the group had been recognised informally previously (Dowe 2009). The most pronounced morphological character of *Saribus* is the trifurcate (rarely bifurcate) inflorescence in which three (rarely two) equal axes are basally united within a single prophyll. The species of *Saribus* in New Guinea have orange, orange-brown, or red mature fruit colour, whereas the fruits of New Guinea *Livistona* are blue or black. All then known species were treated taxonomically within a monograph of *Livistona* (Dowe 2009) published shortly before *Saribus* was recircumscribed.

Key to the species of *Saribus* in New Guinea

1. Inflorescence composed of a single axis in basal 10–30 cm, then trifurcate, with the three equal axes lacking peduncular bracts .... **8. *S. woodfordii***
1. Inflorescence mostly trifurcate, or infrequently bifurcate, at base, with the 3 (or 2) equal axes more or less similar, and all sharing a common prophyll but with each axis bearing its own peduncular bract(s) 2
2. Mature leaf blade with 25 or fewer segments; depth of apical cleft 1–4% of the length of the segment .... **1. *S. brevifolius***
2. Mature leaf blade with 45 or more segments; depth of apical cleft 4–40% of the length of the segment 3

3. Fruit >35 mm long 4
3. Fruit <35 mm long 6

4. Inflorescence pendulous, axes (near basalmost first-order branch) thin, < 20 mm wide .... **4. *S. pendulinus***
4. Inflorescence arching, axes (near basalmost first-order branch) robust > 30 mm wide 5

5. Leaf blade with lower surface silvery glaucous; segment apices rigid; rachillae 6–12 cm long, glabrous; flowers solitary; fruit 35–43 mm long .... **7. *S. tothur***
5. Leaf blade with lower surface green; segment apices pendulous, hanging vertically; rachillae 14–24 cm long, pubescent; flowers in clusters of 2–4; fruit 50–65 mm long.................. **6. *S. surru***

6. Inflorescences branched to 4 orders .... **5. *S. rotundifolius***
6. Inflorescences branched to 3 orders 7

7. Sepals and petals yellow; rachillae reddish-brown pubescent, most dense underneath and immediately outside of enclosing bracts, whitish-green on exposed parts .... **3. *S. papuanus***
7. Sepals and petals red; rachillae with dense chocolate-coloured tomentum mostly throughout, distally with cream-green tomentum .... **2. *S. chocolatinus***

1. *Saribus brevifolius* (Dowe & Mogea) Bacon & W.J.Baker

Synonym: *Livistona brevifolia* Dowe & Mogea

Robust, single-stemmed tree palm to 22 m. **Stem** ca. 12 cm diam., leaf scars slightly raised, light grey, internodes narrow. **Leaves** 16–40 in a globose crown; *leaf blade circular, moderately undulate, rigid, 55–62 × 45–55 cm,* upper surface mid-green, lower surface light green, divided for more or less than half its length, *with 22–25 segments,* depth of apical cleft 1–4% of the length of the segment, apical lobes rigid; petiole ca. 1.1 m long, slightly arching, green, upper surface slightly concave, glabrous, with deciduous white waxy scales on the upper surface, lower surface rounded, margins lacking spines. **Inflorescences** trifurcate, ca. 60 cm long, each axis branched to 3 orders with 2–3 first-order branches; rachis bracts tightly tubular; rachillae 40–90 mm long, straight, pubescent, red at anthesis. **Flowers** solitary or in clusters of 2–4; *sepals red; petals broadly triangular, thick, fleshy, red*; anthers 0.2 mm long, pink. **Fruit** globose, 10–12 mm diam., red. **Seed** globose, 8–10 mm diam.

Saribus brevifolius. Habit, Gag Island (CDH).

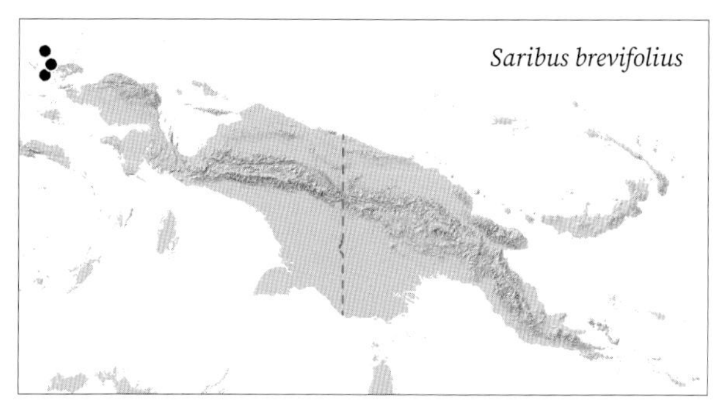

DISTRIBUTION. Raja Ampat Islands.

HABITAT. Forming colonies in open coastal forest on ultrabasic rocks at 10–20 m elevation (Heatubun *et al.* 2014a).

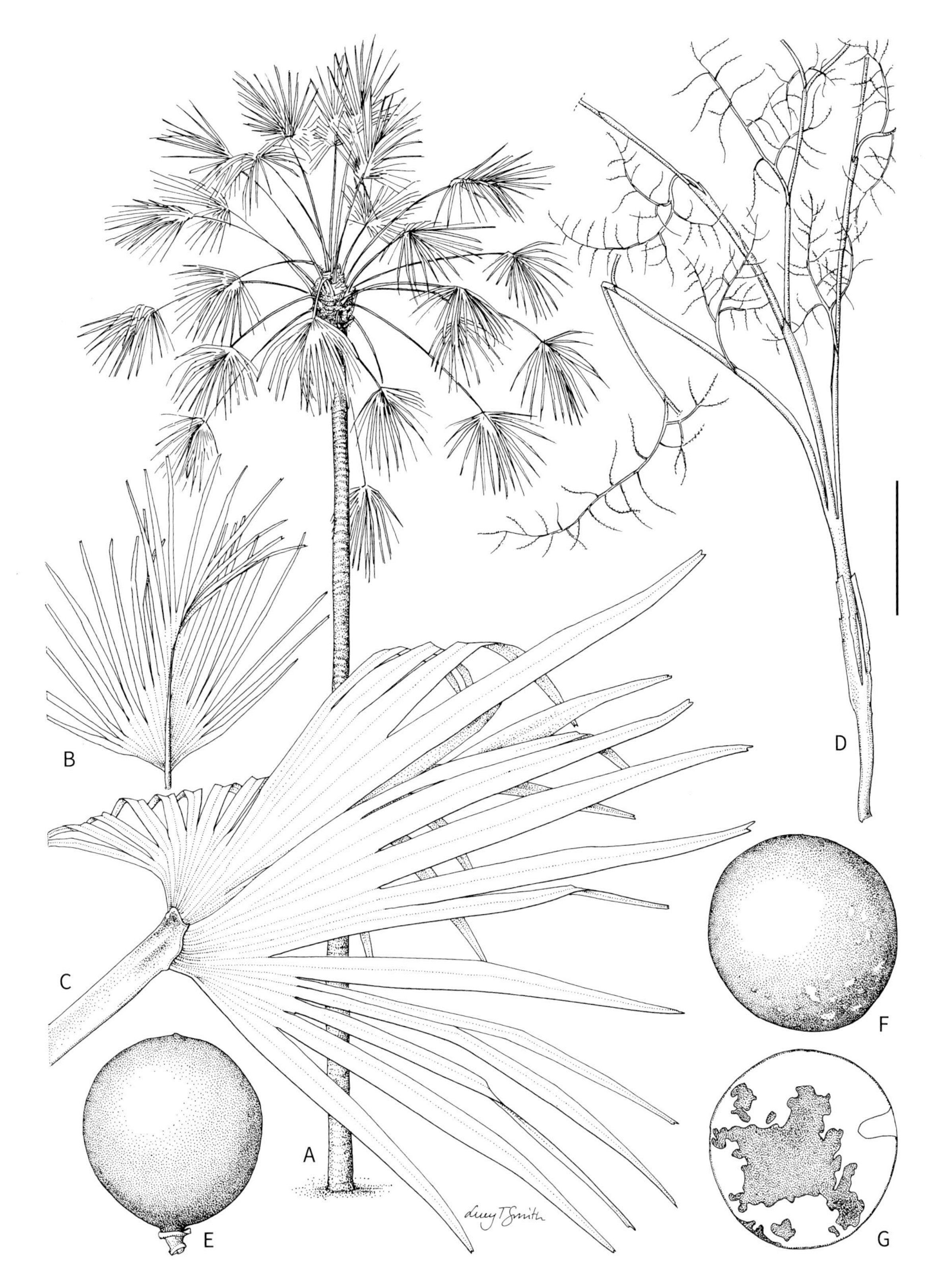

Saribus brevifolius. **A.** Habit. **B.** Leaf, abaxial view. **C.** Leaf, adaxial view. **D.** Inflorescence. **E.** Fruit. **F**, **G.** Seed whole and in longitudinal section. Scale bar: A = 2.6 m; B = 25 cm; C = 8 cm; D = 12 cm; E = 7 mm; F, G = 5 mm. A–C from photographs; D from *Takeuchi et al. 8171*; E–G from *Takeuchi 8224*. Drawn by Lucy T. Smith.

Saribus brevifolius. FROM TOP: crown; juvenile palms. Gag Island (CDH).

LOCAL NAMES. *Ngawan* (Gebe).

USES. None recorded.

CONSERVATION STATUS. Least Concern. Forest clearance and threats in its distribution range are limited. However, monitoring is required due to restricted distribution.

NOTES. *Saribus brevifolius* is a moderate canopy palm. It has the smallest leaves of all the *Saribus* species in New Guinea. They are regularly segmented with the segment apices rigid. Flowers have all parts red, the anthers are pink and the fruit are red at maturity.

2. *Saribus chocolatinus* (Dowe) Bacon & W.J.Baker

Synonym: *Livistona chocolatina* Dowe

Robust, single-stemmed tree palm to 22 m. **Stem** 16–18 cm diam., leaf scars slightly raised, internodes narrow, light grey, petiole stubs not retained. **Leaves** 30–40 in a globose crown; *leaf blade subcircular, flat, rigid, 1–1.2 m long and wide,* upper surface mid-grey green, lower surface light grey green, glaucous waxy, divided for ca. 44% of its length, with 45–60 segments, depth of apical cleft ca. 4% of the segment length, apical lobes rigid; leaf blade moderately undulate; petiole 1.1–1.6 m long, slightly arching, green, glabrous with a cover of deciduous white waxy powder, margins usually spineless in mature plants, or with small single spines to 5 mm long only in the very basal portion in juvenile plants. **Inflorescences** trifurcate but with central axis slightly more robust than the lateral axes all within a common prophyll, 2–2.3 m long; each axis branched to 3 orders with 6–10 first-order branches; peduncular bracts on each axis 2–4; *bases of second-order branches with dense chocolate brown tomentum*; rachillae 8–12 cm long, basally with chocolate brown tomentum, distally with cream-green tomentum. **Flowers** solitary or in clusters of 2–4; *sepals and petals red.* **Fruit** globose, ca. 25 mm diam., shiny orange-red. **Seed** globose, ca. 20 mm diam.

DISTRIBUTION. South-eastern New Guinea.

HABITAT. Grows in isolated colonies, sometimes locally common, on slopes with calcareous or clayey soils at 300–400 m.

LOCAL NAME. *Manganau* (Lababia).

USES. None recorded.

CONSERVATION STATUS. Near Threatened (IUCN 2021). *Saribus chocolatinus* has a somewhat restricted distribution. Deforestation due to mining and logging concessions is a major threat in its distribution range.

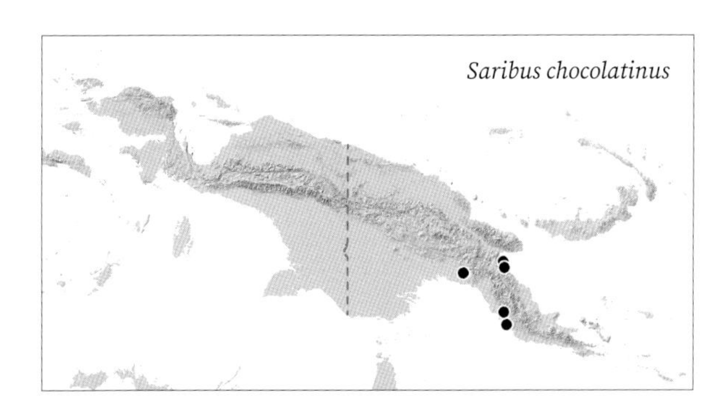

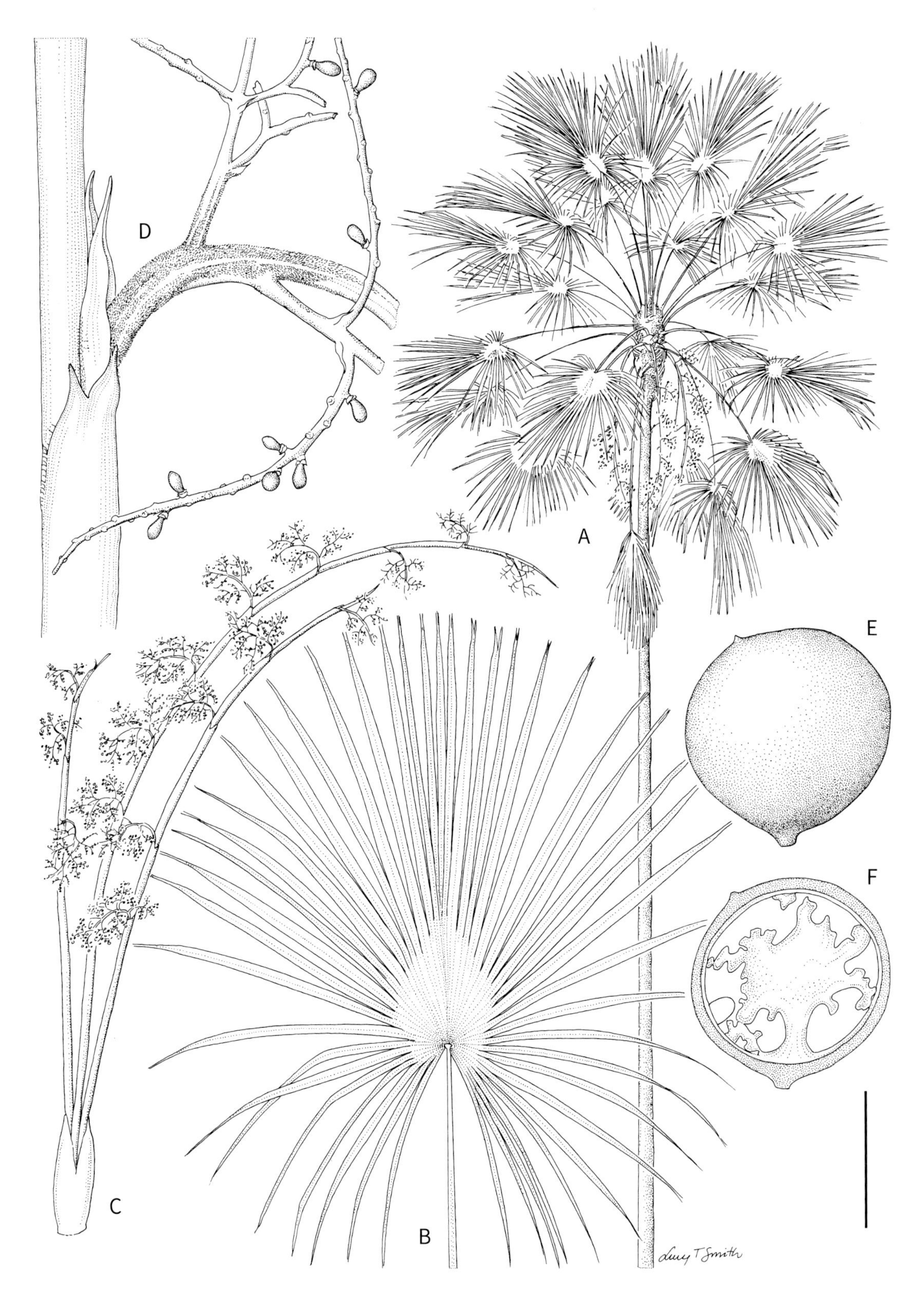

Saribus chocolatinus. **A.** Habit. **B.** Leaf. **C.** Inflorescence. **D.** Portion of inflorescence with fruit and indumentum on rachis. **E**, **F.** Fruit whole and in longitudinal section. Scale bar: A = 2.4 m; B = 23 cm; C = 30 cm; D = 2.5 cm; E, F = 1.5 cm. A, E, F from *Kjaer 514*; B–D from *Barfod 466*. Drawn by Lucy T. Smith.

Saribus chocolatinus. FROM TOP: habit; inflorescences. Kuriva (JLD).

NOTES. *Saribus chocolatinus* is distinguished by the usually unarmed or only lightly armed petiole that initially has a thick coating of white waxy powder. The leaf segments are rigid. The first-order branches and rachillae possess distinctive chocolate brown tomentum and the globose fruit mature orange-red.

3. *Saribus papuanus* (Becc.) Kuntze

Synonym: *Livistona papuana* Becc.

Robust, single-stemmed tree palm to 30 m (50 m?). **Stem** 12–30 cm diam., leaf scars raised, internodes broad, grey. **Leaves** 17–40 in a globose crown; *leaf blade subcircular, 0.9–1.8 × 1–1.5 m*, upper surface shiny or greyish green, lower surface lighter green, rigid, waxy glaucous, divided for 23–69% of its length, with 45–90 segments, depth of apical cleft 5–11% of the segment length; apical lobes acuminate, rigid; *petiole 1.1–2 m long, whitish with a thin, flakily deciduous waxy coating*, margins armed throughout, or with spines infrequently confined to proximal portion or very infrequently lacking; margins sharp when unarmed; spines recurved, to 2 cm long. **Inflorescences** trifurcate with a common prophyll, similar equal axes, 1–2.3 m long, arched-nodding, the central axis slightly longer than the laterals with each axis branched to 3 orders; first-order branches 5–10; peduncular bracts on each axis 1–4; rachillae 3–12 cm long, ca. 1 mm thick, straight, reddish-brown pubescent, most dense underneath and immediately outside of enclosing bracts, whitish-green on exposed parts. **Flowers** solitary or in clusters of 2–4, ca. 1.2 mm long; *sepals and petals yellow*. **Fruit** globose to obovoid-obpyriform, 14–25 × 5–20 mm, orange-red, apex rounded, tapered to a narrow base, stigmatic remains slightly subapical. **Seed** globose, 20–25 mm diam.

DISTRIBUTION. Scattered localities in western New Guinea, including Misool and Yapen.

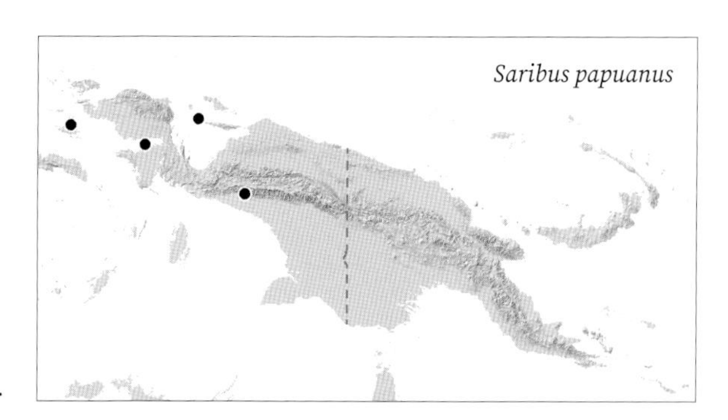

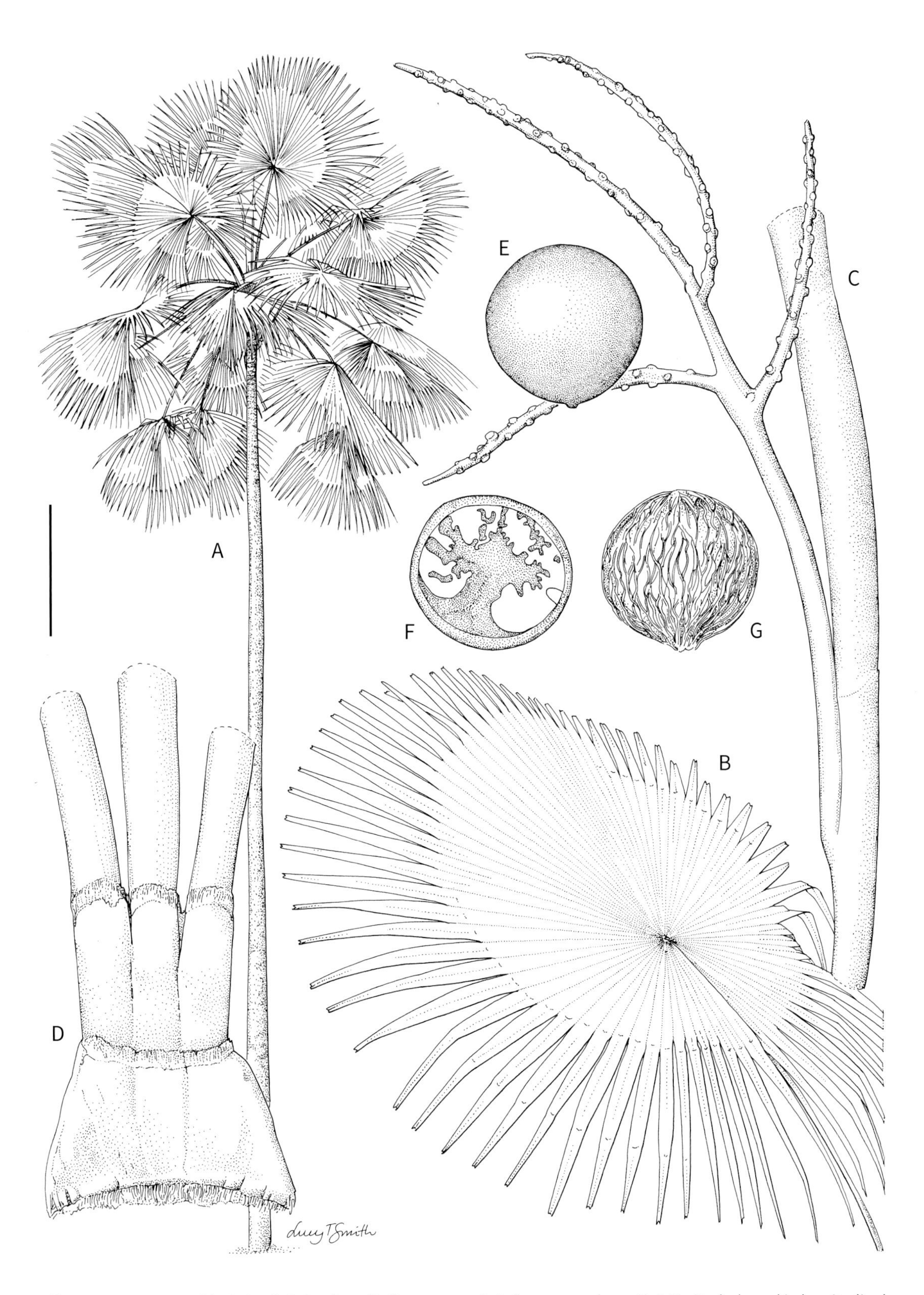

Saribus papuanus. **A.** Habit. **B.** Leaf. **C.** Portion of inflorescence. **D.** Inflorescence base. **E**, **F.** Fruit whole and in longitudinal section. **G.** Seed. Scale bar: A = 3.3 m; B = 30 cm; C, D = 6 cm; E–G = 2.5 cm. All from *Baker et al. 851*. Drawn by Lucy T. Smith.

Saribus papuanus. TOP LEFT: habit, near Jayapura (GP); TOP RIGHT, ABOVE LEFT TO RIGHT: habit; juvenile palm; juvenile palm in forest understorey; leaf sheaths, all near Timika (WB).

HABITAT. Rainforest at 200–540 m.

LOCAL NAMES. *Woka* (Papuan), *Wanna* (Poparo).

USES. None recorded.

CONSERVATION STATUS. Near Threatened. *Saribus papuanus* is known from four scattered sites where it is threatened by logging and mining concessions.

NOTES. *Saribus papuanus* is a tall canopy palm with large leaves that are regularly segmented. The apices are rigid. The inflorescence bracts are tightly tubular. The flowers are yellow and the globose fruit are orange-red at maturity.

4. *Saribus pendulinus* Dowe & S.Venter

Moderately robust, single-stemmed tree palm to 30 m. **Stem** 8–14 cm diam., leaf scars slightly raised, internodes narrow to broad, initially green aging to grey-brown, petiole stubs not retained. **Leaves** 30–40 in a globose to ovoid crown; *leaf blade subcircular, flat, rigid, 1.5–2 m long and wide, upper surface green, lower surface lighter green, divided for ca. 50% of its length,* with 60–80 segments, apex entire acuminate, apical lobes pendulous with age; leaf blade shallowly to moderately costapalmate; petiole 1.3–1.5 m long, straight to arching, green, *readily deciduous papery scales mainly aggregated toward the margins, margins with small single spines to 5 mm long,* sometimes lacking spines in mature individuals or only in the very basal portion. **Inflorescences** bifurcate or trifurcate with a common prophyll, 1.1–2 m long, *pendulous, axes near basalmost first-order branch thin <20 mm wide,* more or less of equal thickness; each axis branched to 3 orders; peduncular bracts on each axis 2–3, tightly sheathing; *second-order branches glabrous or minutely hairy, salmon-red; rachillae finger-like, 30–60 mm long, red at anthesis.* **Flowers** solitary or in clusters of 2–4; *sepals and petals red.* **Fruit** *globose, 40–45 mm diam.,* shiny orange-red, stigmatic remains apical. **Seed** globose, 30–40 mm diam.

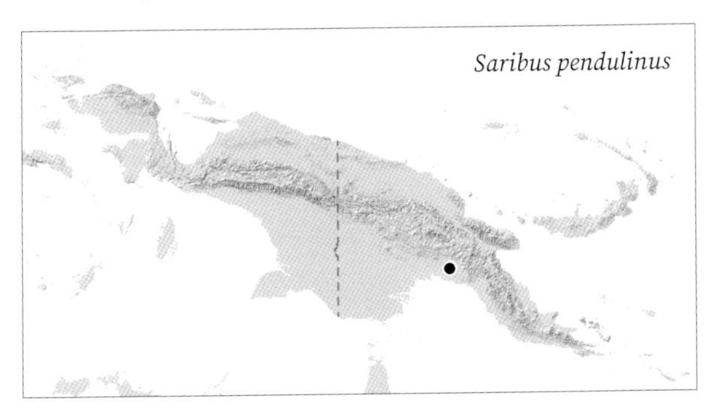

Saribus pendulinus. FROM TOP: habit; juvenile. Purari River (SV).

DISTRIBUTION. Southern Papua New Guinea.

HABITAT. In lowland hill rainforest at 100–150 m, on ridges of conglomerate sandstone.

LOCAL NAMES. None recorded.

USES. None recorded.

CONSERVATION STATUS. Vulnerable. *Saribus pendulinus* populations are small, limited in distribution and threatened by planned developments in the area for liquid natural gas extraction.

NOTES. *Saribus pendulinus* is a moderately tall palm distinguished by the thin pendulous inflorescences and infructescences. Stems are relatively thin and the leaves large. Flowers are red and the fruit are globose, orange to orange-red at maturity.

Saribus pendulinus. BELOW, LEFT TO RIGHT: crown with inforescences; leaf sheaths; inflorescences with fruit; part of inflorescence; flowers; fruit. Purari River (SV).

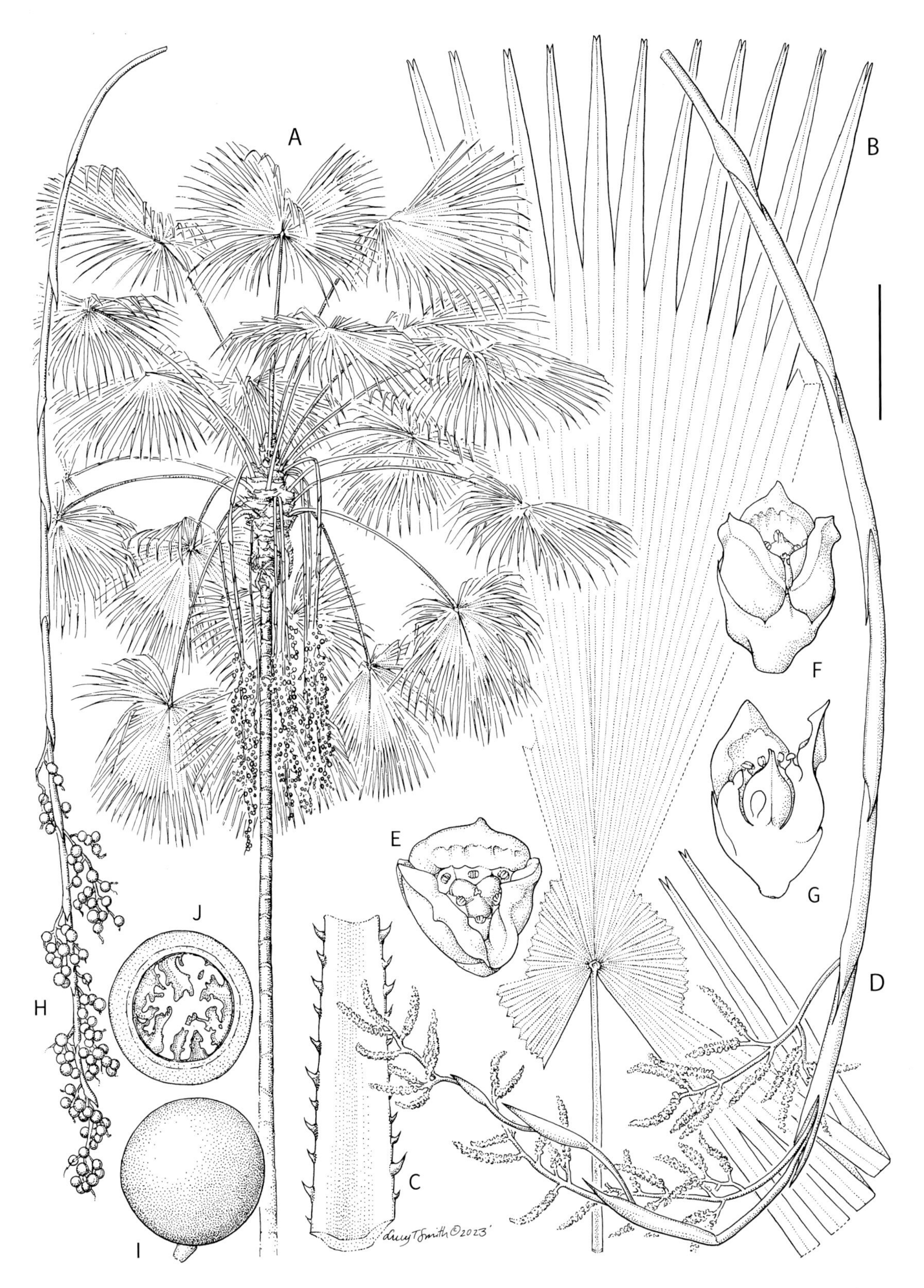

Saribus pendulinus. **A.** Habit. **B.** Leaf. **C.** Portion of petiole. **D.** Inflorescence. E–**G.** Flower in two views and longitudinal section. **H.** Inflorescence with fruit. **I, J.** Fruit whole and in longitudinal section. Scale bar: A = 2 m; B, D, H = 12 cm; E–G = 1.8 mm; C, I, J = 4 cm. All from *Venter 15939*. Drawn by Lucy T. Smith.

5. *Saribus rotundifolius* (Lam.) Blume

Synonyms: *Corypha rotundifolia* Lam., *Licuala rotundifolia* (Lam.) Blume, *Livistona altissima* Zoll., *Livistona microcarpa* Becc., *Livistona mindorensis* Becc., *Livistona robinsoniana* Becc., *Livistona rotundifolia* (Lam.) Mart., *Livistona rotundifolia* var. *luzonensis* Becc., *Livistona rotundifolia* var. *microcarpa* (Becc.) Becc., *Livistona rotundifolia* var. *mindorensis* (Becc.) Becc.

Robust tree palm to 45 m. **Stem** 15–25 cm diam., leaf scars obscure to prominent, light green to white, internodes broad, green to grey, smooth or infrequently with longitudinal fissures. **Leaves** 20–50 in a globose crown; *leaf blade circular to subcircular, 75–150 cm long*, upper surface semi-glossy dark green, lower surface lighter green, divided for 38–62% of its length, with 60–90 segments, depth of apical cleft 4–25% of the segment length, apical lobes usually erect, but pendulous in segments with deeper clefts; petiole slightly arching, 0.9–2.1 m long, margins with recurved black spines 1–20 mm long throughout or proximally only, with largest proximally, distally becoming smaller and more widely spaced, or very infrequently with spines lacking in mature plants. **Inflorescences** trifurcate with a common prophyll, with similar equal axes, 0.9–1.5 m long, arching, with each axis branched to 4 orders; first-order branches ca. 10; peduncular bracts lacking or 1 on each axis; rachillae 3–20 cm long, 1–1.5 mm thick, straight, yellowish, glabrous. **Flowers** solitary or in clusters of 2–4, to 2–3 mm long, *yellowish*. **Fruit** globose to subglobose, 11–25 mm diam., *at first yellow, ripening through to orange-red to red or to dark violet*. **Seed** globose, 10–13 mm diam.

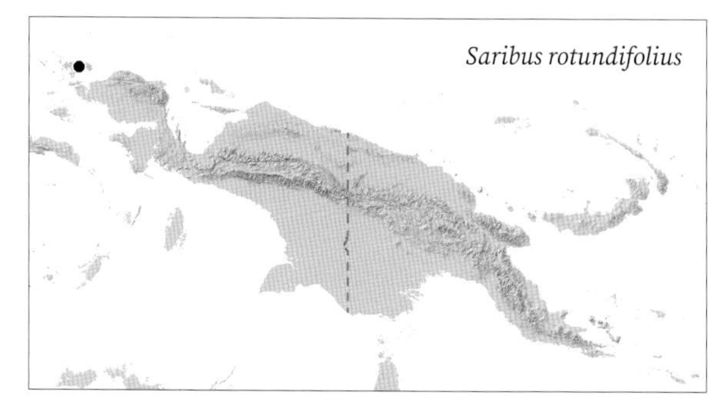

DISTRIBUTION. In New Guinea only recorded from the Raja Ampat Islands. Otherwise widely distributed in Indonesia, Malaysia (Banggi Island only) and the Philippines.

Saribus rotundifolius. LEFT TO RIGHT: crown, cultivated Florida (CL); inflorescence, showing three main axes; base of inflorescence, prophyll removed to show three main axes, cultivated Bogor Botanic Garden, Indonesia (WB).

Saribus rotundifolius. **A.** Habit. **B.** Portion of petiole. **C.** Lower section of leaf, upper surface. **D.** Centre of leaf, lower surface. **E.** Base of inflorescence. **F.** Portion of inflorescence. **G, H.** Fruit whole and in longitudinal section. Scale bar: A = 80 cm; B–D = 4 cm; E, F = 6 cm; G, H = 1 cm. All from *Dransfield JD 3831*. Drawn by Lucy T. Smith.

HABITAT. Locally abundant in swamp-forest, seasonally dry swamp forest, mangrove margins, rainforest, moist evergreen forest, along rivers and secondary forest at sea level to 300 m.

LOCAL NAMES. None recorded (for New Guinea).

USES. The stems are used as flooring material in traditional houses in Waigeo.

CONSERVATION STATUS. Least Concern. Though rare in New Guinea, this species is widespread.

NOTES. *Saribus rotundifolius* is a tall canopy palm with large leaves, tightly tubular inflorescence bracts and yellowish flowers. Certain structures are highly variable, notably the leaf segments, which are rigid to pendulous, and the fruit, which vary in size (1.1–2.5 cm diam.) and colour (at first yellow, then ripening though to orange or red, or to dark violet). These variable characters appear to occur more or less randomly throughout the entire distribution of the species. This is one of the most widespread cultivated palms in the tropics.

6. *Saribus surru* (Dowe & Barfod) Bacon & W.J.Baker

Synonym: *Livistona surru* Dowe & Barfod

Moderately robust, single-stemmed tree palm to 20 m. **Stem** 18–25 cm diam., leaf scars slightly raised, internodes narrow, light grey, petiole stubs not retained. **Leaves** 17–29 in a globose crown; *leaf blade subcircular to ovate, undulate, 1.8–2.3 × 1.4–1.6 m,* upper surface mid-green, lower surface similar green, divided for 45–80% of its length, with 70–90 segments, depth of apical cleft ca. 6% of the segment length; *apical lobes pendulous, hanging vertically;* petiole 1.4–1.8 m long, slightly arching, green, glabrous except for scattered scales, brown in the centre and grey at the margin, more densely so on the lower surface, margins with single or grouped black spines 5–10 mm long, larger and more closely inserted in the proximal portion. **Inflorescences** trifurcate with a common prophyll, with similar equal axes ca. 1.2 m long, each axis branched to 3 orders with 5–7 first-order branches; peduncular bracts lacking; rachillae 14–24 cm long, pubescent with dense long coarse red adpressed scales in the proximal portion, distally with long white scales, less dense to glabrous in the extreme distal portions. **Flowers** in clusters of 2–4. **Fruit** *globose to obovoid, 55–65 × 50–55 mm, orange-red.* Seed globose to subglobose, 42–48 mm diam.

Saribus surru. Habit, Ramu River (WB).

DISTRIBUTION. North-eastern Papua New Guinea.

HABITAT. In rainforest and swamp-forest, 10–1,300 m.

LOCAL NAMES. *Surru, Bop* or *Tim* (Olo, Miwaute).

USES. Preferred species for thatching houses in Sandaun Province. There, the leaves are used as umbrellas, leaf sheath fibers as brooms and sago strainers, and stems as the base of houses and axe handles.

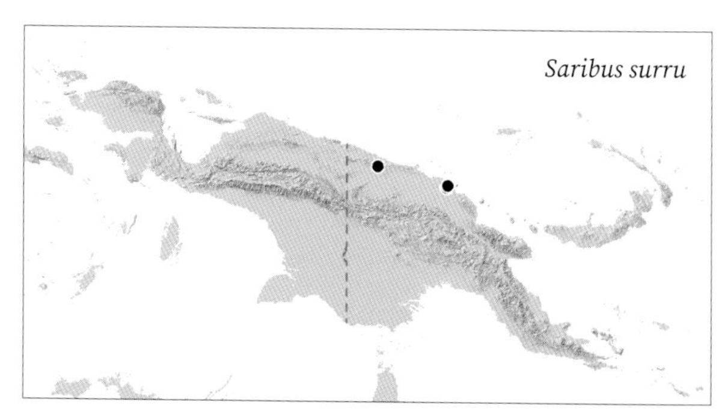

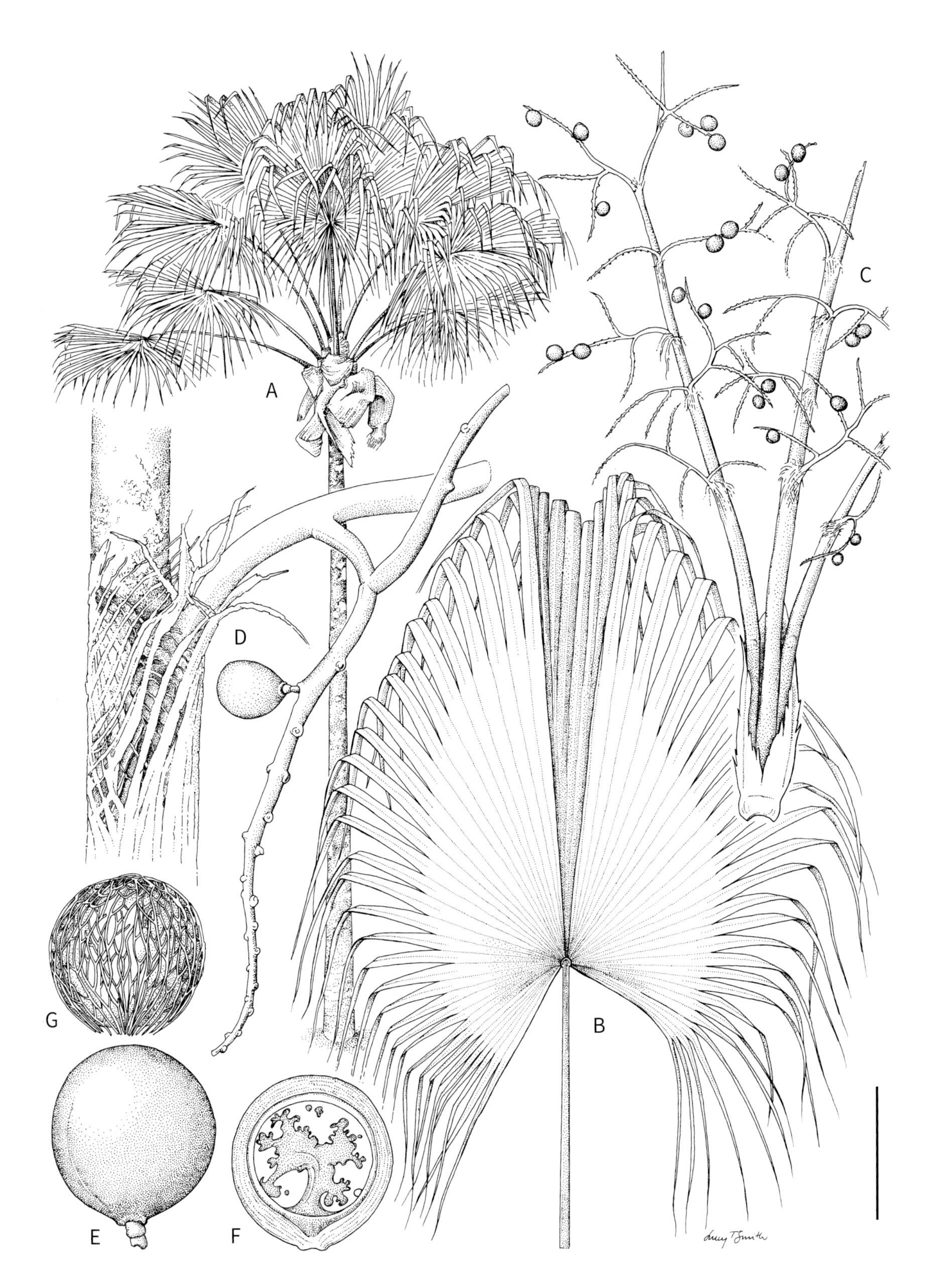

Saribus surru. **A.** Habit. **B.** Leaf. **C.** Inflorescence with fruit. **D.** Detail fibrous rachis with fruit. **E, F.** Fruit whole and in longitudinal section. **G.** Fibrous endocarp. Scale bar: A = 2.7 m; B = 30 cm; C = 24 cm; D = 4 cm; E–G = 3 cm. A, C, D from *Barfod 399*; B, F, G from *Baker & Utteridge 582*; E from *Barfod et al. 390*. Drawn by Lucy T. Smith.

Saribus surru. ABOVE: leaf, Ramu River (WB). RIGHT: fruit, Torricelli Mountains (AB).

CONSERVATION STATUS. Endangered (IUCN 2021). *Saribus surru* has a restricted distribution. Deforestation due to logging concessions and agricultural activities is a major threat in its distribution range.

NOTES. *Saribus surru* is a moderate to large canopy palm. The leaves are large and regularly segmented with pendulous segment apices. The inflorescence bracts are loosely tubular. The globose fruit are the largest in the genus and are orange-red at maturity.

7. *Saribus tothur* (Dowe & Barfod) Bacon & W.J.Baker

Synonym: *Livistona tothur* Dowe & Barfod

Moderately robust palm to 20 m. **Stem** 15–20 cm diam., leaf scars slightly raised, narrow, dark grey, internodes broad, grey. **Leaves** 24–40 in a globose crown; leaf blade subcircular, 1.5–2 ×1.2–1.5 m, upper surface bluish-green, lower surface silvery glaucous, divided for 62–85% of its length, with 60–75 segments, depth of apical cleft 1–3% of the segment length, *apical lobes rigid, but becoming pendulous with age or damage*; petiole 1.5–2 m long, arching, glabrous, green, margins with single, recurved, green spines 1–2 mm long throughout its length, but larger and more closely spaced in the proximal portion. **Inflorescences** trifurcate with a common prophyll, with similar equal axes, each axis branched to 3 orders, ca. 2 m long, with 5–6 first-order branches; each peduncle with one tubular, papery and loosely sheathing bract, rachillae 6–12 cm long, to 3 mm diam., straight, green-red, glabrous. **Flowers** *solitary; sepals and petals red*; anthers cream. **Fruit** globose with a basal constriction, 35–43 mm diam., semi-glossy orange-red. **Seed** *globose, 22–28 mm diam.*, endosperm deeply intruded by orange pulpy tissue.

DISTRIBUTION. Central northern New Guinea.

HABITAT. In rainforest on ridges, limestone and metamorphic rocks at 400–600 m.

LOCAL NAMES. *Tot-hur* and *Yu bbraal* (Bewani).

USES. Bows and roof struts are fashioned from the petioles, umbrellas are made from the leaves, and salt is extracted from the ash of the burnt petioles.

CONSERVATION STATUS. Endangered (IUCN 2021). *Saribus tothur* has a restricted distribution. Deforestation due to logging concessions is a major threat in its distribution range.

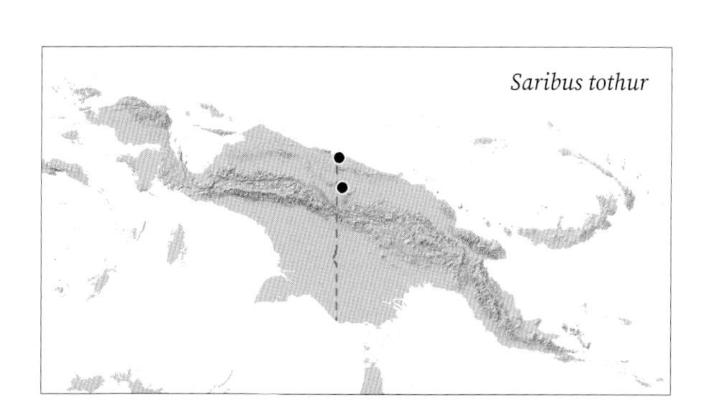

Saribus tothur. **A.** Habit. **B.** Leaf. **C.** Portion of petiole. **D.** Inflorescence. **E.** Portion of rachilla with flowers. **F**, **G.** Flower whole and in longitudinal section. **H**, **I.** Fruit whole and in longitudinal section. Scale bar: A = 1.8 m; B = 36 cm; C, E = 3 cm; D = 21 cm; F, G = 1.5 cm; H, I = 2 cm. A, D–F, G from *Barfod 510*; B from *Dowe & Ferrero 516*; C, H, I from *Damborg & Barfod 418*. Drawn by Lucy T. Smith.

Saribus tothur. CLOCKWISE FROM TOP LEFT: inflorescence; flowers; inflorescence with fruit, Sandaun (AB); leaf, Sandaun (AD).

NOTES. *Saribus tothur* is a moderate canopy palm. The leaves are regularly segmented with the segment apices rigid. The inflorescence bracts are loosely sheathing. The flowers (sepals, petals and carpels) are red and the globose fruit are shiny orange-red at maturity.

8. *Saribus woodfordii* (Ridl.) Bacon & W.J.Baker

Synonyms: *Livistona beccariana* Burret, *Livistona woodfordii* Ridl.

Moderately robust, single-stemmed tree palm to 16 m. **Stem** 12–20 cm diam., leaf scars slightly raised, light grey, internodes broad, greyish-brown to grey with age, petiole stubs not persistent. **Leaves** 30–60 in a globose to broadly conical crown; *leaf blade subcircular to circular, 60–170 × 45–90 cm, rigid,* shiny mid-green on upper surface, glaucous, lower surface lighter green with fine powdery wax, divided for 51–75% of its length, with 60–70 segments, depth of apical cleft 5–23% of the segment length, apical lobes acuminate, semi-pendulous, hanging ca. 45° or more to the vertical; petiole to ca. 1.1 m long, slightly arching, upper surface slightly ridged, covered with a deciduous white powder, margins unarmed or with single, small, curved, green spines, confined to the proximal half. **Inflorescence** *with a single axis basally branched into 3 equal axes 10–30 cm above the base,* each axis branched to 3 orders, 1.2–2.7 m long, slightly curving; first-order branches 5–10; peduncular bracts lacking; rachillae 4–6 cm long, 1 mm thick, straight, basally with brown-purple tomentum. **Flowers** in clusters of 2–6; *sepals and petals red.* **Fruit** globose, 7–12 mm diam., reddish orange to reddish brown. **Seed** globose, 6–9.5 mm diam.; endosperm intruded with brownish tissue.

DISTRIBUTION. Far south-eastern New Guinea and the Louisiade Archipelago (Rossel and Sudest Islands). Elsewhere in the Solomon Islands (Tulagi [Nggela] and San Cristobal Islands).

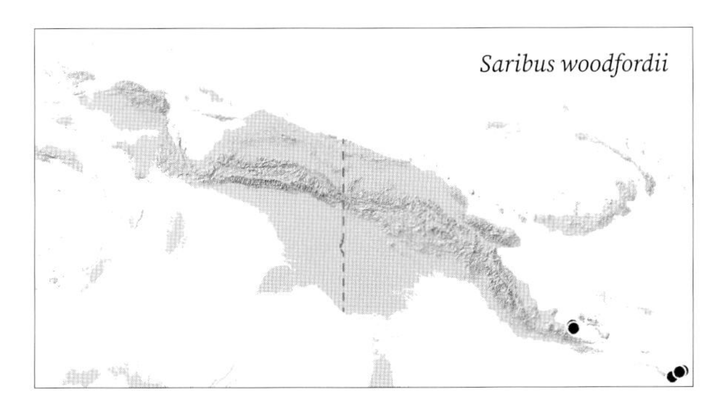

HABITAT. Coastal forest on limestone or lateritic soils at sea level to 120 m.

LOCAL NAMES. *Boda* (Wedau), *Filu* (Kakabai).

USES. None recorded.

CONSERVATION STATUS. Least Concern. Though relatively restricted in New Guinea, *Saribus woodfordii* is more widespread beyond.

NOTES. *Saribus woodfordii* is a moderate canopy palm. The leaves are regularly segmented with the segment apices semi-pendulous. The inflorescence is a single axis divided into 3 main branches, with the branches basally adnate. This arrangement is unusual in *Saribus* in that branching occurs away from the base of the inflorescence rather than at the base as in all other species in the genus. The flowers (petals and sepals) are red and the globose fruit are reddish orange to reddish brown at maturity.

Saribus woodfordii. Habit, Rossel Island (JLD).

Saribus woodfordii. Habit, Rossel Island (JLD).

Saribus woodfordii. **A.** Habit. **B.** Leaf. **C.** Inflorescence with fruit. **D.** Portion of inflorescence with fruit. **E.** Inflorescence. **F**, **G.** Fruit whole and in longitudinal section. Scale bar: A = 1.8 m; B, D = 30 cm; C, E = 6 cm; F, G = 1 cm. A–C from photographs; D–G from *Brass 28281*. Drawn by Lucy T. Smith.

Caryota rumphiana. Habit, Tamrau Mountains (WB).

Caryota L.

Synonyms: *Schunda-pana* Adans., *Thuessinkia* Korth. ex Miq.

Tall – leaf bipinnate – no crownshaft – no spines – leaflets jagged

Robust, single-stemmed tree palms, crownshaft absent, *stems dying after flowering*, monoecious. **Leaf** *bipinnate*, horizontal; sheath disintegrating into a mass of dark fibres; petiole short to long; *leaflets wedge-shaped, with jagged tips*, arranged regularly, horizontal or on edge, *V-shaped in section at the base (induplicate)*. **Inflorescence** among the leaves, *maturing from top of stem downwards*, thus the oldest inflorescence at the stem apex, branched 1–2 orders, branches curved, ± pendulous; prophyll small, inconspicuous, peduncular bracts 5–7, soon splitting, densely covered in indumentum, not dropping off as inflorescence expands; peduncle shorter than or about the same length as rachis; rachilla straight, pendulous. **Flowers** *in triads throughout the length of the rachilla*, not developing in pits, male flowers bullet-shaped, female flowers usually globose. **Fruit** red or black, stigmatic remains apical, flesh juicy, filled with irritant needle crystals. **Seeds** 1 or 2, globose or hemispherical, endosperm homogeneous or ruminate.

DISTRIBUTION. About 13 species from South China to India, through South-East Asia to Australia, the Solomon Islands and Vanuatu, two species in New Guinea.

NOTES. *Caryota* is unique in the palm family for its bipinnate leaves with wedge-shaped leaflets. Inflorescences are borne below or between the leaves, with the oldest inflorescences towards the stem apex and younger inflorescences near the base of the stem. The stems die after flowering. The latest systematic treatment is that of Jeanson (2011). An account of the genus in Australia is given in Dowe (2010).

Key to species of *Caryota* in New Guinea

1. Fresh petioles covered in uniform dull grey-brown indumentum; primary leaflets held in the same plane; inflorescence branched to 1 order; endosperm ruminate................ **1. *C. rumphiana***
1. Fresh petioles strikingly zebra-striped with bands of pale and dark indumentum; primary leaflets held in different planes; inflorescence branched to up to 2 orders; endosperm homogeneous .... **2. *C. zebrina***

1. *Caryota rumphiana* Mart.

Synonyms: *Caryota alberti* F.Muell ex H.Wendl., *Caryota rumphiana* var. *alberti* (F.Muell. ex H.Wendl.) F.M.Bailey, *Caryota rumphiana* var. *australiensis* Becc., *Caryota rumphiana* var. *moluccana* Becc., *Caryota rumphiana* var. *papuana* Becc.

Robust, single-stemmed tree palm to 20 m tall. **Stems** to 40 cm diam.; internodes ca. 30 cm long, with thick pale tomentum. **Leaves** 5–15 in crown, to 7 m long, 5 m wide; sheath 2 m or more long, ca. 30 cm wide, fibrous along margins, *densely covered in uniform grey-white tomentum*; petiole to 1.5 m long, *densely covered in grey-white indumentum*; rachis to 5.5 m long; primary leaflets ca. 12–15 on each side of the rachis, held in one plane, mid-leaf primary leaflet ca. 200 cm long; secondary leaflets ca. 25 on each side of the secondary rachis, to ca. 20 cm, irregularly wedge-shaped, leathery, glabrous above, with bands of brown scales below. **Inflorescences** 2–3 m long, *branching to 1 order*; peduncle to 50 cm long, covered in brown tomentum; peduncular bracts tubular, tomentose; rachillae of various lengths, longest to at least 2.5 m, ca. 5 mm diam., with brown indumentum. **Male flower** ca. 12–20 × 8 mm, stamens ca. 22–36. **Female flower** 3–9 × 4–6 mm. **Fruit** ripening dull crimson, then black, subglobose to 22 × 28 mm. **Seeds** 1–2(–3), hemispherical, ca. 25 × 25 mm; *endosperm ruminate*; embryo subapical.

Caryota rumphiana. LEFT: habit, Wau (WB). RIGHT: inflorescences showing basipetal maturation, cultivated Lae (WB).

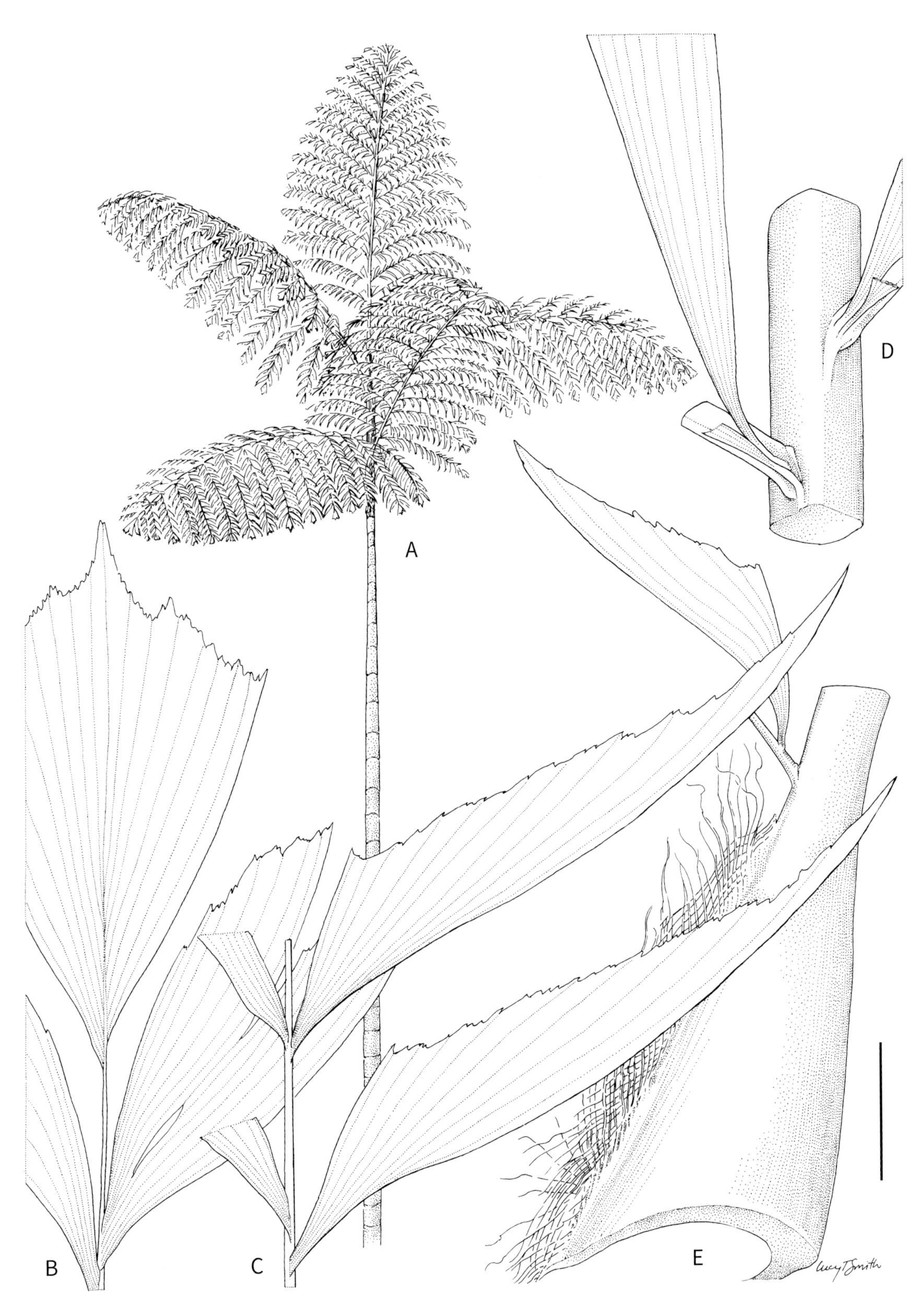

Caryota rumphiana (Plate 1). **A.** Habit. **B.** Leaf apex. **C.** Leaf mid-portion. **D.** Portion of leaf rachis. **E.** Leaf base with sheath. Scale bar: A = 2 m; B–E = 8 cm. A from photograph; B–E from *Baker et al. 646.* Drawn by Lucy T. Smith.

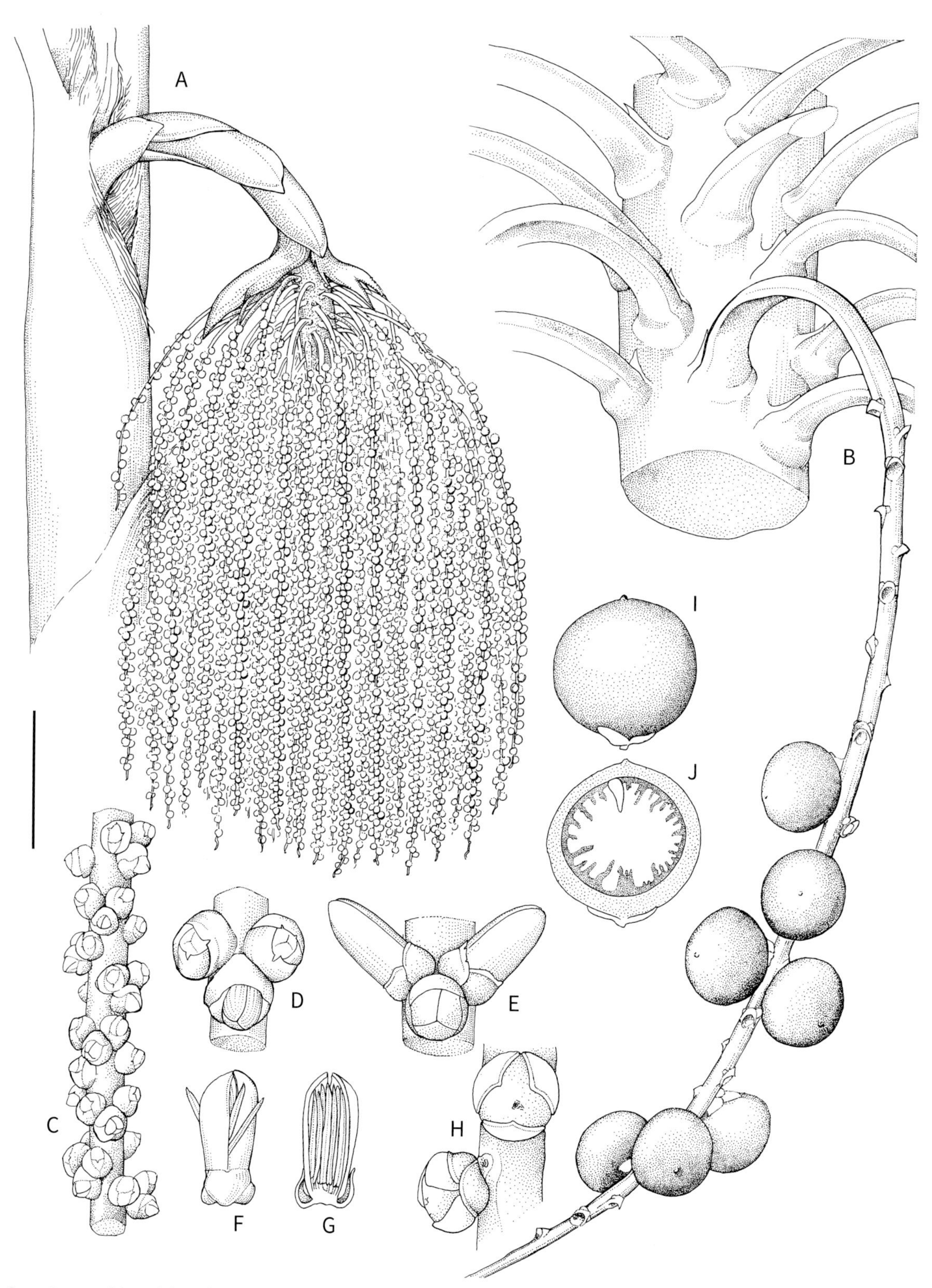

Caryota rumphiana (Plate 2). **A.** Inflorescence with fruit attached to stem. **B.** Rachilla with fruit attached to inflorescence axis. **C.** Portion of rachilla with triads. **D.** Immature triad. **E.** Triad with mature male buds. **F, G.** Male flower whole and in longitudinal section. **H.** Triad with mature female flower, male flowers fallen. **I, J.** Fruit whole and in longitudinal section. Scale bar: A = 50 cm; B = 4 cm; C, I, J = 2 cm; F–H = 1.5 cm; D, E = 1 cm. A, B, I, J from cultivated plant at Kew; C–H from *Baker et al. 646.* Drawn by Lucy T. Smith.

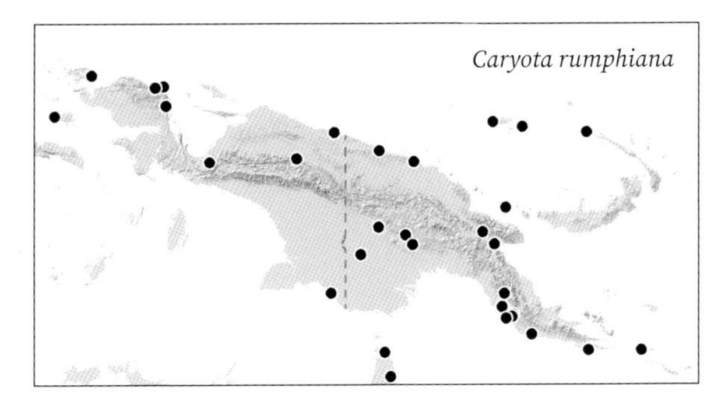

DISTRIBUTION. Widespread throughout the lowlands of New Guinea, including the Raja Ampat Islands and the Bismarck Archipelago. Elsewhere in Philippines, Sulawesi, Maluku, Solomon Islands and Australia (Jeanson 2011).

HABITAT. Lowland to montane forest, often on river banks and disturbed areas from sea level to 1,500 m.

LOCAL NAMES. *Anasi* (Krisa), *Benisri* (Kroy), *Bi* (Biagi), *Biar* (Telefomin), *Kowali* (Kutubu), *Lul* (Matbat), *Moroko* (Sough), *Nandiney* (Marap), *Oa* (Motu), *Sahun* (Kali), *Tere Nyi* (North Cyclops).

USES. Stem for flooring, posts, and rafters, axe handles, spears, and as support for crossbeams of an improvised canoe. Young shoot (heart-of-palm) eaten fresh or cooked. The stem is an inferior source of sago and, when felled, serves to cultivate sago grubs. Stem pith is used to treat coughs. The fruit (and seeds) is edible and is a betel nut substitute. It is eaten by the northern cassowary (Pangau-Adam & Mühlenberg 2014).

CONSERVATION STATUS. Least Concern (IUCN 2018).

NOTES. *Caryota rumphiana* is a very familiar palm of the lowlands on New Guinea. It is variable and, in good fertile soils, can be massive. Immediately recognisable by its bipinnate leaves and fishtail leaflets, the only palm with which it can be confused is *C. zebrina*, a rare palm of montane forest in the central northern part of the island, which is immediately distinguished by its zebra-striped petioles (among other features – see key).

2. *Caryota zebrina* Hambali, Maturb., Heatubun & J.Dransf.

Robust, single-stemmed tree palm to 16 m tall. **Stems** ca. 20–40 cm diam.; internodes 30–40 cm long, cracking longitudinally, with thick brown tomentum. **Leaves** 6–8 in crown, 5–7 m long; sheath 1–2 m long, ca. 15 cm wide, fibrous along margins, in mature leaves covered in dull brown tomentum, *in juveniles zebra-striped with bands of pale and dark indumentum;* petiole ca. 1–2 m long, ca. 2.6 cm diam., in young leaves *zebra-striped as sheath;* rachis to 5 m long; primary leaflets ca. 20 on each side of the rachis, held in different planes, mid-leaf primary leaflets ca. 150 cm long, distal to 100 cm long; *secondary leaflets drying pale coloured,* ca. 7–11 on each side of the secondary rachis, to ca. 26 × 7 cm, irregularly wedge-shaped, *leathery,* glabrous above, with bands of brown scales below. **Inflorescences** 1–2.5 m long, *branching to 2 orders;* peduncle to 30 cm long, covered in brown tomentum; peduncular bracts tubular, tomentose;

Caryota zebrina. Habit, cultivated Hawaii (WB).

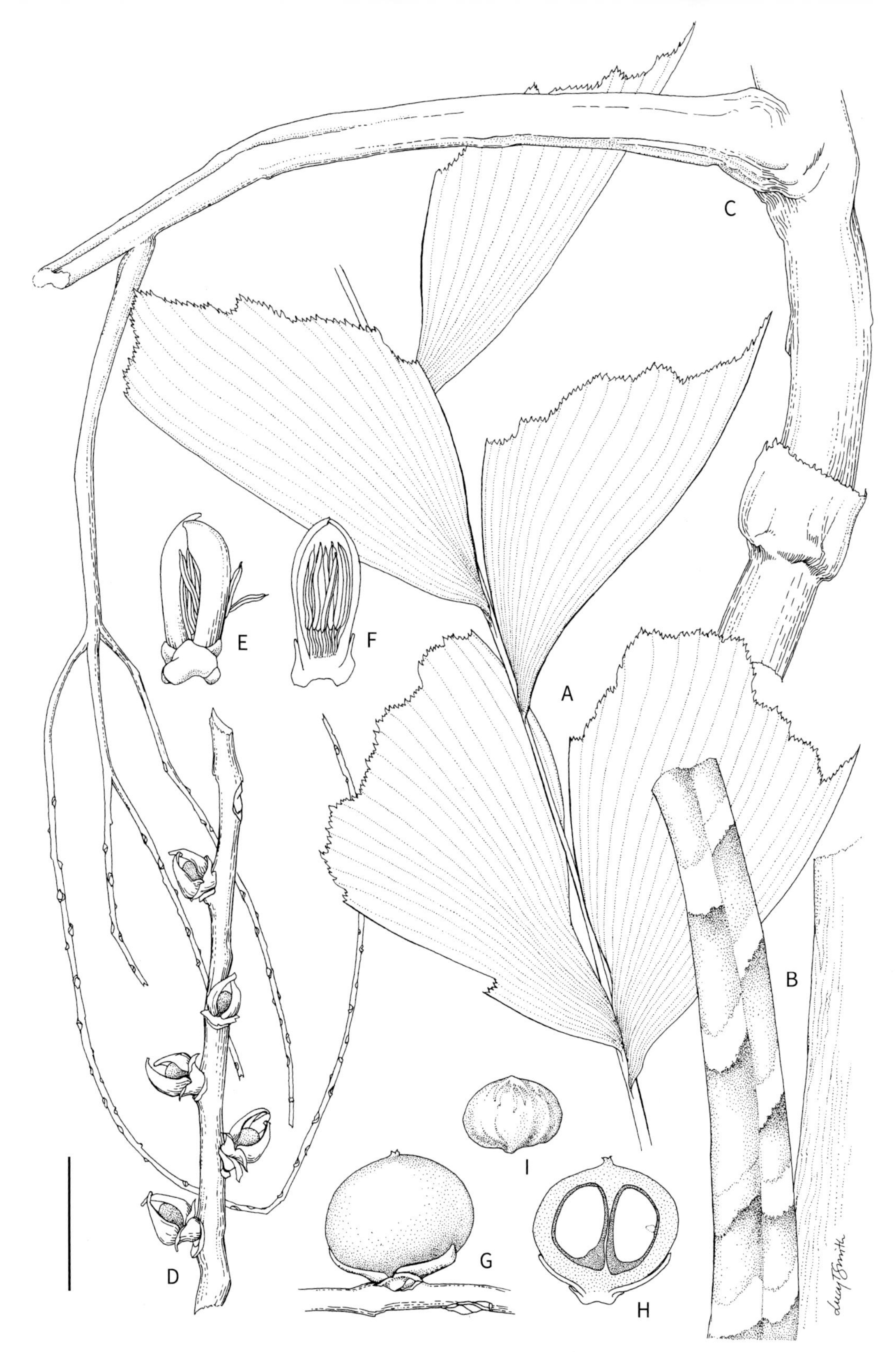

Caryota zebrina. **A.** Leaf mid-portion. **B.** Petiole. **C.** Portion of inflorescence. **D.** Portion of rachilla with female flowers. **E, F.** Male flower whole and in longitudinal section. **G, H.** Fruit whole and in longitudinal section. **I.** Seed. Scale bar: A, B = 3 cm; C = 6 cm; D, G–I = 2 cm; E, F = 1 cm. A from *van Royen & Sleumer 6129*; B–D, G–I from *Heatubun et al. 273*; E, F from *Maturbongs & Wally 586.* Drawn by Lucy T. Smith.

rachillae of various lengths, longest to at least 1 m, 3–4 mm diam., with brown indumentum. **Male flower** ca. 14 × 7 mm, stamens ca. 28. **Female flower** 10.0 × 3.5 mm. **Fruit** ripening white, then red, globose to 15 × 25 mm. **Seeds** 2, hemispherical, ca. 12 × 15 × 8 mm; endosperm *homogeneous*; embryo lateral.

DISTRIBUTION. Cyclops and Torricelli Mountains in northern New Guinea.

HABITAT. In montane forest at 850–1,500 m elevation.

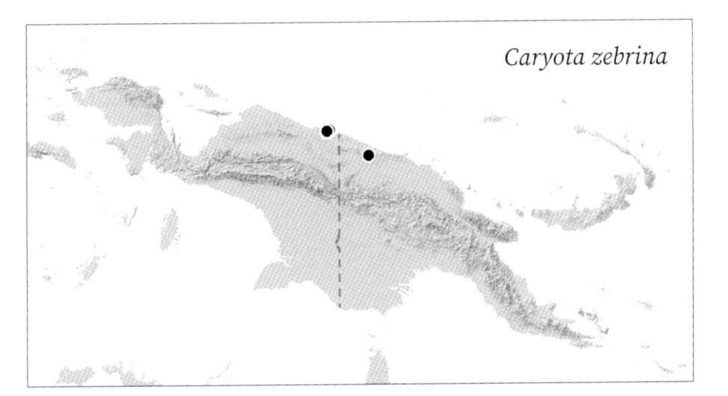

LOCAL NAMES. *Palem belang, Palem tokek* (Bahasa Indonesia names in horticulture).

USES. Apart from horticultural value, probably too scarce to be of use to local people.

CONSERVATION STATUS. Endangered (IUCN 2021). *Caryota zebrina* has a restricted distribution. Deforestation due to logging and mining concessions is a major threat in its distribution range.

NOTES. Immediately distinguished by the extraordinary zebra-striping on young leaves. The primary leaflets are also unusual in being held in different planes, giving the whole leaf a rather untidy appearance. The inflorescence branches to two orders (only one in *C. rumphiana*). The seed has homogeneous endosperm whereas *C. rumphiana* has ruminate endosperm. The unusual, patterned indumentum and the homogeneous endosperm are found elsewhere in the genus only in the very similar *C. ophiopellis* in Vanuatu (Dransfield *et al.* 2000a).

Caryota zebrina. LEFT TO RIGHT: habit, cultivated Bogor, Indonesia (WB); indumentum on petiole, cultivated Royal Botanic Gardens, Kew (WB).

Arenga microcarpa. Inflorescence with fruit, cultivated Montgomery Botanical Centre, Florida (WB).

Arenga Labill. ex DC.

Synonyms: *Blancoa* Blume, *Didymosperma* H.Wendl & Drude ex Hook.f., *Gomutus* Correa, *Saguerus* Steck

Medium – leaf pinnate – no crownshaft – no spines – leaflets jagged

Moderate to massive, single- or multi-stemmed, *untidy*, mid-storey or canopy tree palms, crownshaft absent, *individual stems dying after flowering*, monoecious. **Leaf** pinnate, more or less straight; *sheath disintegrating into a mass of fibres*; petiole long; leaflets numerous, with jagged tips, arranged regularly or irregularly, horizontal, undersurface covered with grey indumentum, *leaflet bases asymmetrical and V-shaped in section (induplicate).* **Inflorescence** among the leaves, *maturing from top of stem downwards*, thus the oldest inflorescence at the stem apex, branched up to 2 orders, branches curved, pendulous; prophyll relatively small, inconspicuous, peduncular bracts 5–7, splitting, densely covered in indumentum, not dropping off as inflorescence expands; peduncle about the same length as inflorescence rachis; rachillae quite robust and straight. **Flowers** male and female flowers borne on separate inflorescences on the same stem, the female inflorescence usually borne at the stem apex, flowers not developing in pits, male flowers bullet-shaped. **Fruit** dark purplish green or red, globose, stigmatic remains apical, flesh juicy, filled with irritant needle crystals. **Seed**s 1–3, globose or angled, endosperm homogeneous.

DISTRIBUTION. About 20 species from South China to India, through South-East Asia to Australia, one species widespread in New Guinea, not recorded from Bismarck Archipelago.

NOTES. There is one native species of *Arenga* in New Guinea, *A. microcarpa,* which is an untidy, multi-stemmed, mid-storey tree palm with conspicuous leaf sheath fibres. The leaflets are jagged at the tip and have asymmetrical bases that are V-shaped (induplicate) in section. Individual stems die after flowering. It is most readily confused with *Orania,* a genus of single-stemmed tree palms with reduplicate leaflets (Λ-shaped [reduplicate] in section), stems that do not die after flowering and quite large fruit. *Arenga pinnata,* the sugar palm, is naturalised in some places. The genus was monographed by Mogea (1991).

Key to the species of *Arenga* in New Guinea

1. Moderate, multi-stemmed palm with leaves to 8 m long, stems ca. 20 cm diam. (lowland forest, riverbanks) . . . . **1. *A. microcarpa***
1. Massive single-stemmed palm with leaves at least 10 m long, stems ca. 40 cm diam. or more (village margins) . . . . **2. *A. pinnata***

1. *Arenga microcarpa* Becc.

Synonyms: *Arenga australasica* (H.Wendl. & Drude) S.T.Blake ex H.E.Moore, *Arenga gracilicaulis* F.M.Bailey, *Arenga microcarpa* var. *keyensis* Becc., *Didymosperma humile* K.Schum. & Lauterb., *Didymosperma microcarpum* (Becc.) Warb. ex K.Schum. & Lauterb., *Didymosperma novoguineense* Warb. ex K.Schum. & Lauterb., *Normanbya australasicus* (H.Wendl. & Drude) Baill., *Saguerus australasicus* H.Wendl. & Drude

Moderate to robust, multi-stemmed tree palm to 4–8 (–14 m). **Stem** ca. 12 cm diam. or more, *dying after flowering and fruiting, obscured by old leaf-sheaths and abundant black fibre,* crownshaft absent. **Leaf** 5–10 in crown, erect to spreading, to ca. 6 m long; petiole and sheath to 25–40(–100) cm long, the petiole with abundant grey indumentum, sheath *disintegrating into an untidy mass of robust black fibres* to 50 cm long; leaflets 40–70 on each side of leaf rachis, to 40–110 × 2–5 cm long, with jagged tips, *arranged irregularly,* grouped in 2s–4s in mid section, dark green on upper surface, covered in pale indumentum on lower surface, leaflet base briefly stalked and with an asymmetric ear-like lobe. **Inflorescences** *branched to 1 (rarely 2) orders,* male and female very similar but female more robust, to 1.7 m long, male inflorescences occasionally multiple; rachillae 45–80 cm long, 6 mm diam. in female, 5 mm diam. in male. **Male flower** bullet-shaped, 6–10 × 4–6 mm; stamens ca. 14–24. **Female flower** spherical, ca. 6 mm diam. **Fruit** *spherical,* dull green, ripening *white then red,* 1.5–1.8 cm diam. **Seed** 1–3, 7–11 × 4 mm, with two flattened sides and one curved side, surrounded by irritant crystals; endosperm homogeneous, embryo lateral.

DISTRIBUTION. Widespread in New Guinea, also in Maluku and Australia.

HABITAT. Lowland forest, particularly along riverbanks and occasionally at higher elevations to 640 m.

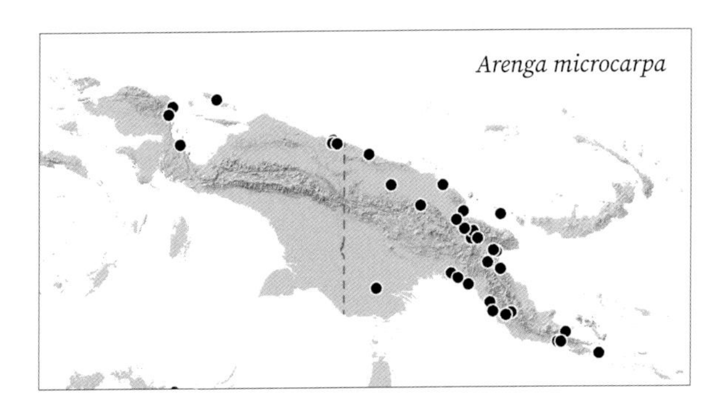

Arenga microcarpa. Habit, Sirebi River (WB).

LOCAL NAMES. *An* (Sough), *Biawar* (Wamesa), *Go-ih* (Matapaili), *Iri Nyi* (North Cyclops).

USES. Leaves are used for thatching. Stem is used for house floors and posts. Stem pith is used medicinally for treating coughs. Elsewhere, in Sangihe-Talaud, the stem is used as a source of sago (Fairchild 1944). The juice of the fleshy fruit is strongly irritant; people cleaning the seeds are likely to suffer painful inflammation of the skin.

CONSERVATION STATUS. Least Concern.

NOTES. *Arenga microcarpa* is a widespread, somewhat weedy palm occurring widely in the lowlands of New Guinea. It is very variable, perhaps related to soil fertility and exposure. In moist sites, particularly along river banks, it can be very robust, but surviving

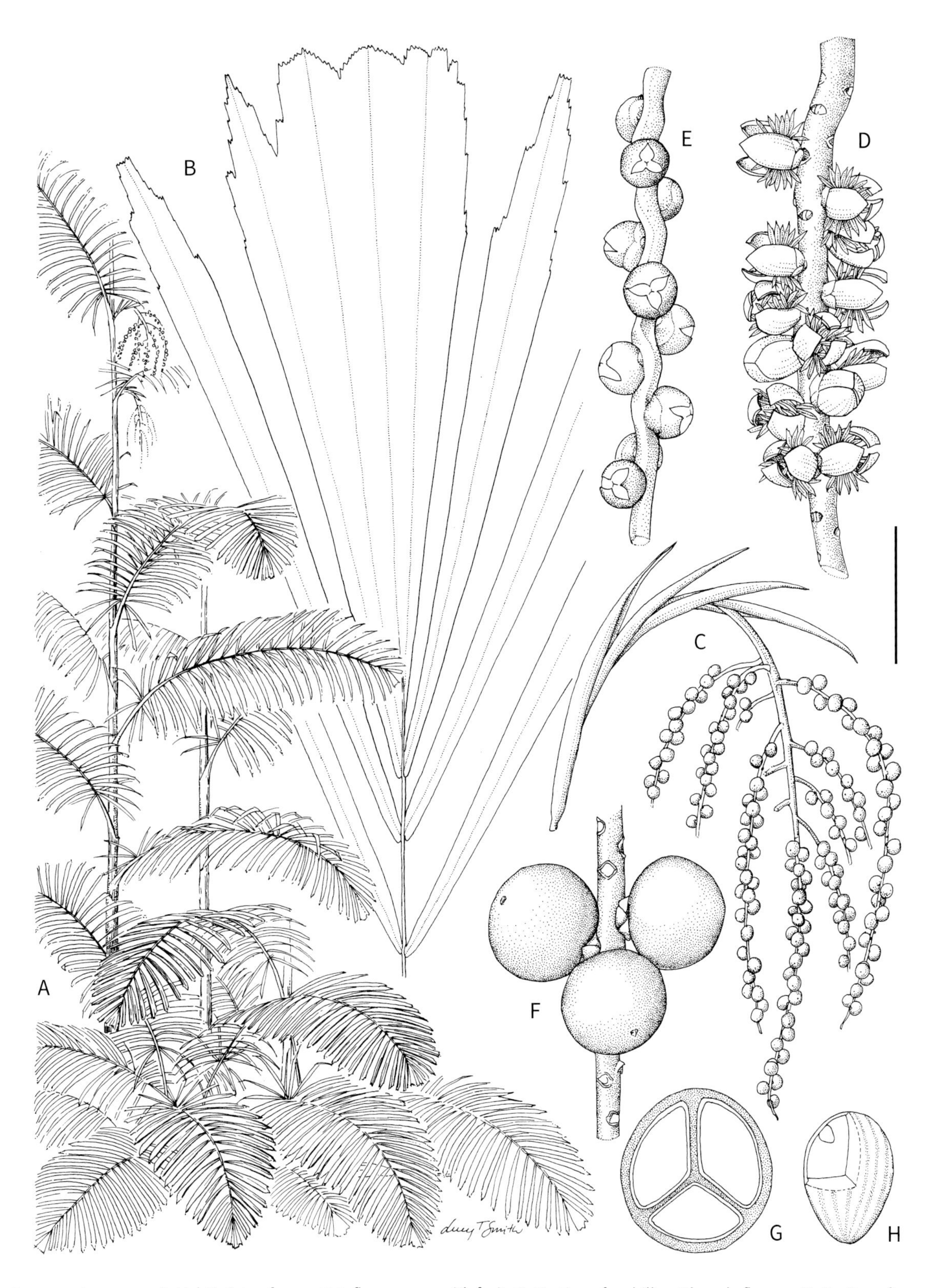

Arenga microcarpa. **A.** Habit. **B.** Leaf apex. **C.** Inflorescence with fruit. **D.** Portion of rachilla with male flowers. **E.** Portion of rachilla with female flowers. **F.** Portion of rachilla with fruit. **G.** Fruit in transverse section. **H.** Seed showing embryo. Scale bar: A = 1 m; B = 8 cm; C = 12 cm; D–F = 2 cm; G, H = 1.5 cm. A, B, E from *Kew cult. 1959-58301*; C, F–H from *Baker & Utteridge 577*; D from *Baker et al. 1044*. Drawn by Lucy T. Smith.

Arenga microcarpa. LEFT TO RIGHT: male flowers; inflorescence, Wandamen Peninsula (WB); fruit, near Madang (WB).

in cleared areas, it can be stunted. Some Australian populations have been separated as a distinct species, *A. australasica* (see Mogea 1991 and discussion in Dowe 2010), but the characters separating the two taxa overlap and phylogenetic analyses conducted by Jeanson (2011) suggest there is but one variable species.

2. *Arenga pinnata* (Wurmb.) Merr.

Synonyms: *Arenga gamuto* Merr., *Arenga griffithii* Seem. ex H.Wendl., *Arenga saccharifera* Labill. ex DC., *Borassus gomutus* Lour., *Gomutus rumphii* Corrêa, *Caryota onusta* Blanco, *Gomutus saccharifer* (Labill. ex DC.) Spreng., *Gomutus vulgaris* Oken, *Saguerus gamuto* Houtt., *Saguerus pinnatus* Wurmb, *Saguerus rumphii* (Corrêa) Roxb. ex Fleming, *Saguerus saccharifer* (Labill. ex. DC.) Blume, *Sagus gomutus* (Lour.) Perr.

Massive, single-stemmed tree palm to 20 m. **Stem** ca. 40 cm diam., *dying after flowering and fruiting, obscured by old leaf-sheaths and abundant black fibre.* **Leaf** 8–12 in crown, *held erect in an untidy shuttlecock,* to 8.2 m long; petiole and sheath to 1.6 m long, petiole with abundant grey indumentum, sheath *disintegrating into an untidy mass of robust black fibres* to 80 cm long; leaflets ca. 150 on each side of leaf rachis, to 120–160 cm long, with jagged tips, *arranged irregularly,* grouped in 3s in mid section, dark green on upper surface, covered in gleaming, pale indumentum on lower surface, the leaflet base often with a stalk to 2 cm long and an asymetric ear-like lobe to 5 cm long. **Inflorescences** *branched to 1 order,* male and female very similar but female more robust, to 1.5 m long; rachillae to ca. 80 cm long, 10 mm diam. in female, 6 mm diam. in male. **Male flower** bullet-shaped, 8 × 5 mm; stamens ca. 60–120. **Female flower** spherical, ca. 14 mm diam. **Fruit** dull dark green, 4–4.5 × 4 cm. **Seeds** 3, 3 × 2.5 cm, with two flattened sides and one curved side, surrounded by irritant crystals; endosperm homogeneous, embryo lateral.

DISTRIBUTION. Native distribution unknown, but widely cultivated in South and South-East Asia. Occasionally naturalised near villages particularly in the lowlands of Indonesian New Guinea.

HABITAT. Lowland to upland cleared areas.

LOCAL NAMES. *Aren* (Bahasa Indonesia).

USES. One of the most important multipurpose palms of the Malesian region. Sheath fibre (*ijuk*) used for thatch, rope and many other purposes, leaves used for temporary packing for fruit, inflorescences tapped for sap, sugar and alcohol production, immature seeds, cooked, eaten as a delicacy.

CONSERVATION STATUS. Least Concern.

NOTES. *Arenga pinnata* has been seen in several areas near towns and villages in Indonesian New Guinea but so far appears not to be well represented in herbaria. It is a very familiar palm in Java, Sumatra and Sulawesi, and elsewhere in South-East Asia.

Arenga pinnata. CLOCKWISE FROM TOP: habit, cultivated Nanga Talong, Sarawak (WB); leaf sheaths with fibre, cultivated Nanga Talong, Sarawak (WB); inflorescence with fruit, cultivated Colombia (WB); inflorescence with flower buds, cultivated Colombia (WB).

Corypha utan. Habit, one stem in fruit, Lakefield National Park, Queensland (CN).

Corypha L.

Synonyms: *Bessia* Raf., *Codda-pana* Adans., *Dendrema* Raf., *Gembanga* Blume, *Taliera* Mart.

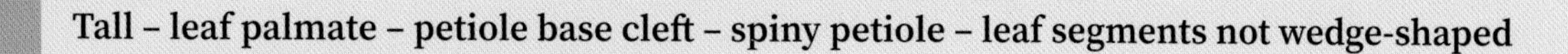
Tall – leaf palmate – petiole base cleft – spiny petiole – leaf segments not wedge-shaped

Massive, single-stemmed fan palm, crownshaft absent, *stems dying after flowering*, hermaphrodite. **Leaf** palmate, very large; sheath splitting to base opposite the petiole; petiole robust, elongate, *with a conspicuous triangular cleft at the base*, the margins with *crowded regularly arranged black spines*; segments numerous, the blade split to ca. one third or more of its radius, with pointed tips. **Inflorescence** *above the leaves, many inflorescences produced simultaneously above the leaves to form a compound, terminal inflorescence*, individual inflorescences branched to 5 orders, branches widely spreading, flowering resulting in the death of the stem; primary bracts tubular, tightly sheathing, not dropping off as inflorescence expands; peduncle shorter than inflorescence rachis; rachillae very slender, short and straight. **Flowers** very small, hermaphrodite, in groups of up to 10. **Fruit** green to black, globose, stigmatic remains basal, flesh rather dry, endocarp thin. **Seed** 1, globose, endosperm homogeneous.

DISTRIBUTION. Six species in India to Philippines, through South-East Asia to Australia, in New Guinea a single species restricted to central southern mainland New Guinea.

NOTES. *Corypha* contains massive, single-stemmed fan palms that produce inflorescences above the leaves on stems that then die after flowering. The petiole bears regular, tooth-like spines and has a distinctive triangular cleft at base. The genus contains approximately six species of which one, *Corypha utan*, occurs in New Guinea in seasonal lowland habitats near to sea level. It is distinguished from other fan palms that do not die after flowering, *Livistona*, *Saribus* and *Licuala*, which lack a cleft in the petiole base and have fibrous leaf sheaths, and *Borassus*, which (in New Guinea) has non-spiny petioles with sharp margins, inflorescences with fat branches and very large fruit (>10 cm long). There has been no critical assessment of the genus since the monograph of Beccari (1931).

Corypha utan Lam.

Synonyms: *Borassus sylvestris* Giseke, *Corypha elata* Roxb., *Corypha gebang* Mart., *Corypha gembanga* (Blume) Blume, *Corypha griffithiana* Becc., *Corypha macropoda* Kurz, *Corypha macrophylla* Roster, *Corypha sylvestris* (Giseke) Mart., *Gembanga rotundifolia* Blume, *Livistona vidalii* Becc., *Taliera elata* (Roxb.) Wall., *Taliera gembanga* Blume, *Taliera sylvestris* (Giseke) Blume

Massive, single-stemmed palm to 25 m tall. **Stem** 50–100 cm diam., sometimes slightly swollen, *often with distinct spiral markings* formed by raised leaf scars, grey. **Leaves** massive, 15–30 or more in the crown; *leaf base with a conspicuous triangular cleft*; petiole curved, 2.5–4 m long, to 15 cm wide, *usually yellowish with conspicuous black bands of relatively small curved triangular spines* to 8 mm long along the margins; blade rounded up to 3.5 m diam., undulate and folded, dull to greyish green above, paler beneath; segments numerous (up to 80) to 1.5 m long, 5–8 cm wide, the segments free for at least 30% of their length. **Inflorescences** compound terminal inflorescence *pyramidal* up to 8 m tall and 6 m wide, individual inflorescences branched to 5 orders, the basalmost inflorescence ca. 2.8 m long, rachillae very crowded ca. 25 cm long. **Flowers** arranged in groups of 2–10, at anthesis ca. 5 mm wide, white. **Fruit** globose, 15–30 mm diam., green to brownish-green, smooth, with scattered brown dots. **Seed** 10–22 mm diam, endosperm homogeneous, embryo apical.

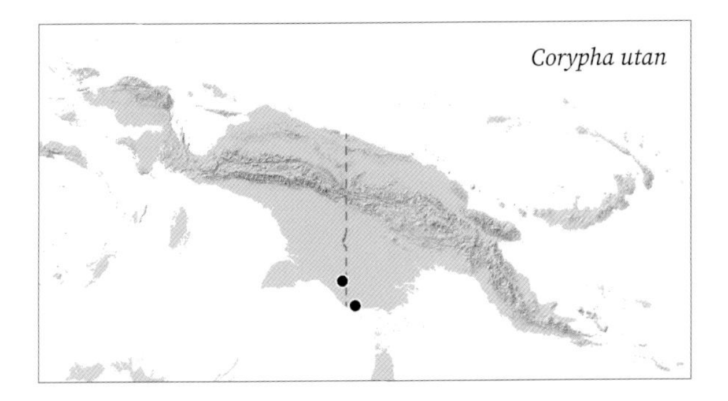

DISTRIBUTION. Central southern New Guinea. Elsewhere widely distributed in disturbed coastal forest from the Bay of Bengal and the Philippines to Australia.

HABITAT. Seasonal vegetation near coast at sea level to 50 m elevation.

LOCAL NAMES. *Ugah* (Marind).

USES. Leaves used for thatching. A reddish edible starch, which is easily digestible, is obtained from the trunk. Stem is used for making hunting tools and traditional weapons. Young shoot (heart-of-palm) and young fruits are edible.

CONSERVATION STATUS. Least Concern (IUCN 2013).

NOTES. When in flower, instantly identifiable by its huge pyramidal compound inflorescence at the top of the stem. When not in flower, the leaf with cleft petiole base is highly distinctive and otherwise known in New Guinea only in *Borassus heineanus*. The latter has smooth sharp petiole margins whereas *C. utan* has neatly arranged black spines throughout the length of the petiole margins. The only other tree fan palms in New Guinea belong to *Livistona* and *Saribus*, but these do not have the cleft petiole base. Juvenile plants might be confused with plants of *Licuala*, but in *Licuala* the leaf blade is divided to the very base into wedge-shaped segments, whereas in *C. utan* the splits in the blade reach no less than one third of the radius. For an account of the species in Australia, see Dowe (2010).

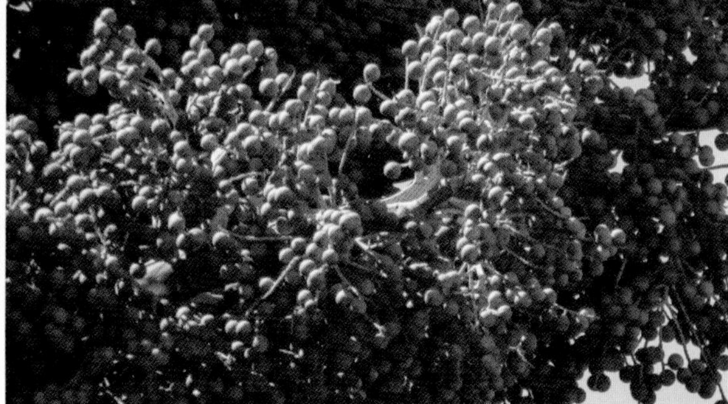

Corypha utan. CLOCKWISE FROM LEFT: habit, stem in full flower, Cape York, Queensland (AB); flowers, Lakefield National Park, Queensland (AB); habit, Lakefield National Park, Queensland (unknown); fruit, Lakefield National Park, Queensland (IC).

Corypha utan. **A.** Habit before and after flowering. **B.** Leaf. **C.** Leaflet tip. **D.** Detail showing transverse veinlets. **E.** Detail spines on petiole. **F.** Inflorescence. **G.** Portion of inflorescence. **H.** Portion of rachilla with flowers. **I.** Flower. **J, K.** Fruit whole and in longitudinal section. Scale bar: A = 5 m; B =1 m; C = 8 cm; D = 5 mm; E = 3 cm; F = 56 cm; G = 12 cm; H = 1 cm; I = 3.3 mm; J, K = 2 cm. A, B from photographs; C from *Bartlett 13577*; D–I from *Maturbongs et al. 660*; J, K from *Bartlett 14138*. Drawn by Lucy T. Smith.

Borassus heineanus. Habit, cultivated National Botanic Garden, Lae (WB).

Borassus L.

Synonym: *Lontarus* Adans.

Tall – leaf palmate – petiole base cleft – no spines – leaf segments not wedge-shaped

Robust, single-stemmed tree fan palm, dioecious. **Leaf** palmate; sheath splitting to the base opposite the petiole; *petiole robust, with sharp edges, with a conspicuous triangular cleft at the base*; segments numerous, the blade split to ca. one third or more of its radius, with pointed tips. **Inflorescence** among the leaves, the male inflorescence branched to 2 orders, the female inflorescence unbranched or with a single branch near the base, pendulous; prophyll and peduncular bract similar, not dropping off as inflorescence expands; peduncle longer than inflorescence rachis; *rachilla very robust*. **Flowers** *male flowers in groups developing in pits hidden by overlapping bracts on a thick sausage-like rachilla*, emerging one at a time, female flowers *borne singly, very much larger, surrounded by overlapping rounded bracts*. **Fruit** *large, black*, stigmatic remains apical, flesh yellow, fibrous. **Seeds** 1–3, each inside its own endocarp, endosperm homogeneous.

DISTRIBUTION. Six species in Africa, Madagascar, India to New Guinea, one known species in northern mainland New Guinea.

NOTES. *Borassus* is a genus of massive, single-stemmed, dioecious fan palms. The petiole of *Borassus* has a distinctive triangular cleft at the base and the fruit are very large. According to the latest monograph (Bayton 2007), there are six species globally, but only one, *B. heineanus*, occurs in New Guinea. It is most easily confused with *Corypha*, which is distinguished by its production of inflorescences above the leaves, stems that die after flowering, spiny petioles and smaller fruit. Other fan palms in New Guinea (*Livistona*, *Saribus* and *Licuala*) do not have the basal triangular cleft to the petiole.

Borassus heineanus Becc.

Robust, single-stemmed tree palm to 25 m. **Stem** ca. 40 cm diam. **Leaf** 20–28 in the crown; petiole and sheath 150–300 cm long; petiole 3.1–5.2 cm wide at midpoint, green with *very sharp black margins, but no spines*; costa 130–150 cm long, hastula conspicuous, to 1.2 cm, blade radius to 180 cm; segments 50–90, 3.9–7.1 cm wide, apices bifid, blade divided to 76–92 cm, transverse veinlets conspicuous. **Inflorescence** male to 1.2 m long, branched to one order, female to 80 cm long, unbranched; *male rachilla cylindrical*, to 70 cm long, 2.7–4.3 cm diam., *rachilla bracts forming pits containing 6–12 flowers*; female rachilla 37–49 cm long with 7–22 flowers arranged spirally. **Male flower** 0.4–1 cm long, stamens 6. **Female flower** 2.5 × 2 cm. **Fruit** *large*, 12–15 × 8–10 cm, ovoid with a slightly pointed apex, greenish black; *pyrenes 1–3*, 9.2–10.5 × 4.8 -7.0 × 4.0–4.5 cm; endocarp sometimes with flanges penetrating seed. **Seed** with homogeneous endosperm; *embryo apical*.

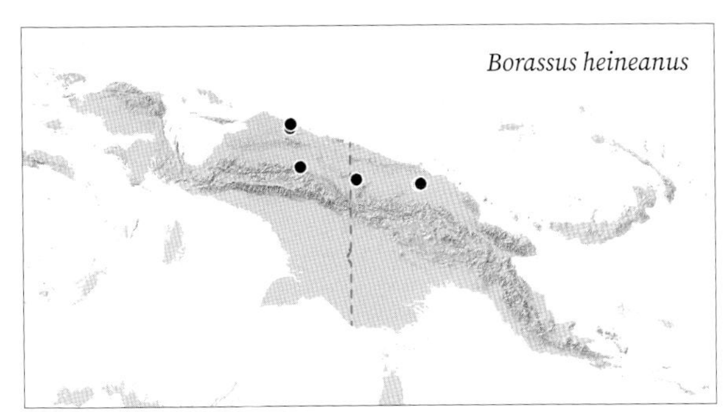

DISTRIBUTION. Restricted to central northern New Guinea.

HABITAT. In lowland forest from sea level to 60 m.

LOCAL NAMES. *Beiwof* (Apau), *Lipmemon* (Kamangauwi dialect).

USES. Leaves used for thatch. Edible young fruits, which are also eaten by the northern cassowary.

CONSERVATION STATUS. Near Threatened. *Borassus heineanus* is threatened at several sites by mining and logging concessions.

NOTES. *Borassus heineanus* is instantly recognisable as the only massive tree palm with fan leaves that have cleft leaf bases and unarmed petioles with very sharp margins. The cleft leaf bases are also known in *Corypha utan*, but that palm has conspicuous teeth along the petiole margins. The fruit of *B. heineanus* is among the largest fruit of any palm in New Guinea, and yet it is known to be consumed by cassowaries, which disperse the sizeable seeds (Pangau-Adam & Mühlenberg 2014).

Borassus heineanus. LEFT TO RIGHT: habit, Sarmi (CDH); leaf, cultivated National Botanic Garden, Lae (WB).

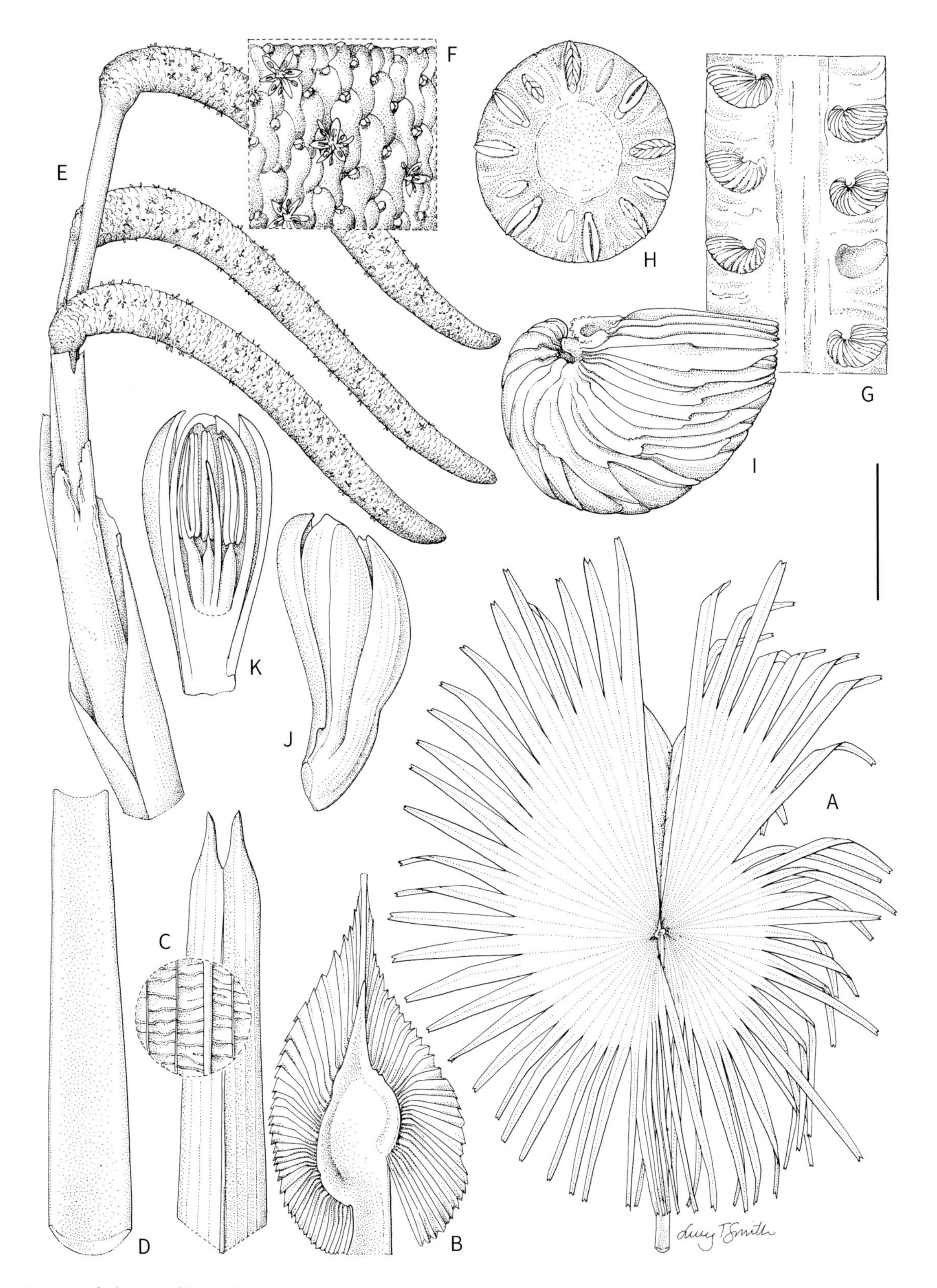

Borassus heineanus (Plate 1). **A.** Leaf. **B.** Hastula. **C.** Segment apex with surface detail. **D.** Petiole. **E.** Male inflorescence. **F.** Surface of rachilla. **G.** Male rachilla in longitudinal section. **H.** Male rachilla in transverse section. **I.** Male flower cluster excised from rachilla. **J, K.** Male flower whole and in longitudinal section. Scale bar: A = 45 cm; B = 6 cm; C = 1.5 cm; D = 12 cm; E = 48 cm; F–H = 3 cm; I = 7 mm; J, K = 4 mm. A, E from photograph; B, C from *Kjaer 525*; D, F–K from *Banka et al. s.n.* Drawn by Lucy T. Smith.

Borassus heineanus. CLOCKWISE FROM TOP LEFT: female inflorescence with fruit, Sarmi (CDH); leaf sheath, cultivated National Botanic Garden, Lae (WB); male flowers, Sarmi (CDH); fruit in transverse section, Sarmi (CDH); male inflorescence, cultivated National Botanic Garden, Lae (WB).

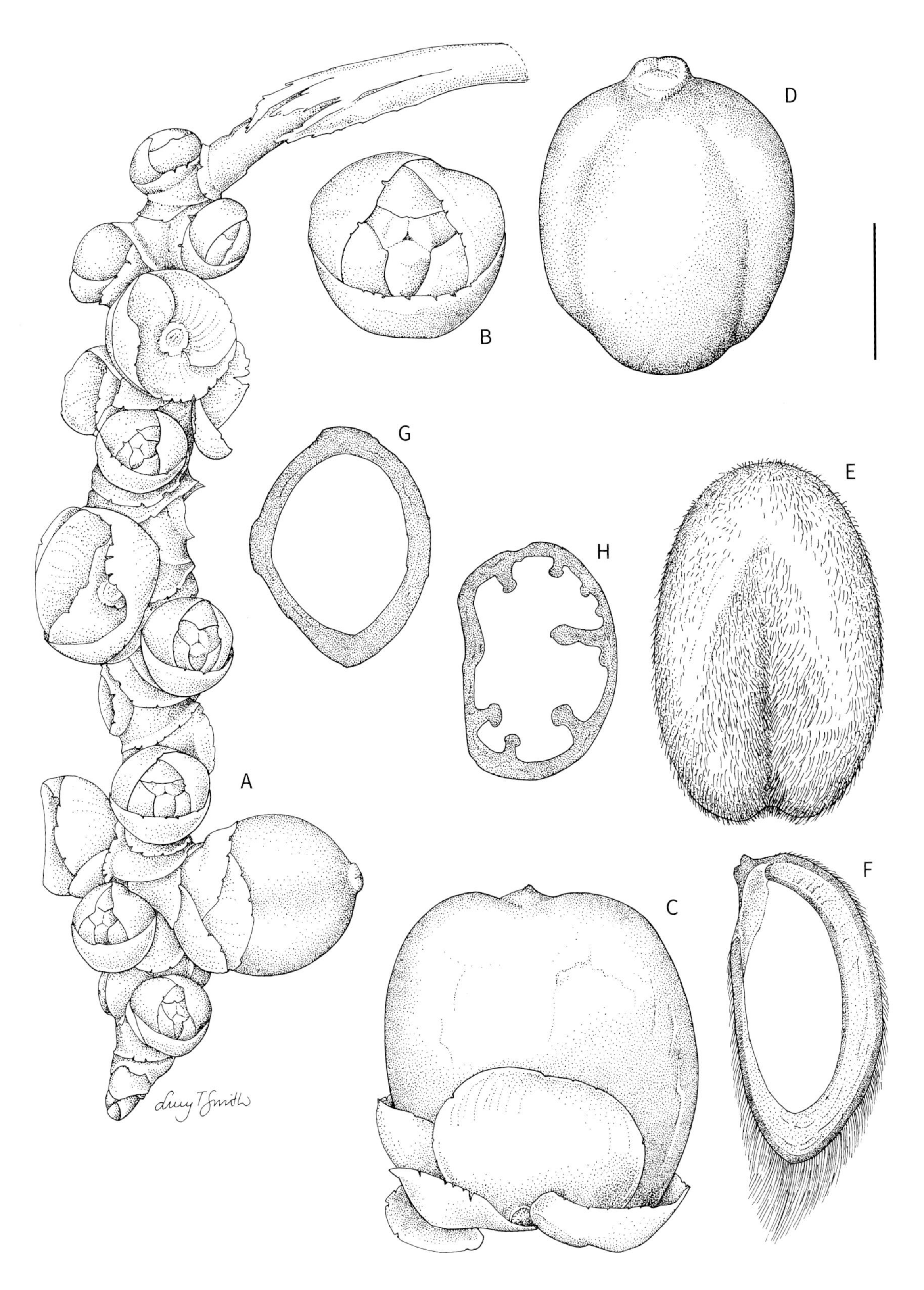

Borassus heineanus (Plate 2). **A.** Female rachilla with flowers and fruit. **B.** Female flower. **C.** Fruit. **D.** Immature fruit. **E, F.** Seed whole and in longitudinal section. **G, H.** Endocarp in transverse section, two different forms. Scale bar: A, C = 6 cm; B, G = 3 cm; D–F, H = 4 cm. A, B, D, G from *Kjaer 525*; C, E, F, H from *Ferrero & Barfod 420*. Drawn by Lucy T. Smith.

Orania regalis. Habit, near Sorong (WB).

Orania Zipp.

Synonyms: *Arausiaca* Blume, *Halmoorea* J.Dransf. & N.Uhl, *Macrocladus* Griff., *Sindroa* Jum.

Medium to tall – leaf pinnate – no crownshaft – no spines – leaflets jagged, pale beneath

Usually robust, single-stemmed tree palms, crownshaft absent, monoecious. **Leaf** pinnate, usually spirally arranged or arranged in a fan (distichous), arching or horizontal; *sheath splitting to base opposite the petiole, fibrous;* petiole elongate; leaflets numerous, *with jagged tips,* arranged regularly or irregularly, sometimes pendulous, covered with whitish indumentum on the undersurface. **Inflorescence** *between the leaves,* branched from 1–3 orders; prophyll hidden among leaf sheaths, *peduncular bract much larger than prophyll,* enclosing the inflorescence in bud, strongly beaked, not dropping off as inflorescence expands, but often decaying later; peduncle usually longer than inflorescence rachis; rachillae generally slender, straight, curved or zig-zag. **Flowers** in triads at the base of the branches, pairs of male flowers towards tip, not developing in pits. **Fruit** green, yellow or orange-red, *large, round, bilobed or trilobed* (depending on number of seeds), fruit with usually 1 but sometimes 2 or 3 seeds, stigmatic remains usually basal (between the lobes in bilobed or trilobed fruit), flesh dry, endocarp smooth *with a basal heart-shaped button.* **Seed** 1 (rarely 2–3), globose, endosperm homogeneous.

DISTRIBUTION. Madagascar, Thailand to New Guinea. Around 30 species, of which 22 occur in New Guinea, mostly on the mainland, not recorded from the Bismarck Archipelago.

NOTES. *Orania* species are usually robust single-stemmed tree palms of mid-storey to canopy with fibrous leaf sheaths that do not form crownshafts and with jagged leaflet apices. Their inflorescences appear between the leaves and bear a peduncular bract that is much longer than the prophyll. The fruits are spherical, quite large (3.5–8 cm), and occasionally bilobed or trilobed. *Orania* occurs throughout New Guinea in lowland to montane rainforest, from sea level to 1,800 m. It can be confused with *Arenga,* which is usually multi-stemmed and has leaflets with asymetrical bases that are V-shaped (induplicate) in section, as well as stems that die after flowering and rather small fruit. *Orania* also somewhat resembles *Cocos,* but this genus has pointed leaflets and very large fruit.

Essig (1980) published a useful synopsis of the genus as it was then understood, including the Australian palm *O. appendiculata,* which is now placed in a completely unrelated genus *Oraniopsis,* and not including the Malagasy species. Several of the species recognised in the latest monograph (Keim & Dransfield 2012) are known from single specimens. Species have sometimes been separated using details of stamen number, as well as number of staminodes and their disposition. The consequence of this is that specimens may be difficult to assign to one of the recognised species without full material. Despite their large size (or perhaps because of it), these majestic palms remain poorly known.

Key to the species of *Orania* in New Guinea

1. Inflorescence branching to one order only 2
1. Inflorescence branching to more than one order 4

2. Large palm, trunk more than 10 cm diam.; number of rachilla in one inflorescence more than 10; bearing triads in the proximal $^{4}/_{5}$ to $^{9}/_{10}$ from the base of rachillae .... **3. *O. dafonsoroensis***
2. Small palm, stem around 10 cm diam. or less; number of rachillae in one inflorescence 10 or fewer; bearing triads in the proximal $^{1}/_{5}$ to $^{1}/_{3}$ from the base of rachillae 3

3. Leaves conspicuously spirally arranged, peduncle of inflorescence glabrous, number of flower clusters 120–128 per rachilla .... **17. *O. parva***
3. Leaves arranged subdistichously; peduncle of inflorescence with dense red-brown tomentum, number of flower clusters 80–85 per rachilla .... **21. *O. timikae***

4. Inflorescence branching to 3 orders 5
4. Inflorescence branching to 2 orders 7

5. Outer leaves deflexed; leaflets arranged irregularly (at least partially) and in more than one plane.. .... **4. *O. deflexa***
5. Outer leaves not deflexed; leaflets arranged regularly in one plane 6

6. Wax absent in inflorescence; rachillae about 30–60 cm long; number of flower clusters per rachilla numerous, 80–90. .... **7. *O. gagavu***
6. Wax abundant in inflorescence; rachillae short-branching in convergent angle, about 13–15 cm long; number of flower clusters per rachilla few, 20 or fewer .... **8. *O. glauca***

7. Leaves in crown not spirally arranged 8
7. Leaves in crown spirally arranged. 10

8. Leaves sub-distichously arranged .... **19. *O. subdisticha***
8. Leaves distichously arranged. 9

9. Small understorey, stem ca. 5 m high, stem ca. 10 cm diam.; leaflets arranged in more than one plane; inflorescence densely covered with red-brown tomentum .... **20. *O. tabubilensis***
9. Taller species, stem 15–20 m high, stem 20–23 cm diam.; leaflets arranged in one plane; inflorescence glabrous .... **5. *O. disticha***

10. Leaflets arranged irregularly in more than one plane giving the whole leaf a plumose appearance .. .. **1. *O. archboldiana***
10. Leaflets arranged regularly in one plane 11

11. Small understorey palm, stem diam. always 15 cm or less 12
11. Medium to large palm, stem diam. always more than 15 cm 14

12. Number of flower clusters per rachilla 100 or more .... **6. *O. ferruginea***
12. Number of flower clusters per rachilla fewer than 100 13

13. Length of rachilla 40 cm or longer; species of coastal and lowland forest, below 1,000 m elevation . . . . **11. *O. littoralis***
13. Length of rachilla 25 cm or shorter; species of highland/mountain forest, 1,000 m or higher elevation . . . . **15. *O. oreophila***

14. Inflorescence congested and brush-like, with branches swept forward; number of stamens 3–5, never more than five . . . . **18. *O. regalis***
14. Inflorescence well expanded; number of stamens more than five 15

15. Filaments always united. . . . . **22. *O. zonae***
15. Filaments always free. 16

16. Staminodes unequal, at least two being different . . . . **12. *O. longistaminodia***
16. Staminodes always similar in size and shape 17

17. Length of petal in female flower 9 mm or more. 18
17. Length of petal in female flower usually less than 9 mm. 20

18. Rachillae 5–6 mm diam., not obviously zigzag, with greyish white indumentum . . . . . . **2. *O. bakeri***
18. Rachillae more robust, 7–8 mm diam., conspicuously zigzag, with grey or brown indumentum . . 19

19. Rachillae densely covered with red-brown tomentum; stamens 6; mature fruits reddish orange, around 5 cm diam. . . . . **9. *O. grandiflora***
19. Rachillae densely covered with greyish white indumentum; stamens 9–14; mature fruits greenish to bright yellow. . . . . **13. *O. macropetala***

20. Inflorescence glabrous or very rarely with red-brown tomentum; length of petal in male flower usually less than 10 mm, very rarely 10–11 mm; mature fruit large, 6.5–7.5 cm diam. . . . . **16. *O. palindan***
20. Inflorescence densely covered with red-brown tomentum; length of petal in male flower always 10 mm or slightly more; mature fruit 3.5–5.0 cm diam. 21

21. Triads borne in the proximal half of rachilla; male flowers blunt ca. 10 × 5 mm; lowland species, found at less than 1,000 m elevation . . . . **10. *O. lauterbachiana***
21. Triads borne up to the proximal $^{4}/_{5}$ part of rachilla; male flowers elongate with 10 × 2–4 mm, only rarely 5 mm; highland species, found at more than 1,000 m elevation. . . . . **14. *O. micrantha***

1. *Orania archboldiana* Burret

Small to medium, single-stemmed tree palm to 15 m. **Stem** 6–10 cm diam. near crown, grey. **Leaves** 6–10 in crown, spirally arranged, to 2.7–3 m long; petiole to 90 cm long, ca. 2 cm diam., glabrescent; rachis to 1.6–2.3 m long, 1.5–2 cm wide, glabrescent, *leaflets numerous, irregularly arranged (at least partially) and displayed in different planes giving the whole leaf a plumose appearance*, 60–90 cm long, 3–4.7 cm wide, upper surface dull green, glabrous or with sparse red-brown tomentum along ribs, lower surface with dense white indumentum and red-brown tomentum along ribs. **Inflorescence** spreading, branched to 2 orders, 90–133 cm long; peduncle 30–45 cm long, with red-brown tomentum; peduncular bract persistent, woody; rachillae numerous, slender, 30–50 cm long, 0.4 cm diam., not zigzag. **Male flowers** ca. 4 mm long, stamens 6. **Female flowers** ca. 3 mm long, staminodes 6, uniform. **Fruits** globose or bilobed, 3.5 cm diam. **Seed** with embryo above seed equator.

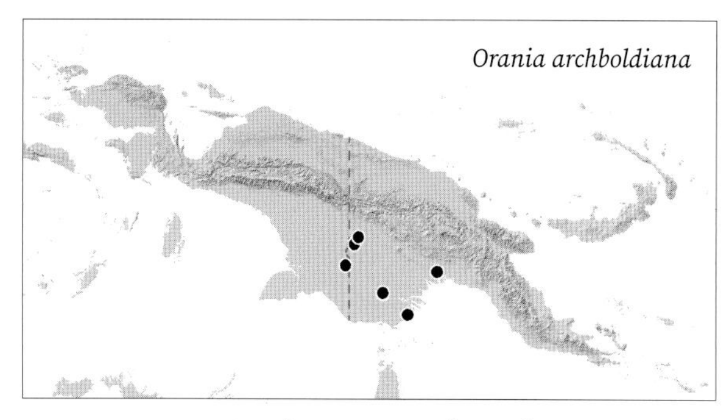

DISTRIBUTION. Southern New Guinea, from near Merauke eastwards.

HABITAT. On ridges in lowland forest, 40–150 m elevation.

LOCAL NAMES. *Kalpo* (Yei).

USES. Outer wood used as arrow tip.

CONSERVATION STATUS. Least Concern. Monitoring is required due to threats from logging concessions.

NOTES. This species is immediately distinguished by its plumose leaves with leaflets arranged in more than one plane throughout the length of the leaf. Leaflets in *Orania deflexa* and *O. tabubilensis* have plumose leaflets in the mid-section of the leaf, but not throughout the leaf length.

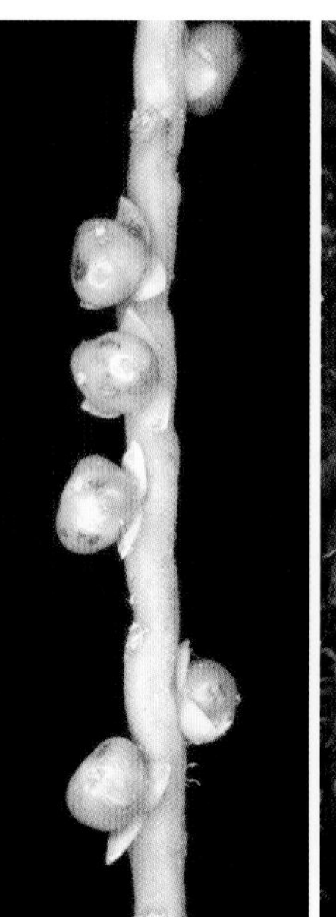

Orania archboldiana. CLOCKWISE FROM TOP RIGHT: habit; leaf showing irregular, plumose leaflets; inflorescence; female flowers, near Kikori (WB).

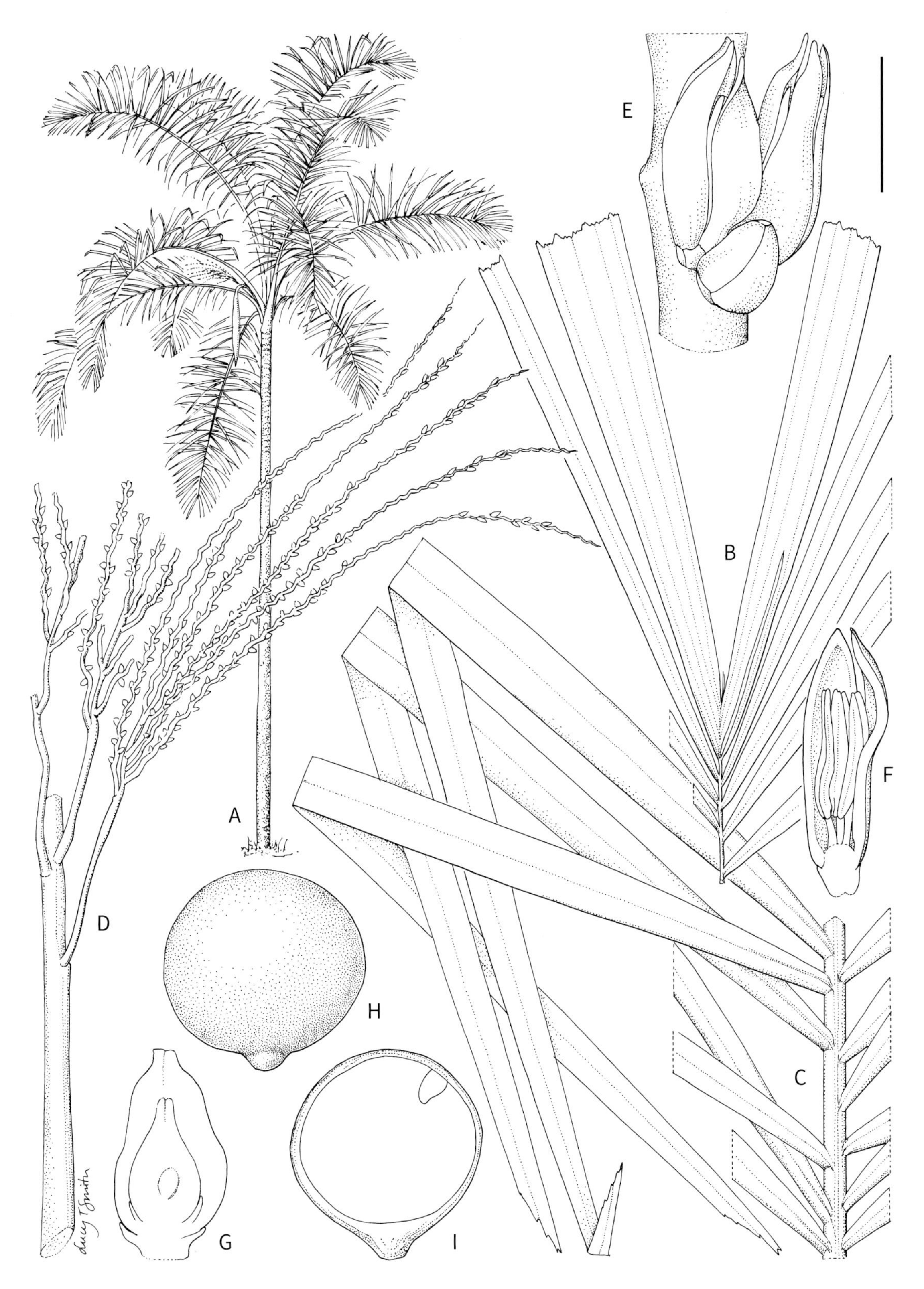

Orania archboldiana. **A.** Habit. **B.** Leaf apex. **C.** Leaf mid-portion. **D.** Portion of inflorescence. **E.** Portion of rachilla with triad. **F.** Male flower in longitudinal section. **G.** Female flower in longitudinal section. **H, I.** Fruit whole and in longitudinal section. Scale bar: A = 1.5 m; B, C = 8 cm; D = 6 cm; E = 4 mm; F = 3.3 mm; G = 2.2 mm; H, I = 3 cm. All from *Baker et al. 1104*. Drawn by Lucy T. Smith.

2. *Orania bakeri* A.P.Keim & J.Dransf.

Large, single-stemmed tree palm to 15 m. **Stem** 20 cm diam., grey. **Leaves** 8 in crown, spirally arranged, to ca. 4–5 m long; petiole to 1.1 m long, ca. 4 cm diam., covered with red-brown tomentum; rachis to 2.6–3.6 m long, ca. 2.5 cm wide in mid-leaf, with red-brown tomentum, leaflets 40–45 on each side of the rachis, regularly arranged in one plane, to 132 cm long, 6–7 cm wide, upper surface dull green, glabrous, lower surface with white indumentum. **Inflorescence** spreading, branched to 2 orders, to 100 cm or more long; peduncle ca. 40 cm long, with thin red-brown tomentum; peduncular bract persistent, woody 1.15–1.40 m long; rachillae numerous, *robust,* 40–55 cm long, 0.8 cm diam., *not zigzag.* **Male flowers** ca. 10 mm long, *stamens 6.* **Female flowers** *ca. 15 mm long,* staminodes 6, uniform. **Fruits** globose or bilobed, 3.8–3.9 cm diam., olive green turning dark brown at maturity. **Seed** with embryo below seed equator.

DISTRIBUTION. North-eastern New Guinea, near the mouth of the Ramu River.

HABITAT. Swampy lowland forest at ca. 50 m elevation.

LOCAL NAMES. None recorded.

USES. None recorded.

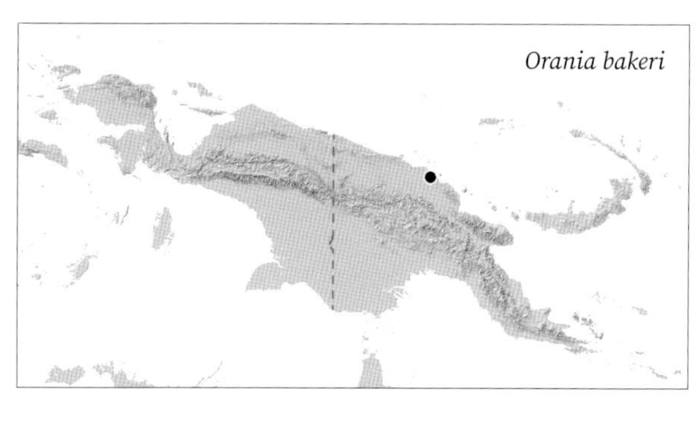

CONSERVATION STATUS. Critically Endangered (IUCN 2021). *Orania bakeri* is known from only one site. Deforestation due to logging is a major threat in its distribution range.

NOTES. In its large size and massive inflorescence this species is very similar to *Orania macropetala* but differs in the much thicker rachillae and 6 rather than 9–14 stamens. It is only known from the type.

Orania bakeri. LEFT TO RIGHT: inflorescence; habit. Ramu River (WB).

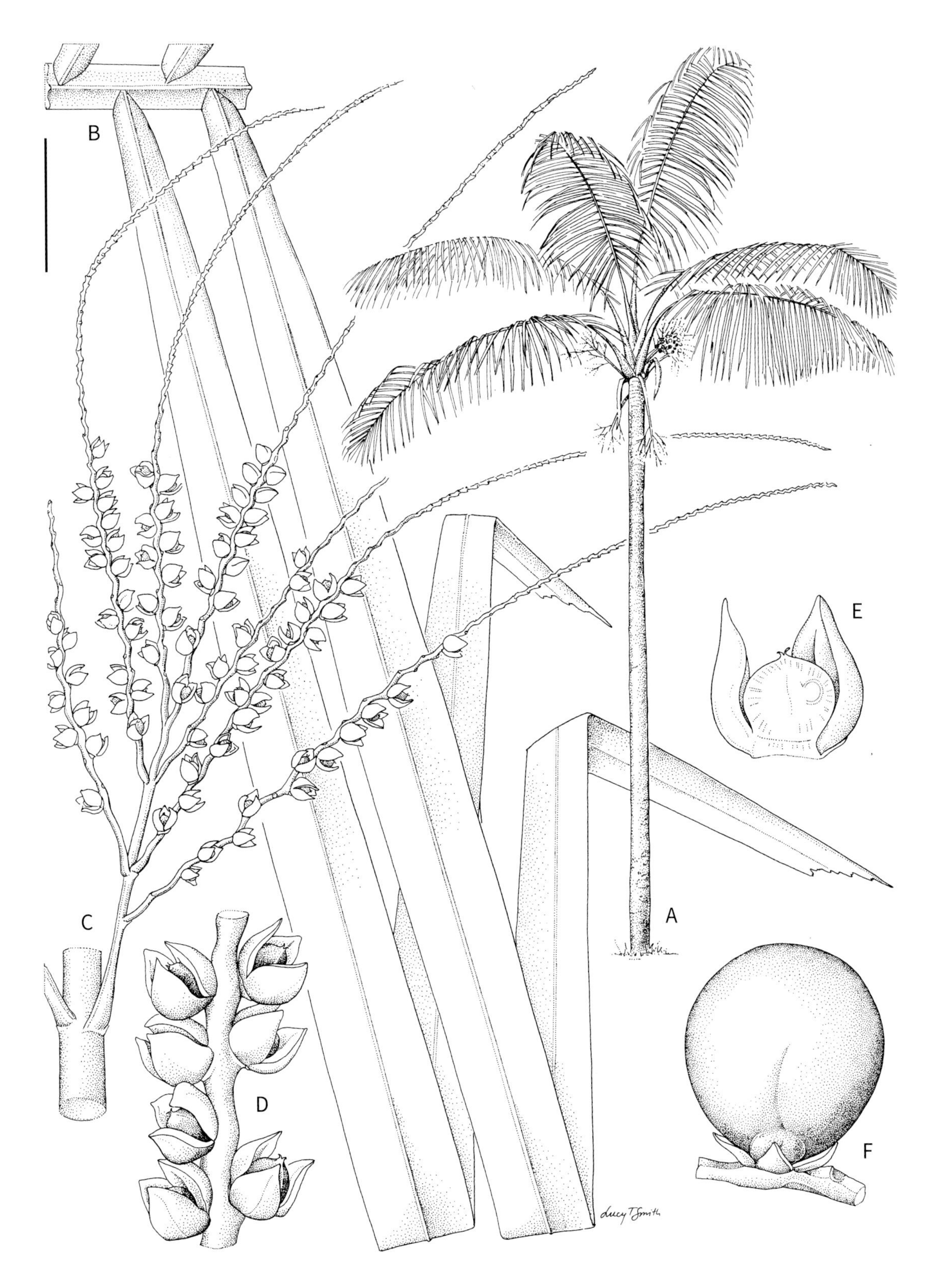

Orania bakeri. **A.** Habit. **B.** Leaf mid-portion. **C.** Portion of inflorescence. **D.** Portion of rachilla with female flowers. **E.** Female flower in longitudinal section. **F.** Fruit. Scale bar: A = 2 m; B = 8 cm; C = 6 cm; D = 2 cm; E = 1 cm; F = 2.5 cm. All from *Baker & Utteridge 581*. Drawn by Lucy T. Smith.

3. *Orania dafonsoroensis* A.P.Keim & J.Dransf.

Large, single-stemmed tree palm to 10 m. **Stem** 10–16 cm diam., grey. **Leaves** 10–11 in crown, *subdistichous* when young, spirally arranged at maturity, to 2.5–4.5 m long; petiole to 70–80 cm long, 2.3–3.5 cm diam., covered with red-brown tomentum; rachis to 1.5–3.3 m long, 2.5 cm wide in mid-leaf, with red-brown tomentum, leaflets 35–40 on each side of the rachis, regularly arranged in one plane, 1.1–1.7 m long, 8.5–9 cm wide, upper surface dark green, glabrous, lower surface with white indumentum and bands of red-brown tomentum along veins and margins. **Inflorescence** spreading, *branched to 1 order, to 2.9 m long*; peduncle ca. 2 m long, glabrous; peduncular bract persistent, woody to 3 m or more long; *rachillae 18–20,* robust, 45–80 cm long, ca. 0.8 cm diam., *not zigzag*. **Male flowers** ca. 6 mm long, stamens 6. **Female flowers** ca. 5 mm long, staminodes 6, uniform. **Fruits** globose or bilobed, 4.0–4.5 cm diam., dull green turning yellow to bright orange at maturity. **Seed** with embryo below seed equator.

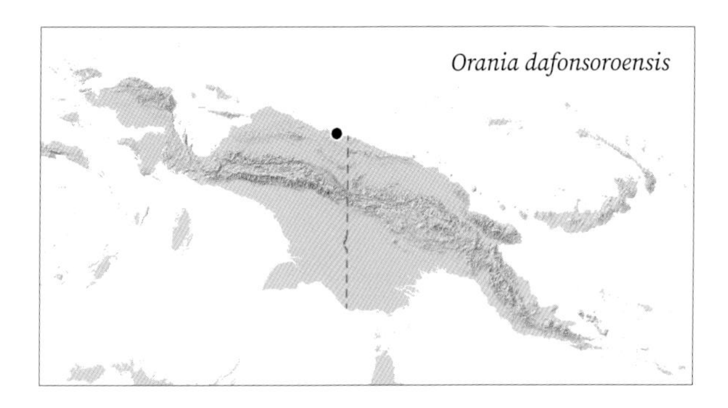

DISTRIBUTION. Cyclops Mountains.

HABITAT. Steep slopes in tropical rainforest at 700–900 m elevation.

LOCAL NAMES. *Hara cho* (Sentani).

USES. None recorded.

CONSERVATION STATUS. Endangered. *Orania dafonsoroensis* has a range restricted to the Cyclops Mountains and, although it occurs in a national park, the habitat is vulnerable to continuing disturbances and degradation.

NOTES. *Orania dafonsoroensis* is a very distinctive palm with its large inflorescence branched to one order only and the subdistichous leaf arrangement in juveniles. It differs from other species with inflorescences branched to one order in its large size and very long inflorescence almost out of proportion to the rest of the palm.

4. *Orania deflexa* A.P.Keim & J.Dransf.

Medium to large, single-stemmed tree palm to 10 m. **Stem** ca. 20 cm diam., grey. **Leaves** ca. 11 in crown, spirally arranged, to 3.3 m long; petiole strongly *decurved, giving crown a drooping appearance,* to 90 cm long, 3 cm diam., covered with red-brown tomentum; rachis ca. 115 cm or more long, 2.5 cm wide in mid-leaf, with red-brown tomentum, leaflets ca. 50 on each side of the rachis, regularly arranged in one plane distally, *in mid third of leaf irregularly arranged in several planes and thus somewhat plumose*, to 90–95 cm long, 5–5.5 cm wide, upper surface dull green, glabrous but with red-brown scales along midrib, lower surface with white indumentum and wax and red-brown tomentum along ribs. **Inflorescence** spreading, *branched to 3 orders,* to 180 cm or more long, densely red-brown tomentose throughout; peduncle 65 cm long, with thin red-brown tomentum; peduncular bract persistent, woody ca. 1 m long; rachillae numerous, slender, 29–32 cm long, 0.5 cm diam., not zigzag. **Male flowers** ca. 8 mm long, stamens 6. **Female flowers** 6 mm long, staminodes 6, uniform. **Fruits** unknown.

DISTRIBUTION. Mountains near Bulolo.

HABITAT. Disturbed montane forest at 1,400 m elevation.

LOCAL NAMES. None recorded.

USES. None recorded.

CONSERVATION STATUS. Critically Endangered (IUCN 2021). *Orania deflexa* is known from only one locality that is disturbed by mining, where its population was estimated to be under 100 individuals in 2006.

NOTES. The drooping leaves are unique to this species. It is also unusual in the inflorescences branched to 3 orders and the somewhat plumose leaves. Inflorescences branched to three orders are also known in *Orania glauca* and *O. gagavu,* but these two species have regularly arranged leaflets.

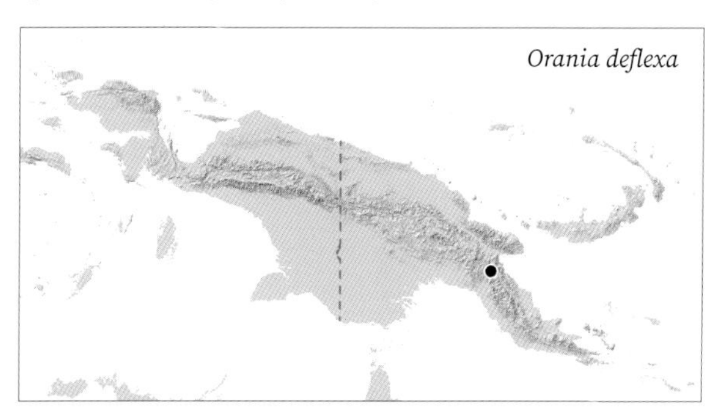

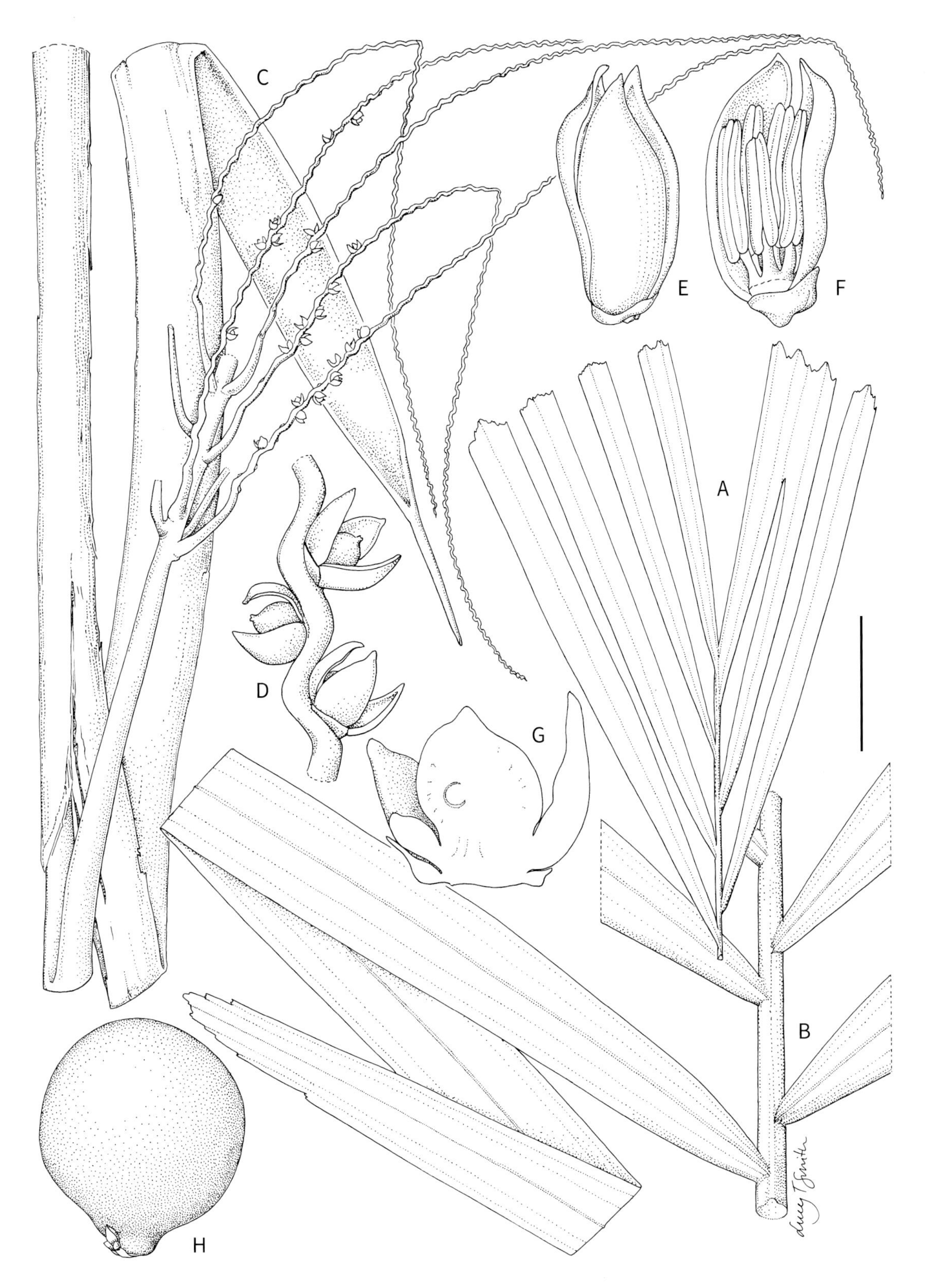

Orania dafonsoroensis. **A.** Leaf apex. **B.** Leaf mid-portion. **C.** Inflorescence. **D.** Portion of rachilla with female flowers. **E, F.** Male flower whole and with one petal removed. **G.** Female flower in longitudinal section. **H.** Fruit. Scale bar: A–C = 8 cm; D = 1 cm; E, F = 4 mm; G = 5 mm; H = 3 cm. All from *Desianto 12*. Drawn by Lucy T. Smith.

Orania deflexa. LEFT TO RIGHT FROM TOP: habit; leaf showing irregular, plumose leaflets; inflorescence; female flowers; male flowers. Near Wau (WB).

***Orania deflexa*.** **A.** Leaf. **B.** Leaf mid-portion. **C.** Portion of inflorescence. **D.** Portion of rachilla with male flowers. **E.** Portion of rachilla with female flowers. **F.** Male flower with one petal removed. **G.** Female flower in longitudinal section. Scale bar: A = 1 m; B = 8 cm; C = 6 cm; D, E = 1.5 cm; F = 5 mm; G = 7 mm. A from photograph; B–G from *Baker et al. 1319*. Drawn by Lucy T. Smith.

5. *Orania disticha* Burret

Large, single-stemmed tree palm to 20 m. **Stem** ca. 23 cm diam., grey. **Leaves** 7–12 in crown, *distichously arranged,* to 4.3 m long; petiole to ca. 1.3 m long, ca. 3 cm diam., covered with red-brown tomentum; rachis to 3 m long, 2.5–3 cm wide in mid-leaf, with red-brown tomentum, leaflets 40–45 on each side of the rachis, regularly arranged in one plane, to 90–150 cm long, 6.5–7.5 cm wide, upper surface dull green, glabrous, shiny green, lower surface with white indumentum and red-brown scales along ribs. **Inflorescence** robust, spreading, branched to 2 orders, to 130 cm or more long; peduncle ca. 50 cm long, ca. 3 cm diam.; peduncular bract persistent, woody to 1.3 m long with dense brown tomentum; rachillae numerous, robust, 41–70 cm long, 0.6 cm diam., conspicuously zigzag. **Male flowers** ca. 12 mm long, stamens 6. **Female flowers** ca. 12 mm long, staminodes 6, uniform. **Fruits** globose or lobed, 5.5–6 cm diam., bright orange. **Seed** with embryo below seed equator.

DISTRIBUTION. Scattered throughout central spine of New Guinea.

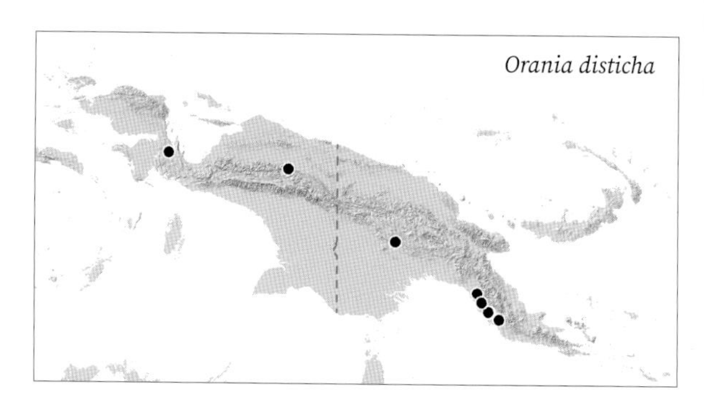

HABITAT. Ridges in lowland and upland rainforest at 10–1,500 m elevation.

LOCAL NAMES. *Pokengeh* (Mekeo-Mapia), *Tidifa* (Kutubu).

USES. Leaves used for thatching.

CONSERVATION STATUS. Least Concern (IUCN 2021).

NOTES. Highly distinctive in the distichously arranged leaves but otherwise similar to *O. palindan*. The only other strictly distichous species in our area is *O. tabubilensis*, but that has leaflets in the middle section of the leaf arranged in several planes and is smaller in stature.

Orania disticha. LEFT TO RIGHT: habit, cultivated National Botanic Garden, Lae (WB); inflorescence and leaf; leaf sheaths, showing distichous arrangement, near Port Moresby (FE).

Orania disticha. **A.** Habit. **B.** Leaf mid-portion. **C.** Portion of inflorescence. **D, E.** Male flower whole and in longitudinal section. **F, G.** Female flower whole and in longitudinal section. **H.** Fruit. **I.** Seed. Scale bar: A = 1 m; B = 9 cm; C = 6 cm; D–G = 1 cm; H, I = 4 cm. A, B, D, E, H, I from *Brass 5599*; C, F, G from *Carr 11705*. Drawn by Lucy T. Smith.

***Orania ferruginea*. A.** Leaf apex. **B.** Inflorescence with peduncular bract. **C.** Immature fruit. **D.** Fruit. Scale bar: A = 9 cm; B = 6 cm; C = 7 mm; D = 2.5 cm. All from *Keim et al. 42*. Drawn by Lucy T. Smith.

6. *Orania ferruginea* A.P.Keim & J.Dransf.

Small, single-stemmed tree palm to 10 m. **Stem** 5–6 cm diam., grey. **Leaves** 10 in crown, spirally arranged, to 2.3–4 m long; petiole to 80 cm long, ca. 2 cm diam., densely covered with red-brown tomentum; rachis to 1.1–2.8 m long, ca. 2 cm wide in mid-leaf, with red-brown tomentum, leaflets 40–45 on each side of the rachis, regularly arranged in one plane, to 87–100 cm long, ca. 5 cm wide, upper surface dull green, glabrous apart from scales along midrib, lower surface with white indumentum and *red-brown tomentum* along ribs. **Inflorescence** spreading, branched to 2 orders, to 1–1.2 m long; peduncle 50–56 cm long, *densely covered with red-brown tomentum*; peduncular bract persistent, woody 1.1–1.4 m long; rachillae numerous, slender, 18–20 cm long, 0.4 cm diam., slightly zigzag. **Male flowers** unknown. **Female flowers** unknown. **Fruits** globose or bilobed, 3.5–4.0 cm diam., dull green turning yellow at maturity. **Seed** with embryo below seed equator.

DISTRIBUTION. Scattered records near Manokwari and Wamena.

HABITAT. Lowland forest at 200–300 m elevation.

LOCAL NAMES. None recorded.

USES. None recorded.

CONSERVATION STATUS. Critically Endangered. *Orania ferruginea* is known from only two sites. The Manokwari site is protected, but very limited. Deforestation is a major threat in its distribution range.

NOTES. *Orania ferruginea* is easily recognized by the dense, thick red-brown tomentum covering sheaths, petioles and inflorescence. It is known from very few collections.

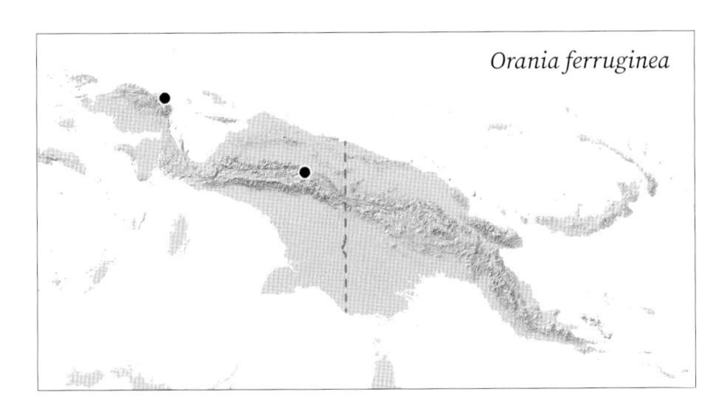

7. *Orania gagavu* Essig

Large, single-stemmed tree palm to 7 m. **Stem** 20–25 cm diam., grey. **Leaves** ca. 9 in crown, spirally arranged, to 4 m long; petiole to 1 m long, ca. 2.5 cm diam., covered with red-brown tomentum; rachis to 3 m long, ca. 2 cm wide in mid-leaf, with red-brown tomentum, leaflets numerous at least 60 on each side of the rachis, regularly arranged in one plane, to 125 cm long, 4.5–5.5 cm wide, upper surface dark green, glabrous apart from midrib with red-brown tomentum, lower surface with white indumentum and thin red-brown tomentum along ribs. **Inflorescence** spreading, *branched to 3 orders*, to 2.5 m or more long, glabrous; peduncle ca. 1 m long; peduncular bract not seen; *rachillae slender, very numerous*, 55–65 cm long, 0.4 cm diam., distally zigzag. **Male flowers** unknown. **Female flowers** unknown. **Fruits** globose, immature fruit ca. 1 cm diam. **Seed** with embryo below seed equator.

DISTRIBUTION. Far south-eastern Papua New Guinea.

HABITAT. Rainforest on mountain slopes at 400 m elevation.

Orania gagavu. Leaf and inflorescence, Alotau (FE).

***Orania gagavu*.** **A.** Leaf mid-portion. **B.** Inflorescence. **C.** Portion of inflorescence. **D.** Fruit. Scale bar: A = 9 cm; B = 83 cm; C = 9 cm; D = 1 cm. A, C, D from *Essig LAE 74096*; B from photograph. Drawn by Lucy T. Smith.

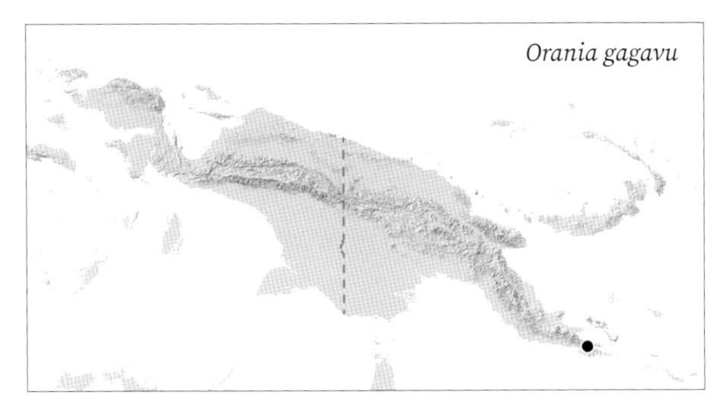

LOCAL NAMES. *Gagavu* (Kaporika).

USES. None recorded.

CONSERVATION STATUS. Critically Endangered (IUCN 2021). *Orania gagavu* is known from only one site where it is threatened by a logging concession.

NOTES. *Orania gagavu* is very similar to *O. lauterbachiana* but differs in the massive inflorescence branched to three rather than two orders. The rachillae are very slender and numerous.

8. *Orania glauca* Essig.

Robust, single-stemmed tree palm to ca. 20 m. **Stem** 30 cm diam., grey. **Leaves** 8–10 in crown, spirally arranged, to 4.6 m long; petiole to 1.4 m long, ca. 4 cm diam., covered with thin red-brown tomentum; rachis to 3.2 m long, 3 cm wide in mid-leaf, with red-brown tomentum, leaflets to 60 on each side of the rachis, regularly arranged in one plane, 80–84 cm long, 6.8–7 cm wide, upper surface dark green, glabrous with conspicuous white wax, lower surface with white indumentum and *abundant wax*. **Inflorescence** spreading, *branched to 3 orders*, to 260 cm or more long, the whole with abundant white wax and appearing glaucous; peduncle ca. 80 cm long; peduncular bract persistent, woody 3 m long, with *abundant white wax*; rachillae numerous, stiff, arranged more or less parallel to each other, very slender, 13.5–0.2 cm. **Male flowers** unknown. **Female flowers** unknown. **Fruits** globose bilobed or trilobed, 3.4–4.8 cm diam., dull

Orania glauca. LEFT TO RIGHT: habit; inflorescence and peduncular bract. Amanab (FE).

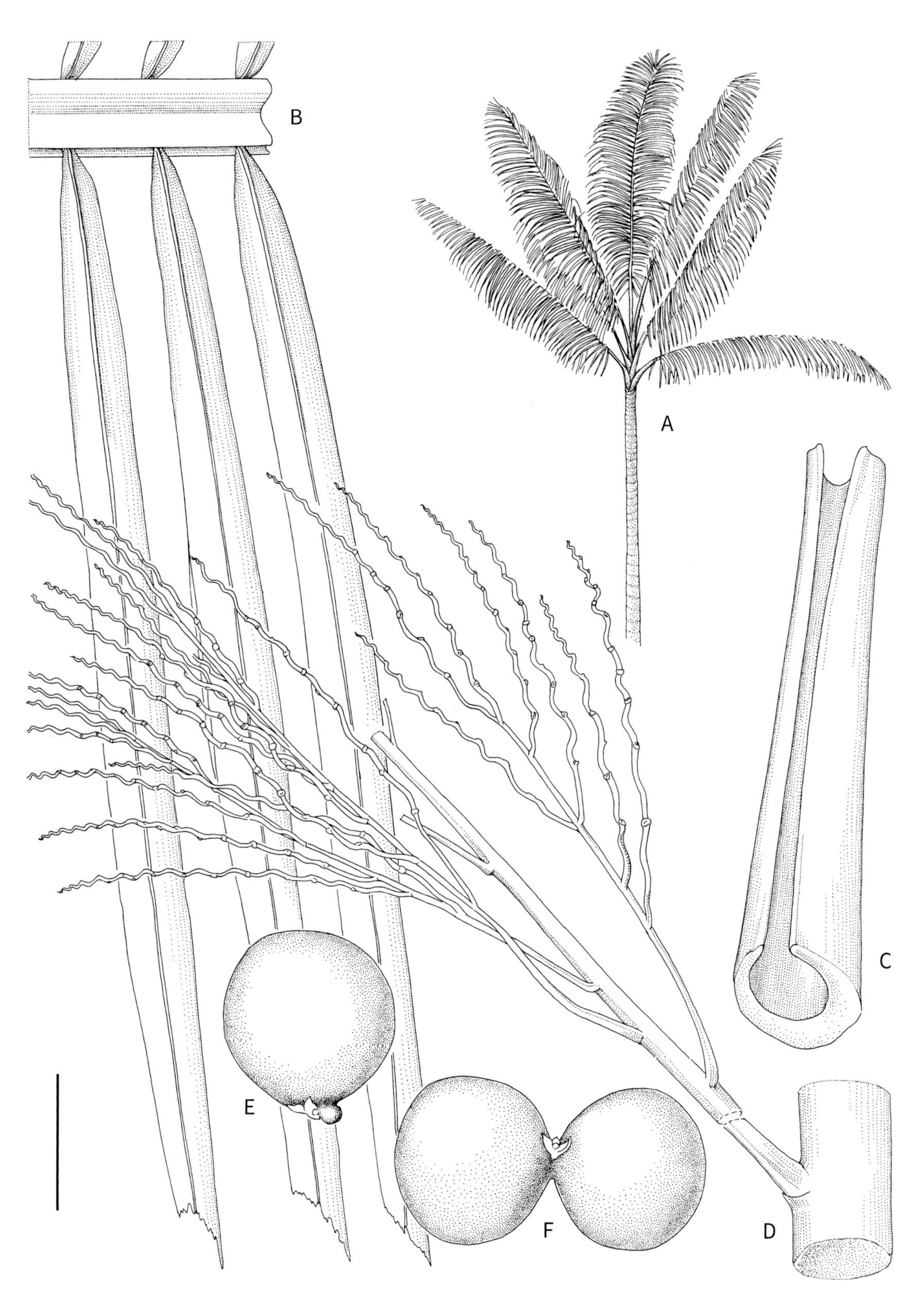

Orania glauca. **A.** Habit. **B.** Leaf mid-portion. **C.** Petiole. **D.** Portion of inflorescence. **E.** Fruit. **F.** Bilobed fruit. Scale bar: A = 2 m; B = 9 cm; C, D = 4.5 cm; E, F = 4 cm. All from *Essig LAE 55089*. Drawn by Lucy T. Smith.

green turning yellowish red or pale orange at maturity. **Seed** with embryo below seed equator.

DISTRIBUTION. Near Amanab, central northern New Guinea.

HABITAT. Disturbed forest on red soils, ca. 400 m elevation.

LOCAL NAMES. None recorded.

USES. None recorded.

CONSERVATION STATUS. Data Deficient (IUCN 2021). *Orania glauca* is known from only one site where it is threatened by logging. Ambiguity in the assessment published on the IUCN Red List (IUCN 2021) implies that the assessor was uncertain and may have intended to assess it as Critically Endangered.

NOTES. *Orania glauca* is among the largest species in the genus, distinctive in its glaucous appearance and inflorescence branched to three orders with many stiff, slender, glaucous rachillae, arranged almost parallel to each other.

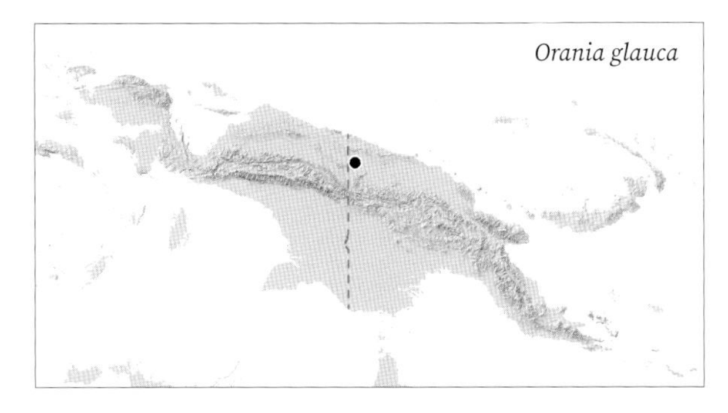

9. *Orania grandiflora* A.P.Keim & J.Dransf.

Large, single-stemmed tree palm to 25 m. **Stem** 30 cm diam., grey. **Leaves** ca. 10 in crown, spirally arranged, to 6 m long; petiole to 2 m long, 4–4.5 cm diam., covered with red-brown and white tomentum; rachis to 4 m long, ca. 3 cm wide in mid-leaf, with red-brown tomentum, leaflets to 60 on each side of the rachis, regularly arranged in one plane, to 1.5 m long, ca. 7 cm wide, upper surface dull green, glabrous, lower surface with white indumentum, wax and red-brown tomentum along ribs. **Inflorescence** spreading, branched to 2 orders, to 1.4 m or more long, with *abundant red-brown tomentum*; peduncle 80 cm long, 10 cm diam. at base; peduncular bract persistent, woody, 1.1 m long; rachillae numerous, very robust, 54–63 cm long, 0.7 cm diam., slightly zigzag. **Male flowers**

Orania grandiflora. FROM TOP: habit; inflorescence. Near Timika (JD).

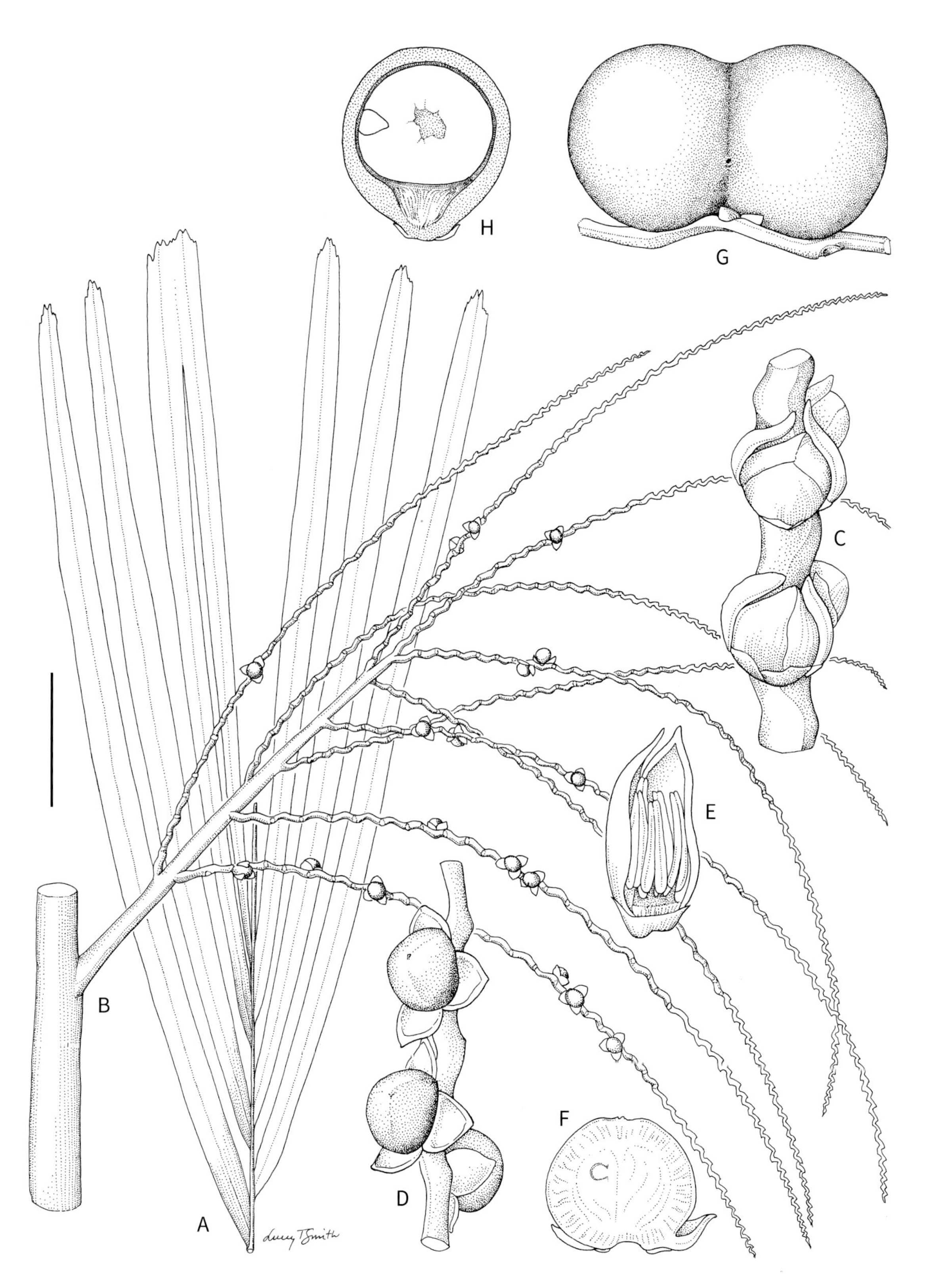

Orania grandiflora. **A.** Leaf apex. **B.** Portion of inflorescence. **C.** Portion of rachilla with triads. **D.** Portion of rachilla with female flowers. **E.** Male flower, one petal removed. **F.** Female flower in longitudinal section. **G.** Bilobed fruit. **H.** Fruit in longitudinal section. Scale bar: A, B = 8 cm; C, F = 1.5 cm; D = 2 cm; E = 7 mm; G, H = 4 cm. All from *Heatubun et al. 179*. Drawn by Lucy T. Smith.

Orania grandiflora. LEFT TO RIGHT: fruit; flowers. Near Timika (JD).

ca. 10 mm long, *stamens* 6. **Female flowers** ca. 10 mm long, staminodes 6, uniform. **Fruits** globose or bilobed, 4.8–5 cm diam., dull green turning brilliant orange at maturity. **Seed** with lateral embryo.

DISTRIBUTION. Near Timika, south-western New Guinea.

HABITAT. Disturbed alluvial forest at ca. 30 m elevation.

LOCAL NAMES. None recorded.

USES. None recorded.

CONSERVATION STATUS. Critically Endangered. *Orania grandiflora* is known from only one site where it is threatened by a mining concession. The type locality is already destroyed by tailings from the Freeport gold and copper mine.

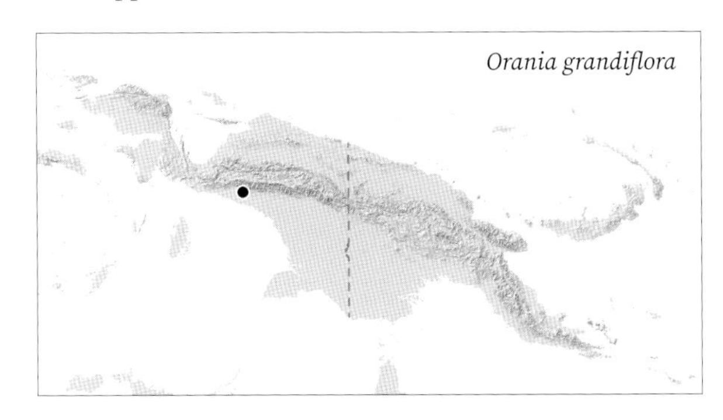

NOTES. In habit, this spectacular palm is very similar to *Orania macropetala* but differs in having six rather than 9–14 stamens and staminodes, and in the abundant tomentum in the inflorescence. The brilliant orange fruits at maturity are particularly striking.

10. *Orania lauterbachiana* Becc.

Synonym: *Orania brassii* Burret

Large, single-stemmed tree palm to 10–20 m. **Stem** 18–32 cm diam., grey. **Leaves** 10 in crown, spirally arranged, to 3.5–5.1 m long; petiole to 1.2–2.2 m long, ca. 2.5 cm diam., covered with red-brown tomentum; rachis to 3 m long, ca. 2.5 cm wide in mid-leaf, with red-brown tomentum, leaflets 18–21 on each side of the rachis, regularly arranged in one plane, to 70–132 cm long, 5.5–11.5 cm wide, upper surface dull green, glabrous or with tomentum along ribs, lower surface with white indumentum and red-brown tomentum. **Inflorescence** spreading, robust, *branched to 2 orders*, to 95–192 cm or more long; peduncle to 65 cm long, with thin red-brown tomentum; peduncular bract not seen; rachillae numerous, robust, 35–60 cm long, 0.8 cm diam., distally zigzag, glabrous or with thin brown tomentum. **Male flowers** ca. 11 mm long, stamens 6. **Female flowers** 8 mm long, staminodes 6, uniform. **Fruits** globose or bilobed or trilobed, 3.3–5 cm diam.,

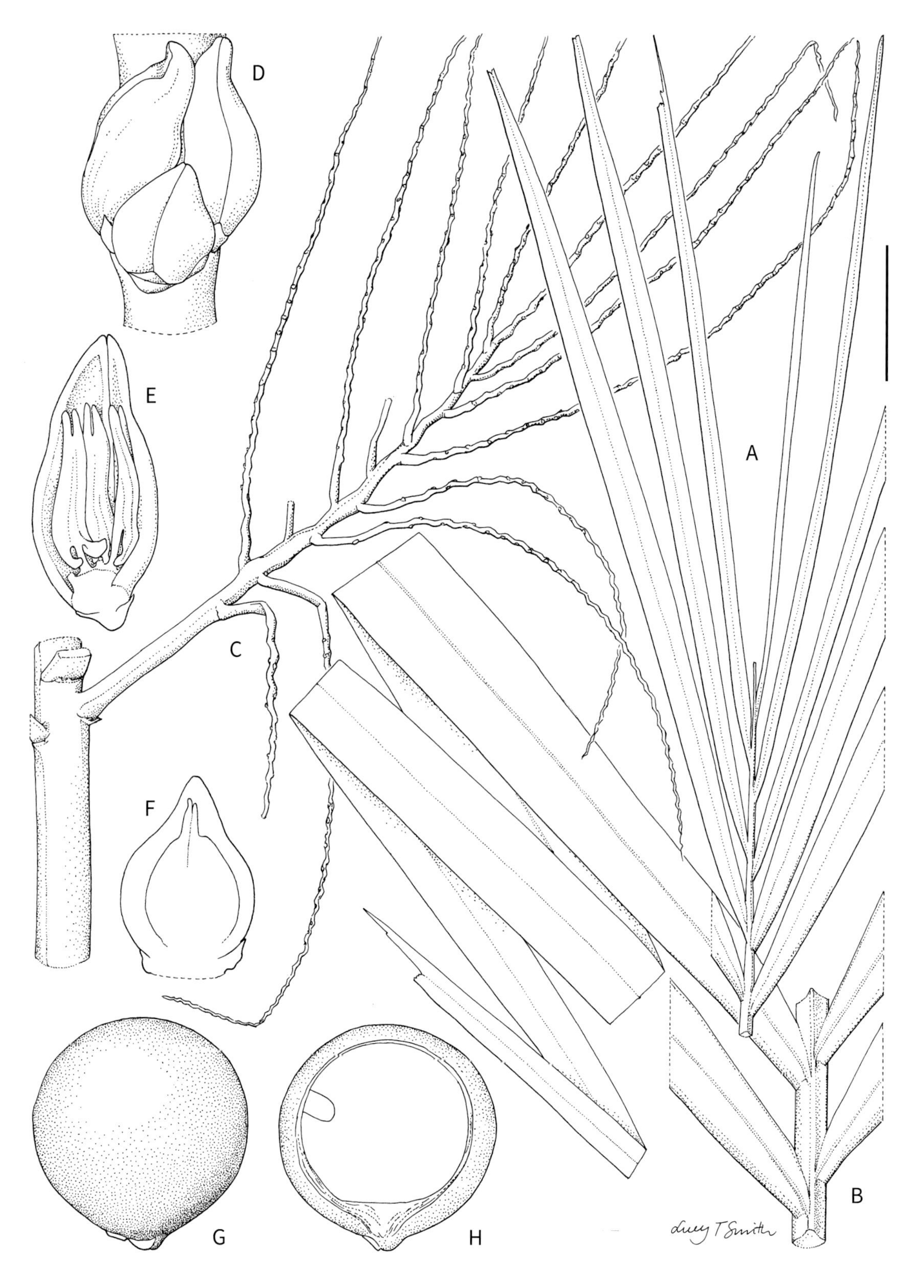

***Orania lauterbachiana*.** **A.** Leaf apex. **B.** Leaf mid-portion. **C.** Portion of inflorescence. **D.** Triad. **E.** Male flower in longitudinal section. **F.** Female flower in longitudinal section. **G, H.** Fruit whole and in longitudinal section. Scale bar: A–C = 8 cm; D = 5 mm; E, F = 3 mm; G, H = 3 cm. All from *Baker et al. 1107*. Drawn by Lucy T. Smith.

Orania lauterbachiana. FROM TOP: habit; inflorescences. Near Kikori (WB).

brownish green turning bright or brownish orange at maturity. **Seed** with lateral embryo.

DISTRIBUTION. Throughout New Guinea, including D'Entrecasteaux Islands (Normanby).

HABITAT. Rainforest from sea level to 1250 m elevation.

LOCAL NAMES. *Bananak* (Amele), *Bu-lowe* (Lababia), *Haoh* (Ayawasi), *Kolu* (Mokian), *Kunakwan* (Jal), *Mamvuiti* (Kairi), *Omoo* (Yamur).

USES. Trunk used for floorboards.

CONSERVATION STATUS. Least Concern.

NOTES. This species is very similar in general appearance to *Orania gagavu* but the latter has inflorescences branched to three orders (rather than two) and the rachillae bear up to 90 flower clusters rather than over 95 in *O. lauterbachiana*.

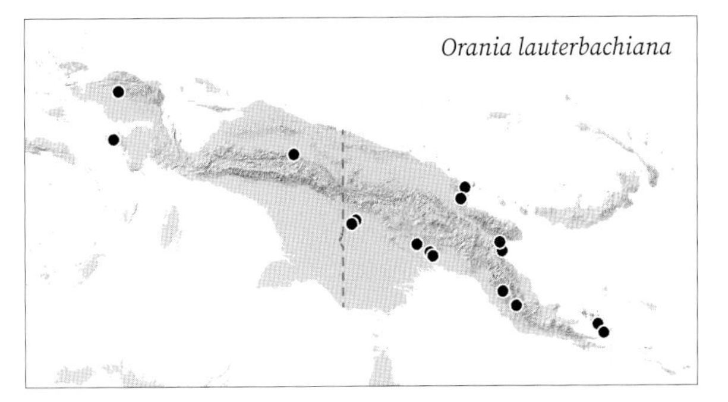

11. *Orania littoralis* A.P.Keim & J.Dransf.

Small, single-stemmed tree palm to 6 m. **Stem** 14 cm diam., grey. **Leaves** ca. 12 in crown, spirally arranged, to 6 m long; petiole short, to 2 cm diam., covered with red-brown tomentum; rachis to 4 m long, ca. 2 cm wide in mid-leaf, with red-brown tomentum, leaflets up to 70 on each side of the rachis, regularly arranged in one plane, to 120 cm long, ca. 7 cm wide, upper surface bright green, shiny, glabrous, lower surface with white indumentum and red-brown tomentum along veins. **Inflorescence** spreading, branched to 2 orders, to 2.5 m or more long; peduncle ca. 85 cm long, ca. 7 cm diam., with dense red-brown tomentum; peduncular bract persistent, woody to 2.4 m long; *rachillae long*, numerous, thick, 40–55 cm long, 0.6 cm diam., not zigzag. **Male flowers** ca. 9–10 mm long, stamens 6. **Female flowers** 4 mm long, staminodes 6, uniform. **Fruits** globose or bilobed, 4.5–5 cm diam., turning orange and red at maturity. **Seed** with embryo above seed equator.

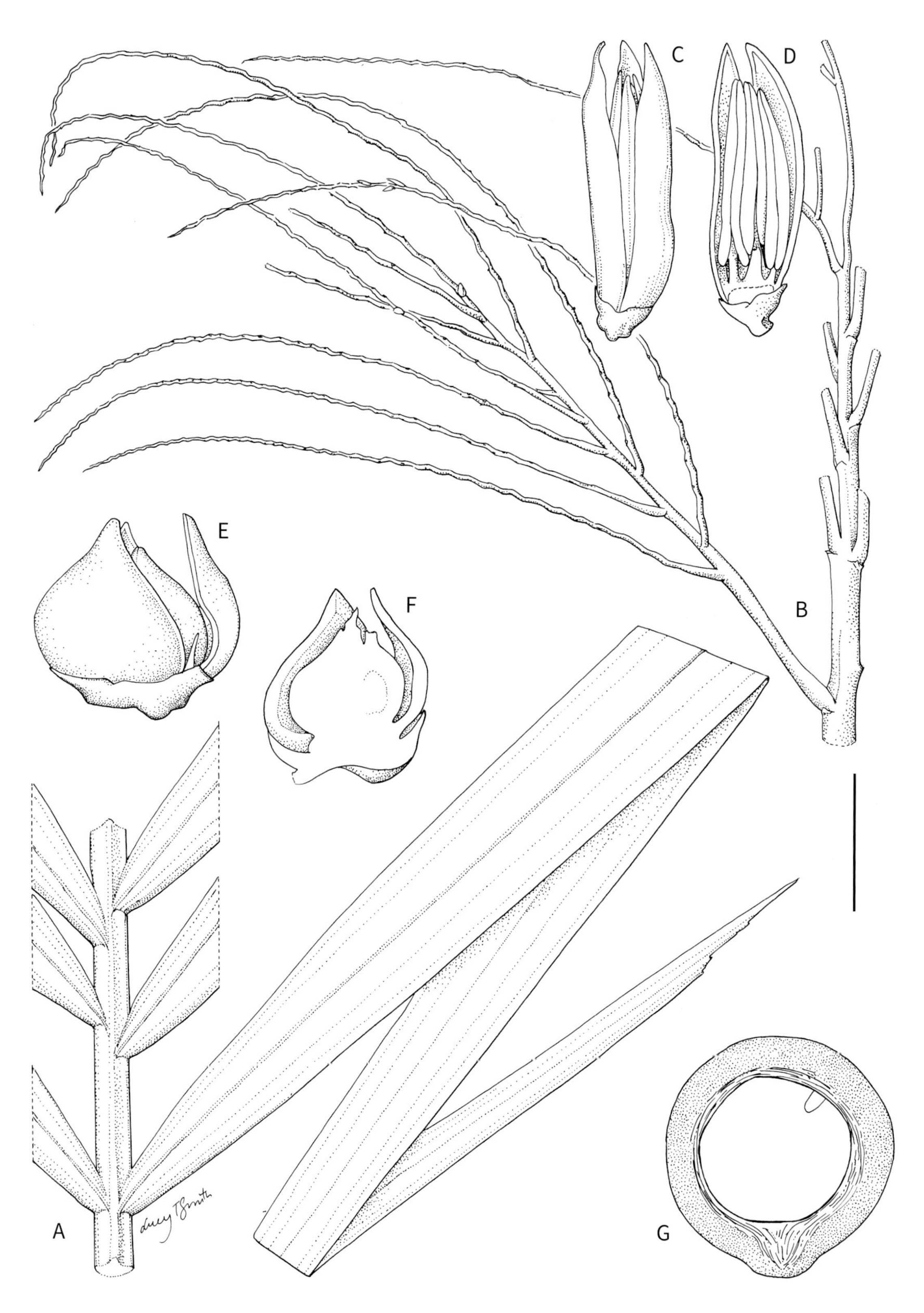

Orania littoralis. **A.** Leaf mid-portion. **B.** Portion of inflorescence. **C, D.** Male flower whole and with one petal removed. **E, F.** Female flower whole and in longitudinal section. **G.** Fruit in longitudinal section. Scale bar: A, B = 8 cm; C–F = 4 mm; G = 2.5 cm. All from *Barfod et al. 456*. Drawn by Lucy T. Smith.

Orania littoralis. FROM TOP, LEFT TO RIGHT: habit; female flowers; male flowers. Alotau (AB).

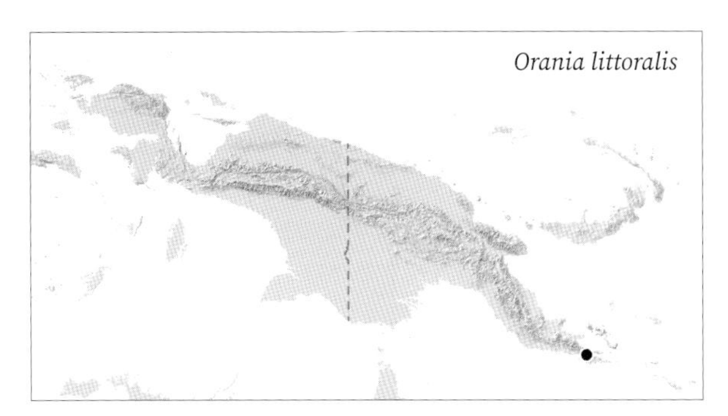

DISTRIBUTION. Far south-eastern New Guinea.

HABITAT: Lowland rainforest at sea level.

LOCAL NAMES. None recorded.

USES. None recorded.

CONSERVATION STATUS. Critically Endangered (IUCN 2021). *Orania littoralis* is known from only one site where it is threatened by logging.

NOTES. *Orania littoralis* shares many similarities with both *O. ferruginea* and *O. oreophila*. It can be distinguished by the much longer rachillae with relatively fewer flower clusters.

12. *Orania longistaminodia* A.P.Keim & J.Dransf.

Large, single-stemmed tree palm to 6 m. **Stem** ca. 22 cm diam., grey. **Leaves** 12 in crown, spirally arranged, to ca. 6 m long; petiole to 2.6–2.8 m long, ca. 4 cm diam., covered with thin red-brown tomentum; rachis to 2.6–3.6 m long, ca. 2.5 cm wide in mid-leaf, with thin red-brown tomentum, leaflet number not recorded, regularly arranged in one plane, to 130 cm long, ca. 10 cm wide, upper surface dull green, glabrous apart from thin red-brown tomentum along midrib, lower surface with white indumentum and red-brown tomentum along margins and main veins. **Inflorescence** spreading, branched to 2 orders, 53–94 cm or more long; peduncle 33–58 cm long, with red-brown tomentum; peduncular bract persistent, woody 80–90 cm long; rachillae numerous, robust, 18–20 cm long, 0.6 cm diam., somewhat zigzag. **Male flowers** ca. 5 mm long, *stamens 5–7*. **Female flowers** ca. 8 mm long, staminodes 6, *unequal*, varying from flower to flower, with a *mixture of small (1 mm long) and large* (5 mm long). **Fruits** globose or bilobed, 4.8–7 cm diam., bright orange at maturity. **Seed** with embryo below seed equator.

DISTRIBUTION. Two widely separated records on the north and south coasts of east New Guinea.

Orania longistaminodia. **A.** Leaf apex. **B.** Inflorescence. **C, D.** Male flower whole and with one petal removed. **E.** Female flower. **F.** Female flower, one petal removed. **G.** Petal from female flower with staminode. Scale bar: A = 6 cm; B = 4 cm; C, D = 5 mm; E–G = 7 mm. All from *Barfod & Damborg 374*. Drawn by Lucy T. Smith.

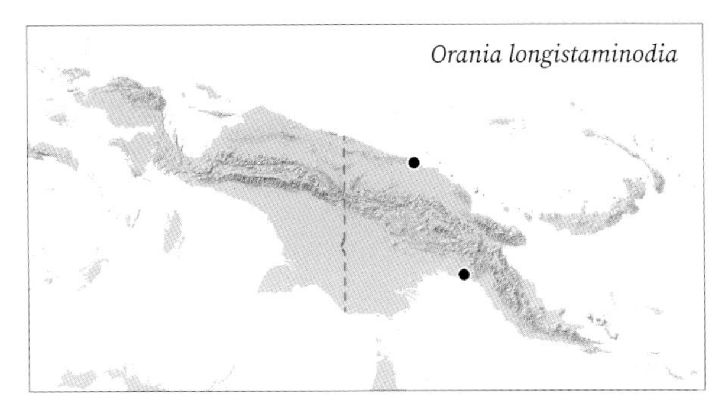

HABITAT. Disturbed coastal forest at sea level.

LOCAL NAMES. None recorded.

USES. Sap is said to be used by local tribes as a medicine for sore throats.

CONSERVATION STATUS. Critically Endangered (IUCN 2021). *Orania longistaminodia* is known from only two sites where it is threatened by logging concessions.

NOTES. Difficult to distinguish from other large species without female flowers with their distinctive unequal staminodes and male flowers with their variable stamen number.

13. *Orania macropetala* K.Schum. & Lauterb.

Large, single-stemmed tree palm to 10–20 m. **Stem** ca. 20 cm diam., grey. **Leaves** ca. 12 in crown, spirally arranged, to 3.9–5 m long; petiole ca. 1 m long, ca. 4 cm diam., covered with red-brown tomentum; rachis to 2.9–4 m long, 2.5 cm wide in mid-leaf, with red-brown tomentum, leaflets numerous, regularly arranged in one plane, 79–140 cm long, 5–8.5 cm wide, upper surface dull green, glabrous, lower surface with white indumentum and thin red-brown tomentum along margins and main veins. **Inflorescence** spreading, branched to 2 orders, to 200 cm or more long; peduncle ca. 100 cm long, with thin red-brown tomentum; peduncular bract no recorded; rachillae numerous, robust, 30–60 cm long, 0.8 cm diam., *conspicuously zigzag*. **Male flowers** *very large, ca. 17 mm long, stamens 9–14*. **Female flowers** 9–13 mm long, staminodes 6 or 10, uniform. **Fruits** globose, rarely bilobed, to 7.5 cm diam., colour not recorded. **Seed** with embryo below seed equator.

DISTRIBUTION. Mostly reported from north-eastern New Guinea, with an outlying record in the Bird's Head Peninsula.

HABITAT. Lowland to upland tropical rainforest, 100–900 m elevation.

Orania macropetala. Inflorescences with fruit and floral triads in bud, Usino (FE).

LOCAL NAMES. *Gamain* (Usino).

USES. Ripe fruits eaten by cassowaries.

CONSERVATION STATUS. Least Concern. Surveys are required as most collections were made at least 40 years ago.

NOTES. Distinguishing the large New Guinea species of *Orania* with certainty requires male and female flowers so that stamen and staminode number can be counted. In *Orania macropetala*, the rachillae are conspicuously zigzag and the male flowers are strikingly large with 9–14 stamens.

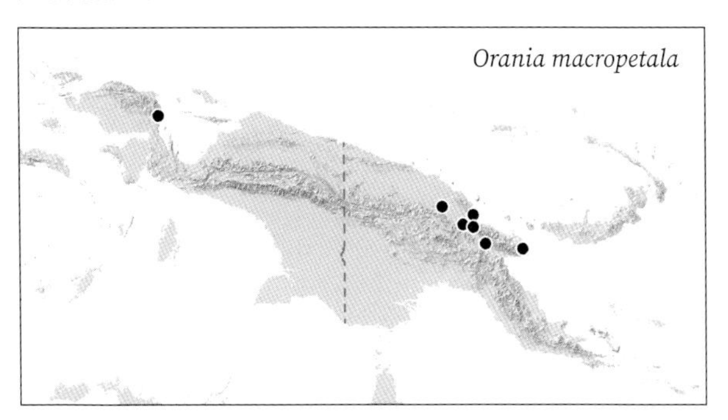

Orania macropetala. **A.** Leaf mid-portion. **B.** Portion of inflorescence. **C.** Portion of rachilla with male flowers. **D.** Portion of rachilla with female flowers. **E.** Male flower in longitudinal section. **F.** Female flower in longitudinal section. Scale bar: A = 9 cm; B = 6 cm; C, D = 1 cm; E, F = 7 mm. All from *Essig & Young LAE 74001*. Drawn by Lucy T. Smith.

14. *Orania micrantha* Becc.

Synonym: *Orania clemensiae* Burret

Large, single-stemmed tree palm to 17 m. **Stem** 25–30 cm diam., grey. **Leaves** spirally arranged, to 2.6 m long; petiole not recorded; rachis fragment seen 1.5 cm wide, with dense red-brown tomentum, leaflets ca. 30 on each side of the rachis, regularly arranged in one plane, to 80–100 cm long, 4.7–6.5 cm wide, upper surface dull green, glabrous apart from thin red-brown tomentum along midrib, lower surface with white indumentum and red-brown tomentum along margins and main veins. **Inflorescence** spreading, branched to 2 orders, to 100–270 cm or more long; peduncle 45–130 cm long, with dense red-brown tomentum; peduncular bract not seen; *rachillae numerous, slender*, 37–60 cm long, ca. 0.4 cm diam., not zigzag, glabrous or with thin brown tomentum. **Male flowers** ca. 10 mm long, *elongate*, stamens 6. **Female flowers** ca. 5 mm long, staminodes 6, uniform. **Fruits** globose or bilobed, 2.8 cm diam. (young), colour not recorded. **Seed** with embryo below seed equator.

DISTRIBUTION. Scattered records in the eastern half of the Central Range.

HABITAT. Upland rainforest at 1,000–1,400 m elevation.

LOCAL NAMES. None recorded.

USES. None recorded.

CONSERVATION STATUS. Vulnerable. *Orania micrantha* has a restricted distribution. Deforestation due to mining is a major threat in its distribution range.

NOTES. With only very few collections, this palm remains very poorly known. It seems closest to *Orania lauterbachiana* but is distinguished by its more slender rachillae and elongate rather than blunt male flowers and smaller female flowers. Further collections of these two taxa are required to establish whether they are distinct or not.

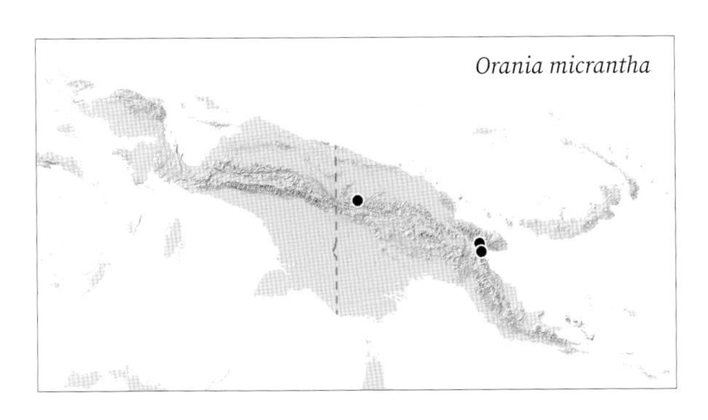

Orania micrantha. FROM TOP: inflorescence; floral triads in bud. Frieda River (FE).

Orania micrantha. **A.** Leaf mid-portion. **B.** Portion of inflorescence. **C.** Portion of rachilla with male flowers. **D**, **E.** Male flower whole and in longitudinal section. **F**, **G.** Female flower whole and in longitudinal section. Scale bar: A = 6 cm; B = 8 cm; C = 7 mm; D–G = 3.3 mm. All from *Essig & Young LAE 74064.* Drawn by Lucy T. Smith.

15. *Orania oreophila* Essig

Small, single-stemmed tree palm *to 4 m*. **Stem** ca. 20 cm diam., grey. **Leaves** 8 in crown, spirally arranged, to 2.8–3.1 m long; petiole to 1.2 m long, 4 cm diam., covered with red-brown tomentum; rachis to 2.6–3.6 m long, 1 cm wide in mid-leaf, with red-brown tomentum, leaflets 22–24 on each side of the rachis, regularly arranged in one plane, to 60–78 cm long, 4–4.5 cm wide, upper surface dark green, glabrous apart from tomentum along midrib, lower surface with white indumentum and red-brown tomentum along ribs and margins. **Inflorescence** spreading, branched to *2 orders*, length not recorded; peduncle dimensions not recorded; peduncular bract persistent, woody, 59 cm long; rachillae slender, 18–24 cm long, 0.2 cm diam., not zigzag. **Male flowers** ca. 8 mm long, stamens 6. **Female flowers** ca. 5 mm long, staminodes 6, uniform. **Fruits** globose or bilobed, 3.0–4.5 cm diam., colour not recorded. **Seed** with embryo below seed equator.

DISTRIBUTION. Highlands of eastern New Guinea.

HABITAT. Montane forest of *Castanopsis* at 1,300–1,600 m elevation.

LOCAL NAMES. None recorded.

USES. None recorded.

CONSERVATION STATUS. Endangered (IUCN 2021). Known only from two localities where habitat quality is declining.

NOTES. A distinctive small montane palm, occurring at the highest elevation recorded for the genus.

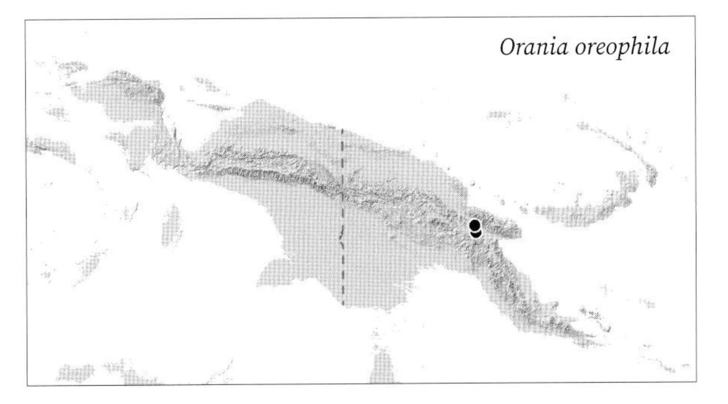

Orania oreophila. LEFT TO RIGHT: habit, Kassam Pass (FE); inflorescence, Eastern Highlands (MC).

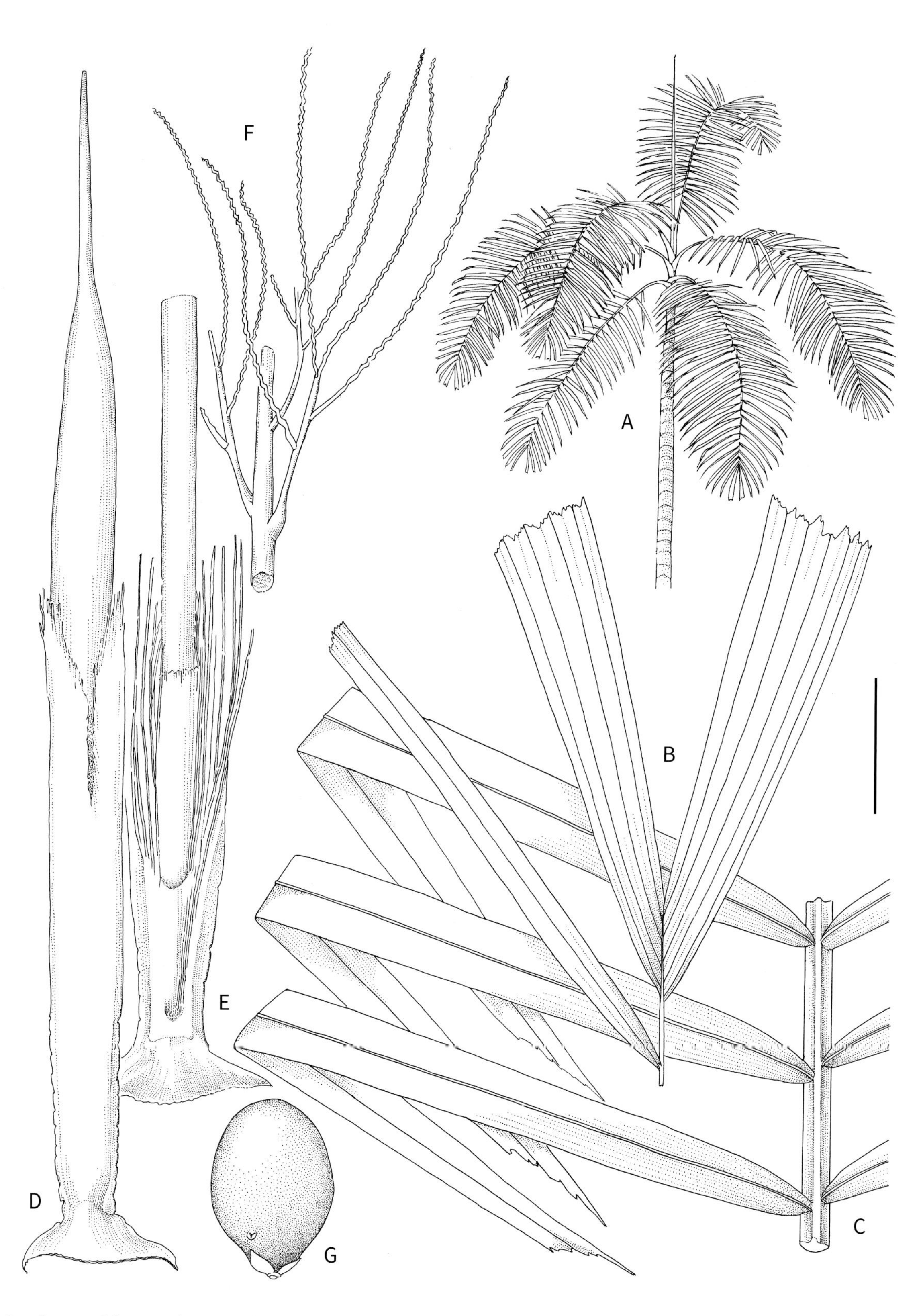

Orania oreophila. **A.** Habit. **B.** Leaf apex. **C.** Leaf mid-portion. **D.** Inflorescence bud. **E.** Inflorescence base. **F.** Portion of inflorescence. **G.** Fruit. Scale bar: A = 1.5 m; B, C, F = 8 cm; D, E = 6 cm; G = 3 cm. All from *Essig LAE 55147*. Drawn by Lucy T. Smith.

16. *Orania palindan* (Blanco) Merr.

Synonyms: *Caryota palindan* Blanco, *Orania decipiens* var. *montana* Becc., *Orania moluccana* Becc., *Orania philippinensis* Scheff. ex Becc., *Orania regalis* Naves non Zipp.

Large, single-stemmed tree palm to 30–40 m. **Stem** 15–40 cm diam., grey. **Leaves** 10–15 in crown, spirally arranged, to 6 m long; petiole to 1.5 m long, ca. 6 cm diam., covered with red-brown tomentum; rachis to 4 m or more long, 2.5 cm wide in mid-leaf, with red-brown tomentum, leaflets to 60 or more on each side of the rachis, regularly arranged in one plane, to 200 cm long, 4–7 cm wide, upper surface dull green, glabrous apart from red-brown tomentum along midrib, lower surface with dense white indumentum and red-brown tomentum along ribs and margins. **Inflorescence** spreading, robust, *branched to 2 orders*, 1.1–2 m or more long; peduncle 80–150 cm long, with thin red-brown tomentum; peduncular bract persistent, woody 53–160 cm long; rachillae numerous, robust, *glabrous*, 30–66 cm long, 0.8 cm diam., straight or conspicuously zigzag. **Male flowers** to 11 mm long, stamens 6. **Female flowers** ca. 6 mm long, staminodes 6, *uniform*. **Fruits** globose or bilobed, *6.5–7.5 cm diam.*, olive green turning greenish yellow to bright yellow at maturity. **Seed** with embryo above seed equator.

Orania palindan. Habit, Wandamen Peninsula (WB).

Orania palindan. Habit, Tamrau Mountains (WB).

DISTRIBUTION. Widespread from the Bird's Head Peninsula to northern-central New Guinea. Elsewhere in the Philippines, Sulawesi and Maluku.

HABITAT. Lowland and hill rainforest, 10–1050 m elevation.

LOCAL NAMES. *Ajabu* (Sumuri), *Hiyaub* (Siwi), *Kok* (Madik), *Nibun, Kelapa hutan* (Malay-Sentani dialect), *Saser* (Wandama), *Wroh* (Marap).

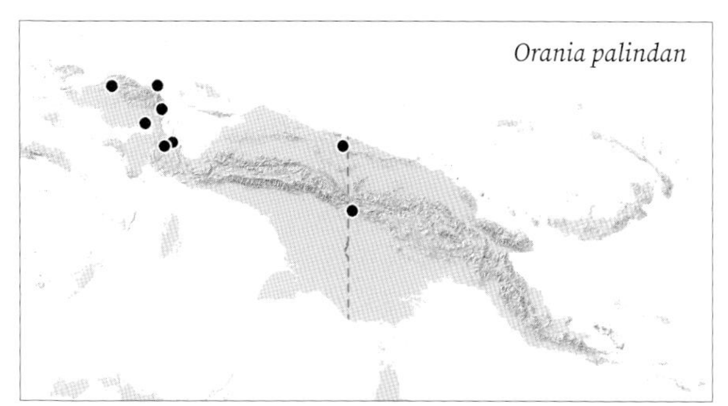

Orania palindan**. A.** Habit. **B.** Leaf apex. **C.** Leaf mid-portion. **D.** Portion of inflorescence. **E**, **F.** Male flower whole and in longitudinal section. **G**, **H.** Female flower whole and in longitudinal section. **I**, **J.** Fruit whole and in longitudinal section. **K.** Seed. Scale bar: A = 3.3 m; B, C = 9 cm; D = 6 cm; E–H = 7 mm; I–K = 4 cm. A from photograph; B–D from *Keim et al. 37*; E–H from *de Vogel 4402*; I–K from *Keim et al. 44.* Drawn by Lucy T. Smith.

Orania palindan. FROM TOP: inflorescence; fruit. Wandamen Peninsula (WB).

USES. Leaves used for thatching, trunk for house and bridge construction, harpoons and bows. Fruit and trunk used to make poison. Fruit, especially the endosperm, eaten by bats. People do not eat the fruit, as they say it is bitter and poisonous.

CONSERVATION STATUS. Least Concern.

NOTES. The largest species in New Guinea, *Orania palindan* can be distinguished from other robust species in the genus by the inflorescences branched to two orders, the more or less glabrous inflorescence, the male flowers with petals usually less than 10 mm long and the very large fruit at least 6.5 cm diam.

17. *Orania parva* Essig

Small, single-stemmed tree palm to 4 m. **Stem** 8 cm diam., grey. **Leaves** number in crown not recorded, *spirally arranged*, to 2 m or more long; petiole to 1 m long, 2 cm diam., densely covered in red-brown tomentum; rachis to 1 m long, 1.5–2 cm wide, with red brown tomentum, leaflets ca. 23 on each side of the rachis, regularly arranged, 90–106 cm long, 5.5–6 cm wide, upper surface dull green, glabrous apart from sparse red-brown tomentum along midrib, lower surface with dense white indumentum and red-brown tomentum along ribs. **Inflorescence** spreading, *branched to 1 order*, ca. 1.2 m long; peduncle ca. 80 cm long, *glabrous*, peduncular bract persistent, woody, to 128–130 cm long; rachillae *few, ca. 8–10, slender*, to 35–36 cm long, 3–3.5 mm diam. **Male flowers** ca. 8 mm long, stamens 6. **Female flowers** ca. 7 mm long, staminodes 6, uniform. **Fruits** globose or bilobed, 1.3 cm diam. (young). **Seed** with embryo below equator of seed.

DISTRIBUTION. Central New Guinea, near Telefomin.

HABITAT. Well-drained slopes at 60 m elevation.

LOCAL NAMES. None recorded.

USES. None recorded.

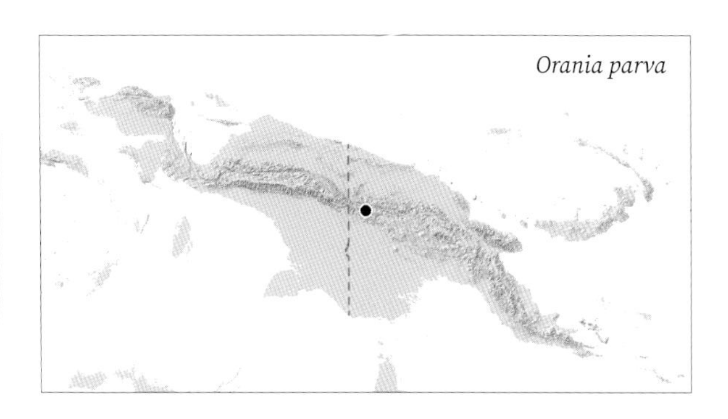

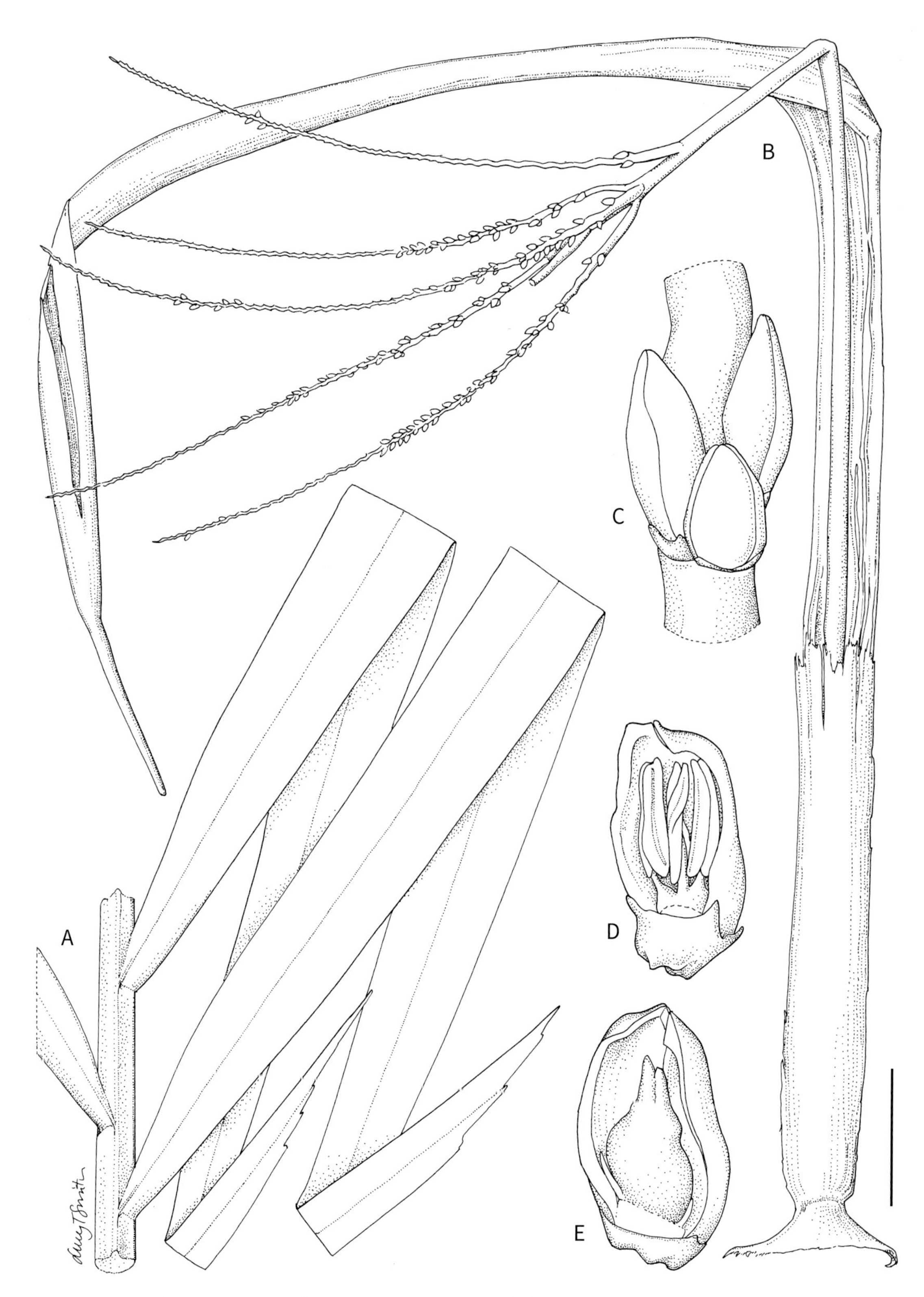

***Orania parva*. A.** Leaf mid-portion. **B.** Inflorescence. **C.** Portion of rachilla with triad. **D.** Male flower, one petal removed. **E.** Female flower in longitudinal section. Scale bar: A, B = 8 cm; C = 5 mm; D, E = 2.5 mm. All from *Essig & Young LAE 74046*. Drawn by Lucy T. Smith.

Orania parva. Leaf and inflorescences, Frieda River (FE).

CONSERVATION STATUS. Data Deficient (IUCN 2021). This species is known only from the type locality, where there is limited information on ongoing threats. Due to insufficient evidence, an extinction risk category cannot be selected.

NOTES. The smallest species in the genus, *Orania parva* is distinguishable by its spirally arranged leaves and inflorescence branched to one order only with glabrous peduncle. The only other species with inflorescences branched to one order are *Orania dafonsoroensis, O. tabubilensis* and *O. timikae*. The first of these is a large palm with subdistichous leaves as a juvenile and disproportionately large inflorescences. Of the remaining two, the former is larger than *O. parva* and has distichous leaves and the latter, although of more or less similar stature as *O. parva*, has a subdistichous crown and a densely red-brown tomentose peduncle.

18. *Orania regalis* Zipp.

Synonyms: *Arausiaca excelsa* Blume, *Orania aruensis* Becc.

Large, single-stemmed tree palm to 11 m. **Stem** 20–25 cm diam., grey. **Leaves** 8–10 in crown, spirally arranged, to 2.5–4.2 m long; petiole to 60–100 cm long, 4–4.5 cm diam., covered with red-brown tomentum; rachis to 3.2 m long, ca. 2 cm wide in mid-leaf, with red-brown tomentum, leaflets 50–60 on each side of the rachis, regularly arranged in one plane, to 100–160 cm long, 6.5–7.2 cm wide, upper surface dark green, glabrous, apart from scattered red scales along midrib, lower surface with white indumentum and red-brown tomentum along margins and main veins. **Inflorescence** *conspicuously congested,* branched to 2 orders, to 120–125 cm or more long; peduncle ca. 62 cm long, glabrous; peduncular bract persistent, woody 80–150 cm long; rachillae numerous, swept forward, slender, 12–20 cm long, 0.2 cm diam., conspicuously zigzag. **Male flowers** ca. 7 mm long, *stamens 3* (occasionally 5), filaments free or united. **Female flowers** ca. 7 mm long, *staminodes 3,* uniform. **Fruits** globose or bilobed, 4.5–8 cm diam., olive green turning bright orange at maturity. **Seed** with embryo above seed equator.

DISTRIBUTION. North-western New Guinea from the Bird's Head Peninsula to Nabire, and in Misool and Aru Islands.

HABITAT. Lowland rainforest, sea level to 150 m elevation.

LOCAL NAMES. *Fai* (Matbat), *Kabet* (Matbat), *Wowi* (Wandama).

USES. Trunk used in house construction, for example cut into strips and used for tying roofs.

CONSERVATION STATUS. Least Concern. However, monitoring is required due to mining, oil palm plantations and logging concessions in the distribution range of this species.

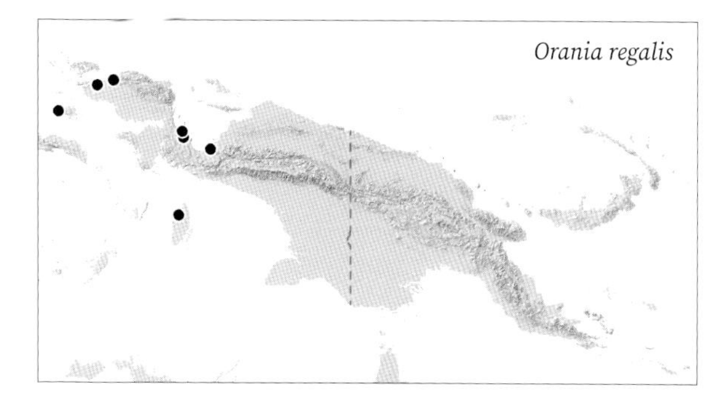

Orania regalis. CLOCKWISE FROM TOP LEFT: habit, near Sorong (WB); floral triads in bud, Wandamen Peninsula (WB); fruit, near Sorong (WB); seedling, Wandamen Peninsula (WB); inflorescence, Wandamen Peninsula (WB).

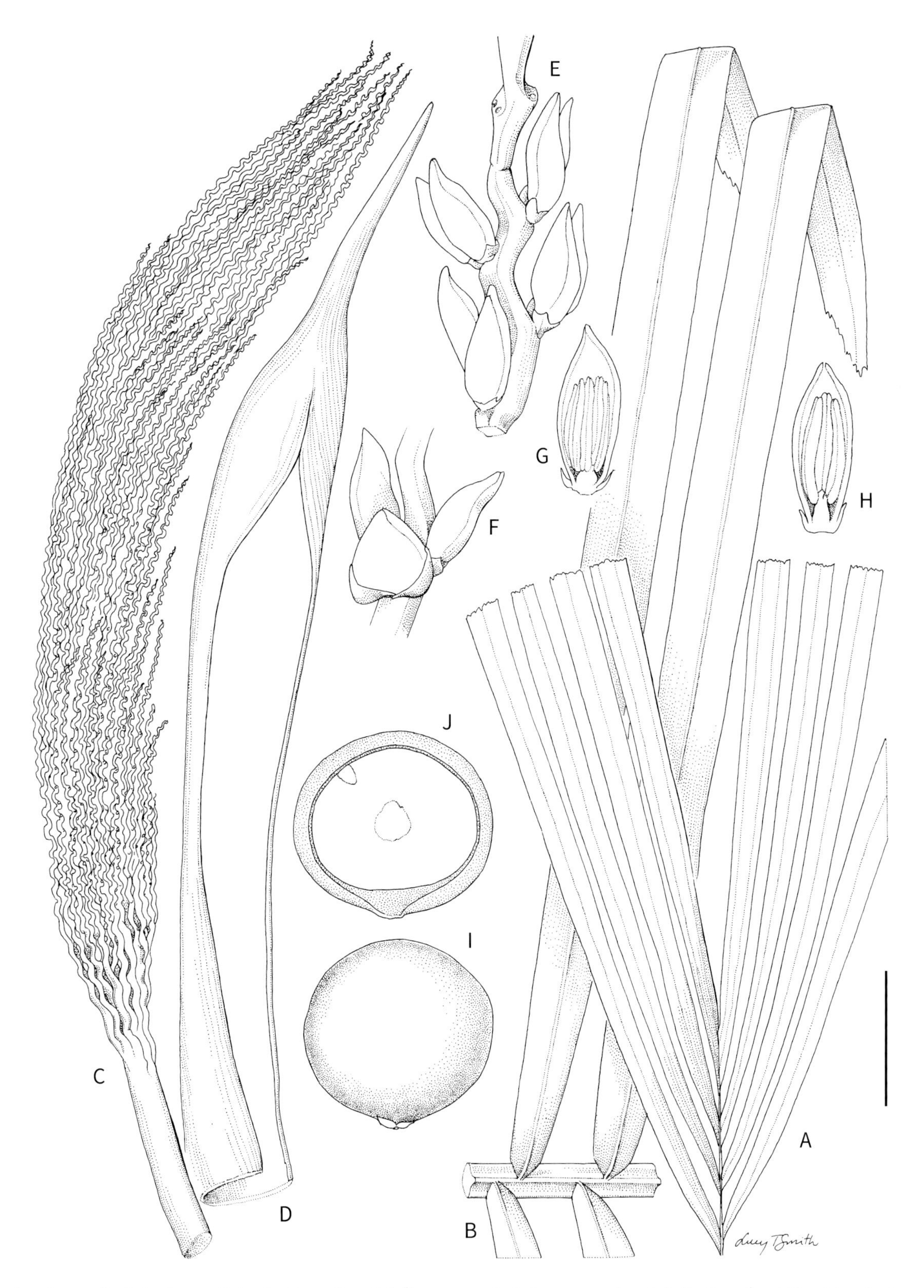

Orania regalis. **A.** Leaf apex. **B.** Leaf mid-portion. **C.** Inflorescence. **D.** Inflorescence bract. **E.** Portion of rachilla with paired male flowers. **F.** Portion of rachilla with triad. **G.** Male flower with five stamens in longitudinal section. **H.** Male flower with three stamens and fused filaments in longitudinal section. **I, J.** Fruit whole and in longitudinal section. Scale bar: A, B = 9 cm; C, D = 6 cm; E, F = 5 mm; G = 4 mm; H = 3.3 mm; I, J = 4 cm. A–D, I, J from *Keim et al. 32*; E–G from *Keim 33*; H from *Beccari s.n.* Drawn by Lucy T. Smith.

NOTES. Instantly recognizable by the highly congested inflorescence, *Orania regalis* is also remarkable for having three (rarely more) stamens and staminodes. For a long time, known from just a few herbarium specimens and cultivated individuals in the Kebun Raya in Bogor, it was rediscovered in the wild on the Wandamen Peninsula in Indonesian New Guinea (Baker *et al.* 2000a). Details of typification, etymology and a brief history of the cultivation of *O. regalis* in Kebun Raya, Bogor have been explored by Dowe & Latifah (2020).

19. *Orania subdisticha* A.P.Keim & J.Dransf.

Small, single-stemmed tree palm to 8 m. **Stem** ca. 8 cm diam., grey. **Leaves** 7 in crown, *subdistichously arranged*, to 3 m long; petiole to 1.5 m long, 2–2.5 cm diam., densely covered in red-brown tomentum; rachis to 1.5 m long, ca. 2 cm wide, with red-brown tomentum, leaflets ca. 22 on each side of the rachis, regularly arranged, 70–80 cm long, 5–5.5 cm wide, upper surface dull green, glabrous apart from red-brown scales along midrib, lower surface with dense white indumentum and red-brown tomentum along ribs and margins. **Inflorescence** spreading, *branched to 2 orders*, ca. 1.2 m long; peduncle covered in red-brown tomentum, peduncular bract persistent, woody, to 150 cm long; rachillae slender, 33–50 cm long, ca. 0.5 cm diam., *not conspicuously zigzag*. **Male flowers** ca. 7 mm long, stamens 6. **Female flowers** 6 mm long, staminodes 6, uniform. **Fruits** globose, ca. 4 cm diam., bright red. **Seed** with embryo below seed equator.

DISTRIBUTION. One record to the south of the Central Range in East New Guinea.

HABITAT. Montane forest on karst limestone, 1050 m elevation.

LOCAL NAMES. None recorded.

USES. None recorded.

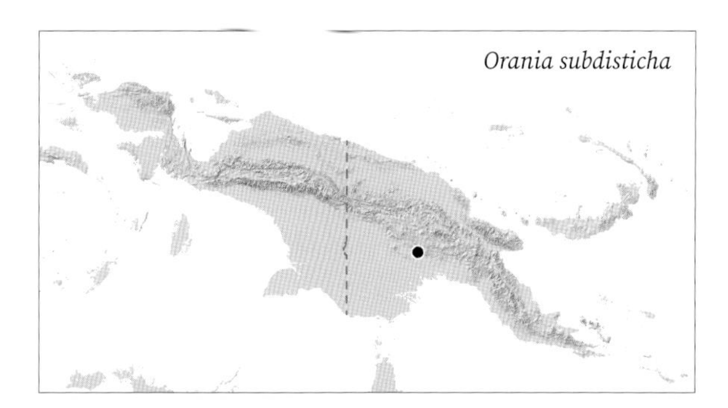

Orania subdisticha. FROM TOP: habit; crown with inflorescence. Southern Highlands (WB).

Orania subdisticha. **A.** Habit. **B.** Leaf apex. **C.** Leaf mid-portion. **D.** Inflorescence. **E.** Portion of rachilla with male flowers. **F.** Portion of rachilla with female flowers. **G.** Male flower with one petal removed. **H.** Female flower in longitudinal section. Scale bar: A = 1 m; B–D = 8 cm; E = 7 mm; F = 1 cm; G = 2.5 mm; H = 5 mm. All from *Baker et al. 1116*. Drawn by Lucy T. Smith.

Orania subdisticha. LEFT TO RIGHT: male flowers; female flowers. Southern Highlands (WB).

CONSERVATION STATUS. Vulnerable (IUCN 2021). Known only from one locality, but specific threats are not reported. It occurs in some abundance in montane forest on rocky limestone slopes that are not likely to be used for agriculture.

NOTES. *Orania subdisticha* is easily distinguished from other small species with subdistichous leaves by the inflorescence branched to two rather than a single order. *Orania disticha* is a much larger palm, with conspicuously zigzag rachillae and much larger fruit.

20. *Orania tabubilensis* A.P.Keim & J.Dransf.

Small, single-stemmed tree palm to 5 m. **Stem** ca. 10 cm diam., grey. **Leaves** 8 in crown, *distichously arranged,* to 3.8 m long; petiole and sheath to 1.5 m long, 1.5–2.0 cm diam. towards the tip, densely covered in red-brown tomentum; rachis to 2.1 m long, ca. 2 cm wide, with red-brown tomentum, leaflets ca. 35 on each side of the rachis, *regularly arranged near the base, irregularly arranged in mid-leaf and displayed in different planes,* to 90 cm long, 4–4.5 cm wide, upper surface dull green, glabrous apart from red-brown scales along midrib, lower surface with dense white indumentum and red-brown tomentum along ribs and margins. **Inflorescence** spreading, branched to 2 orders, ca. 1.8 m long; peduncle to 60 cm long, covered in red-brown tomentum, peduncular bract persistent, woody, to 150 cm long; rachillae rather thick, 20–50 cm long, 0.5 cm diam., slightly zigzag. **Male flowers** ca. 6 mm long, stamens 6. **Female flowers** 2.5 mm long, staminodes 6, uniform. **Fruits** globose, 2 cm diam. (immature), mature colour unknown. Seed with embryo above seed equator.

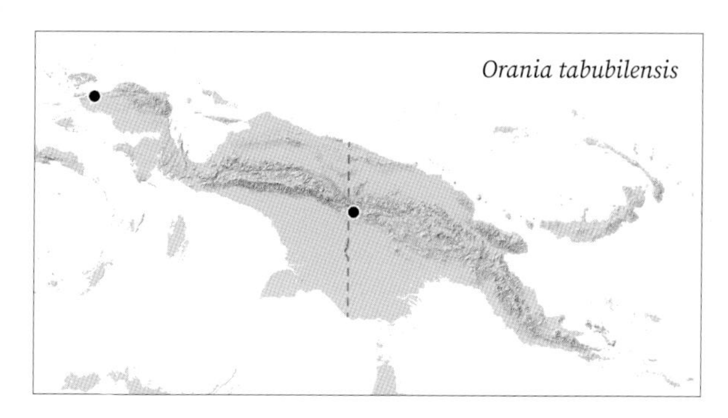

DISTRIBUTION. Known from two widely separated sites near Sorong and Tabubil.

HABITAT. In Indonesian New Guinea, *Orania tabubilensis* occurs in forest at sea level whereas in Papua New Guinea, the collection was made in montane forest on limestone at 1,000 m elevation.

LOCAL NAMES. *Lai* (Matbat).

USES. None recorded.

Orania tabubilensis. Habit, near Tabubil (WB).

Orania tabubilensis. **A.** Habit. **B.** Whole leaf diagram. **C.** Leaf apex. **D.** Leaf mid-portion. **E.** Portion of inflorescence. **F.** Portion of rachilla with triad. **G.** Male flower with one petal removed. **H.** Female flower in longitudinal section. **I, J.** Fruit whole and in longitudinal section. Scale bar: A = 1.6 m; B = 1.3m; C–E = 8 cm; F = 7 mm; G, H = 4 mm; I, J = 2 cm. All from *Baker et al. 1129*. Drawn by Lucy T. Smith.

Orania tabubilensis. Inflorescences, near Tabubil (WB).

CONSERVATION STATUS. Critically Endangered (IUCN 2021). *Orania tabubilensis* is known from only two sites, which are threatened by mining concessions.

NOTES. Easily distinguished from all other species by the combination of distichously arranged leaves and leaflets arranged in more than one plane. That the two localities, widely separated, should be at such different elevations suggests that further study is required to determine whether the two populations truly represent the same species.

21. *Orania timikae* A.P.Keim & J.Dransf.

Small, single-stemmed tree palm to 4 m. **Stem** 10 cm diam., grey. **Leaves** 6 in crown, *subdistichously* arranged, to 3 m long; petiole to 1 m long, ca. 2 cm diam., densely covered in red-brown tomentum; rachis to 2 m long, 1.5–2 cm wide, with *red-brown tomentum*, leaflets ca. 23 on each side of the rachis, regularly arranged, 88–90 cm long, 5–6 cm wide, upper surface dull green, glabrous, lower surface with dense white indumentum and red-brown tomentum along ribs. **Inflorescence** spreading, *branched to 1 order*, ca. 1.5 m long; peduncle covered in red-brown tomentum, peduncular bract persistent, woody, to 1.6 m long; rachillae few, ca. 10, to 45 cm long, 0.5 cm diam. **Male flowers** ca. 8 mm long, stamens 6. **Female flowers** 6 mm long, staminodes 6, uniform. **Fruits** and **seeds** unknown.

DISTRIBUTION. Recorded between Timika and Tembagapura, western New Guinea.

HABITAT. Heath forest and the area transitional between heath forest and lowland forest, 435–540 m elevation.

LOCAL NAMES. None recorded.

USES. None recorded.

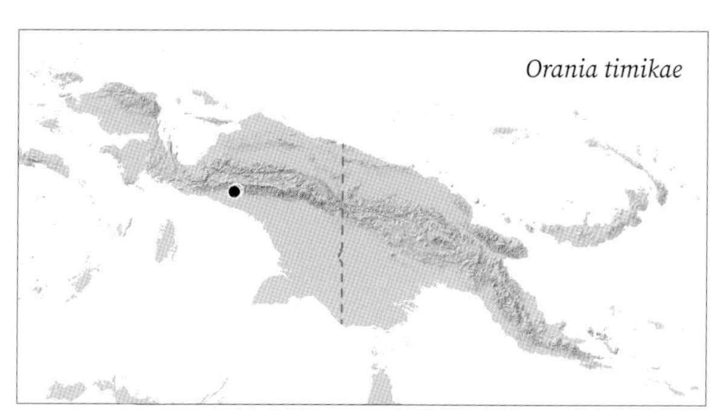

CONSERVATION STATUS. Critically Endangered. *Orania timikae* is known from only two adjacent sites where it is threatened by mining concessions.

NOTES. *Orania timikae* is distinctive in its small size, subdistichous crown, inflorescence branched to only one order and the unusual habitat (heath forest). It is the only species with persistent subdistichous leaf arrangement. Reminiscent of *O. dafonsoroensis* and *O. parva* with its inflorescence branched to one order, it is immediately distinguished by the red-brown tomentum on the peduncle.

Orania timikae. Habit, Timika-Tembagapura road (WB).

Orania timikae. **A.** Habit. **B.** Leaf apex. **C.** Leaf mid-portion. **D.** Inflorescence. **E.** Portion of rachilla with triads. **F.** Male flower in longitudinal section. **G**, **H.** Female flower whole and in longitudinal section. Scale bar: A = 1 m; B–D = 8 cm; E = 1 cm; F–H = 5 mm. All from *Dransfield et al. JD 7667*. Drawn by Lucy T. Smith.

22. *Orania zonae* A.P.Keim & J.Dransf.

Large, single-stemmed tree palm to 20 m. **Stem** ca. 27 cm diam., grey. **Leaves** 12 in crown, spirally arranged, to 4.3 m long; petiole and sheath to 2 m long, 4–4.5 cm diam., covered with red-brown and white tomentum; rachis to ca. 2 m long, ca. 3 cm diam. in mid-leaf, with red-brown tomentum, leaflets to 60 on each side of the rachis, regularly arranged in one plane, to 150–190 cm long, 8.5–11 cm wide, upper surface dull green, glabrous, lower surface with white indumentum, wax and red-brown tomentum along ribs. **Inflorescence** spreading, branched to 2 orders, to 1.5 m or more long, with abundant red-brown tomentum; peduncle 90 cm long, ca. 3 cm diam.; peduncular bract persistent, woody 1.4 m long; rachillae numerous, thick, 45–50 cm long, 0.7 cm diam., somewhat zigzag. **Male flowers** ca. 3 mm long, *stamens 6, 9 or 12, filaments united.* **Female flowers** 4 mm long, *staminodes 6, uniform.* **Fruit** globose, ca. 4 cm diam., dull green (immature), colour at maturity not known. **Seed** with embryo below seed equator.

DISTRIBUTION. Vicinity of Manokwari, Bird's Head Peninsula.

HABITAT. Lowland forest on coralline limestone at 165 m elevation.

LOCAL NAMES. *Motuaga* (Meyah).

USES. Stem used for arrow heads.

CONSERVATION STATUS. Critically Endangered. *Orania zonae* is known from only one site. Deforestation is a major threat in its distribution range.

NOTES. *Orania zonae* is a poorly known species, distinctive in the variable number of stamens in the same inflorescence, with filaments united. Filaments in *O. regalis* are also sometimes united, but that species is instantly distinguishable by the congested inflorescence and low stamen number.

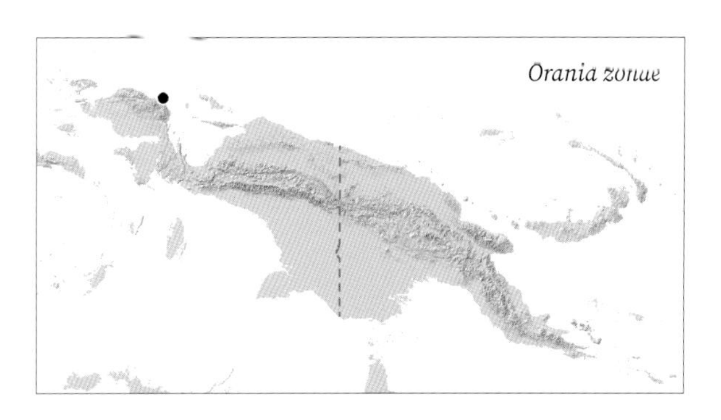

Orania zonae. FROM TOP: habit; inflorescence. Nuni (JD).

***Orania zonae*. A.** Habit. **B.** Portion of leaf sheath. **C.** Portion of inflorescence. **D.** Tip of peduncular bract. **E, F.** Male flower whole and in longitudinal section. **G.** Anthers, filaments fused. Scale bar: A = 1.5 m; B, D = 9 cm; C = 6 cm; E, F = 4 mm; G = 3 mm. All from *Zona et al. 674*. Drawn by Lucy T. Smith.

Cocos nucifera. Coconuts in village garden, Morobe (WB).

Cocos L.

Synonyms: *Calappa* Steck, *Coccos* Gaertn., *Coccus* Mill.

Tall – leaf pinnate – no crownshaft – no spines – leaflets pointed

Robust, single-stemmed tree palm, crownshaft absent, monoecious. **Leaf** pinnate, arching; sheath forming a fibrous network; petiole absent until *sheath has disintegrated into fibres*; leaflets reduplicate, numerous, with pointed tips, arranged regularly. **Inflorescence** *between the leaves,* branched to 1 order, branches widely spreading; prophyll short, hidden among leaf sheaths, peduncular bract much longer, *woody,* deeply grooved, enclosing inflorescence in bud and then splitting along its length, forming a cowl over the flowers, not dropping off as inflorescence expands; peduncle about the same length as inflorescence rachis; rachillae quite robust and ± straight. **Flowers** in triads at the base of the rachillae, pairs of male flowers towards tip, not developing in pits, female flowers much larger than the male flowers. **Fruit** *very large,* yellow or green to brown, globose or ovoid, stigmatic remains apical, *husk fibrous, endocarp thick, spherical, with three eyes.* **Seed** 1, globose, endosperm homogeneous, with a *central hollow partially filled with fluid.*

DISTRIBUTION. Pantropical, one species. Distribution extensively modified by humans.

NOTES. An iconic and economically important species, *Cocos nucifera,* the coconut, occurs throughout coastal vegetation in the tropics and further inland, and is widely cultivated for the numerous useful products it provides. It is a robust, single-stemmed tree palm with fibrous leaf sheaths bearing inflorescences among the leaves. It is instantly recognised by its massive edible fruit with liquid contents. It is unlikely to be confused with any other palm. *Orania* is similar in overall form, but it has jagged leaflet tips and smaller round fruits that lack a fibrous husk.

Cocos nucifera L.

Synonyms: *Calappa nucifera* (L.) Kuntze, *Cocos indica* Royle, *Cocos mamillaris* Blanco, *Cocos nana* Griff., *Diplothemium henryanum* F.Br., *Palma cocos* Mill. In addition, a very large number of varietal names have been published.

Robust, *emergent palm to 30 m tall,* bearing 20–40 leaves. **Stem** 30–35 cm diam. often curving. **Leaf** up to 6 m long; petiole 50–100 cm long; rachis ca. 5.5 m long, *twisted 90° in its distal half*; 90–120 leaflets per side, mid-leaf leaflet 100 × 2–5 cm. **Inflorescence** *among the leaves,* 1–1.5 m long; peduncular bract ca. 1 m long, woody; peduncle 40 cm long; rachillae up to 35 cm long. **Male flower** ca. 13 × 4 mm in bud, slightly asymmetrical, creamy white; stamens 6; anthers ca. 8 mm long; pistillode short, trifid. **Female flower** ca. 30 mm diam., globose. **Fruit** ca. *30 × 25 cm,* ovoid, *weakly to strongly 3-sided,* green, yellow, orange or brown (depending on cultivar); *husk fibrous and air-filled* (giving buoyancy), endocarp 10–15 cm, globose, endocarp wall ca. 5 mm thick, brown, very hard, with *three basal "eyes" (germination pores).* **Seed** ca. 10–13 cm, globose; endosperm homogeneous with large, *liquid-filled cavity.*

DISTRIBUTION. Pantropical and throughout New Guinea; extensively cultivated.

HABITAT. A common palm around human habitations. It is widely cultivated in home gardens and plantations and it naturalizes along the beach strand, sea level to 1,000 m (Bourke & Vlassak 2004).

LOCAL NAMES. *Coconut* (English), *Kelapa* (Bahasa Indonesia), *Kokonas* (Papua New Guinea Tok Pisin).

USES. Arguably, the most useful of all palms. Various parts of this palm are used for construction, fibre, containers, food, drink, etc. Copra (dried endosperm), coir (mesocarp fibre), coconut water (seed liquid) and coconut oil (seed oil) are commodities that have international markets. A full account of the coconut's uses is beyond the scope of this book (see Westphal & Jansen (1989) for a summary).

CONSERVATION STATUS. Least Concern.

NOTES. The origin and domestication of the coconut have been topics of speculation for decades. Gunn *et al.* (2011), using molecular methods, identified two potential areas of domestication, namely the southern margins of the Indian subcontinent and islands of South-East Asia between the Malay Peninsula and New Guinea.

Cocos nucifera. RIGHT: habit, near Madang (WB). BELOW: crown with fruit, Mission Beach, Queensland (JD).

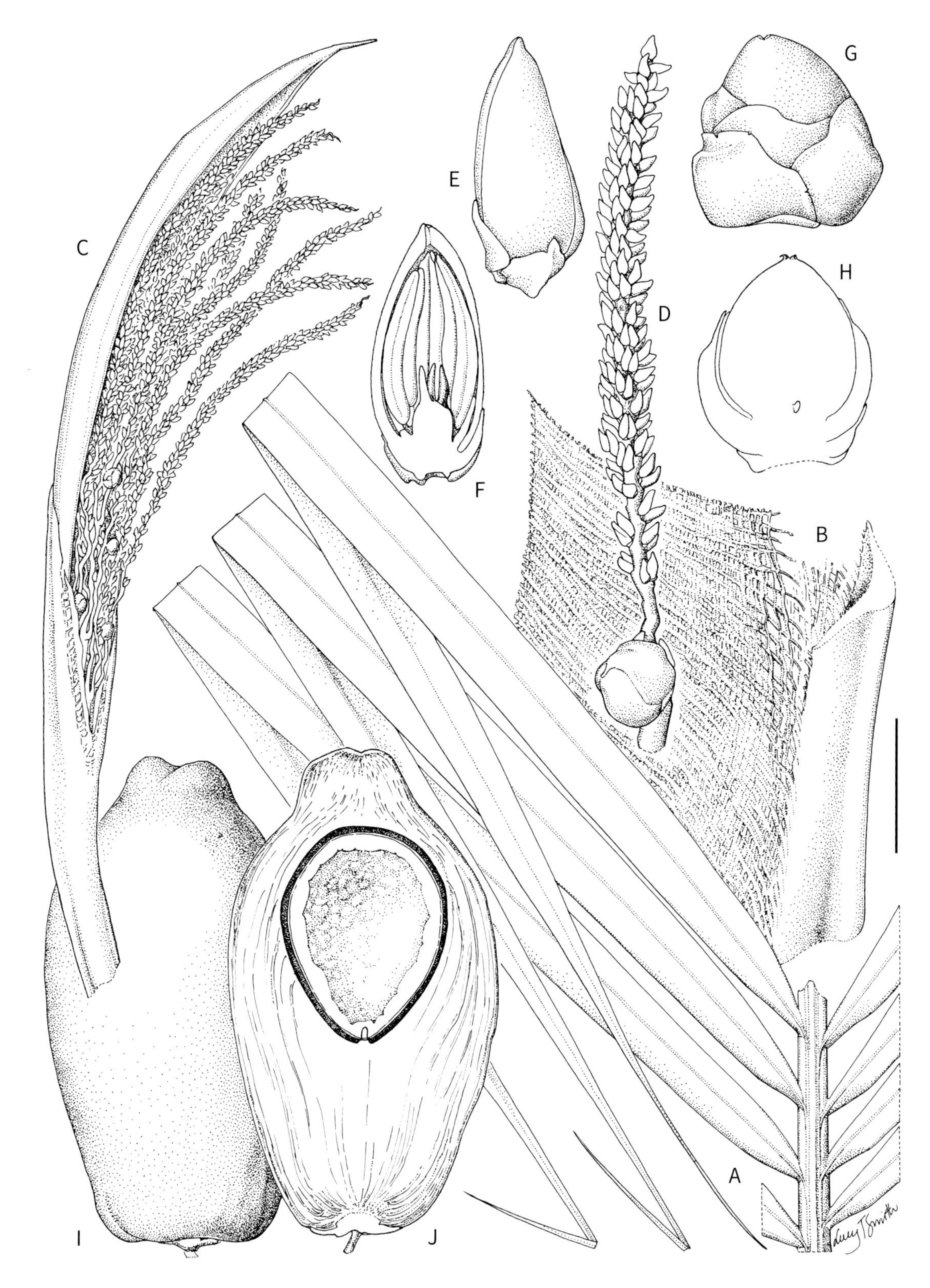

Cocos nucifera. **A.** Leaf mid-portion. **B.** Leaf sheath. **C.** Inflorescence. **D.** Rachilla with flowers. **E, F.** Male flower whole and in longitudinal section. **G, H.** Female flower whole and in longitudinal section. **I, J.** Fruit whole and in longitudinal section. Scale bar: A–C, I, J = 8 cm; D = 4 cm; E, F = 5 mm; G, H = 2.5 cm. All from *Kew cult. 1969–4480*. Drawn by Lucy T. Smith.

Sommieria leucophylla. Habit, near Sorong (WB).

ARECOIDEAE | PELAGODOXEAE

Sommieria Becc.

Small – leaf pinnate – no crownshaft – no spines – leaflets toothed or pointed

Slender, single-stemmed palm, sometimes stemless, crownshaft absent, monoecious. **Leaf** *entire or pinnate with few broad leathery leaflets,* arching; *sheath fibrous,* splitting almost to base opposite the petiole; petiole short; leaflets (where blade divided) 2–3 on each side of leaf rachis, multifold, with pointed or toothed tips, upper surface dark green, *lower surface chalky white.* **Inflorescence** *between the leaves, branched to 1 order with few branches, flowers developing within pits;* prophyll hidden among leaf sheaths, peduncular bract enclosing the inflorescence in bud, borne at the tip of the peduncle, splitting and usually dropping as the inflorescence expands; peduncle longer than inflorescence rachis; rachillae slender and straight. **Flowers** in triads throughout the length of the rachilla, developing in shallow pits. **Fruit** *yellow brown to pink, round, the surface cracked into corky warts,* flesh soft, endocarp smooth. **Seed** 1, round, endosperm homogeneous, embryo basal.

DISTRIBUTION: One species endemic to New Guinea, widespread in the western half of the island.

NOTES: This New Guinea endemic is a slender, single-stemmed, understorey palm with leathery leaves that are usually entire-bifid (or divided into a few broad leaflets) and chalky white below. Leaf sheaths are fibrous and do not form a crownshaft. The inflorescence develops between the leaves and has rather few branches clustered at the tip of a long peduncle. The fruit are notable for their corky warts. *Sommieria* was once thought to be related in some way to *Heterospathe* (Uhl & Dransfield 1987, Stauffer *et al.* 2004), but molecular evidence places it in tribe Pelagodoxeae, together with the enigmatic genus *Pelagodoxa* from the western Pacific islands, but of uncertain origin (Dransfield *et al.* 2008, Hodel *et al.* 2019).

Though highly distinctive, *Sommieria* can be confused with *Calyptrocalyx,* which has unbranched inflorescences and leaves that are not chalky white beneath, or juveniles of the rattan genus *Korthalsia,* with leaves that are white beneath, but spiny. *Heterospathe elegans* shares similar inflorescence morphology with *Sommieria* (Trudgen & Baker 2008), but its leaves are usually finely divided (never white below) and the fruits are smooth, not warty. *Sommieria* was monographed most recently by Heatubun (2002).

Sommieria leucophylla Becc.

Synonyms: *Sommieria affinis* Becc., *Sommieria elegans* Becc.

Slender *single-stemmed acaulescent or short stemmed palm* of forest undergrowth, up to 40 leaves in crown. **Stem** 3–4 cm diam. **Leaf** 90–180 cm long; sheaths splitting to the base and with sparsely fibrous margins; petiole 10–40 cm long, *blade undivided or divided into 2–4 leaflets* each side of the rachis, *blade dark green on upper surface, dense chalky white on undersurface.* **Inflorescence** to 160 cm long, branched to 1 order; peduncle to 136 cm; *rachillae fewer than 20, radiating from the tip of the peduncle,* 11–27 × ca. 0.5 cm, flowers borne in pits. **Male flower** 2.5 mm diam. **Female flower** 2.5–3 mm diam. **Fruit** 9–15 × 8–15 mm, rounded to somewhat ellipsoidal, *covered in corky warts, bright pink at maturity, fleshy.* **Seed** 5–9 × 5–8 mm, rounded.

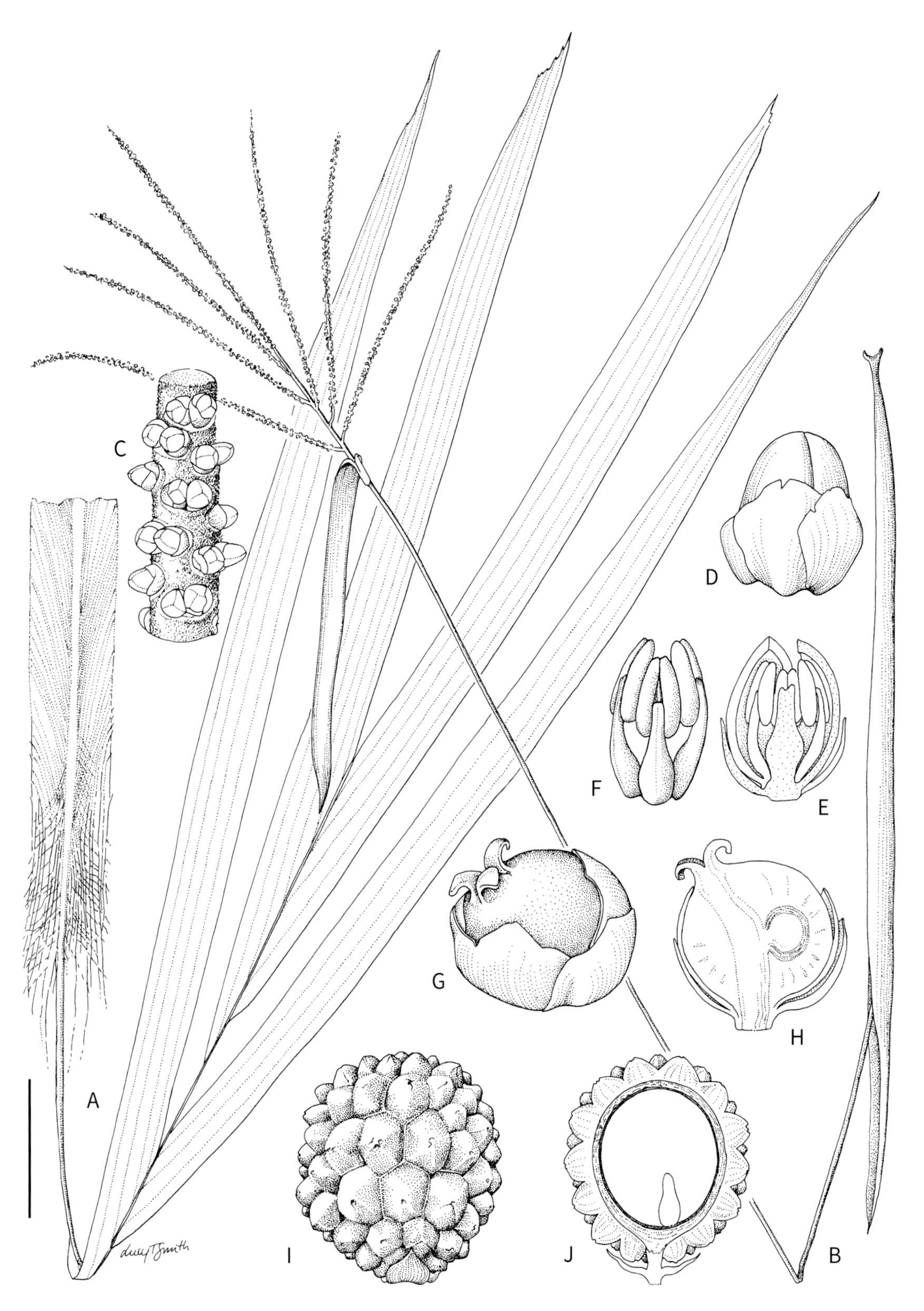

Sommieria leucophylla (Plate 1). **A.** Leaf. **B.** Inflorescence. **C.** Portion of rachilla with male flowers. **D**, **E.** Male flower whole and in longitudinal section. **F.** Stamens, abaxial view. **G**, **H.** Female flower whole and in longitudinal section. **I**, **J.** Fruit whole and in longitudinal section. Scale bar: A = 9 cm; B = 6 cm; C = 7 mm; D, E, G, H = 2 mm; F = 1.6 mm; I, J = 1 cm. All from *Dransfield et al. JD 7535*. Drawn by Lucy T. Smith.

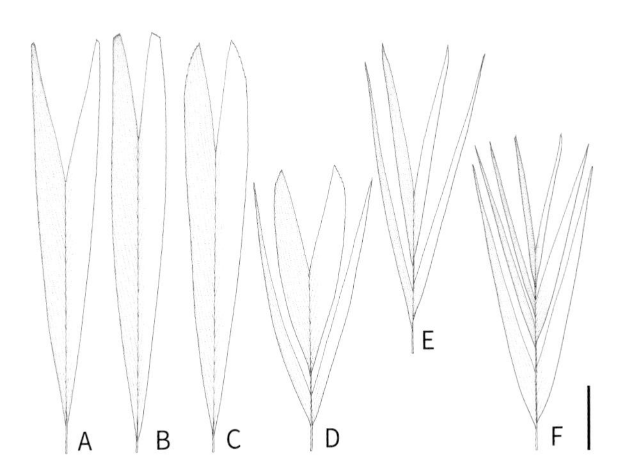

Sommieria leucophylla (Plate 2). A–**F.** Leaf forms from different specimens. Scale bar: A–F = 25 cm. A from *Heatubun 96*; B from *Heatubun et al. 196*; C from *Barfod 404*; D from *Barrow et al. 130*; E from *Dransfield et al. JD 7535*; F from *Heatubun 151*. Drawn by Lucy T. Smith.

DISTRIBUTION. Widespread from the Raja Ampat Islands and the Bird's Head Peninsula to the vicinity of Vanimo.

HABITAT. Lowland forest from sea level to 500 m elevation.

LOCAL NAMES. *Amkaulas* (Moi), *Are* (Sayal), *Man* (Bewani), *Mbebmega* (Hatam), *Ovenatenae* (Sumuri), *Som* (Biak), *Sunggumi* (Wandamen), *Yet* (Marap).

USES. Leaves used for wrapping food (e.g. fish), as adornments in traditional dances and as thatching.

CONSERVATION STATUS. Least Concern.

NOTES. This is a very beautiful and distinctive palm. Until relatively recently it was known only from a few fragmentary specimens. In the 1980s it was found in northwestern Papua New Guinea. Since then many herbarium collections have been made in Indonesian New Guinea allowing a much better appreciation of variation. Beccari distinguished three separate species on the basis of leaf dissection and size, but as more material became available for study, Heatubun (2002) was able to show that variation in leaf dissection and size overlaps, and that within-population variation is considerable.

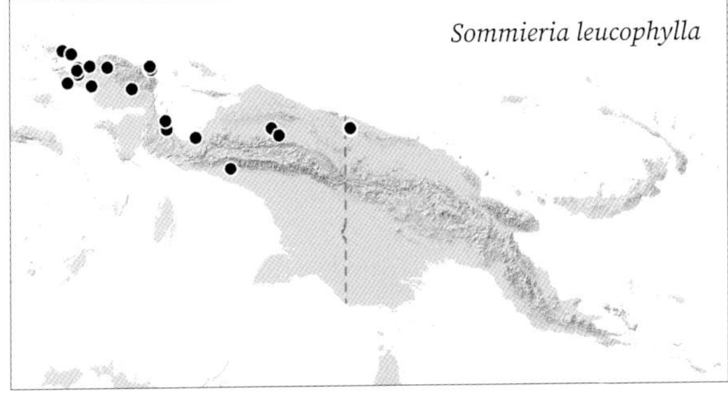

Sommieria leucophylla. LEFT TO RIGHT FROM TOP: inflorescence, near Timika (JD); fruit, near Sorong (WB); leaf undersurface showing pale indumentum, near Sorong (GP); female flowers, Wandamen Peninsula (WB).

Actinorhytis calapparia. Habit, near Lae (WB).

Actinorhytis H.Wendl. & Drude

Tall — leaf pinnate — crownshaft — no spines — leaflets pointed

Robust, single-stemmed tree palm, *slender crownshaft present*, the same diameter as stem, monoecious. **Leaf** pinnate, *strongly arching*; sheath tubular, pale green; petiole short; leaflets numerous, with pointed tips, arranged regularly, ascending. **Inflorescence** *below the leaves, branched to 2–3 orders, branches widely spreading*; prophyll and peduncular bract similar, enclosing inflorescence in bud, dropping off as inflorescence expands; peduncle shorter than inflorescence rachis, *grossly swollen at base*; rachillae slender and straight. **Flowers** in triads at the base of the branches, pairs of male flowers towards tip, not developing in pits, female flowers much larger than the male flowers. **Fruit** *large, red when ripe, ovoid*, stigmatic remains apical, flesh fibrous, endocarp thin, closely adhering to seed. **Seed** 1, globose, *endosperm strongly ruminate.*

DISTRIBUTION. One species from New Guinea to Solomon Islands (Bougainville and adjacent islands). Frequently cultivated in South-East Asia.

NOTES. *Actinorhytis* is a robust, single-stemmed tree palm with strongly arching leaves and a slender crownshaft. The inflorescence is borne below the leaves, and is widely spreading and branched to three orders. The fruit are large and contain a seed with endosperm strongly ruminate that is often used as an alternative to betel nut (*Areca catechu*).

Actinorhytis is highly distinctive, but could be confused with some of the larger species of *Hydriastele*. These would be distinguished by their jagged rather than pointed leaflet tips, horse's tail-like inflorescences and small fruit. The genus lacks a recent monograph.

Actinorhytis calapparia (Blume) H.Wendl. & Drude ex Scheff.

Synonyms: *Actinorhytis poamau* Becc., *Areca calapparia* Blume, *Areca cocoides* Griff., *Pinanga calapparia* (Blume) H.Wendl., *Ptychosperma calapparia* (Blume) Miq., *Seaforthia calapparia* (Blume) Mart.

Very robust, single-stemmed palm to 30 m or more, bearing 9–15 leaves in crown. **Stem** 18–34 cm diam., internodes 5–25 cm. **Leaf** 2–5 m long including petiole; sheath 0.8–1.8 m long, pale green; petiole 10–50 cm long; leaflets 55–109 each side of rachis, linear, with grey-brown ramenta on lower surface of mid-rib near base; mid-leaf leaflets 62–92 × 3.5–5 cm wide; apical leaflets 23–33 × 0.3–3 cm wide, not united. **Inflorescence** 75–130 cm long including 5–20 cm peduncle, *widely spreading*; primary branches 25–40, 70–95 cm long, with up to ca. 24 rachillae each; rachillae 16–70 cm long. **Male flower** 3.5–4 × ca. 2.5 mm in bud; stamens 24–33. **Female flower** 8–8.5 × 6–7 mm. Fruit 6–8.2 × 3.7–5 cm. Seed 3.4–3.7 × 3.5–3.8 cm.

DISTRIBUTION. Widespread in New Guinea, including the Bismarck Archipelago, and Bougainville and adjacent islands. Also cultivated in New Guinea and elsewhere in South-East Asia.

HABITAT. Varied habitats from lowland swamp forest to montane forest, 0–1,800 m.

LOCAL NAMES. *Manpung* (Amungme), *Tilitili* (Huli), *Wowi* (Wandama).

Actinorhytis calapparia, Areca catechu and *Cocos nucifera* cultivated among bananas and cassava, near Lae (WB).

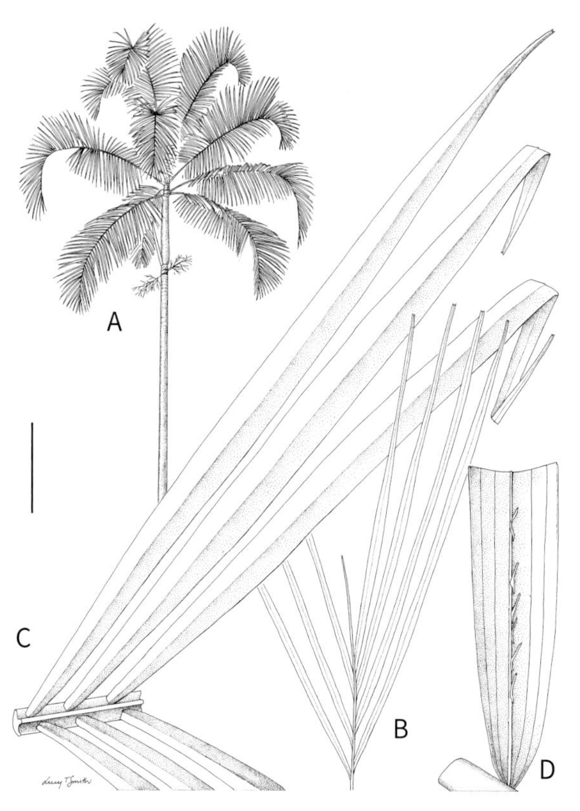

Actinorhytis calapparia (Plate 1). **A.** Habit. **B.** Leaf apex. **C.** Leaf mid-portion. **D.** Section of leaflet abaxial side with ramenta. Scale bar: A = 2 cm; B, C = 9 cm; D = 3 cm. All from *Baker & Utteridge 583*. Drawn by Lucy T. Smith.

USES. Stems are used to make bows and flooring in houses. Shoots are often used as a vegetable. The fruit is sometimes used in a lotion to treat scurf. It is eaten by the northern cassowary (*Casuarius unappendiculatus*; Pangau-Adam & Mühlenberg 2014). The fruit kernel can be eaten, although it is tough. Seed are used as a substitute for betel. Powdered seeds are occasionally used as baby powder. Often cultivated for its seeds (betel substitute) or as an ornamental.

CONSERVATION STATUS. Least Concern.

NOTES. Here, we follow Dransfield *et al.* (2008) in treating *A. poamau* from the Treasury and Shortland Islands as a synonym of *A. calapparia*.

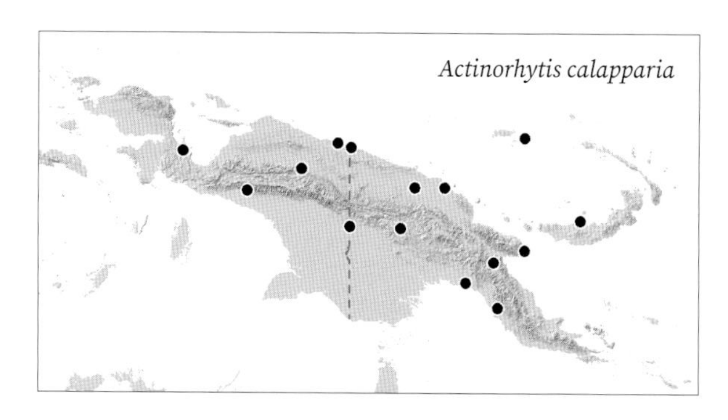

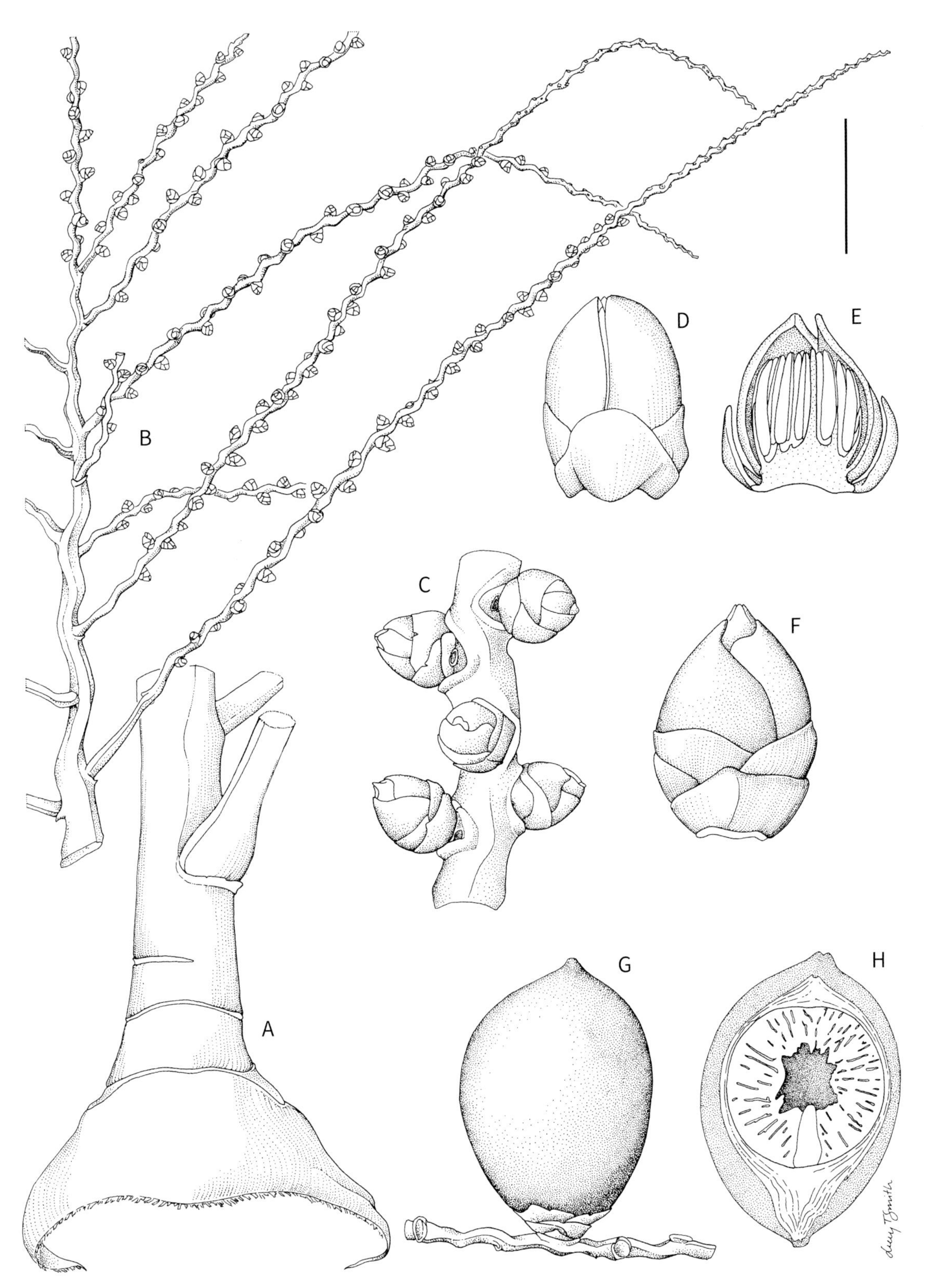

Actinorhytis calapparia (Plate 2). **A.** Inflorescence base. **B.** Portion of inflorescence. **C.** Portion of rachilla with female flowers. **D**, **E.** Male flower whole and in longitudinal section. **F.** Female flower. **G**, **H.** Fruit whole and in longitudinal section. Scale bar: A, B = 6 cm; C = 1.5 cm; D = 3.3 mm; E = 2.5 mm; F = 5 mm; G, H = 3 cm. A–D, F from *Baker & Utteridge 583*; E from *Nur s.n.*; G, H from *Maturbongs & Wally 587*. Drawn by Lucy T. Smith.

Areca catechu. Habit, Morobe (WB).

Areca L.

Synonyms: *Gigliolia* Becc., *Mischophloeus* Scheff., *Pichisermollia* H.C.Monteiro.

Small to medium — leaf pinnate — crownshaft — no spines — leaflets lobed or pointed

Single-stemmed tree palms, crownshaft present, variously coloured, monoecious. **Leaf** pinnate, straight; sheath tubular, variously coloured; petiole very short to elongate; leaflets few to numerous, rarely single-fold, *usually composed of many folds*, with *pointed or lobed tips*, arranged regularly or irregularly, horizontal. **Inflorescence** *below the leaves*, branched 2–3 orders, branches congested or widely spreading; prophyll thin, enclosing inflorescence in bud, dropping off as inflorescence expands, peduncular bract absent; peduncle shorter than inflorescence rachis; rachillae slender and usually straight. **Flowers** *in triads at the base of the branches, pairs or solitary male flowers towards tip*, not developing in pits, female flowers much larger than the male flowers. **Fruit** green ripening orange to red, stigmatic remains apical, flesh usually fibrous, endocarp thin, closely adhering to seed. **Seed** 1, globose or ellipsoid, endosperm strongly ruminate.

DISTRIBUTION. About 55 species from South China to India, through South-East Asia to Solomon Islands, six species in New Guinea.

NOTES. *Areca* contains understorey to mid-storey, single-stemmed tree palms. The inflorescences are presented below the leaves and female flowers occur only towards the base of the inflorescence branches. The fruiting inflorescence is club-like in the most widespread species, *Areca macrocalyx*. Six species occur in New Guinea, though one of these, *Areca catechu* (betel nut palm), is only known from cultivation. *Areca* is most easily confused with the closely related *Pinanga*, but this genus bears female flowers and fruit throughout the entire length of the inflorescence branches. The genus east of Wallace's Line has recently been monographed by Heatubun *et al.* (2012a) with later additions (Heatubun *et al.* 2013, Heatubun 2016).

Key to the species of *Areca* in New Guinea

1. Palm with stilt roots; floral triads spirally arranged on the rachillae; complete triads with female flowers occurring from the base to half the length of each rachilla; fruit with juicy flesh .... **5. *A. novohibernica***
1. Palm without stilt roots; floral triads uniseriate or distichously (or subdistichously) arranged on the rachillae; complete triads with female flowers occurring only at the base and/or along the lower third of each rachilla; fruits with fibrous flesh 2

2. Leaflets single-fold, held in different planes (plumose)....................... **4. *A. mandacanii***
2. Leaflets mostly composed of several folds, the leaflets held in one plane and the leaf not plumose .. .. 3.

3. Small, undergrowth to robust, emergent tree palm; inflorescence compact with rachillae congested along main axis, branched to one order (rarely two orders), most of rachillae shrivelling after pollination, inflorescence then club-like in fruit. . . . . **3. *A. macrocalyx***
3. Moderate to robust, emergent tree palm; inflorescence spreading, branched from 2–3 orders; rachillae persistent, scarcely shrivelling, inflorescence not club-like in fruit. 4

4. Inflorescences branched to 4 orders; triads (and hence fruit) solitary at the base of the rachilla, but many rachillae lacking complete triads . . . . **2. *A. jokowi***
4. Inflorescences branched to 2–3 orders 5

5. Leaf rachis slender; leaflets few (9–10 on each side of the rachis), sigmoid, at least 3 most distal leaflets broadly wedge-shaped and deeply lobed. . . . .**6. *A. unipa***
5. Leaf rachis robust; leaflets many (more than 20 on each side of the rachis), crowded, not sigmoid, not wedged-shaped; widely cultivated . . . . **1. *A. catechu***

1. *Areca catechu* L.

Synonyms: *Areca catechu* var. *alba* Blume, *Areca catechu* var. *batanensis* Becc., *Areca catechu* f. *communis* Becc., *Areca catechu* var. *longicarpa* Becc., *Areca catechu* var. *nigra* Giseke, *Areca catechu* var. *silvatica* Becc., *Areca cathechu* Burm.f., *Areca faufel* Gaertn., *Areca himalayana* Griff., *Areca hortensis* Lour., *Areca macrocarpa* Becc., *Areca nigra* Giseke ex. H.Wendl., *Sublimia areca* Comm. ex Mart.

Moderate to robust, single-stemmed tree palm, to 25(–30) m tall, 8–12 leaves in the crown, *crown shuttlecock-shaped to arching*. **Stem** 15–25(–40) cm diam. **Leaf** 1.5–2.7 m long (including petiole); crownshaft 100–175 × 15–20 cm, green; petiole *lacking or short*, to 15 cm long; leaflets 20–35 on each side of rachis, *not sigmoid, each comprising more than one fold, more or less regularly arranged, in one plane*. **Inflorescence** 29–80 cm long, divaricate, protandrous, *branched two or three orders*; peduncle ca. 6 cm long; rachillae ca. 10–40 cm long, *numerous, triads distichously arranged with 1–3 complete triads occurring at the base of each rachilla*. **Male flower** 4.0–7.5 × 2–5 mm, stamens 6. **Female flower** 12–15 × 7–10 mm. **Fruit** 5–7 × 2–4 cm, ovoid to ellipsoid; mesocarp fibrous. **Seed** 3.0–3.5 × 2.5–3.0 cm, variously subglobose to ovoid, more or less flat at base.

DISTRIBUTION. Of unknown wild origin, the betel palm is widely cultivated throughout New Guinea.

HABITAT. Village settings from sea level to over 600 m.

LOCAL NAMES. *Areca* or *Areca-nut palm*, *Betel nut* (English), *Buai* (Papua New Guinea Tok Pisin), *Buei* (Pala), *Mala'chu* (Gebe), *Malalolef* (Gebe), *Pinang* (Bahasa Indonesia), *Pinang Nau* (Yapen), *Sawu* (Wandamen), *Vua* (Lamekot).

USES. The most important use is as a masticatory, extremely popular throughout New Guinea and elsewhere in the Old World Tropics. The scale of betel nut use is enormous, used by around 200–400 million people, making it the fourth most widely "abused" substance after caffeine, nicotine and alcohol (Norton 1998, Gupta & Warnakulasuriya 2002). Medicinally, in New Guinea the betel nut is used to treat toothache, dysentery, upset stomach and body ache, among other ailments. The red oral mixture of chewed betel nut and lime is applied to tropical ulcers. The nut is heated over a fire and pressed on sores caused by sea urchins. The leaf sheaths and prophylls are used as wrapping material. The scraped bark is mixed with sea water and drunk to treat asthma.

CONSERVATION STATUS. Least Concern.

NOTES. The betel palm is one of the most familiar palms in New Guinea. The origin and dispersal of the betel nut palm and the chewing habit have been discussed for many years (reviewed by Zumbroich 2008). The discovery of *A. mandacanii*, *A. jokowi* and *A. unipa*, apparently all close relatives of *A. catechu*, in western New Guinea is important because it suggests that New Guinea should be considered alongside the Philippines as a potential area of origin for *A. catechu* (see Heatubun *et al.* 2012a).

Areca catechu. CLOCKWISE FROM TOP: betel nut garden, near Lae (WB); betel nut for sale, Sorong (LG); inflorescence, Morobe (WB).

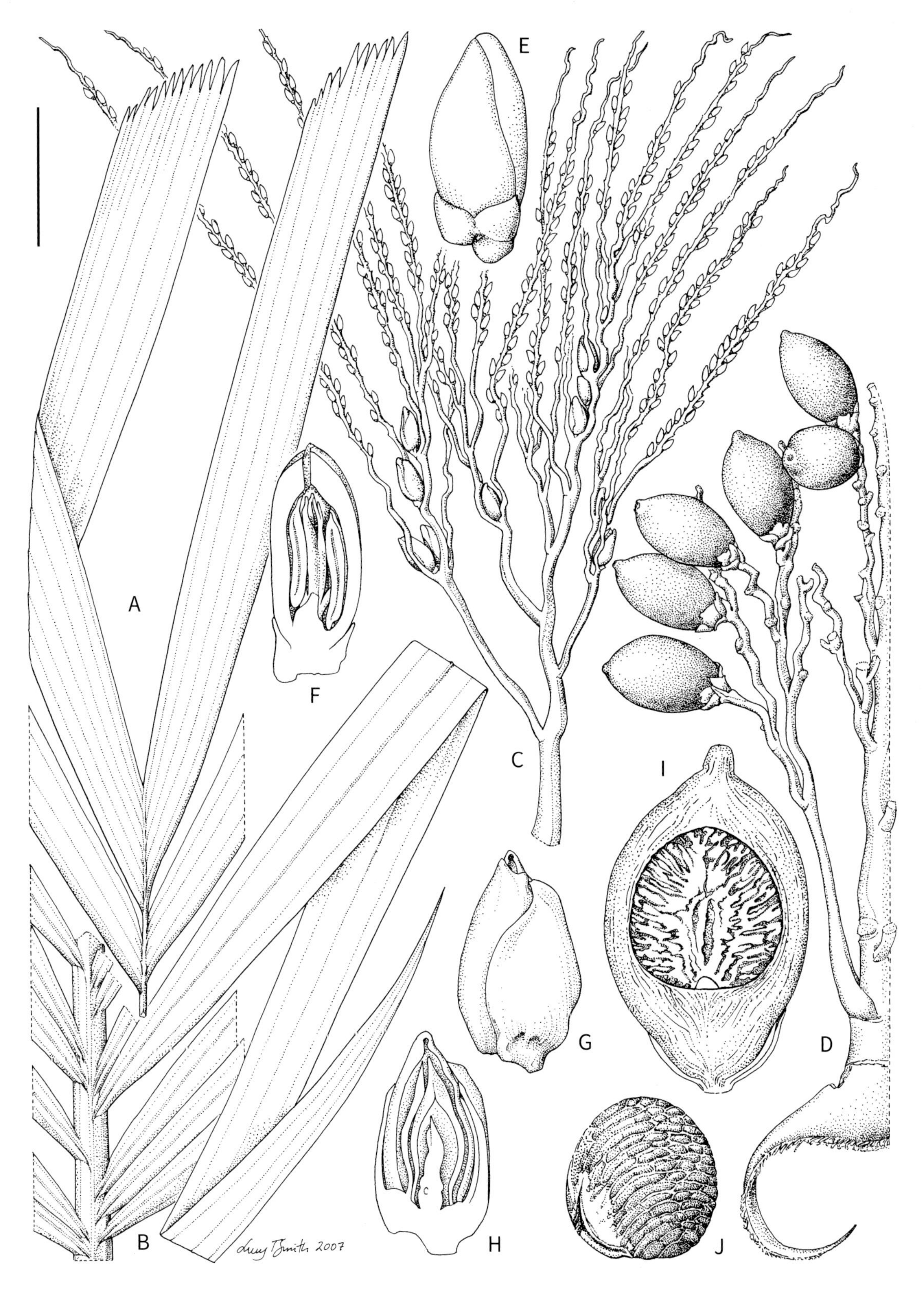

Areca catechu. **A.** Leaf apex. **B.** Leaf mid-portion. **C.** Inflorescence. **D.** Portion of inflorescence with fruit. **E, F.** Male flower whole and in longitudinal section. **G, H.** Female flower whole and in longitudinal section. **I.** Fruit in longitudinal section. **J.** Seed. Scale bar: A, B = 8 cm; C, D = 6 cm; E, F = 3 mm; G, H = 1 cm; I, J = 2 cm. A–C, E–J from *Noblick et al. 5180*; D from *de Vogel 3266*. Drawn by Lucy T. Smith.

2. *Areca jokowi* Heatubun

Slender, single-stemmed palm to 15 m, 9 leaves in crown. **Stem** 7–8 cm diam. **Leaf** 90–93 cm long (including petiole); crownshaft up to 40 cm long, light to dull green; petiole 5–6 cm long; leaflets ca. 11 on each side of rachis, *more or less regularly arranged*, comprising 1–10 folds, sigmoid. **Inflorescence** 30–37 cm long, ca. 46 cm wide, divaricate, *yet somewhat congested*, protandrous, *branching to 4 orders*; peduncle 3 cm long; rachillae 14–17 × 0.1–0.2 cm, *numerous, crowded, not expanding widely*, green, elongate, sinuous especially in the distal ⅔, *triads uniseriate, becoming distichous near the tip of rachillae, complete floral triads always solitary at the base of rachillae* or absent, lower ca. ⅓ to ½ of rachilla naked (with exception of solitary complete triad if present). **Male flower** 3.2– 4.3 × 2.0–2.5 mm, stamens 6. **Female flower** ca. 11 × 9.1 mm. **Fruit** 3.2–3.5 × 2.5–2.8 cm, ovoid, mesocarp fibrous, mature fruits not seen. **Seed** not seen.

DISTRIBUTION. Known only from two individual palms cultivated on the shores of Lake Yamur, Indonesian New Guinea, said to be grown from seeds collected in hill forest in the headwaters of the Ima River in Gunung Daweri.

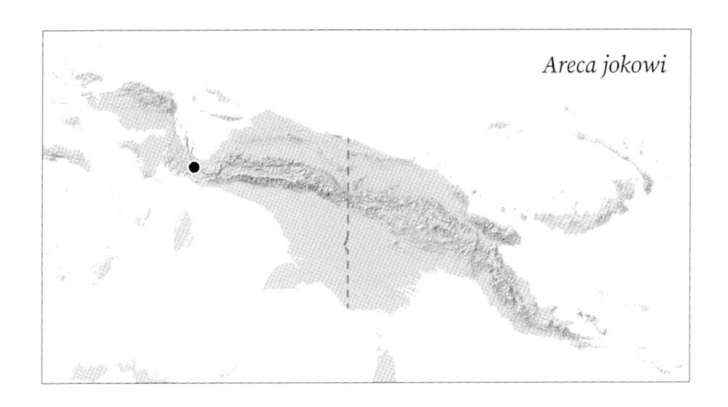

HABITAT. Hill forest on soils derived from sandstones at ca. 300 m.

LOCAL NAMES. *Siaku'* (Yamur).

USES. Fruits chewed as a betel nut substitute. The palm has potential as an ornamental.

CONSERVATION STATUS. Critically Endangered. The only known site of *Areca jokowi* is located within a logging concession.

NOTES. Similar to *Areca catechu*, *A. mandacanii* and *A. unipa* in habit and inflorescence structure, *A. jokowi* differs in the inflorescence branched to four orders, the crowded rachillae, borne very close together and not expanding widely, sinuous especially in the

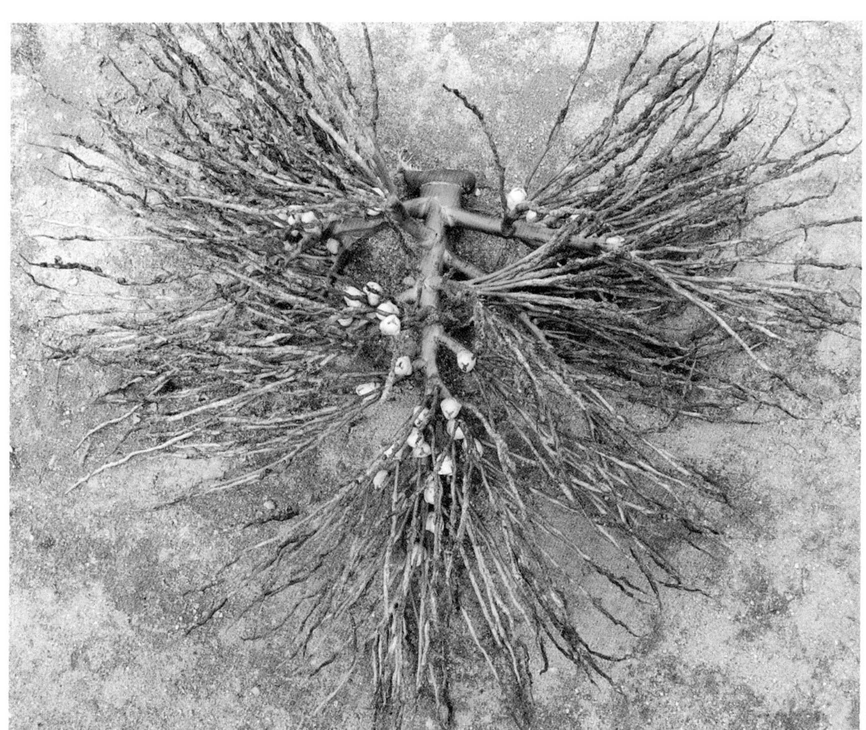

Areca jokowi. LEFT TO RIGHT: inflorescence; female flower. Yamur Lake (CDH).

Areca jokowi. **A.** Habit. **B.** Leaf apex. **C, D.** Leaf mid-portion. **E.** Leaf base. **F.** Portion of inflorescence. **G.** Rachilla with female flower attached. **H.** Male flower. **I.** Female flower in longitudinal section. Scale bar: A = 2.4 m; B–E = 8 cm; F = 4 cm; G = 1.5 cm; H = 3 mm; I = 7 mm. All from *Heatubun et al. 1252*. Drawn by Lucy T. Smith.

distal two thirds and the floral triads uniseriate in arrangement (although distichous near the tip of the rachilla). Complete floral triads are always solitary at the base of rachillae, but absent from many rachillae, the remaining floral triads consisting of paired or solitary male flowers. The calyx of the female flowers at anthesis is conspicuously bright white.

3. *Areca macrocalyx* Zipp. ex Becc.

Synonyms: *Areca jobiensis* Becc., *Areca macrocalyx* var. *aruensis* Becc., *Areca macrocalyx* var. *conophyla* Becc., *Areca macrocalyx* var. *intermedia* Becc., *Areca macrocalyx* var. *waigheuensis* Becc., *Areca macrocalyx* var. *zippeliana* Becc., *Areca multifida* Burret, *Areca nannospadix* Burret, *Areca nigasolu* Becc., *Areca rechingeriana* Becc., *Areca rostrata* Burret, *Areca torulo* Becc., *Areca warburgiana* Becc.

Slender to robust, single-stemmed palm to 25 m tall, 6–10 leaves in crown. **Stem** 2.5–25 cm diam. **Leaf** to 2.5 m long (including petiole); crownshaft to 150 cm long, green to dark green (reddish-green to bright red in some populations); petiole lacking or to 10 cm long; leaflets 6–75 on each side of rachis, *regularly to irregularly arranged, single to multifold*, linear to sigmoid. **Inflorescence** 10–65 cm long, erect to pendulous, *protogynous, branched to 1 order* (sometimes basal-most rachillae branched again), *branches spirally arranged, more congested distally*; peduncle to 10 cm long; rachillae to 41 cm long, to 3 mm wide, *much less robust than rachis, numerous* (12–600), cream to green, sinuous to zigzag, *triads distichously arranged, 1–5 complete triads at the very base of each rachilla,* male portion of rachilla *very slender, and drying and falling after anthesis,* the inflorescence becoming *congested, club-like in fruit.* **Male flower** ca. 14 × 7 mm, stamens 6. **Female flower** ca. 20 × 15 mm. **Fruit** to 5 × 3 cm, usually obovoid, with short beak, mesocarp fibrous. **Seed** 3 × 2 cm, ovoid, rounded apically and flattened basally.

DISTRIBUTION. The most widespread species of *Areca* in East Malesia, distributed from Maluku through New Guinea to the Solomon Islands.

HABITAT. In forest from sea-level to 1,500 m, sometimes cultivated around villages.

LOCAL NAMES. *Are* (Sayal), *Ariki* (Onate), *Aupmo* (Keroom), *Kasimya* (Gebe), *Kasmai* (Waigeo), *Men* (Mianmin), *Mon* (Matbat), *Monbat* (Matbat), *Muncu sirbi* (Arfak), *Owee* (Yamur), *Puaxau* (Krisa), *Piawan* (Wandamen), *Pinang Hutan* (Indonesia), *Rigi* (Kotte), *Ripafe* (Sumeri), *Rofero* (Irarutu), *Sasoro* (Sumuri), *Sias* (Karon), *Sung-geri* (Amungkal), *Sunggeri-Piawan* (Wandamen), *Torheru Nyi* (North Cyclops), *Wauneb* (Amungkal), *Wissara* (Waskuk).

USES. Often used as a substitute for betel chewing when *A. catechu* is unavailable. Stems and leaves are used for building materials (flooring and thatch for huts or temporary shelters) and fruits are used for medicine. The red crownshaft form from the Finschhafen area of Papua New Guinea is a much-prized ornamental. Young shoot (heart-of-palm) is edible. The fruit is eaten by the northern cassowary (Pangau-Adam & Mühlenberg 2014).

CONSERVATION STATUS. Least Concern (IUCN 2013).

NOTES. *Areca macrocalyx* is easy to distinguish from other *Areca* species in New Guinea by the congested, club-like fruiting inflorescence. The male-only portions of the rachillae are thin and dry after anthesis and fall off as the fruits mature leaving a spike-like structure. It is very well represented in herbaria. Flynn (2004) analysed *Areca* in New Guinea and the Solomon Islands using a morphometric approach, and concluded that seven species (*A. congesta, A. jobiensis, A. ledermanniana, A. multifida, A. nannospadix, A. rostrata* and *A. warburgiana*) should be included within *A. macrocalyx* (two of these *A. congesta* and *A. ledermanniana* are now considered names of uncertain application, along with *A. glandiformis*; see below). Morphological variation is complex, perhaps as a response to different habitat types, occurring as it does in a wide range of ecological conditions from littoral and swampy areas in lowlands to heath forest in lower montane vegetation, and from rainforest to drier areas in savannah lands. Specimens from different localities can appear very different. However, morphology overlaps and it has proved impossible to accept reliable separate taxa. The narrow species concept used in the past reflects limited information obtained from single

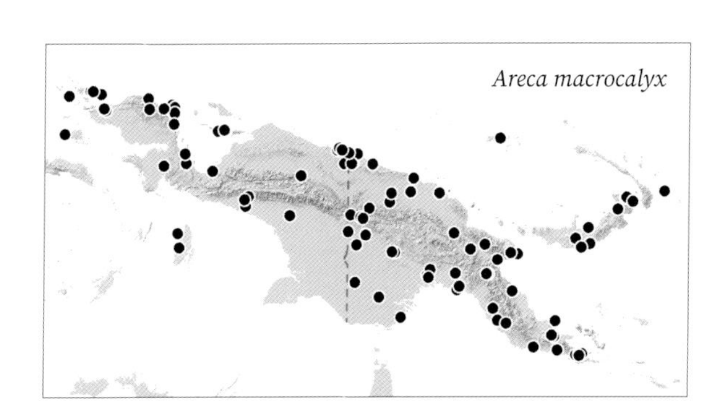

Areca macrocalyx (slender form). TOP LEFT TO RIGHT: habit, Sirebi River (WB); male and female flowers, Sirebi River (WB); inflorescence with fruit, near Sorong (WB). *Areca macrocalyx* (robust, red crownshaft form). BOTTOM LEFT TO RIGHT: habit ; inflorescence; male and female flowers. Jivewaneng (WB).

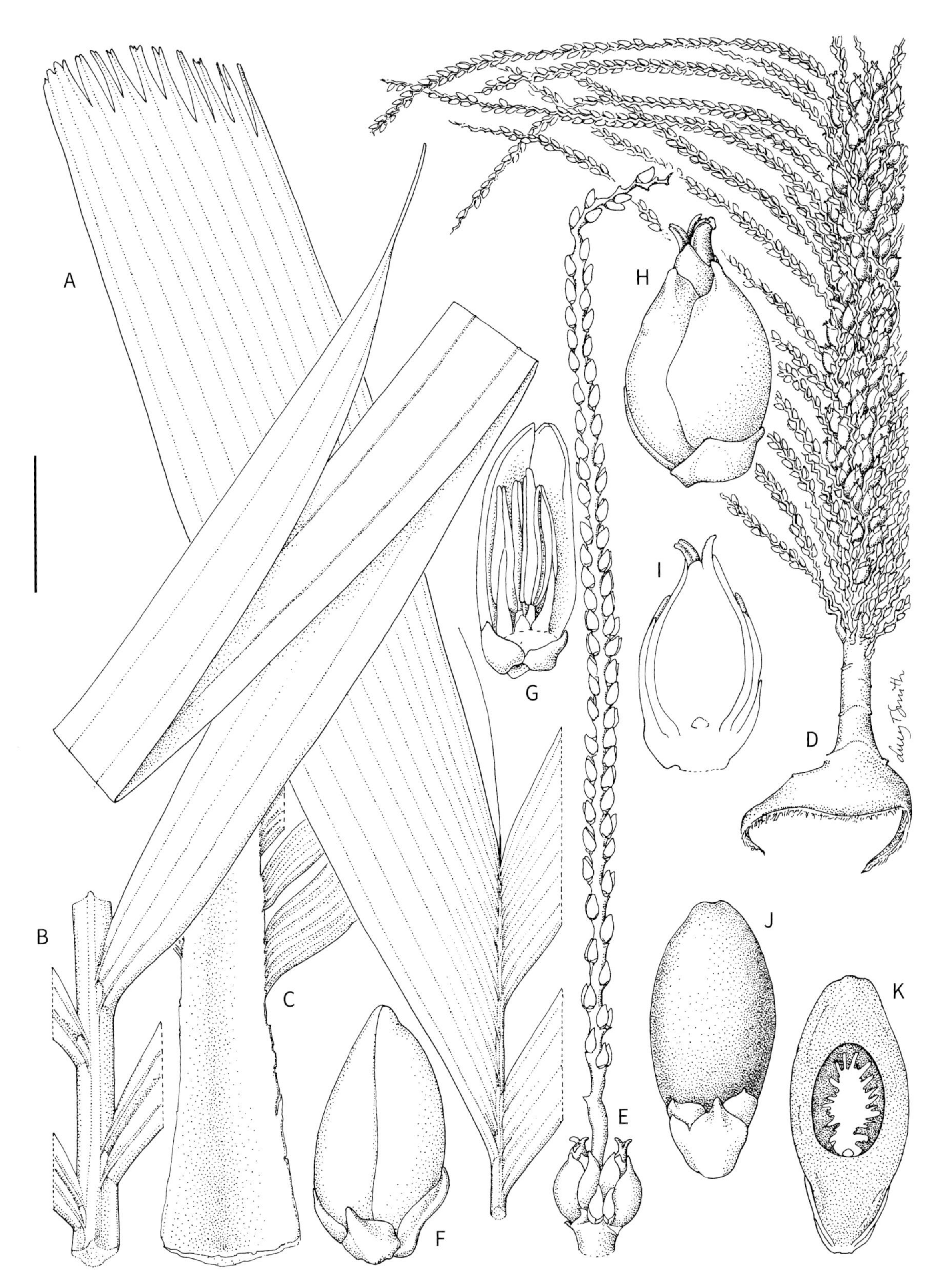

Areca macrocalyx. **A.** Leaf apex. **B.** Leaf mid-portion. **C.** Leaf base. **D.** Inflorescence. **E.** Rachilla with flowers. **F**, **G.** Male flower whole and in longitudinal section. **H**, **I.** Female flower whole and in longitudinal section. **J**, **K.** Fruit whole and in longitudinal section. Scale bar: A–C = 8 cm; D = 6 cm; E = 2.5 cm; F, G = 3 mm; H, I = 1 cm; J, K = 2 cm. All from *Baker et al. 1100*. Drawn by Lucy T. Smith.

Areca mandacanii. Crown, cultivated Nong Nooch Tropical Garden, Thailand (ZJ).

collections and, with many more recent specimens, proves to be unworkable. For a full discussion, see Heatubun *et al.* (2012a).

4. *Areca mandacanii* Heatubun

Moderate, single-stemmed tree palm, ca. 8 leaves in crown. **Stem** up to 15 m tall, 8–10 cm diam. **Leaf** 2–2.5 m long (including petiole), *plumose*; crownshaft to 152 × 15 cm; petiole to 6 cm long; ca. 60 leaflets on each side of the rachis, *single-fold, linear, irregularly arranged in 13–15 groups, held in several planes*, terminal ca. 13–14 leaflets regularly arranged, leaflets somewhat arching. **Inflorescence** ca. 60 cm long, protandrous, *laxly branched, mostly branching to 2 (rarely 3) orders*; peduncle ca. 5 cm long; rachillae 37–50 cm long, numerous; *triads distichously arranged, only one complete triad occurring at very base of each rachilla*. **Male flower** 4.5 × 2.5 mm, stamens 6. **Female flower** 10–13 mm diam. **Fruit** 65–70 × 42–45 mm, ellipsoid, with *conspicuous, woody, shallow, disc-shaped depression at apex*; mesocarp fibrous. **Seed** ca. 28 × 25 mm, subglobose.

DISTRIBUTION. One known locality in the Bird's Head Peninsula, Indonesian New Guinea.

HABITAT. Forest transition between swamp forest and lowland forest in areas temporarily flooded at ca. 10 m elevation.

LOCAL NAMES. *Ngafa* (Sayal).

USES. Fruits chewed as a betel-nut substitute. Stems used for flooring.

CONSERVATION STATUS. Critically Endangered. The only known site of *Areca mandacanii* is used heavily by local people. Deforestation due to oil palm plantation and logging concessions is a major threat in the region.

NOTES. The plumose leaves are unique in *Areca*. It is most similar to *A. catechu*, but the inflorescence of *A. mandacanii* is more slender and laxly branched to 2 (rarely 3) orders, as opposed to congested or crowded and branched mostly to 3 orders in *A. catechu*.

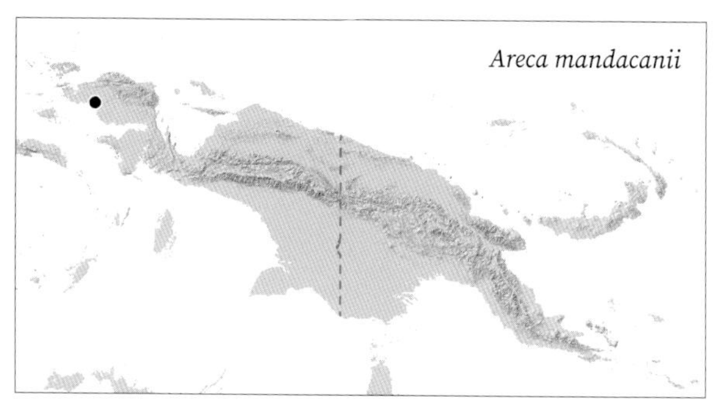

5. *Areca novohibernica* (Lauterb.) Becc.

Synonyms: *Areca guppyana* Becc., *Areca novohibernica* var. *salomonensis* Burret, *Areca salomonensis* (Burret) Burret ex A.W.Hill & E.Salisb., *Nenga novohibernica* Lauterb.

Slender, single-stemmed tree palm, to 4(–5) m, *with stilt-roots*, 5–8 leaves in crown. **Stem** to 10 cm diam. **Leaf** 1.2–1.5 m long (including petiole); crownshaft 60–90 cm long, pale to mid-green; petiole 30–50 cm long; leaflets *more or less regularly arranged, ca. 5 leaflets on each side of rachis*, slightly sigmoid, *each comprising 3–15 folds*. **Inflorescence** 15–36 × 14–30 cm, protandrous, *branched to 1 order* (sometimes basally branched to 2 orders), divaricate, erect at anthesis, pendulous in fruit; rachilla 7–16 cm long, 2–4 mm wide, 10–21, greenish-cream to green, *triads spirally*

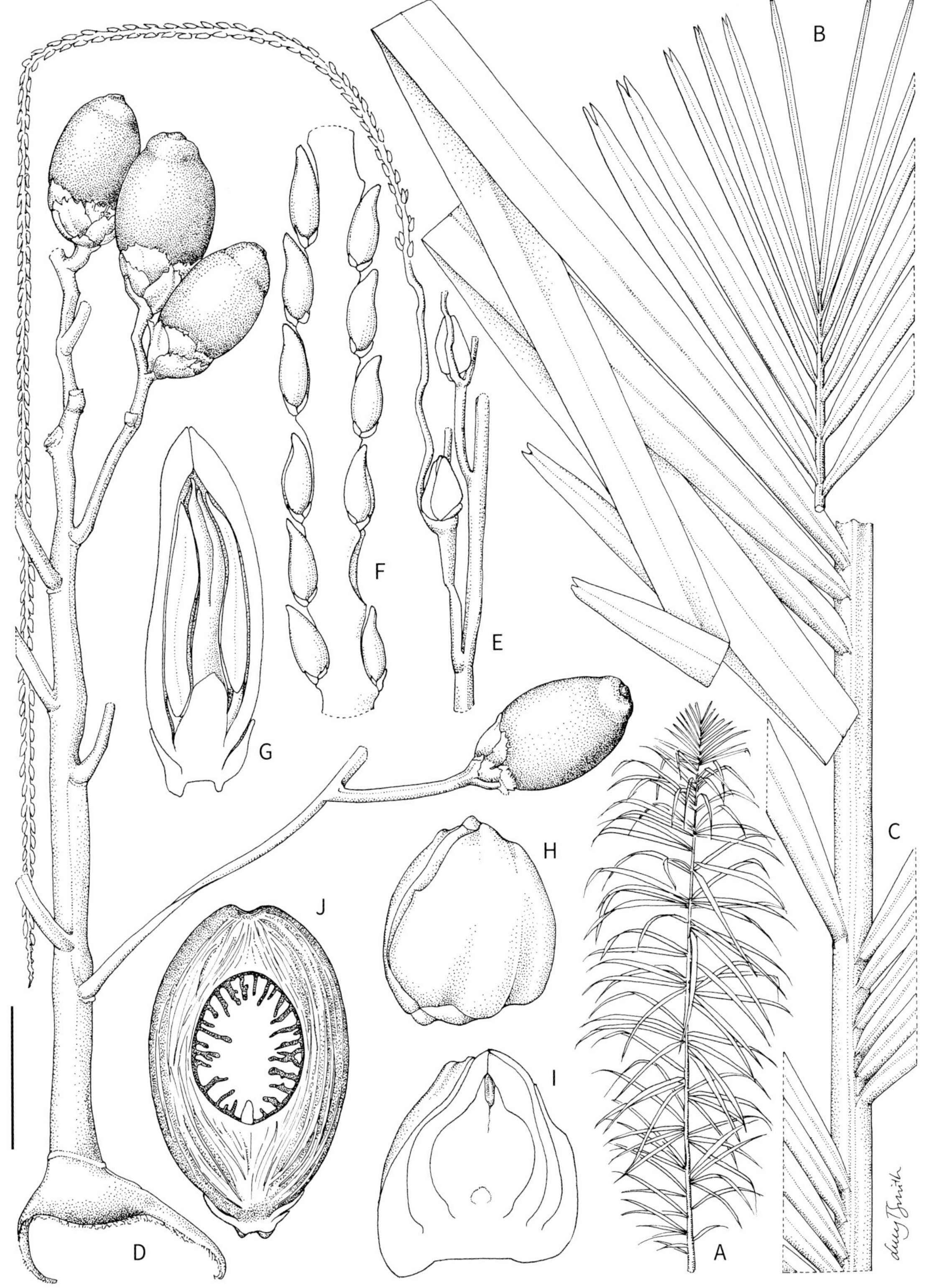

***Areca mandacanii*. A.** Whole leaf diagram. **B.** Leaf apex. **C.** Leaf mid-portion. **D.** Portion of inflorescence with fruit. **E.** Rachilla. **F.** Portion of rachilla with flowers. **G.** Male flower in longitudinal section. **H, I.** Female flower whole and in longitudinal section. **J.** Fruit in longitudinal section. Scale bar: A = 50 cm; B–D = 4 cm; E, J = 3 cm; F = 7 mm; G = 1.6 mm; H, I = 1 cm. A–E, H, I from *Heatubun et al. 423*; F, G, J from *Heatubun et al. 413*. Drawn by Lucy T. Smith.

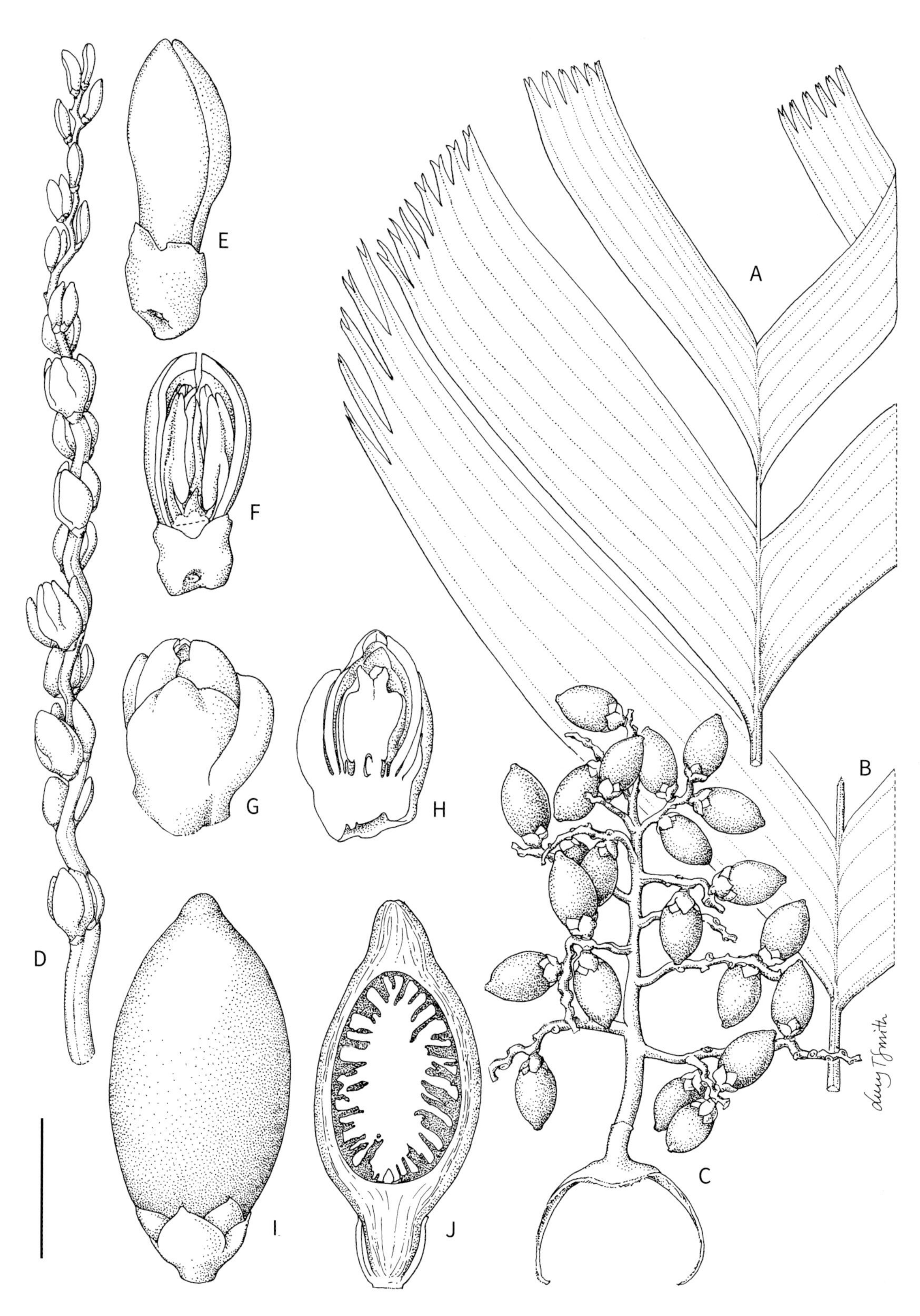

***Areca novohibernica*.** **A.** Leaf apex. **B.** Leaf mid-portion. **C.** Inflorescence with fruit. **D.** Rachilla with flowers. **E, F.** Male flower whole and in longitudinal section. **G, H.** Female flower whole and in longitudinal section. **I, J.** Fruit whole and in longitudinal section. Scale bar: A, B = 8 cm; C = 6 cm; D, I, J = 1.5 cm; E, F = 3 mm; G, H = 7 mm. A, B from *Sands 726*; C from *Takeuchi 16802*; D–J from *Sands 2124*. Drawn by Lucy T. Smith.

Areca novohibernica. FROM TOP: inflorescences, cultivated National Botanic Garden, Lae (WB); stilt roots, cultivated Carlsmith Estate, Hawaii (WB).

arranged on the rachilla, complete triads occurring from base up to half the length of the rachilla. **Male flower** 5.2–6.5 × 2.5–2.7 mm, stamens 6. **Female flower** 8–9 × 7.0–7.5 mm. **Fruit** 3.2–4.0 × 1.6–2.0 × 1.5–2.0 cm, ellipsoid to slightly ovoid, *mesocarp fleshy and juicy*. **Seeds** 1.2–1.5 × 1.0–1.5 cm, globose to ellipsoid, rounded apically, flattened basally.

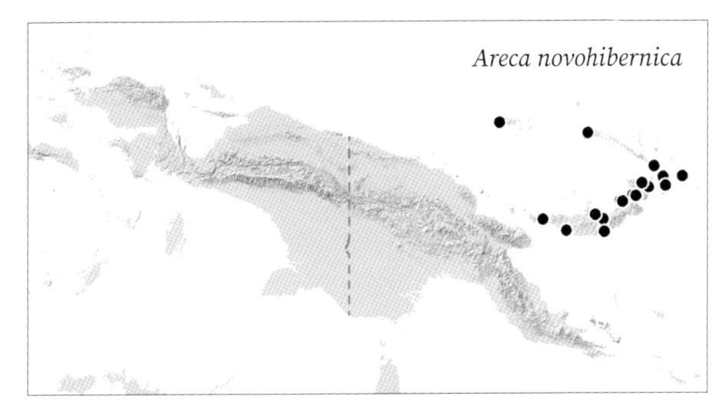

DISTRIBUTION. Bismarck Archipelago (Manus, New Britain and New Ireland) and the Solomon Islands.

HABITAT. Forest on volcanic and limestone soils at 10–1350 m.

LOCAL NAMES. None recorded.

USES. The fruits are used as a betel nut substitute. The species is an attractive ornamental.

CONSERVATION STATUS. Endangered. *Areca novohibernica* is an island species and logging concessions are a major threat within its range.

NOTES. *Areca novohibernica* is immediately distinguished from other New Guinea species in the consistent presence of stilt roots, in the spirally arranged triads on the rachillae and in the juicy fleshy fruit. It is most similar to *A. vestiaria* Giseke of Sulawesi and Maluku.

6. *Areca unipa* Heatubun

Slender, single-stemmed palm to 12 m, leaves 7 in crown. Stem ca. 7.5 cm diam. Leaf ca. 1.2 m long (including petiole); crownshaft up to 75 cm long, 7 cm diam., pale to dull green; petiole ca. 16.5 cm long; leaflets *9–10 on each side of rachis,* sigmoid, each comprising 3–6 folds, *more or less regularly arranged,* distant. Inflorescence 30–40 cm long, 10–15 cm wide, slender, protandrous, *branched to 2 orders*; peduncle 1–4.5 cm long; rachillae to 21 cm long, 1–4 mm wide, pale green, sinuous near the base; *triads distichous, only one complete triad occurring at the base of each rachilla,* remaining triads comprising very few paired and solitary male flowers. Male flower 4.5–6 × 2.1–2.5 mm, stamens 6. Female

Areca unipa. **A.** Habit. **B.** Leaf apex. **C.** Leaf mid-portion and basal portion. **D.** Portion of inflorescence. **E.** Portion of rachilla with male flowers. **F**, **G.** Male flower whole and in longitudinal section. **H**, **I.** Female flower whole and in longitudinal section. **J.** Detail of ovary and staminode tube. **K.** Fruit. Scale bar: A = 2 m; B, C = 8 cm; D = 4 cm; E = 1.5 cm; F, G = 2.5 mm; H, I = 1 cm; J = 7 mm; K = 3 cm. All from *Iwanggin & Simbiak 138*. Drawn by Lucy T. Smith.

flower to 2 × 1 cm. Fruit 5.5–6 × 3.5–3.8 cm (unripe fruits), obovoid or ovoid, beaked, mesocarp fibrous. Seed ca. 3 × 2.2 cm, obovoid, slightly flattened at base.

DISTRIBUTION. One known locality close to Ayata village in East Maybrat District, Maybrat Regency in the central part of the Bird's Head Peninsula.

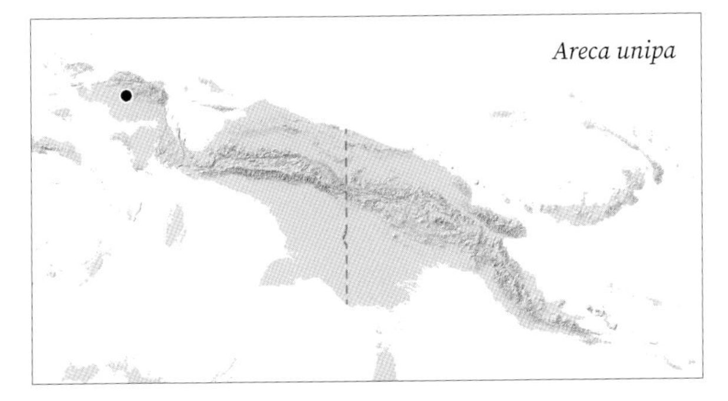

HABITAT. Primary lowland peat forest at an elevation of about 200 m above sea level. It appears to be adapted to extreme conditions on underlying coal beds, which sometimes lack any apparent soil, except for leaf litter over the coal outcrop.

Areca unipa. Habit, Aifat (CDH).

Areca unipa. LEFT TO RIGHT: inflorescence; inflorescence with fruit. Aifat (CDH).

LOCAL NAMES. *Srah owei knu* (Aifat dialect, Mai Brat language).

USES. Fruits are chewed as a betel nut substitute.

CONSERVATION STATUS. Critically Endangered. The only known population of *A. unipa* is estimated to comprise fewer than 250 mature palms. Deforestation, due to mining, oil palm plantations and logging concessions, is a major threat in its distribution range.

NOTES. *Areca unipa* is most similar to *A. catechu* and *A. mandacanii* in its single-stemmed, moderate tree palm habit and inflorescence structure, but it can immediately be distinguished by the small, slender leaves with relatively long petiole and very few multifold, sigmoid and broadly wedge-shaped leaflets.

NAMES OF UNCERTAIN APPLICATION

Areca congesta Becc.

The types of *Areca congesta* and *A. ledermanniana*, both collected by Ledermann, were destroyed in Berlin. Though likely to be synonyms of *A. macrocalyx*, we regard the names as being of uncertain application, following Heatubun *et al.* (2012a).

Areca glandiformis Lam.

This very early name from 1783 was rejected by Heatubun *et al.* (2014b; Applequist 2016). Though a likeness to *A. macrocalyx* is clear in the illustration that serves as the type, too many uncertainties prevail for it to be used with confidence. See Heatubun *et al.* (2014b) for further discussion.

Areca ledermanniana Becc.

See note under *Areca congesta*.

Pinanga rumphiana. Habit, Biak Island (WB).

Pinanga Blume

Synonyms: *Cladosperma* Griff., *Ophiria* Becc., *Pseudopinanga* Burret

Medium – leaf pinnate – crownshaft – no spines – leaflets lobed or pointed

Moderately robust, single-stemmed (in New Guinea) tree palm; crownshaft present, monoecious. **Leaf** pinnate, arching; sheath tubular; petiole short; leaflets numerous, *mostly composed of 2 folds*, with pointed tips, arranged regularly, *apical pair much broader, multifold with lobed tips*. **Inflorescence** *below leaves, branched to 1 order*; prophyll papery, enclosing the inflorecence in bud, soon dropping off when inflorescence expands, *peduncular bract absent*; peduncle shorter than inflorescence rachis; rachillae curved at the base then straight, pendulous. **Flowers** *in triads throughout the length of the rachilla*, spirally arranged, male flowers fleshy, asymmetrical, much larger than the spherical female flowers. **Fruit** red to black, ovoid, stigmatic remains apical, flesh thin, endocarp fibrous. **Seed** 1, globose, endosperm ruminate.

DISTRIBUTION. Around 140 species from South China to India, through South-East Asia to New Guinea. Not recorded from the Bismarck Archipelago.

NOTES. *Pinanga* is a highly variable genus of understorey to mid-storey palms. The inflorescences are produced below the leaves and are branched to one order. Female flowers and fruit are present from the base to the tip of the inflorescence branches. It can be confused with *Areca*, which, like *Pinanga*, lacks a peduncular bract, but in contrast *Areca* bears female flowers only at the base of inflorescence branches. It may also be confused with *Hydriastele*, which has jagged leaflet tips (rather than pointed or lobed tips) and inflorescences enclosed within prophyll and peduncular bract in bud, rather than a prophyll only.

Pinanga has not been the subject of a monograph treatment for more than a century (Beccari 1886). A comprehensive taxonomic account is long overdue.

Pinanga rumphiana (Mart.) J.Dransf. & Govaerts

Synonyms: *Areca gigantea* H.Wendl., *Areca punicea* Zipp. ex Blume, *Areca sanguinea* Zipp. ex Blume, *Drymophloeus puniceus* (Zipp. ex Blume) Becc., *Drymophloeus rumphianus* Mart., *Pinanga caudata* Becc., *Pinanga punicea* (Zipp. ex Blume) Merr., *Pinanga ternatensis* Scheff., *Pinanga ternatensis* var. *papuana* Becc., *Ptychosperma caudatum* Becc., *Ptychosperma puniceum* (Zipp. ex Blume) Miq., *Saguaster puniceus* (Zipp. ex Blume) Kuntze, *Seaforthia rumphiana* Mart.

Moderately robust, *single-stemmed palm to 15 m*, bearing 6–8 leaves in crown. Stilt roots sometimes reported. **Stem** 7–9 cm diam., internodes 8–20 cm. **Leaf** 2–3.5 m long including petiole; sheath 0.7–1.2 m long; petiole 10–46 cm long; leaflets 29–44 each side of rachis, *arranged regularly or somewhat irregularly*, linear, comprising 1–5 folds (mostly 2 folds), with fine brown indumentum on undersurface of major ribs; mid-leaf leaflet 70–112 × 2.5–7.5 cm wide; apical leaflets 14–35 × 0.5–6 cm wide, free or united to 1/5 of their length. **Inflorescence** 45–75 cm long including 5–10 cm peduncle, pendulous, with 20–30 rachillae in total; rachillae 25–46 cm long. **Male flower** 7–10 × 4–6 mm in bud, white; stamens 26–31. **Female flower** ca. 5 × 4 mm, yellow-green. **Fruit** *15–19 × 9–12 mm diam., ellipsoid, red or purple*. **Seed** 9–11 × 7.5–10 mm, globose to ovoid, shallow depression basal or laterally at base.

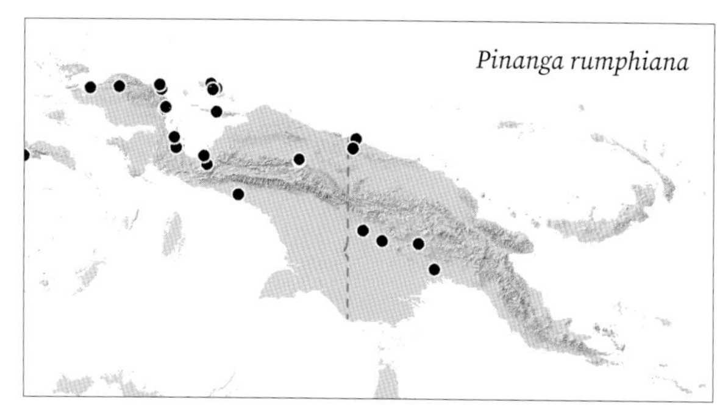

DISTRIBUTION. Widespread in New Guinea, reaching the eastern recorded limit of the genus near Kikori. Elsewhere in South-East Sulawesi and Maluku.

HABITAT. Widespread in lowland rainforest from 10–880 m.

LOCAL NAMES. *Ansansup* (Biak), *Ansirep* (Biak), *Owe* (Miyah), *Sutu* (Kaimana).

USES. Leaves used as thatch and to wrap meat or rice for carrying. Young leaves used to obtain a thread to weave rough textiles. Slender stems used as walking sticks, spear shafts, rafters, and for walls. Endosperm used as a dye in the batik industry. Kernel sometimes used as a substitute for betel nut. Palm cabbage of young plants is edible.

CONSERVATION STATUS. Least Concern.

NOTES. *Pinanga rumphiana* is the most easterly occurring species in the genus.

Pinanga rumphiana. Leaf apex, Tamrau Mountains (WB).

Pinanga rumphiana. LEFT TO RIGHT: inflorescence, Wandamen Peninsula (SB); floral triads in bud, Tamrau Mountains (WB); female flowers, Kikori River (WB).

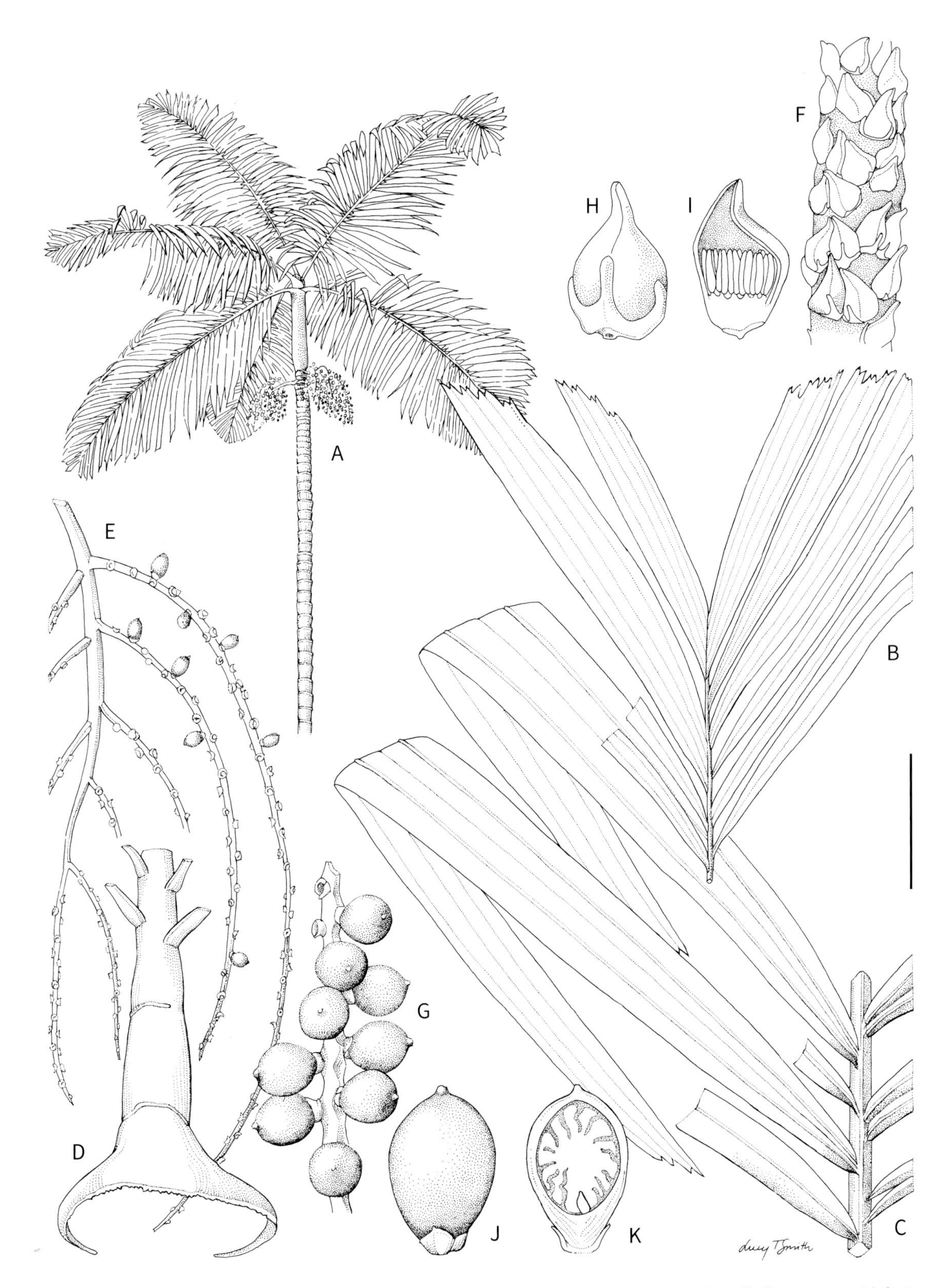

Pinanga rumphiana. **A.** Habit. **B.** Leaf apex. **C.** Leaf mid-portion. **D.** Inflorescence base. **E.** Portion of inflorescence with fruit. **F.** Portion of rachilla with male flowers. **G.** Portion of rachilla with fruit. **H, I.** Male flower whole and in longitudinal section. **J, K.** Fruit whole and in longitudinal section. Scale bar: A = 1.5 m; B, C = 9 cm; D = 4 cm; E = 6 cm; F, J, K = 1.5 cm; G = 3 cm; H, I = 7 mm. A from photograph; B–E, G, J, K from *Baker et al. 850*; F, H, I from *Baker et al. 1047*. Drawn by Lucy T. Smith.

Unidentified *Physokentia* species (possibly *P. insolita*).
LEFT TO RIGHT: habit; stilt roots, Bougainville (JP).

Physokentia Becc..

Synonyms: *Goniocladus* Burret, *Goniosperma* Burret

Small – leaf pinnate – crownshaft – no spines – leaflets pointed

Slender, single-stemmed tree palm with stilt roots, crownshaft present, monoecious. **Leaf** pinnate, few in crown, arching; sheath tubular; petiole short; leaflets reduplicate, numerous, with pointed tips, arranged regularly. **Inflorescence** *below the leaves,* branched to 3 orders; *prophyll not encircling peduncle completely,* peduncular bract much longer than the prophyll, enclosing the inflorescence in bud, dropping off with the prophyll as the inflorescence expands; *peduncle longer than inflorescence rachis;* rachillae moderately slender and curved. **Flowers** in triads at the base of the branches, pairs of male flowers towards tip, not developing in pits, female flowers much larger than the male flowers. **Fruit** black, *globose, stigmatic remains subapical,* flesh thin, *endocarp shallowly angled and ridged.* **Seed** 1, globose, endosperm ruminate.

DISTRIBUTION. Seven species from the Bismarck Archipelago, Solomon Islands, Fiji to Vanuatu. In the New Guinea region, *Physokentia* has been recorded only in New Britain.

NOTES. In our region, *Physokentia* is a rare, slender, single-stemmed tree palm with stilt roots, bearing inflorescences below the leaves with the prophyll incompletely encircling the peduncle. The fruit is globose and the stone (endocarp) enclosing the seed has shallow, angular ridges. It occurs in submontane to montane rainforest at 450–1,800 m. *Physokentia* might be confused with *Areca,* but that genus has inflorescences with a completely encircling prophyll, and no peduncular bract.

No modern taxonomic monograph is available for the genus. The taxonomic account of Moore (1977) is followed here.

Physokentia avia H.E. Moore

Single-stemmed, emergent, *stilt-rooted* palm ca. 15 m tall. **Stem** ca. 10–15 cm diam., prominently ringed with leaf scars. **Leaf** ca. 2 m long; sheath 50–80 cm long, with dense woolly-scaly indumentum at the apex; petiole 60–70 cm long, also woolly-scaly; ca. 26 pairs of leaflets, evenly spaced, mid-leaf leaflet 50–65 × 3.5–4.5 cm wide, lacking chaffy scales on the underside. **Inflorescence** 45–55 cm long; *prophyll incompletely circling the peduncle;* peduncular bract 33 cm long; peduncle 7–8.5 cm long; rachillae 25–28.5 cm long, ca. 2–3 mm diam. **Male flower** 5–8 mm long, markedly asymmetrical; sepals ovate, 2–3 mm long, acute to rounded at apex; petals linear-elliptical, ovate or asymmetrically ovate, 5–7 × 3–4 mm wide, grooved internally to match anthers; stamens 6; pistillode ovoid to columnar, half as long as stamens, deeply trifid. **Female flower** not seen. **Fruit** 13–15 mm diam., depressed-globose with subapical stigmatic residue, black; *epicarp minutely pebbled;* endocarp ca. 9 × 10 mm, *globose with irregular, slight depressions and bumps,* brown, *fragile,* operculum basal. **Seed** 9 × 10 mm, depressed globose, brown, with prominent, pale raphe; *endosperm ruminate.*

DISTRIBUTION. Central New Britain.

HABITAT. Montane forest dominated by *Nothofagus* at 1,500–1,830 m.

LOCAL NAMES. None recorded.

USES. None recorded.

CONSERVATION STATUS. Endangered (IUCN 2021). *Physokentia avia* has a restricted distribution. Deforestation is a major threat to this island species.

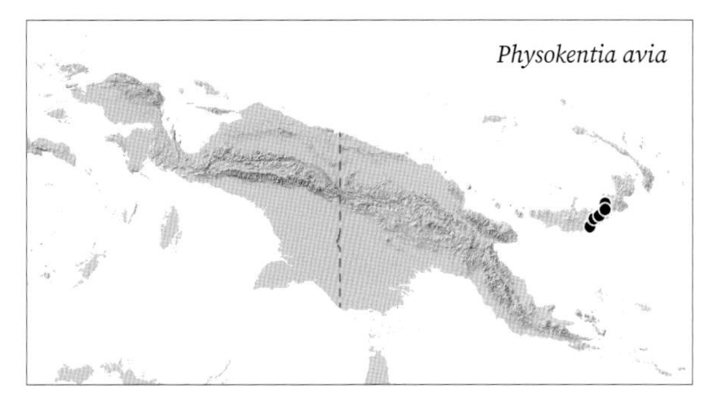

NOTE. *Physokentia avia* is unusual in the genus in having only slightly sculptured endocarps (Moore 1977). The other species (from the Solomon Islands, Vanuatu and Fiji) have elaborately ridged and angled endocarps. Limited numbers of specimens of this species are available; very few botanists have surveyed the montane forests of New Britain. We illustrate the species here with photographs of an unknown species from nearby Bougainville; to our knowledge *P. avia* has not been photographed in the wild.

In the absence of flowers and fruits, *Physokentia avia* might be confused with *Areca novohibernica*, as both have stilt roots, but *A. novohibernica* has leaflets with multiple folds. By contrast, in *P. avia*, each leaflet has only a single fold.

Physokentia avia. **A.** Leaf apex. **B.** Leaf mid-portion. **C.** Portion of inflorescence with fruit. **D.** Portion of rachilla with triads. **E**, **F.** Male flower whole and in longitudinal section. **G**, **H.** Female flower whole and in longitudinal section. **I.** Fruit. **J.** Endocarp. **K**, **L.** Seed whole and in longitudinal section. Scale bar: A–C = 6 cm; D, I = 1 cm; E, F = 4 mm; G, H = 3.3 mm; J–L = 7 mm. A, B, J–L from *Clunie LAE 63260*; C–I from *Stevens & Isles LAE 58394*. Drawn by Lucy T. Smith.

Calyptrocalyx yamutumene . Habit, cultivated Merwin Conservancy, Hawaii (WB).

Calyptrocalyx Blume

Synonyms: *Linospadix* Becc. ex Hook.f. (not *Linospadix* H. Wendl.), *Paralinospadix* Burret

Very small to medium – leaf pinnate – no crownshaft – no spines – leaflets lobed, toothed or pointed

Slender to moderate, single- or multi-stemmed tree palms, usually small, crownshaft absent, monoecious. **Leaf** pinnate, straight, leaf blade sometimes not divided into leaflets; sheath splitting to the base opposite the petiole, *fibrous at the margins;* petiole very short to elongate; leaflets few to numerous each side of leaf rachis, with lobed, toothed or pointed tips, arranged regularly or irregularly, horizontal. **Inflorescence** *between or below the leaves, unbranched (spicate), some species producing more than one spike at a node;* prophyll and peduncular bract similar, enclosing inflorescence in bud, not dropping off as inflorescence expands, *peduncular bract borne just above the prophyll,* projecting from prophyll; peduncle longer than rachilla; rachilla usually slightly thicker than peduncle, usually straight. **Flowers** *in triads throughout the length of the rachilla, developing in pits,* female flowers emerging from pits some time after the male flowers drop off. **Fruit** red, pink or purple to black, globose to ellipsoid, stigmatic remains apical, flesh thin to quite thick, endocarp thin, closely adhering to seed. **Seed** 1, globose to ellipsoid, endosperm ruminate or homogeneous.

DISTRIBUTION. Twenty-eight species from Maluku to New Guinea and the Bismarck Archipelago, 27 species endemic to New Guinea, one in both Maluku and New Guinea.

NOTES. *Calyptrocalyx* species are small understorey to moderately robust mid-storey palms that are single- or multi-stemmed, with fibrous leaf sheaths. Inflorescences are spicate and appear between the leaves, although may later be presented below the leaves; in many species the inflorescences are multiple in the leaf axil. The peduncular bract is inserted near the base of the peduncle and is persistent. Flowers develop within shallow to deep pits in the inflorescence spike.

Calyptrocalyx is a common genus throughout New Guinea from sea level to 2,000 m. It is most easily confused with rarer *Linospadix,* which also bears spicate inflorescences, but the peduncular bract is inserted at the top of the peduncle and falls at maturity. It can resemble *Sommieria* from western New Guinea, but this genus can be distinguished by its leaves being chalky white beneath and the branched inflorescence bearing corky warted fruits.

Many of the taxa accepted here are poorly understood and there may well be several taxa yet to be described. Species are widely cultivated by horticulturists and untangling the identity of species in cultivation has proved to be challenging. This account draws heavily on the latest monograph (Dowe & Ferrero 2001) with some updates and modifications.

Key to the species of *Calyptrocalyx* in New Guinea

1. Flower-bearing part of inflorescence the same diameter or only slightly wider than the peduncle 2
1. Flower-bearing part of inflorescence at least 1.5 times wider than the peduncle 12

2. Leaves entire bifid, occasionally with one or two basal leaflets 3
2. Leaves divided with irregularly or regularly arranged leaflets, or with leaflets of variable width, but not entire bifid 5

3. Petiole absent; leaves somewhat hooded, sometimes mottled . . . . **19. *C. micholitzii***
3. Petiole exceeding 5 cm long, leaves not hooded. 4

4. Inflorescence with a single spike. . . . . **24. *C. pusillus***
4. Inflorescence with 3–7 spikes . . . . **26. *C. sessiliflorus***

5. Leaves with irregularly arranged leaflets, mostly comprising several folds, not single-fold. . . . . **17. *C. leptostachys***
5. Leaflets mostly consisting of single folds, occasionally terminal leaflets multifold 6

6. Leaflets grouped. 7
6. Leaflets regularly arranged, not grouped 8

7. Leaflets grouped into two distant groups, fanned within the groups; inflorescence densely dark scaly. . . . . **16. *C. lepidotus***
7. Leaf with slightly grouped narrow divaricate or reflexed leaflets; inflorescence sparsely scaly, becoming glabrous. . . . . **25. *C. reflexus***

8. Leaflets sigmoid . . . . **18. *C. merrillianus***
8. Leaflets linear or lanceolate, not sigmoid. 9

9. Small, slender palms; peduncle very slender, 3 mm or less diam. 10
9. Robust palms; peduncle greater than 3 mm diam.. 11

10. Floral pits congested, 1–5 mm apart; fruit 9–13 mm long . . . . **2. *C. amoenus***
10. Floral pits well-spaced, 10–15 mm apart; fruit 19–21 mm long. . . . . **15. *C. laxiflorus***

11. Leaves with numerous distinctly grouped leaflets with the leaflets displayed in several planes, the whole leaf thus plumose . . . . **14. *C. lauterbachianus***
11. Leaflets regularly arranged . . . . **1. *C. albertisianus***

12. Robust single-stemmed palm with stems at least 15 cm diam. and regularly arranged pinnate leaves at least 2.5 m long; leaflets strongly plicate or ribbed. . . . . **27. *C. spicatus***
12. Smaller palms with stems less than 15 cm diam. and smaller leaves 13

13. Leaves with regularly arranged linear or sigmoid leaflets. 14
13. Leaves entire bifid or irregularly pinnate 20

14. Leaflets sigmoid, sometimes narrowly so, with conspicuous drip tips. 15
14. Leaflets linear or lanceolate, not conspicuously sigmoid, lacking conspicuous drip tips 19

15. Inflorescence with a single spike 16
15. Inflorescence with 2 or more spikes . . . . 6. *C. caudiculatus*

16. Stamen filaments basally united to form a tube with anthers borne on short free filaments 17
16. Stamen filaments free 18

17. Single or muti-stemmed moderate palm; leaflets mottled, broadly sigmoid; stamens with filaments united to form a conspicuous tube with inflexed anthers . . . . 7. *C. doxanthus*
17. Short single-stemmed palm; leaflets not mottled, narrowly sigmoid or lanceolate; stamens united in a cup with erect free filaments, anthers not inflexed. . . . . 5. *C. calcicola*

18. Single- or multi-stemmed palm; inflorescence with deep floral pits; endosperm homogeneous. . . . . 21. *C. pachystachys*
18. Multi-stemmed palm; inflorescence with shallow floral pits; endosperm ruminate . . . . 23. *C. polyphyllus*

19. Inflorescence with 2 or more spikes . . . . 10. *C. forbesii*
19. Inflorescence consisting of a single spike. . . . . 20. *C. multifidus*

20. Leaves entire bifid or with irregular segments, the veins strongly curved towards the tip . . . . 11. *C. geonomiformis*
20. Leaves various but the veins not strongly curved towards the tip 21

21. Male flowers closely appressed, appearing like rodent teeth . . . . 12. *C. hollrungii*
21. Male flowers not closely appressed, not appearing like rodent teeth 22

22. Endosperm ruminate 23
22. Endosperm homogeneous 25

23. Inflorescence consisting of a single spike. . . . . 8. *C. elegans*
23. Inflorescence with 2–4 spikes (rarely 1 spike) 24

24. Fruit globose or subglobose . . . . 3. *C. arfakianus*
24. Fruit ovoid or ellipsoid . . . . 22. *C. pauciflorus*

25. Fruit globose . . . . 4. *C. awa*
25. Fruit ellipsoid 26

26. Petiole present 8–12 cm long . . . . 9. *C. flabellatus*
26. Petiole absent or very short, not exceeding 3 cm in length 27

27. Leaves entire bifid, narrow, 11–25 cm wide, split to ca. 1/3 length . . . . 28. *C. yamutumene*
27. Leaves irregularly segmented, rarely bifid, if bifid the blade at least 30 cm wide split to 1/2 length . . . . 13. *C. julianettii*

1. *Calyptrocalyx albertisianus* Becc.

Synonyms: *Calyptrocalyx albertisianus* var. *minor* Burret, *Calyptrocalyx clemensiae* Burret, *Calyptrocalyx minor* Burret (invalidly published), *Ptychosperma normanbyi* Becc.

Single- or rarely multi-stemmed, *moderate to robust* palm to 2–15 m, crown with 6–18 leaves. **Stem** erect, to 4–15 cm diam., internodes 2–10 cm long. **Leaf** *regularly pinnate*, 1.5–4.5 m long, usually emerging reddish brown; sheath 40–110 cm long, brownish green with scattered brown scales, margins fibrous; petiole 20–73 × 2–3 cm, with scattered dark brown scales; leaflets 14–37 on each side of the rachis, held more or less horizontally, linear, 60–90 × 3–8 cm, plicate, upper surface dark green, lower surface paler, lower surface with ramenta along major veins. **Inflorescence** *robust*, 60–260 cm long, becoming pendulous, with 1–6 equal spikes; prophyll 30–40 cm long; peduncular bract 35–65 cm long; rachilla 32–110 cm long, 10–14 mm diam. with crowded deep floral pits. **Male flower** with 22–60 stamens. **Fruit** ellipsoid to globose, 15–42 × 10–26 mm, red; stigmatic remains apical on a short beak. **Seed** 10–30 × 8–18 mm diam., globose to ellipsoid; *endosperm ruminate*.

DISTRIBUTION. Widespread throughout New Guinea, including New Britain and the Louisiade Archipelago.

HABITAT. In rainforest at sea level to 1,800 m.

LOCAL NAMES. *Kerekere* (Milne Bay), *Korakh* (Daga), *Kuwei sinii* (Baiamo Sani), *Panjawing* (Ndu), *Sasep* (Wandama), *Wokoton hokilibe* (Wasisi).

Calyptrocalyx albertisianus. LEFT TO RIGHT: habit, Wau (WB); inflorescence, Eastern Highlands (MC).

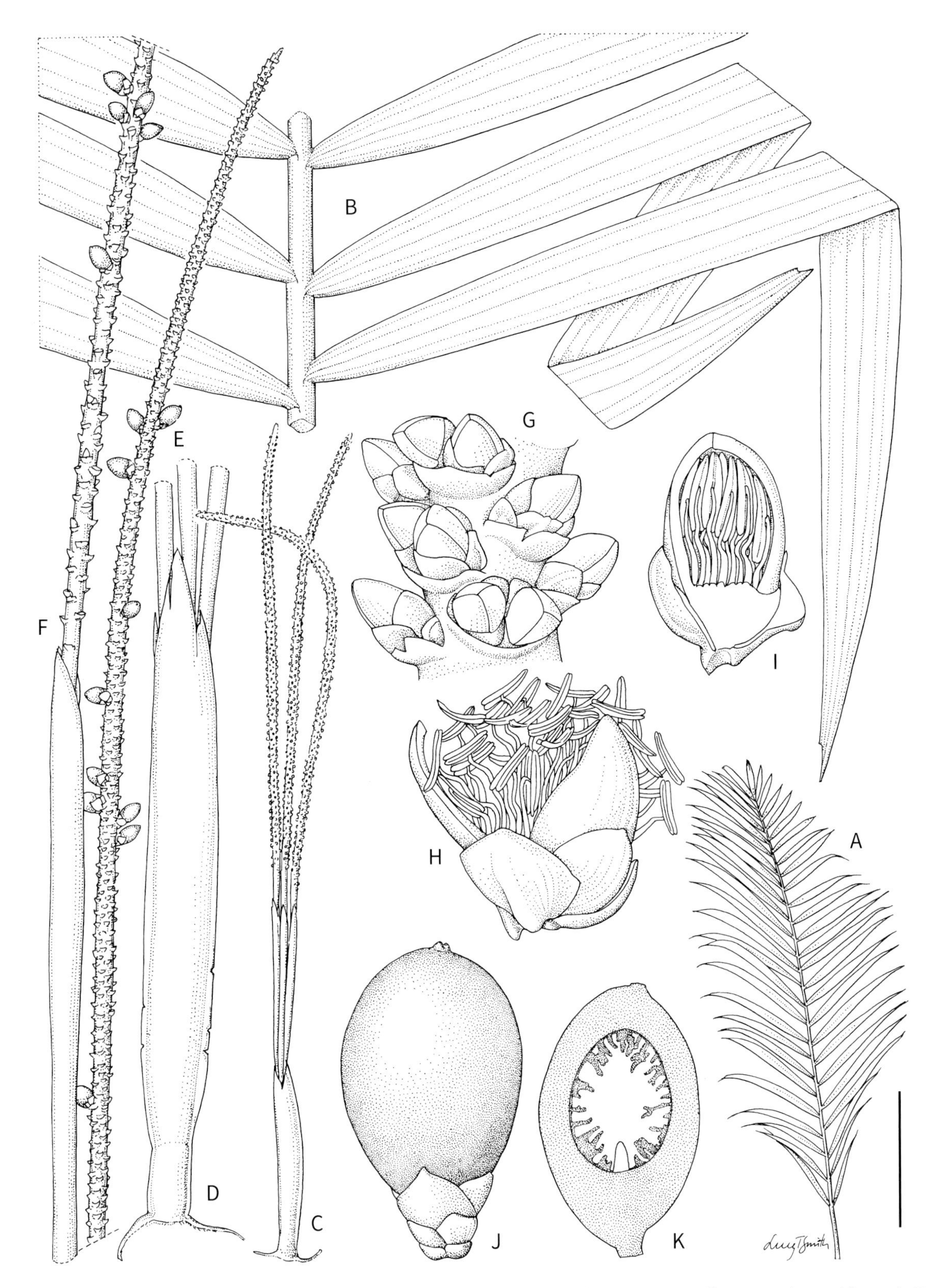

Calyptrocalyx albertisianus. **A.** Whole leaf diagram. **B.** Leaf mid-portion. **C.** Inflorescence. **D.** Inflorescence base with prophyll. **E, F.** Inflorescence spike. **G.** Portion of rachillla with flowers. **H, I.** Male flower whole and in longitudinal section. **J, K.** Fruit whole and in longitudinal section. Scale bar: A = 1 m; B, E, F = 8 cm; C = 25 cm; D = 12 cm; G = 1.5 cm; H, I = 7 mm; J, K = 2 cm. A, C, D, H from *Baker et al. 1101*; B, E–G, I–K from *Baker et al. 1109*. Drawn by Lucy T. Smith.

Calyptrocalyx albertisianus. LEFT TO RIGHT: leaf sheaths and base of Inflorescence; male flowers; fruit. Kikori River (WB).

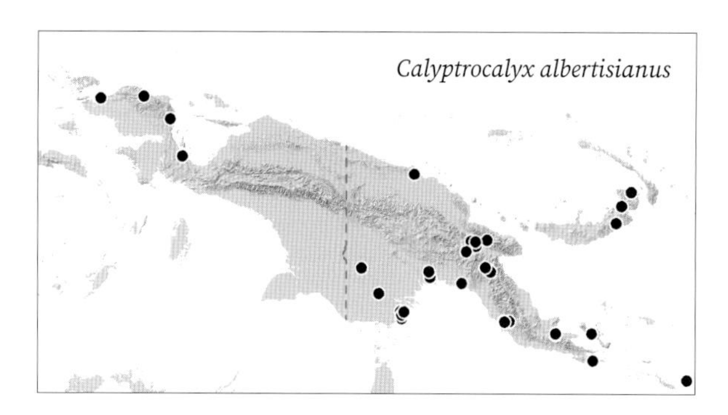

USES. Seed used as a betel nut substitute. Mature fruits eaten by cassowaries.

CONSERVATION STATUS. Least Concern (IUCN 2021).

NOTES. *Calyptrocalyx albertisianus* is widespread in New Guinea where it is the largest species of the genus. It is closely similar to *C. spicatus*, known only from Gag Island in our area, but differs in the smaller stature, smaller lip of the floral pits and the smaller fruit. One collection (*Davis 745*) from Kebar Valley, Manokwari, seems to be a particularly slender form of this widespread species.

2. *Calyptrocalyx amoenus* Dowe & M.D.Ferrero

Single- or multi-stemmed, *small to moderate* palm to 2.6–5 m, crown with 8–12 leaves. **Stem** erect, to 2–5 cm diam., internodes 3–6 cm long. **Leaf** *regularly pinnate*, 1.4–1.9 m long, emerging crimson-red; sheath 30–40 cm long, light green, glabrous, margins fibrous; petiole 8–10 × 0.3–0.5 cm, glabrous; leaflets 9–24 on each side of the rachis, held more or less horizontally, 2–22 cm or more apart, broadly lanceolate, hooded, 13–44 × 1.2–12 cm, acuminate in a long drip tip 4–10 × 0.2 cm, upper surface glossy dark green, lower surface dull paler green. **Inflorescence** *slender rigid*, 70–115 cm long, becoming pendulous, with 2 equal spikes or single spiked; prophyll not seen; peduncular bract not seen; rachilla 30–35 cm long, 3 mm diam. with congested floral pits. **Male flower** with 6 stamens. **Fruit** ellipsoid to globose, 9–13 × 6–10 mm, red; stigmatic remains apical. **Seed** 7–10 × 4–6mm diam., ellipsoid; *endosperm homogeneous*.

DISTRIBUTION. Eastern central New Guinea.

HABITAT. In rainforest, 120–700 m.

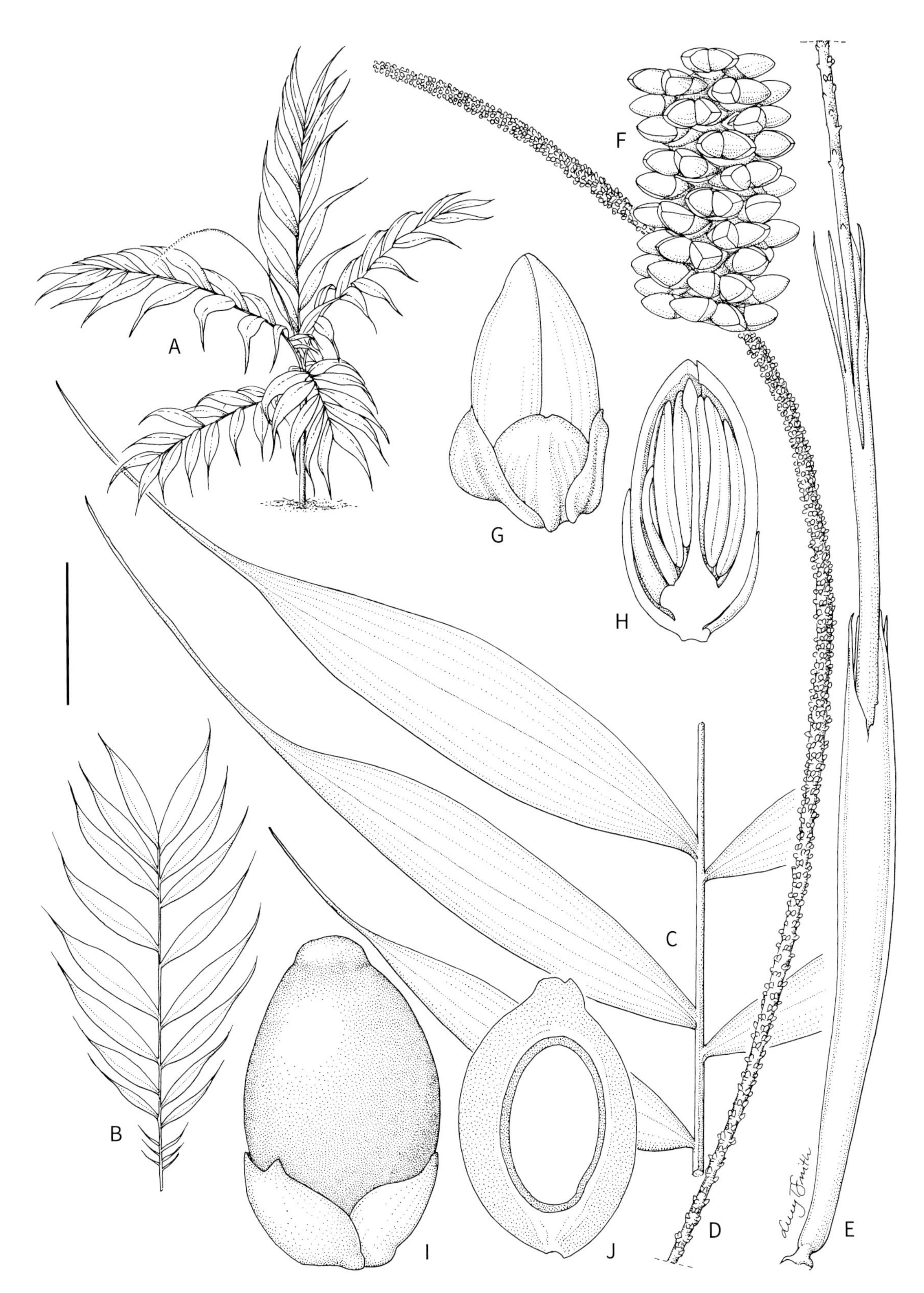

Calyptrocalyx amoenus. **A.** Habit. **B.** Whole leaf diagram. **C.** Leaf mid-portion. **D**, **E.** Inflorescence. **F.** Portion of rachilla with male flowers. **G**, **H.** Male flower whole and in longitudinal section. **I**, **J.** Fruit whole and in longitudinal section. Scale bar: A = 1 m; B = 24 cm; C = 6 cm; D, E = 4 cm; F = 4 mm; G, H = 1.5 mm; I, J = 3.3 mm. A–E from *Baker et al. 618*; F–H from *Baker et al. 619*; I, J from *Baker et al. 620*. Drawn by Lucy T. Smith.

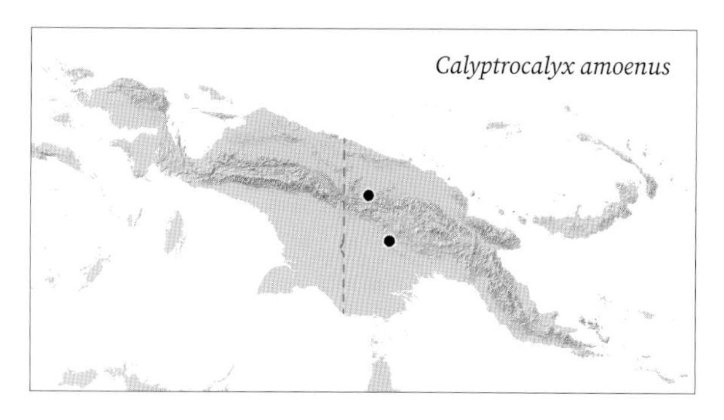

LOCAL NAMES. *Yurrimak* (Sandaun).

USES. Stem used to make spears, spearheads and practice bows and cross-beams in houses.

CONSERVATION STATUS. Critically Endangered. *Calyptrocalyx amoenus* is know from only two sites that are threatened by deforestation.

NOTES. This species is distinctive in its crimson-flushed emerging leaves, broad hooded leaflets with long drip tips and homogeneous endosperm. It can be distinguished from the similar *C. pachystachys* by the slender inflorescence. Specimens from Mount Bosavi (*Baker et al. 618, 619, 620*) match *C. amoenus* in all respects, except for the the inflorescences being single rather than paired at each node. It is included here as *C. amoenus*.

Calyptrocalyx amoenus. Habit, Mount Bosavi (WB).

3. *Calyptrocalyx arfakianus* (Becc.) Dowe & M.D. Ferrero

Synonyms: *Bacularia arfakiana* (Becc.) Burret, *Linospadix arfakianus* Becc., *Linospadix pachystachys* Burret, *Paralinospadix arfakianus* (Becc.) Burret, *Paralinospadix pachystachys* (Burret) Burret.

Single-stemmed, *small* palm to 1–2 m, crown with 6–10 leaves. **Stem** erect, to 10–14 mm diam., internodes to 3 cm long. **Leaf** *irregularly pinnate*, 75–100 cm long; sheath 15–18 cm long, green striate, margins lacerate; petiole 15–25 × 2–3 cm; leaflets *irregularly arranged, 2 on each side of the rachis, or 1–4 on one side and up to 12 on the other side*, basal leaflets ca. 12 mm wide, distal pair united to form a deeply bifid apex, main veins prominent upper surface, not so on lower surface. **Inflorescence** to 75 cm long, with *2–4 equal spikes*, pendulous; prophyll 12 cm long; peduncular bract to 20 cm long; rachilla 18–20 cm long, 3–5 mm diam. with shallow pits. **Male flower** greenish to white, with 6 or 7 stamens. **Fruit** *globose*, 10–12 mm diam., orange; stigmatic remains apical on a short beak. **Seed** 7–10 mm diam., globose; *endosperm ruminate*.

DISTRIBUTION. Bird's Head Peninsula and Waigeo.

HABITAT. In rainforest at 250 to 1,500 m.

LOCAL NAMES. *Mbep* (Anggi, Arfak Mountains).

USES. None recorded.

CONSERVATION STATUS. Critically Endangered. *Calyptrocalyx arfakianus* is known from only two sites. Deforestation due to oil palm plantations and logging concessions is a major threat in its distribution range.

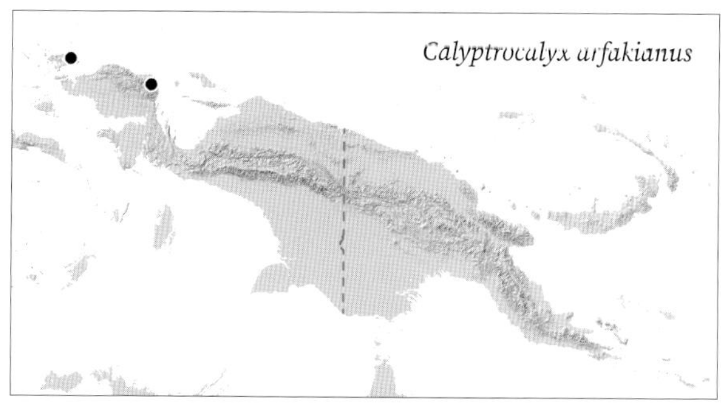

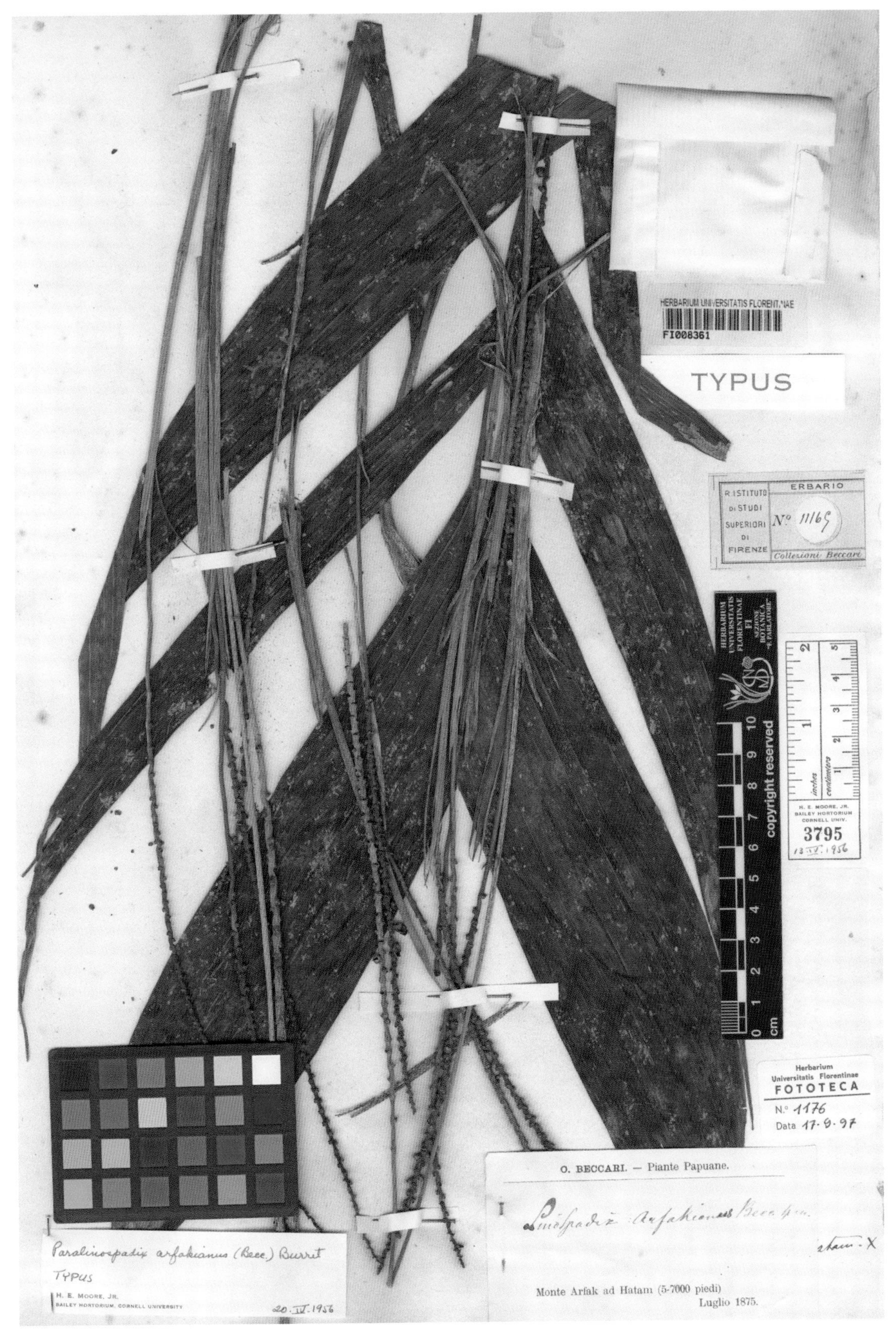

Calyptrocalyx arfakianus. Holotype specimen of *C. arfakianus* (*Beccari s.n.*) at the herbarium of the Museum of Natural History, University of Florence (FI). The holotype gathering comprises three herbarium sheets.

NOTES. Confined to Indonesian New Guinea, *C. arfakianus* is distinguishable by its small size, multiple inflorescences and irregularly pinnate leaves. It is known from very few collections. It could be confused with *Linospadix* but the lack of a scar at the base of the rachilla portion of the inflorescence will immediately distinguish it as a species of *Calyptrocalyx*, and *Linospadix* lacks multiple spikes at each node.

4. *Calyptrocalyx awa* Dowe & M.D.Ferrero

Multi-stemmed, *small* palm to 2–3 m, crown with 8–11 leaves. **Stem** erect, to 20–30 mm diam., internodes to 50 mm. **Leaf** *entire-bifid or irregularly pinnate* (rarely with one or two divisions each side), 90–130 × 30–60 cm, emerging bronze-coloured; sheath 16–30 cm long, margins smooth; petiole 5–18 cm long; rachis 50–70 cm long; leaflets 2–6 on each side of the rachis, 80 × 15 cm. **Inflorescence** *with 2 equal spikes*, 40–70 cm long; prophyll 17 cm long; peduncular bract to 26 cm long; rachilla 30–50 cm long, 35–45 mm diam., floral pits shallow. **Male flower** with *red-tinged petals* and 6–7 stamens. **Fruit** globose, 10–12 mm diam., *orange*, stigmatic remains apical. **Seed** 7–10 mm diam., globose; *endosperm homogeneous*.

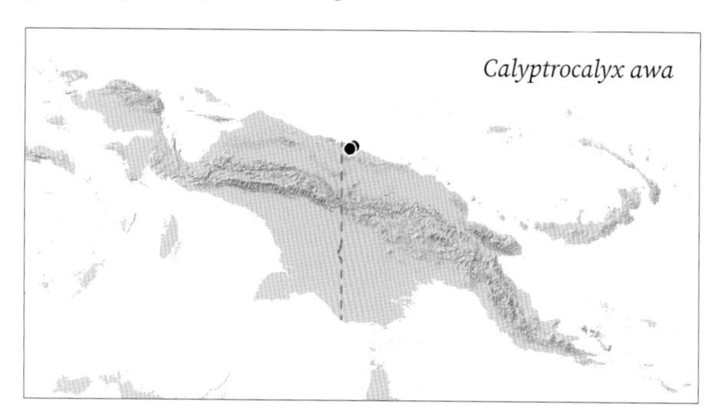

DISTRIBUTION. Central northern New Guinea.

HABITAT. In lowland rainforest at 10–60 m.

LOCAL NAMES. *Awa* (Osima).

USES. Stems used to make practice bows for small children.

CONSERVATION STATUS. Critically Endangered. *Calyptrocalyx awa* is known from only two adjacent sites that are threatened by logging concessions.

Calyptrocalyx awa. FROM TOP: habit; leaf. Cultivated Floribunda Palms and Exotics, Hawaii (WB). BOTTOM LEFT TO RIGHT: Inflorescence with male flowers, Bewani (JLD); fruit, cultivated Floribunda Palms and Exotics, Hawaii (WB).

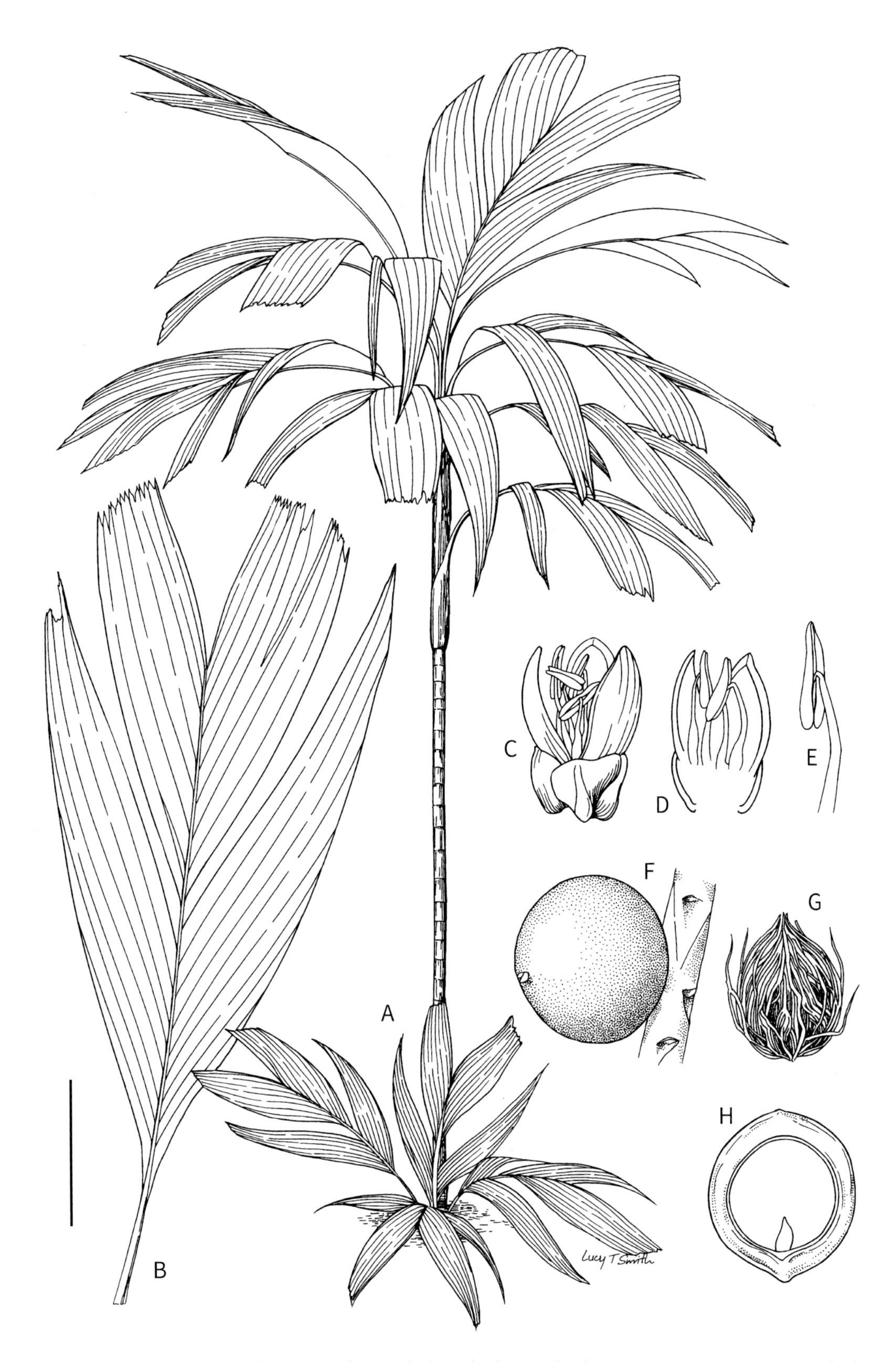

Calyptrocalyx awa. **A.** Habit. **B.** Leaf. C, **D.** Male flower whole and in longitudinal section. **E.** Stamen. **F.** Fruit attached to rachilla. **G.** Seed with attached mesocarp fibres. **H.** Fruit in longitudinal section. Scale bar: A = 30 cm; B = 15 cm; C, D = 1.7 mm; E = 0.7 mm; F–H = 10 mm. All from *Dowe & Ferrero 507*. Drawn by Lucy T. Smith.

NOTES. *Calyptrocalyx awa* is a distinctive multi-stemmed small palm with 2-spiked inflorescences bearing male flowers with red-tinged petals, and orange fruit.

5. *Calyptrocalyx calcicola* J.Dransf. & L.T.Sm.

Multi-stemmed, *small* palm to 1.2–1.5 m, crown with 6 leaves. **Stems** erect, to 1.6 cm diam., internodes 3–5 cm long. **Leaf** *regularly pinnate*, 80–85 cm long; sheath to 13 cm long, dull green, with scattered dark scales, margins fibrous; petiole 34–40 × 0.3 cm, with abundant dark scales; leaflets 5–7 on each side of the rachis, held more or less horizontally, lanceolate, 20–30 × 2–2.5 cm, including the long drip tips, both surfaces with scattered dark scales, apical leaflet pair multifold, to 45 × 5 cm, irregularly lobed. **Inflorescence** 50–70 cm long, becoming pendulous, single spiked; prophyll 15 × 1–1.5 cm with abundant pale scales; peduncular bract 30 × 1.5 cm with abundant dark scales; rachilla to 10 cm long, 5 mm diam. with congested floral pits and *abundant mid brown scales.* **Male flower** with *12 stamens, the filaments joined to form a conspicuous fleshy tube, the anthers erect at first then spreading.* **Fruit** (immature) ellipsoid, 12–15 × 5 mm, dull green, covered with abundant scales; stigmatic remains apical. **Seed** 10–11 × 5 mm, ellipsoid; *endosperm homogeneous.*

DISTRIBUTION. Southern Highlands, Papua New Guinea.

HABITAT. Rainforest on karst limestone, 860–900 m.

LOCAL NAMES. None recorded.

USES. None recorded.

CONSERVATION STATUS. Critically Endangered. *Calyptrocalyx calcicola* is known from only three nearby sites. Deforestation due to logging concessions is a major threat in its distribution range.

NOTES. *Calyptrocalyx calcicola* is known from two collections from karst limestone in the Southern Highlands of Papua New Guinea. It is distinctive in its lanceolate leaflets with very long drip tips and especially in the conspicuously thick rachilla portion of the inflorescence. The inflorescence and flowers are densely covered with overlapping scales. The remarkable androecial tube is otherwise known in the genus only in *C. doxanthus*.

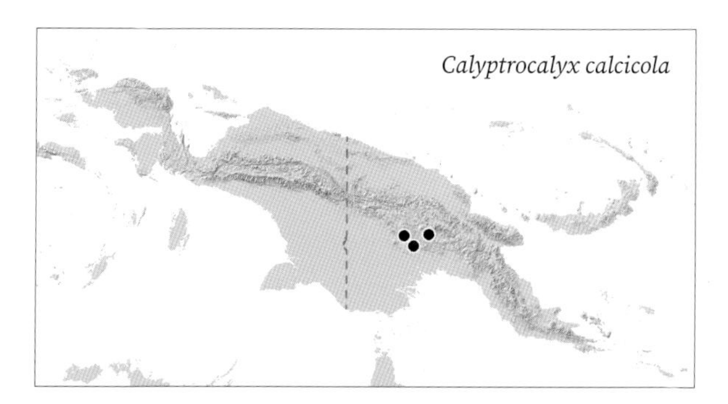

Calyptrocalyx calcicola. Inflorescence, Lake Kutubu (WB).

Calyptrocalyx calcicola. **A.** Leaf and leaf sheath. **B.** Inflorescence. **C.** Inflorescence with fruit. **D.** Portion of rachilla with male flower buds. **E, F.** Male flower in bud and open. **G, H.** Staminodial tube removed and opened up. Scale bar: A = 6 cm; B, C = 4 cm; D = 1 cm; E, F = 4 mm; G, H = 3.3 mm. A, B, D–H from *Baker et al. 668*; C from *Baker et al. 660*. Drawn by Lucy T. Smith.

6. *Calyptrocalyx caudiculatus* (Becc.) Dowe & M.D.Ferrero

Synonyms: *Linospadix caudiculatus* Becc., *Paralinospadix caudiculatus* (Becc.) Burret

Multi-stemmed, *small* palm to 3 m, crown with 9–11 leaves. **Stems** 2–10, erect, to 2 cm diam., internodes 2.5–4 cm long. **Leaf** *regularly pinnate*, 55–70 cm long, leaf emerging crimson-red; sheath to 17 cm long, light green, with scattered brown scales, margins fibrous; petiole very short to 35 × 0.7 cm, with scattered brown scales; leaflets 8–9 on each side of the rachis, rather distant, *sigmoid to lanceolate, hooded*, 27–28 × 4.5–5 cm, acuminate in a drip tip 7–8 × 0.1 cm, upper surface dark green, blotched with dark veins, lower surface pale green. **Inflorescence** *erect*, 90–120 cm long, with *2–8 equal spikes*; prophyll to 22 × 0.4 cm; peduncular bract to 64 × 0.4 cm; rachilla 40–50 cm long, 4 mm diam. with shallow floral pits. **Male flower** with 6–8 stamens. **Fruit** ellipsoid to globose, 10–12 × 5–6 mm, purple-black; stigmatic remains apical. **Seed** ovoid, 8–10 × 6–8 mm, ellipsoid; endosperm homogeneous.

DISTRIBUTION. Western and north-western New Guinea.

HABITAT. In rainforest at 120–760 m elevation.

LOCAL NAMES. *Saube* (Kaimana), *Serahmut* (Miyah).

USES. None recorded.

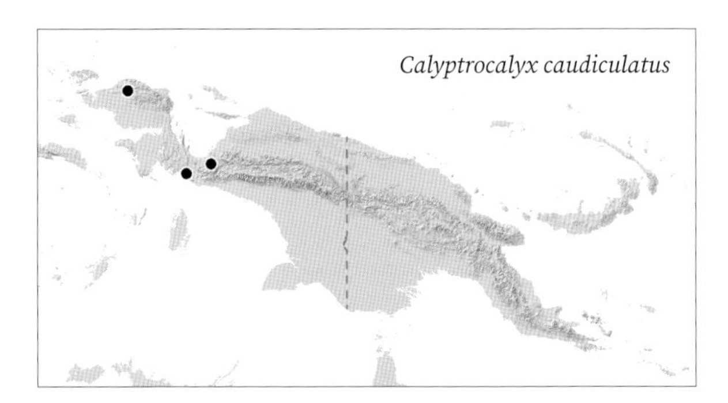

Calyptrocalyx caudiculatus. LEFT TO RIGHT: inflorescence; habit. Tamrau Mountains (WB).

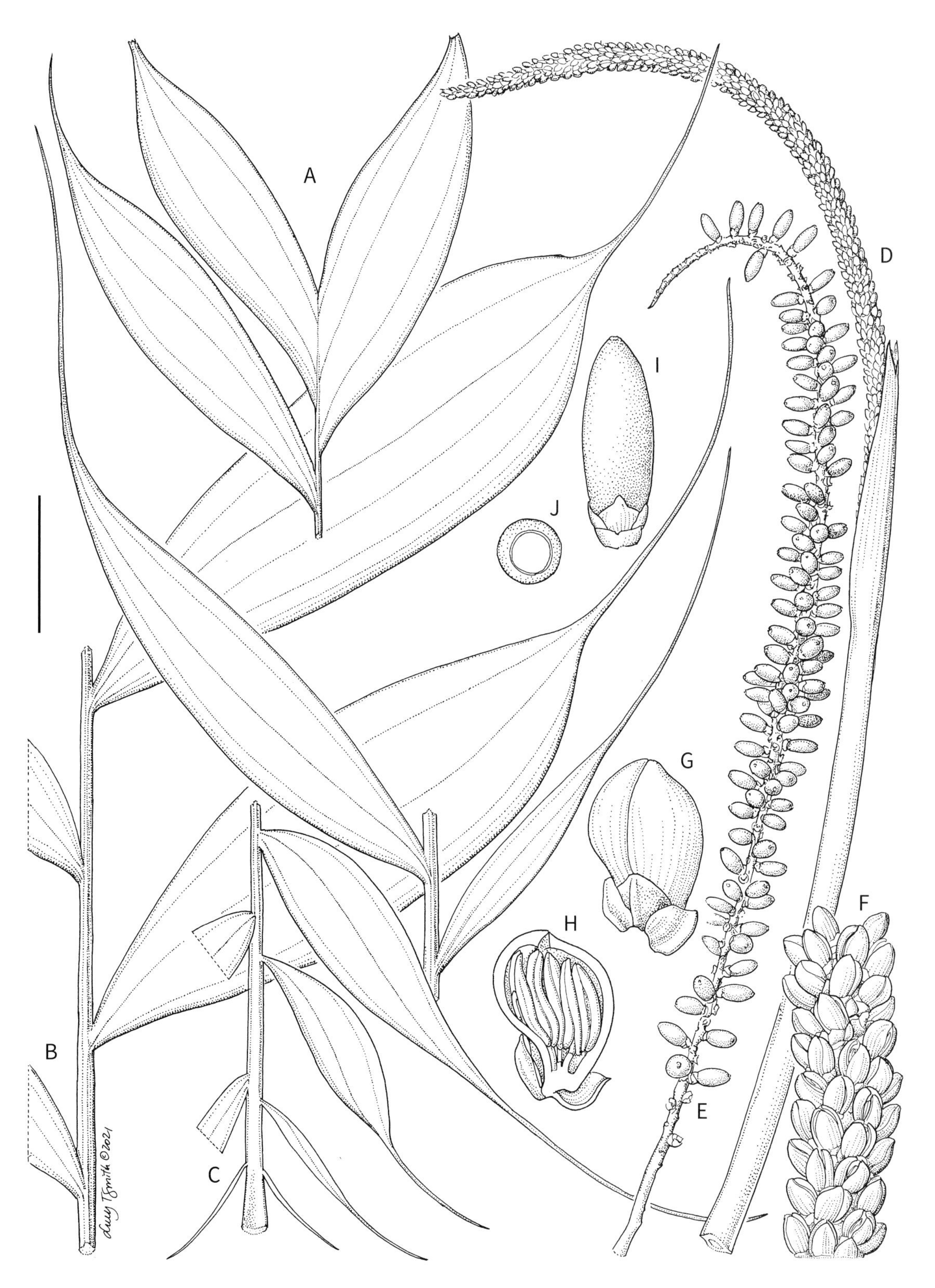

Calyptrocalyx caudiculatus. **A.** Leaf apex. **B.** Leaf mid-portion. **C.** Leaf base. **D.** Inflorescence with flower buds. **E.** Portion of inflorescence with fruit. **F.** Portion of rachilla with male flowers. **G**, **H.** Male flower whole and in longitudinal section. **I**, **J.** Fruit whole and in transverse section. Scale bar: A–C = 6 cm; D, E = 4 cm; F = 1 cm; G, H = 3.3 mm; I, J = 1 cm. All from *Mehen 4*. Drawn by Lucy T. Smith.

CONSERVATION STATUS. Near Threatened. *Calyptrocalyx caudiculatus* is known from three sites, but deforestation is a major threat in its distribution range.

NOTES. *Calyptrocalyx caudiculatus* is distinguished by its multi-stemmed habit and broadly lanceolate, hooded leaflets, the young leaf emerging crimson red.

7. *Calyptrocalyx doxanthus* Dowe & M.D.Ferrero

Single- or multi-stemmed, small to moderate palm to 1.5–2 m, crown with 7 leaves. **Stems** 1–4, erect, to 2 cm diam., internodes 2.5–3 cm long. **Leaf** *regularly pinnate*, 77–120 cm long, emerging crimson-red; sheath 16–17 cm long, light green, glabrous, margins fibrous; petiole 28–30 × 1 cm, glabrous; leaflets 14–17 on each side of the rachis, held more or less horizontally, *sigmoid to broadly lanceolate, hooded*, 11–25 × 3–10 cm, acuminate in a drip tip 5–10 × 0.2 cm, upper surface matt light to yellow green, *blotched with dark veins*, lower surface greyish white to pale green. **Inflorescence** *slender rigid*, 54–55 cm long, becoming pendulous, single spiked; prophyll 14–17 × 2.5 cm; peduncular bract 32–37 × 1.2–1.3 cm; rachilla 13–15 cm long, 4–10 mm diam. with congested floral pits. **Male flower** with 8–10 bright *lavender* stamens, the *filaments joined to form a conspicuous fleshy cup, the anthers purple, pendulous inside*. **Fruit** ellipsoid, 9–11–15 × 9–12 mm, red; stigmatic remains apical. **Seed** 10–11 × 7–10 mm, ellipsoid; *endosperm homogeneous*.

DISTRIBUTION. Vicinity of Jayapura, Indonesian New Guinea (Cyclops Mountains, Tami River).

HABITAT. On slopes above creeks in seasonally dry forest at 100–700 m.

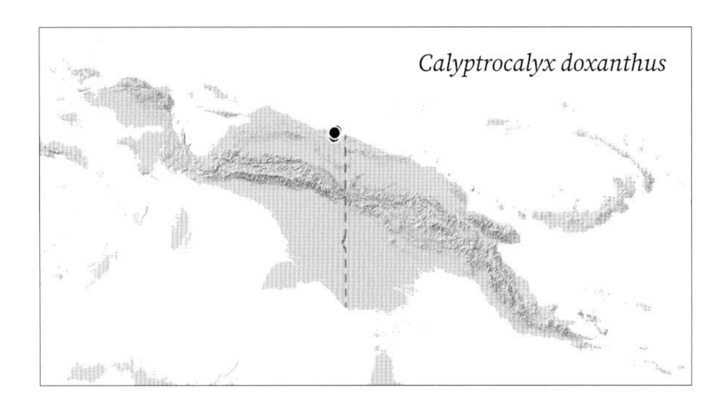

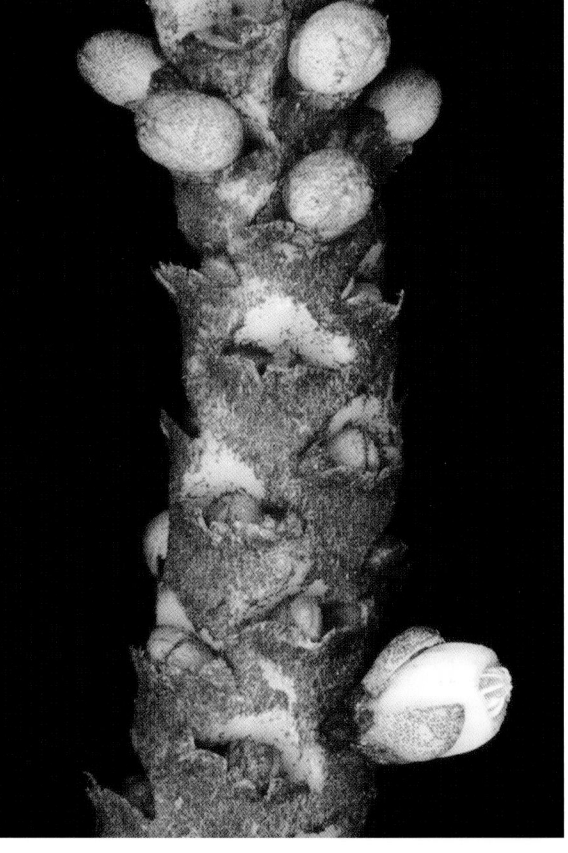

Calyptrocalyx doxanthus. FROM TOP: habit, cultivated Bogor, Indonesia (JD); male flowers, open flower showing androecial cup, cultivated Bogor, Indonesia (WB).

Calyptrocalyx doxanthus. **A.** Habit. **B.** Leaf apex. **C.** Leaf mid-portion. **D.** Leaf base. **E.** Inflorescence with fruit. **F**, **G.** Male flower whole and in longitudinal section. **H.** Female flower. **I.** Fruit. **J**, **K.** Seed (with attached mesocarp fibres)whole and in longitudinal section. Scale bar: A = 50 cm; B–D = 6 cm; E = 4 cm; F, H = 5 mm; G = 4.5 mm; I, J, K = 13 mm. A–D, F–K = *Hambali s.n.*; E from *Desianto 2*. Drawn by Lucy T. Smith.

LOCAL NAMES. *Demah kupei* (Cyclops), *Yanyasa Nyi* (North Cyclops), *Yet* (Marap).

USES. Stem used to make spears, spearheads and practice bows and cross-beams in houses.

CONSERVATION STATUS. Data Deficient. This species is known from only one area, where there is limited information on ongoing threats. There is insufficient evidence to select a single extinction risk category.

NOTES. *Calyptrocalyx doxanthus* is a most unusual species. It has broad lanceolate blotched hooded leaflets, the young leaf emerging crimson red, but most distinctive is the extraordinary androecium; the stamens have filaments united to form a mauve-coloured fleshy cup, with the purple anthers inserted on the inside. This stamen structure was thought to be unique to this species within the genus, but another taxon has been shown to display an androecial tube. *Calyptrocalyx calcicola* has a similar androecial tube but with erect rather than pendulous anthers.

8. *Calyptrocalyx elegans* Becc.

Synonyms: *Calyptrocalyx bifurcatus* Becc., *Calyptrocalyx moszkowskianus* Becc., *Calyptrocalyx schultzianus* Becc.

Single- or multi-stemmed, *small to moderate* palm to 1–5 m, crown with 6–15 leaves. **Stems** erect or leaning, to 2.5–60 mm diam., internodes to 6 cm long. **Leaf** *entire-bifid, or regularly or irregularly pinnate*, 30–120 cm long; sheath to 35 cm long, green striate, margins lacerate-fibrous; petiole 12–30 × 0.2–0.5 cm; leaflets few to numerous, regularly or irregularly arranged and grouped, close or distant, 40–60 × 2–5 cm, if bifid, leaf blade to 30 cm long, 19–25 main veins prominent on upper surface, not so on lower surface. **Inflorescence** 50–90 cm long, *with 1 spike with abundant dark scales*; prophyll 12 cm long; peduncular bract to 30 cm long; rachilla 9–30 cm long, 5–10 mm diam. with congested pits. **Male flower** greenish to cream, with 7–30 stamens. **Fruit** ellipsoid, 12–14 × 10–12 mm, *red*; stigmatic remains apical. **Seed** ellipsoid 12 × 7 mm; *endosperm ruminate.*

DISTRIBUTION. Northern New Guinea.

HABITAT. In rainforest from sea level to 1,000 m.

LOCAL NAMES. *Boalak* (Kaka), *Kel keiyik* (Mai), *Kohili* (Madang), *Malu* (Bifrau), *Mara* (Ndu), *Sanumb* (Bewani), *Siterarum* (Olo, Sandaun).

USES. Stems used to make spearheads and shafts and

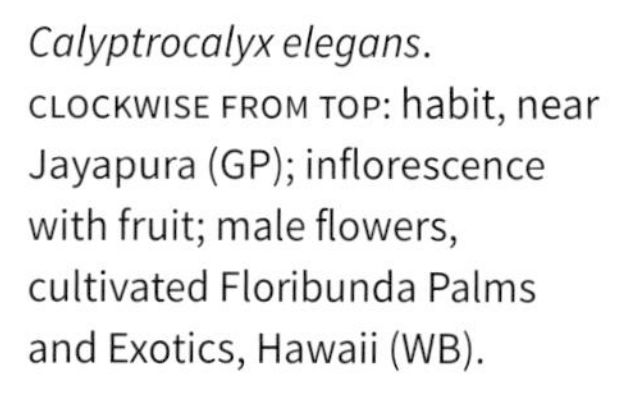

Calyptrocalyx elegans. CLOCKWISE FROM TOP: habit, near Jayapura (GP); inflorescence with fruit; male flowers, cultivated Floribunda Palms and Exotics, Hawaii (WB).

Calyptrocalyx elegans**. A.** Habit. **B, C.** Whole leaf diagram. **D.** Leaf apex. **E.** Inflorescence. **F, G.** Male flower whole and in longitudinal section. **H.** Female flower. **I, J.** Fruit whole and in longitudinal section. Scale bar: A = 1 m; B, C = 48 cm; D = 12 cm; E = 6 cm; F = 4 mm; G = 3.3 mm; H = 7 mm; I, J = 1.5 cm. A, C–E, H–J from *Baker & Utteridge 588*; B from *Heatubun et al. 289*; F, G from *Dowe 511*. Drawn by Lucy T. Smith.

leaves to wrap food. Palm heart occasionally eaten. Stems sometimes intentionally damaged to encourage beetle larvae for food.

CONSERVATION STATUS. Least Concern.

NOTES. *Calyptrocalyx elegans* is widespread in northern New Guinea. It is distinctive in its single- or multi-stemmed habit, variously dissected leaves and solitary-spiked inflorescence with ellipsoid fruit and ruminate endosperm.

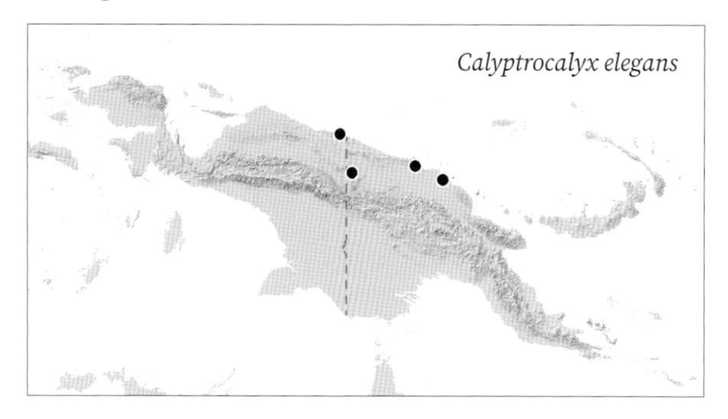

9. *Calyptrocalyx flabellatus* (Becc.) Dowe & M.D.Ferrero

Synonyms: *Bacularia flabellata* (Becc.) F.Muell., *Linospadix flabellatus* Becc., *Paralinospadix flabellatus* (Becc.) Burret

Single-stemmed, small palm to 1–5 m, crown with 6–9 leaves. **Stems** erect, to 7 cm diam., internodes 2–3 cm long, green. **Leaf** *bifid, irregularly or regularly pinnate,* 35–150 cm long; sheath to 7–8 cm long, margins fibrous; petiole 8–12 cm long, 10 mm wide at the base; leaflets 3–5 on each side of the rachis, *mostly very broad,* 30–50 × 6–11 cm, upper surface dark green, lower surface paler, main vein prominent. **Inflorescence** with *2–4 equal spikes,* 35–70 cm long; peduncle 18–50 cm long, 1.5–2 mm diam., glabrescent; prophyll to 7–12 cm; peduncular bracts to 14–20 cm long; rachilla 17–20 cm long, 2.3 mm wide, slightly spindle-shaped, *wider than the peduncle,* floral pits not densely crowded, drying with prominent lips. **Male flower** with 9 stamens. **Fruit** ellipsoid, 11–12 × 7–8 mm, *orange to red,* stigmatic remains apical. **Seed** ellipsoid, 8–10 × 5–6 mm; *endosperm homogeneous.*

DISTRIBUTION. Bird's Head Peninsula, Indonesian New Guinea, with one outlier near Jayapura (Tami River).

HABITAT. In rainforest from 65 to 2,000 m.

LOCAL NAMES. *Kiligata* (Amoi), *Meahrah* (Wariori River), *Owe* (Maibrat), *Sunggem* (Wondama), *Wube* (Marap).

Calyptrocalyx flabellatus. Habit, Arfak Plains (JD).

USES. Stem used for spears.

CONSERVATION STATUS. Least Concern.

NOTES. This species is distinctive in its crown of crowded, relatively short, wide leaves, bifid or irregularly divided, many spiked inflorescences and orange to red fruit.

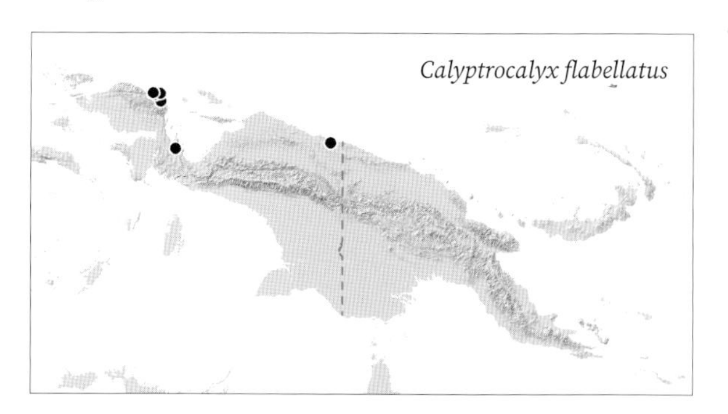

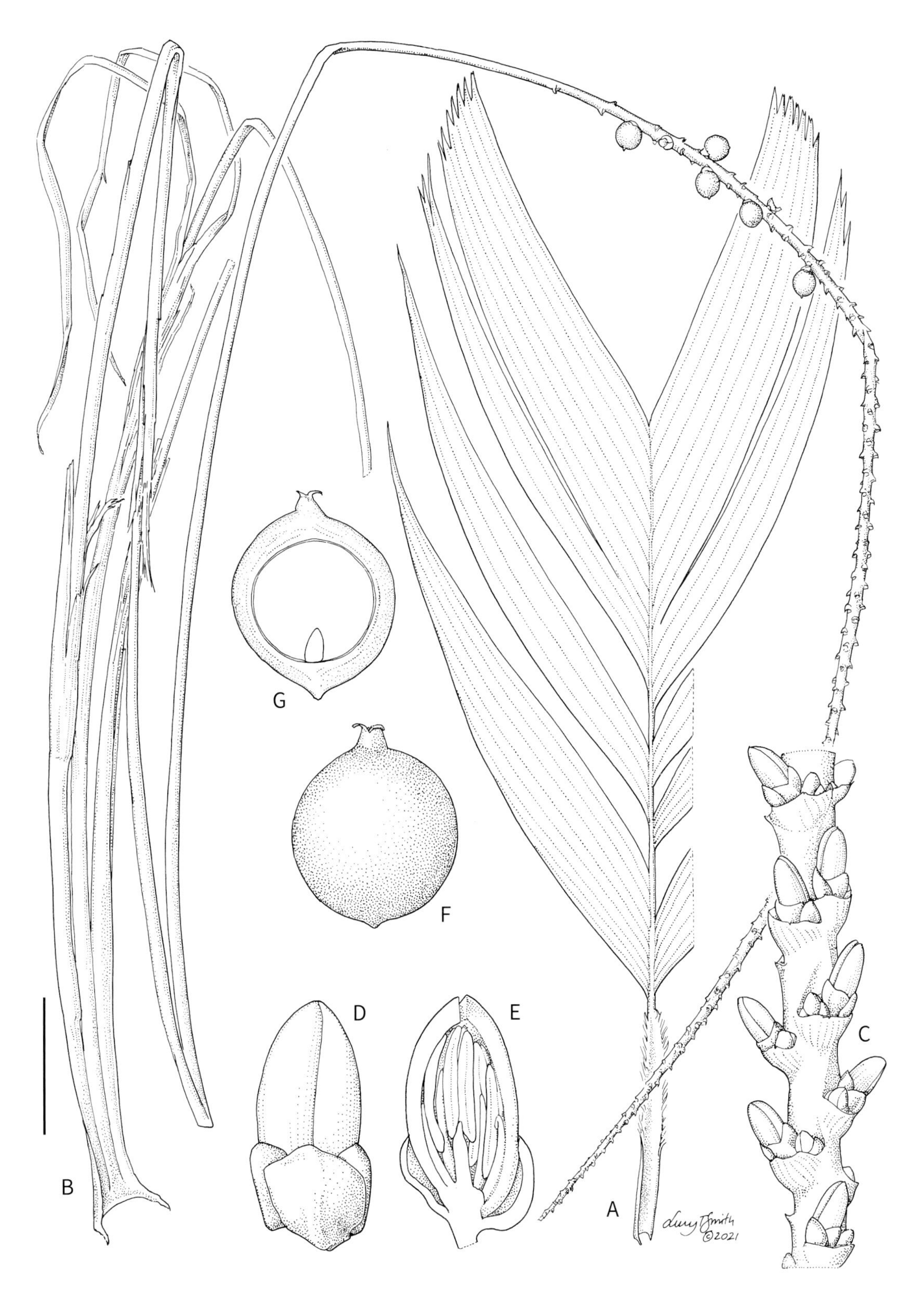

Calyptrocalyx flabellatus. **A.** Leaf. **B.** Inflorescence with fruit. **C.** Portion of rachilla with male flowers. **D, E.** Male flower whole and in longitudinal section. **F, G.** Fruit whole and in longitudinal section. Scale bar: A = 15 cm; B = 4 cm; C, F, G = 7 mm; D, E = 2.5 mm. All from *Dransfield et al JD 7597*. Drawn by Lucy T. Smith.

10. *Calyptrocalyx forbesii* (Ridl.) Dowe & M.D.Ferrero

Synonyms: *Linospadix forbesii* Ridl., *Linospadix petrickianus* Sander, *Paralinospadix forbesii* (Ridl.) Burret, *Paralinospadix petrickianus* (Sander) Burret, *Paralinospadix stenoschistus* Burret

Single- or multi-stemmed, *small to moderate* palm to 5 m, crown with 8–10 leaves. **Stems** 1–10, erect or leaning, 1.3 to 9 cm diam., internodes 1–12 cm long, dark green. **Leaf** *regularly pinnate,* 50–200 cm long, *emerging purple-brown;* sheath to 14–18 cm long, green, margins fibrous; petiole 2–10 cm long, 4–10 mm diam. with abundant dark scales; leaflets 10–30 on each side of the rachis, upper surface dark glossy green, lower surface paler, *narrow linear to slightly sickle-shaped,* acuminate, 15–40 cm × 0.4–0.6 cm. **Inflorescence** with 1–4 equal spikes, 30–130 cm long; peduncle 10–65 cm long, 3–4 mm diam.; prophyll to 10–23 cm long; peduncular bract to 20–42 cm long, papery with abundant dark scales; rachilla 20–54 cm long, 4.5 mm wide, slightly spindle-shaped, with abundant velvety tomentum, floral pits not congested. **Male flower** with 6–12 stamens. **Fruit** *ellipsoid,* 7–20 × 4–12 mm, *orange to red,* stigmatic remains more or less apical, epicarp smooth, coarsely granular when dry. **Seed** ellipsoid, 15 × 8 mm.; *endosperm homogeneous.*

DISTRIBUTION. South-eastern New Guinea and the D'Entrecasteaux Islands.

HABITAT. In rainforest from sea level to 1,200 m.

LOCAL NAMES. None recorded.

USES. None recorded.

CONSERVATION STATUS. Least Concern (IUCN 2021).

NOTES. *Calyptrocalyx forbesii* is a seemingly widespread palm in eastern New Guinea, distinctive in its regularly pinnate leaves with conspicuously narrow leaflets, the usually many spiked inflorescence, and orange to red ellipsoid fruit.

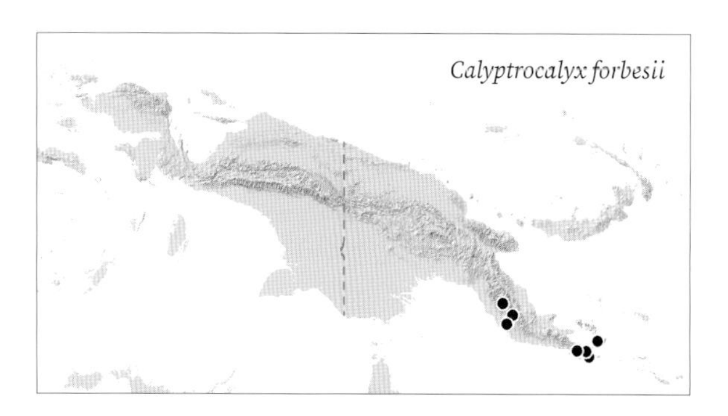

Calyptrocalyx forbesii. FROM TOP: habit; leaves. Cultivated Lae (WB).

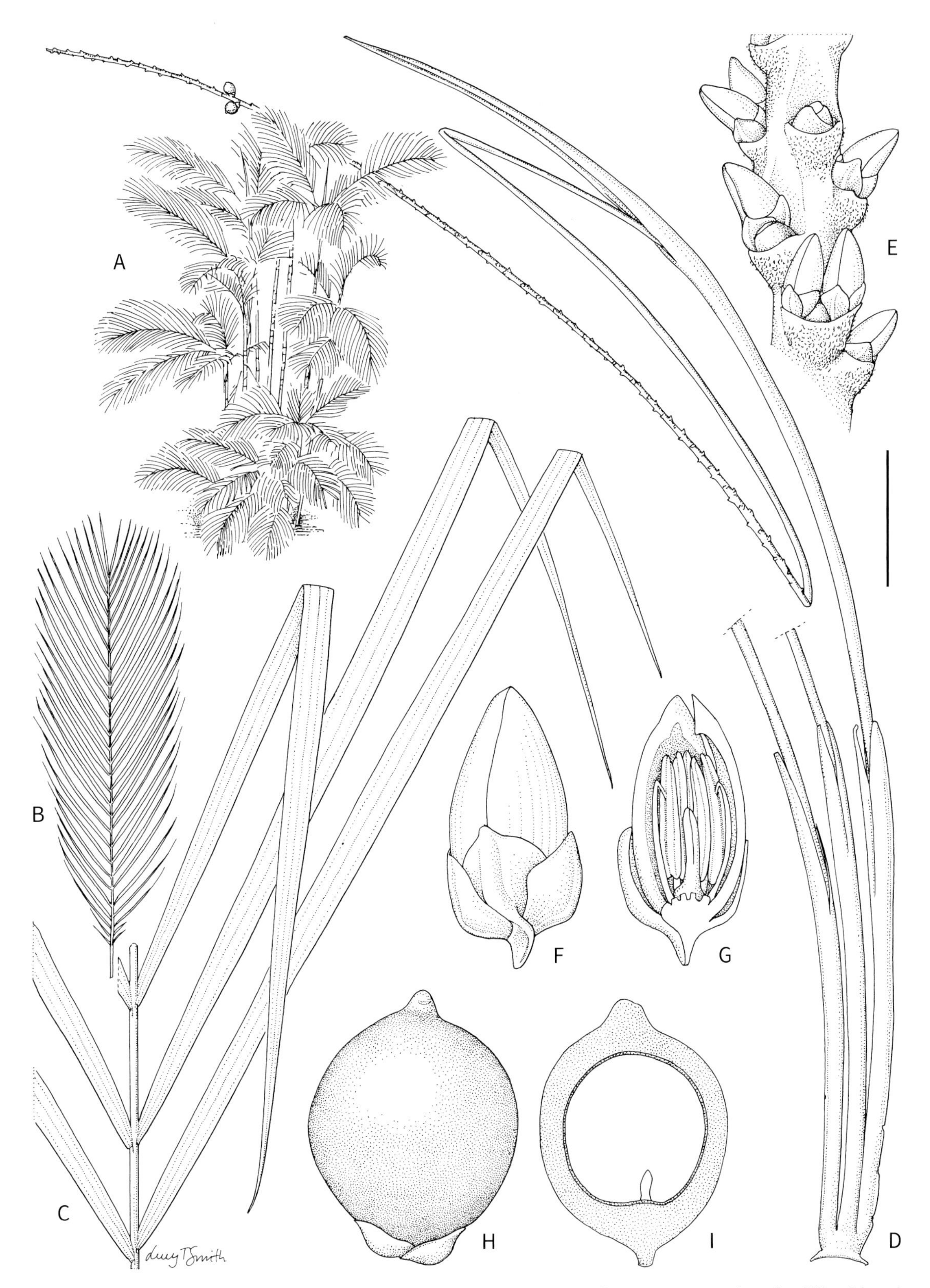

***Calyptrocalyx forbesii*. A.** Habit. **B.** Whole leaf diagram. **C.** Leaf mid-portion. **D.** Inflorescence. **E.** Portion of rachilla with male flowers. **F, G.** Male flower whole and in longitudinal section. **H, I.** Fruit whole and in longitudinal section. Scale bar: A = 1.6 m; B = 30 cm; C, D = 4 cm; E = 3.3 mm; F, G = 1.6 mm; H, I = 4 mm. A from photograph; B–D from *Baker et al. 1179*; E–H from *Gideon 77079*. Drawn by Lucy T. Smith.

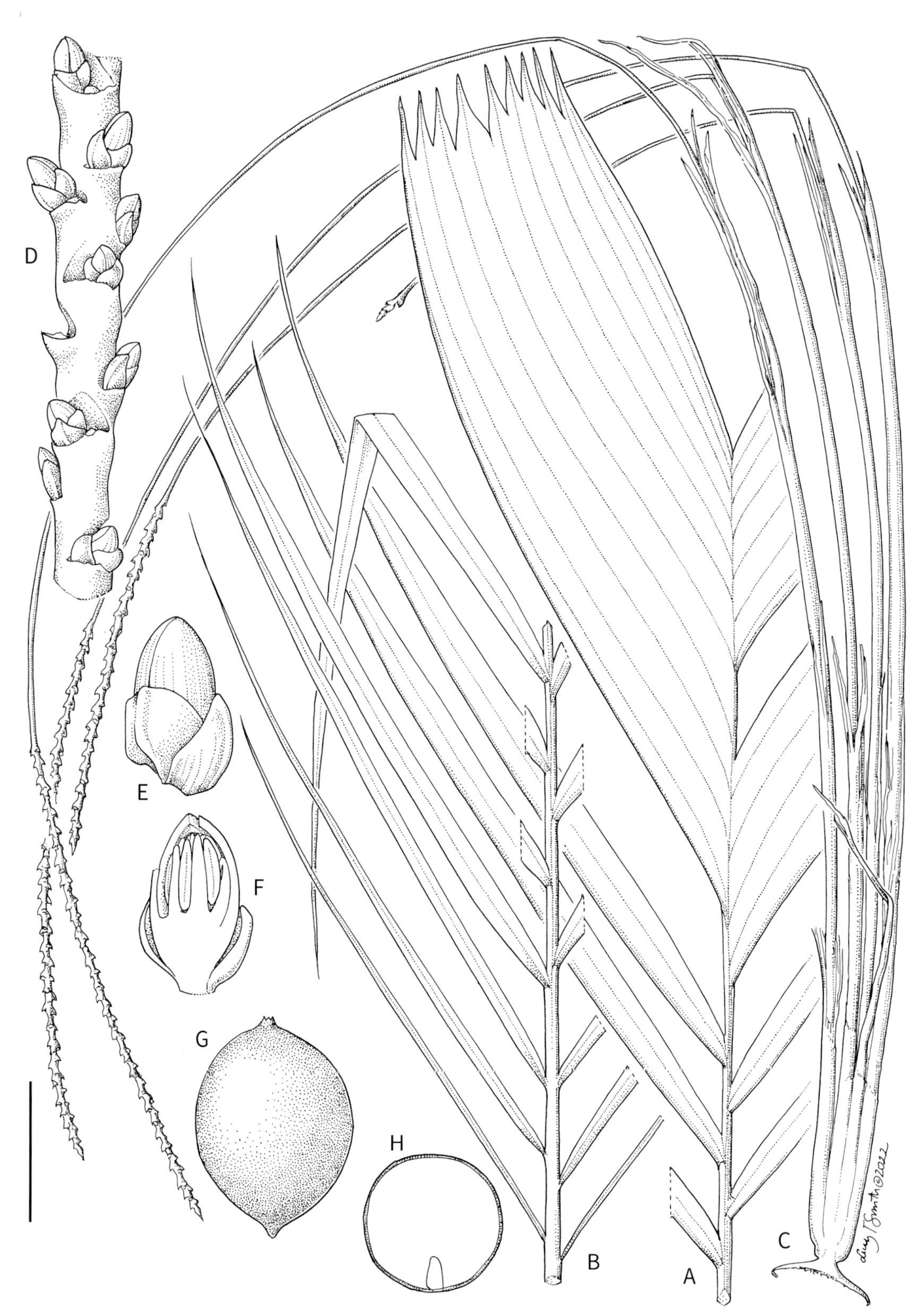

Calyptrocalyx geonomiformis. **A.** Leaf apex. **B.** Leaf mid-portion. **C.** Inflorescence. **D.** Portion of rachilla with male flowers. **E, F.** Male flower whole and in longitudinal section. **G.** Fruit. **H.** Seed in longitudinal section. Scale bar: A, B = 6 cm; C = 4 cm; D = 7 mm; E, F = 3 mm; G, H = 7 mm. A–C from *Baker et al. 834*; D–F from *Baker et al 833*; G, H from *Baker et al. 816*. Drawn by Lucy T. Smith.

11. *Calyptrocalyx geonomiformis* (Becc.) Dowe & M.D.Ferrero

Synonyms: *Linospadix geonomiformis* Becc., *Paralinospadix geonomiformis* (Becc.) Burret

Single- or multi-stemmed, *small* palm to 1 m, crown with 8 leaves. **Stems** erect, 1.2–1.4 cm diam., internodes 2–2.5 cm long, green. **Leaf** bifid or irregularly pinnate, 45–55 cm long, 28–30 cm wide; sheath to 27 cm long, with scattered orange-brown scales, margins fibrous; petiole 4–11 cm long; leaf when bifid with 16–18 upper ribs, *curved and converging on the leaf tip*, when pinnate, with 2–3 broad leaflets on each side of the rachis, or with numerous linear leaflets proximally and broad flabellate apical pair, leaflets 25–40 × 0.7–8 cm, sickle-shaped, acuminate, upper surface dark green, lower surface paler. **Inflorescence** with 1–5 equal spikes, 40–60 cm long; peduncle 36–48 cm long, 1.5–2 mm diam., glabrescent; prophyll to 8 cm long; peduncular bract to 21–26 cm long; rachilla spindle-shaped, wider than the peduncle, 18–20 cm long, 0.5–0.6 cm wide, floral pits congested, densely reddish scaly. **Male flower** with 6 stamens. **Fruit** *orange*, more or less ellipsoid, 10 × 9 mm, stigmatic remains apical. **Seed** globose or oblate, 5 × 6 mm; *endosperm homogeneous.*

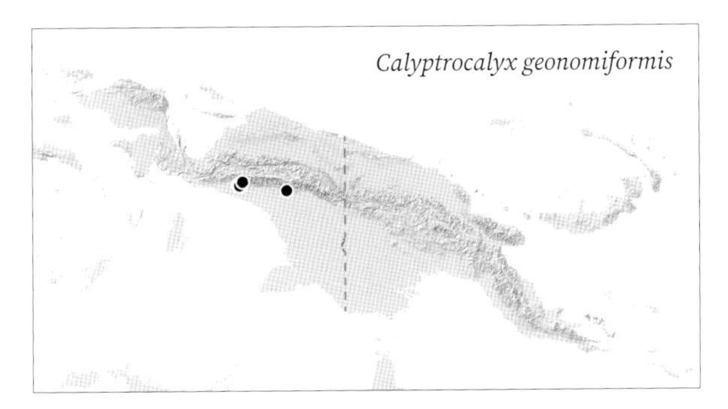

DISTRIBUTION. South-western New Guinea.

HABITAT. In rainforest from 250 to 540 m.

LOCAL NAMES. None recorded.

USES. None recorded.

CONSERVATION STATUS. Endangered. *Calyptrocalyx geonomiformis* has a restricted distribution. Deforestation due to mining concessions is a major threat in its distribution range.

NOTES. This small palm from Indonesian New Guinea is distinctive in the main ribs in the leaf, which are strongly curved and converge at the leaf tip, a character otherwise known only in *C. julianettii* from eastern New Guinea. The latter, however, lacks a petiole.

12. *Calyptrocalyx hollrungii* (Becc.) Dowe & M.D.Ferrero

Synonyms: *Linospadix hellwigianus* Warb. ex Becc., *Linospadix hollrungii* Becc., *Paralinospadix hollrungii* (Becc.) Burret, *Linospadix schlechteri* Becc., *Paralinospadix clemensiae* Burret, *Paralinospadix schlechteri* (Becc.) Burret

Single- or multi-stemmed, *small* palm to 3 m, crown with 6–11 leaves. **Stems** 1–20, erect or leaning, 1.2 to 2 cm diam., internodes 1.5–8 cm long, pale brown. **Leaf** bifid, irregularly or regularly pinnate, 45–180 × 16–29 cm, *emerging dark red*, crimson or purple; sheath to 11–22 cm long, green, margins fibrous; petiole 4–30 cm long, 4–7 mm diam. with abundant dark scales; leaf when bifid deeply so, with 12–13 upper ribs; when

Calyptrocalyx hollrungii. Habit, Jivewaneng (WB).

***Calyptrocalyx hollrungii*.** **A.** Habit. B–**D.** Whole leaf diagram. **E.** Leaf apex. **F**, **G.** Inflorescence. **H.** Portion of rachilla with male flowers. **I.** Male flower in longitudinal section. **J**, **K.** Fruit whole and in longitudinal section. Scale bar: A = 50 cm; B–D, F, G = 6 cm; E = 8 cm; H = 5 mm; I = 2 mm; J, K = 1 cm. A, D–G from *Zona & Smith 867*; B from *Ferrero s.n.*; C, J, K from *Baker & Utteridge 571*; H, I from *Baker et al. 1176*. Drawn by Lucy T. Smith.

Calyptrocalyx hollrungii. LEFT TO RIGHT: emerging leaf, Jivewaneng (WB); inflorescence with male flowers, near Finschhafen (AB); inflorescence with fruit, Jivewaneng (WB) .

irregularly pinnate, leaflets 3–14 on each side, with 3–7 upper ribs; when regularly pinnate, leaflets 13–16 on each side of the rachis, mid-leaf leaflets 22 cm long, 0.6 cm wide, lower surface of blade scaly. **Inflorescence** with 1–3 equal spikes, 68–120 cm long; peduncle 40–59 cm long, 2–3 mm diam.; prophyll to 18–24 cm × 0.6 cm, with *dense dark scales;* peduncular bract to 30–48 × 0.8 cm long, with abundant dark scales; rachilla 20–30 × 2–3 cm, more or less terete, slightly thicker than the peduncle, with abundant scales, floral pits congested or not. **Male flower** *elongate, closely pressed against each other in pairs, appearing like rodent teeth,* with 6–12 stamens. **Fruit** ovoid to ellipsoid, 12–16 × 6–9 mm, *orange to red,* stigmatic remains more or less apical, epicarp smooth, coarsely granular when dry. **Seed** globose to pyriform, 9–10 × 6–7 mm; endosperm homogeneous.

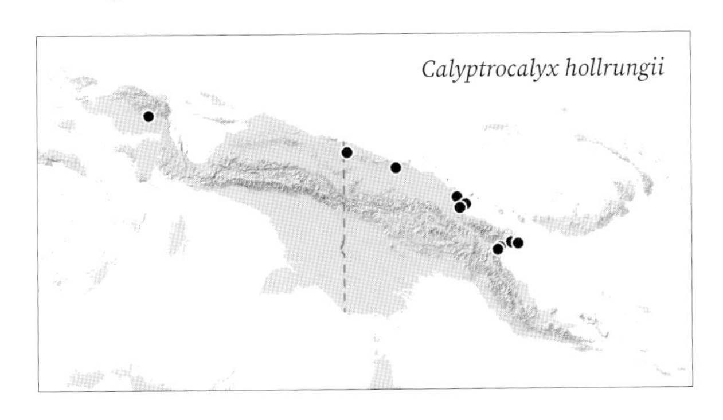

DISTRIBUTION. North-eastern New Guinea, with one outlier record reported from the Bird's Head Peninsula.

HABITAT. In rainforest 60 to 600 m.

LOCAL NAMES. *Hulameng* (Kotte), *Kising* (Kotte), Kumdu (*Samu Kundi*).

USES. Leaves used for wrapping food. Stem sometimes used for spear and arrow tips. Used in *haus tambaran* roof construction.

CONSERVATION STATUS. Least Concern.

NOTES. This widespread multi-stemmed relatively small palm in Papua New Guinea is unique in the nature of its male flowers, which are elongate and closely pressed against each other in pairs, appearing like rodent teeth.

13. *Calyptrocalyx julianettii* (Becc.) Dowe & M.D.Ferrero

Synonyms: *Linospadix julianettii* Becc., *Paralinospadix amischus* Burret, *Paralinospadix julianettii* (Becc.) Burret

Single- or multi-stemmed, *small* palm to 1–2 m, crown with ca. 10 leaves. **Stems** erect, 1.2–2.4 cm diam. **Leaf** irregularly pinnate, 60–90 × 25–30 cm; sheath to 14–18 cm long, with *dark brown scales,* margins fibrous; petiole absent or very short, to 3 cm long, 0.3–0.4 diam.; leaflets 5–12 on each side of the rachis, *sickle-shaped to sigmoid,* rigid, upper surface dark green,

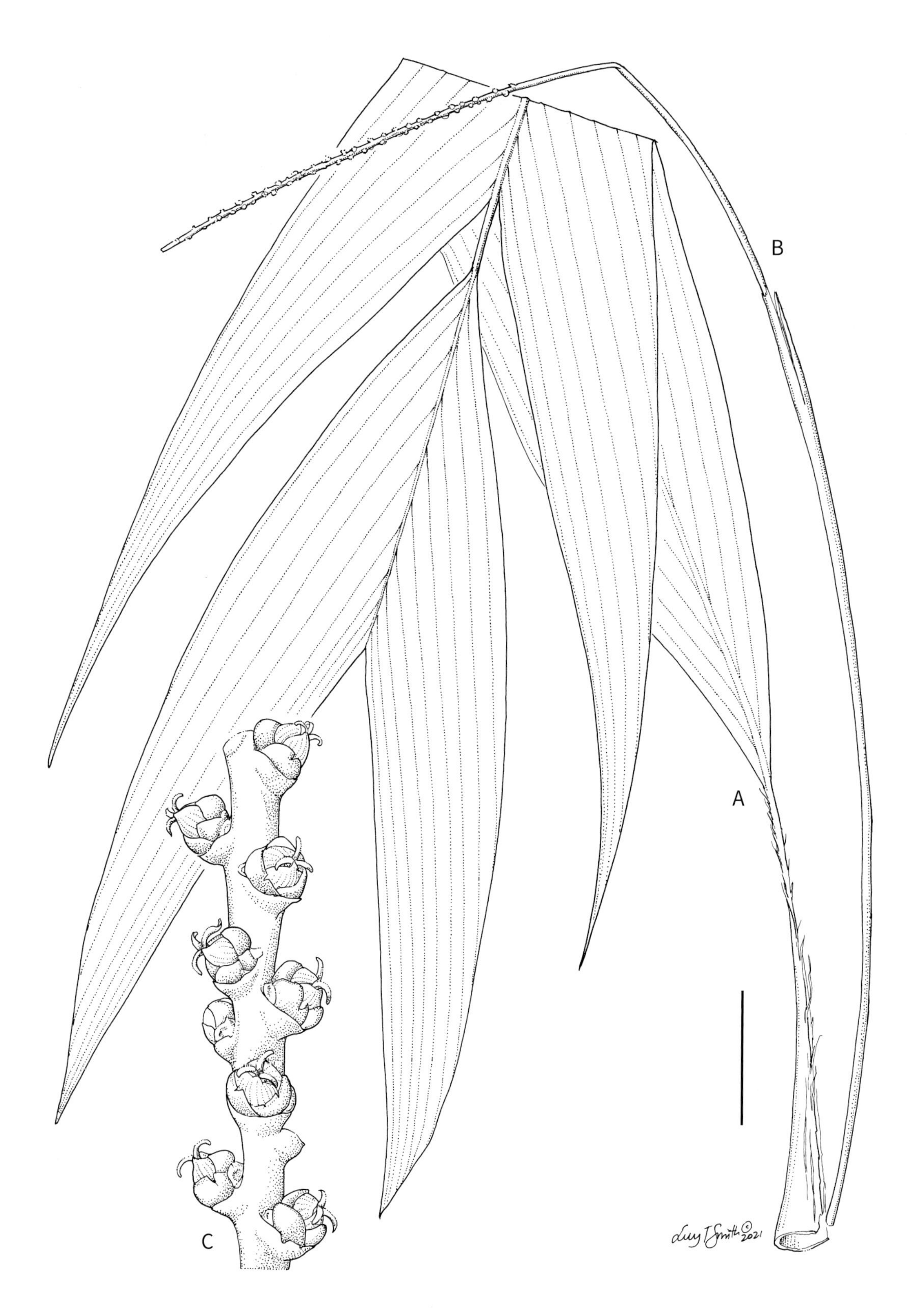

Calyptrocalyx julianettii. **A.** Leaf and leaf sheath. **B.** Inflorescence. **C.** Portion of rachilla with female flowers. Scale bar: A, B = 6 cm; C = 5 mm. All from *Carr 16193.* Drawn by Lucy T. Smith.

lower surface paler, mid-leaf leaflets usually narrower than apical or basal leaflets, 15–20 × 2–4.5 cm, *leaf ribs strongly curved and converging at the leaf apex.* **Inflorescence** with *1–4 equal spikes,* 40–90 cm long; peduncle 20–40 cm long, 1.5–3 mm diam.; prophyll to 16–17 cm long; peduncular bract to 24–38 cm long, *with abundant dark scales;* rachilla spindle-shaped, wider than the peduncle, 15–30 cm long, 3–3.5 mm diam., floral pits congested, with dark scales. **Male flower** with 9 stamens. **Fruit** ellipsoid 11–13 × 6–9 mm, *red.* **Seed** globose, 5–8 mm diam.; *endosperm homogeneous.*

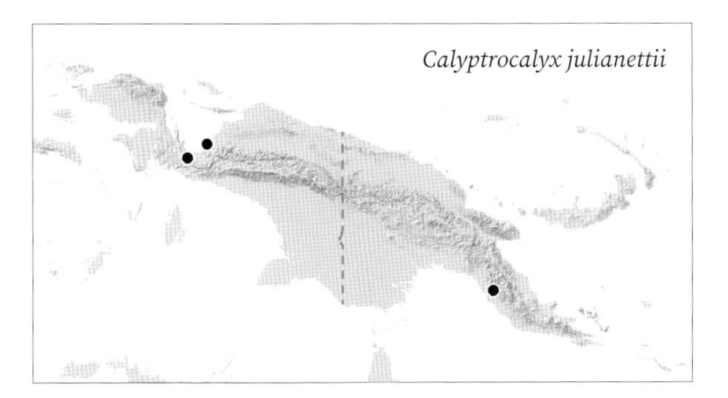

DISTRIBUTION. Disjunct records in north-western and south-eastern New Guinea.

HABITAT. In rainforest from 100 to 1,200 m.

LOCAL NAMES. None recorded.

USES. None recorded.

CONSERVATION STATUS. Least Concern.

NOTES. This small undergrowth palm shares with *C. geonomiformis* the character of leaf ribs strongly curved and converging at the leaf apex. For differences between the two species, see under *C. geonomiformis.*

14. *Calyptrocalyx lauterbachianus* Warb. ex Becc.

Synonyms: *Calyptrocalyx archboldianus* Burret, *Calyptrocalyx stenophyllus* Becc., *Laccospadix lauterbachianus* (Warb. ex Becc.) Burret, *Linospadix lauterbachianus* (Warb. ex Becc.) Becc.

Single- or multi-stemmed, *moderate* palm to 9 m, crown with 7–9 leaves. **Stem** erect, 3.5–8 cm diam. **Leaf** pinnate, 160–300 cm long, emerging reddish brown; sheath 40–60 cm long, densely brown tomentose, margins fibrous, apex with two auricles to 10 cm; petiole 20–50 cm long, 9–11 mm wide, leaflets 20–60 each side of rachis, *single-fold, arranged in groups of 2–5 and held in various planes, whole leaf thus plumose,* ensiform, acuminate, mid-leaf leaflet 45–60 × 3.5–5 cm, ramenta present along lower midrib. **Inflorescence** usually *multiple,* 80–200 cm long with *1–6 equal spikes;* prophyll 30–50 cm long; peduncular bracts 60–70 cm long; rachilla 19–60 × 3.5–9.5 cm, floral pits crowded. **Male flower** with 12–24 stamens. **Fruit** more or less globose, 12–40 × 9–25 mm, *red,* stigmatic remains more or less eccentric. **Seed** 7–11 mm diam., globose; *endosperm ruminate.*

DISTRIBUTION. Widespread in south-eastern Papua New Guinea.

HABITAT. In rainforest from sea level to 1,800 m.

LOCAL NAMES. *Gi* (Khotte), *Koiya* (Anga), *Tofe* (Okapa).

Calyptrocalyx lauterbachianus. Leaf, near Finschhafen (WB).

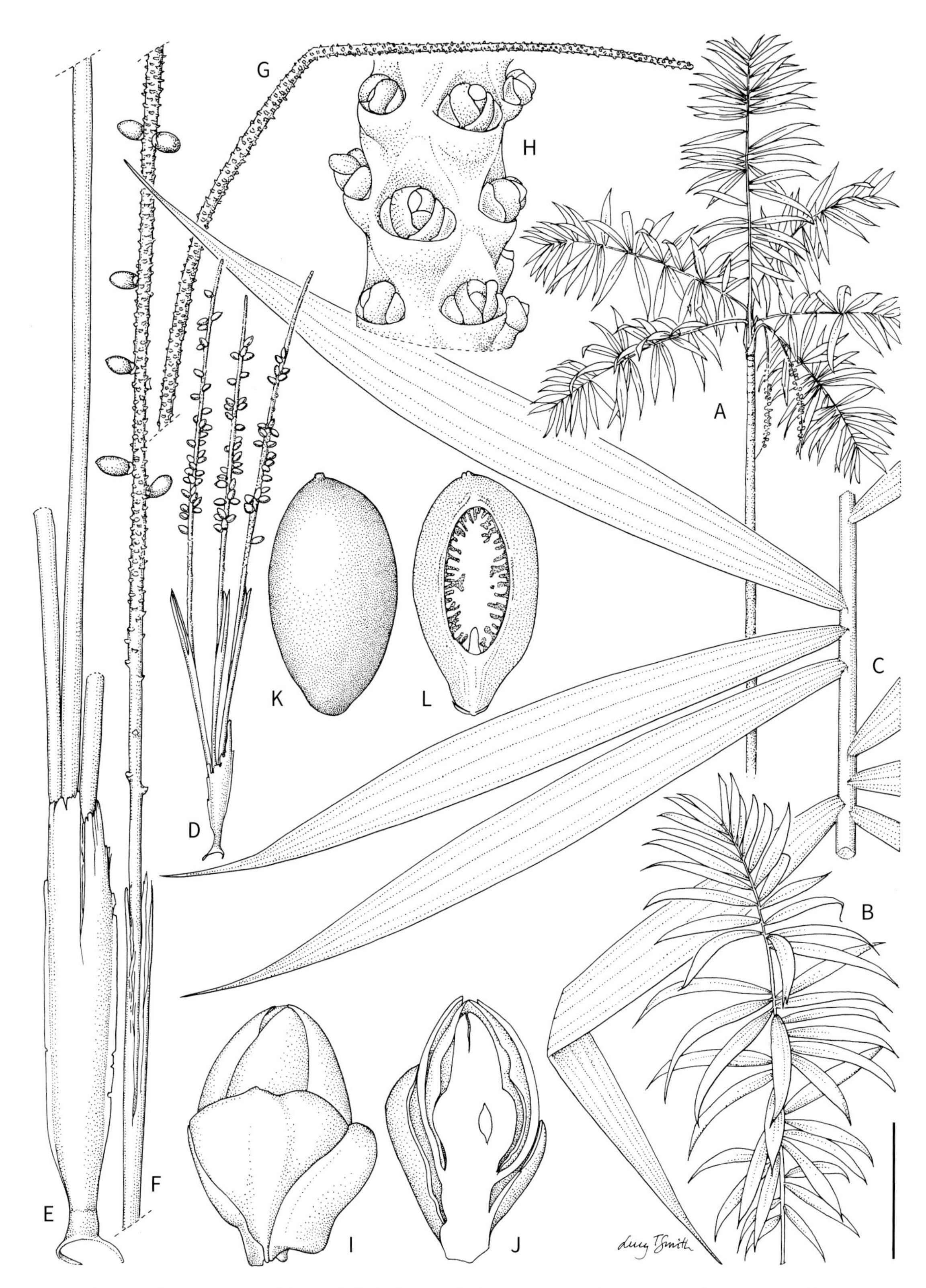

Calyptrocalyx lauterbachianus. **A.** Habit. **B.** Whole leaf. **C.** Leaf mid-portion. **D.** Inflorescence with fruit. **E–G.** Inflorescence showing prophyll and spike with fruit. **H.** Portion of rachilla with female flowers. **I, J.** Female flower whole and in longitudinal section. **K, L.** Fruit whole and in longitudinal section. Scale bar: A = 1.6 m; B = 60 cm; C = 8 cm; D = 40 cm; E–G = 6 cm; H, K, L = 7 mm; I, J = 3 mm. A–G from *Banka et al. 2009*; H–J from *Jermy 5126*. Drawn by Lucy T. Smith.

Calyptrocalyx lauterbachianus. Inflorescence with fruit, near Finschhafen (WB).

USES. Mature fruits used as a betel substitute by Aseki and Menyamya people of the Morobe Highlands. The wood is used for spears and arrows. The leaves are burnt to make salt and the inner flesh of the fruit is eaten raw (Adelbert Mountains).

CONSERVATION STATUS. Least Concern (IUCN 2021).

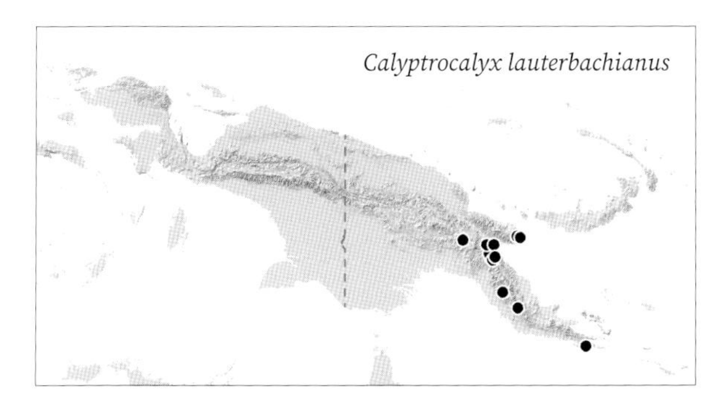

NOTES. Immediately distinguishable by the leaflet arrangement and robust inflorescences, *C. lauterbachianus* is one of only three species with strongly grouped leaflets. In *C. lauterbachianus* the leaflets are arranged in several planes that give the whole leaf a plumose appearance. The leaflets of *C. lepidotus* are similarly grouped but the whole palm is much smaller and the inflorescences very slender, in comparison with those of *C. lauterbachianus*. *Calyptrocalyx reflexus* also has grouped leaflets in different planes, but is a much smaller palm, with fewer much narrower leaflets that are strongly reflexed at the leaf tip.

15. *Calyptrocalyx laxiflorus* Becc.

Stemless or very short-stemmed, single- or multi-stemmed palm, crown with 4–9 leaves. **Stem** 1.5–2 cm diam., internodes very short. **Leaf** *regularly pinnate,* 64–90 cm long, emerging pink to bronze; sheath to

Calyptrocalyx laxiflorus. Habit, cultivated Floribunda Palms and Exotics, Hawaii (WB).

Calyptrocalyx laxiflorus. **A.** Habit with leaf and inflorescences. **B.** Portion of rachilla. Scale bar: A = 4.5 cm; B = 7.5 mm. All from *Barfod 396*. Drawn by Lucy T. Smith.

Calyptrocalyx laxiflorus. LEFT TO RIGHT: leaf; inflorescences; immature fruit. Cultivated Floribunda Palms and Exotics, Hawaii (WB).

12–24 cm long, glabrous, yellowish to green, margins lacerate-fibrous; petiole 9–70 cm long; leaflets *regularly arranged,* 2–11 on each side of the rachis, 15–20 × 3–4 cm in mid-leaf, *broadly lanceolate to sigmoid, long acuminate,* upper surface with a metallic sheen, lower surface paler. **Inflorescence** to 60–80 cm long, with 1–8 equal filiform spikes; prophyll and peduncular bract not seen; peduncle 30–40 cm, densely scaly; peduncular bract papery, very densely scaly; rachilla 26–70 cm long, 2 mm diam., densely scaly, floral pits c. 10–15 mm distant. **Male flower** not seen. **Fruit** *ellipsoid,* 19–21 × 9 mm., *red;* stigmatic remains apical; epicarp finely pilose. **Seed** spinde-shaped to sickle-shaped, 15 × 7 mm; endosperm homogeneous.

DISTRIBUTION. Torricelli Mountains.

HABITAT. In rainforest at 1,000 m.

LOCAL NAMES. *Timenum* (Olo).

USES. None recorded.

CONSERVATION STATUS. Critically Endangered.

Calyptrocalyx laxiflorus is known from only one site that is threatened by logging concessions. Additional localities have not been found through subsequent fieldwork.

NOTES. This species is distinctive in its acaulescent habit and regularly pinnate leaves with sigmoid leaflets with long drip tips. The leaves of juvenile plants apparently have a silvery sheen that is lost in the adult palm. The inflorescence is multi-spiked and the red fruit ellipsoid.

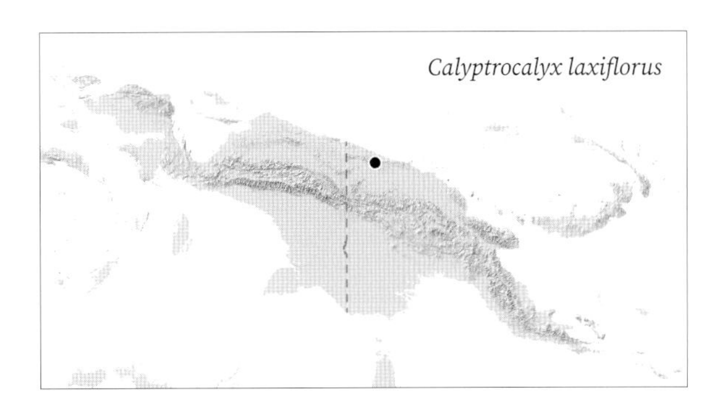

***Calyptrocalyx lepidotus*. A.** Leaf. **B.** Inflorescence. **C**, **D.** Portion of rachilla with triads. **E.** Portion of rachilla with female flowers. **F.** Fruit whole and in transverse section. Scale bar: A = 8 cm; B = 6 cm; C–G = 5 mm. A–E from from *Brass 7316*; F, G from *Brass 6716*. Drawn by Lucy T. Smith

16. *Calyptrocalyx lepidotus* (Burret) Dowe & M.D.Ferrero

Synonym: *Paralinospadix lepidotus* Burret

Single-stemmed, moderate palm to 4–10 m, crown with 10–12 leaves. **Stem** erect, to 25–40 mm diam., internodes elongate. **Leaf** *irregularly pinnate,* 80–160 cm long, new leaf emerging brown; sheath to 24 cm long, with *patchy dark tomentum,* margins lacerate-fibrous; petiole 44–64 × 0.6–0.7 cm; leaflets 6–12 on each side of the rachis, *arranged in two distant groups,* 17–40 cm long, to 5 cm wide in mid-leaf, lanceolate, slightly sickle-shaped, long acuminate, upper surface dark green, *lower surface with brown scales.* **Inflorescence** to 125–140 cm long, with 1–7 equal spikes, pendulous; prophyll 30 cm long; peduncle ca. 70 × 0.2–0.3 cm, *densely scaly;* peduncular bract papery, very densely scaly; rachilla 26–70 cm long, 2–3.2 mm diam., *densely scaly and with congested pits.* **Male flower** with 6 stamens. **Fruit** ellipsoid, 8–9 × 4–4.5 mm, *red;* stigmatic remains apical. **Seed** subglobose, ca. 4 mm diam.; endosperm homogeneous.

DISTRIBUTION. Central New Guinea.

HABITAT. In rainforest at 80–1,400 m.

LOCAL NAMES. None recorded.

USES. None recorded.

CONSERVATION STATUS. Near Threatened (IUCN 2021). *Calyptrocalyx lepidotus* has a restricted distribution. At least one site where it is found is threatened by a mining concession. Our own verified occurrence records suggest that it may in fact be Endangered.

NOTES. This small single-stemmed species is immediately distinguishable by the grouped leaflets and dense dark scales on the petioles, leaf rachises and inflorescence. For differences from other species with grouped leaflets, see under *C. lauterbachianus.*

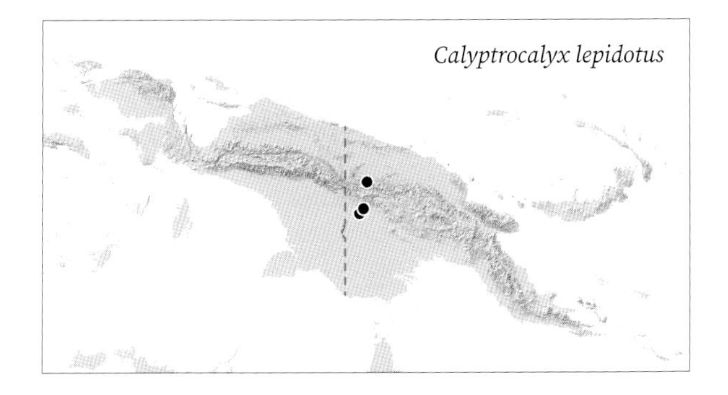

17. *Calyptrocalyx leptostachys* Becc.

Single-stemmed, *small* palm to 2 m, crown with 7–9 leaves. **Stem** erect, to 10–20 mm diam., internodes to 5 cm long. **Leaf** *irregularly pinnate,* 50–70 cm long; sheath to 25 cm long, margins fibrous; petiole 8–10 × 0.4–1.0 cm; rachis with sparse orange-brown scales; leaflets 3–10 on each side of the rachis, irregularly arranged, narrowly lanceolate, acuminate, upper surface green, lower surface paler, consisting of 1–7 folds, largest 30 × 3 cm. **Inflorescence** *with 2 equal spikes,* 60–70 cm long, *very slender;* prophyll not seen; peduncle 4 mm diam.; peduncular bract not seen; rachilla to 20 cm long, 2 mm diam., *floral pits shallow,* lip rounded. **Male flower** with ca. 15 stamens. **Fruit** globose to obovoid, 16–17 × 12 mm, colour unknown, epicarp smooth minutely granular when dry. **Seed** globose, ca. 9 mm diam.; *endosperm ruminate.*

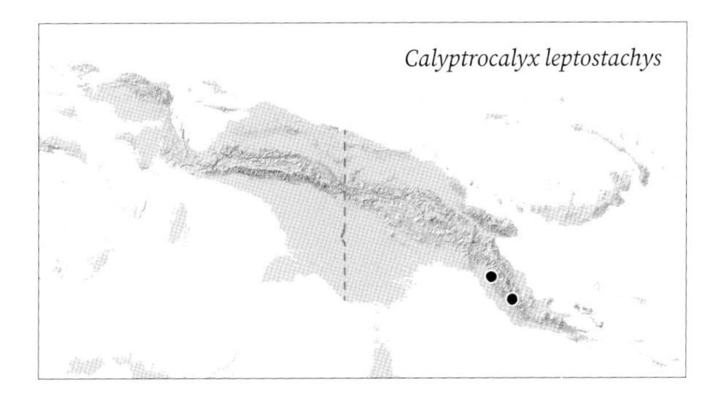

DISTRIBUTION. Owen Stanley Range, south-eastern New Guinea.

HABITAT. Montane forest at 1,500 m.

LOCAL NAMES. None recorded.

USES. None recorded.

CONSERVATION STATUS. Critically Endangered. *Calyptrocalyx leptostachys* is known from only one site that is threatened by logging concessions.

NOTES. A poorly known species with very few collections. The inflorescences are thread-like and comprise two spikes. The endosperm is ruminate.

Calyptrocalyx leptostachys. Holotype specimen of *C. leptostachys* (*Loria s.n.*) at the herbarium of the Museum of Natural History, University of Florence (FI).

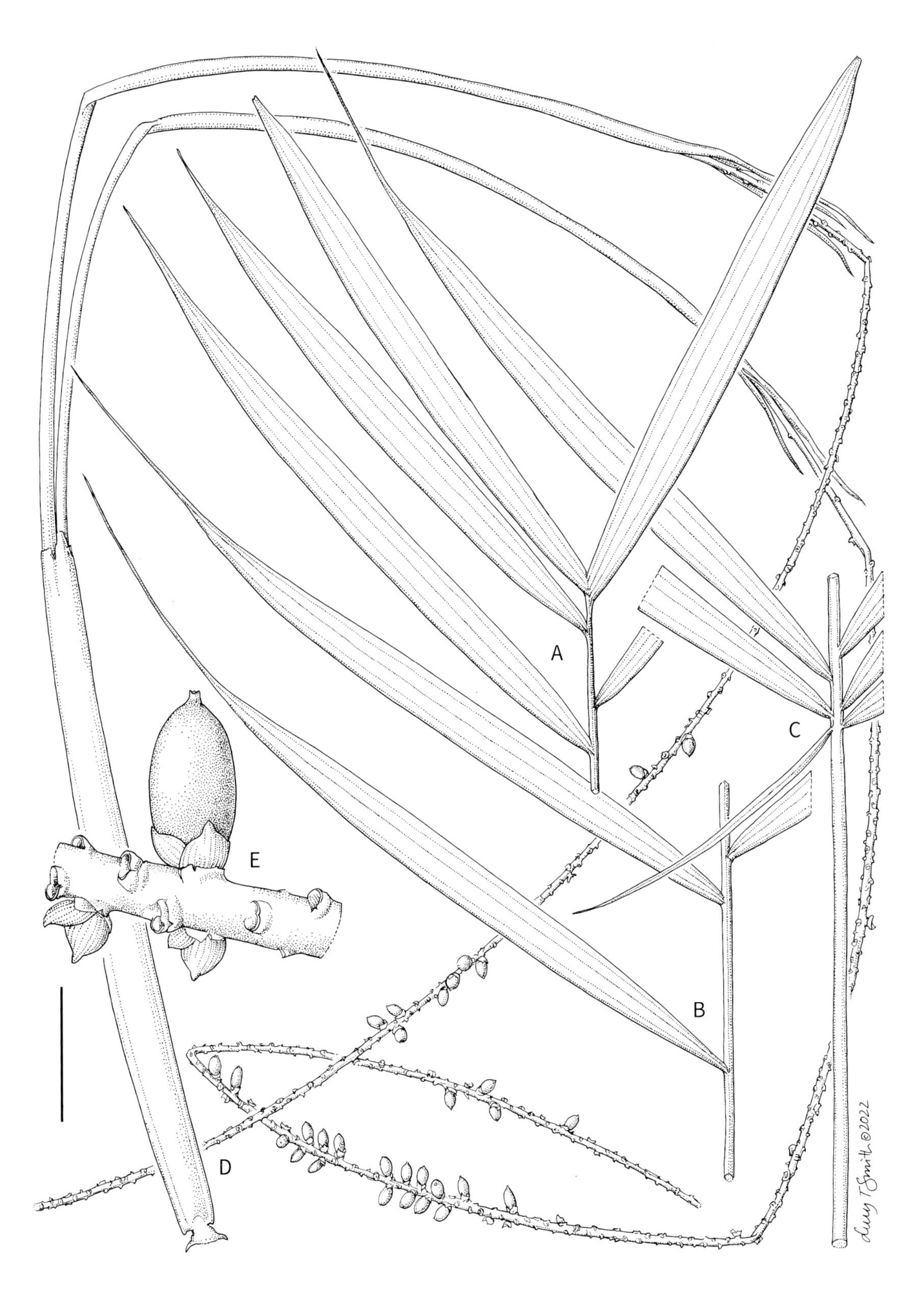

Calyptrocalyx merrillianus. **A.** Leaf apex. **B.** Leaf mid-portion. **C.** Leaf base. **D.** Inflorescence with fruit. **E.** Portion of rachilla with fruit. Scale bar: A–C = 8 cm; D = 4 cm; E = 7 mm. All from *Baker et al. 649*. Drawn by Lucy T. Smith.

18. *Calyptrocalyx merrillianus* (Burret) Dowe & M.D.Ferrero

Synonym: *Paralinospadix merrillianus* Burret

Multi-stemmed, *small* palm to 2–4 m, crown with 7–9 leaves. **Stems** 2–3, erect, to 13 mm diam., internodes to 5 cm long. **Leaf** *regularly pinnate*, 80–200 cm long; sheath to 30 cm long, margins fibrous; petiole 14–24 × 1.0–1.2 cm; rachis with dark scales; leaflets to 15 on each side of the rachis, regularly arranged, basal 1 or 2 very narrow, the rest linear lanceolate or sigmoid, acuminate, upper surface dark green, lower surface paler, mid-leaf leaflets to 38 × 3.3 cm. **Inflorescence** with *2–4 equal spikes*, 80–90 cm long, *thread-like*; prophyll ca. 21 cm long; peduncle to 60 cm long, 3–4 mm diam.; peduncular bract 30–33 cm long, covered with dark scales; rachilla to 30 cm long, 3–4 mm diam., floral pits well-spaced proximally, congested distally. **Male flower** with 8 stamens. **Fruit** ovoid, beaked, 18–20 × 8–10 mm, *red to purple-black*, stigmatic remains apical on a short beak, epicarp smooth when fresh, coarsely granular when dry. **Seed** globose, ca. 10 mm diam.; *endosperm ruminate*.

DISTRIBUTION. Southern central New Guinea.

HABITAT. In rainforest at 80–450 m.

LOCAL NAMES. *Gurrinem* (Drimskai, Upper Fly River).

USES. Leaves used to wrap fish and other food; stems used to make spears.

CONSERVATION STATUS. Data Deficient (IUCN 2021). This species is known from only two sites, where there is limited information on ongoing threats. Owing to insufficient evidence, an extinction risk category cannot be selected.

NOTES. This multi-stemmed palm is distinctive in the regularly pinnate leaves with linear to sigmoid leaflets, inflorescences with many spikes and fruit with ruminate endosperm. That it occurs on the banks of the upper Fly River in full sun suggests that it may be a rheophyte.

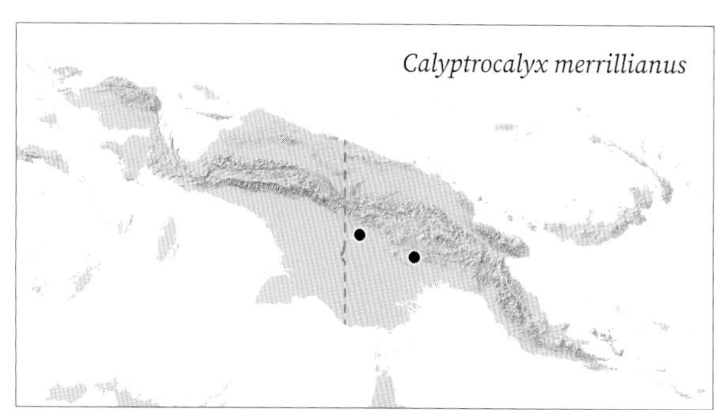

19. *Calyptrocalyx micholitzii* (Ridl.) Dowe & M.D.Ferrero

Synonyms: *Linospadix micholitzii* Ridl., *Linospadix pauciflorus* Ridl., *Paralinospadix micholitzii* (Ridl.) Burret, *Paralinospadix pauciflorus* (Ridl.) Burret

Single-stemmed (or multi-stemmed), *dwarf* palm to 0.5–1 m, crown with 10–12 leaves. **Stem** erect, to 10 mm diam. **Leaf** *entire-bifid* (rarely with one or two divisions each side), 16–100 × 8–18 cm, somewhat hooded, *leathery, metallic-coloured and often mottled paler green on dark green*, leaf emerging pale green; sheath 6–8 cm long, densely brown tomentose, margins fibrous; *petiole absent*. **Inflorescence** *thread-like, usually multiple*, 35–90 cm long with *1–5 equal spikes*; prophyll 2.5–13 cm long; peduncular bract 20–25 cm long; rachilla 18–22 × 0.3–0.4 cm, floral pits very shallow, distant. **Male flower** with 8–25 stamens. **Fruit** ellipsoid to globose, 11–12 × 6–8 mm, *red*, stigmatic remains more or less eccentric. **Seed** 7 mm diam., globose; *endosperm homogeneous*.

DISTRIBUTION. Western and central northern New Guinea.

HABITAT. In rainforest at 30–600 m.

Calyptrocalyx micholitzii. Habit, near Timika (JD).

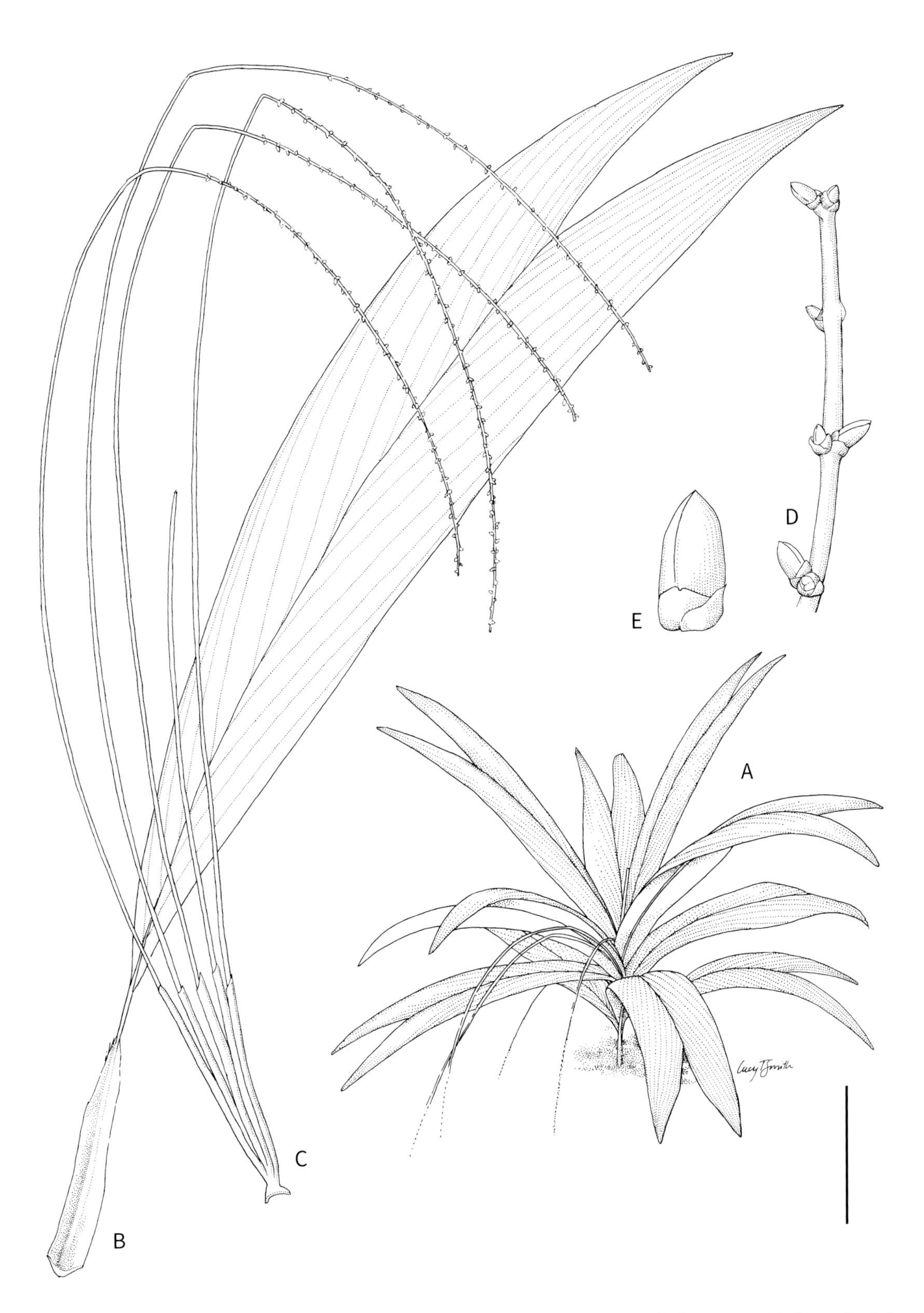

Calyptrocalyx micholitzii. **A.** Habit. **B.** Leaf. **C.** Inflorescence. **D.** Portion of rachilla with male flowers. **E.** Male flower. Scale bar: A = 40 cm; B = 6 cm; C = 9 cm; D = 5 mm; E = 1.8 mm. A from photograph; B, D, E from *Heatubun et al. 176*; C from *Heatubun et al. 181*. Drawn by Lucy T. Smith.

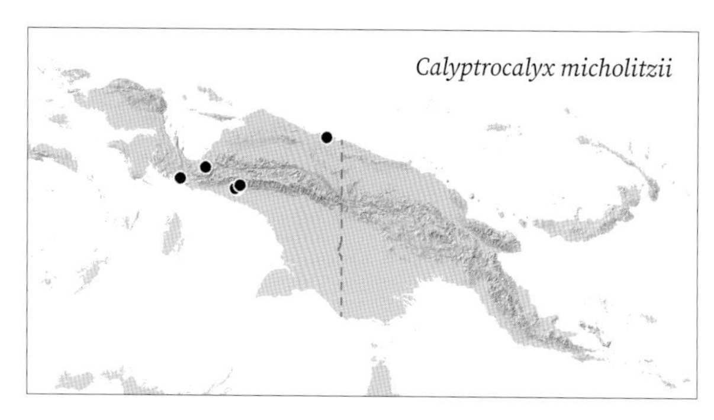

LOCAL NAMES. *Sawat net, Sasif* (Kroy).

USES. None recorded.

CONSERVATION STATUS. Near Threatened. *Calyptrocalyx micholitzii* has a relatively restricted distribution. Deforestation due to mining concessions is a threat in its distribution range.

NOTES. *Calyptrocalyx micholitzii* is a distinctive and beautiful dwarf species with dark metallic green leaves mottled paler green and thread-like multiple spikes.

Calyptrocalyx aff. *micholitzii*. Habit, Tamrau Mountains (WB).

Two collections (*Baker et al. 1372, Gardiner et al. 429*) from the Tamrau mountains in the western Bird's Head Peninsula, seem very close to *C. micholitzii* if not conspecific, but the leaf shape is subtly different, it dries dull dark chocolate brown in the herbarium and the ripe fruit are black rather than red.

20. *Calyptrocalyx multifidus* Becc.

Synonyms: *Linospadix multifidus* Becc., *Paralinospadix multifidus* (Becc.) Burret

Multi-stemmed, *small* palm to 1–2 m. **Stems** 2–5, erect, 8–40 mm diam., internodes 2–4 cm. **Leaf** *irregularly or regularly pinnate*, 70–125 cm long, emerging reddish brown; sheath 8–35 cm long, glabrous, margins fibrous; petiole 20–60 cm long, 3–10 mm wide, glabrous; leaflets 4–18 on each side of the rachis, regularly arranged, or one side of leaf not split into leaflets, or split into few irregular leaflets, mid-leaf leaflets 20–60 × 10–25 cm, ribs prominent. **Inflorescence** erect, stiff, *1-spiked*, 40–105 cm long; peduncle 20–85 cm long, 2–6 mm diam., scaly; prophyll to 15–28 × 1.4–2.5 cm; peduncular bract 30–40 × 1.3 cm; rachilla 10–28 × 0.4– 0.9 cm, wider than the peduncle, densely scaly, floral pits crowded, deep. **Male flower** with 8–10 stamens. **Fruit** ellipsoid, 10–16 × 10–12 mm, *red*, stigmatic remains apical. **Seed** globose to ellipsoid, 11 × 7 mm; *endosperm homogeneous*.

DISTRIBUTION. North-western New Guinea.

HABITAT. In rainforest from 300 to 900 m.

LOCAL NAMES. *Puah* (Maibrat)

USES. None recorded.

CONSERVATION STATUS. Data Deficient. This species is known from only three sites, where there is limited information on ongoing threats. Due to insufficient evidence, an extinction risk category cannot be selected.

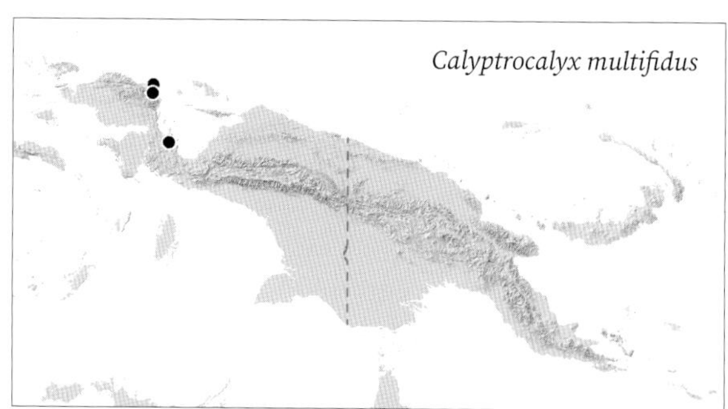

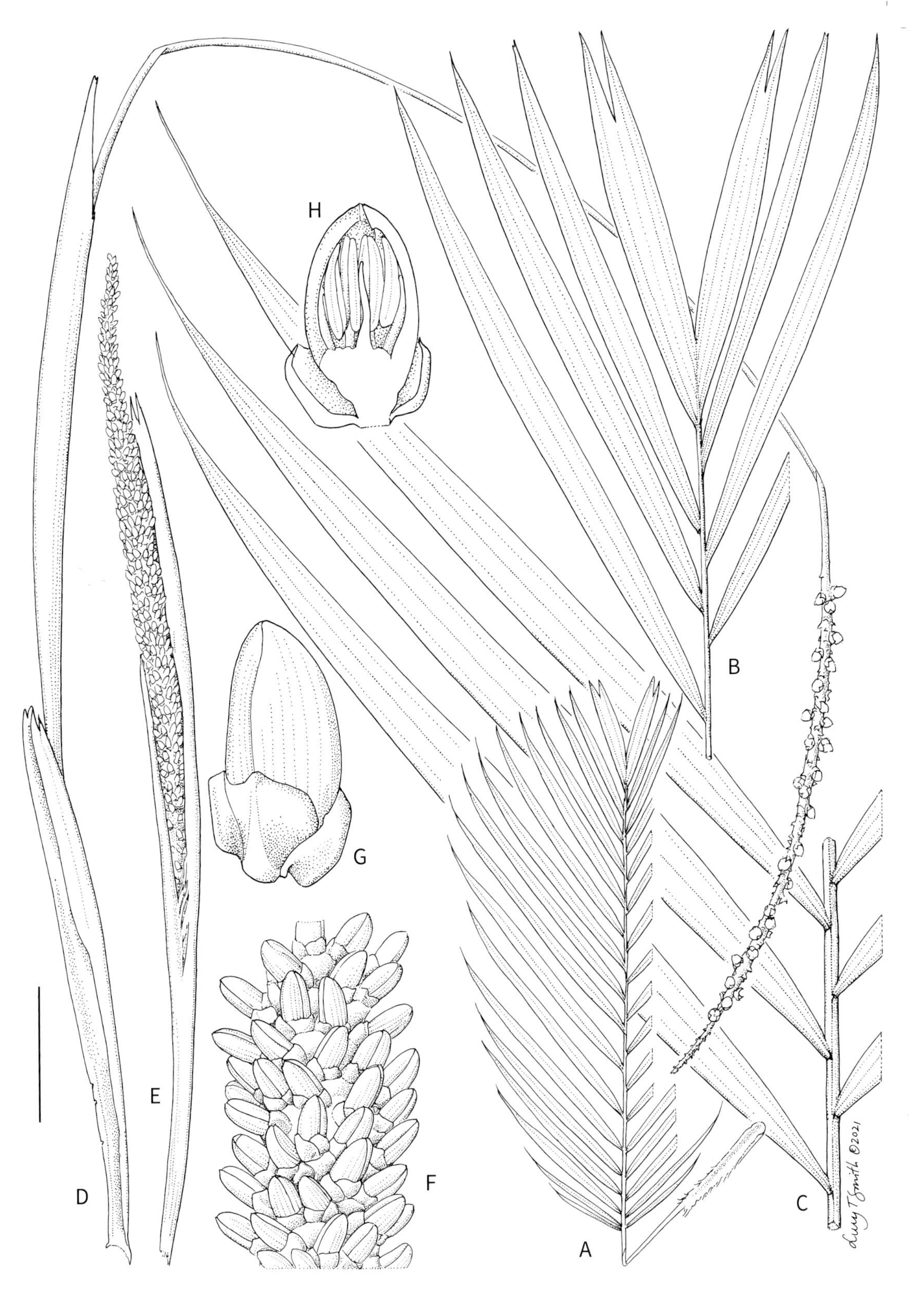

Calyptrocalyx multifidus. **A.** Whole leaf diagram. **B.** Leaf apex. **C.** Leaf mid-portion. **D.** Inflorescence with fruit. **E.** Rachilla with male flowers. **F.** Portion of rachilla with male flowers. **G, H.** Male flower whole and in longitudinal section. Scale bar: A = 15 cm; B–E = 4 cm; F = 7 mm; G, H = 2.5 mm. All from *Dransfield et al. JD 7607*. Drawn by Lucy T. Smith.

Calyptrocalyx multifidus. FROM TOP: habit; inflorescence with fruit. Arfak Mountains (SZ).

NOTES. *Calyptrocalyx multifidus* is distinctive in its multi-stemmed habit, robust 1-spiked inflorescence and ellipsoid fruit with homogeneous endosperm; the leaf varies in dissection from regularly pinnate through to leaves with few, broad leaflets of varying widths. A specimen from East Sepik, Papua New Guinea, *Kjaer 518*, seems to match *C. multifidus* but is far from the range of this species

21. *Calyptrocalyx pachystachys* Becc.

Synonym: *Calyptrocalyx schlechterianus* Becc.

Single- or multi-stemmed, *small to moderate* palm to 1–5 m, crown with 8–10 leaves. **Stems** 1–5, erect, to 22–40 mm diam., internodes 30–45 mm long. **Leaf** *regularly pinnate*, 80–140 cm long, emerging reddish brown; sheath to 26 cm long, glabrous, margins fibrous; petiole 8–25 cm long, 3–5 mm diam., glabrous or tomentose; leaflets 9–17 on each side of the rachis, slightly irregularly arranged, sometimes grouped, *more or less sigmoid, hooded*, 13–44 × 1.2–11 cm, with drip tips 4–10 cm long, upper surface glossy dark green, lower surface paler, main vein prominent. **Inflorescence** *stiff, 1-2-spiked*, 50–90 cm long; peduncle 30–65 cm long, 3–5 mm diam.; prophyll to 18–24 × 1–2.5 cm; peduncular bracts to 30 cm long; rachilla 16–39 × 0.5–1.2 cm, wider than the peduncle, covered in reddish-brown tomentum, floral pits crowded, deep. **Male flower** with 7–19 stamens. **Fruit** ellipsoid, 16–18 × 7–8 mm, *orange to red*, stigmatic remains apical. **Seed** ovoid, 9–11 × 4–6 mm.; *endosperm homogeneous*.

DISTRIBUTION. Scattered records in central northern and eastern New Guinea.

HABITAT. In rainforest 300–900 m.

LOCAL NAMES. *Angop* (Ningera, East Sepik), *Apantsj* (Waskuk, East Sepik), *Awa* (Osima, Sandaun), *Briau* (Bewani, Sandaun), *Giagiau* (Waskuk, East Sepik), *Gurrinem* (Nu, Western), *Yagiyas* (Waskuk, East Sepik).

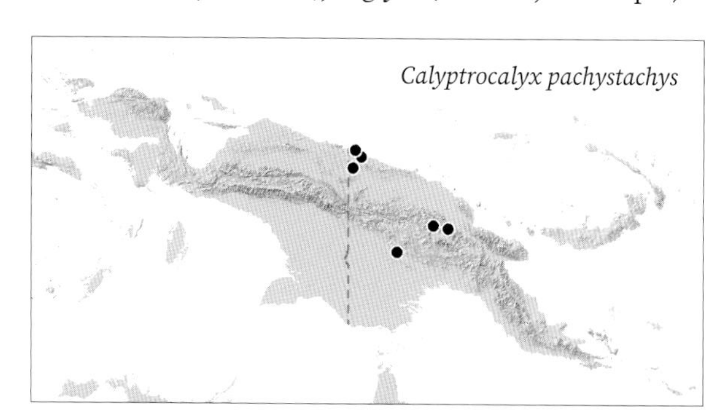

Calyptrocalyx pachystachys. **A.** Leaf apex. **B.** Leaf mid-portion. **C.** Leaf base. **D.** Inflorescence. **E.** Portion of rachilla with male flowers. **F.** Male flower bud in longitudinal section. **G.** Open male flower. **H, I.** Female flower whole and in longitudinal section. **J.** Fruit. **K, L.** Seed whole and in longitudinal section. Scale bar: A–D = 8 cm; E = 7 mm; F, H, I = 3.3 mm; G = 4 mm; J–L = 1 cm. All from *Marcus s.n.* Drawn by Lucy T. Smith.

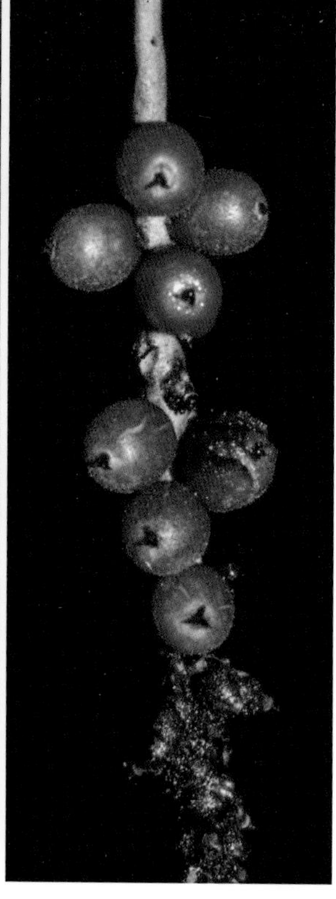

Calyptrocalyx pachystachys. LEFT TO RIGHT: habit, cultivated Floribunda Palms and Exotics, Hawaii (JSM); inflorescence, cultivated Merwin Conservancy (WB); inflorescence with fruit, Bewani (JLD).

USES. Stems used to make spears.

CONSERVATION STATUS. Least Concern (IUCN 2021).

NOTES. *Calyptrocalyx pachystachys* is distinctive in its regularly pinnate leaves with sigmoid leaflets with drip tips; the inflorescence is single or rarely 2-spiked, with the rachilla portion much wider than the peduncle. The fruit is ellipsoid and the endosperm homogeneous.

22. *Calyptrocalyx pauciflorus* Becc.

Synonym: *Calyptrocalyx angustifrons* Becc.

Single- or multi-stemmed, *small* palm to 1–2 m, crown with 5–9 leaves. **Stem** erect, to 15–50 mm diam., internodes to 2 cm long. **Leaf** *entire bifid or irregularly pinnate*, when bifid to 65 cm long, when pinnate tending to be longer; sheath 16–24 cm long, green, glabrous, margins lacerate; petiole 9–80 cm long; pinnate leaves with narrow basal leaflets; main veins prominent, upper surface dark green, lower surface paler. **Inflorescence** to 45–100 cm long, with 1 or 2 equal filiform spikes; prophyll 14 cm long; peduncular bract to 26 cm long; peduncle 30–60 cm long, 1.5–2 mm diam. rachilla 10–40 cm long, 3–3.5 mm diam. with rather distant floral pits. **Male flower** greenish to white, with 9 stamens. **Fruit** ovoid to ellipsoid, 13×10 mm diam., *red*; stigmatic remains apical. **Seed** globose 9 mm diam., globose; *endosperm ruminate.*

DISTRIBUTION. Scattered records in the eastern half of the Central Range.

HABITAT. In rainforest at sea level to 1,500 m.

LOCAL NAMES. *Basini* (Wagu), *Beisini* (Baihinimo, East Sepik).

USES. Stems used to make spears for catching fish by people in Hunstein Mountains, East Sepik. Leaves used for wrapping sago.

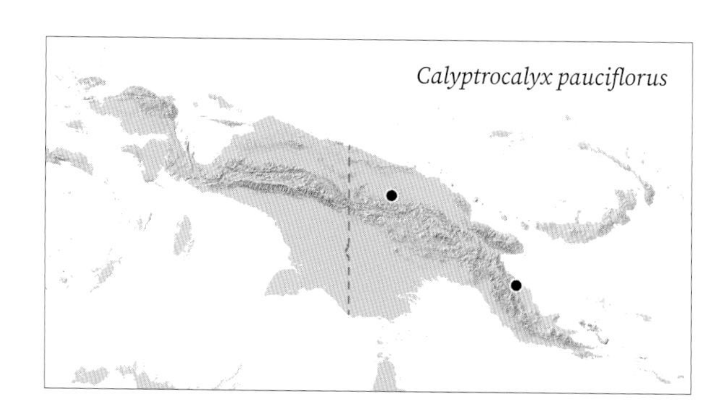

Calyptrocalyx pauciflorus. **A.** Leaf. **B.** Inflorescence. **C.** Rachilla with flowers. Scale bar: A = 6 cm; B = 4 cm; C = 2 cm. All from *Hoogland & Craven 10998*. Drawn by Lucy T. Smith.

CONSERVATION STATUS. Critically Endangered. *Calyptrocalyx pauciflorus* is known from only two sites that are threatened by logging concessions.

NOTES. *Calyptrocalyx pauciflorus* is distinguishable by its small size, 1- or 2-spiked inflorescence, bifid or irregularly pinnate leaves and ruminate endosperm.

23. *Calyptrocalyx polyphyllus* Becc.

Multi-stemmed, *small* palm to 1–4 m, crown with 7–10 leaves. **Stems** 2–5, erect, to 15–20 mm diam., internodes elongate to 35 mm long. **Leaf** *regularly pinnate,* 60–100 cm long, emerging reddish brown; sheath 16–20 cm long, tomentum thick, reddish brown, margins fibrous; petiole 2–15 cm, glabrous; leaflets 10–21 on each side of the rachis, regularly arranged, mid-leaf leaflets 40–45 × 14–18 cm, with drip tips, midribs prominent. **Inflorescence** *stiff, 1-spiked,* 75–100 cm long, more or less arching; prophyll to 13 cm long; peduncular bracts to 30 cm long; rachilla 10–35 × 1.2 cm, wider than the peduncle, floral pits widely spaced, shallow. **Male flower** with 10–16 stamens. **Fruit** ovoid, beaked, 9–11 × 6 mm, *red,* stigmatic remains apical on short beak. **Seed** ovoid, 8 × 5 mm; *endosperm ruminate.*

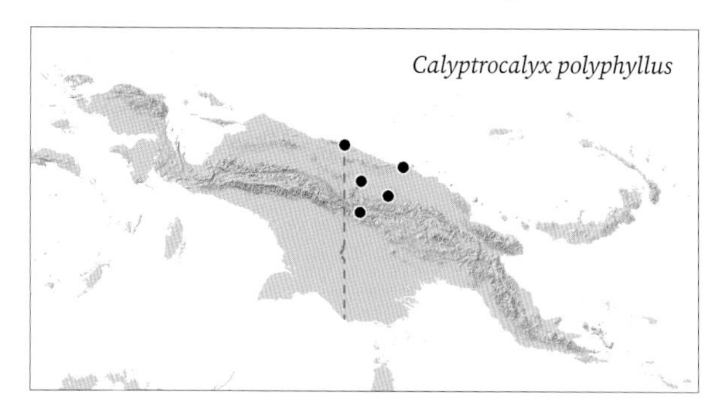

DISTRIBUTION. Central northern New Guinea.

HABITAT. In rainforest at 60–1,000 m.

LOCAL NAMES. *Apolop* (Waskuk), *Belin eddi* (Wasissi), *Giagiau* (Baihinemo), *Gongok* (Mianmin), *Peliah* (Brrinhimo), *Pirare* (Waskuk).

USES. Stems are used to make spears and arrows in the Hunstein Mountains and April River, and for making children's bows. Stems are used as digging sticks. Near

Calyptrocalyx polyphyllus. CLOCKWISE FROM LEFT: leaves; inflorescence with fruit; fruit; female flowers. Cultivated Floribunda Palms and Exotics, Hawaii (WB).

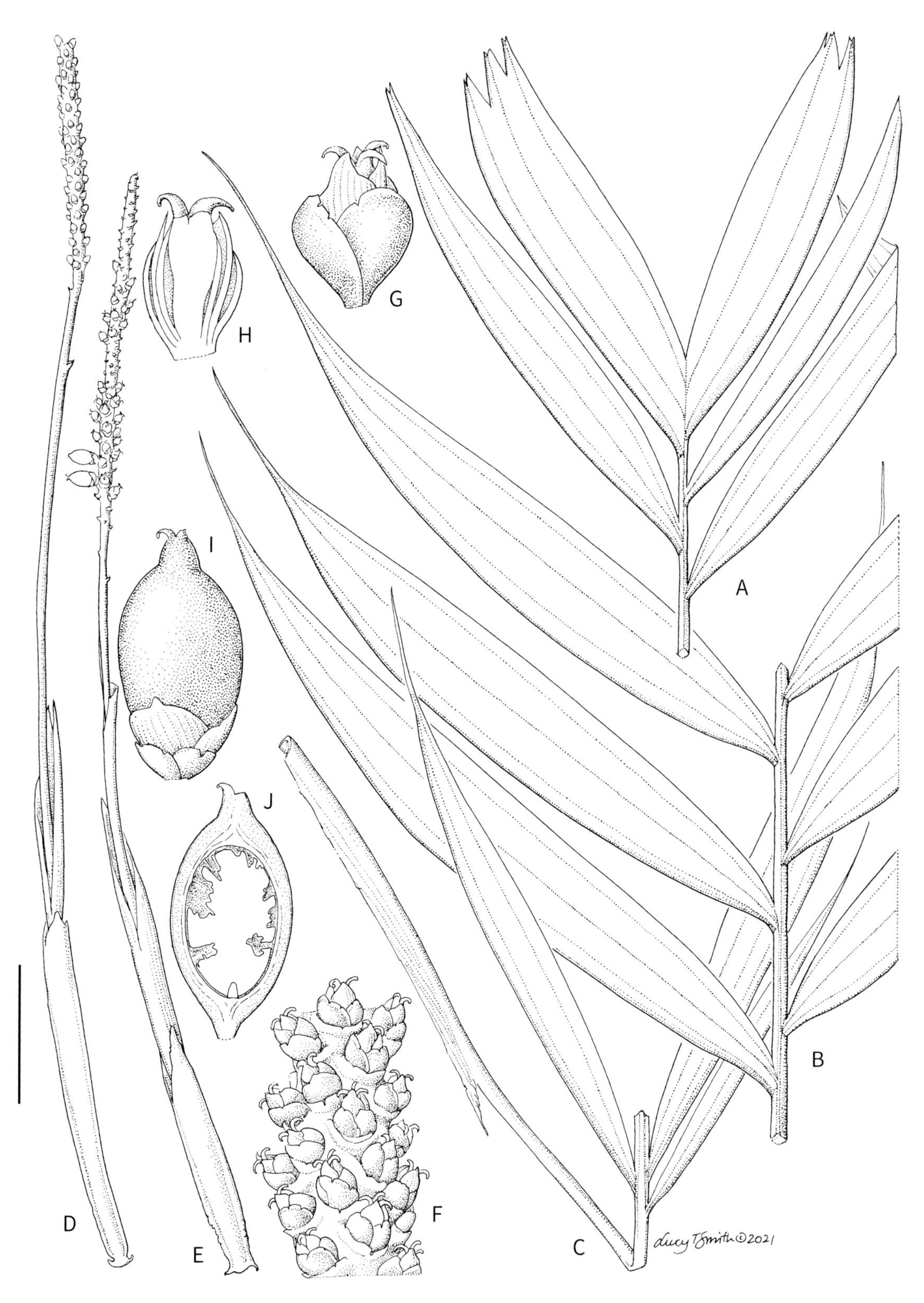

Calyptrocalyx polyphyllus. **A.** Leaf apex. **B.** Leaf mid-portion. **C.** Leaf base. **D.** Inflorescence. **E.** Inflorescence with fruit. **F.** Portion of rachilla with female flowers. **G**, **H.** Female flower whole and in longitudinal section. **I**, **J.** Fruit whole and in longitudinal section. Scale bar: A–C = 8 cm; D, E = 6 cm; F = 1 cm; G, H = 5 mm; I, J = 7.5 mm. All from *Hoogland & Craven 10356*. Drawn by Lucy T. Smith.

Calyptrocalyx pusillus. **A.** Habit. **B.** Inflorescence emerging from crown. **C.** Portion of rachilla with male flowers and peduncular bract. **D.** Portion of rachilla with female flowers. **E, F.** Male flower whole and in longitudinal section. Scale bar: A = 4 cm; B = 2.5 cm; C, D = 5 mm; E, F = 1.5 mm. All from *Carr 13881*. Drawn by Lucy T. Smith.

Wagu, it is said that betel nuts and coconuts planted in holes dug with this species will produce earlier. Used in roof construction.

CONSERVATION STATUS. Near Threatened. *Calyptrocalyx polyphyllus* has a relatively restricted distribution. Deforestation due to logging concessions is a major threat in its distribution range.

NOTES. This species is distinctive in its multi-stemmed habit, regularly pinnate leaves, ovoid fruit and ruminate endosperm.

24. *Calyptrocalyx pusillus* (Becc.) Dowe & M.D.Ferrero

Synonyms: *Linospadix parvulus* Becc., *Linospadix pusillus* Becc., *Paralinospadix pusillus* (Becc.) Burret

Single-stemmed (or multi-stemmed), *dwarf* palm to 1 m, crown with 4–7 leaves. **Stem** erect or decumbent, 5–6 mm diam., internodes 12–20 mm long. **Leaf** *entire-bifid* or irregularly segmented, 20–50 × 7–12 cm, thin, *upper surface dark green, lower surface paler,* 7–10 ribs on each side of the rachis, if segmented, leaflets 6–12 mm wide, lower surface with ribs silvery scaly; sheath 5–10 cm long, glabrous, margins fibrous; petiole 5–18 cm long, 2–3 mm diam. **Inflorescence** *thread-like,* single, 15–60 cm long; prophyll 4–13 cm long; peduncular bracts 10–28 cm long; flower bearing part 4–21 × 0.3–0.7 cm, *densely scaly,* floral pits shallow, distant. **Male flower** with 8 stamens. **Fruit** *ellipsoid,* 8–11 × 3–5 mm, beaked, *red.* **Seed** 7 × 4 mm diam., ellipsoid; *endosperm homogeneous.*

DISTRIBUTION. South-eastern New Guinea.

HABITAT. In montane forest at 1,250–1,800 m.

LOCAL NAMES. None recorded.

USES. None recorded.

CONSERVATION STATUS. Endangered. *Calyptrocalyx pusillus* has a restricted distribution. Deforestation due to logging concessions is a major threat in its distribution range.

NOTES. The smallest species in the genus, *Calyptrocalyx pusillus,* is distinguished by its usually entire bifid leaf, single spike and small ellipsoid fruit. It could be confused with a species of *Linospadix,* but the position of the peduncular bract or its scar just below the flower-bearing portion of the inflorescence rather than towards its base would immediately indicate the latter genus.

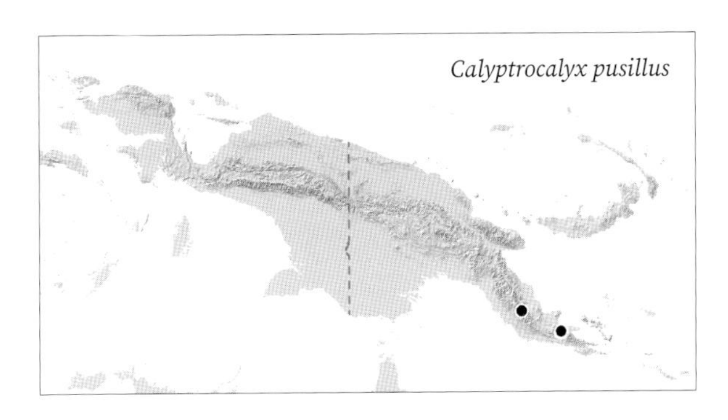

A specimen (*Takeuchi 10712*), collected on Mount Oibo in the Bismarck Range, is very similar to *C. pusillus* but has narrower and stiffer leaves; unfortunately flowers and fruit are lacking in the single specimen. It is tentatively assigned to *C. pusillus,* representing a significant range extension.

25. *Calyptrocalyx reflexus* J.Dransf. & J.Marcus

Single-stemmed, small palm to 0.6 m, crown with ca. 13 leaves. **Stems** erect, to 3 cm diam., internodes *very short.* **Leaf** *irregularly pinnate,* to 80 cm long, leaf *emerging purplish-tinged;* sheath to 15 cm long, light green, with scattered brown scales, margins fibrous; petiole 17 × 0.8 cm, with abundant brown scales; leaflets 13–15 on each side of the rachis, *arranged singly*

Calyptrocalyx reflexus. Habit, cultivated Floribunda Palms and Exotics, Hawaii (WB).

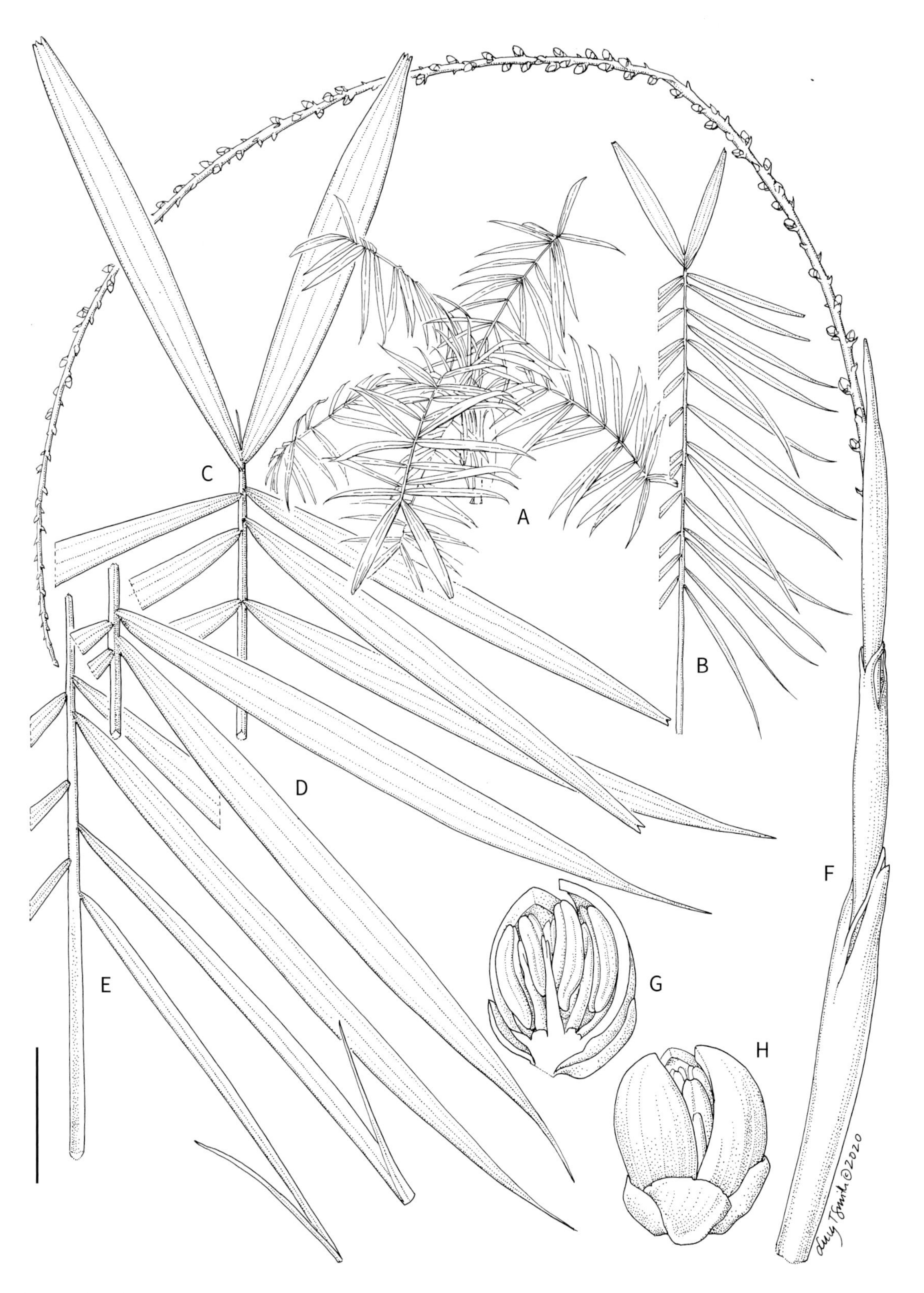

Calyptrocalyx reflexus. **A.** Habit. **B.** Whole leaf diagram. **C.** Leaf apex. **D.** Leaf mid-portion. **E.** Leaf base. **F.** Inflorescence. **G, H.** Male flower whole and in longitudinal section. Scale bar: A = 40 cm; B = 15 cm; C–E = 8 cm; F = 4 cm; G, H = 2.5 mm. All from *Baker 1453*. Drawn by Lucy T. Smith.

Calyptrocalyx reflexus. LEFT TO RIGHT: leaf showing reflexed leaflets; inflorescence; inflorescence with male flowers. Cultivated Floribunda Palms and Exotics, Hawaii (WB).

or in pairs, sometimes in different planes, linear, reflexed or divaricate in pairs, 27–43 × 0.6–3 cm, long acuminate, upper surface mid green, lower surface pale green, upper surface glabrous, lower surface with scattered ramenta. **Inflorescence** arching or pendulous, 90–120 cm long, with 1–3 equal spikes; prophyll 13 × 1.6 cm disintegrating into fibres; peduncular bract 18 × 1 cm; rachilla 37–65 cm long, 2 mm diam. with shallow floral pits. **Male flower** with 9 stamens. **Fruit** globose, 14 × 13 mm, *purple-black*; stigmatic remains apical. **Seed** ellipsoid, 12 × 8 mm; endosperm ruminate.

DISTRIBUTION. Reported from one locality in central northern New Guinea.

HABITAT. In rainforest.

LOCAL NAMES. None recorded.

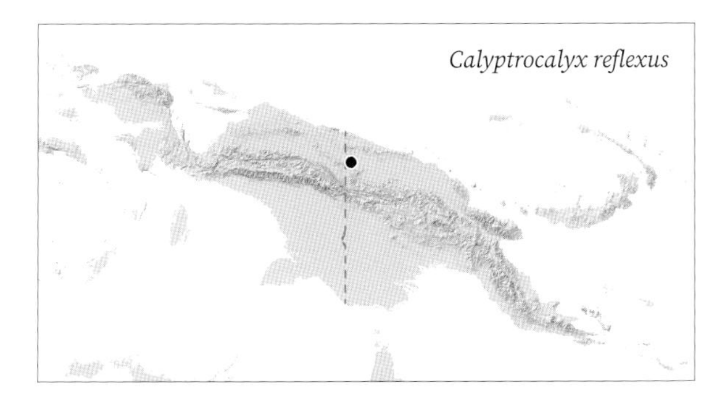

USES. None recorded.

CONSERVATION STATUS. Data Deficient. The distribution and status of this species is insufficiently known.

NOTES. *Calyptrocalyx reflexus* is a distinctive species with its narrow, reflexed leaflets unlike the leaflets of any other species; the young leaf emerging crimson red. It was described from cultivation and we have seen no specimens collected in the wild. It is somewhat similar to *C. lepidotus* but that species has strongly grouped leaflets that are not reflexed and the inflorescences are sparsely scaly, becoming glabrescent.

26. *Calyptrocalyx sessiliflorus* Dowe & M.D.Ferrero

Synonyms: *Linospadix leptostachys* Burret, *Paralinospadix leptostachys* (Burret) Burret

Single-stemmed, *small* palm to 0.5–3 m, crown with 6–15 leaves. **Stem** erect, to 10–20 mm diam. **Leaf** *entire-bifid*, 50–70 × 20–30 cm, somewhat *leathery*, emerging reddish brown; sheath 12–18 cm long, densely brown tomentose, margins fibrous; petiole very short 2–5 cm long, with dense brown tomentum. **Inflorescence** *thread-like, multiple*, 35–90 cm long with *2–5 equal spikes*; prophyll 5–10 cm long; peduncular bracts 15–20 cm long; rachilla 12–20 × 0.2–0.3 cm,

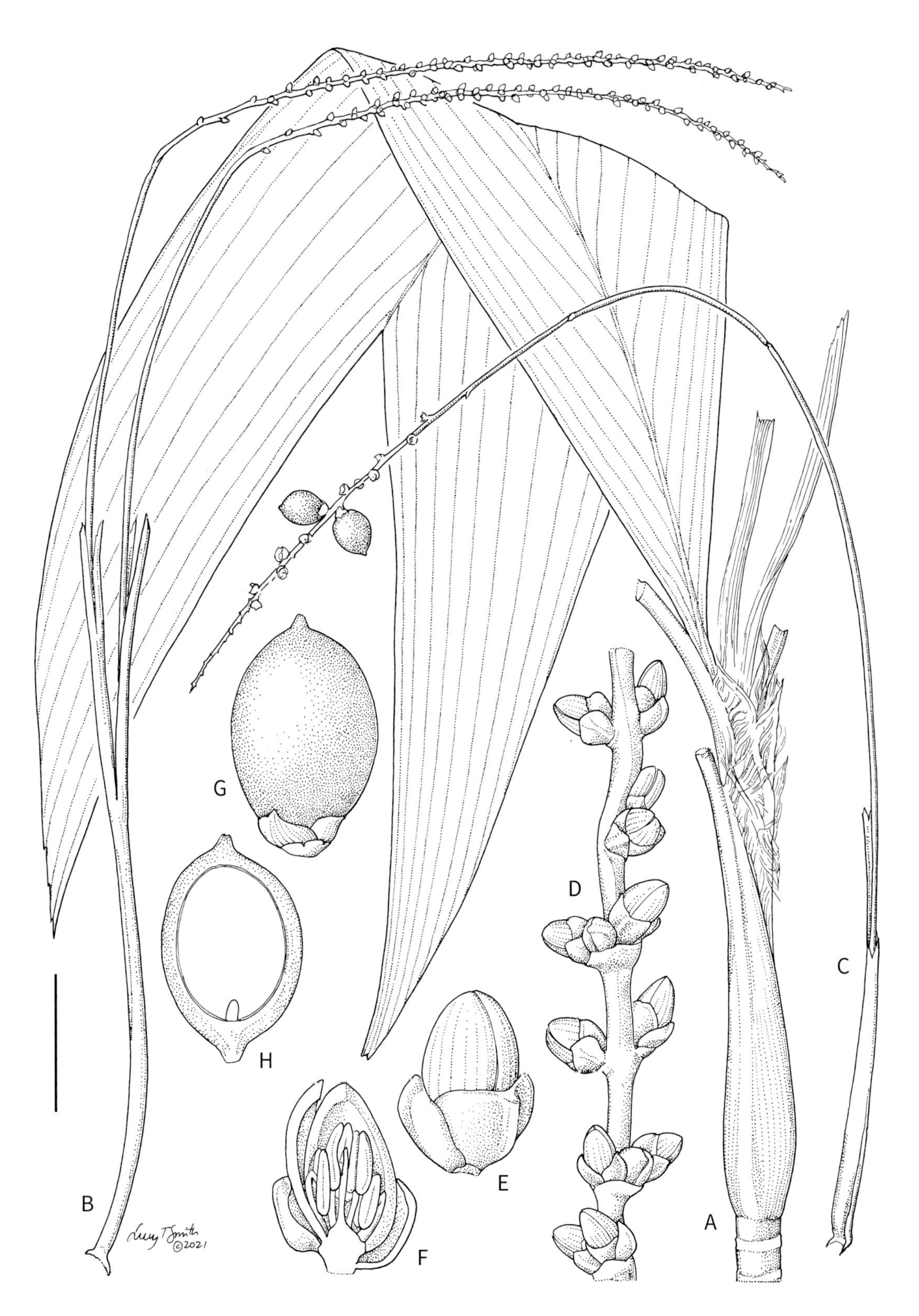

Calyptrocalyx sessiliflorus. **A.** Crown with leaf and leaf sheath. **B.** Inflorescence. **C.** Inflorescence with fruit. **D.** Portion of rachilla with triads. **E, F.** Male flower whole and in longitudinal section. **G, H.** Fruit whole and in longitudinal section. Scale bar: A–C = 4 cm; D = 5 mm; E, F = 2 mm; G, H = 7 mm. All from *Maturbongs 612*. Drawn by Lucy T. Smith.

more or less the same diam. as the peduncle, floral pits very shallow, rather distant. **Male flower** with 9–10 stamens. **Fruit** ellipsoid to globose, 10–12 × 8 mm, *red*, stigmatic remains apical on short beak. **Seed** 7 mm diam., globose; *endosperm homogeneous.*

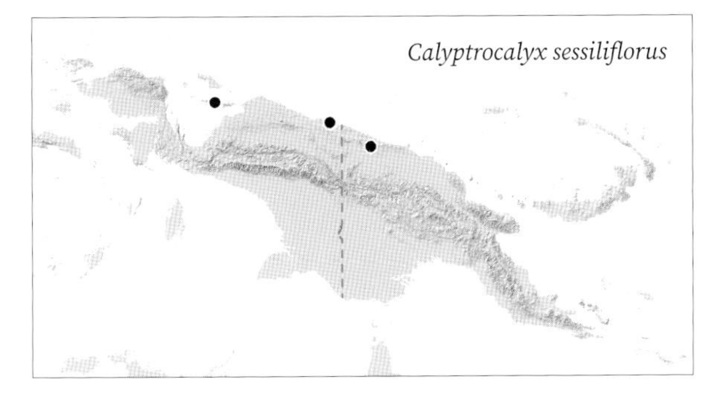

DISTRIBUTION. Northern New Guinea, including Yapen Island.

HABITAT. In rainforest at 500–1,200 m.

LOCAL NAMES. *Ansuni* (Unate), *Dami dami demah* (Wambena), *Siterarum* (Olo).

USES. None recorded.

CONSERVATION STATUS. Vulnerable. *Calyptrocalyx sessiliflorus* has a relatively restricted distribution. Deforestation due to logging and mining concessions is a major threat in its distribution range.

NOTES. *Calyptrocalyx sessiliflorus* is distinctive in its leathery entire bifid leaves, emerging leaves reddish tinged, and thread-like multiple spikes.

27. *Calyptrocalyx spicatus* (Lam.) Blume

Synonyms: *Areca spicata* Lam., *Pinanga globosa* G.Nicholson

Single-stemmed, *moderate to robust* palm to 5–14 m, crown with 9–12 leaves. **Stem** erect, to 15–25 cm diam., internodes 1–10 cm long. **Leaf** *regularly pinnate*, 2.5–4 m long, leaf usually emerging reddish brown; sheath 35–40 cm long, brownish green with scattered brown tomentum, margins fibrous; petiole 15–30 × 2–4 cm, with dense brown tomentum; leaflets 25–40 on each side of the rachis, held more or less horizontally or pendulous, linear, 35–70 × 2–6.5 cm, plicate, upper surface dark green, lower surface paler. **Inflorescence** *robust*, 1.8–2.5 m long, becoming pendulous, with (1)2–3 equal spikes; prophyll 35–40 cm long; peduncular bracts 40–50 cm long; rachilla 1.4–1.9 m long, 2.5–3 cm diam. with crowded deep floral pits. **Male flower** with 60–140 stamens. **Fruit** ellipsoid to

Calyptrocalyx spicatus. FROM TOP: male flowers; fruit. Gag Island (CDH).

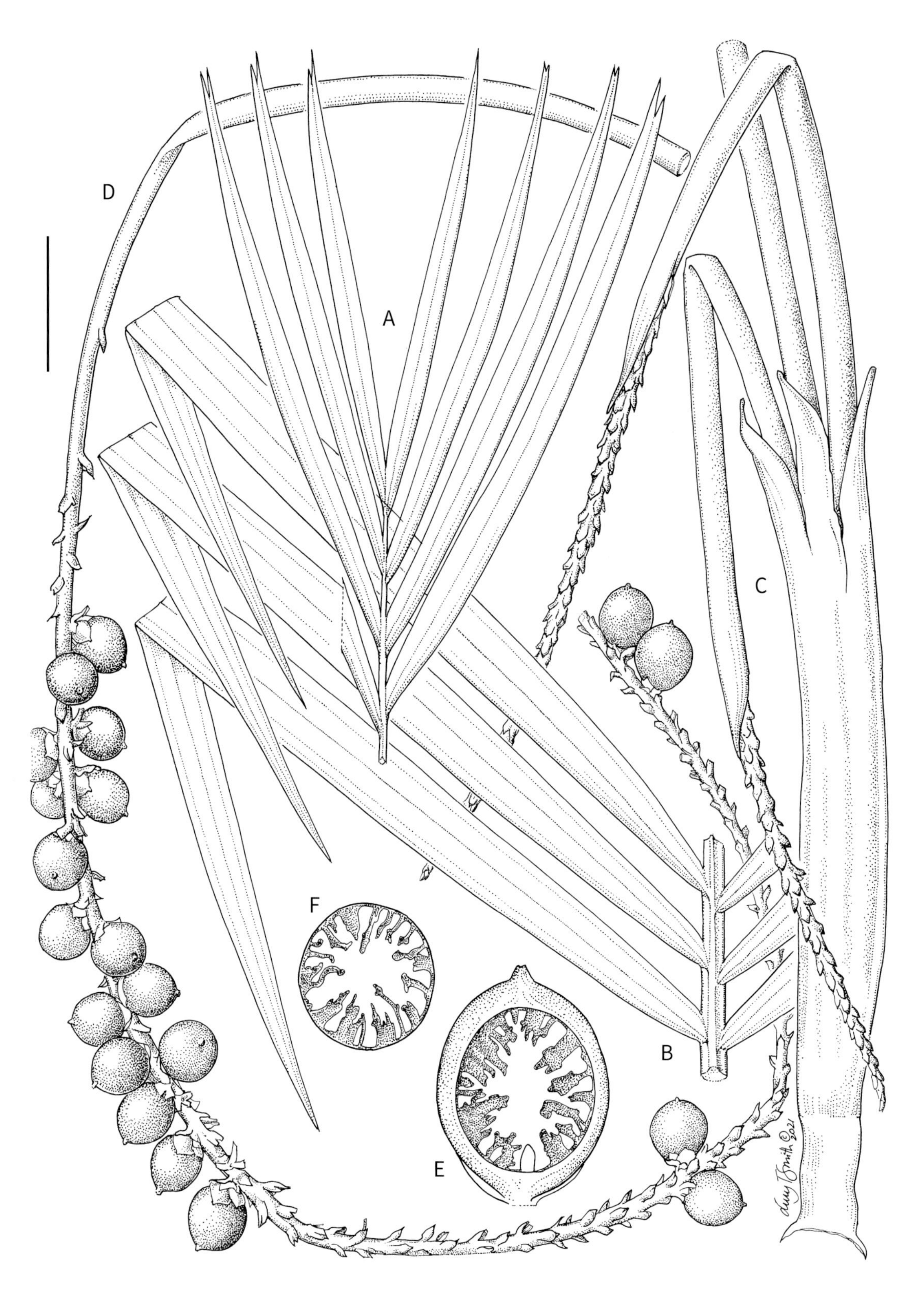

Calyptrocalyx spicatus. **A.** Leaf apex. **B.** Leaf mid-portion. **C.** Inflorescence. **D.** Portion of inflorescence with fruit. **E.** Fruit in longitudinal section. **F.** Seed in cross section. Scale bar: A, B = 12 cm; C, D = 6 cm; E, F = 1.5 cm. A, B, D–F from *Heatubun et al. 749*; C from *Heatubun et al. 750*. Drawn by Lucy T. Smith.

globose, 40–60 × 20–30 mm, *brownish cream or red-brown*, sometimes with brown tomentum; stigmatic remains apical on short beak. **Seed** 20–30 mm diam., globose; *endosperm ruminate*.

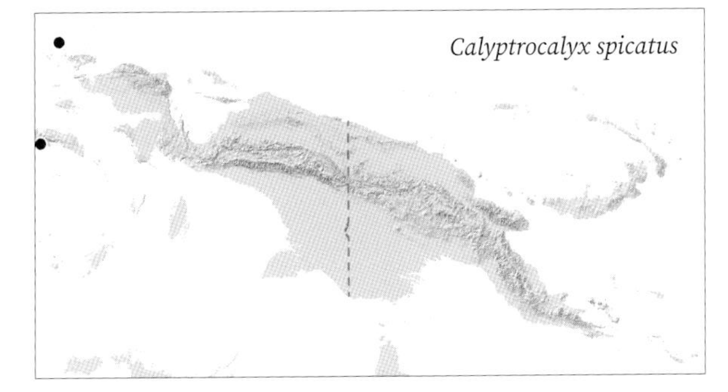

DISTRIBUTION. In New Guinea, known only from the Raja Ampat Islands (Gag Island). Elsewhere in Maluku (Ternate, Halmahera, Buru, Seram, Ambon).

HABITAT. In rainforest at sea level to 60 m.

LOCAL NAMES. None recorded.

USES. None recorded.

CONSERVATION STATUS. Least Concern. This species, though widespread, may be threatened in New Guinea.

NOTES. *Calyptrocalyx spicatus* has been recorded in our area only from Gag Island (*Heatubun 749, 750*). Heatubun *et al.* (2014a) mention that the Gag specimen differs from material from Maluku in the inflorescence always consisting of 3 spikes (as opposed to 1 or 2), the fruit being creamy white and covered with thick brown indumentum (as opposed to being dull red-brown and not covered in dense brown indumentum) (Dowe & Ferrero 2001). However, the number of spikes in *C. spicatus* from Maluku varies from 1–3, not 1–2 (e.g. *Mogea 3133* from Seram has three spikes) and the difference in fruit colour may not be significant. Otherwise the Gag material is a good match for *C. spicatus*. Within *Calyptrocalyx*, the species can only be confused with *C. albertisianus*; see under that species for a comparison.

28. *Calyptrocalyx yamutumene* Dowe & M.D.Ferrero

Multi-stemmed, *small* palm to 2.5 m, crown with 11–14 leaves. **Stems** erect, to 2.5 m, 10–30 mm diam., internodes 2–3 cm long, green. **Leaf** *entire, deeply bifid*, 95–120 × 11–25 cm, somewhat hooded, *leathery*, emerging pale green; sheath 14–17 cm long, margins fibrous; petiole *absent or very short*. **Inflorescence**

Calyptrocalyx yamutumene. FROM TOP: crown with inflorescence; inflorescence with fruit. Cultivated Floribunda Palms and Exotics, Hawaii (WB).

slender, single, 55–96 cm long; prophyll 6.5–12 × 1–1.2 cm; peduncular bract 13–27 × 1.5–4 cm; rachilla 15–34 × 0.2–0.3 cm, floral pits well-spaced, shallow. **Male flower** with 8–9 stamens. **Fruit** ellipsoid, 16–18 × 10–12 mm, *red*, stigmatic remains apical on a very short beak. **Seed** ovoid 9–11 × 6–7 mm; *endosperm homogeneous*.

DISTRIBUTION. Bewani Mountains; sight records from Cyclops Mountains (Dowe & Ferrero 2001).

HABITAT. In rainforest at 100–900 m.

LOCAL NAMES. None recorded.

USES. Crushed endosperm tastes salty and is used to enhance the taste of food.

CONSERVATION STATUS. Critically Endangered. Known with certainty from only one site in an area with intensifying deforestation.

NOTES. This small multi-stemmed species is distinctive in its narrow bifid leaf lacking a petiole, and the inflorescence with a single spike. The fruits are red and the endosperm homogeneous.

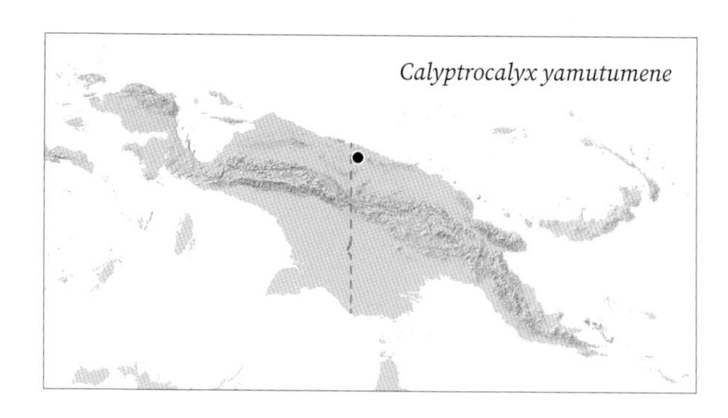

NAMES OF UNCERTAIN APPLICATION

Paralinospadix microspadix (Becc.) Burret

Synonym: *Linospadix microspadix* Becc.

The holotype of this species in Berlin was destroyed and it is not possible to reconstruct what the species looks like from the photograph of the holotype in Singapore and the fragments in Florence.

Calyptrocalyx yamutumene. LEFT TO RIGHT: male flowers; Inflorescence with female flowers. Cultivated Floribunda Palms and Exotics, Hawaii (WB).

***Calyptrocalyx yamutumene*. A.** Habit. **B.** Leaf. **C.** Portion of rachilla. **D.** Male flower. **E**, **F.** Fruit whole and in longitudinal section. **G.** Seed. Scale bar: A = 64 cm; B = 9 cm; C = 9 mm; D = 2.2 mm; E–G = 10 mm. All from *Dowe & Ferrero 508*. Drawn by Lucy T. Smith.

Linospadix albertisianus. Habit, Tamrau Mountains (WB).

Linospadix H. Wendl.

Synonym: *Bacularia* F.Muell.

Very small – leaf pinnate – no crownshaft – no spines – leaflets slightly toothed or pointed

Slender, multi-stemmed dwarf palms, crownshaft absent, monoecious. **Leaf** pinnate, straight, *leaf blade bifid or irregularly divided into leaflets*; sheath splitting to the base opposite the petiole, fibrous at the margins; petiole very short to elongate; leaflets, when leaf split, few, irregularly arranged, with lobed, or pointed tips, arranged regularly. **Inflorescence** *between the leaves, unbranched (spicate), not multiple at the node*; prophyll and peduncular bract similar, enclosing inflorescence in bud, the prophyll persistent, the *peduncular bract borne at the tip of the peduncle just below the flower-bearing portion,* deciduous leaving a scar; peduncle usually longer than the flower-bearing portion (rachilla); rachilla more or less the same diameter as the peduncle. **Flowers** arranged spirally in triads throughout the length of the rachilla, developing in *very shallow pits*; male flowers with 6–12 stamens; female flowers emerging from pits some time after the male flowers drop off. **Fruit** red, ellipsoid to fusiform, stigmatic remains apical, flesh thin, endocarp thin, closely adhering to seed. **Seed** 1, ellipsoid to falcate, endosperm *homogeneous.*

DISTRIBUTION. Six species, one in New Guinea mainland (not known from surrounding islands) and five in Australia.

NOTES. *Linospadix* species are diminutive multi-stemmed understorey palms with fibrous leaf sheaths. Inflorescences are spicate, appearing between the leaves, and are never multiple in the leaf axil as occurs in *Calyptrocalyx*. The peduncular bract is inserted at the tip of the peduncle, just below the flower-bearing portion of the spike, sheaths tightly and is later deciduous at anthesis, leaving a circular scar, which is diagnostically useful. Flowers develop within shallow pits in the inflorescence spike.

In New Guinea, while *Linospadix* may be found in lowland forest it is more common in submontane and montane forest up to 1350 m. It is most easily confused with *Calyptrocalyx*, which also bears spicate inflorescences, but in *Calyptrocalyx* the peduncular bract is inserted very near the prophyll on the peduncle and is persistent. The genus in New Guinea was revised by Dowe & Ferrero (2001).

1. *Linospadix albertisianus* (Becc.) Burret

Synonyms: *Bacularia albertisiana* Becc., *Bacularia angustisecta* Becc., *Bacularia canina* Becc., *Bacularia longicruris* Becc., *Linospadix angustisectus* (Becc.) Burret, *Linospadix caninus* (Becc.) Burret, *Linospadix elegans* Ridley, *Linospadix longicruris* (Becc.) Burret

Clustering, understorey palm to 2 m. **Stems** 2–10, erect or arching to 4–10 mm diam.; internodes to 8 cm long; crown with 4–12 leaves. **Leaf** bifid, or irregularly or regularly pinnate, 20–90 cm long; sheath 10–12 cm long, slightly swollen, green, glabrous, margins smooth; petiole 5–12 × 0.3 cm, glabrous; leaflets 1–13 on each side of the rachis, held more or less horizontally, 20–30 cm long, width various, upper surface glossy mid to dark green, lower surface paler, ribs and veins prominent on both surfaces. **Inflorescence** very slender, 20–65 cm long; prophyll to 12 cm long; peduncular bract 9–25 cm long; rachilla

8–20 cm long, 2–3 mm diam. **Male flower** elongate, curved or subcylindrical, 3.5–4 mm long in bud, with 6–12 stamens. **Fruit** ellipsoid to spindle-shaped, sometimes slightly curved, 13–28 × 3.5–6 mm, red; stigma remains apical on a short beak. **Seed** to 10–12 × 2–3 mm, fusiform ellipsoid; endosperm homogeneous.

DISTRIBUTION. Widespread throughout mainland New Guinea.

HABITAT. In rainforest from lowlands to 1350 m.

LOCAL NAMES. *Apalop* (Waskuk), *Hara Cho* (North Cyclops), *Ibu* (Wagu), *Nana* (Sentani), *Niyo* (Orne), *Ubo* (Wagu), *Yamu* (Bewani), *Yamu Ten* (Bewani), *Yalilim* (Olo).

USES. Stems used for stirring and serving sago.

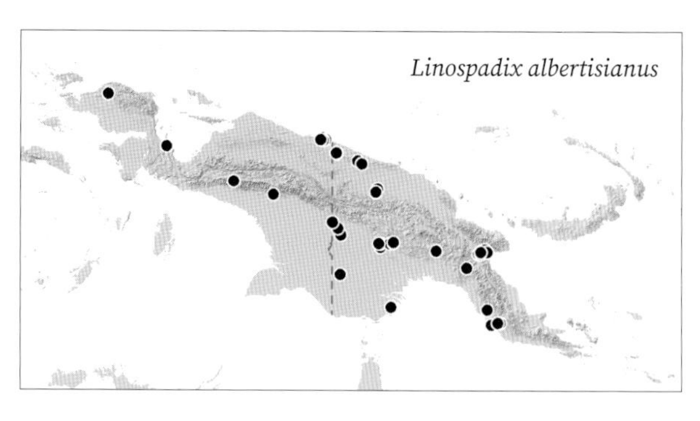

CONSERVATION STATUS. Least Concern.

NOTES: In all five different taxa have been recognised in *Linospadix* in New Guinea. In the latest revision (Dowe & Ferrero 2001) only two of these are accepted – *L. albertisianus* and *L. caninus*. The former is said to be distinguished by the prominent veins on both leaf surfaces and the asymmetrical fruit, whereas the latter has prominent veins only on the upper leaf surface and slightly curved fruit.

We have examined many specimens of *Linospadix* from New Guinea and can only recognise one variable species. The leaf character used to separate the two taxa by Dowe and Ferrero seems not to work and, in fact, the leaf of the type of *L. caninus* (*Versteegh 1635*) has veins prominent on both surfaces. The slightly curved fruit in *L. caninus* seems to us to be insufficient to recognise a separate taxon when all other features are part of a continuum of variation.

Linospadix albertisianus. LEFT TO RIGHT FROM TOP: inflorescences, Tamrau Mountains (WB); inflorescence with splitting peduncular bract, Mount Bosavi (WB); male flowers, Bewani (AB); fruit, Tamrau Mountains (WB).

Linospadix albertisianus. **A.** Habit. **B.** Stem with leaf sheath, leaves and inflorescence. **C.** Leaf. **D.** Inflorescence prophyll and peduncular bract. **E**, **F.** Portion of inflorescence with peduncular bract detail, two views. **G.** Detail of peduncular bract scar. **H.** Portion of rachilla with male flower buds. **I.** Male flower in longitudinal section. **J.** Female flower. **K.** Fruit. Scale bar: A = 42 cm; B = 4 cm; C = 8 cm; D = 3 cm; E–G = 1 cm; H = 3.3 mm; I = 1.8 mm J = 2 mm; K = 7 mm. A from *Baker et al. 629*; B, D–G from *Katik LAE 70786*; C from *Baker & Kage 663*; H, I from *Baker et al. 615*; J from *Craven & Schodde 1402*; K from *van Royen & Sleumer 5721*. Drawn by Lucy T. Smith.

Ptychosperma keiense. Habit, Manokwari-Ransiki road (JD).

Ptychosperma Labill.

Synonyms: *Actinophloeus* (Becc.) Becc., *Romanovia* Sander ex André, *Seaforthia* R.Br., *Strongylocaryum* Burret

Small to medium – leaf pinnate – crownshaft – no spines – leaflets jagged

Slender or moderately slender, single- or multi-stemmed tree palms, crownshaft present, monoecious. **Leaf** pinnate, slightly arching; sheath tubular; petiole present; leaflets few to numerous, reduplicate, linear to broadly wedge-shaped, with jagged tips, arranged regularly or irregularly, horizontal. **Inflorescence** *below the leaves,* branched 2–4 orders, branches widely spreading; prophyll and peduncular bract similar, enclosing inflorescence in bud, dropping off as inflorescence expands; peduncle shorter than inflorescence rachis (rarely longer); rachillae slender to quite robust, *green or reddish at anthesis,* straight or curving. **Flowers** in triads throughout the length of the rachillae, not developing in pits, male flowers larger than female flowers, bullet-shaped, purplish or whitish. **Fruit** *red, orange, purple or black,* globose to ellipsoid, stigmatic remains apical, flesh quite thick and juicy, sometimes with irritant needle-shaped crystals, exocarp wrinkled or smooth (and conforming to the endocarp) when dry, *endocarp pale, thin, grooved and lobed* (obscurely so in *P. ramosissimum*). **Seed** 1, *grooved and lobed, lobes rounded, sharp or blocky (angular),* endosperm homogeneous or ruminate.

DISTRIBUTION. Around 25 species from New Guinea to Australia and the Solomon Islands, with 22 species in the New Guinea region.

NOTES. *Ptychosperma* species are slender, single- or multi-stemmed, mid-storey tree palms, with inflorescences borne below the leaves that sometimes have colourful branches. The fruits are red, purple, orange or black, small (ca. 1 cm), each with a pale, hard endocarp enclosing the seed that is angled or grooved. They occur in lowland to montane rainforest from sea level to 1,900 m. *Ptychosperma* can be confused with a number of genera: *Brassiophoenix,* distinguished by its broad, wedge-shaped leaflets, orange, medium-sized fruit (ca. 3.5 cm long), and pale endocarp with conspicuous grooves and ridges; *Dransfieldia,* with pointed leaflet tips and a persistent prophyll and terete seed; *Drymophloeus,* which is distinctive for its stilt roots and a usually persistent prophyll, and *Ptychococcus,* which is a moderate to robust palm bearing quite large fruit (4–6 cm long) containing a black or dark brown endocarp with conspicuous grooves and ridges.

The taxonomic history of this genus is complex, particularly as many species were admired for their ornamental qualities and introduced into cultivation before they were named by botanists. The first modern revision of the genus is that of Essig (1978), who undertook fieldwork in Papua New Guinea, described eight new species and recognized four subgenera. The species comprising two of his subgenera have been returned to *Ponapea,* which Moore and Fosberg (1956) had subsumed in *Ptychosperma*. Essig's revision was the basis for this treatment, along with new data from living collections, molecular analyses (Zona *et al.* 2011, Alapetite *et al.* 2014) and fieldwork in Papua. Additional fieldwork, however, is still much needed in the Bismarck Archipelago, D'Entrecasteaux Islands, and Louisiade Archipelago. Our knowledge of *Ptychosperma* outside of New Guinea (Bougainville and the Solomon Islands) is far from complete.

Key to the species of *Ptychosperma* in New Guinea

1. Mature fruits orange or red 2
1. Mature fruits purple or black. 13

2. Endosperm ruminate 3
2. Endosperm homogeneous 8

3. Leaflets strongly cuneate to narrowly wedge-shaped, widest toward the apex of the leaflet, apices truncate to notched . . . . **3. *P. caryotoides***
3. Leaflets lanceolate to linear (rarely narrowly wedge-shaped), widest in the middle, leaflet apices oblique or obliquely notched 4

4. Inflorescence with 3 or 4 orders of branching; rachillae ≤10 cm long 5
4. Inflorescence with 2 (rarely 3) orders of branching; rachillae >10 cm long 6

5. Peduncle >20 cm long; fruit conforming to the endocarp when dry; seed 10–11 mm long, strongly lobed or angled . . . . **20. *P. tagulense***
5. Peduncle ca. 7 cm long; pericarp not conforming to the endocarp when dry; seed 6–8 mm long, nearly circular in cross section, lobes indistinct . . . . **16. *P. ramossisimum***

6. Fruit smooth, conforming to the endocarp when dry; stigma curved on dried fruit. . . . **6. *P. gracile***
6. Fruit wrinkled and not conforming smoothly to endocarp when dry; stigma erect on dried fruit. . 7

7. Largest bracts subtending inflorescence branches present as low ridges <1 cm long or absent; rachillae (dried) ≤2 mm diam.; seed (9–)10–13 mm long; endosperm strongly ruminate . . . . **17. *P. rosselense***
7. Largest bracts subtending inflorescence branches >1 cm long, leathery; rachillae (dried) >2 mm diam.; seed 6–10 mm long; endosperm weakly and shallowly ruminate . . . . **8. *P. lauterbachii***

8. Leaflets in middle third of leaf regularly arranged. 9
8. Leaflets in middle third of leaf irregularly arranged 12

9. Leaflets 4.0–7.5 cm wide; distance between adjacent leaflets ≥6 cm; apices oblique; seed lobes usually unequal in size (three deep grooves and two shallow grooves) . . . . **7. *P. keiense***
9. Leaflets 1.8–3.3(–4.0) cm wide; distance between leaflets <6 cm; apices notched or obliquely notched; seed lobes more or less equal. 10

10. 30 or more female flowers per 5 cm of rachilla . . . . **18. *P. sanderianum***
10. 10–18 female flowers per 5 cm of rachilla. 11

11. Terminal leaflets not wider than subterminal leaflets ..... **5. *P. furcatum***

11. Terminal leaflets wider than subterminal leaflets ..... **13. *P. nicolai***

12. Leaflets arranged in more than one plane; male flower buds <6 mm long; stamens 27 or 28 ..... **11. *P. microcarpum***

12. Leaflets arranged in one plane; male flower buds >6 mm long; stamens 42–51 ..... **14. *P. propinquum***

13. Endosperm ruminate 14

13. Endosperm homogeneous 16

14. Leaflets in middle third of leaf irregularly arranged; fruits <12 mm long; endosperm profoundly and strongly ruminate ..... **12. *P. mooreanum***

14. Leaflets in middle third of leaf regularly arranged; fruits >12 mm long; endosperm weakly ruminate 15

15. Palm of mangrove swamps; ca. 27 female flowers per 5 cm of rachilla; fruit ca. 14 mm long ..... **10. *P. mambare***

15. Palm of rainforests; 11–18 female flowers per 5 cm of rachilla; fruit 14.7–19.8 mm long ..... **19. *P. schefferi***

16. Leaflets linear-lanceolate 17

16. Leaflets wedge-shaped or narrowly wedge-shaped with deeply notched apices 18

17. Leaflets irregularly arranged throughout, apex oblique; 6–11 female flowers per 5 cm of rachilla ..... **1. *P. ambiguum***

17. Leaflets on middle third of leaf regularly arranged, apex obliquely excavate; 15–21 female flowers per 5 cm of rachilla ..... **9. *P. lineare***

18. Leaflets irregularly arranged throughout 19

18. Leaflets regularly arranged in the middle of leaf but irregularly arranged (sometimes appearing grouped) at base or apex 20

19. Fruit ca. 13 mm long; seed ca. 8 mm long ..... **4. *P. cuneatum***

19. Fruit ca. 20 mm long; seed ca. 15 mm long ..... **2. *P. buabe***

20. Lower surfaces of the leaflets completely covered with conspicuous stellate or ragged scales; ramenta absent ..... **21. *P. vestitum***

20. Lower surfaces of the leaflets smooth except for scattered ramenta along the midveins and/or at the base of the leaflets 21

21. Peduncle <6 cm; inflorescence with 1 or 2 orders of branching; largest bracts subtending inflorescence branches >1 cm long; stamens ca. 55 ..... **22. *P. waitianum***

21. Peduncle ≥6 cm long, inflorescence with 2 or 3 orders of branching; largest bracts subtending inflorescence branches absent or obscure; stamens ca. 17 ..... **15. *P. pullenii***

Ptychosperma ambiguum. **A.** Whole leaf diagram. **B.** Leaf apex. **C.** Leaf mid-portion. **D.** Leaf base. **E.** Inflorescence with fruit. **F.** Male flower bud. **G.** Female flower. **H.** Fruit. **I.** Endocarp. **J.** Seed in transverse section. Scale bar: A = 30 cm; B–D = 8 cm; E = 4 cm; F = 5 mm; G = 4 mm; H–J = 1 cm. A–G from *Vink 12128*; H–J = *Beccari s.n.* Drawn by Lucy T. Smith.

1. *Ptychosperma ambiguum* (Becc) Becc. ex Martelli

Synonyms: *Actinophloeus ambiguus* (Becc.) Becc., *Drymophloeus ambiguus* Becc., *Saguaster ambiguus* (Becc.) Kuntze

Single-stemmed, understorey palm 2–5 m tall. **Stem** ca. 2.1 cm diam. **Leaf** ca. 1.5 m long; sheath ca. 24.7 cm long; petiole ca. 37.5 cm long; rachis length not recorded; ca. 22 pairs of leaflets, mid-leaf leaflet ca. 57 × ca. 2.6 cm, *linear-lanceolate, irregularly arranged, in one plane*. **Inflorescence** 21–30 cm long, branched to 2 orders; peduncle 3.2–4.8 × 0.5–0.6 cm wide; rachillae 16.5–20.5 cm long, ca. 2 mm diam., colour not recorded, with 6–11 female flowers per 5 cm. **Male flower** ca. 6 × 3 mm in bud; colour not recorded, stamens ca. 26. **Female flower** ca. 5 × ca. 6 mm, globose. **Fruit** 15–18 × 8–10 mm, *black*; exocarp wrinkled when dry; endocarp ca. 17 × 6–8 mm. **Seed** 11–12 × 7–8 mm, *lobes blocky; endosperm homogeneous.*

DISTRIBUTION. Fakfak (Onin) Peninsula.

HABITAT. Rainforest on limestone, at ca. 50 m elevation.

LOCAL NAMES. None recorded.

USES. None recorded.

CONSERVATION STATUS. Critically Endangered. *Ptychosperma ambiguum* has a restricted distribution. Deforestation due to mining and logging concessions is a major threat in its distribution range.

NOTE. As the species epithet suggests, this species is confusing and not easily differentiated from others. Nevertheless, the combination of black fruit, homogeneous endosperm, and irregularly arranged, linear-lanceolate leaves is sufficient to diagnose the species. It is, however, poorly known and not in cultivation.

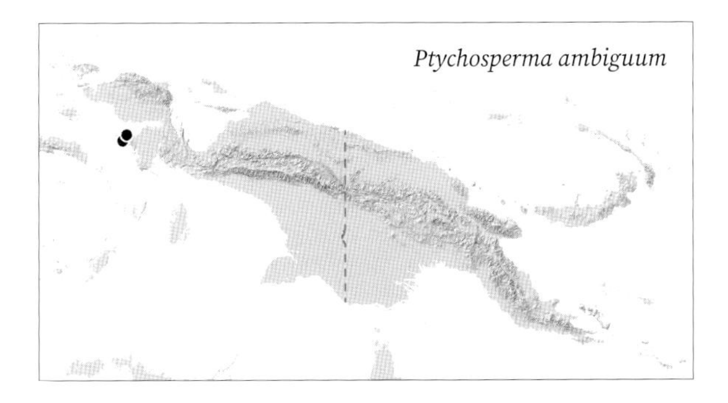

2. *Ptychosperma buabe* Essig

Single-stemmed, understorey palm ca. 3 m tall bearing ca. 6 leaves. **Stem** 4.2–6.0 cm diam. **Leaf** 2–3.5 m long; sheath 50–60 cm long; petiole ca. 68 cm long; rachis 136–155 cm long; 16–18 pairs of leaflets, *leaflets arranged irregularly in one plane*, mid-leaf leaflet 30–41 × 10–11 cm, *wedge-shaped, terminal leaflets wider than subterminals*. **Inflorescence** ca. 40 cm long, branched to 2 orders; peduncle 5.0–7.5 × ca. 1.4 cm wide; rachillae 32–43 cm long, 3–4 mm diam., colour not recorded, with 8 or 9 female flowers per 5 cm. **Male flower** ca. 8 × 3–5 mm in bud; colour not recorded, stamens ca. 47. **Female flower** not seen. **Fruit** ca. 20 × 10–11 mm, *black*; exocarp wrinkled when dry; endocarp not seen. **Seed** not seen at maturity; *endosperm homogeneous.*

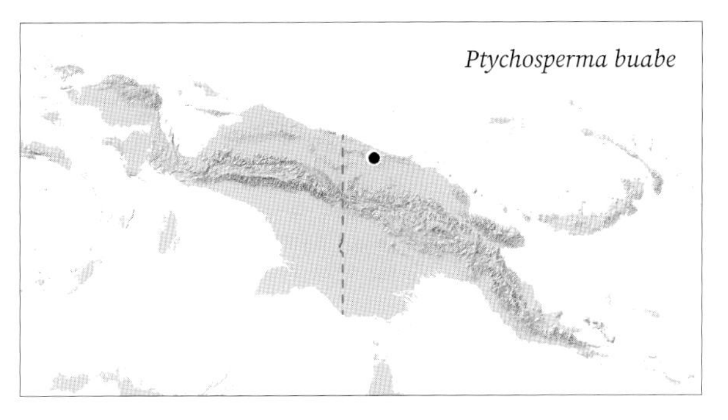

DISTRIBUTION. Torricelli Mountains.

HABITAT. Rainforest at 760–1,000 m in elevation.

LOCAL NAMES. *Buabe* (Wapi).

USES. The wood is used for making spears.

CONSERVATION STATUS. Critically Endangered (IUCN 2021). *Ptychosperma buabe* is known from only two sites that are threatened by logging concessions.

NOTE. Among the black-fruited species with homogeneous endosperm, this relatively large *Ptychosperma* is recognised by its wedge-shaped leaflets, arranged irregularly in one plane throughout the leaf, terminal leaflets larger than subterminals, and fruits ca. 20 mm long.

3. *Ptychosperma caryotoides* Ridl.

Synonyms: *Actinophloeus montanus* (K.Schum. & Lauterb.) Burret, *Drymophloeus montanus* K.Schum. & Lauterb., *Ptychosperma discolor* Becc., *Ptychosperma josephense* Becc., *Ptychosperma leptocladum* Burret, *Ptychosperma montanum* (K.Schum. & Lauterb.) Burret, *Ptychosperma polyclados* Becc., *Ptychosperma ridleyi* Becc., *Ptychosperma sayeri* Becc.

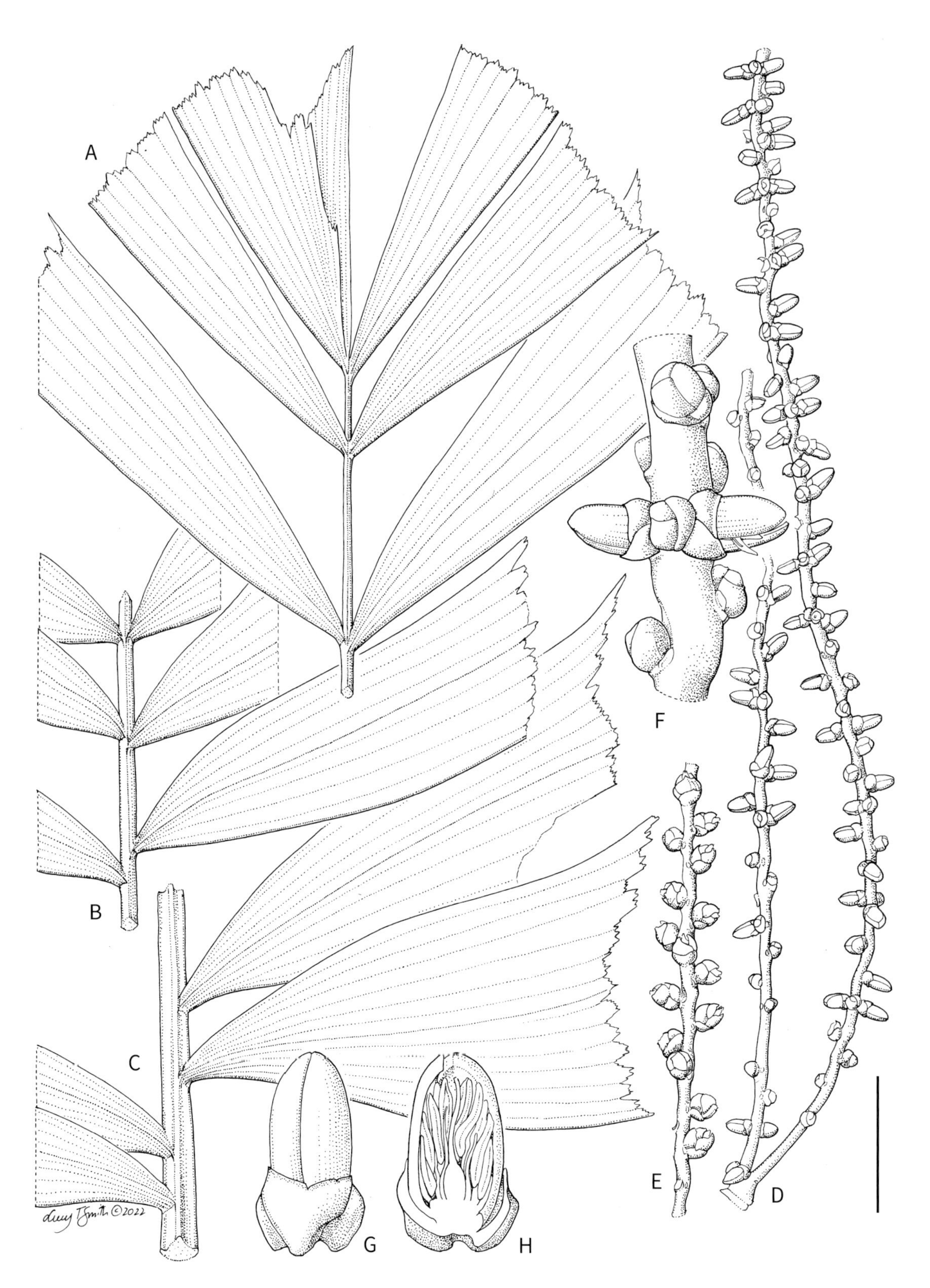

Ptychosperma buabe. **A.** Leaf apex. **B.** Leaf mid-portion. **C.** Leaf base. **D.** Portion of inflorescence with male and female flower buds. **E.** Portion of inflorescence with female flowers. **F.** Triad. **G, H.** Male flower whole and in longitudinal section. Scale bar: A–C = 8 cm; D, E = 4 cm; F = 1 cm; G, H = 7 mm. All from *Darbyshire 352*. Drawn by Lucy T. Smith.

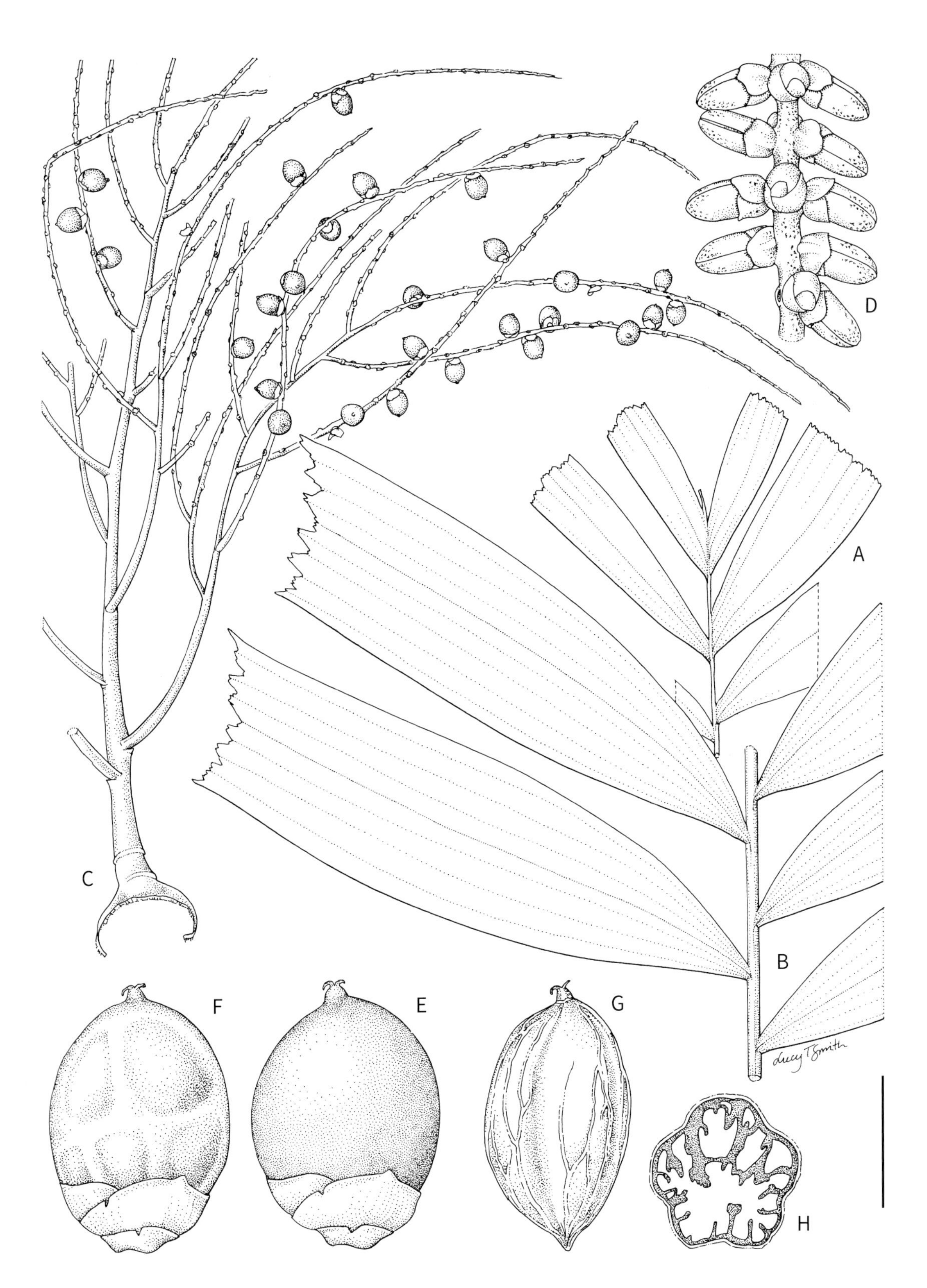

Ptychosperma caryotoides. **A.** Leaf apex. **B.** Leaf mid-portion. **C.** Portion of inflorescence with fruit. **D.** Portion of rachilla with flowers. **E.** Fruit. **F.** Fruit, dried. **G.** Endocarp. **H.** Seed in transverse section. Scale bar: A–C = 6 cm; D = 1 cm; E– H = 7 mm. A–C, E–H from *Zona 611*; D from *Brass 23783*. Drawn by Lucy T. Smith.

Single-stemmed, understorey to mid-storey palm 2–9 m tall, bearing 5 or 6 leaves. **Stem** 1–10 cm diam. **Leaf** 1–2.5 m long; sheath 18–45 cm long; petiole 10–50 cm long; rachis 35–153 cm long; 5–39 pairs of leaflets, *arranged irregularly basally, regularly distally, and in one plane,* mid-leaf leaflet 17–45 × 4–16 cm, *broadly to narrowly wedge-shaped,* borne 8–18 cm apart; *terminal leaflets larger than subterminal leaflets.* **Inflorescence** 10–50 cm long, branched to 2 or 3 orders (rarely 1); peduncle 3–9 × 0.5–1.4 cm; rachillae 6–34 cm long, 1–2 mm diam., green at anthesis, with 8–45 female flowers per 5 cm. **Male flower** ca. 9 mm long in bud, whitish; stamens 30–34. **Female flowers** not seen. **Fruit** 12–21 × 7–13 mm, *orange to red*; exocarp wrinkled when dry; endocarp 12–18 × 5–8 mm. **Seed** 8–12 × 6–8 mm, *lobes rounded; endosperm ruminate.*

Ptychosperma caryotoides

DISTRIBUTION. South-eastern Papua New Guinea.

HABITAT. Rainforest in mountains, usually at elevations of 50–1,200 m.

LOCAL NAMES. None recorded.

USES. None recorded.

CONSERVATION STATUS. Least Concern (IUCN 2021). Monitoring is necessary due to the presence of logging and mining concessions in its distribution range.

NOTES. This is one of the most polymorphic species, as evidenced by the numerous names that have been applied to this species in all its guises. Consequently, it is one of the most difficult to identify and may, in fact, comprise a complex of closely related taxa. More detailed field studies, coupled with molecular genetic studies, are needed to resolve the taxonomy of this species. Alapetite *et al.* (2014) found that it is phylogenetically close to *P. gracile, P. salomonense* (Solomon Islands) and *P. elegans* (Australia), all of which have red fruits and ruminate endosperms. *Ptychosperma caryotoides* is distinguished from these and other species by its wedge-shaped leaflets, terminal leaflets larger than subterminals, finely rugose exocarp (when dry), and rounded lobes on the seed.

4. *Ptychosperma cuneatum* (Burret) Burret

Synonyms: *Actinophloeus cuneatus* Burret, *Actinophloeus hospitus* Burret, *Actinophloeus macarthurii* var. *hospitus* (Burret) L.H.Bailey, *Ptychosperma hospitum* (Burret) Burret, *Ptychosperma tenue* Becc.

Multiple-stemmed, understorey palm bearing 5–8 leaves per stem. **Stem** 3–5 m tall, diam. 2–3 cm. **Leaf** ca. 1.3 m, sheath 34–43 cm long; petiole ca. 34 cm long; rachis length not recorded, ca. 21 pairs of leaflets, *arranged irregularly in one plane,* mid-leaf leaflet 33–49 × 4.0–8.5 cm, *wedge-shaped;* terminal leaflets not larger than subterminal leaflets. **Inflorescence** 43–51 cm long, branched to 2 orders; peduncle 5–6 × ca. 0.9 cm wide; rachillae 22.5–24.5 cm long, ca. 2 mm diam., colour not recorded, with 8–9 female flowers per 5 cm. **Male flower** 4–5 × ca. 3 mm in bud, colour not

Ptychosperma cuneatum. Leaf, cultivated Fairchild Tropical Botanic Garden, Florida (SZ).

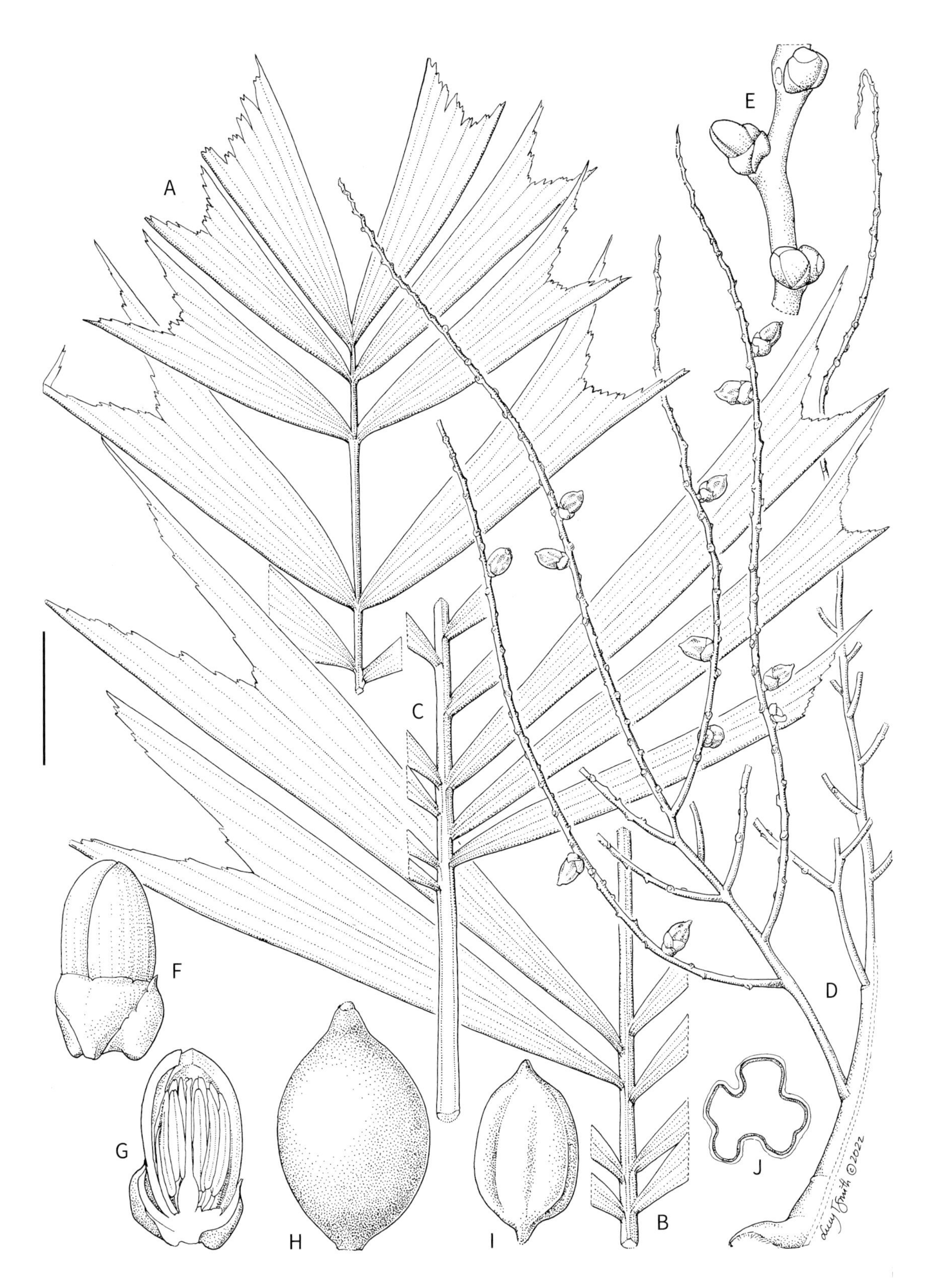

Ptychosperma cuneatum. **A.** Leaf apex. **B.** Leaf mid-portion. **C.** Leaf base. **D.** Portion of inflorescence with fruit. **E.** Portion of rachilla with flower buds. **F**, **G.** Male flower whole and in longitudinal section. **H.** Fruit. **I.** Endocarp. **J.** Seed in transverse section. Scale bar: A–C = 6 cm; D = 4 cm; E = 7 mm; F, G = 3.3 mm, H–J = 5 mm. All from *Heatubun et al. 282*. Drawn by Lucy T. Smith.

recorded; stamens ca. 22. **Female flowers** not seen. **Fruit** 11–13 × 6–7 mm, *black*; exocarp wrinkled when dry; endocarp not seen. **Seed** ca. 8 × ca. 4 mm, *lobes rounded; endosperm homogeneous.*

DISTRIBUTION. Central northern New Guinea, near Jayapura.

HABITAT. Rainforest, recorded at 80–290 m elevation.

LOCAL NAMES. None recorded.

USES. None recorded.

CONSERVATION STATUS. Endangered. *Ptychosperma cuneatum* has a restricted distribution. Deforestation due to land conversion for oil palm and logging concessions is a major threat in its distribution range.

NOTE. This species was first described from cultivated material in the Kebun Raya, Bogor, Indonesia. According to Essig (1978), the source of the material was the vicinity of Lake Sentani near Jayapura. It is phylogenetically close to *P. pullenii* (Alapetite *et al.* 2014). *Ptychosperma cuneatum* is similar in appearance to *P. caryotoides*, but *P. cuneatum* has black fruits (vs. orange to red in *P. caryotoides*). Furthermore, the leaflet arrangement differs: leaflets are regularly arranged throughout the leaf in *P. cuneatum*, and irregularly arranged at the base of the leaf of *P. caryotoides*. The female flowers are sweetly fragrant.

The name *Ptychosperma hospitum*, which is a synonym of *P. cuneatum*, has been widely misapplied to palms now identified as *P. propinquum*.

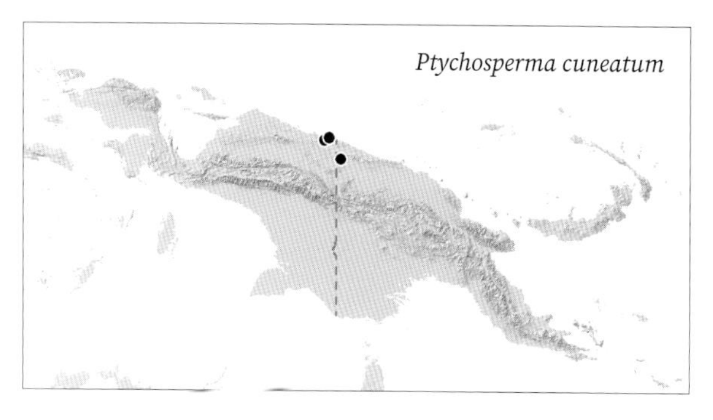

5. *Ptychosperma furcatum* (Becc.) Becc. ex Martelli

Synonym: *Actinophloeus furcatus* Becc.

Single-stemmed or multiple-stemmed, understorey to mid-storey palm 0.6–4.0 m tall. **Stem** ca. 4 cm diam. **Leaf** 1.5–2.5 m long; sheath 30–40 cm long; petiole 15–41 cm long; rachis 98–120 cm long; 27–35 pairs of leaflets, *leaflets arranged regularly in one plane*, mid-leaf leaflet 29–48 × 1.6–4.2 cm wide, linear, borne 3–5 cm apart; terminal leaflets not larger than subterminal leaflets. **Inflorescence** 30–58 cm long, branched to 2 orders; peduncle 3–6 × 0.8–1.1 cm; rachillae 15–35 cm long, 1–2 mm diam., colour not recorded, with 11–18 female flowers per 5 cm. **Male flower** ca. 4 × ca. 2 mm in bud, colour not recorded; stamens 16–23. **Female flowers** ca. 5 × 3 mm, ovoid. **Fruit** 11–12 × 5–8 mm, *orange to red*; exocarp wrinkled when dry; endocarp ca. 9 × ca. 5 mm. **Seed** 7–8 × 4–6 mm, deeply lobed, *lobes rounded; endosperm homogeneous.*

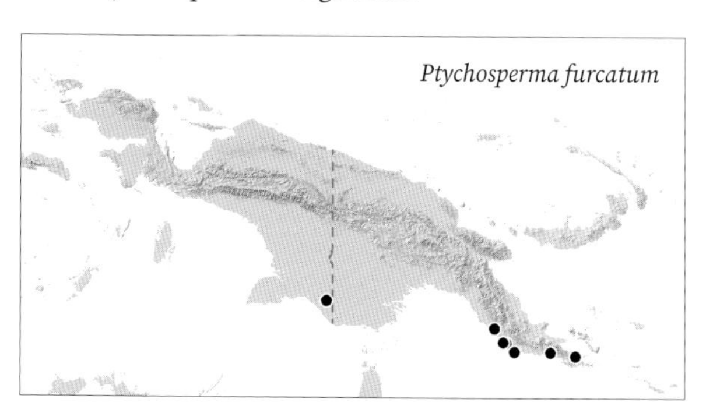

DISTRIBUTION. South and south-eastern New Guinea.

HABITAT. Rainforest below 20–200 m elevation.

LOCAL NAMES. None recorded.

USES. None recorded.

CONSERVATION STATUS. Least Concern (IUCN 2021).

NOTE. *Ptychosperma furcatum* might be confused with the poorly known *P. nicolai*, but the former species has terminal leaflets not wider than the subterminals (vs. subterminals larger in *P. nicolai*).

6. *Ptychosperma gracile* Labill.

Synonyms: *Saguaster gracilis* (Labill.) Kuntze, *Seaforthia ptychosperma* Mart., *Ptychosperma hartmannii* Becc.

Single-stemmed, mid-storey palm 3–8 m tall. **Stem** 3–8 cm diam. **Leaf** 2.5–4 m long; sheath 47–64 cm long; petiole 27–60 cm long; rachis 1.4–2.2 m long; 20–27 pairs of leaflets, arranged regularly in one plane, mid-leaf leaflet 22–68 × 2.5–7.6 cm, linear-lanceolate; terminal leaflets not larger than subterminal leaflets. **Inflorescence** ca. 71 cm long, branched to 2 (or 3) orders; peduncle 2.7–4.0 × 2.4–4.0 cm; rachillae 12–44 cm long, 1–3 mm diam., colour not recorded, with 8–12 female flowers per 5 cm. **Male flower** ca. 5 × ca. 3 mm in bud, whitish; stamens ca. 24. **Female flowers** not seen. **Fruit** 14–18 × 7–13 mm, *orange to red; exocarp*

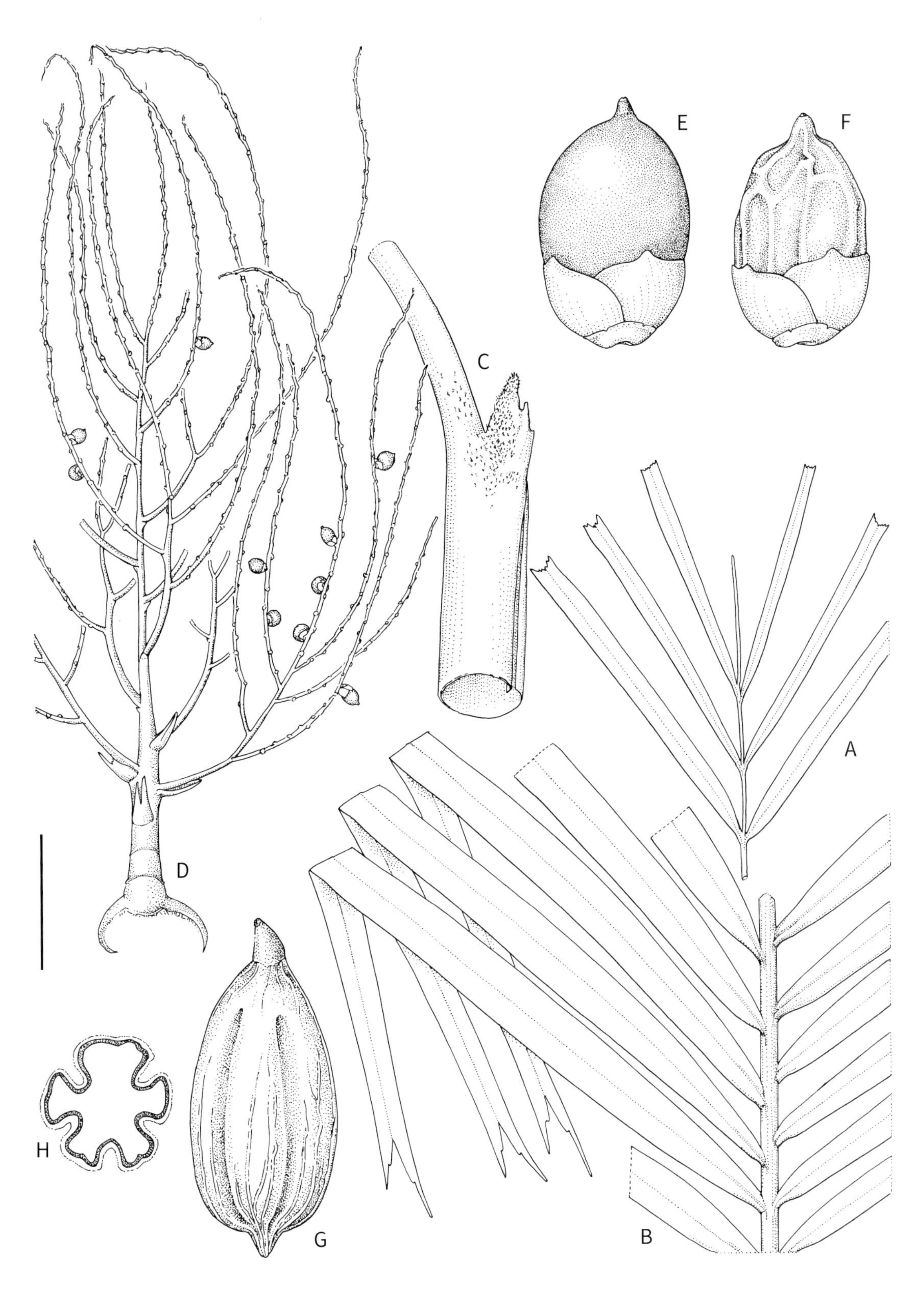

Ptychosperma furcatum. **A.** Leaf apex. **B.** Leaf mid-portion. **C.** Leaf sheath with ligule. **D.** Inflorescence with fruit. **E.** Fruit. **F.** Fruit, dried. **G.** Endocarp. **H.** Seed in transverse section. Scale bar: A, B, D = 6 cm; C = 3 cm; E, F = 7 mm; G, H = 5 mm. All from *Schodde 2720*. Drawn by Lucy T. Smith.

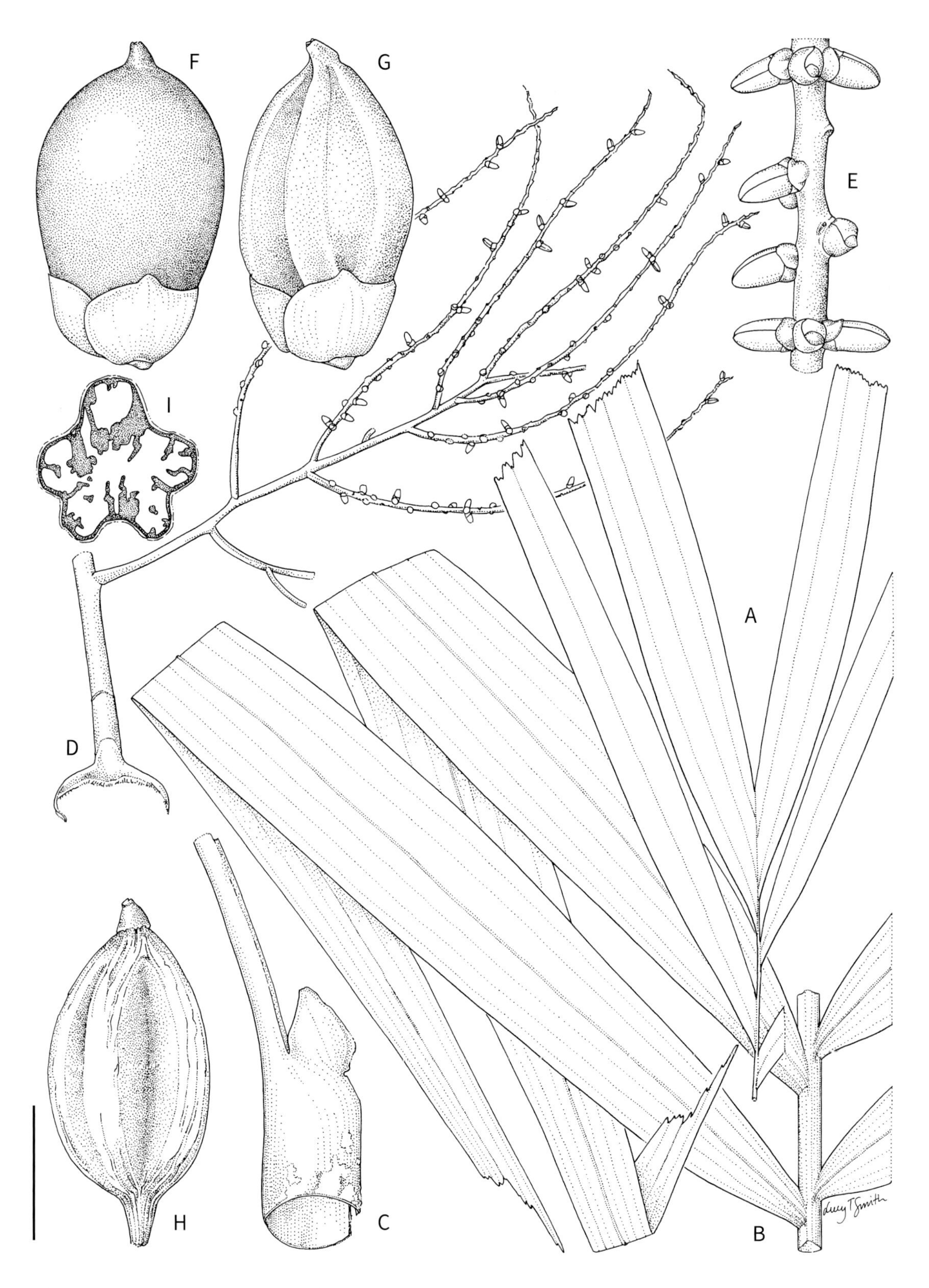

Ptychosperma gracile. **A.** Leaf apex. **B.** Leaf mid-portion. **C.** Leaf sheath with ligule. **D.** Portion of inflorescence. **E.** Portion of rachilla with flowers. **F.** Fruit. **G.** Fruit, dried. **H.** Endocarp. **I.** Seed in transverse section. Scale bar: A–D = 6 cm; E–I = 7 mm. A, B, F–I from *Womersley NGF 43626*; C from *Zona & Dransfield 615*; D, E from *Coode et al. 29766*. Drawn by Lucy T. Smith.

smooth, adhering tightly and conforming to the endocarp when dry, stigmatic remains apical but curving to one side; endocarp 14–15 × 7–10 mm. **Seed** 10–11 × 7–10 mm; *endosperm ruminate.*

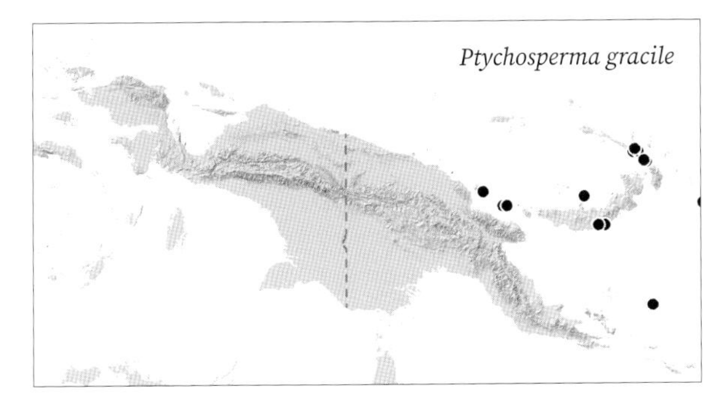

DISTRIBUTION. Islands of the Bismarck Sea (New Ireland, New Britain, Long and Bagabag Islands), Woodlark Island and Bougainville. Not recorded from the mainland New Guinea.

HABITAT. Rainforest at elevations of sea level to 1,400 m.

LOCAL NAMES. *Kiskis* (Long Island), *Mameli* (Manda).

USES. None recorded.

CONSERVATION STATUS. Least Concern. However, close monitoring is required due to heavy logging of the forests within the distribution of this island species.

NOTE: Among red-fruited species with ruminate endosperm, it is distinguished by its dried fruits, in which the epicarp and mesocarp conform closely to the endocarp and the stigma curves to one side. It is phylogenetically close to *P. caryotoides* and two other species from outside New Guinea (Alapetite *et al.* 2014).

7. *Ptychosperma keiense* (Becc.) Becc. ex Martelli

Synonym: *Drymophloeus propinquus* Becc. var. *keiensis* Becc.

Multiple-stemmed, understorey to mid-storey palm 3–7 m tall, bearing 6–8 leaves per crown. **Stem** 2.5–6.0 cm diam. **Leaf** 1.5–3 m long; sheath 24–42 cm long; petiole 15–63 cm long; rachis 83–137 cm long; 9–26 pairs of leaflets, *arranged regularly in one plane,* mid-leaf leaflet 30–55 × *3.0–7.5 cm, lanceolate,* borne 5–9 cm apart; *terminal leaflets slightly to much larger than subterminal leaflets.* **Inflorescence** 17–62 cm long, branched to 2 orders; peduncle 3.5–8.5 × 0.4–2.0 cm; rachillae 13–34 cm long, 2–3 mm diam., green at anthesis, with 8–14 female flowers per 5 cm. **Male flower** 6–8 × 3–4 mm in bud, whitish; stamens ca. 36. **Female flowers** ca. 6 × ca. 4 mm, ovoid. **Fruit** 13–18 × 7–13 mm, *orange to red;* exocarp wrinkled when dry; endocarp 13–16 × 7–9 mm. **Seed** 9–10 × 6–9 mm, *unevenly lobed (three deep grooves and two shallow grooves), lobes rounded; endosperm homogeneous.*

DISTRIBUTION. Widely distributed in western New Guinea, the Kai and Aru Islands.

HABITAT. Lowland rainforest and riverbanks, from sea level to 650 m.

LOCAL NAMES. *Guafat* (Kai Islands), *Kirepi* (Kanum).

USES. Young shoot (heart-of-palm) edible.

CONSERVATION STATUS. Least Concern.

NOTE. This species is widely cultivated, but it has been confused with *P. propinquum* (more commonly known by the synonym *P. macarthurii*). *Ptychosperma keiense* is a more robust species, with wider, evenly arranged leaflets, and may even have more invasive tendencies (Dowe 2007). The distinctive features of *P. keiense,* the

Ptychosperma keiense. Crown, Manokwari-Ransiki road (JD).

regularly arranged leaflets and the unevenly lobed seed, are not always obvious, especially in herbarium specimens. Further study of *P. propinquum* and *P. keiense,* both of which have similar distributions, may reveal that they are conspecific, in which case, *P. propinquum* is the name with priority.

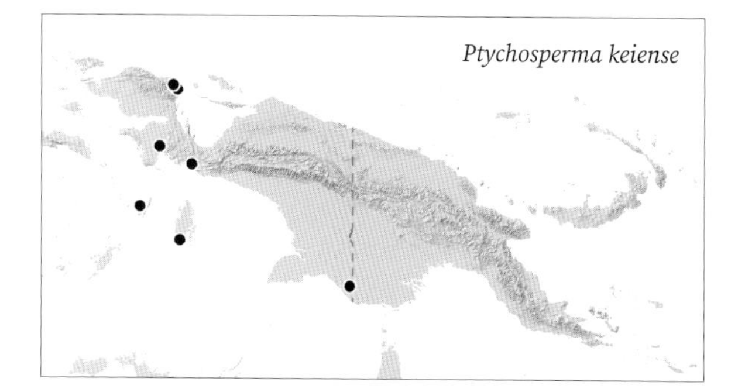

Ptychosperma keiense. RIGHT: leaf, Manokwari-Ransiki road (JD). BELOW, LEFT TO RIGHT: inflorescences; male flowers; fruit, Manokwari-Ransiki road (JD).

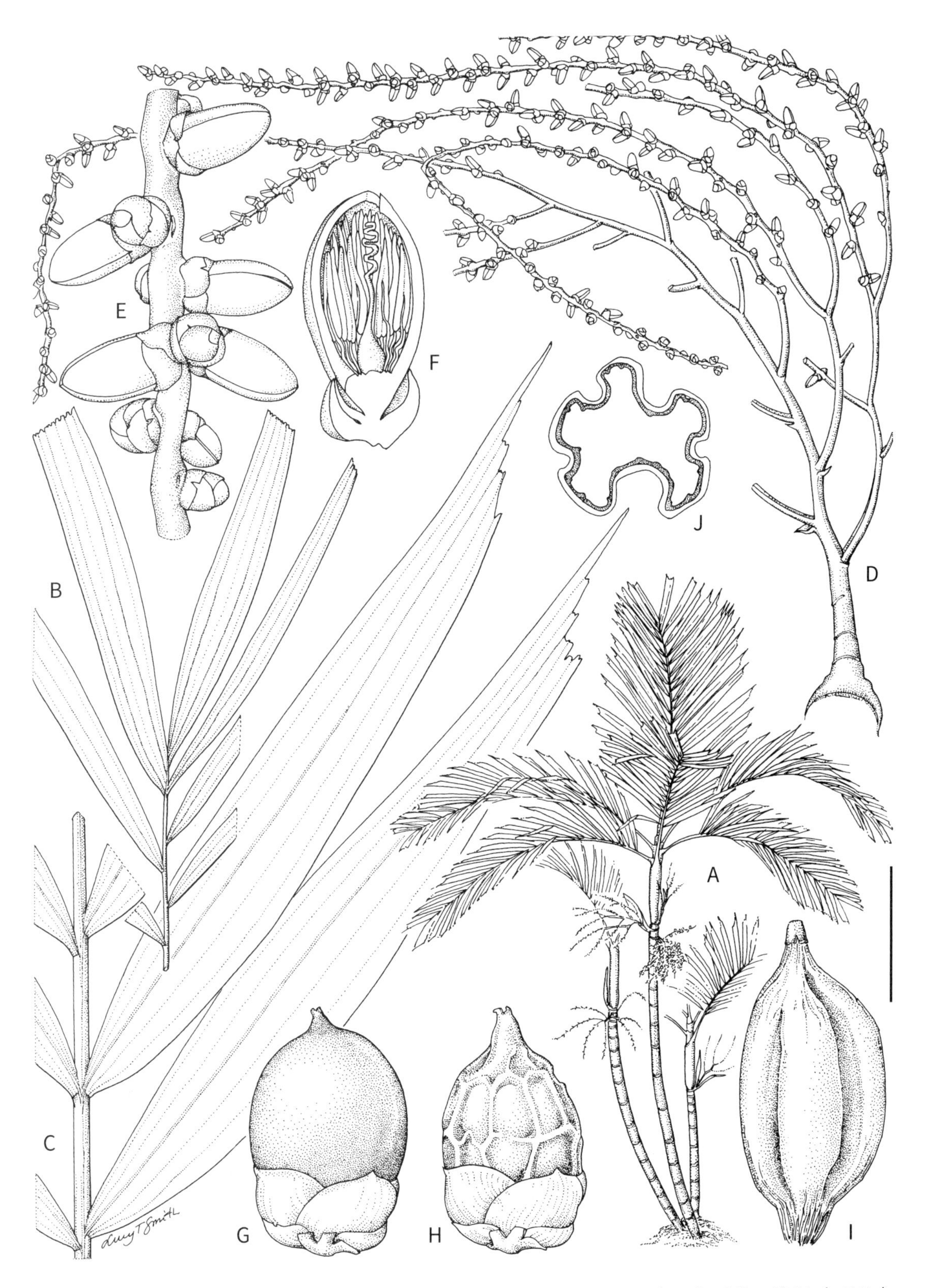

***Ptychosperma keiense*. A.** Habit. **B.** Leaf apex. **C.** Leaf mid-portion. **D.** Inflorescence. **E.** Portion of rachilla with triads. **F.** Male flower in longitudinal section. **G.** Fruit. **H.** Fruit, dried. **I.** Endocarp. **J.** Seed in transverse section. Scale bar: A = 1 m; B, C = 8 cm; D = 6 cm; E, G, H = 1 cm; F = 5 mm; I, J = 7 mm. All from *Zona et al. 693*. Drawn by Lucy T. Smith.

8. *Ptychosperma lauterbachii* Becc.

Synonyms: *Actinophloeus punctulatus* Becc., *Ptychosperma hollrungii* Warb. ex Burret, *Ptychosperma punctulatum* (Becc.) Becc. ex Martelli

Single-stemmed or multiple-stemmed, mid-storey palm bearing 6–8 leaves. **Stem** 4–12 m tall, 4–8 cm diam. **Leaf** 2–3.5 m long; sheath 26–62 cm long; petiole 15–61 cm long; rachis 160–250 cm long; 19–42 pairs of leaflets, *arranged regularly in one plane*, mid-leaf leaflet 32–72 × 3.0–7.5 cm, linear-lanceolate, borne 4–10 cm apart; terminal leaflets larger or not than subterminal leaflets. **Inflorescence** 33–60 cm long, branched to 2 orders; *largest bract subtending inflorescence branches >1 cm long*; peduncle 3.5–6.8 × 1.0–1.6 cm wide; rachillae 21–36 cm long, 2–5 mm diam., green at anthesis, with 11–15(–21) female flowers per 5 cm. **Male flower** 6–7 × ca. 3 mm in bud; whitish, stamens ca. 32. **Female flowers** not seen. **Fruit** 12–18 × 6–10 mm, *orange to red*; *exocarp wrinkled when dry*; endocarp 13–15 × 5–7 mm. **Seed** 6–10 × 5–7 mm; *endosperm weakly ruminate.*

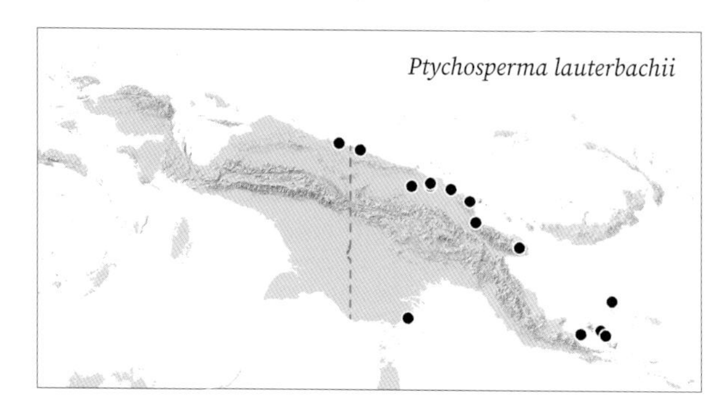

Ptychosperma lauterbachii. Habit, cultivated Fairchild Tropical Botanic Garden, Florida (SZ).

DISTRIBUTION. Widespread in north-eastern, southern and south-eastern New Guinea, including the D'Entrecasteaux and Trobriand Islands.

HABITAT. Rainforest and edges of swamps and mangrove forest close to sea level.

LOCAL NAMES. *Bugorr* (Sempi, Madang), *Kaka* (Gavien, Angoram), *Kikimotar* (Wanigela), *Tetai* (Gabobora), *Waanguh* (Maprik).

USES. The wood is used for spears.

CONSERVATION STATUS. Least Concern (IUCN 2021).

NOTES. Along with *Ptychosperma caryotoides*, this taxon is polymorphic and difficult to characterise. *Ptychosperma lauterbachii* differs from *P. caryotoides* in having leaflets regularly arranged throughout the length of the leaf (vs. basal leaflets irregularly arranged in *P. caryotoides*). Detailed field and genetic studies may reveal that *P. lauterbachii*, as circumscribed here, is a complex of two or more genetically distinct taxa.

9. *Ptychosperma lineare* (Burret) Burret

Synonyms: *Actinophloeus linearis* Burret, *Ptychosperma streimannii* Essig

Multiple-stemmed, understorey to mid-storey palm 5–15 m tall, bearing 8 or 9 leaves per crown. **Stem** 3–6 cm diam. **Leaf** 3–3.5 m long; sheath 50–60 cm long; petiole 10–50 cm long; rachis 197–226 cm long; 24–48 pairs of leaflets, arranged *irregularly basally, regularly distally, and in one plane*, mid-leaf leaflet 34–60 × 2.3–5.5 cm, *linear-lanceolate*, borne 2.4–6.3 cm apart; terminal leaflets not larger than subterminal leaflets. **Inflorescence** 50–80 cm long, branched to 2 or 3

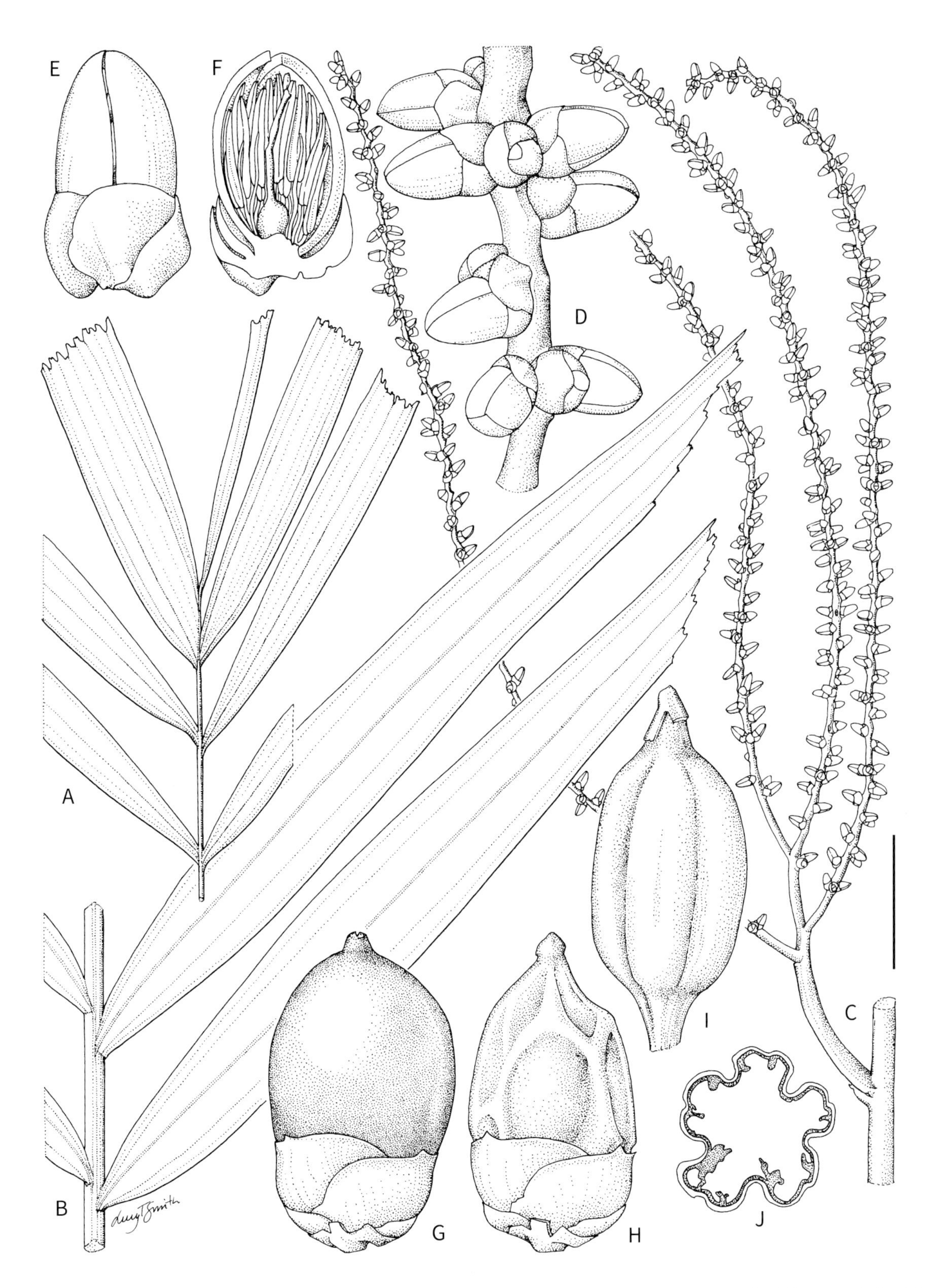

Ptychosperma lauterbachii. **A.** Leaf apex. **B.** Leaf mid-portion. **C.** Portion of inflorescence. **D.** Portion of rachilla with triads. **E**, **F.** Male flower whole and in longitudinal section. **G.** Fruit. **H.** Fruit, dried. **I.** Endocarp. **J.** Seed in transverse section. Scale bar: A, B = 6 cm; C = 4 cm; D = 7 mm; E, F = 4 mm; G–J = 5 mm. All from *Brass 27351*. Drawn by Lucy T. Smith.

Ptychosperma lineare. Crowns, cultivated Fairchild Tropical Botanic Garden, Florida (SZ).

orders; peduncle 3.9–8.0 × 1.5–4.0 cm wide; rachillae 23–35 cm long, ca. 3 mm diam., colour not recorded, with 13–21 female flowers per 5 cm. **Male flower** 6–7 × 2–3 mm in bud, colour not recorded; stamens 25–35. **Female flowers** ca. 6 × ca. 4 mm, ovoid. **Fruit** 14–15 × 8–10 mm, *black*; exocarp wrinkled when dry; endocarp 13–14 × 5–6 mm. **Seed** 7–9 × 5–6 mm; *endosperm homogeneous*.

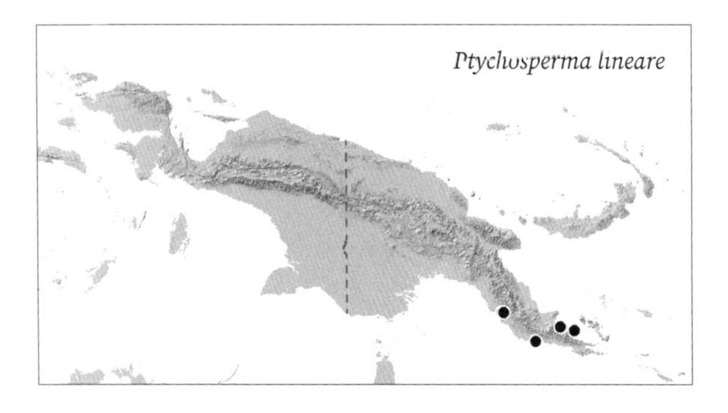

DISTRIBUTION. South-eastern New Guinea.

HABITAT. This species grows in rainforest, forested flood plains, swampy forest, and even coastal mangroves at sea level to 70 m.

LOCAL NAMES. None recorded.

USES. None recorded.

CONSERVATION STATUS. Vulnerable (IUCN 2021). *Ptychosperma lineare* has a restricted distribution. Deforestation due to logging and mining concessions is a major threat in its distribution range.

NOTE. *Ptychosperma lineare* bears a superficial resemblance to *P. furcatum*, but the former species has black fruits, and the latter species has red fruits. The characteristics used to distinguish *P. streimannii* (leaflet number and arrangement, male flower length, fruit length) are readily accommodated within *P. lineare*.

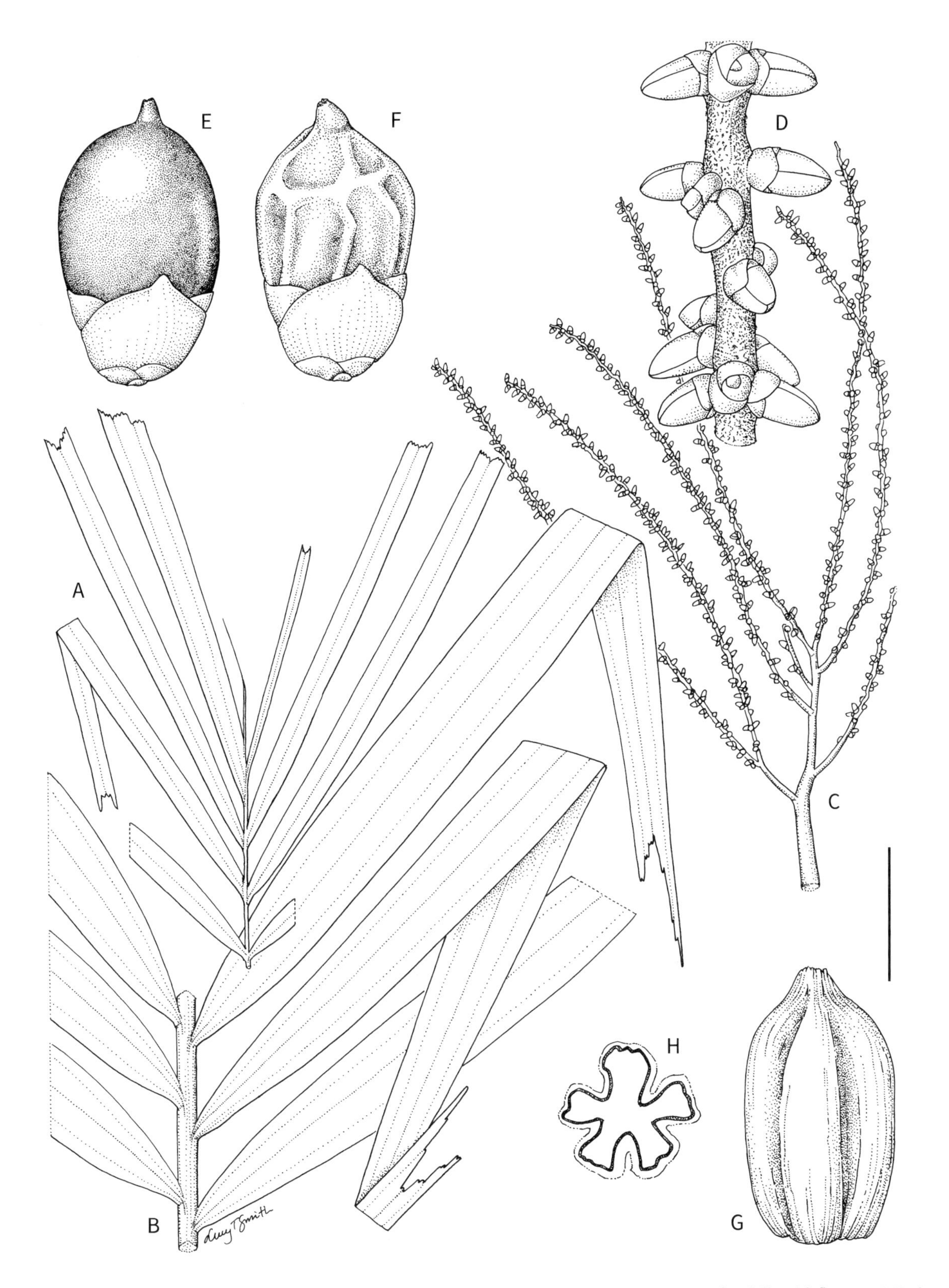

Ptychosperma lineare. **A.** Leaf apex. **B.** Leaf mid-portion. **C.** Portion of inflorescence. **D.** Portion of rachilla with flowers. **E.** Fruit. **F.** Fruit, dried. **G.** Endocarp. **H.** Seed in transverse section. Scale bar: A–C = 6 cm; D, G, H = 5 mm; E, F = 7 mm. A–D from *Pullen 8157*; E–H from *Brass 24250*. Drawn by Lucy T. Smith.

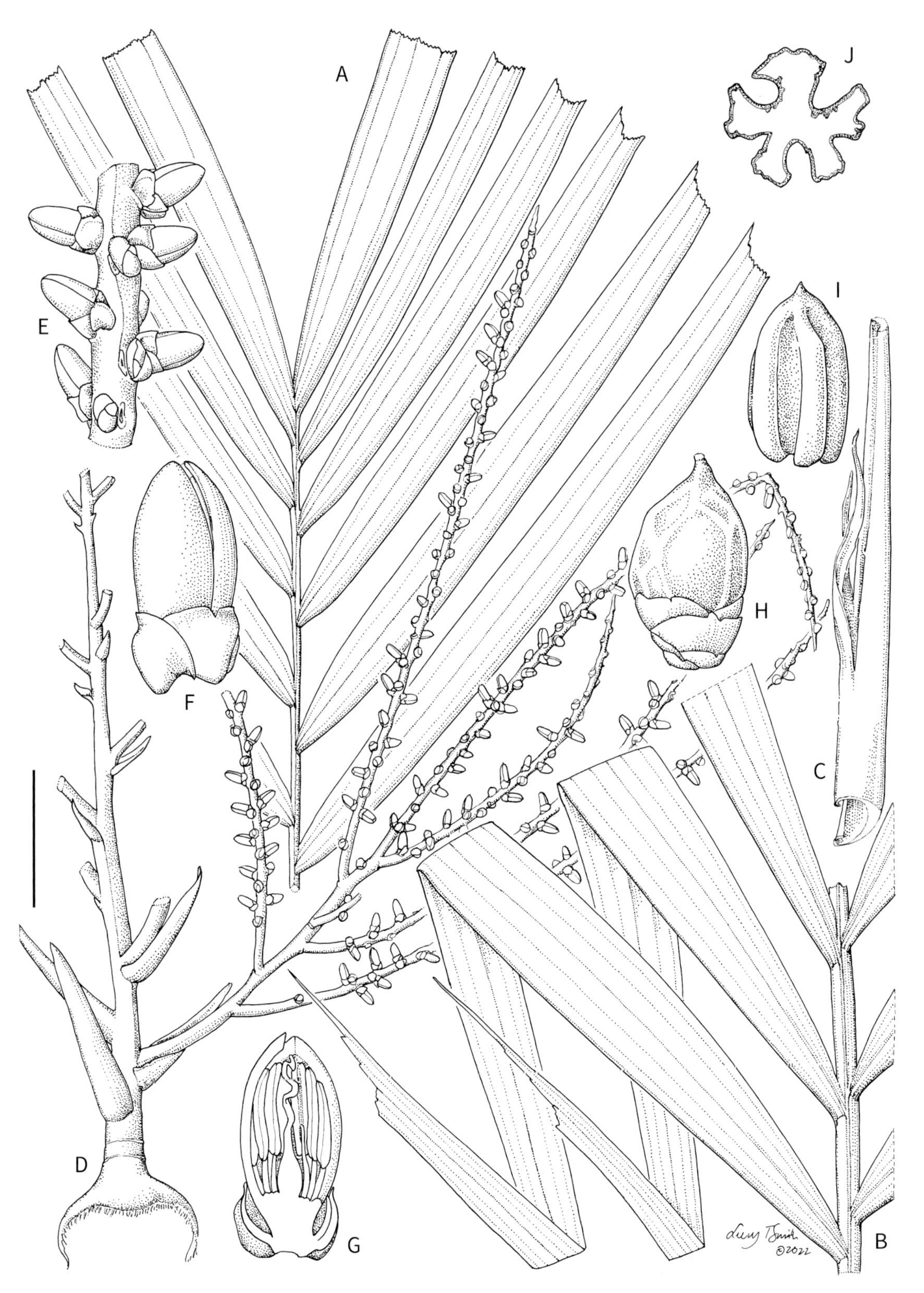

Ptychosperma mambare. **A.** Leaf apex. **B.** Leaf mid-portion. **C.** Leaf sheath with ligule. **D.** Portion of inflorescence. **E.** Portion of rachilla with flowers. **F, G.** Male flower whole and in longitudinal section. **H.** Fruit, dried. **I.** Endocarp. **J.** Seed in transverse section. Scale bar: A–C = 6 cm; D = 4 cm; E, H = 1 cm; F, G, J = 4 mm; I = 5 mm. A–G from *Essig LAE 55151*; H–J from *Essig LAE 55152*. Drawn by Lucy T. Smith.

10. *Ptychosperma mambare* (F.M.Bailey) Becc. ex Martelli

Synonym: *Drymophloeus mambare* F.M.Bailey

Multiple-stemmed, mid-storey palm ca. 6 m tall. **Stem** diam. not recorded. **Leaf** ca. 2.8 m long; sheath 34–55 cm long; petiole 19–34 cm long; rachis 100–200 cm long; 28–37 pairs of leaflets, *arranged regularly in one plane*, mid-leaf leaflet 50–60 × 3.2–3.5 cm, *linear*, borne 7–20 cm apart; terminal leaflets not larger than subterminal leaflets. **Inflorescence** branched to 2 orders; peduncle ca. 4 cm × 1.1 cm wide; rachillae 21–26 cm long, 3–4 mm diam., colour not recorded, with 17–27 female flowers per 5 cm. **Male flower** ca. 6 × ca. 3 mm in bud, colour not recorded; stamens ca. 37. **Female flowers** not seen. **Fruit** 13–15 × ca. 8 mm, *black*; exocarp wrinkled when dry; endocarp not seen. **Seed** 7–8 × ca. 5 mm; *endosperm weakly ruminate.*

DISTRIBUTION. North coast of south-eastern New Guinea.

HABITAT. Mangrove swamp, in brackish water, at sea level.

LOCAL NAMES. None recorded.

USES. None recorded.

CONSERVATION STATUS. Critically Endangered. *Ptychosperma mambare* is known from only two sites which are threatened by mining and logging concessions. It occupies the mangrove habitat which may be threatened by rising sea levels and coastal development.

NOTE. This species is restricted to mangrove swamp habitat, where its tolerance of brackish water is unusual. The only other species to grow in such a wet habitat is *P. lineare*, which grows in rainforest, forested flood plains, swampy forest and occasionally in coastal mangroves.

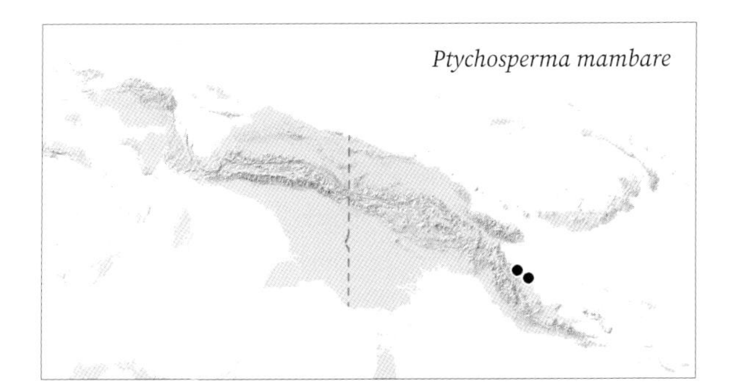

11. *Ptychosperma microcarpum* (Burret) Burret

Synonyms: *Actinophloeus macrospadix* Burret, *Actinophloeus microcarpus* Burret, *Ptychosperma macrospadix* (Burret) Burret

Multiple-stemmed (very rarely single-stemmed), understorey to mid-storey palm 3–9 m tall. **Stem** 4–6 cm diam. **Leaf** 2.5–3 m long; sheath 25–40 cm long; petiole 9–33 cm long; rachis 90–220 cm long; 32–45 pairs of leaflets, *arranged irregularly in many planes*, mid-leaf leaflet 30–59 × 1.6–5.7 cm, linear to narrowly cuneate; terminal leaflets not larger than subterminal leaflets. **Inflorescence** 31–90 cm long, branched to 2 orders; peduncle 2.5–6.0 × 1.1–1.9 cm; rachillae 18–33 cm long, ca. 2 mm diam., green at anthesis, with 10–20 female flowers per 5 cm. **Male flower** 4–5 × 2–3 mm in bud, whitish; stamens 27 or 28. **Female flowers** ca. 5 × 3 mm; ovoid. **Fruit** ca. 12 × 7–8 mm, *orange to red*; exocarp wrinkled when dry; endocarp not seen. **Seed** ca. 7 × ca. 5 mm; *endosperm homogeneous.*

Ptychosperma microcarpum. Habit, cultivated Fairchild Tropical Botanic Garden, Florida (SZ).

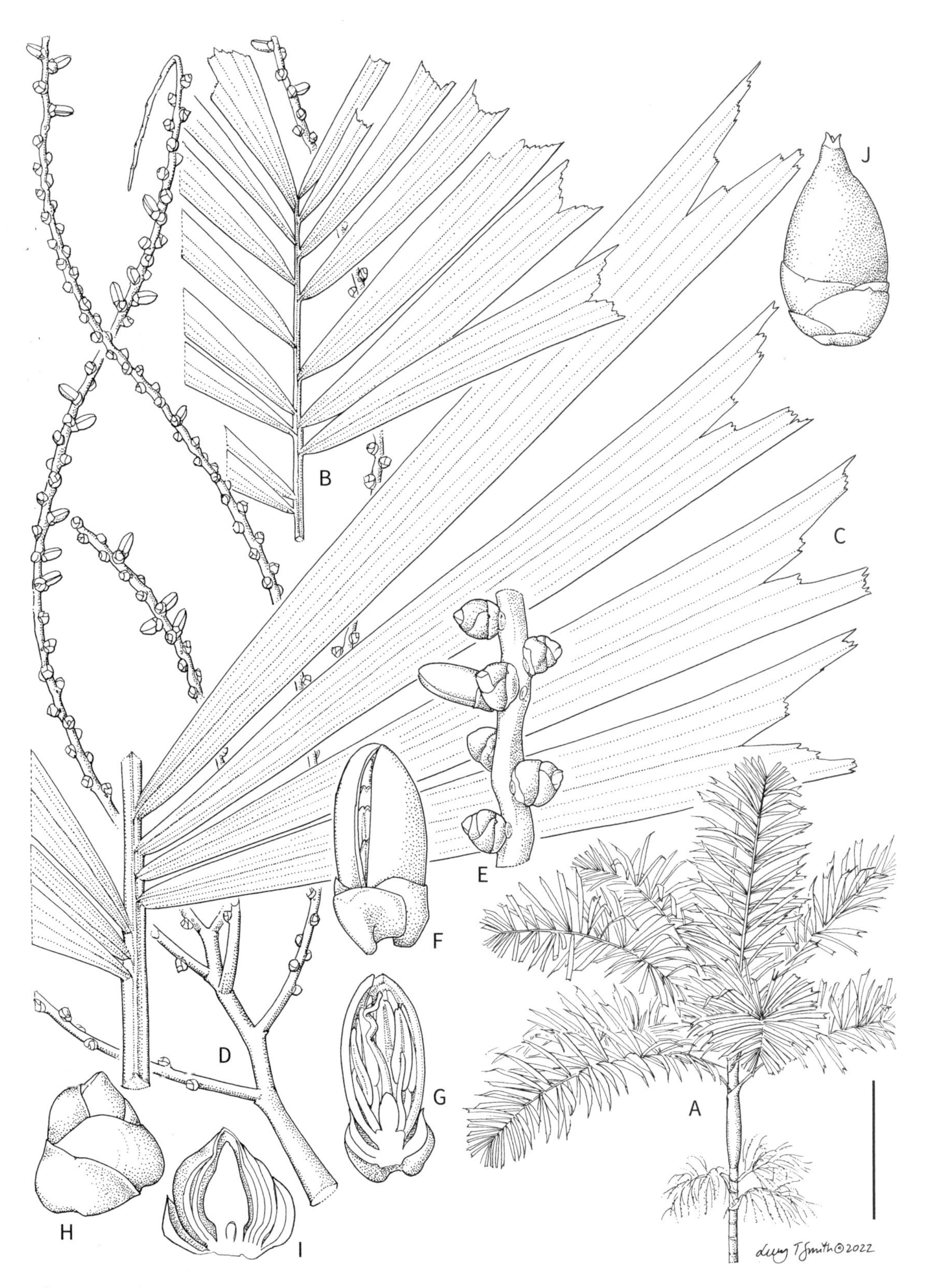

Ptychosperma microcarpum. **A.** Habit. **B.** Leaf apex. **C.** Leaf mid-portion. **D.** Portion of inflorescence with male and female flower buds. **E.** Portion of rachilla with male and female flower buds. **F**, **G.** Male flower whole and in longitudinal section. **H**, **I.** Female flower whole and in longitudinal section. **J.** Immature fruit. Scale bar: A = 1 m; B–C = 6 cm; D = 4 cm; E = 1 cm; F–I = 4 mm; J = 7 mm. All from *Darbyshire* 667. Drawn by Lucy T. Smith.

Ptychosperma microcarpum. CLOCKWISE FROM ABOVE: habit; inflorescence with fruit; female flowers. Cultivated Fairchild Tropical Botanic Garden, Florida (SZ).

DISTRIBUTION. South-eastern New Guinea, vicinity of Port Moresby.

HABITAT. Low elevation rainforest, often along riverbanks at sea level to 490 m.

LOCAL NAMES. *Aiaba* (Roro), *Paiva* (Matapaili).

USES. The wood is used for bows.

CONSERVATION STATUS. Endangered. *Ptychosperma microcarpum* has a restricted distribution. Deforestation due to logging concessions is a major threat in its distribution range.

NOTE. The most distinctive feature of this species is its irregularly arranged, clustered leaflets, which diverge from the rachis in more than one plane. The species epithet means "small fruit", although the fruits of this species are within the size range typical for the genus. The female flowers of this species, which are white, have a pleasant fragrance not unlike jasmine.

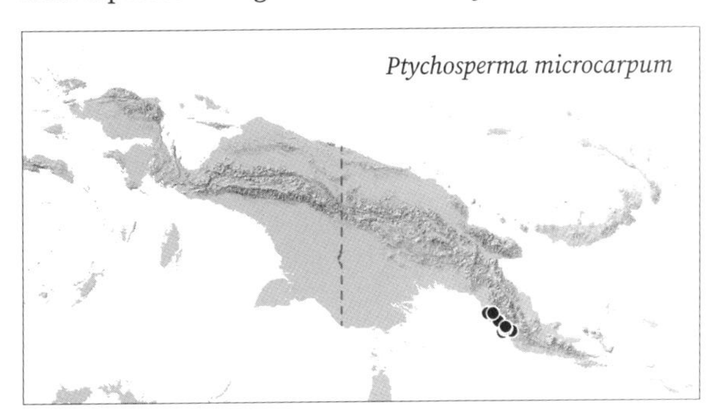

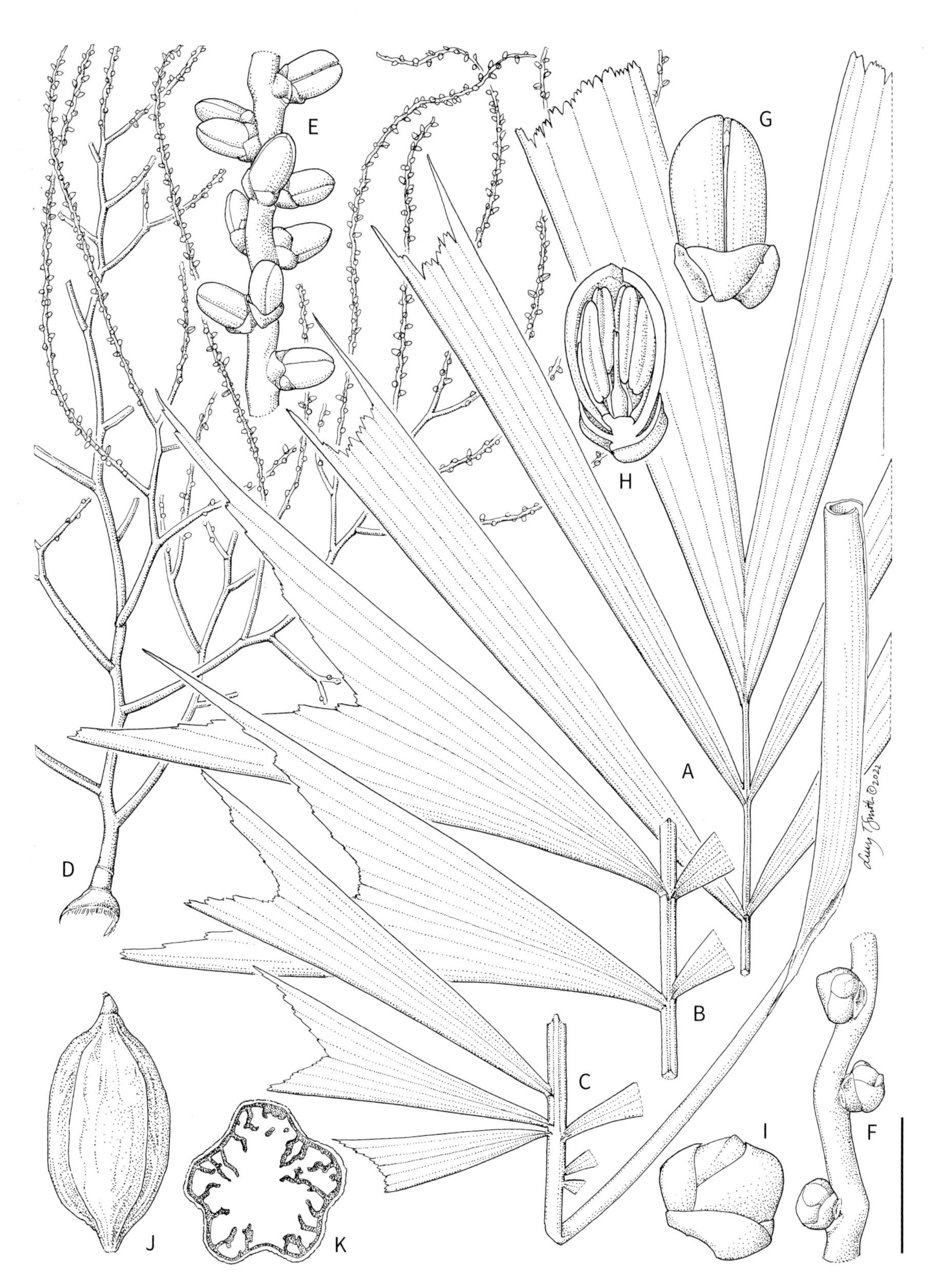

Ptychosperma mooreanum. **A.** Leaf apex. **B.** Leaf mid-portion. **C.** Leaf base with leaf sheath. **D.** Inflorescence. **E.** Portion of rachilla with male flower buds. **F.** Portion of rachilla with female flower buds. **G**, **H.** Male flower whole and in longitudinal section. **I.** Female flower. **J.** Endocarp. **K.** Seed in transverse section. Scale bar: A–C = 6 cm; D = 4 cm; E, F = 4 mm; G, H = 2 mm; I = 1.2 mm; J, K = 5 mm. A–I from *Barfod et al. 455*; J, K from *Brass 28883*. Drawn by Lucy T. Smith.

12. *Ptychosperma mooreanum* Essig

Single-stemmed, understorey to mid-storey palm 3–8 m tall, bearing 5–8 leaves. **Stem** 2–8 cm diam. **Leaf** 1.2–2.7 m long; sheath 23–30 cm long; petiole 29–60 cm long; rachis 120–200 cm long; 14–20 pairs of leaflets, *arranged irregularly in one plane*, mid-leaf leaflet 21–51 × 3–12 cm, *cuneate*; *terminal leaflets larger than subterminals.* **Inflorescence** 33–145 cm long, branched to 2 orders; peduncle ca. 6.5 × ca. 0.9 cm; rachillae 13–24 cm long, 1–2 mm diam., colour not recorded, with 11–19 female flowers per 5 cm. **Male flower** 4–6 × ca. 2 mm in bud, colour not recorded; stamens ca. 20. **Female flowers** ca. 3 × ca. 2 mm, ovoid. **Fruit** 9–12 × 7–9 mm, *black*; exocarp wrinkled when dry; endocarp 10–11 × 5–7 mm. **Seed** 6–7 × 5–7 mm; *endosperm ruminate.*

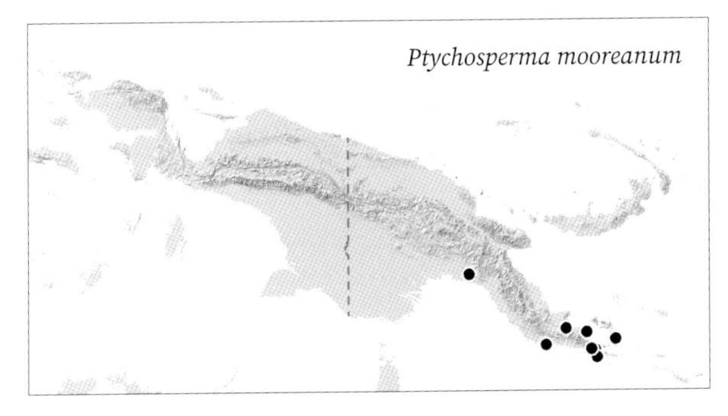

DISTRIBUTION. South-eastern New Guinea, including the D'Entrecasteaux Islands and the Louisiade Archipelago, with an outlier in Gulf Province.

HABITAT. Low elevation rainforest at sea level to 810 m.

LOCAL NAMES. None recorded.

USES. None recorded.

CONSERVATION STATUS. Least Concern (IUCN 2021).

NOTE. *Ptychosperma mooreanum* is one of only three species possessing black fruits and ruminate endosperm (the other two are *P. mambare* and *P. schefferi*). It differs from these species in having fruits <12 mm long and deeply ruminate endosperm. It differs from *P. mambare* in habitat and fewer female flowers per length of rachillae, and from *P. schefferi* in the arrangement of leaflets.

13. *Ptychosperma nicolai* (Sander ex André) Burret

Synonyms: *Actinophloeus nicolai* (Sander ex André) Burret, *Ptychosperma macrocerum* Becc., *Romanovia nicolai* Sander ex André

Stem not seen. **Leaf** not seen; leaflet ca. 43 cm long, 4 cm wide, linear, *terminal leaflets wider than subterminals.* **Inflorescence** not seen; rachillae ca. 35 cm long, ca. 2.1 mm diam., colour not recorded, with ca. 10 female flowers per 5 cm, *black, crustose indumentum.* **Male flower** not seen. **Female flower** not seen. **Fruit** 13–14 × 7–8 mm, *orange to red*; exocarp wrinkled when dry. **Seed** ca. 7 × ca. 5 mm; *endosperm homogeneous.*

DISTRIBUTION. Unknown, but perhaps southern Indonesian New Guinea.

HABITAT. Unknown.

LOCAL NAMES. None recorded.

USES. None recorded.

CONSERVATION STATUS. Data deficient. The native distribution of *P. nicolai* is unknown. It may in fact be extinct in the wild.

NOTE. This puzzling species is known only from the type collection of *Ptychosperma macrocerum*, a sparse description and an illustration of a juvenile plant, all of these from cultivated plants. Essig (1978) reported Beccari's belief that the species may have originated in southern Indonesian New Guinea; however, no wild-collected material from that area answers to this species. The type specimen is noteworthy in having praemorse, deeply notched leaflets, with dark ramenta on the lower side along the midvein, and broad terminal leaflets. The inflorescence axes are heavily vested with a crustose, black indumentum. Further research may show that this species is conspecific with *P. furcatum*, which is morphologically similar.

14. *Ptychosperma propinquum* (Becc.) Becc. ex Martelli

Synonyms: *Actinophloeus bleeseri* (Burret) Burret, *Actinophloeus macarthurii* (H.Wendl. ex H.J.Veitch) Becc., *Actinophloeus propinquus* (Becc.) Becc., *Carpentaria bleeseri* (Burret) Burret, *Drymophloeus propinquus* Becc., *Kentia macarthurii* H.Wendl. ex H.J.Veitch, *Ptychosperma bleeseri* Burret, *Ptychosperma julianetti* Becc., *Ptychosperma macarthurii* (H.Wendl. ex H.J.Veitch) H.Wendl. ex Hook.f., *Saguaster macarthurii* (H.Wendl. ex H.J.Veitch) Kuntze, *Saguaster propinquus* (Becc.) Kuntze

Multiple-stemmed, mid-storey palm 3–8 m tall, bearing 8–10 leaves per stem. **Stem** 2–8 cm diam. **Leaf** 1.5–2.8 m long; sheath 29–61 cm long; petiole 20–58 cm long; rachis up to 2 m long; 19–25 pairs of leaflets, *arranged irregularly in one plane*, mid-leaf leaflet 26–61

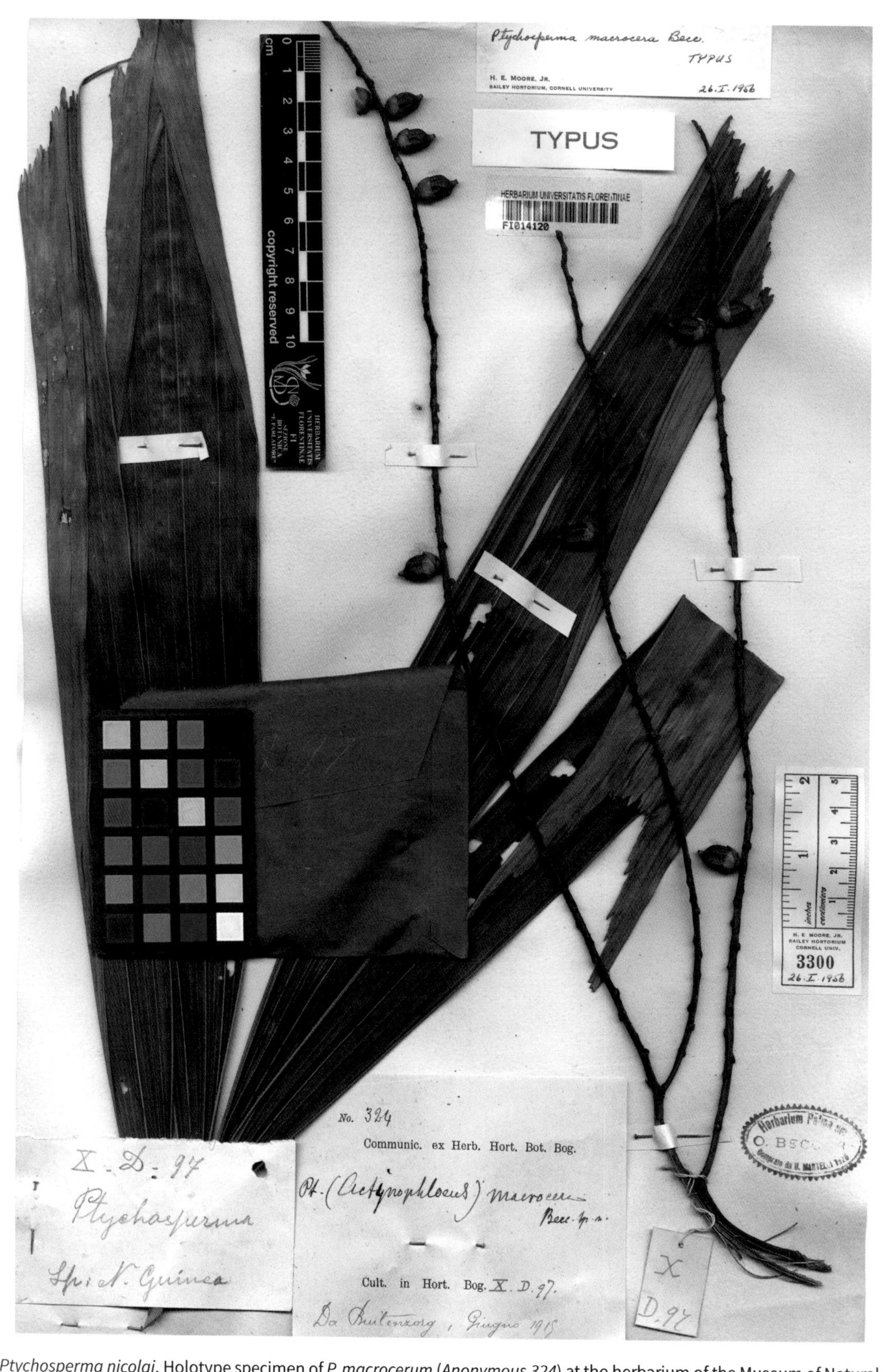

Ptychosperma nicolai. Holotype specimen of *P. macrocerum* (*Anonymous 324*) at the herbarium of the Museum of Natural History, University of Florence (FI). *Ptychosperma macrocerum* is treated here as a synonym of the poorly known *P. nicolai*.

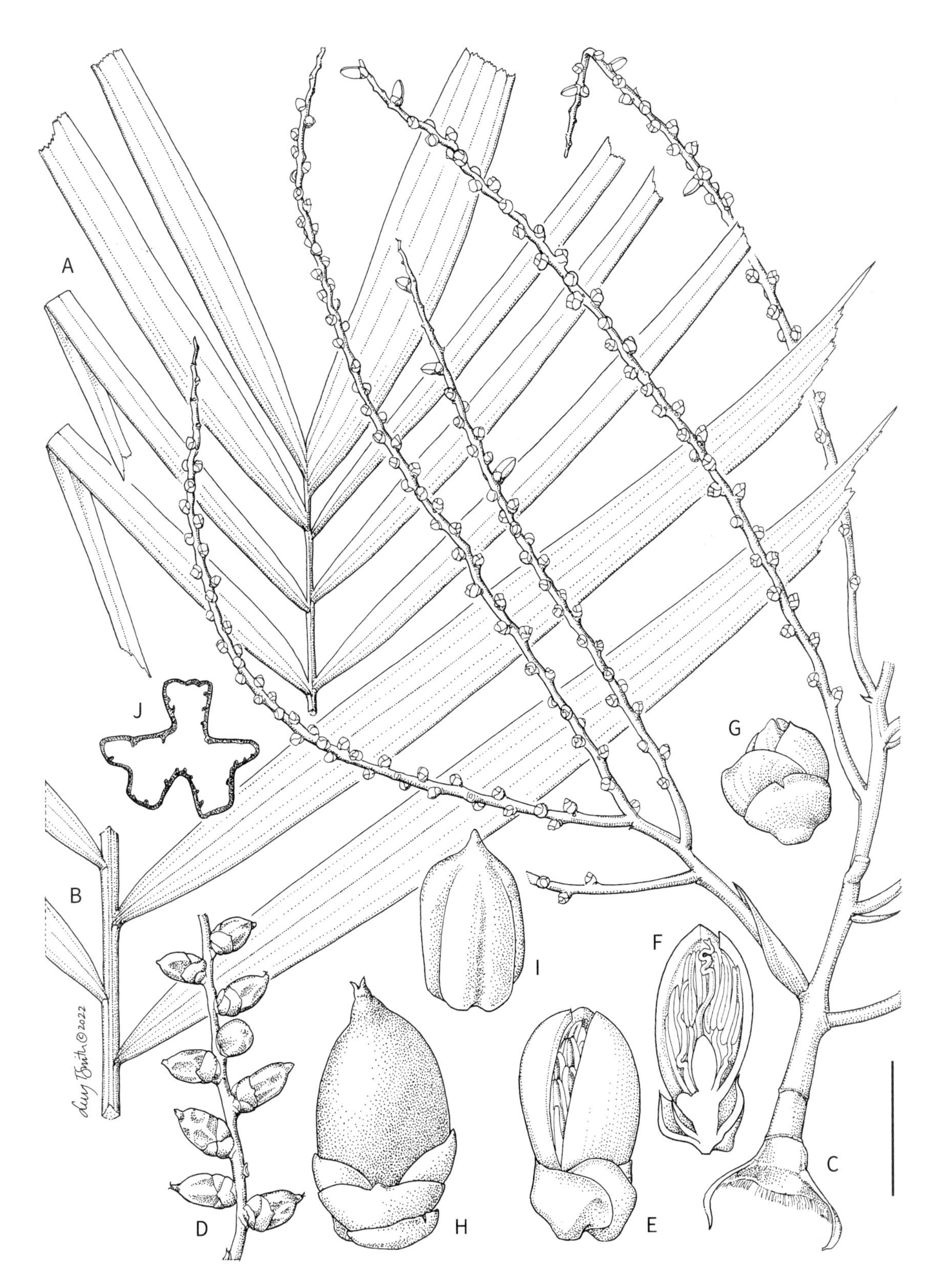

Ptychosperma propinquum. **A.** Leaf apex. **B.** Leaf mid-portion. **C.** Inflorescence. **D.** Portion of rachilla with fruit. **E, F.** Male flower whole and in longitudinal section. **G.** Female flower bud. **H.** Fruit. **I.** Endocarp. **J.** Seed in transverse section. Scale bar: A, B = 8 cm; C = 4 cm; D = 3 cm; E, F, H, I = 7 mm; G, J = 5 mm. All from *Pullen 7065*. Drawn by Lucy T. Smith.

Ptychosperma propinquum. LEFT TO RIGHT: habit, naturalised Singapore (WB); male flowers, cultivated Forest Research Institute Malaysia (WB); inflorescence with fruit, cultivated Forest Research Institute Malaysia (WB).

× 1.9–7.5 wide, linear-lanceolate, borne ca. 6 cm apart; terminal leaflets not larger than subterminal leaflets. **Inflorescence** 25–44 cm long, branched to 2 orders; peduncle 2.4–9.5 × 0.3–1.3 cm; rachillae 16–40 cm long, 2–4 mm diam., green at anthesis, with 8–14 female flowers per 5 cm. **Male flower** 6–8 × 3–4 mm in bud, whitish; stamens 42–51. **Female flowers** ca. 4 × ca. 4 mm diam., globose. **Fruit** 13–16 × 7–11 mm, *orange to red*; exocarp wrinkled when dry; endocarp 14–15 × ca. 7 mm. **Seed** 8–10 × 5–7 mm, *5-lobed, rounded to blocky, unequal in size and spacing; endosperm homogeneous.*

DISTRIBUTION. Widespread from Raja Ampat Islands and the Bird's Head Peninsula to the Fly River catchment. Also occurs in the Aru Islands and Australia.

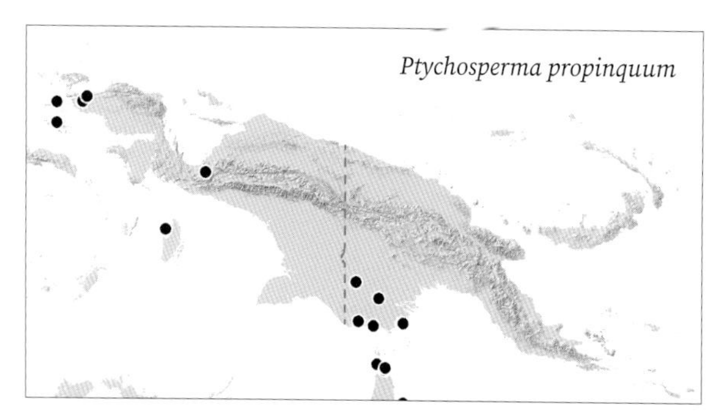

HABITAT. Gallery or riverine rainforest, swamp forest or savannahs, at sea level to 70 m.

LOCAL NAMES. *Gosura* (Etmahona-Kambrauw).

USES. Spears and arrowheads.

CONSERVATION STATUS. Least Concern.

NOTE. This species is well known, having been in cultivation under the name *Ptychosperma macarthurii* for over a century (Dowe 2007). Beccari noted its similarity to the black-fruited *P. ambiguum*, but did not have fruits to examine and did not know that *P. propinquum* is red-fruited. Many plants cultivated as *P. macarthurii* are *P. keiense*. *Ptychosperma propinquum* can be distinguished from *P. keiense* by the regular arrangement of its leaflets (vs. irregular in *P. keiense*).

15. *Ptychosperma pullenii* Essig

Single-stemmed, understorey to mid-storey palm ca. 6 m tall, bearing ca. 6 leaves. **Stem** ca. 2.4 cm diam. **Leaf** 1.9–3.5 m long; sheath ca. 61 cm long; petiole 39–68 cm long; rachis 91–193 cm long; 10–12 pairs of leaflets, *arranged irregularly basally, regularly distally, and in one plane*, mid-leaf leaflet 33–49 × 14–18 cm wide, *cuneate*, borne ca. 12 cm apart; *terminal leaflets larger than subterminal leaflets*. **Inflorescence** 45–61 cm long,

Ptychosperma pullenii. **A.** Leaf apex. **B.** Leaf mid-portion. **C.** Inflorescence. **D.** Portion of rachilla with flowers. **E.** Fruit. **F.** Fruit, dried. **G.** Endocarp. **H.** Seed in transverse section. Scale bar: A–C = 6 cm; D–F = 7 mm; G, H = 5 mm. A–D from *Zona 888*; E–H from *Zona 969*. Drawn by Lucy T. Smith.

Ptychosperma pullenii. LEFT TO RIGHT: leaf and inflorescence with fruit, East Sepik (AB); fruit, cultivated Fairchild Tropical Botanic Garden, Florida (SZ).

branched to 3 orders; peduncle 8.0–11.5 × 0.6–1.0 cm wide; rachillae 23–33 cm long, 1–2 mm diam., reddish at anthesis, with 7–9 female flowers per 5 cm. **Male flower** ca. 5 × ca. 2 mm in bud, purple; stamens ca. 17. **Female flowers** not seen. **Fruit** 12–13 × 7–8 mm, *black*; exocarp wrinkled when dry; endocarp not seen. **Seed** 6–7 × ca. 5 mm, lobes blocky; *endosperm homogeneous.*

DISTRIBUTION. North coast of New Guinea in the vicinity of Wewak.

HABITAT. Rainforest on rocky substrates at 150 m.

LOCAL NAMES. None recorded.

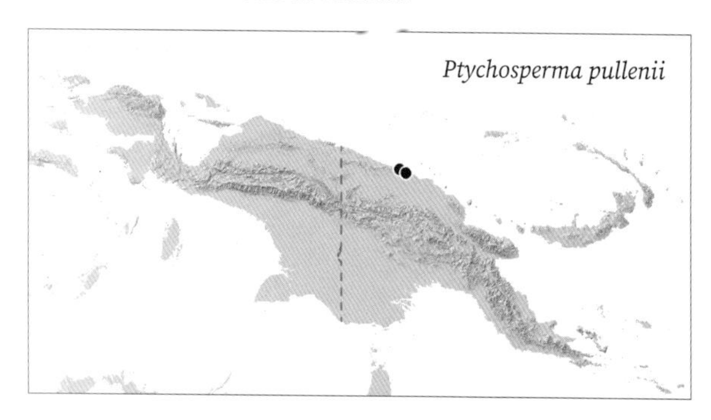

USES. None recorded.

CONSERVATION STATUS. Critically Endangered (IUCN 2021). *Ptychosperma pullenii* is known from only three sites. Deforestation due to logging concessions and other human activity is a major threat in its distribution range.

NOTE. *Ptychosperma pullenii* is phylogenetically close to *P. cuneatum* (Alapetite *et al.* 2014). It is morphologically similar to *P. waitianum* but has a longer peduncle and obscure bracts subtending the inflorescence branches.

16. *Ptychosperma ramosissimum* Essig

Single-stemmed understorey to mid-storey palm 2–8 m tall. **Stem** ca. 2.5 cm diam. **Leaf** ca. 1.5 m long; sheath not seen; petiole ca. 40 cm long; rachis not seen; leaflets arranged regularly in one plane, mid-leaf leaflet 32–62 × 3.0–3.6 cm, linear-lanceolate, borne ca. 6 cm apart; terminal leaflets not larger than subterminal leaflets. **Inflorescence** ca. 60 cm long, *branched to 4 orders*; peduncle ca. 7 cm × 4.9 cm; *rachillae 3.0–8.0 cm long, ca. 1 mm diam.*, colour not

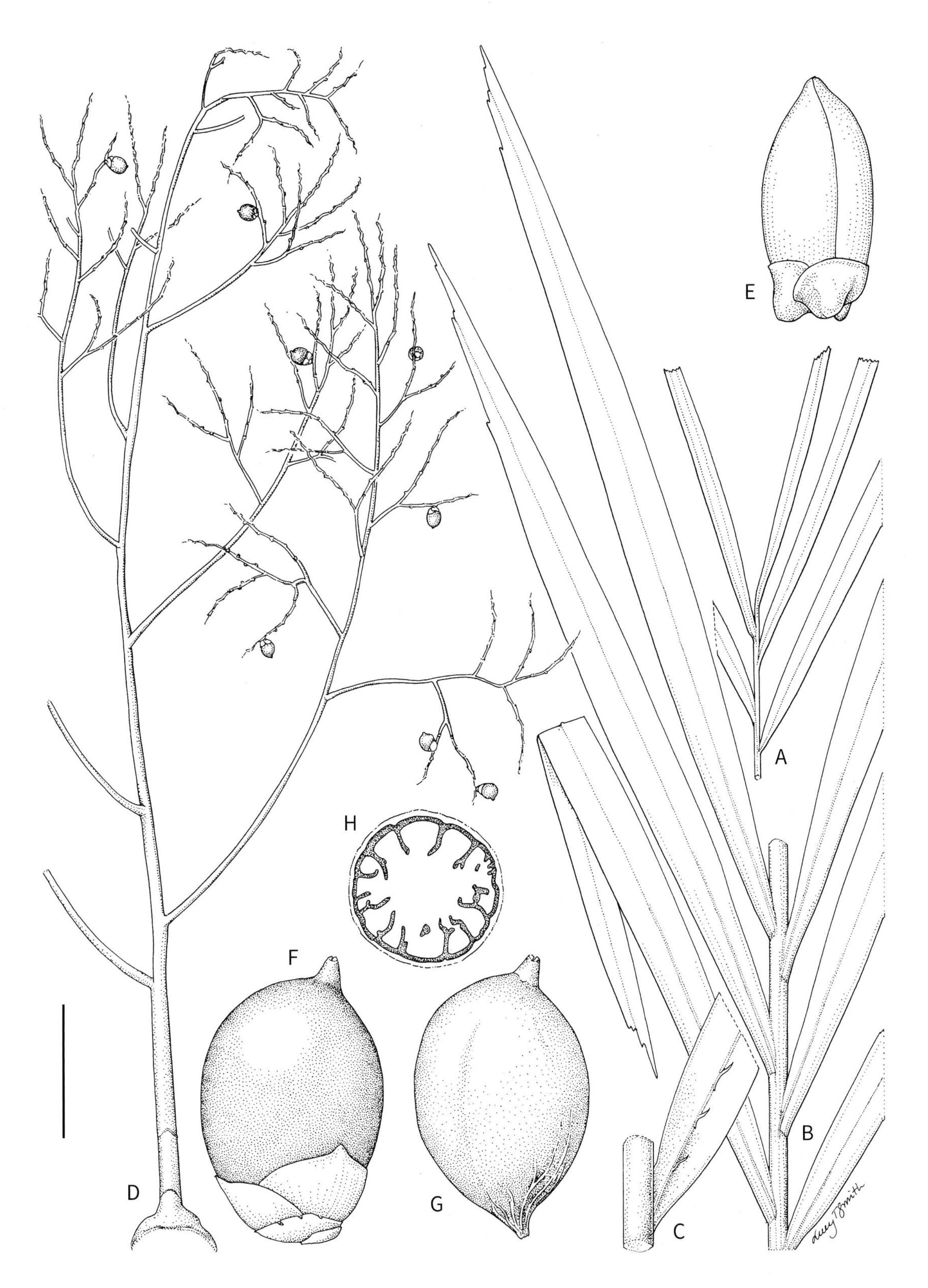

Ptychosperma ramosissimum. **A.** Leaf apex. **B.** Leaf mid-portion. **C.** Leaflet abaxial side detail. **D.** Inflorescence with fruit. **E.** Male flower. **F.** Fruit. **G.** Endocarp. **H.** Seed in transverse section. Scale bar: A, B = 6 cm; C = 3 cm; D = 4 cm; E = 2 mm; F–H = 4 mm. All from *Gillison NGF 25399*. Drawn by Lucy T. Smith.

Ptychosperma ramosissimum. LEFT TO RIGHT: habit; inflorescence with fruit. Rossel Island (JLD).

recorded, with 10–25 female flowers per 5 cm. **Male flower** not seen. **Female flower** not seen. **Fruit** 8–14 × 5–9 mm, *orange to red*; exocarp smooth when dry; endocarp not seen. **Seed** 6–8 × 5–7 mm, *scarcely lobed and nearly circular in cross-section*; endosperm *ruminate*.

DISTRIBUTION. Rossel Island in the Louisiade Archipelago.

HABITAT. Low forest on exposed ridges and hillsides, at 180–700 m.

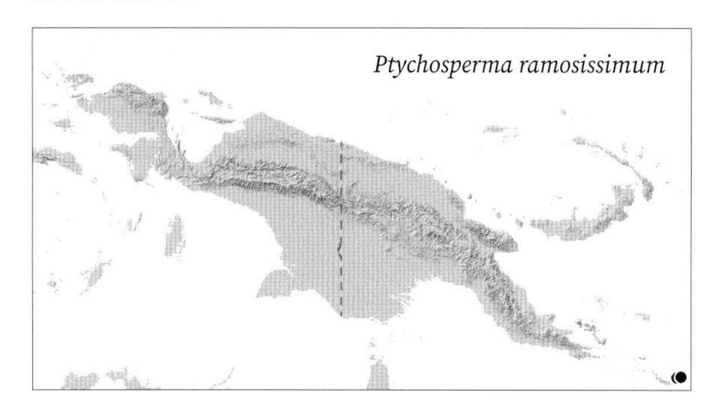

LOCAL NAMES. None recorded.

USES. None recorded.

CONSERVATION STATUS. Endangered (IUCN 2021). This narrowly endemic island species is at high risk from future changes in global climate, being found at high elevation on a single island where the highest peak, Mount Rossel, rises only slightly higher (to 838 m).

NOTE. The species is noteworthy for the inflorescence branched to four orders, its very short rachillae and seeds that are nearly lacking in ridges. These characteristics are useful in separating it from *P. rosselense*, which also occurs on Rossel Isand.

17. *Ptychosperma rosselense* Essig

Single-stemmed, understorey to mid-storey palm 3–8 m tall. **Stem** 4–6 cm diam. **Leaf** ca. 2.4 m long; sheath 30–43 cm long; petiole ca. 61 cm long; rachis 115–130 cm long; ca. 15 pairs of leaflets, arranged regularly in

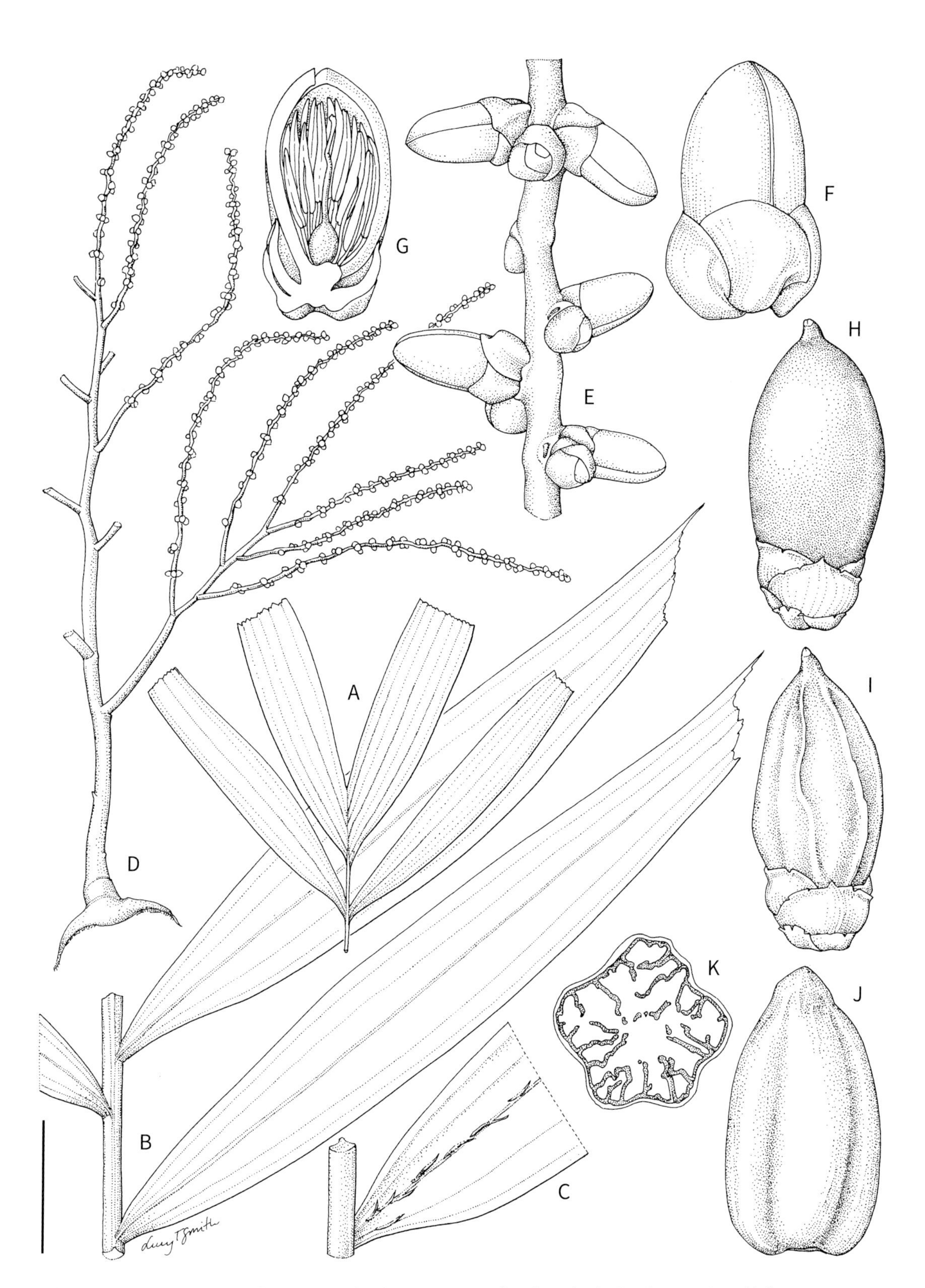

Ptychosperma rosselense. **A.** Leaf apex. **B.** Leaf mid-portion. **C.** Leaflet abaxial side detail. **D.** Portion of inflorescence. **E.** Portion of rachilla with triads. **F, G.** Male flower whole and in longitudinal section. **H.** Fruit. **I.** Fruit, dried. **J.** Endocarp. **K.** Seed in transverse section. Scale bar: A, B = 8 cm; C = 5 cm; D = 6 cm; E, J, K = 7 mm; F, G = 4 mm; H, I = 1 cm. A–C from *Brass 28408*; D–K from *Katik LAE 70919*. Drawn by Lucy T. Smith.

one plane, mid-leaf leaflet 53–60 × 7.0–7.7 cm, linear-lanceolate, borne 8–11 cm apart; terminal leaflets not larger than subterminal leaflets. **Inflorescence** 40–41 cm long, branched to 2 orders; peduncle 5.5–9.5 × 0.7–6.9 cm; *rachillae 13–24 cm long*, ca. 2 mm diam., colour not recorded, with 7–10 female flowers per 5 cm. **Male flower** 7–9 mm long in bud, colour not recorded; stamens 46 or 47, 3–6 mm long. **Female flowers** not seen. **Fruit** 19–23 × 8–11 mm, *orange to red*, exocarp wrinkled when dry; endocarp 18–19 × 7–8 mm. **Seed** 11–14 × 7–8 mm, *lobes low, rounded; endosperm ruminate.*

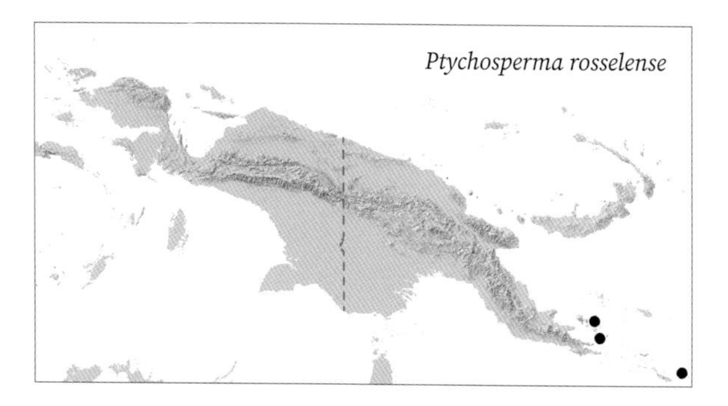

DISTRIBUTION. The Louisiade Archipelago (Rossel Island) and the D'Entrecasteaux Islands (Fergusson and Normanby Islands).

HABITAT. Low, exposed forest on ridges at 700–1,500 m.

LOCAL NAMES. None recorded.

USES. None recorded.

CONSERVATION STATUS. Near Threatened (IUCN 2021). *Ptychosperma rosselense* has a restricted distribution, known only from the peaks of a few small islands. Deforestation due to logging and mining concessions is a threat in its distribution range. Global climate change may be detrimental to this narrowly endemic mountain palm.

NOTE. *Ptychosperma rosselense* is noteworthy for growing at high elevation, in somewhat exposed ridge-line habitats. It can be distinguished from *P. ramosissimum*, which also occurs on Rossel Island, by its inflorescence branched to two orders, rachillae 13–24 cm long, and seeds with low, rounded lobes.

18. *Ptychosperma sanderianum* Ridl.

Synonym: *Actinophloeus sanderianus* (Ridl.) Burret

Stem not seen. **Leaf** not seen; leaflets arranged regularly in one plane, mid-leaf leaflet ca. 35.5 × ca. 0.9 cm wide, narrowly linear-lanceolate, borne ca. 2.8 cm apart. **Inflorescence** not seen; peduncle not seen; rachillae ca. 21.5 cm long, ca. 4 mm diam., colour not recorded, *with ca. 32 female flowers per 5 cm.* **Male flower** ca. 5 × ca. 3 mm in bud, colour not recorded; stamens ca. 21. **Female flower** not seen. **Fruit** 15–17 × 8–9 mm, *orange to red*; exocarp smooth when dry. **Seed** ca. 9 × ca. 6 mm, deeply grooved with blocky lobes; *endosperm homogeneous.*

DISTRIBUTION. Papua New Guinea, possibly the vicinity of Port Moresby.

HABITAT. Unknown.

LOCAL NAMES. None recorded.

USES. None recorded.

CONSERVATION STATUS. Extinct in the Wild. As no wild-collected material has come to light since the original collection in 1898, we infer that it is probably extinct in the wild.

NOTE. This species is well known from cultivated sources; the only wild-collected specimen is that of the type, collected by Micholitz in Papua New Guinea, probably in the vicinity of Port Moresby (Essig 1978). No additional material of this species has been collected from the wild, giving rise to the possibility that urban growth may have extirpated this species. The species is easily distinguished from all others by the dense arrangement of flowers on the rachillae, completely obscuring the axis of the rachilla when in fruit.

Ptychosperma sanderianum. Lectotype specimen of *P. sanderianum* (*Micholitz 9807*) at Singapore Botanic Garden herbarium (SING).

19. *Ptychosperma schefferi* Becc. ex Martelli

Multiple-stemmed, understorey to mid-storey palm ca. 6 m tall. **Stem** ca. 7 cm diam. **Leaf** ca. 1.8 m long; sheath ca. 31 cm long; petiole 12–26 cm long; rachis ca. 99 cm long; 17–25 pairs of leaflets, *arranged regularly in one plane*, mid-leaf leaflet 36–50 × 4.7–6.0 cm, linear-lanceolate, borne 7–9 cm apart; terminal leaflets not larger than subterminal leaflets. **Inflorescence** 32–41 cm long, branched to 1 or 2 orders; peduncle 4–6 × 1–2 cm wide; rachillae 27–37 cm long, 2–3 mm diam., reddish at anthesis, with 11–18 female flowers per 5 cm. **Male flower** 6–7 × 3–4 mm in bud, purple; stamens 30–36. **Female flowers** ca. 6 × ca. 4 mm in bud, ovoid. **Fruit** *black, 14–20 mm long*, 7–12 mm diam.; exocarp wrinkled when dry; endocarp not seen. **Seed** ca. 9 mm long, ca. 6 mm diam., *lobes low, rounded; endosperm ruminate (often weakly so).*

DISTRIBUTION. Central northern New Guinea, from the vicinity of Jayapura to the vicinity of Aitape.

HABITAT. Coastal forest, near sea level.

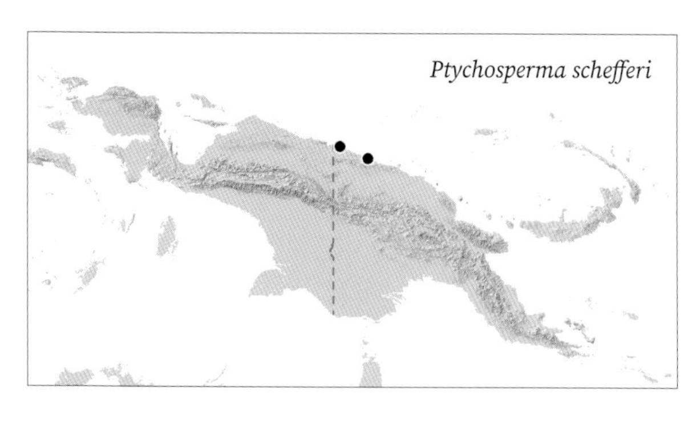

LOCAL NAMES. *Tanai* (Wapi).

USES. None recorded.

CONSERVATION STATUS. Critically Endangered. *Ptychosperma schefferi* is known from only two sites where it is threatened by logging concessions.

NOTE. This species is sometimes confused with *P. lineare*, which differs primarily by having a homogeneous endosperm. *Ptychosperma schefferi* might be confused with *P. mambare*, but the latter is a species of mangroves and has smaller fruits.

Ptychosperma schefferi. LEFT TO RIGHT: habit, cultivated Fairchild Tropical Botanic Garden, Florida (SZ); habit, unusual entire-leaved form, cultivated Gardens by the Bay, Singapore (WB).

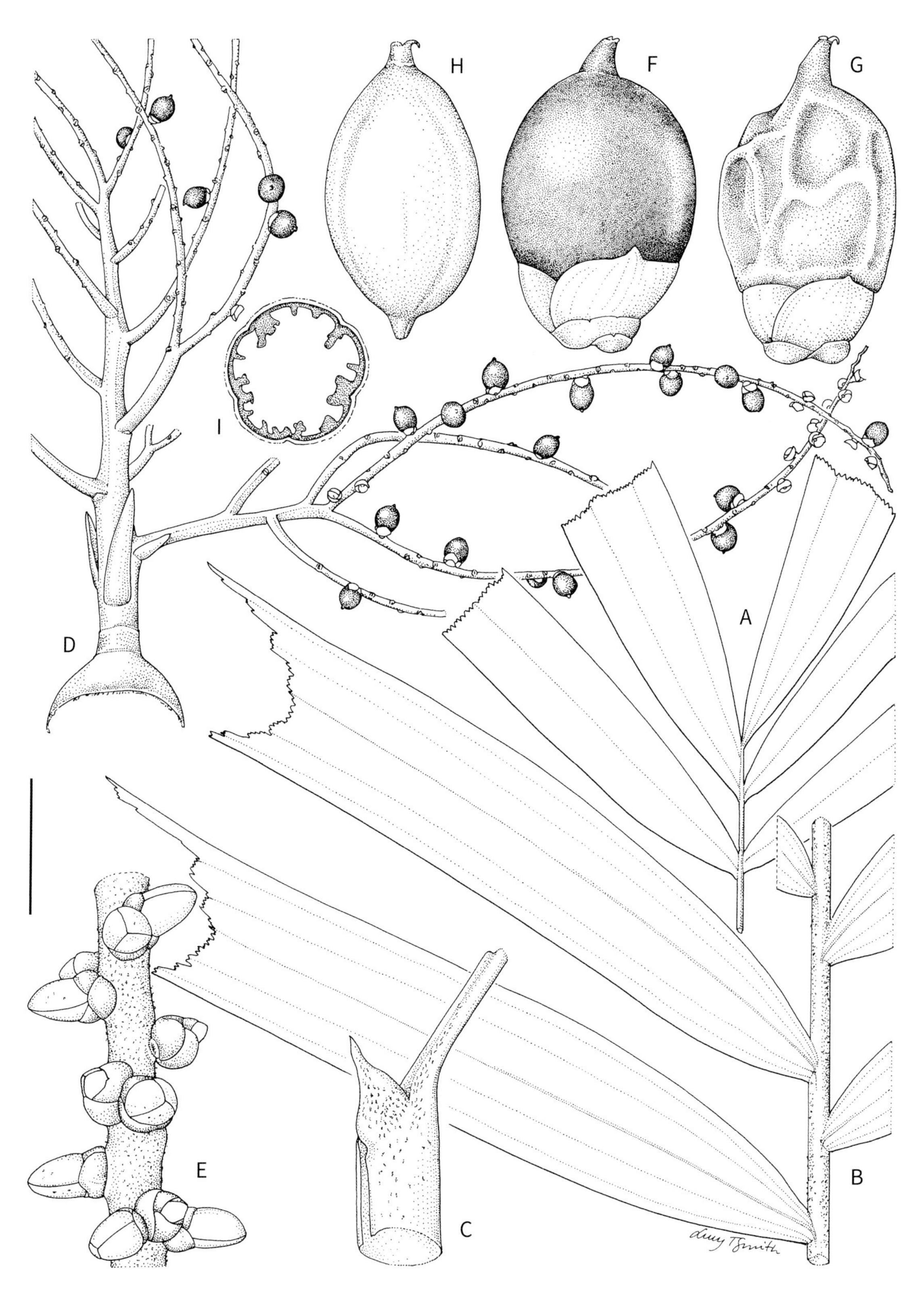

Ptychosperma schefferi. **A.** Leaf apex. **B.** Leaf mid-portion. **C.** Leaf sheath with ligule. **D.** Portion of inflorescence with fruit. **E.** Portion of rachilla with flowers. **F.** Fruit. **G.** Fruit, dried. **H.** Endocarp. **I.** Seed in transverse section. Scale bar: A–D = 6 cm; E = 1 cm; F–I = 7 mm. A–D, F–I from *Zona & Smith 968*; E from *Zona 970*. Drawn by Lucy T. Smith.

Ptychosperma schefferi. LEFT TO RIGHT: habit; inflorescence with fruit. Near Jayapura (GP).

20. *Ptychosperma tagulense* Essig

Single-stemmed, understorey to mid-storey palm ca. 2 m tall. **Stem** ca. 3 cm diam. **Leaf** not seen; sheath 32–35 cm long; petiole not seen; rachis not seen; leaflets arranged regularly in one plane, mid-leaf leaflet 24–28 × 4.7–5.1 cm wide, linear to narrowly cuneate, borne ca. 8 cm apart; terminal leaflets not larger than subterminal leaflets. **Inflorescence** 53–55 cm long, branched to 3 orders; *peduncle 23–24 × ca. 0.5 cm; rachillae ca. 4 cm long*, ca. 1 mm diam., colour not recorded, with ca. 7 female flowers per 2.5 cm (=14 female flowers per 5 cm). **Male flower** 4–5 × ca. 2 mm in bud, colour not recorded; stamens not seen. **Female flowers** not seen. **Fruit** 18–20 × 8–11 mm, *orange to red; exocarp smooth and conforming to endocarp when dry.* **Seed** 10–11 × 7–9 mm diam., *grooves shallow and lobes rounded; endosperm ruminate.*

DISTRIBUTION. Sudest Island in the Louisiade Archipelago.

HABITAT. Montane forest at ca. 700 m elevation.

LOCAL NAMES. None recorded.

USES. None recorded.

CONSERVATION STATUS. Data Deficient (IUCN 2021). This species is known only from one island, where there is limited information on ongoing threats, though global climate change is a concern given its montane habitat. Due to insufficient evidence, an extinction risk category cannot be selected.

NOTE. This species is distinguished from all other species by its long peduncle, short rachillae, dried fruits in which the exocarp and mesocarp conform to the endocarp, and by its montane habitat.

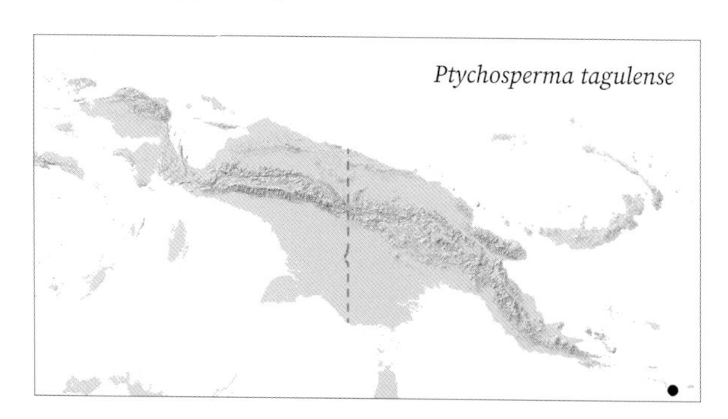

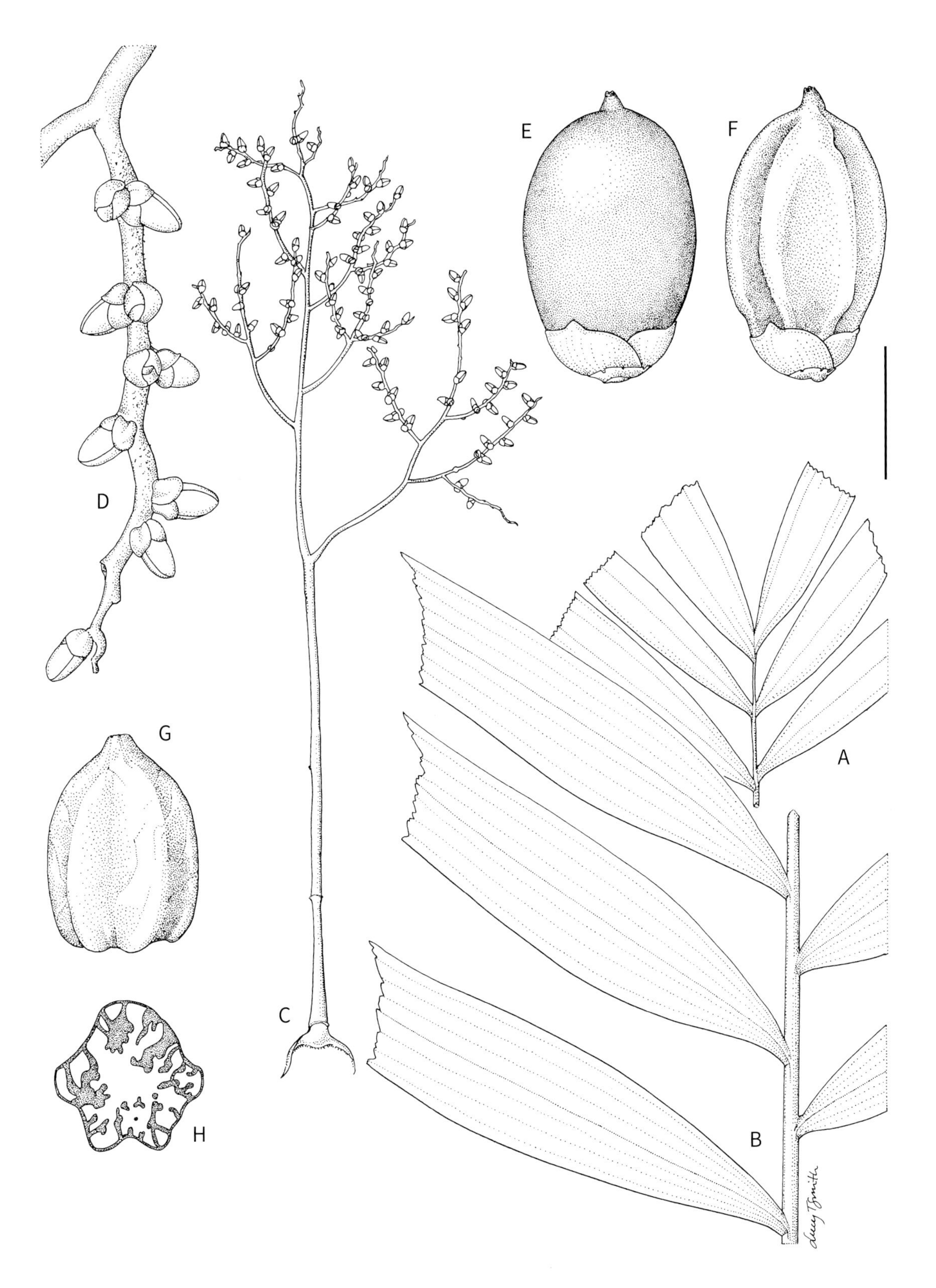

Ptychosperma tagulense. **A.** Leaf apex. **B.** Leaf mid-portion. **C.** Inflorescence. **D.** Rachilla with flowers. **E.** Fruit. **F.** Fruit, dried. **G.** Endocarp. **H.** Seed in transverse section. Scale bar: A, B = 6 cm; C = 4 cm; D, G, H = 7 mm; E, F = 1 cm. All from *Katik LAE 70867*. Drawn by Lucy T. Smith.

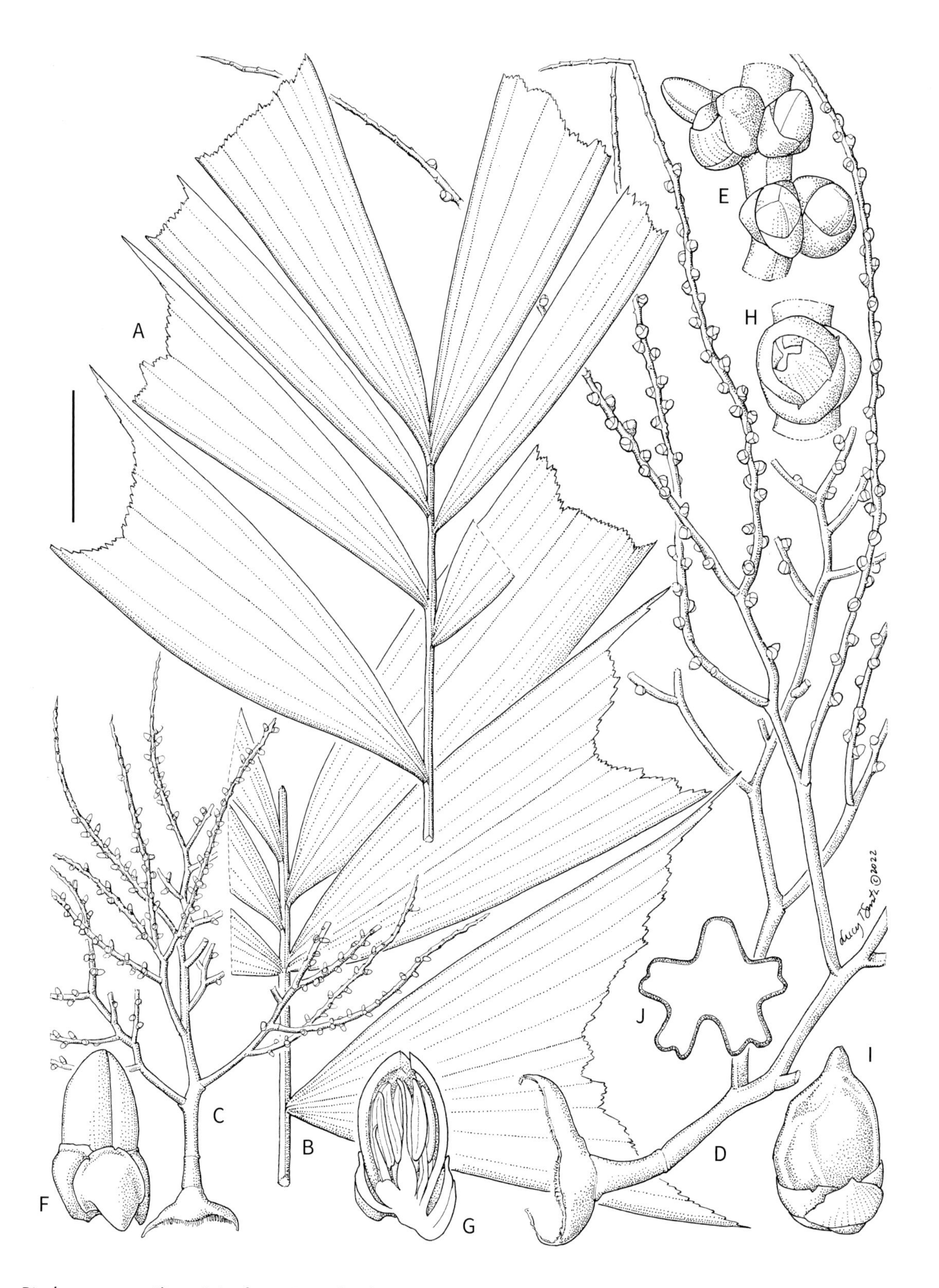

Ptychosperma vestitum. **A.** Leaf apex. **B.** Leaf mid-portion. **C.** Inflorescence with mature male flower buds. **D.** Inflorescence with mature female flower buds. **E.** Portion of rachilla with triads. **F**, **G.** Male flower whole and in longitudinal section. **H.** Female flower. **I.** Fruit, dried. **J.** Seed in transverse section. Scale bar: A, B = 6 cm; C, D = 4 cm; E, J = 4 mm; F, G = 3 mm; H = 7 mm; I = 1 cm. All from *Essig LAE 53132*. Drawn by Lucy T. Smith.

21. *Ptychosperma vestitum* Essig

Single-stemmed, understorey palm. **Stem** not seen. **Leaf** ca. 1.9 m long including petiole; sheath ca. 30 cm long; petiole 30–45 cm long; rachis 100–120 cm long; 9–11 pairs of leaflets, mid-leaf leaflet ca. 25 × ca. 11 cm wide, *cuneate, arranged irregularly and in one plane; lower surface of leaflets covered with conspicuous stellate or ragged scales.* **Inflorescence** ca. 44 cm long, branched to 2 or 3 orders; peduncle ca. 5 × 0.8 cm wide; rachillae ca. 32 cm long, ca. 2 mm diam., colour not recorded, with ca. 15 female flowers per 5 cm. **Male flower** not seen. **Female flower** ca. 5 × ca. 4 mm, ovoid. **Fruit** ca. 12 × ca. 8 mm, *black*; exocarp wrinkled when dry; endocarp not seen. **Seed** ca. 6 × ca. 5 mm diam., lobes blocky; *endosperm homogeneous.*

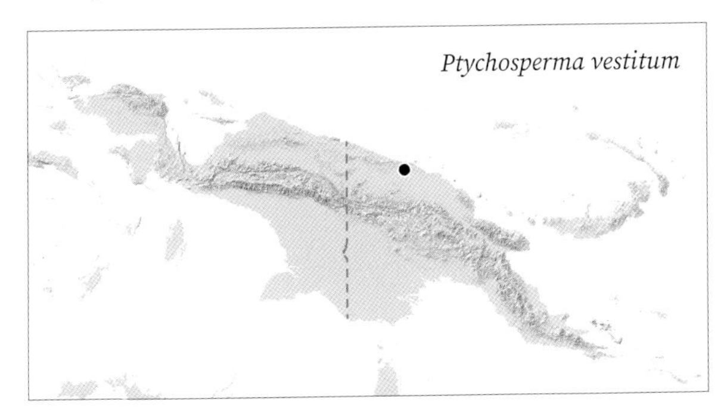

DISTRIBUTION. Near Wewak, north-eastern New Guinea.

HABITAT. Said to grow on "small mounds of black humus" within swamp forest, at ca. 150 m elevation.

LOCAL NAMES. None recorded.

USES. None recorded.

CONSERVATION STATUS. Data Deficient (IUCN 2021). This species is known from only one site, where there is limited information on ongoing threats. Due to insufficient evidence, an extinction risk category cannot be selected.

NOTE. This poorly known species is remarkable for the dense covering of ragged scales on the lower surface of the leaves.

22. *Ptychosperma waitianum* Essig

Synonym: *Ptychosperma burretianum* Essig

Multiple-stemmed (rarely solitary), understorey palm 2–8 m tall, bearing 5–7 leaves per stem. **Stem** 2–8 cm diam. **Leaf** 93–245 cm long including petiole; sheath 18–60 cm long; petiole 10–50 cm long; rachis 46–190 cm long; 7–17 pairs of leaflets, *arranged irregularly basally, regularly distally, and in one plane*, mid-leaf leaflet 19–38 × 7–15 cm, *cuneate*, borne 11–14 cm apart; *terminal leaflets not larger than subterminal leaflets; leaves are reddish upon emergence.* **Inflorescence** 18–43 cm long, branched to 2 orders; peduncle 1.9–4.7 × 0.5–1.8 cm wide; rachillae 12–35 cm long, 2–4 mm diam., reddish at anthesis, with 7–13 female flowers per 5 cm. **Male flower** 7–8 × 3–4 mm in bud, purple; stamens ca. 55. **Female flowers** not seen. **Fruit** 13–18 × 7–13 mm, *black*; exocarp wrinkled when dry; endocarp 14–18 × 7–8 mm. **Seed** 8–9 × 5–7 mm, lobes blocky; *endosperm homogeneous.*

DISTRIBUTION. South-eastern tip of New Guinea, and the D'Entrecasteaux Islands (Normanby Island) and Louisade Archipelago (Sudest Island).

HABITAT. Rainforest on hillsides and ridge slopes sea level to 430 m.

LOCAL NAMES. *Keleh* (Amele), *Kamuntua* (Daga).

USES. None recorded.

CONSERVATION STATUS. Near Threatened (IUCN 2021). *Ptychosperma waitianum* has a relatively restricted distribution. Deforestation due to logging concessions is a threat in its distribution range.

NOTE. This species is widely cultivated and noteworthy for its wedge-shaped leaflets and reddish emerging leaf. It may be distinguished from *P. pullenii* by its short peduncle and prominent bracts subtending the inflorescence branches. Moreover, *P. waitianum* has ca. 55 stamens as opposed to ca. 17 in *P. pullenii.*

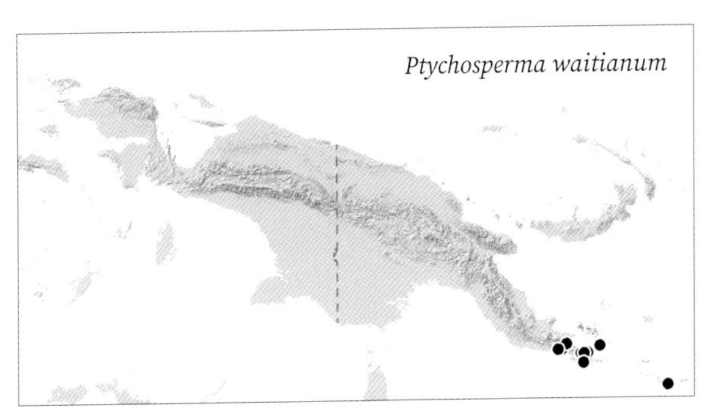

Ptychosperma waitianum. FROM TOP: emerging leaf, cultivated Royal Botanic Gardens, Kew (WB); male flowers, cultivated Fairchild Tropical Botanic Garden, Florida (SZ).

NAMES OF UNCERTAIN APPLICATION

Kentia sanderiana André

André described a juvenile palm shown at an exhibition in Ghent, Belgium, by the English nursery Sander & Co. The description is inadequate to fix this name to any known species and no illustration was published. Although this name is sometimes associated with *Ptychosperma sanderianum* Ridl., the two names cannot with certainty be shown to be conspecific. André's name remains a *nomen obscurum*.

Ptychosperma advenum Becc.

The source of this collection was surely a cultivated or naturalized palm (hence the specific epithet meaning "foreign"). The type specimen is insufficient for identification; however, as noted by Essig (1978), it bears a strong resemblance to *P. propinquum*, which is widely naturalised in Singapore (usually reported under the synonym *P. macarthurii*).

Ptychosperma angustifolium Blume (as "angustifolia")

The description omits critical details of fruits and seeds, and the type, an illustration of the habit, is not useful in fixing the identity of this name. It remains a *nomen obscurum*.

Ptychosperma praemorsum hort. ex Becc.

The type specimen is insufficient for identification. As noted by Essig (1978), it resembles either *P. furcatum* or *P. lineare*.

Ptychosperma *warletii* Sander ex M.T. Masters

Synonym: *Ptychosperma warteliana* André

This species is based on a juvenile palm, whose generic identity cannot be determined with certainty. These two names are typified by the same engraving of a juvenile palm exhibited by Sander's nursery of St Albans. The plant is described as having come from Seram (André 1898: 263) and having leaflets that are "silvery beneath" (Masters 1898: 242). While this name is often listed as a synonym of *Drymophloeus oliviformis*, its generic placement, in the absence of flowers and fruits, is equivocal. The spelling of the epithet is questionable; it may be a typographic error for *wartelii*.

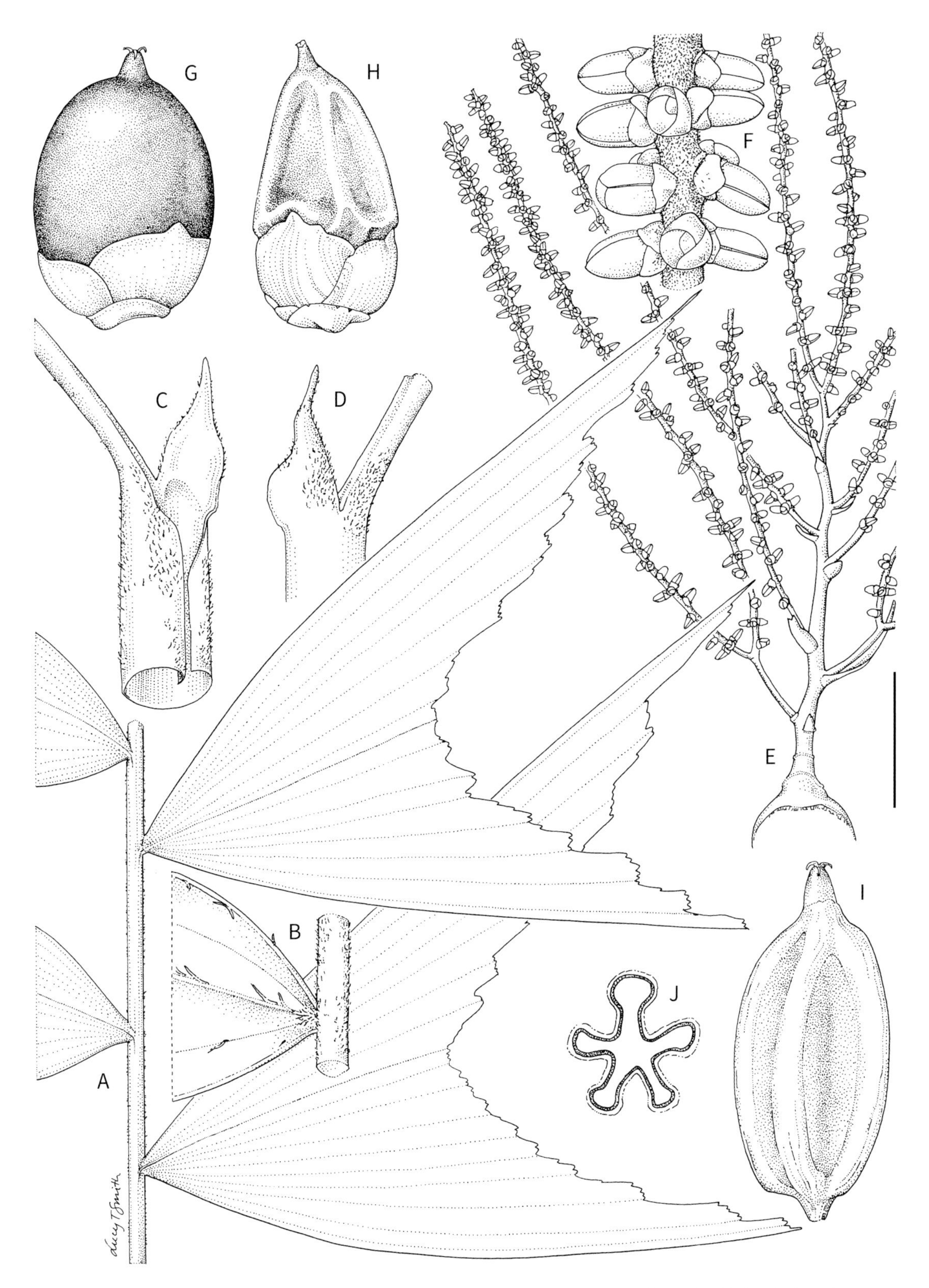

Ptychosperma waitianum. **A.** Leaf mid-portion. **B.** Leaflet abaxial side detail. **C, D.** Leaf sheath with ligule, two views. **E.** Inflorescence. **F.** Portion of rachilla with flowers. **G.** Fruit. **H.** Fruit, dried. **I.** Endocarp. **J.** Seed in transverse section. Scale bar: A, E = 6 cm; B–D = 3 cm; F = 1 cm; G, H = 7 mm; I, J = 5 mm. A–D, G–J from *Zona & Smith 964*; E, F from *Zona 867*. Drawn by Lucy T. Smith.

Ponapea hentyi. Habit, cultivated Ho'omaluhia Botanical Garden, Hawaii (JD).

Ponapea Becc.

Small to medium – leaf pinnate – crownshaft – no spines – leaflets jagged

Slender, single-stemmed tree palm, crownshaft present, monoecious. **Leaf** pinnate, *strongly arching;* sheath tubular; petiole present; leaflets few to numerous, reduplicate, *wedge-shaped,* with jagged tips, arranged regularly, horizontal. **Inflorescence** *below the leaves,* branched to 2 orders, branches spreading; peduncular bract projecting from prophyll, enclosing inflorescence in bud, prophyll (and sometimes peduncular bract) usually not dropping off as inflorescence expands; peduncle sometimes longer than inflorescence rachis; rachillae slender to quite robust, straight to curving. **Flowers** in triads throughout the length of the rachillae, not developing in pits, male flowers larger than female flowers, bullet-shaped. **Fruit** red, ellipsoid, stigmatic remains apical, flesh quite thick, *endocarp pale to black, thin, weakly 5-grooved.* **Seed** 1, conforming to endocarp, endosperm ruminate.

DISTRIBUTION. One species in New Britain and three species in the Caroline Islands.

NOTES. *Ponapea* is represented in the New Guinea region by the New Britain endemic species *P. hentyi,* a slender palm with strongly recurving leaves, wedge-shaped leaflets and an endocarp with a distinctive black, glassy inner surface. It is known from elevations up to around 700 m. It resembles *Drymophloeus,* although the ranges of the two genera do not overlap. *Drymophloeus* is immediately distinguished by its stilt roots and is generally less robust with less strongly recurving leaves. It is also similar to *Ptychosperma* and *Brassiophoenix,* which are distinguishable by their deeply, sharply lobed or angled endocarps (*P. hentyi* endocarps are only weakly lobed) and seeds and leaves that do not strongly recurve.

Ponapea hentyi (Essig) C.Lewis & Zona

Synonyms: *Drymophloeus hentyi* (Essig) Zona, *Ptychosperma hentyi* Essig

Solitary, understory palm bearing 10(–13) leaves. **Stem** 5–8 cm diam. **Leaf** 1.8–3 m long; sheath 45.5–75 cm long; petiole 15–21 cm long; rachis 157–261 cm long; 12–21 leaflets per side, mid-leaf leaflet 29–54 × 9.5–30.0 cm, *broadly wedge-shaped, gradually diminishing in size toward the tip of the leaf; terminal leaflets not enlarged.* **Inflorescence** ca. 75 cm long; peduncular bract 9.0–24.5 × ca. 4.5 cm; peduncle 12–17 cm long; rachillae 11–20 cm long, ca. 2 mm diam. **Male flower** 6–7(–10) mm × 4 mm; stamens 25–36; *pistillode conical,* up to 2 mm long, style absent. **Female flower** not seen. **Fruit** 15–16 × 9–13 mm, broadly fusiform to nearly globose, red; exocarp slightly striate when dry; endocarp 15–16 × 10 mm, fusiform, *inner wall black, glassy.* **Seed** 8–9 × 8 mm, globose or slightly oblate spheroid, sometimes weakly five-lobed, brown; *endosperm ruminate.*

DISTRIBUTION. New Britain.

HABITAT. Rainforest from sea level to 820 m.

LOCAL NAMES. None recorded.

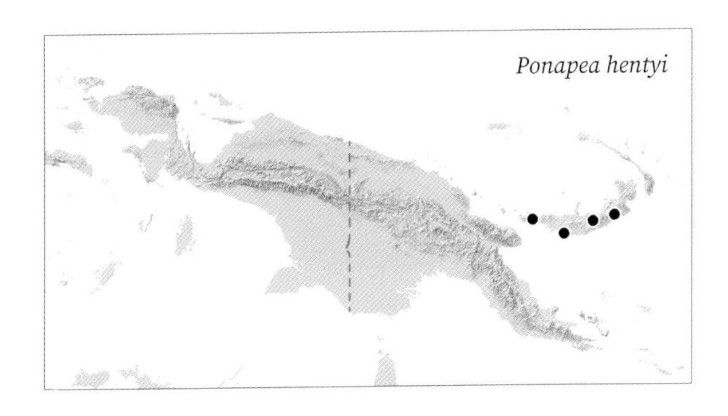

USES. None recorded.

CONSERVATION STATUS. Endangered (IUCN 1998). *Ponapea hentyi* has a restricted distribution. Deforestation due to land conversion for oil palm is a major threat in its distribution range.

NOTE. This species was originally decribed in *Ptychosperma* and then included in *Drymophloeus* by Zona (1999), but molecular evidence (Zona *et al.* 2011, Alapetite *et al.* 2014) suggested its inclusion in the morphologically heterogeneous *Ponapea*.

Ponapea hentyi. LEFT TO RIGHT FROM TOP: inflorescences, cultivated National Botanic Garden, Lae (WB); male flowers, cultivated Nong Nooch Tropical Garden, Thailand (SZ); inflorescence with fruit, New Britain (FE).

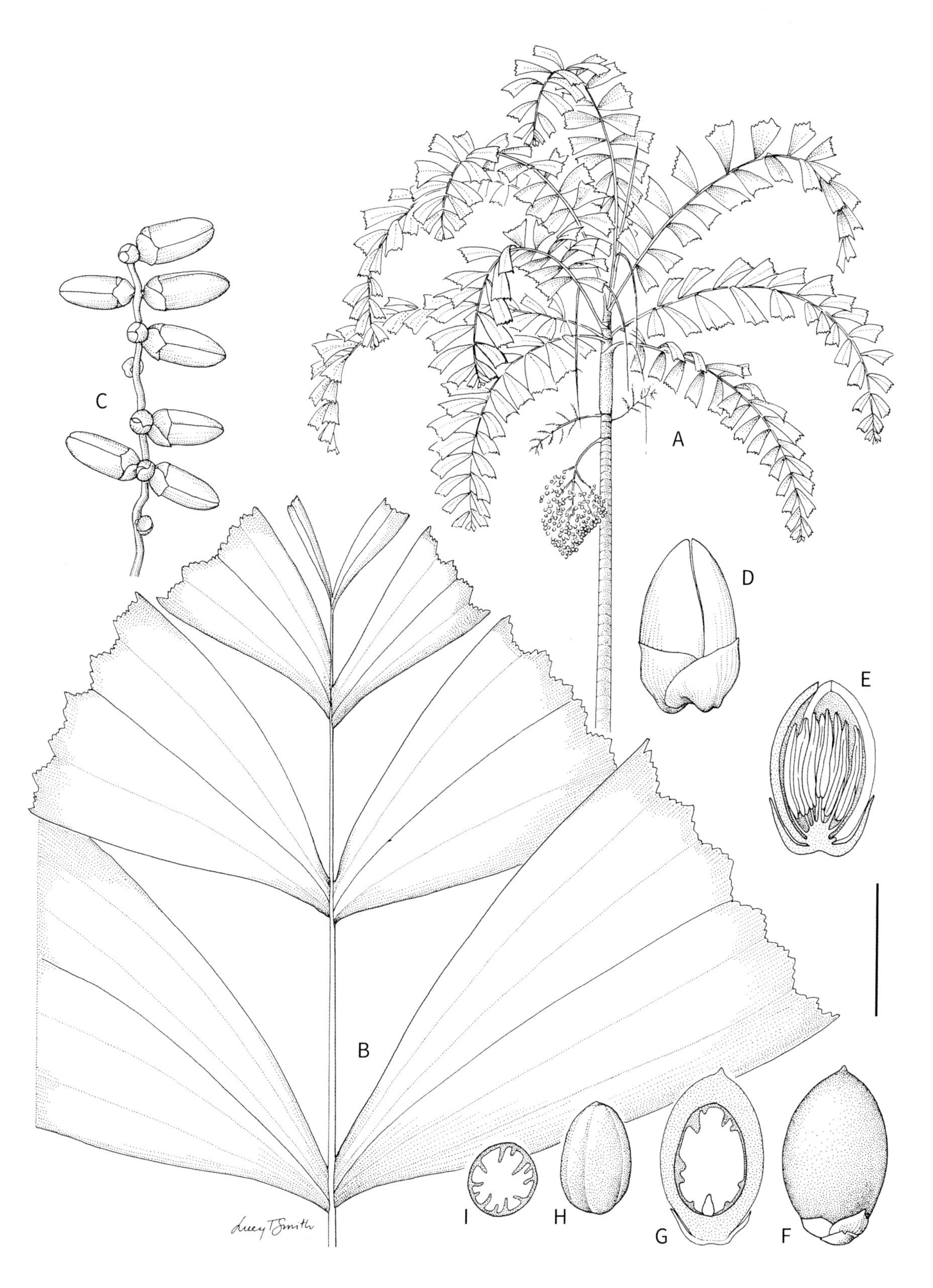

Ponapea hentyi. **A.** Habit. **B.** Leaf apex. **C.** Portion of rachilla with flowers. **D**, **E.** Male flower whole and in longitudinal section. **F**, **G.** Fruit whole and in longitudinal section. **H**, **I.** Seed whole and in transverse section. Scale bar: A = 1 m; B = 4 cm; C, F–I = 1.5 cm; D, E = 5 mm. A from photograph; B–I from *Millar NGF 40558*. Drawn by Lucy T. Smith.

Drymophloeus litigiosus. Habit, Tamrau Mountains (WB).

Drymophloeus Zipp.

Synonyms: *Coleospadix* Becc, *Rehderophoenix* Burret, *Saguaster* Kuntze

Small to medium – leaf pinnate – crownshaft – no spines – leaflets jagged

Slender, usually single-stemmed tree palm, usually with a cone of stilt roots, crownshaft present, monoecious. **Leaf** pinnate, straight to arching; sheath tubular; petiole present; leaflets narrowly to broadly wedge-shaped, with jagged tips, reduplicate, arranged regularly, horizontal. **Inflorescence** *below the leaves,* branched 1 or 2 orders, branches spreading, green; peduncular bract longer than and projecting from prophyll, enclosing inflorescence in bud, *prophyll (and sometimes peduncular bract) usually not dropping off as inflorescence expands, disintegrating in place;* peduncle sometimes longer than inflorescence rachis; rachillae slender to quite robust, straight to curving. **Flowers** in triads throughout the length of the rachillae, not developing in pits, male flowers larger than female flowers, bullet-shaped. **Fruit** *red, ovoid,* stigmatic remains apical, flesh quite thick, juicy, filled with irritant needle crystals, *endocarp pale, thin, smooth.* **Seed** 1, conforming to endocarp, endosperm homogeneous or ruminate.

DISTRIBUTION. Two species from Maluku to New Guinea, both in New Guinea, but only in the far north-west.

NOTES. *Drymophloeus* species are small to medium-sized palms distinguished readily by the cone of stilt roots, well-defined crownshaft and pinnately arranged leaflets with jagged apices. The genus occurs in lowland and montane forest up to 1,200 m. *Ponapea hentyi* is similar in habit, but differs in its strongly reflexed leaves and the absence of stilt roots, and is known only from New Britain. Other genera, usually lacking stilt roots, that might be confused with *Drymophloeus* are *Brassiophoenix,* which has very broad wedge-shaped leaflets, larger, orange fruit (ca. 3.5 cm long) and endocarps with conspicuous grooves and ridges, *Dransfieldia,* which has pointed leaflet tips and black fruit, and *Ptychosperma,* which has a grooved endocarp.

A taxonomic account of *Drymophloeus* was published by Zona (1999), but the genus has since undergone considerable contraction in recent years as a result of molecular analyses (Zona *et al.* 2011). These changes affect one other species from our area, *Ponapea hentyi,* and three species from the Solomon Islands now placed in *Veitchia* H.Wendl. We follow the taxonomy of Zona (1999) for the two species that remain in *Drymophloeus.*

Key to the species of *Drymophloeus* in New Guinea

1. Terminal leaflet usually bifid, endosperm ruminate. . . . . **1. *Drymophloeus litigiosus***
1. Terminal leaflet broadly fan-like, not cleft or with an apical notch, endosperm homogeneous . . . . **2. *Drymophloeus oliviformis***

1. *Drymophloeus litigiosus* (Becc.) H.E.Moore

Synonyms: *Coleospadix beguinii* Burret, *Coleospadix litigiosa* (Becc.) Becc., *Coleospadix oninensis* (Becc.) Becc., *Coleospadix porrectus* Burret, *Drymophloeus beguinii* (Burret) H.E.Moore, *Drymophloeus oninensis* (Becc.) H.E.Moore, *Drymophloeus porrectus* (Burret) H.E.Moore, *Ptychosperma litigiosum* Becc., *Ptychosperma litigiosum* var. *oninense* Becc., *Saguaster oninensis* (Becc.) Kuntze

Small, single- or multiple-stemmed palm to 6 m bearing 7–9 leaves. **Stem** 1.3–5.0 cm diam.; stilt root cone 30–100 m tall. **Leaf** 1.5–3 m long; sheath 20–79.5 cm long; petiole 22–48 cm long; rachis 85–200 cm long; 7–16 pairs of leaflets, mid-leaf leaflet 23–64 cm long, 4.3–20.5 cm wide; *terminal leaflets not united.* **Inflorescence** 47–75 cm long; peduncular bract 17.5–41.5 × 1.1–2.5 cm wide; peduncle 13.5–37.5 cm long, 0.3–1.2 cm wide; rachillae 12–40 cm long, slender, 1–3 mm diam. **Male flower** 4–6 × 2–3 mm; stamens 24–32; pistillode 2–4 × ca. 1 mm, *style absent.* **Female flowers** 3–5 × 3–4 mm, flattened globose. **Fruit** 14–23 × 6–11 mm; endocarp 15–16 × 7–8 mm. **Seed** 7–14 × 4–9 mm, globose, sometimes flattened at the base, brown; *endosperm ruminate.* **Eophyll** *elliptical, broadly notched at the apex.*

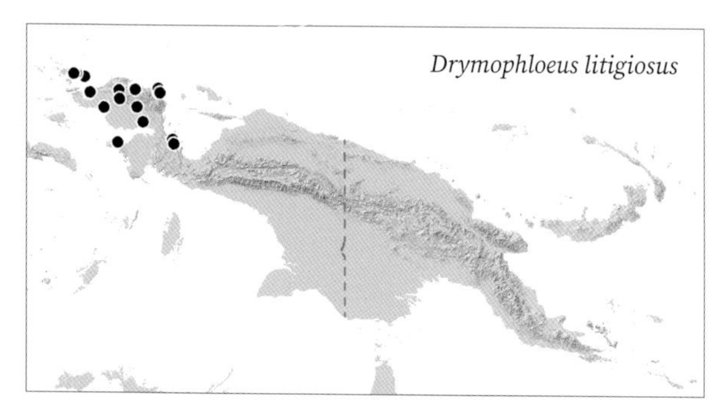

DISTRIBUTION. North-western New Guinea and Waigeo. Also Maluku (Morotai, Halmahera, Obi).

HABITAT. Hill forest or alluvial forest at sea level to 1,200 m.

LOCAL NAMES. *Kiligata* (Moi), *Maiterate* (Sumury), *Meraningga Afok* (Meyah), *Taupu* (Yamur) and many others (see Zona 1999).

USES. Stems used for spears and arrowheads.

CONSERVATION STATUS. Least Concern.

NOTES. This palm is highly variable in its habit (usually single-stemmed but occasionally multiple-stemmed), leaf indument density, and depth of the cleft in the bifid terminal leaflet.

Drymophloeus litigiosus (narrow leaflet form).
FROM TOP: crown; inflorescence. Near Sorong (WB).

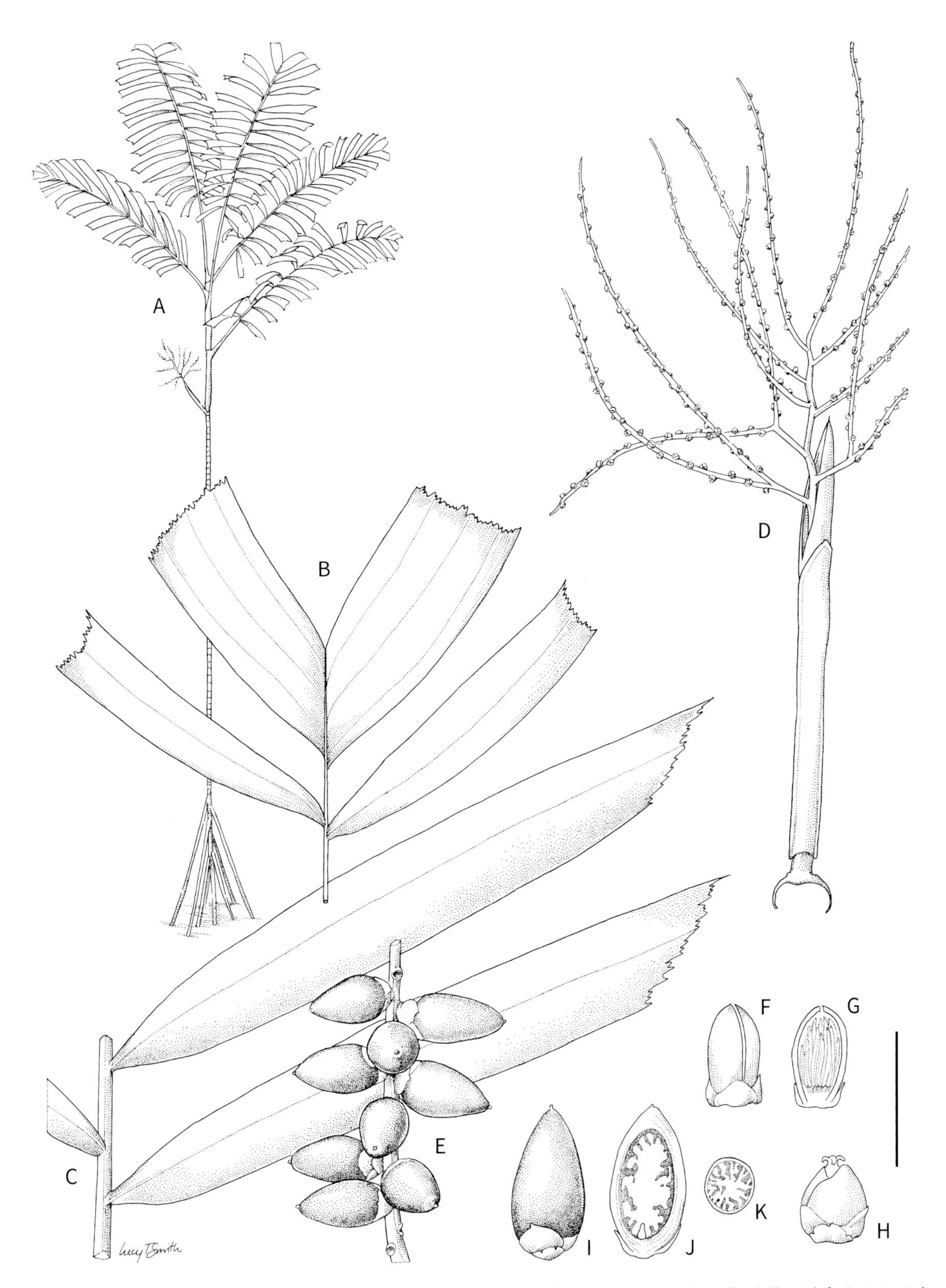

Drymophloeus litigiosus. **A.** Habit. **B.** Leaf apex. **C.** Leaf mid-portion. **D.** Inflorescence. **E.** Portion of rachilla with fruit. **F**, **G.** Male flower whole and in longitudinal section. **H.** Female flower. **I**, **J.** Fruit whole and in longitudinal section. **K.** Seed in transverse section. Scale bar: A = 85 cm; B, C = 9 cm; D = 8 cm; E = 2.5 cm; F–H = 5 mm; I–K = 2 cm. A from photograph; B–D, H from *Zona et al. 680*; E, I–K from *Sands et al. 6483*; F, G from *Zona et al 696.* Drawn by Lucy T. Smith.

Drymophloeus litigiosus. ANTI-CLOCKWISE FROM TOP: leaf apex, Wandamen Peninsula (WB); stilt roots, Waigeo Island (GP); fruit, Arfak Plains (JD); male flowers, Wandamen Peninsula (WB).

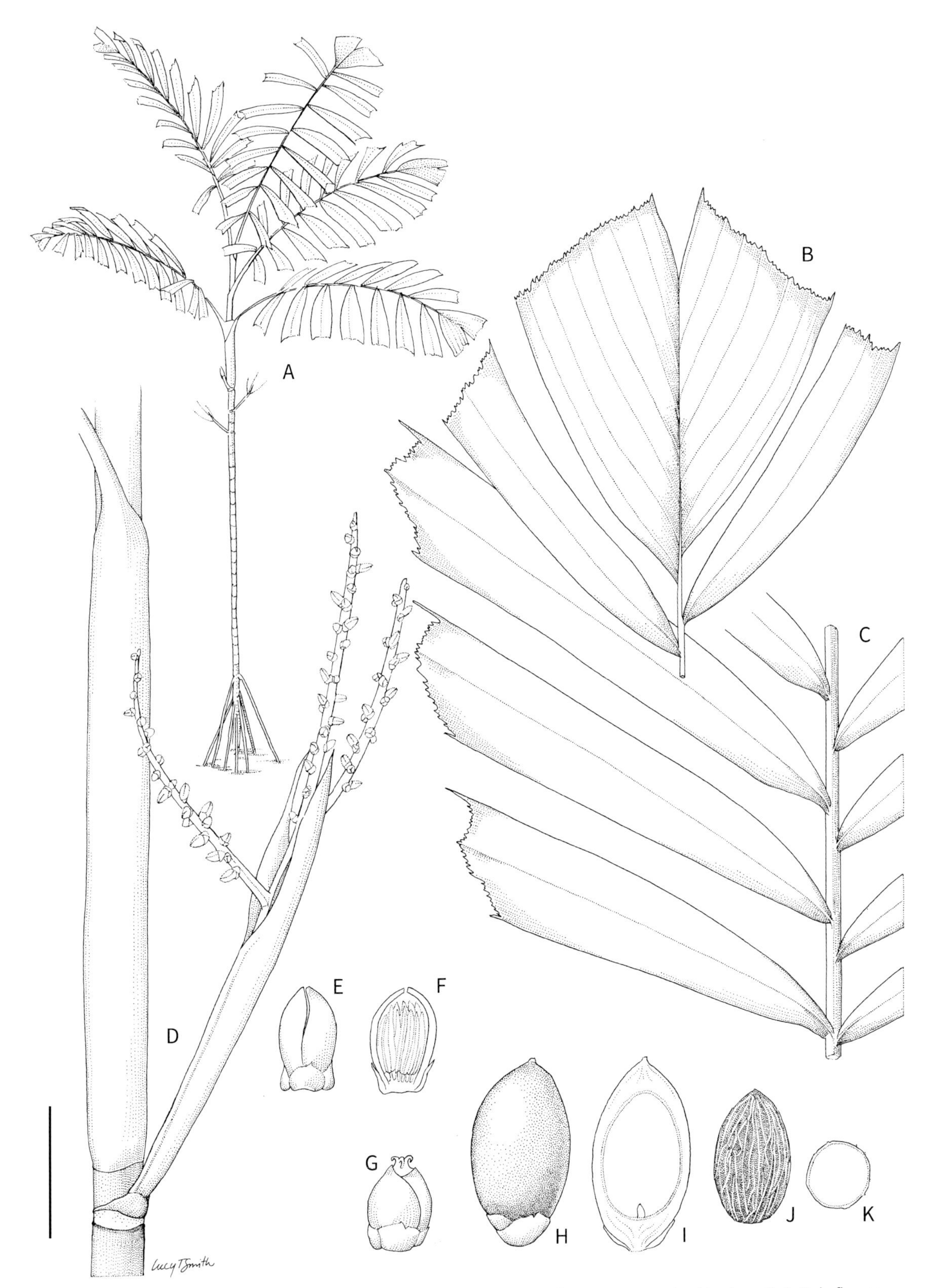

Drymophloeus oliviformis. **A.** Habit. **B.** Leaf apex. **C.** Leaf mid-portion. **D.** Inflorescence attached to stem. **E**, **F.** Male flower whole and in longitudinal section. **G.** Female flower. **H**, **I.** Fruit whole and in longitudinal section. **J**, **K.** Seed whole and in transverse section. Scale bar: A = 1.2 m; B, C = 9 cm; D = 4 cm; E–K = 1 cm. A from photograph; B, C, G from *Zona et al. 684*; D–F from *Sands 6199*; H–K from *van Balgooy 4714*. Drawn by Lucy T. Smith.

Drymophloeus oliviformis. TOP LEFT TO RIGHT: habit, Wandamen Peninsula (WB); stilt roots, Wosimi River (WB); crown with inflorescences, Arfak Plains (JD); Inflorescence with fruit, Wosimi River (WB).

2. *Drymophloeus oliviformis* (Giseke) Mart.

Synonyms: *Areca elaeocarpa* Reinw. ex Kunth, *Areca oliviformis* Giseke, *Areca oliviformis* var. *gracilis* Giseke, *Areca vaginata* Giseke, *Coleospadix gracilis* (Giseke) Burret, *Drymophloeus appendiculatus* (Blume) Miq., *Drymophloeus bifidus* Becc., *Drymophloeus ceramensis* Miq., *Drymophloeus jaculatorius* Mart., *Drymophloeus leprosus* Becc., *Drymophloeus rumphii* Blume ex Scheff., *Harina rumphii* (Blume) Mart., *Ptychosperma appendiculatum* Blume, *Ptychosperma oliviforme* (Giseke) Schaedtler, *Ptychosperma rumphii* Blume, *Saguaster appendiculatus* (Blume) Kuntze, *Saguaster bifidus* (Becc.) Kuntze, *Saguaster leprosus* (Becc.) Kuntze, *Saguaster oliviformis* (Giseke) Kuntze, *Seaforthia blumei* Kunth, *Seaforthia jaculatoria* Mart., *Seaforthia oliviformis* (Giseke) Mart.

Small, usually single-stemmed, sometimes multiple-stemmed palm to 7 m, bearing 6 or 7 leaves. **Stem** 2–7 cm diam.; stilt root cone 30 m tall. **Leaf** 1–2.5 m long; petiole 21–61 cm long; sheath 15.5–70 cm long; rachis 60–150 cm long; 8–19 pairs of leaflets, *rubbery in texture*, mid-leaf leaflet 18–71 × 3–24 cm wide; *terminal leaflets usually united to form a single fan-like leaflet*, which may be slightly cleft. **Inflorescence** 18–30 cm long; peduncular bract 8–28 × 1.5–3.5 cm wide; peduncle 7–26 cm long, 0.3–0.7 cm wide; rachillae 7.5–40.5 cm long, 1–4 mm diam. **Male flower** 5–10 × 2–5 mm; stamens 30–66; pistillode 4–6 × ca. 1 mm, *style 2–3 mm long*. **Female flowers** 3–6 × 3–6 mm, globose. **Fruit** 11–24 × 6–12 mm; endocarp 9–18 × 7–8 mm. **Seed** 5–11 × 4–8 mm, globose to ovoid, sometimes flattened at the base, brown; *endosperm homogeneous*. **Eophyll** *elliptical, rarely shallowly notched at the apex.*

DISTRIBUTION. North-western New Guinea. Also Maluku (Ambon, Buru, Seram, Sula Islands).

HABITAT. Lowland or hillside rainforest or alluvial forest, often over limestone, at 10–600 m elevation.

LOCAL NAMES. *Songgomi* (Wandama).

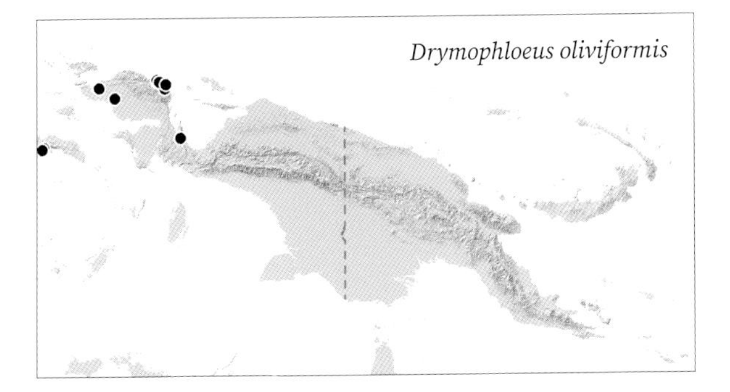

USES. Stems used for spears, harpoons; leaves used for wrapping sago.

CONSERVATION STATUS. Least Concern.

NOTES. In the field, this palm is easily recognised by the texture and shape of the leaflets. The leaflets have a firm, almost rubbery texture, and the terminal pair of leaflets are united into a single fan-like leaflet, sometimes with an inconspicuous apical notch.

NAMES OF UNCERTAIN APPLICATION

Drymophloeus angustifolius (Blume) Mart.

Synonyms: *Actinophloeus angustifolius* (Blume) L.H.Bailey, *Coleospadix angustifolius* (Blume) Burret, *Ptychosperma angustifolium* Blume, *Saguaster angustifolius* (Blume) Kuntze

This species was described on vegetative characters and inflorescence branches; mature flowers, fruits and seeds were not described. The type illustration, the shorter, multi-stemmed palm in the foreground, is inadequate for determining the generic identity of this taxon: it could be *Drymophloeus* or *Ptychosperma*. The latter is more likely, given that the illustration does not show a persistent peduncular bract, as it does for *D. oliviformis* (as *P. rumphii*) in the background.

Drymophloeus communis (Zipp. ex Blume) Miq.

Synonyms: *Areca communis* Zipp. ex Blume, *Seaforthia communis* (Zipp. ex Blume) Mart., *Ptychosperma commune* (Zipp. ex Blume) Miq.

The description for this taxon is vague and could correspond to any of several species of *Ptychosperma*, as well as *Drymophloeus*. Although Blume (1839) originally described it in *Areca*, he believed it might well belong to the genus *Ptychosperma*. This conclusion is further bolstered by the stated distribution of this entity, in south-eastern New Guinea, where *Drymophloeus* is not known to occur.

Brassiophoenix schumannii. Habit, East Sepik (AB).

Brassiophoenix Burret

Medium – leaf pinnate – crownshaft – no spines – leaflets jagged

Slender, single-stemmed, tree palm, crownshaft present, monoecious. **Leaf** pinnate, reduplicate; sheath tubular; petiole present; leaflets arranged regularly, horizontal, distinctively *wedge-shaped with a broad, jagged apical margin formed into 2 or 3 prongs.* **Inflorescence** *below the leaves,* branched to 2 or 3 orders, branches spreading, green; peduncular bract longer than and projecting from prophyll, enclosing inflorescence in bud, prophyll and peduncular bract dropping off as inflorescence expands; peduncle shorter or longer than rachis; rachillae slender and ± curved. **Flowers** in triads throughout the length of the rachillae, not developing in pits, male flowers larger than female flowers, bullet-shaped. **Fruit** orange, globose, stigmatic remains apical, flesh thick and juicy, *endocarp pale, thick with conspicuous grooves and sharp ridges, closely adhering to seed.* **Seed** 1, conforming to endocarp shape, endosperm homogeneous.

DISTRIBUTION. Two species endemic to eastern mainland New Guinea, not recorded from Bismarck Archipelago or from Indonesian New Guinea.

NOTES. *Brassiophoenix* is a New Guinea endemic genus with two species, both slender, single-stemmed, mid-storey tree palms with broad wedge-shaped leaflets and deeply ridged, straw-coloured endocarps. The genus is known only from lowland rainforest in eastern New Guinea up to 800 m. *Brassiophoenix* can be confused with *Drymophloeus,* which has stilt roots and an endocarp lacking conspicuous ridges, *Ptychococcus,* a more robust palm, with linear leaflets and red fruit with a black or dark brown endocarp, or *Ptychosperma,* which has small fruit and endocarps with shallow ridges and grooves. This account follows the most recent taxonomic synopsis of *Brassiophoenix* (Zona & Essig 1999).

Key to the species of *Brassiophoenix*

1. Endocarp with five ridges . . . . **1. *B. drymophloeoides***
1. Endocarp with nine ridges . . . . **2. *B. schumannii***

1. *Brassiophoenix drymophloeoides* Burret

Slender, single-stemmed palm to 10 m. **Stem** 4–7 cm diam. **Leaf** to ca. 3 m long, 5–14 leaves in crown; sheath 30–60 cm long; petiole 3–20 cm long; rachis 210–256 cm long; leaflets 13–18 each side of rachis; mid-leaf leaflet 41–67 × 22–33 cm. **Inflorescence** 50–94 cm long; peduncular bract ca. 31 × 4 cm wide; peduncle 6–10 cm long, ca. 1 cm wide; rachillae 22–30 cm long, 1–2 mm diam., densely scaly. **Male flower** ca. 10 × 5 mm in bud; stamens ca. 170; pistillode up to 1 mm long, short conical. **Female flower** globose, ca. 8 mm long, 6 mm diam. **Fruit** 31–38 × 18–29 mm; endocarp ca. 33 × 23 mm, strongly 5-ridged longitudinally. Seed 15–16 × 11–12 mm.

DISTRIBUTION. South-eastern New Guinea.

HABITAT. Moist lowland forest to hill forest on limestone at sea level to 360 m.

LOCAL NAMES. *Pawa* (Mekeo), *Kitat* (Daga)

USES. The hard wood is sometimes used.

CONSERVATION STATUS. Least Concern (IUCN 2021). Monitoring is required due to threats from logging.

NOTES. In overall appearance, the two species of *Brassiophoenix* are similar, and they lack any obvious differences in their flowers. They are readily distinguished by their endocarps: five-ridged in *B. drymophloeoides* and nine-ridged in *B. schumannii*. Moreover, their ranges scarcely overlap.

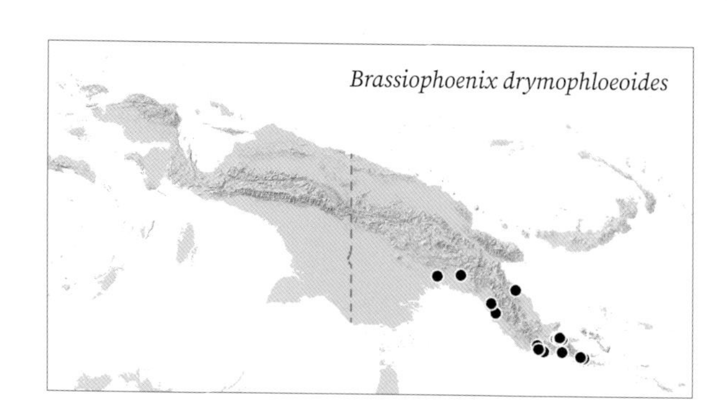

Brassiophoenix drymophloeoides. LEFT TO RIGHT: inflorescence with fruit, cultivated Fairchild Tropical Botanic Garden, Florida (SZ); leaf, near Kikori (WB).

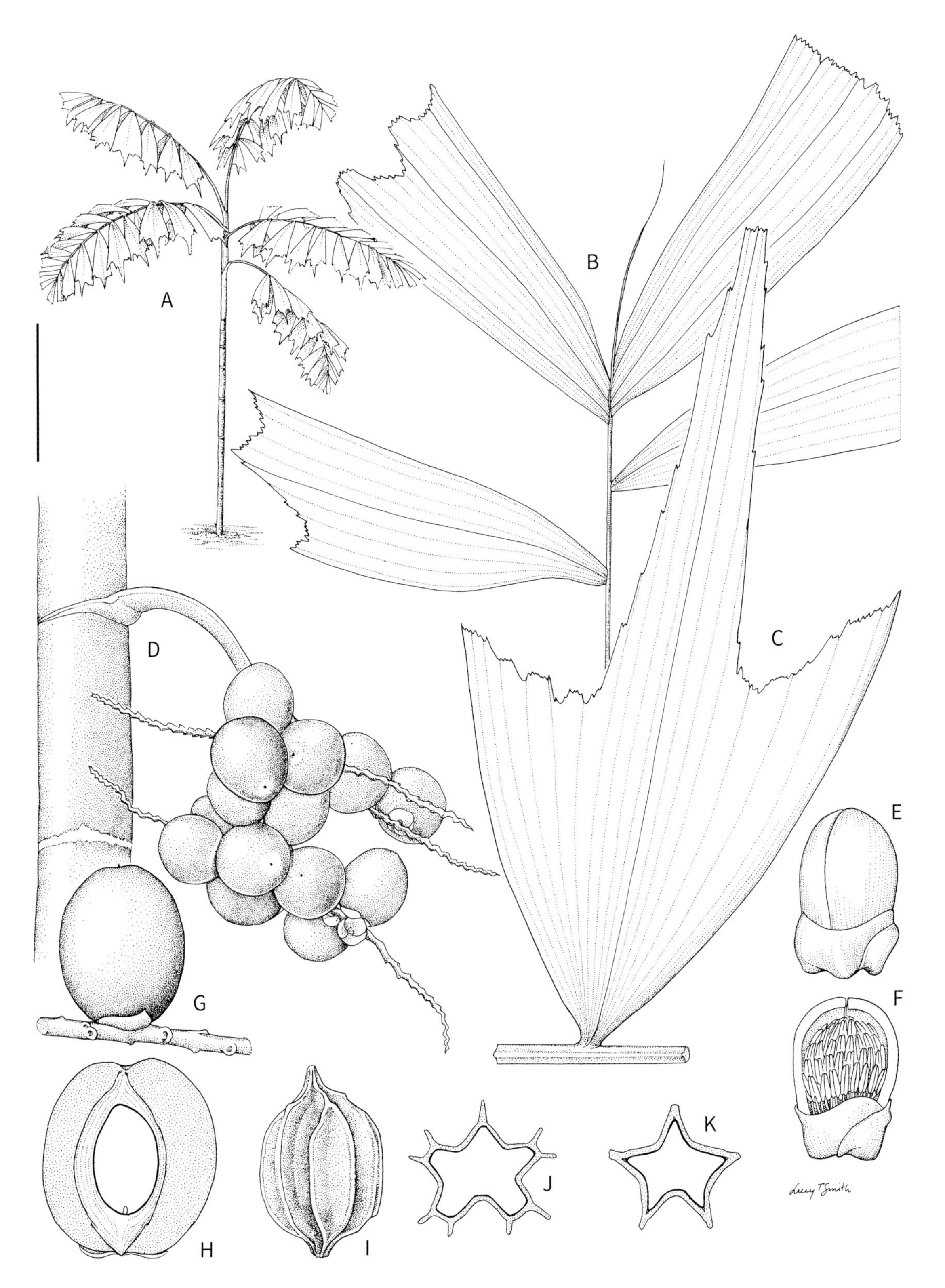

Brassiophoenix schumannii (**A–J**) and ***B. drymophloeoides*** (**K**). **A.** Habit. **B.** Leaf apex. **C.** Leaf mid-portion. **D.** Inflorescence with fruit. **E.** Male flower. **F.** Male flower, one petal removed. **G, H.** Fruit whole and in longitudinal section. **I, J.** Endocarp whole and in transverse section (*B. schumannii*). **K.** Endocarp in transverse section (*B. drymophloeoides*). Scale bar: A = 3.6 m; B, C = 8 cm; D = 6 cm; E, F = 1 cm; G = 2.5 cm; H–K = 2 cm. A–J from *Baker & Utteridge 570*; K from *Pullen 7645*. Drawn by Lucy T. Smith.

Brassiophoenix drymophloeoides. TOP: female flowers, cultivated Fairchild Tropical Botanic Garden, Florida (SZ); BOTTOM LEFT TO RIGHT: fruit; fruit in transverse section. Near Kikori (WB).

2. *Brassiophoenix schumannii* (Becc.) Essig

Synonyms: *Actinophloeus schumannii* Becc., *Drymophloeus schumannii* (Becc.) Warb. ex K.Schum. & Lauterb., *Ptychococcus schumannii* (Becc.) Burret

Slender palm to 12 m, bearing ca. 10 leaves in crown. **Stem** 6–7 cm diam. **Leaf** to ca. 2 m long; sheath 30–60 cm long; petiole ca. 22 cm long; rachis ca. 120 cm long; leaflets 9 each side of rachis; mid-leaf leaflet 41–64 × 11–30 cm wide. **Inflorescence** ca. 25 cm long; peduncular bract not seen; peduncle 11–28 cm long, 1–2 cm wide; rachillae 27–31 cm long, 3–4 mm diam. **Male flower** 7–8 × 5 mm in bud; stamens 50–175; pistillode ca. 1 mm long. **Female flower** not seen. **Fruit** 30–50 × 17–27 mm; endocarp ca. 28 × 20 mm, *ridged and grooved longitudinally, with 9 ridges (5 deep grooves and 4 shallow grooves)*. **Seed** 16–23 × 9–15 mm.

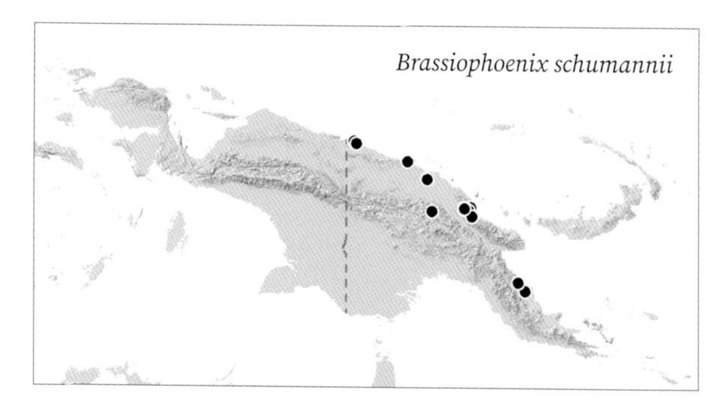

Brassiophoenix schumannii. CLOCKWISE FROM ABOVE: crown, Ioma (MC); inflorescence with fruit, Ioma (MC); male flowers, Finschhafen (AB).

DISTRIBUTION. North-eastern New Guinea, with outlying records in the south-east. It is widespread, but populations are never large.

HABITAT. Moist lowland forest at sea level to 800 m.

LOCAL NAMES. *Chiram Pui* (Angoram), *Nablok Glan* (Kakal), *Pewi* (Kilmeri, Eastern Kilmeri or Ossima dialect), and others (see Ferrero & Dowe 1996).

USES. Ferrero & Dowe (1996) recorded that the species is little used, with leaflets used occasionally as food wrappers and the wood used for combs, forks and spear heads. The fruit are used as a substitute for betel. In Madang, the stems are used for flooring and made into implements for sharpening bows and arrows (Takeuchi 2000).

CONSERVATION STATUS. Least Concern (IUCN 2021). Monitoring is required due to threats from human habitation and from forest degradation through logging and agriculture.

NOTES. See notes under *B. drymophloeoides*.

Ptychococcus lepidotus. Habit, Wau (WB).

Ptychococcus Blume

Medium to tall – leaf pinnate – crownshaft – no spines – leaflets jagged

Moderate to robust, single-stemmed tree palms, crownshaft present, monoecious. **Leaf** pinnate, slightly arching, *typically with some leaves tilted 90° to one side;* sheath tubular; petiole absent to elongate; leaflets numerous, with jagged tips, reduplicate, arranged regularly, horizontal or tilted. **Inflorescence** *below the leaves,* branched to 3 orders, *branches widely spreading;* prophyll and peduncular bract similar, enclosing inflorescence in bud, dropping off as inflorescence expands; peduncle shorter than inflorescence rachis; rachillae robust and straight. **Flowers** in triads throughout the length of the rachillae, not developing in pits, male flowers larger than female flowers, bullet-shaped. **Fruit** red, ovoid, stigmatic remains apical, *flesh thick and juicy, endocarp thick, black or dark brown, with conspicuous grooves and sharp ridges.* **Seed** 1, conforming to endocarp, endosperm ruminate to homogeneous.

DISTRIBUTION. Two species restricted to New Guinea, the Bismarck Archipelago and Bougainville.

NOTES. *Ptychococcus* is readily recognized by its large, emergent stature, leaves that are often tilted 90° to one side, lanceolate leaflets, large, white inflorescence axes, fleshy, red fruits, and black or brown, ridged and lobed endocarps. The genus is phylogenetically close to *Brassiophoenix* (Zona *et al.* 2011, Alapetite *et al.* 2014). It is widespread in lowland to montane rainforest, from sea level to 1,800 m. *Ptychococcus* can be confused with three related, but more slender genera: *Brassiophoenix,* which has broad wedge-shaped leaflets and a ridged but pale endocarp; *Drymophloeus,* which is smaller and has stilt roots and a pale endocarp lacking ridges; and *Ptychosperma,* which has small fruit with pale endocarps bearing shallow ridges and grooves.

The taxonomic revision of Zona (2005) is followed here.

Key to the species of *Ptychococcus*

1. Male flower perianth scaly on the lower surface; anthers smooth; endocarp brown with embedded black fibres, 2–4 mm thick .... **1. *Ptychococcus lepidotus***
1. Male flower perianth not scaly on the lower surface; anthers bullate; endocarp black, ca. 1 mm thick .... **2. *Ptychococcus paradoxus***

1. *Ptychococcus lepidotus* H.E.Moore

Robust, emergent palm 6–19 m tall, bearing 10 or 11 leaves. **Stem** 10–30 cm diam. **Leaf** 3.3–8 m long; sheath 60–150 cm long, green with silvery indumentum; petiole 10–150 cm long; rachis 260–500 cm long; 44–46 leaflets per side, mid-leaf leaflet 60–85 × 5–8 cm. **Inflorescence** 43–90 cm long; peduncular bract ca. 39.5 × ca. 5.5 cm; peduncle 10–16 cm long, 1.9–3.5 cm wide; rachillae 15–25 cm long, 3–3 mm diam., *brown scaly.* **Male flower** 12–14 × 7–8 mm in bud; *densely brown scaly on the lower surface;* stamens 69–138; anthers 4–6 × 1 mm wide, *smooth;* pistillode 8–12 ×1 mm. **Female**

flower ca. 7 mm, globose. **Fruit** 39–49 × 25–34 mm, red; endocarp 39–40 × 26–32 mm, *endocarp wall 2–4 mm thick, brown with embedded black fibres*. **Seed** ca. 23 ×19 mm; endosperm *homogeneous* (to ruminate?).

DISTRIBUTION. Upland areas of north-eastern and eastern New Guinea, with one outlying in the West.

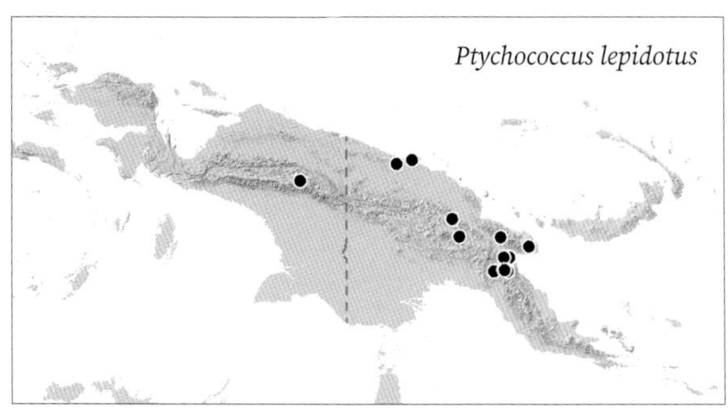

Ptychococcus lepidotus. Habit, Menyama (FE).

HABITAT. Montane rainforest, but widely cultivated in gardens, villages and cemeteries. It is found at 760–1620 m, but said to occur as high as 3,000 m (Ferrero 1996).

LOCAL NAMES. *Tewi* (Hube), *Suhunin* (Yali), *Ba'ha, Mbuna, Val, Wakal* (languages not specified).

USES. Wood is used for spears, bows and arrow points. Highly valued by local people for various construction purposes.

CONSERVATION STATUS. Least Concern (IUCN 2021). However, monitoring is necessary at it is vulnerable to overexploitation for its wood.

NOTES. This palm is readily distinguished by its dense chestnut brown scales found on its leaves, inflorescences and flowers and by its large fruits with thick-walled, brown endocarps. The specific epithet refers to the scales on the inflorescence and flowers.

2. *Ptychococcus paradoxus* (Scheff.) Becc.

Synonyms: *Actinophloeus guppyanus* Becc., *Actinophloeus kraemerianus* Becc., *Drymophloeus paradoxus* Scheff., *Ptychococcus archboldianus* Burret, *Ptychococcus archboldianus* var. *microchlamys* Burret, *Ptychococcus arecinus* (Becc.) Becc., *Ptychococcus elatus* Becc., *Ptychococcus guppyanus* (Becc.) Burret, *Ptychococcus kraemerianus* (Becc.) Burret, *Ptychosperma arecinum* Becc., *Ptychosperma paradoxum* Scheff.

Robust, emergent palm 6–26 m tall, bearing 6–13 leaves. **Stem** 9–25 cm diam. **Leaf** 3.1–6.5 m long; sheath 63–150 cm long, green with silvery indumentum; petiole 0–33 cm long; rachis 250–470 cm long; 32–95 leaflets per side; mid-leaf leaflet 33-102 × 3–12 cm. **Inflorescence** 50–150 cm long; peduncular bract 33–58 × ca. 7 cm; peduncle 8–19 cm long, 2–10 cm wide; rachillae 9–28 cm long, 1–4 mm diam, *glabrous*. **Male flower** 11–17 × 5–10 mm in bud; stamens 100–213; anthers 2–7 mm × 1 mm, *bullate*; pistillode 7–16 × 1–3 mm. **Female flower** ca.

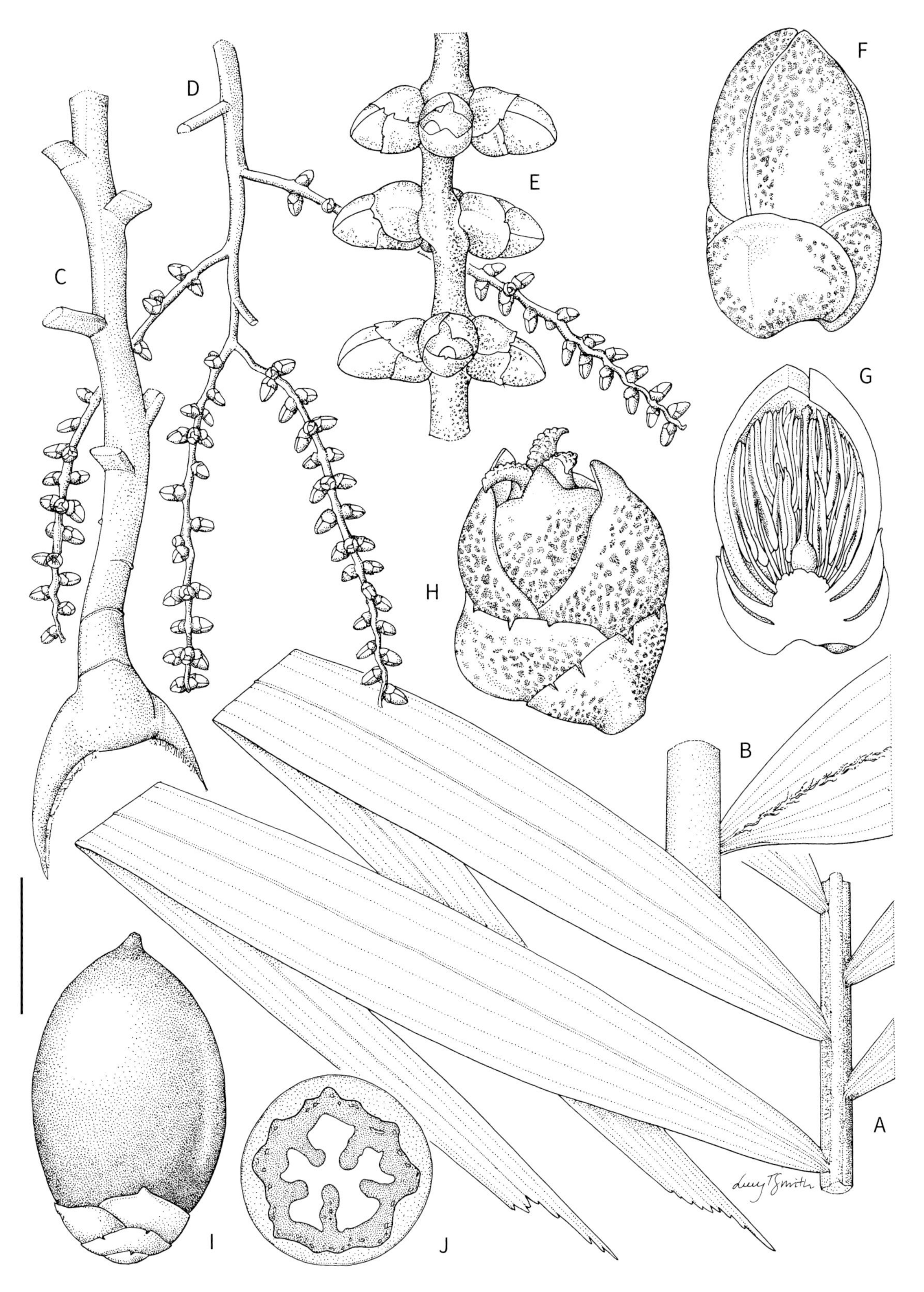

Ptychococcus lepidotus. **A.** Leaf mid-portion. **B.** Leaflet abaxial side detail. **C.** Inflorescence base. **D.** Portion of inflorescence. **E.** Portion of rachilla with triads. **F, G.** Male flower whole and in longitudinal section. **H.** Female flower. **I, J.** Fruit whole and in transverse section. Scale bar: A = 8 cm; B = 4 cm; C, D = 6 cm; E = 1.5 cm; F–H = 7 mm; I, J = 2 cm. All from *Hoogland 9033*. Drawn by Lucy T. Smith.

8 × 9 mm, flattened-globose. **Fruit** 39–60 mm × 22–45 mm; endocarp 31–55 × 20–33 mm, *endocarp wall ca. 1 mm thick, black*. **Seed** 21–28 × 15–25 mm; *endosperm homogeneous to deeply ruminate*.

DISTRIBUTION. Widespread in mainland New Guinea, Bismarck Archipelago and Bougainville.

HABITAT. Widespread throughout lowland New Guinea from sea level to 1,000 m in elevation. In lowland rainforest, alluvial forest and wet or swampy habitats, often over limestone. Widely cultivated in gardens and villages.

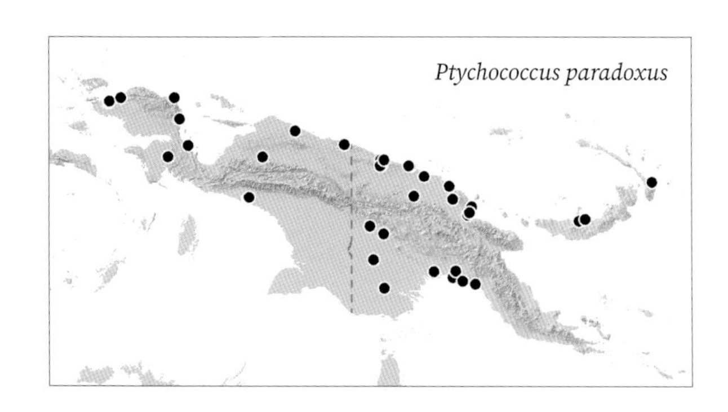

Ptychococcus paradoxus. LEFT TO RIGHT: habit; juvenile palms. Kikori River (WB).

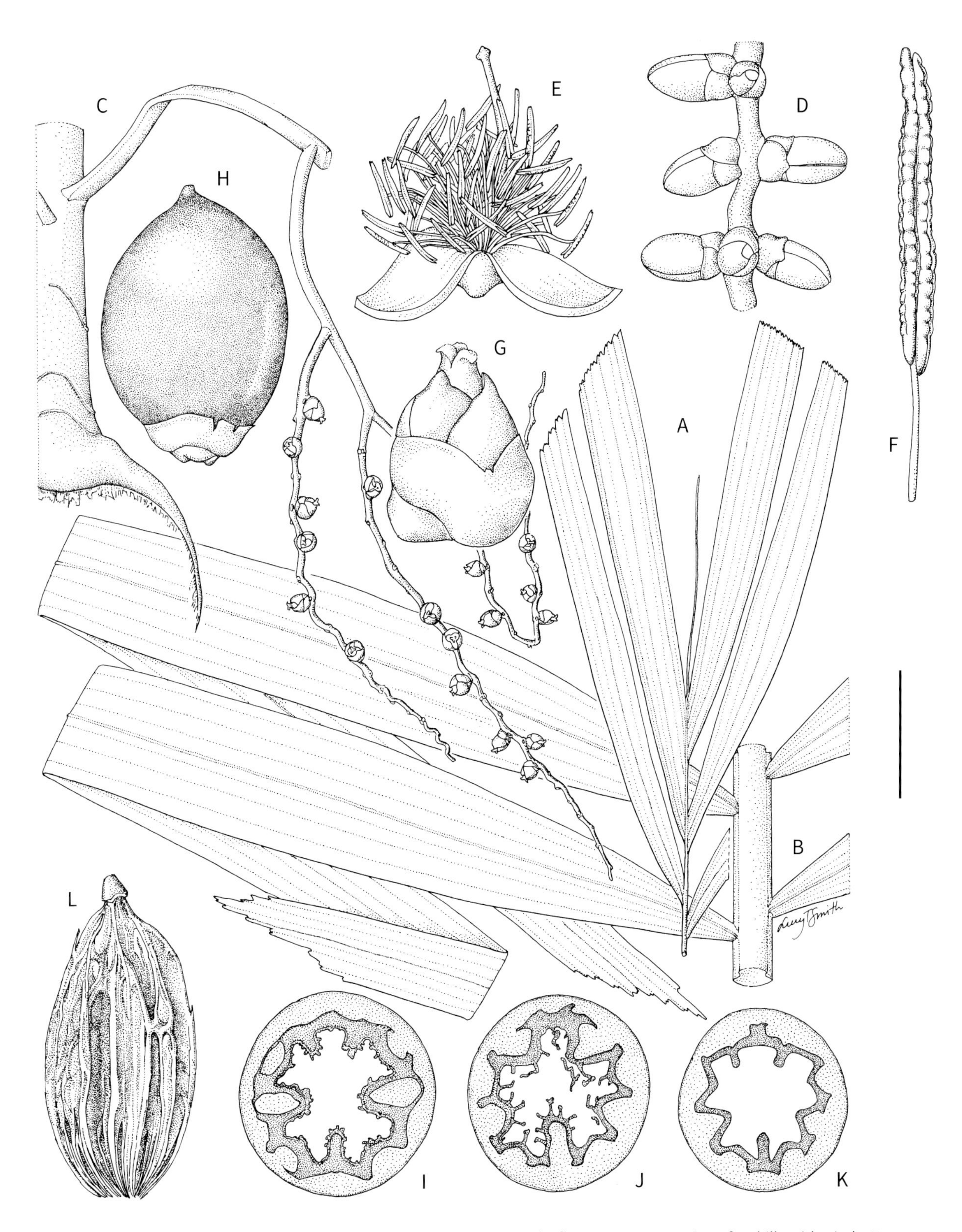

Ptychococcus paradoxus. **A.** Leaf apex. **B.** Leaf mid-portion. **C.** Portion of inflorescence. **D.** Portion of rachilla with triads. **E.** Male flower. **F.** Stamen. **G.** Female flower. **H.** Fruit. I–**K.** Fruit in transverse section. **L.** Endocarp. Scale bar: A–C = 8 cm; D, I–L = 2 cm; E = 1 cm; F = 2 mm; G = 3 cm; H = 2.5 cm. A–I, L from *Barrow et al. 124*; J from *Heatubun et al. 195*; K from *Baker & Utteridge 579*. Drawn by Lucy T. Smith.

LOCAL NAMES. *Apa Imo* (Bahasa Wamesa), *Wilau* (Wapi), *Nu* (Wapi), *Nongrow* (Orne), *Faia Fili* (Tuaripi), *Kem* (Jal), *Mesigef Akta* (Meyha), *Berbar* (Irarutu), *Tari, Nium* (languages not specified).

USES. The wood is used for building houses, spears, bows, axe handles, digging sticks, and arrow points. The leaf sheath is occasionally used as a food platter and as sleeping mats. The terminal bud (heart-of-palm) is edible. Fruit is used as a substitute for betel nut in some parts of Papua New Guinea and by the Barriai people in the Bismarck Archipelago.

CONSERVATION STATUS. Least Concern (IUCN 2013).

NOTES: This palm is widespread in New Guinea. The extent to which its present distribution is the result of human activities is not known. The species is found in the diet of the northern cassowary (Pangau-Adam & Mühlenberg 2014). It is readily distinguished by its large, intricately lobed endocarps that are nearly black. The species epithet was not explained but may refer to the perplexing variability of the endosperm condition, from homogeneous to ruminate (Zona 2003).

Ptychococcus paradoxus. Inflorescence, near Madang (WB).

Ptychococcus paradoxus. CLOCKWISE FROM TOP LEFT: habit, Kikori River (WB); fruit, Wandamen Peninsula (WB); fruit in transverse section, Wandamen Peninsula (WB); male flowers, Madang (WB).

Manjekia maturbongsii. Habit, Biak Island (WB).

Manjekia W.J.Baker & Heatubun

Medium to tall – leaf pinnate – crownshaft – no spines – leaflets jagged

Moderately robust, single-stemmed tree palm, crownshaft present, monoecious. **Leaf** pinnate, *arching;* sheath tubular; petiole present; leaflets reduplicate, with jagged, concave tips, arranged regularly, *pendulous.* **Inflorescence** *below the leaves,* branched to 4 orders, branches spreading, *white;* prophyll and peduncular bract similar, enclosing inflorescence in bud, dropping off as inflorescence expands; peduncle shorter than inflorescence rachis; rachillae sinuous. **Flowers** in triads more or less throughout the length of the rachillae, not developing in pits, male flowers longer than female flowers, bullet-shaped. **Fruit** ripening through orange to red, ellipsoid, stigmatic remains apical, flesh fibrous, *endocarp thin, straw-coloured, with few, thick, longitudinal fibres interspersed with numerous, fine fibres.* **Seed** 1, endosperm ruminate.

DISTRIBUTION. One species, endemic to Biak Island.

NOTES. *Manjekia maturbongsii* is an elegant palm of lowland forest on limestone in Biak. It is distinguished by its moderate habit, well-defined crownshaft, arching leaves, broadly lanceolate, pendulous leaflets with wide, concave, jagged apices, white inflorescences, and a mixture of thick and fine, straw-coloured fibres covering the endocarp. It could be confused with *Wallaceodoxa,* which has more numerous linear leaflets and thick woolly indumentum on the leaf rachis and sheath, or *Ptychococcus,* which has linear leaflets and large (4–6 cm), red fruits each with a black/brown endocarp enclosing the seed with conspicuous grooves and ridges.

The taxonomic treatment of Heatubun *et al.* (2014c) is followed here.

Manjekia maturbongsii (W.J.Baker & Heatubun) W.J.Baker & Heatubun

Synonym: *Adonidia maturbongsii* W.J.Baker & Heatubun

Single-stemmed palm, 10–15 m tall, with ca. 10 leaves in crown. **Stem** 10–20 cm diam. **Leaves** 3.4–4.2 m long; sheath 60–70 cm long, petiole 26–45 cm long, rachis 2.5–3 m long, *arching;* leaflets 25–30 each side of the rachis, *regularly arranged, pendulous;* mid-leaf leaflets 40–49 cm × 9–12 cm. **Inflorescence** 60–70 cm long; peduncular bract 31–35 cm × 5–7 cm; peduncle 8–14 cm long, 2–2.5 cm wide; rachillae 8–19 cm long, 2–4 mm diam., *white.* **Male flower** 7–8 × 3 mm in bud; stamens 30–32; pistillode ca. 5 mm long, ca. 1 mm diam, *bottle-shaped.* **Female flower** ca. 5 × 5 mm, globose. **Fruit** 24–31 × 14–16 mm, red; endocarp 23–30 × 12–13 mm. **Seed** 14–20 × 10–12 mm, *ellipsoid, round in cross-section; endosperm ruminate.*

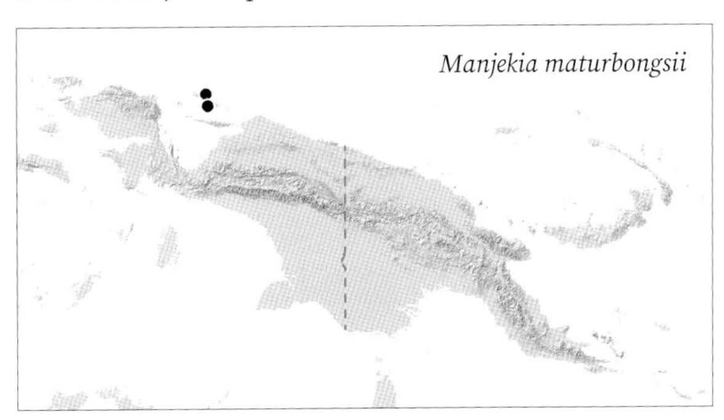

DISTRIBUTION. Biak Island.

HABITAT. Lowland forest on limestone with thin soils and many sink holes at 80–170 m elevation.

LOCAL NAME. *Manjek* (Biak).

USES. The stems are used for flooring and pillars in traditional houses.

CONSERVATION STATUS. Endangered. *Manjekia maturbongsii* has a restricted distribution and is threatened by ongoing forest degradation.

NOTES. This rather robust, handsome palm was first described as a species of *Adonidia,* a genus from the western Philippines and nearby Borneo (Baker & Heatubun 2012). The two genera are strikingly similar in inflorescence and fruit morphology. However, molecular data (Alapetite *et al.* 2014) suggested that this species cannot be accommodated in *Adonidia* as currently delimited, and hence a new genus was required (Heatubun *et al.* 2014c).

Manjekia maturbongsii. LEFT TO RIGHT FROM TOP: crown; inflorescence; male flowers; inflorescence; fruit. Biak Island (WB).

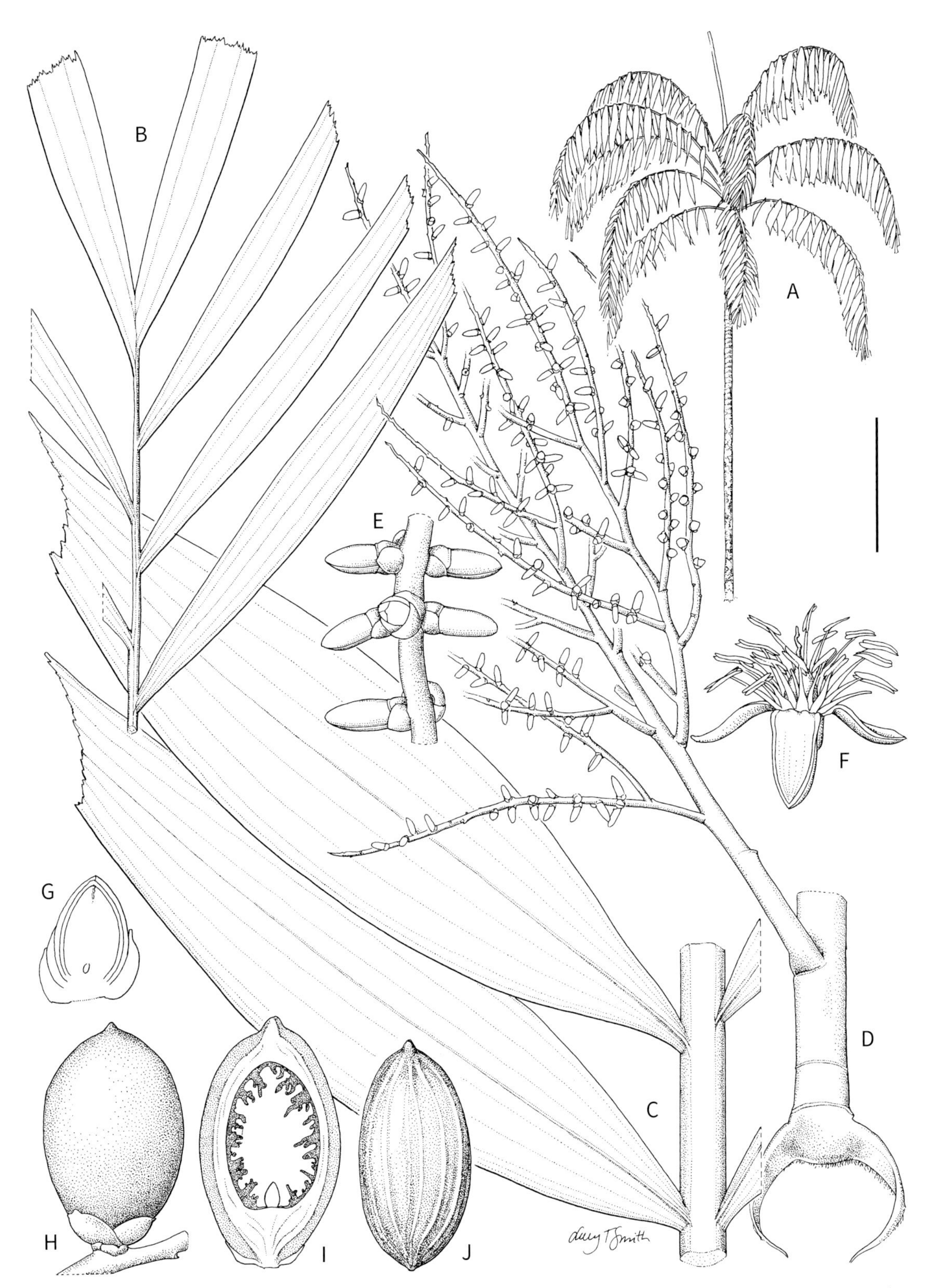

Manjekia maturbongsii. **A.** Habit. **B.** Leaf apex. **C.** Leaf mid-portion. **D.** Portion of inflorescence. **E.** Portion of rachilla with triads. **F.** Male flower. **G.** Female flower in longitudinal section. **H**, **I.** Fruit whole and in longitudinal section. **J.** Endocarp. Scale bar: A = 2 m; B–D = 6 cm; E = 1.5 cm; F, G = 7 mm; H–J = 15 mm. A from photograph; B–F, H–J from *Heatubun et al. 971*; G from *Baker et al. 1338*. Drawn by Lucy T. Smith.

Wallaceodoxa raja-ampat. Habit, Waigeo Island (CDH).

Wallaceodoxa Heatubun & W.J.Baker

Tall – leaf pinnate – crownshaft – no spines – leaflets inconspicuously jagged

Moderately robust, single-stemmed tree palm, crownshaft present, monoecious. **Leaf** pinnate, *arching,* with *woolly, white indumentum with brown hairs* on sheath, petiole and rachis; sheath tubular; petiole present; leaflets numerous, *linear,* reduplicate, with pointed, *inconspicuously jagged tips,* arranged regularly, *pendulous.* **Inflorescence** below the leaves, branched to 3 orders, branches spreading, white; prophyll and peduncular, enclosing inflorescence in bud, dropping off as inflorescence expands; peduncle shorter than inflorescence rachis; rachillae straight, sinuous at tips. **Flowers** in triads more or less throughout the length of the rachillae, not developing in pits, male flowers longer than female flowers, bullet-shaped. **Fruit** ripening through orange to red, ellipsoid, stigmatic remains apical, flesh fibrous, *endocarp thin, straw-coloured, with numerous, flat adherent fibres.* **Seed** 1, endosperm ruminate.

DISTRIBUTION. One species, endemic to Gag and Waigeo in the Raja Ampat Islands.

NOTES. *Wallaceodoxa raja-ampat* is a tall, moderately robust palm, with arching leaves, pendulous, linear leaflets, thick, woolly indumentum on the sheath, petiole and rachis, white inflorescences, and fruit with rounded, straw-coloured endocarp with adherent, longitudinal fibres. The genus was discovered as recently as 2006 (Heatubun *et al.* 2014a, 2014b) and has been recorded only from forest on limestone up to 50 m elevation on Gag and Waigeo Islands. It is most readily confused with *Manjekia*, a Biak endemic, which has broader, conspicuously jagged leaflets and which lacks woolly indumentum, or *Ptychococcus*, which also lacks woolly indumentum and can be immediately distinguished by its large (4–6 cm), red fruits, each with a black/brown endocarp with conspicuous grooves and ridges. *Wallaceodoxa* also superficially resembles other robust New Guinea palms with pendulous leaflets, such as *Cyrtostachys loriae, Hydriastele costata, H. procera* and *Rhopaloblaste* species, but only the last two co-occur in the same region and are likely to cause any confusion.

The taxonomic treatment of Heatubun *et al.* (2014c) is followed here.

Wallaceodoxa raja-ampat Heatubun & W.J.Baker

Solitary palm up to 30 m tall with 11–19 leaves in crown. **Stem** 9–30 cm diam. **Leaves** 2.8–4.1 m long, *woolly, white indumentum and brown chaffy hairs* on sheath, petiole and rachis; sheath 76–115 cm long; petiole 28–50 cm long, rachis 172–245 cm long; leaflets 50–85 on each side, *linear, pendulous*; mid-leaf leaflets 70–114 cm × 1–4.5 cm. **Inflorescence** 50–100 cm long; peduncle 15–20 cm long, 4–6 cm wide; peduncular bract 50–60 × 7–12 cm, with *woolly, white indumentum*; rachillae numerous, 11–18 cm long. **Male flowers** 7–8 × 3–4 mm; stamens 58–64; pistillode 1–2 × ca. 1 mm, 2- or 3-lobed, cream to brown. **Female flowers** 5–6 × ca. 4 mm, globose to ovoid. **Fruits** 18–20 × 10–12 mm, ripening through orange to red; endocarp with *flat, adherent, straw-coloured fibres.* **Seed** 10–12 × 9–10 mm, *endosperm ruminate.*

DISTRIBUTION. Raja Ampat Islands (Gag and Waigeo Islands).

HABITAT. Lowland forest on limestone up to 50 m elevation. It has been found in secondary and heavily disturbed forest, and old gardens, where it appears to grow as a relict from times when the forest was not disturbed.

LOCAL NAME. *Gulbotom* (Wayaf or Gebe)

USES. The stem of this palm is used for flooring, and the fruit is chewed as a substitute for betel nut (*Areca catechu* L.)

CONSERVATION STATUS. Conservation status. Critically Endangered. *Wallaceodoxa raja-ampat* is known from only two subpopulations that are under threat from land clearance for slash-and-burn agriculture, urbanisation and the expansion of coconut plantation.

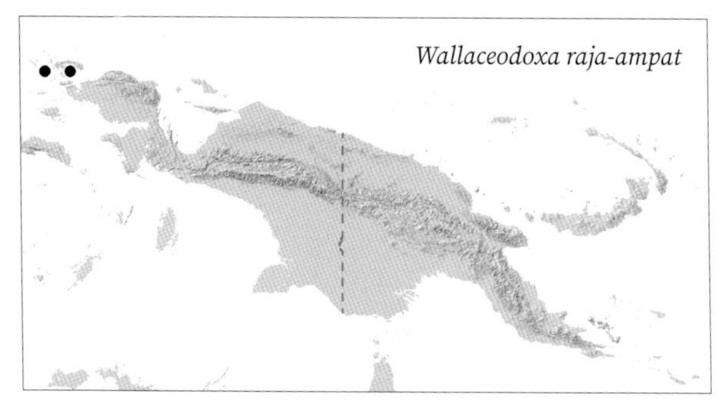

NOTES. *Wallaceodoxa raja-ampat* is an impressive, solitary canopy palm that is recognised by its arching leaves with narrow, linear leaflets, the woolly, white indumentum and brown hairs throughout the sheath, petiole and leaf rachis, and the white inflorescence, branched to three orders, with thick rachillae crowded with floral triads.

Wallaceodoxa raja-ampat. LEFT: habit, Waigeo Island (GP); LEFT TO RIGHT FROM TOP MIDDLE: leaf bases showing indumentum, Waigeo Island (CDH); indumentum on leaf sheath, Gag Island (CDH); inflorescence; fruit; floral triads in bud; fruit, endocarps and seed in transverse section, all Waigeo Island (CDH).

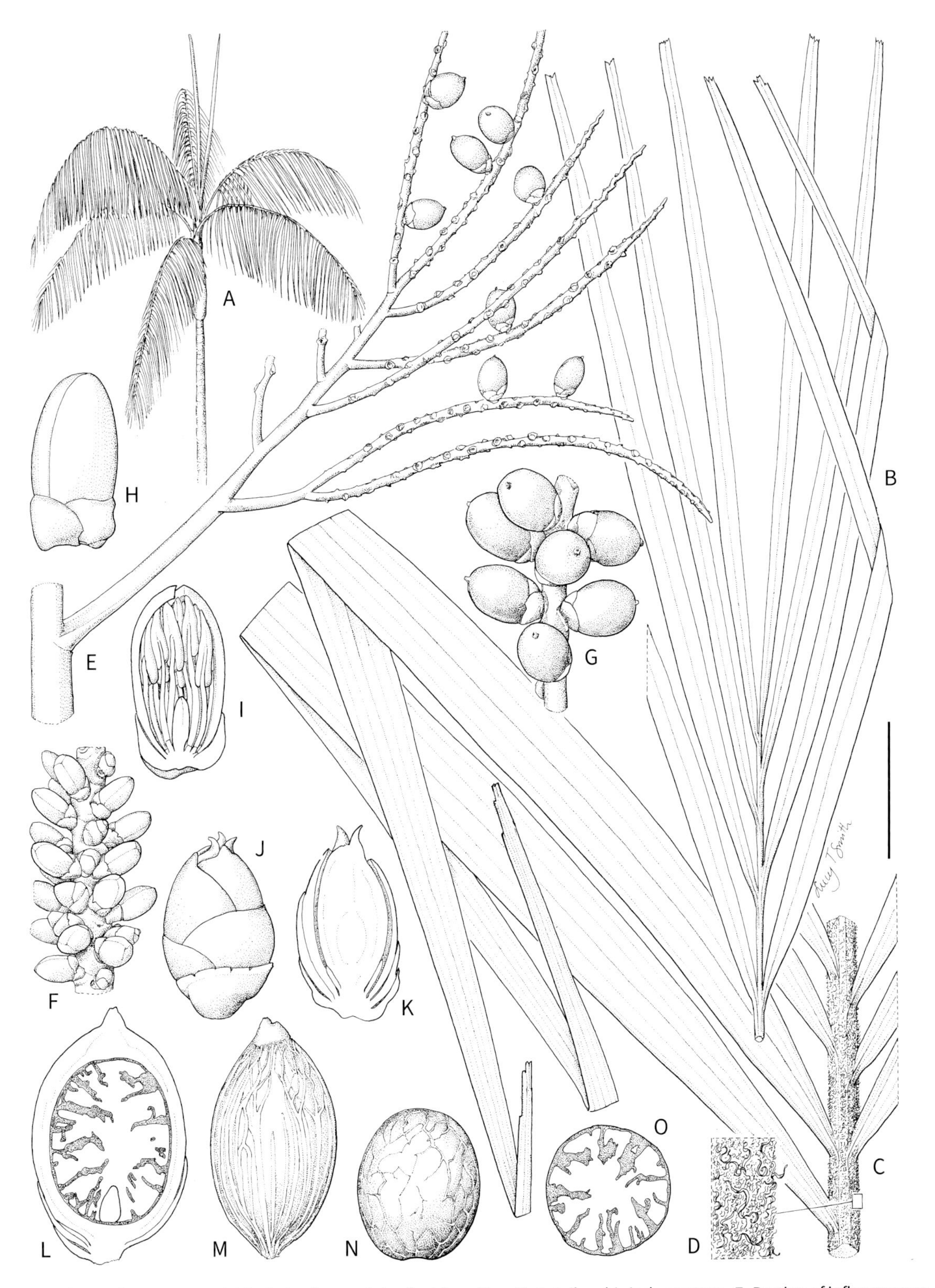

Wallaceodoxa raja-ampat. **A.** Habit. **B.** Leaf apex. **C.** Leaf mid-portion. **D.** Detail rachis indumentum. **E.** Portion of inflorescence with fruit. **F.** Portion of rachilla with triads. **G.** Portion of rachilla with fruit. **H, I.** Male flower whole and in longitudinal section. **J, K.** Female flower whole and in longitudinal section. **L.** Fruit in longitudinal section. **M.** Endocarp with fibres. **N, O.** Seed whole and in transverse section. Scale bar: A = 2 m; B, C = 8 cm; D, J, K = 7 mm; E = 6 cm; F = 1.5 cm; G = 2.5 cm; H, I = 5 mm; L–O = 1 cm. A from photograph; B–D from *Heatubun et al. 1126*; E from *Heatubun et al. 741*; F from photograph; G–O from *Heatubun et al. 746*. Drawn by Lucy T. Smith.

Clinostigma collegarum. Habit, New Ireland (JW).

Clinostigma H. Wendl.

Synonyms: *Bentinckiopsis* Becc., *Clinostigmopsis* Becc., *Exorrhiza* Becc.

Tall – leaf pinnate – crownshaft – no spines – leaflets pointed

Robust, single-stemmed tree palm, slender crownshaft present, monoecious. **Leaf** pinnate; sheath tubular, pale green; petiole short; leaflets numerous, reduplicate, with pointed tips, arranged regularly, pendulous. **Inflorescence** *below the leaves,* branched to 3 orders, *resembling a horse's tail;* prophyll and peduncular bract similar, enclosing inflorescence in bud, dropping off as inflorescence expands; peduncle shorter than inflorescence rachis, grossly swollen at base; rachillae slender and straight. **Flowers** in triads throughout most of the length of the rachilla, pairs of male flowers towards the tip, not developing in pits, male flower asymmetrical, slightly larger than the female. **Fruit** black, ellipsoid, *asymmetric, stigmatic remains to one side of the apex,* flesh juicy, endocarp thin, closely adhering to seed. **Seed** 1, ellipsoid, endosperm homogeneous.

DISTRIBUTION. About 11 species in the western Pacific from the Bonin and Caroline Islands to Samoa, Fiji Islands, Vanuatu and the Solomon Islands, with one species in our region confined to New Ireland. Not recorded from the New Guinea mainland.

NOTES. *Clinostigma* is a robust, single-stemmed, canopy tree palm distinguished by its inflorescence borne below the leaves, resembling a horse's tail, which bears asymmetric fruit with the stigma to one side of apex. Of the 11 species, none occurs on the New Guinea mainland, but one, *C. collegarum* (Dransfield 1982), is endemic to New Ireland. A complete taxonomic account of the genus throughout its range is lacking.

Clinostigma is potentially confused with three other tree palm genera: *Cyrtostachys,* which has inflorescences with widely spreading branches and flowers that usually develop in pits; *Hydriastele,* which has leaflets with almost always jagged tips; and *Rhopaloblaste,* which has dark hairs on the leaf rachis, inflorescences with widely spreading branches and rounded fruit. All three of these genera have fruit with apical or nearly apical stigmatic remains, in contrast with the eccentric position of this structure in *Clinostigma.*

Clinostigma collegarum J.Dransf.

Robust, single-stemmed palm to 20 m, bearing 13 leaves in the crown. **Stem** 20 cm diam., arising from a *cone of roots.* **Leaf** ca. 4 m long; sheath ca. 200 cm long; petiole 60–70 cm long; rachis 330–340 cm; ca. 56 pairs of leaflets, *pendulous,* mid-leaf leaflet ca. 106 × 2.5 cm wide, with chestnut brown, forked, chaffy scales along the upper mid-vein, densest near the base of the leaflet. **Inflorescence** ca. 75 cm long; peduncular bract ca. 83 cm long; peduncle 9–10 cm long; rachillae ca. 43–45 cm long, ca. 3 mm diam. **Male flower** ca. 6 mm long in bud; petals ca. 5 × 3 mm, asymmetric, strongly nerved; stamens 6; pistillode up to 2 mm long, conical, style absent. **Female flower** ca. 5 mm diam., globose. **Fruit** ca. 12 × 8 mm, ovoid, red, subtended by enlarged perianth parts; *stigmatic remains subapical and beak-like;* epicarp smooth, endocarp thin. **Seed** ca. 9 × 6 mm, ovoid, brown; *endosperm homogeneous.*

DISTRIBUTION. New Ireland.

HABITAT. Montane forest on limestone at 1350 m elevation.

LOCAL NAMES. None recorded.

USES. None recorded.

CONSERVATION STATUS. Data Deficient. This species has been collected only once, in 1975 (Dransfield 1982). Its remote, high-elevation habitat may protect it from human disturbance, but high-elevation plants may be especially vulnerable to climate change.

NOTE. One of the most poorly known species in the flora, it appears from photos taken by the original collectors to be strikingly beautiful. Because of its size, pendulous leaflets and well-defined crownshaft, it is most likely to be confused with *Hydriastele costata* or *Cyrtostachys loriae*. The former is known at low to mid-elevations in New Guinea (sight records from the Bismarck Archipelago, but no confirmed herbarium specimens from New Ireland) and has symmetrical fruits with longitudinal stripes and apical stigmatic remains. The latter is known from New Ireland (as well as mainland New Guinea and the Solomon Islands), though from elevations up to 400 m, and is distinguished by its robust, widely spreading inflorescence, bearing flowers in pits on thick rachillae, which later form small, black, ellipsoid fruits with apical stigmatic remains.

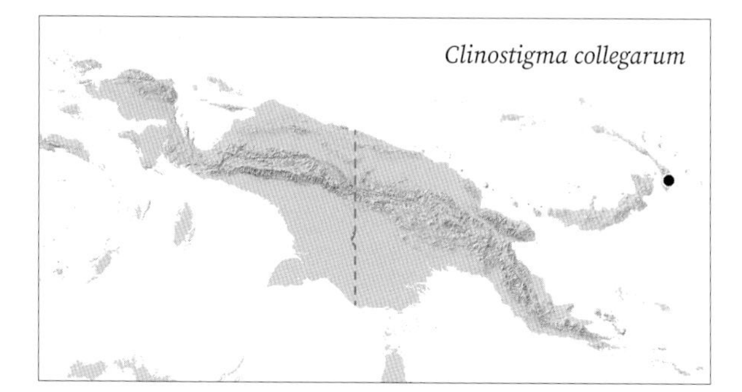

Clinostigma collegarum. Stilt roots, New Ireland (MS).

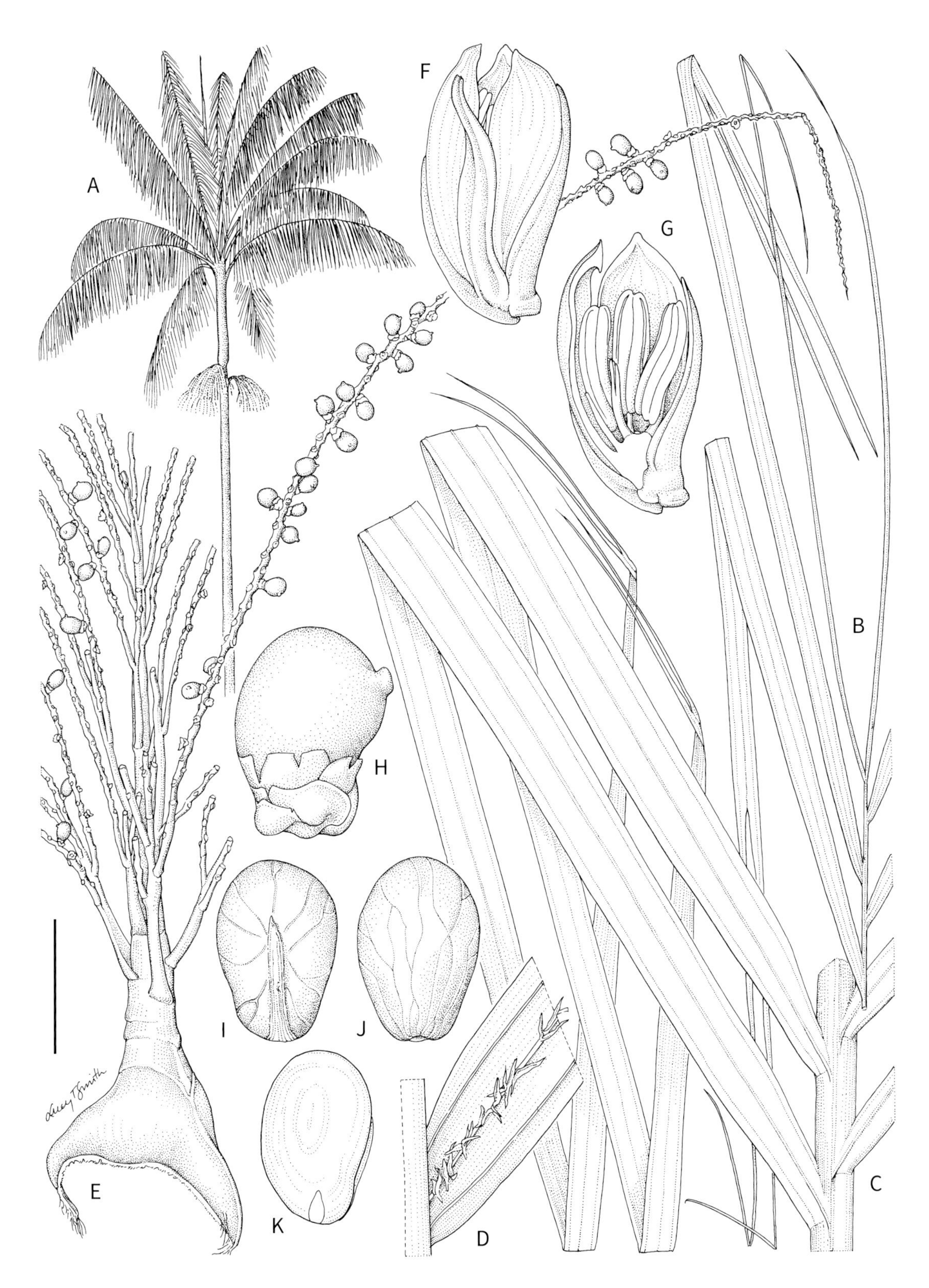

Clinostigma collegarum. **A.** Habit. **B.** Leaf apex. **C.** Leaf mid-portion. **D.** Leaflet abaxial side with ramenta. **E.** Inflorescence with fruit. **F**, **G.** Male flower whole and in longitudinal section. **H.** Fruit. **I**, **J.** Seed in two views. **K.** Seed in longitudinal section. Scale bar: A = 3 m; B, C, E = 6 cm; D = 3 cm; F, G = 3.3 mm; H–K = 1 cm. All from *Sands et al. 2552*. Drawn by Lucy T. Smith.

Cyrtostachys loriae. Habit, near Timika (WB).

Cyrtostachys Blume

Medium to tall – leaf pinnate – crownshaft – no spines – leaflets pointed

Robust, single- or multi-stemmed tree palm, slender crownshaft present, monoecious. **Leaf** pinnate, straight or slightly arching; sheath tubular, pale green; petiole short; leaflets numerous, with pointed tips, arranged regularly, rarely grouped, *often pendulous.* **Inflorescence** *below the leaves,* branched to 3 (rarely 4) orders, *branches widely spreading;* prophyll and peduncular bract similar, enclosing inflorescence in bud, dropping off as inflorescence expands; peduncle shorter than inflorescence rachis, swollen at base; rachillae slender to robust. **Flowers** in triads at the base of the branches, pairs of male flowers towards tip, *usually developing in pits.* **Fruit** *small, black, ellipsoid, stigmatic remains apical,* flesh thin and juicy, endocarp thin, closely adhering to seed. **Seed** 1, ellipsoid, endosperm homogeneous.

DISTRIBUTION. Malay Peninsula, Sumatra and Borneo, New Guinea, Bismarck Archipelago to Solomon Islands, seven species in total, of which six occur in New Guinea.

NOTES. New Guinea species of *Cyrtostachys* are robust, single- or multi-stemmed tree palms, often with pendulous leaflets. They bear inflorescences below the leaves with widely spreading branches and flowers that usually develop within pits. The fruits are very small (≤1.2 cm) and are black and ellipsoid. The New Guinea species occur in lowland to submontane rainforest. *Cyrtostachys* can be confused easily with three other genera: *Clinostigma* (New Ireland only, in our area), which has an inflorescence resembling a horse's tail, flowers not developing in pits and asymmetric fruit with stigmatic remains to one side of apex; *Hydriastele,* which has jagged leaflet tips, inflorescences resembling a horse's tail and flowers not developing in pits; and *Rhopaloblaste,* which has dark hairs on the leaf rachis, flowers not developing in pits and larger fruit with deeply ruminate endosperm. The genus has been monographed by Heatubun *et al.* (2009).

The only species to grow outside New Guinea, *Cyrtostachys renda* Blume, the Sealing Wax Palm, native to Borneo, Sumatra and the Malay Peninsula, is widely cultivated in gardens throughout the tropics and immediately recognised by its bright red crownshaft.

Key to the species of *Cyrtostachys* in New Guinea

1. Crownshaft and leaf sheath glaucous; petiole elongate (25–100 cm long); inflorescence slender, whitish when dried, rachillae with superficial pits .... **5. *C. glauca***
1. Crownshaft and leaf sheath green; petiole short (less than 20 cm long); inflorescence relatively large to robust, not whitish when dried, rachillae with deep pits 2

2. Leaflets irregularly arranged (at least partially), major inflorescence branches pink when young... .... **4. *C. excelsa***
2. Leaflets regularly arranged, inflorescence branches cream, yellow to green when young 3

3. Leaflets crowded, arranged very close to each other, with beard-like ramenta continuous along mid-vein on lower surface of leaflets . . . . 2. *C. barbata*
3. Leaflets regularly pinnate but leaflets not crowded, ramenta rare on lower surface of leaflets. 4

4. Single-stemmed; crown spherical in outline . . . . 6. *C. loriae*
4. Multi-stemmed; crown not spherical in outline 5

5. Crown shuttlecock-shaped in outline; inflorescence branched to 3 orders; stamens 12 . . . 1. *C. bakeri*
5. Crown hemispherical in outline; inflorescence branched to 4 orders; stamens 9 . . . . . . 3. *C. elegans*

1. *Cyrtostachys bakeri* Heatubun

Robust, *multi-stemmed* tree palm to 15–25 m, with up to 3–7 adult stems, leaves ca. 8 in crown, *crown shuttlecock-shaped*. **Stem** 15–25 cm diam. **Leaf** 4–4.5 m long, ascending, with pendulous leaflets; crownshaft ca. 180 cm long, 15–25 cm diam.; petiole 4–10 cm long; 78–90 leaflets on each side of rachis, regularly arranged, mid-leaf leaflets ca. 92 × 5 cm, green, with *fine sparse ramenta* on mid-vein on lower surface. **Inflorescence** 90–120 cm long, branched to 3 orders; peduncle ca. 16 cm; rachillae ca. 55–57 cm long, ca. 9 mm diam., *20–22 pits per 1 cm of rachilla length* (at fruiting stage). **Male flower** 2.3 × 2.3 mm; *stamens 12*. **Female flower** ca. 3 × 2.8 mm. **Fruit** 9–25 × 3–5 mm. **Seed** 7–8 × 4–4.5 mm.

DISTRIBUTION. One record from near Tabubil in central New Guinea.
HABITAT. Rainforest at ca. 750 m above sea level.

Cyrtostachys bakeri. LEFT TO RIGHT: habit, near Tabubil (WB); crown with inflorescences, cultivated National Botanic Garden, Lae (WB).

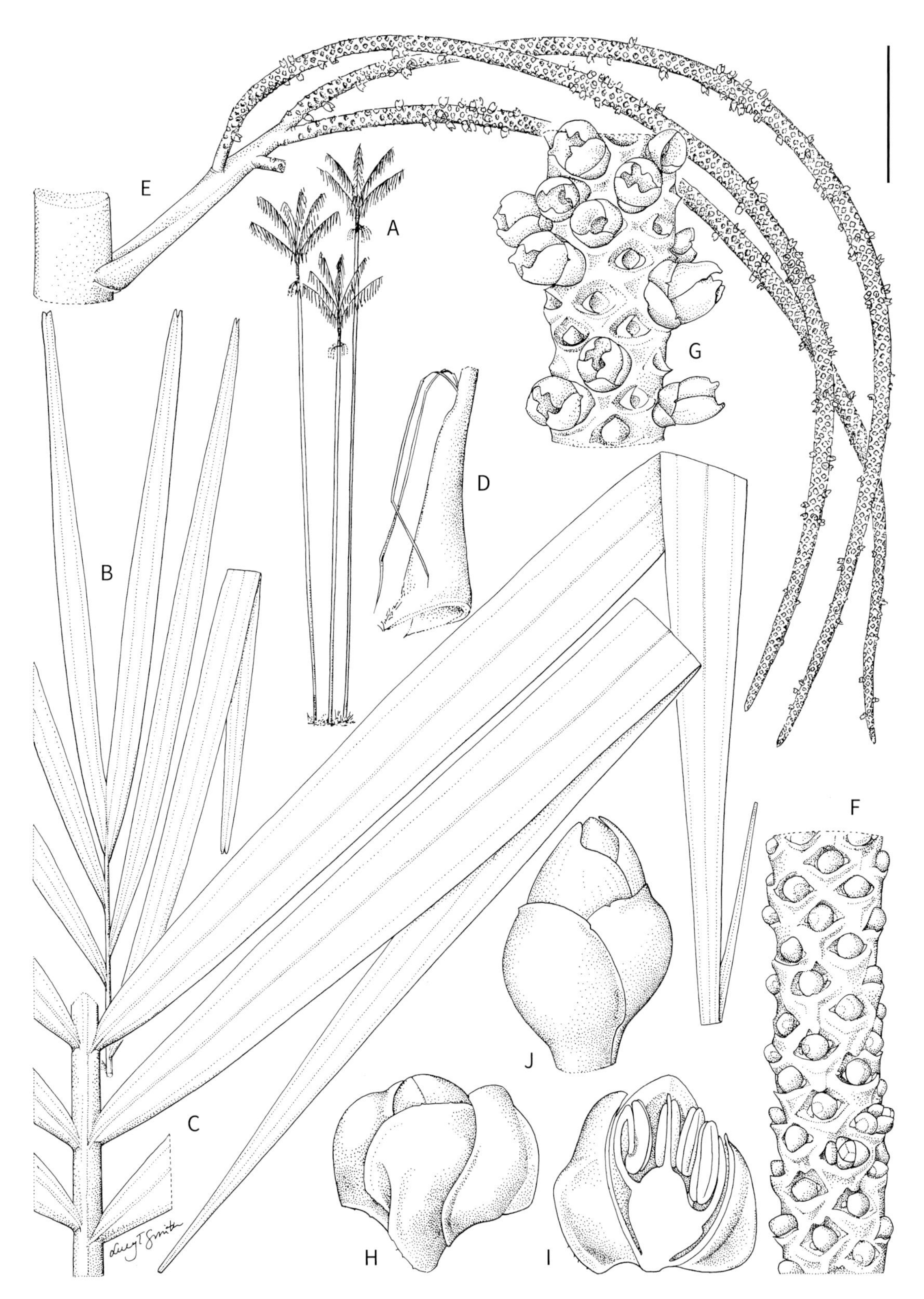

Cyrtostachys bakeri. **A.** Habit. **B.** Leaf apex. **C.** Leaf mid-portion. **D.** Petiole. **E.** Inflorescence with fruit. **F.** Portion of rachilla with flowers. **G.** Portion of rachilla with remnant calyxes. **H**, **I.** Male flower whole and in longitudinal section. **J.** Female flower. Scale bar: A = 5 m; B, C = 8 cm; D = 18 cm; E = 6 cm; F, G = 1 cm; H, I = 1.6 mm; J = 3 mm. All from *Baker et al. 1138*. Drawn by Lucy T. Smith.

Cyrtostachys bakeri. LEFT, FROM TOP: inflorescence; female flowers; fruit. Cultivated National Botanic Garden, Lae (WB).

LOCAL NAMES. None recorded.

USES. None recorded.

CONSERVATION STATUS. Critically Endangered (IUCN 2021). *Cyrtostachys bakeri* is known from only one site where it is threatened by encroaching mine tailings.

NOTES. *Cyrtostachys bakeri* differs from all other species in its rather widely spaced leaflets and congested pits on the rachilla. It is also the only species that has both a shuttle-cock shaped crown and pendulous leaflets. It is most similar to *C. elegans* in the robust, multi-stemmed habit and pendulous leaflets, but differs in the shuttle-cock-shaped crown, the inflorescence branched to 3 orders, the 12 stamens and the habitat at relatively high elevation.

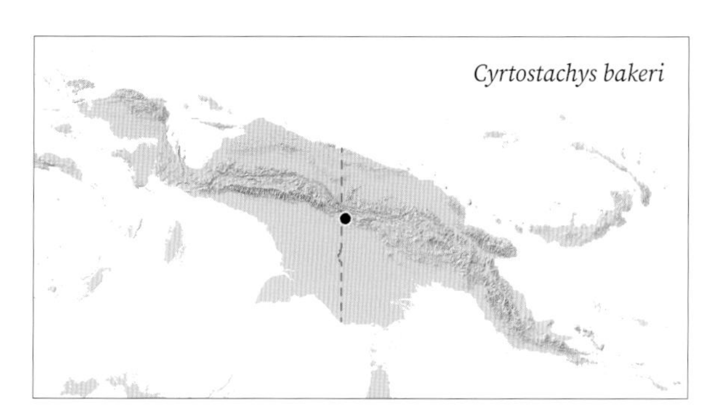

2. *Cyrtostachys barbata* Heatubun

Large, *single-stemmed tree* palm to 25 m. **Stem** 13–20 cm diam. **Leaf** ca. 3.5 m long (including petiole); crownshaft ca. 150 × 40 cm, green, sheath margins fibrous near petiole; petiole 16–20 cm long; leaflets *closely regularly arranged, comb-like,* number on each side unknown, mid-leaf leaflets ca. 105 × 6.5 cm, *with thick, membranous brown ramenta, inflexed, beard-like and continuous along lower surface of mid-vein.* **Inflorescence** branched to 3 orders, light brown when dried; rachillae 43–49 cm long, ca. 4 mm diam.; pits superficial to shallow, 13 pits per 1 cm length of rachilla (at fruiting stage), pits 2–3 mm diam. **Male flower** ca. 3.5 × 2.6 mm; *stamens 8–10.* **Female flower** ca. 3 × 3.3 mm. **Fruit** ca. 12 × 5 mm. **Seed** ca. 5 × 4 mm.

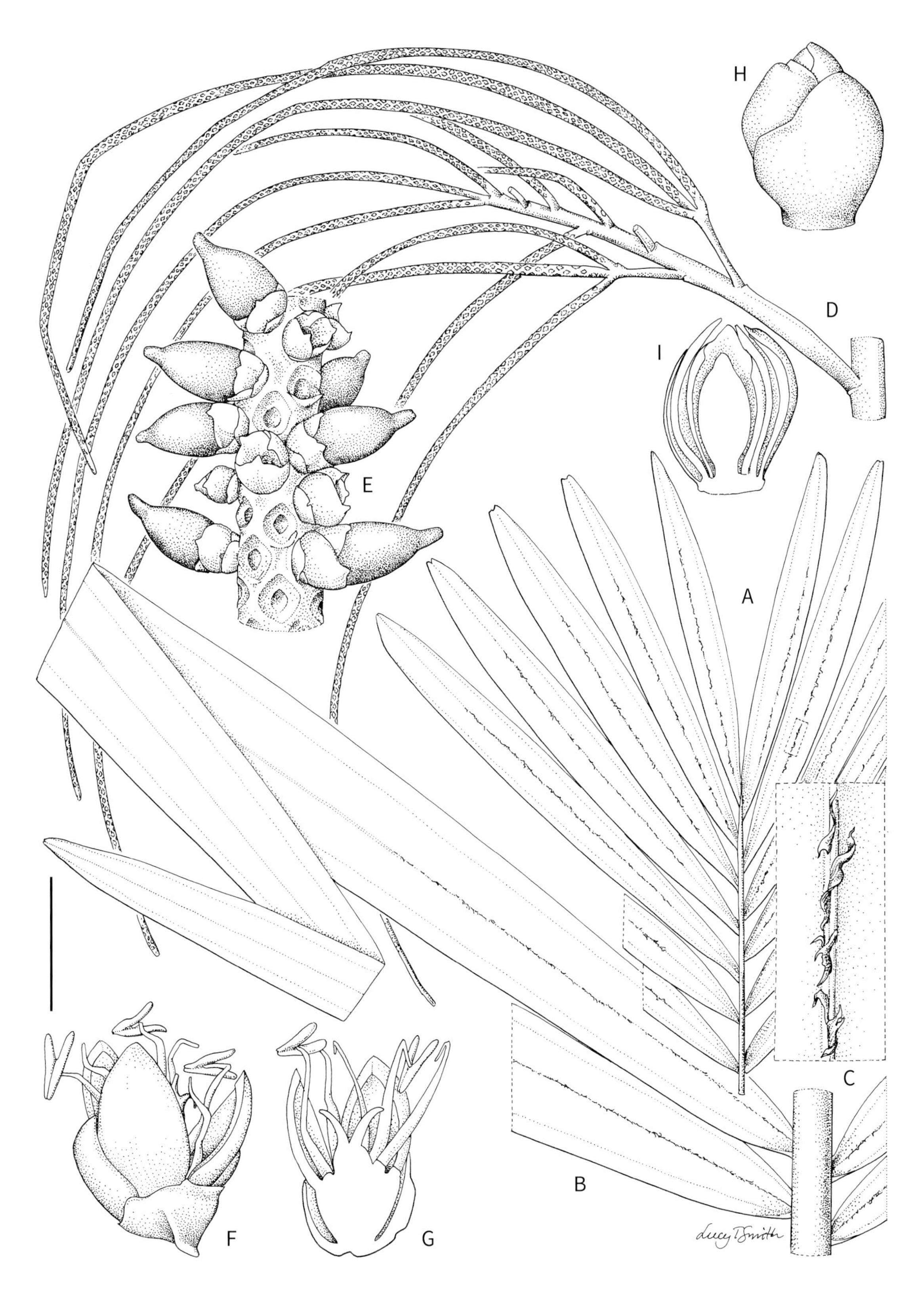

Cyrtostachys barbata. **A.** Leaf apex. **B.** Leaf mid-portion. **C.** Ramenta detail. **D.** Inflorescence. **E.** Portion of rachilla with fruit. **F**, **G.** Male flower whole and in longitudinal section. **H**, **I.** Female flower whole and in longitudinal section. Scale bar: A, B = 8 cm; C = 5 mm; D = 6 cm; E = 1 cm; F–I = 2.5 mm. All from *Brass 13707*. Drawn by Lucy T. Smith.

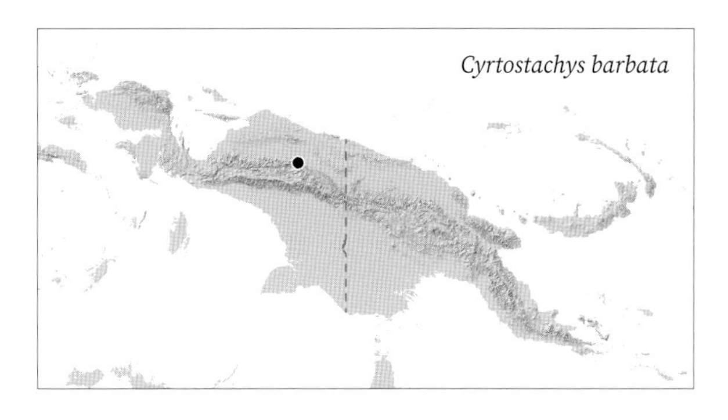

DISTRIBUTION. Idenburg River in the Mamberamo basin.

HABITAT. In *Agathis* forest at 900 m above sea level.

LOCAL NAME. *Nibung* (Bahasa Indonesia, also used for other palms).

USES. None recorded.

CONSERVATION STATUS. Data Deficient. This species is known only from the type locality, where there is limited information on ongoing threats. Due to insufficient evidence, a single extinction risk category cannot be selected.

NOTES. *Cyrtostachys barbata* is similar to *C. loriae* in habit, but differs in the presence of continuous beard-like brown ramenta along the mid-veins on the lower surface of leaflets, the very short space between leaflets giving the leaf a comb-like appearance and the slender rachillae with congested superficial pits. The habitat in *Agathis* forest is unusual.

3. *Cyrtostachys elegans* Burret

Robust, *multi-stemmed* tree palm to 15(–20) m, with up to ca. 3 adult stems and 4–6 or more suckers at the base, 9 leaves in crown, *crown hemispherical in outline*. **Stem** 15–30 cm diam. **Leaf** 3–3.5 m long; *curved*; crownshaft ca. 2.5 cm long; petiole to 10 cm long; leaflets ca. 100 on each side of rachis, *pendulous*, regularly arranged, mid-leaf leaflets 100–127 × 3.5–4.5 cm, *fine brown ramenta discontinuous along mid-vein on lower surface*. **Inflorescence** 75–100 cm long, ca. 160 cm wide, branched to 4 orders, creamy to yellowish green, brown when dried; peduncle very short to 7.5 cm; rachillae 34–62 cm long, *17–19 pits per 1 cm rachilla length* (at fruiting stage), pits 2–4 mm diam. **Male flower** 2–2.5 × 2–2.7 mm; *stamens 9*. **Female flower** 2–5.2 × 1.3–4.5 mm. **Fruit** 12–17 × 5–6 mm. **Seed** 6–7 × 4–5 mm.

DISTRIBUTION. Vicinity of Nabire and Timika, western New Guinea.

HABITAT. Swampy areas in lowland rainforest at up to 300 m above sea level.

LOCAL NAMES. *Nibung* (Bahasa Indonesia, also used for other tree palms).

USES. None recorded.

Cyrtostachys elegans. FROM TOP: habit; near Timika (CDH); inflorescence, near Timika (WB).

Cyrtostachys elegans. **A.** Leaf apex. **B.** Leaf mid-portion. **C.** Portion of inflorescence. **D.** Portion of rachilla with female flowers. **E.** Portion of rachilla with fruit. **F, G.** Male flower whole and in longitudinal section. **H, I.** Female flower whole and in longitudinal section. **J.** Fruit in longitudinal section. Scale bar: A, B = 8 cm; C = 6 cm; D = 1 cm; E = 1.5 cm; F, G = 1.6 mm; H–J = 2.5 mm; J = 5 mm. A–E, H, I from *Heatubun et al. 194*; F, G from *Heatubun et al. 341*. Drawn by Lucy T. Smith.

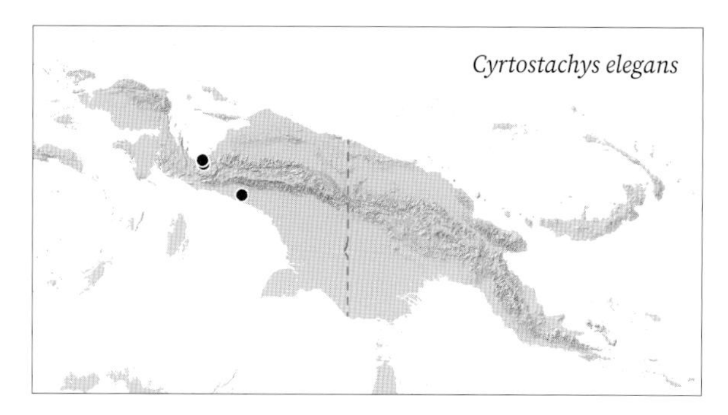

CONSERVATION STATUS. Near Threatened (IUCN 2013). *Cyrtostachys elegans* has a restricted distribution. Mining activity and human settlements are a threat within its distribution range.

NOTES. *Cyrtostachys elegans* is similar to *C. loriae* in being a robust tree palm with pendulous leaflets and short petioles, but differs from the latter in its multi-stemmed habit, the hemispherical crown and elongate inflorescence branched to 4 orders. It also differs from *C. bakeri* in the presence of curved leaves and the hemispherical crown.

4. *Cyrtostachys excelsa* Heatubun

Robust, *single-stemmed* tree palm to 35 m, ca. 7 in crown, *crown appearing shuttle-cock-shaped in outline.* **Stem** ca. 15–18 cm diam. **Leaves** 2.5–3.3 m long, *stiff, ascending*; crownshaft ca. 1.8 m long; petiole 11–15(–27) cm; leaflets 70–80 on each side of rachis, *at least partly irregularly arranged (e.g. in middle of leaf), rather stiff*, mid-leaf leaflets ca. 112 × 4 cm; *rachis with thin, caducous, fine scales, interspersed with white scales.* **Inflorescence** ca. 100 cm long, *branched to 4 orders, pink to reddish-purple when young (at least major branches) to greenish-grey when mature and dark brown when dried*; peduncle ca. 5 cm; rachillae 28–75 cm long, ca. 15 pits per 1 cm rachilla length (before anthesis). **Male flower** 2.5 × 2.5 mm; *stamens 12*. **Female flower** ca. 3.1 × 2.5 mm. **Fruit** not seen. **Seed** not seen.

Cyrtostachys excelsa. BELOW, LEFT TO RIGHT: crown; leaf showing irregularly arranged leaflets. Tamrau Mountains (WB).

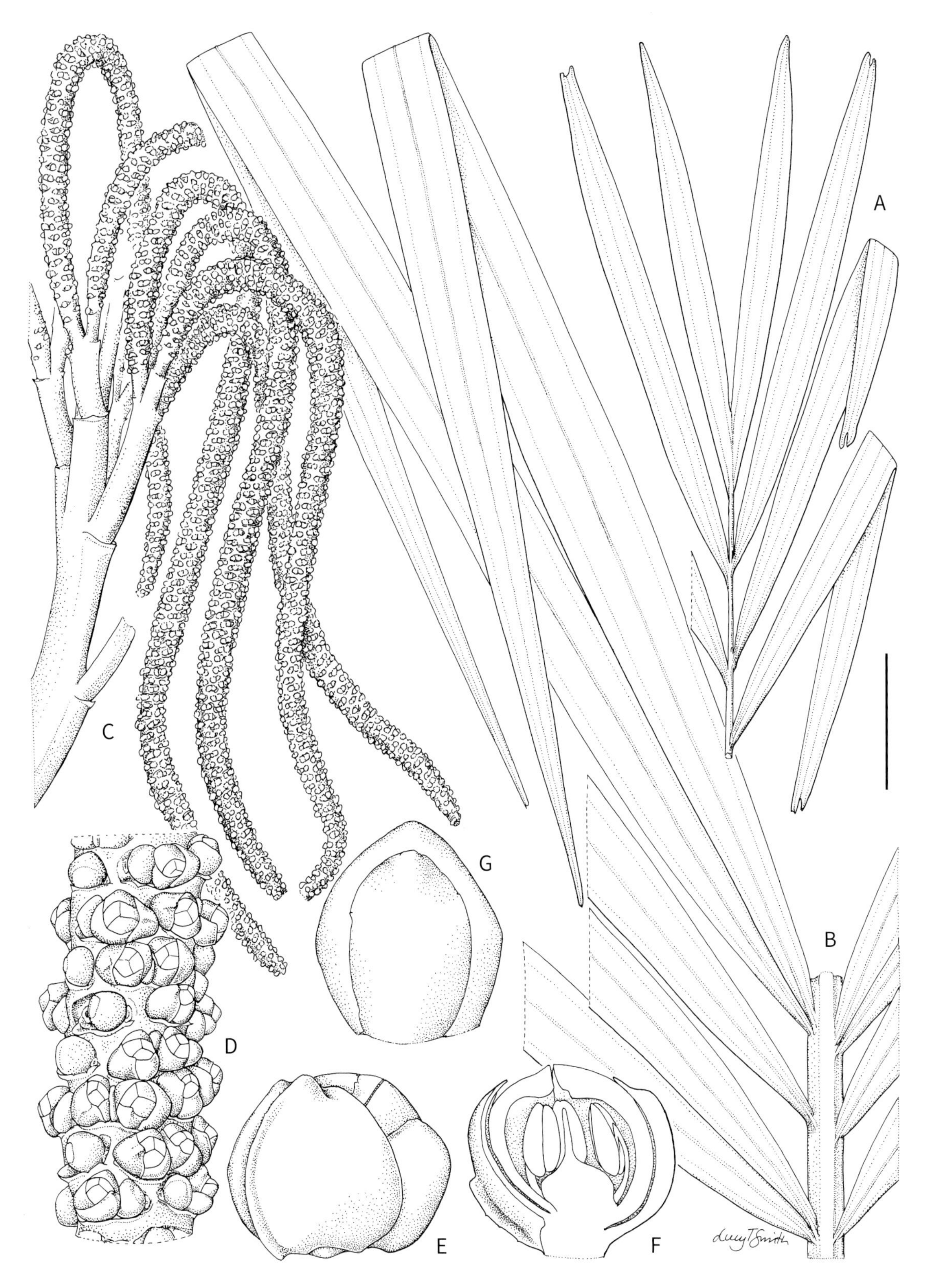

Cyrtostachys excelsa. **A.** Leaf apex. **B.** Leaf mid-portion. **C.** Portion of inflorescence. **D.** Portion of rachilla with flowers. **E, F.** Male flower whole and in longitudinal section. **G.** Female flower. Scale bar: A–C = 8 cm; D = 7 mm; E–G = 1.5 mm. All from *Heatubun et al. 330*. Drawn by Lucy T. Smith.

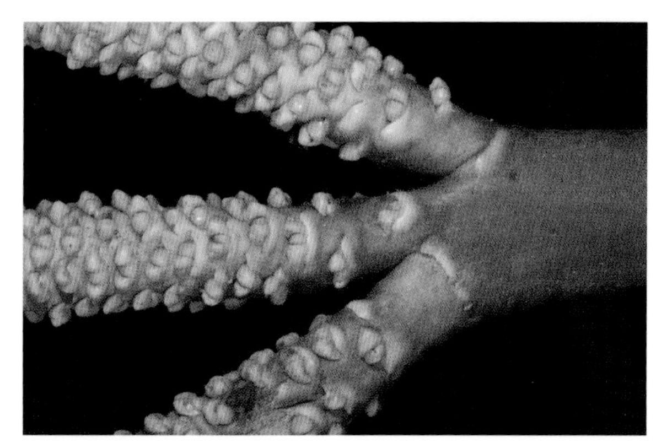
Cyrtostachys excelsa. Inflorescence branch with flower buds, Tamrau Mountains (WB).

DISTRIBUTION. Bird's Head Peninsula and Kwatisore, north-western New Guinea.

HABITAT. Limestone hill forest at ca. 500–700 m.

LOCAL NAMES. *Warita* (Jamur), *Mojigre* (Kwatisore), *Nibung* (Bahasa Indonesia, also used for other tree palms).

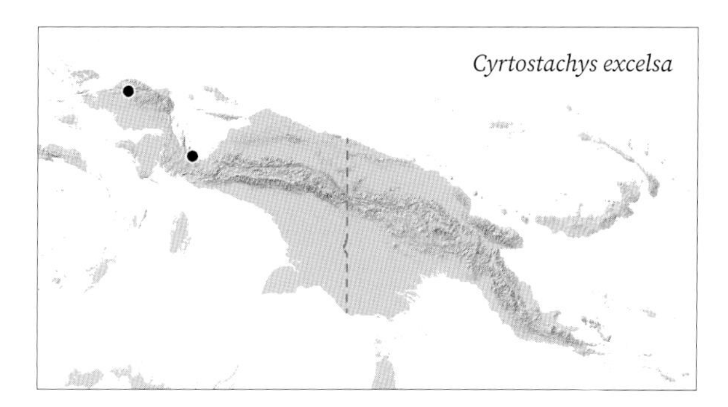

USES. Stems and leaves are used as materials to construct traditional houses (flooring and thatch).

CONSERVATION STATUS. Endangered. *Cyrtostachys excelsa* is known from only a few sites in which deforestation due to logging activities and road building is a major threat.

NOTES. *Cyrtostachys excelsa* is the only species in the genus with irregular leaflets. Together with *C. elegans*, it has inflorescences branched to 4 orders, the highest number in the genus, all others being branched to 2 and 3 orders. Also, *C. excelsa* is the only species from the New Guinea region that has pink to reddish purple inflorescences. The crown of *C. excelsa* is quite distinct, being composed of a small number of ascending leaves, forming a shuttle-cock shape, with stiff leaflets.

5. *Cyrtostachys glauca* H.E.Moore

Slender to moderate, multi-stemmed tree palm to 5.5–15(–21.6) m, with up to ca. 3 adult stems and up to 4 or more suckers at the base, *crown spherical in outline*, 6–8 leaves in crown. **Stem** 5–15 cm diam. **Leaf** 1.7–4.1 m long, *spreading; crownshaft 70–240 cm long, glaucous*; petiole *elongate, 25–88(–100) cm long*; leaflets regularly arranged, *papery*, 45–67 on each side of rachis, mid-leaf leaflets 58–99 × 3–5 cm, with thick, brown membranous ramenta, discontinuous along mid-vein on lower surface. **Inflorescence** 50–130 cm long, 100–200 cm wide, branched to 3–4 (mostly 3) orders, green to pale yellow, pale brown to whitish-green when dried; peduncle to 5 cm; rachillae 25–50 cm long, 2–4 mm diam., white, *pits superficial*, 3–5 per 1 cm rachilla length. **Male flower** 2.5–3.2 × 2.5–2.8 mm; stamens 9–10. **Female flower** 3–3.5 × 2.5–3 mm. **Fruit** 9–12 × 4–6 mm. **Seed** 5–7 × 3–4 mm.

Cyrtostachys glauca. Habit, cultivated National Botanic Garden, Lae (WB).

***Cyrtostachys glauca*. A.** Leaf apex. **B.** Leaf mid-portion. **C.** Portion of inflorescence. **D.** Portion of rachilla with flowers. **E.** Portion of rachilla with fruit. **F, G.** Male flower whole and in longitudinal section. **H, I.** Female flower whole and in longitudinal section. **J.** Fruit in longitudinal section. Scale bar: A, B = 8 cm; C = 6 cm; D, J = 5 mm; E = 1 cm; F, G = 2 mm; H, I = 2.5 mm. A, B, H, I from *Kjaer 512*; C, D, F, G, J from *Moore 9272*; E from *Essig & Katik LAE 55009*. Drawn by Lucy T. Smith.

Cyrtostachys glauca. CLOCKWISE FROM LEFT: crown with inflorescences; Inflorescence; fruit. Cultivated National Botanic Garden, Lae (WB).

DISTRIBUTION. South-eastern New Guinea.

HABITAT. Primary and secondary forest in lowlands or hill forest at 30–400 m.

LOCAL NAMES. *Hek* (Madang), *Vekintambu* (Lababia).

USES. Stems for building purposes.

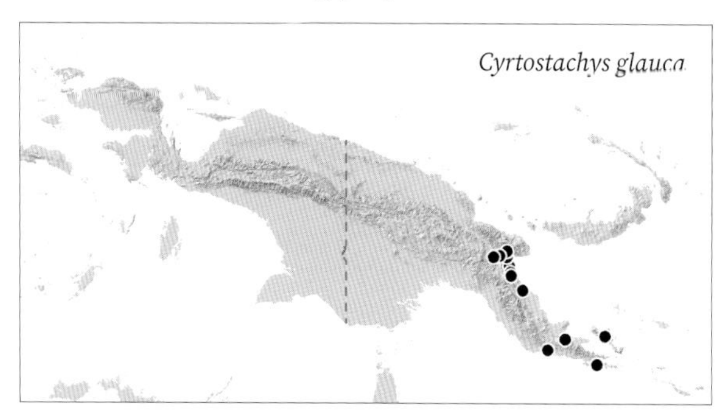

CONSERVATION STATUS. Least Concern (IUCN 2013).

NOTES. *Cyrtostachys glauca* is easily distinguished from all other species by its multiple stems, spherical crown, glaucous leaf sheath and crownshaft, slender and elongate petioles, thin leaflets, slender inflorescence and rachillae with sparse superficial pits. The superficial pits are very peculiar in *Cyrtostachys*. The pits are very shallow or even lacking in a few specimens (*Kjaer & Magun 512* and *Barfod 454*). Together with the tiny flowers (in bud) the species may confused with *Heterospathe* or *Rhopaloblaste*. However, the generic characters such as the presence of a crownshaft, united petals and stamens in male flowers, and the fruit and seed morphology still clearly indicate that it is a species of *Cyrtostachys*.

6. *Cyrtostachys loriae* Becc.

Synonyms: *Cyrtostachys brassii* Burret, *Cyrtostachys kisu* Becc., *Cyrtostachys microcarpa* Burret, *Cyrtostachys peekeliana* Becc., *Cyrtostachys phanerolepis* Burret

Robust, *single-stemmed* tree palm to 10–30 m, 8–14 leaves in crown, *crown hemispherical in outline*. **Stem** 11.5–30 cm diam. **Leaf** 2.5–4.8 m long (including petiole), *spreading*; crownshaft 125–200 cm long; *petiole short to almost absent* (1–10 cm long); leaflets 76–189 on each side of rachis, regularly arranged, mid-leaf leaflets 80–152 × 3.4–6.8 cm, *fine brown ramenta discontinuous along mid-vein on lower surface*. **Inflorescence** 43–150 cm long, up to 250 cm wide, *branched to 3 orders*, green to pale yellow, light brown to black when dried; peduncle to 10 cm; rachillae 25–88.5 cm long, 6–9 mm diam., 8–16 pits per 1 cm rachilla length, pits 2–6 mm diam., deep. **Male flower** 2.5–4 × 2–3 mm; *stamens 9–13*. **Female flower** 2–6 × 2.2–5.5 mm. **Fruit** 8–16 × 4–5 mm. **Seed** 5–8 × 2–5 mm.

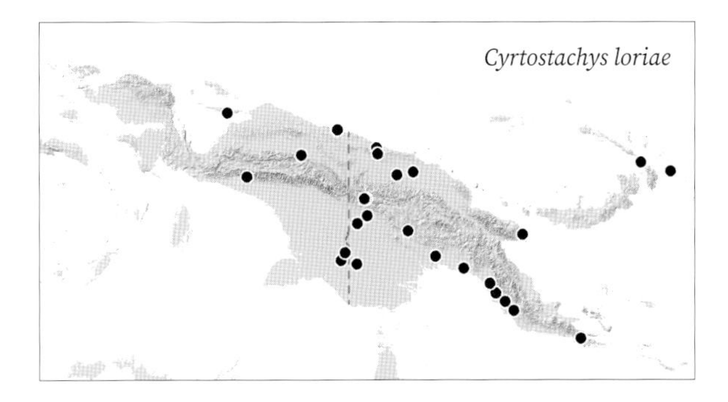

Cyrtostachys loriae. LEFT TO RIGHT: habit, near Timika (JD); inflorescence, Torricelli Mountains (AB).

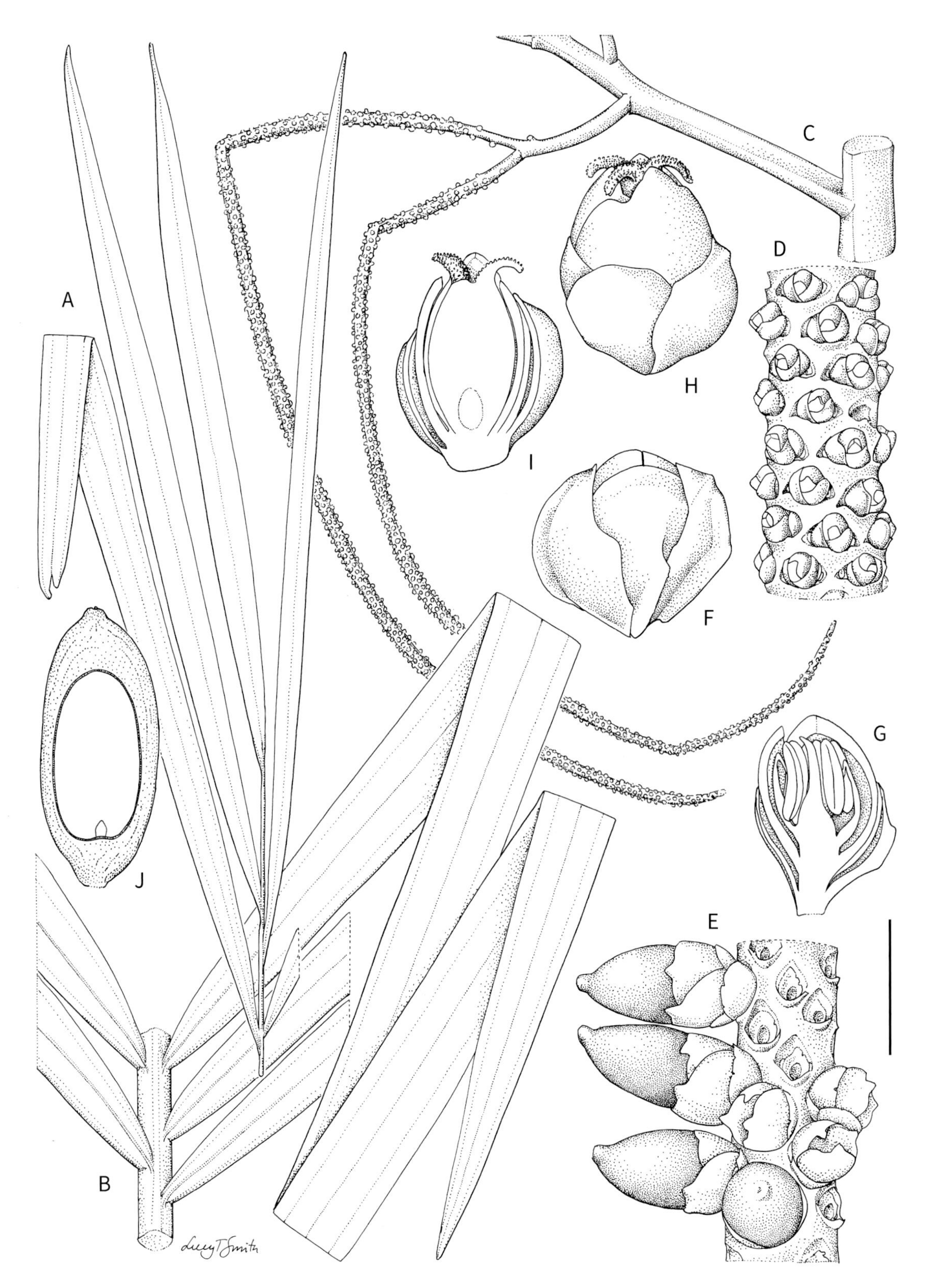

Cyrtostachys loriae. **A.** Leaf apex. **B.** Leaf mid-portion. **C.** Portion of inflorescence. **D.** Portion of rachilla with male flowers. **E.** Portion of rachilla with fruit. **F**, **G.** Male flower whole and in longitudinal section. **H**, **I.** Female flower whole and in longitudinal section. **J.** Fruit in longitudinal section. Scale bar: A–C = 8 cm; D, E = 1 cm; F, G = 2 mm; H, I = 3 mm; J = 7 mm. A–D from *Baker et al. 1110*; E–G, J from *Hoogland & Craven 10144*; H, I from *Heatubun et al. 208*. Drawn by Lucy T. Smith.

DISTRIBUTION. Widespread throughout mainland New Guinea, including Yapen Island and New Ireland, to the Solomon Islands.

HABITAT. Primary or secondary forest in the lowlands or hill forest at an altitude of 30–400 m above sea level.

LOCAL NAMES. *Aikul* (New Ireland), *Apaku* (Mekeo), *Flim* (Mianmin), *Gap* (Marap), *Hek/He-ek* (Amele), *Terep* (Jal), *Lobu* (Wapi), *Mun* (Orme), *Nibung* (Bahasa Indonesia, also used for other tree palms), *Terep/ Terrip* (Yei), *Tnang nyi* (Sentani), *Toono-i* (Bougainville), *Wai'eba* (Kutubu), *Yomberi* (Timbunke), *Yowoh* (Waskuk).

USES. The stems and leaves are used as building materials for traditional houses, e.g., piles, flooring, water pipes, thatch and mattresses. The palm heart is also eaten fresh or cooked.

CONSERVATION STATUS. Least Concern.

NOTES. This very beautiful palm is easily distinguished by its single-stemmed and robust habit, spherical crown, pendulous leaflets, very short (to 10 cm long) or absent petiole and an inflorescence that is more robust than those of other species, branched to 3 orders, with robust rachillae bearing large and deep pits.

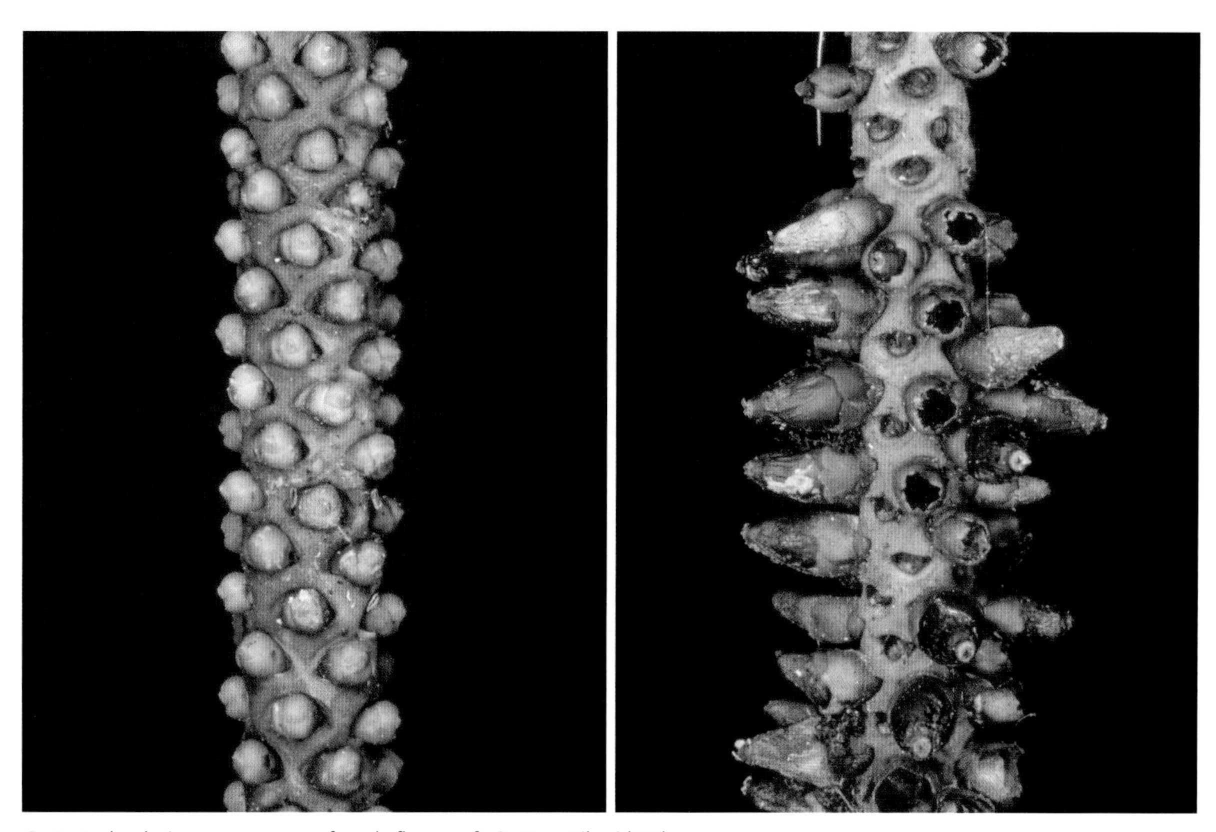

Cyrtostachys loriae. LEFT TO RIGHT: female flowers; fruit. Near Kikori (WB).

Dransfieldia micrantha. Habit, Wandamen Peninsula (WB).

Dransfieldia W.J.Baker & Zona

Small – leaf pinnate – crownshaft – no spines – leaflets pointed

Slender, multi-stemmed (rarely single-stemmed) tree palm, narrow crownshaft present, monoecious. **Leaf** pinnate, more or less straight; sheath tubular, green with dark indumentum; petiole present; leaflets with pointed tips, arranged regularly, horizontal. **Inflorescence** *below the leaves,* branched to 2 (rarely 3) orders, branches spreading; prophyll and peduncular bract similar, but peduncular bract projecting from prophyll and enclosing inflorescence in late bud, *prophyll and sometimes peduncular bract not dropping off as inflorescence expands*; peduncle same length as or slightly longer than inflorescence rachis; rachillae quite robust and curving. **Flowers** in triads throughout the length of the rachillae (rarely inflorescences lacking female flowers), not developing in pits, *male flowers bullet-shaped.* **Fruit** *black, ellipsoid,* stigmatic remains apical, flesh thin, endocarp thin, closely adhering to seed. **Seed** 1, ovoid with flattened base, endosperm ruminate.

DISTRIBUTION. One species endemic to north-western New Guinea and the Raja Ampat Islands.

NOTES. *Dransfieldia* is a slender, multi-stemmed (rarely single-stemmed), mid-storey tree palm with inflorescences borne below the leaves. The prophyll and sometimes the peduncular bract are persistent and the ripe fruit are black. *Dransfieldia* is most easily confused with three genera: *Drymophloeus,* which has jagged leaflet tips and stilt roots; *Heterospathe,* which almost always lacks a crownshaft and has fibrous leaf sheaths and *Ptychosperma,* which has jagged leaflet tips and inflorescences with deciduous major bracts.

Although the sole species of *Dransfieldia* has been known to science for more than a century, the genus itself was not established until 2006, when molecular evidence shed new light on the morphology and relationships of the species (Baker *et al.* 2006, Norup *et al.* 2006).

Dransfieldia micrantha (Becc.) W.J.Baker & Zona

Synonyms: *Heterospathe micrantha* (Becc.) H.E.Moore, *Ptychosperma micranthum* Becc., *Rhopaloblaste micrantha* (Becc.) Hook.f. ex Salomon (not *Rhopaloblaste micrantha* Burret)

Slender, multi-stemmed (rarely single-stemmed) tree palm to 10 m tall, with 4–7 leaves in crown. **Stem** 2–5 cm diam., internodes 4.0–19.5 cm. **Leaf** 1–2 m long including petiole, emerging reddish; sheath 30–45 cm long; petiole 10–20 cm; leaflets 12–27 on each side of rachis, *with raised ridges on upper surface of major veins, linear to narrowly elliptic, attenuate in a narrow acute apex,* with brown ramenta on lower surface of major veins, mid-leaf leaflet 52–76 × 2–5 cm, apical leaflets 18.0–36.0 × 0.8–1.7 cm, apical pair not united at base. **Inflorescence** 34–60 cm long including peduncle, *all axes red to purple at anthesis,* spreading; peduncle 12–26 cm long; primary branches 11–14, to 35 cm; rachillae 8.5–29.0 cm long. **Male flower** 4.5–5.5 mm long in bud, purple; *stamens 15–19.* **Female flower** 3.8–4.3 mm in bud, purple. **Fruit** 15.0–15.9 × 7.6–9.5 mm; black when ripe, flesh ca. 0.7 mm thick. **Seed** 8.9–11.0 × 6.1–7.0 mm.

DISTRIBUTION. Restricted to north-western New Guinea, in the Bird's Head Peninsula, adjacent areas to the east and Waigeo.

HABITAT. Lowland forest and on slopes and ridge tops at 10–250 m elevation.

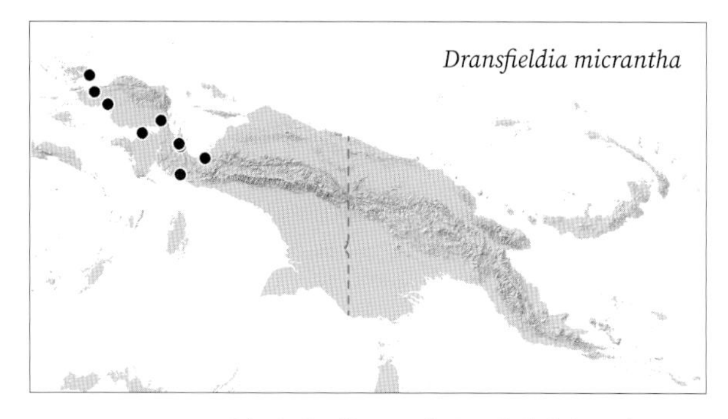

LOCAL NAMES. *Ititohoho* (Yamur), *Kapis* (Biak-Raja Ampat), *Tama'e* (Wondama).

USES. Stems used for harpoons, leaves used for thatch, unspecified parts used for sewing thatch.

Dransfieldia micrantha. LEFT: habit, cultivated Floribunda Palms and Exotics, Hawaii (WB). RIGHT, CLOCKWISE FROM TOP: inflorescences; male flowers; young fruit. Wosimi River (WB).

CONSERVATION STATUS. Least Concern. However, monitoring is required due to mining, oil palm plantations and logging concessions in the range of this species.

NOTES. *Dransfieldia micrantha* is distinct among New Guinea palms in its slender, usually multi-stemmed habit with crownshaft and few leaves in the crown, leaflets with pointed (not jagged) tips and raised ridges along the main veins, and red-purple inflorescences borne below the leaves with relatively long peduncles. For a full discussion of *Dransfieldia* and its similarities to other genera see Baker *et al.* (2006) and Baker & Zona (2006).

Dransfieldia micrantha has a complex taxonomic history, having been placed in *Heterospathe, Ptychosperma* and *Rhopaloblaste* in the past. Molecular evidence (e.g. Norup *et al.* 2006) has helped significantly to clarify the boundaries of these genera and to provide the evidence for the recognition of *Dransfieldia*.

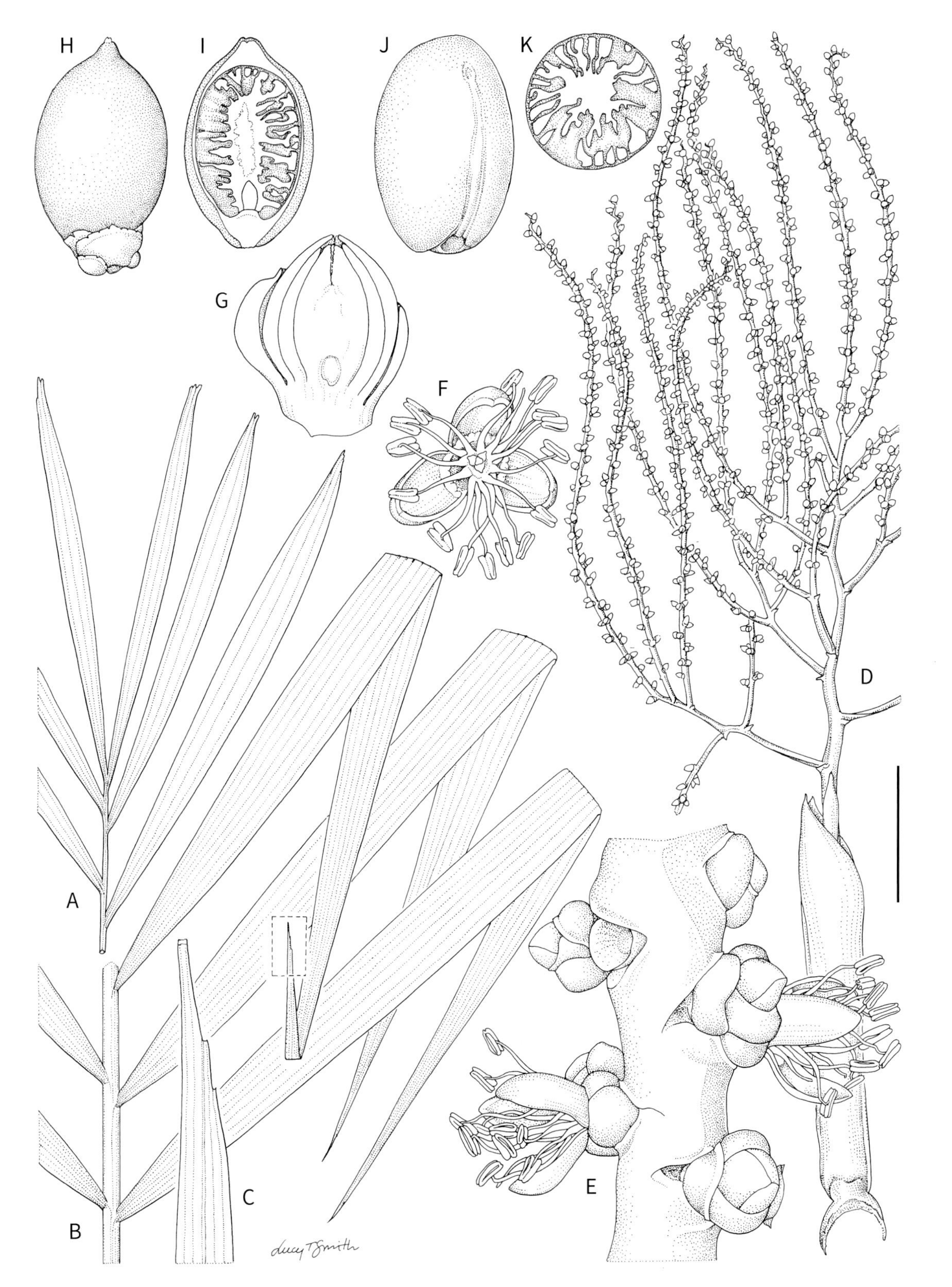

***Dransfieldia micrantha*.** **A.** Leaf apex. **B.** Leaf mid-portion. **C.** Detail of leaflet tip. **D.** Inflorescence. **E.** Portion of rachilla with flowers. **F.** Male flower. **G.** Female flower in longitudinal section. **H**, **I.** Fruit whole and in longitudinal section. **J**, **K.** Seed whole and in transverse section. Scale bar: A, B = 6 cm; C, J, K = 7 mm; D = 4 cm; E, F = 5 mm; G = 3 mm; H, I = 1 cm. A–C, H–K from *Baker et al. 1067*; D from *Heatubun et al. 321*; E–G from *Baker et al. 1066*. Drawn by Lucy T. Smith.

Heterospathe macgregorii. Habit, Kikori River (WB).

ARECOIDEAE | ARECEAE

Heterospathe Scheff.

Synonyms: *Alsmithia* H.E.Moore, *Barkerwebbia* Becc., *Ptychandra* Scheff.

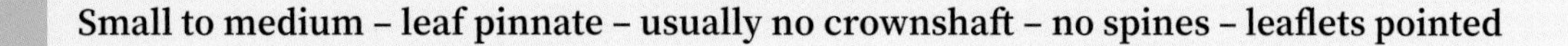

Small to medium – leaf pinnate – usually no crownshaft – no spines – leaflets pointed

Moderate to quite robust single- and multi-stemmed tree palms, some species lacking aerial stem, crownshaft absent (rarely present), monoecious. **Leaf** pinnate, straight; sheath splitting to the base opposite the petiole, *fibrous at the margins*; petiole present; leaflets few to numerous, with pointed tips, arranged regularly (rarely somewhat irregularly), horizontal. **Inflorescence** *between or below the leaves, branched 1–3 (rarely 4) orders,* branches widely spreading; *peduncular bract larger than and projecting from the prophyll,* peduncular bract alone enclosing inflorescence in late bud, splitting, but *usually remaining attached as inflorescence expands*; peduncle as long as to much longer than the inflorescence rachis; rachillae slender and straight or curving. **Flowers** in triads throughout the length of the rachilla, not developing in pits. **Fruit** orange to red, ellipsoid to globose, stigmatic remains apical to lateral, flesh thin, endocarp thin, closely adhering to seed. **Seed** 1 (very rarely 2 or 3), ellipsoid to globose, endosperm ruminate.

DISTRIBUTION. Philippines to Maluku, through New Guinea, the Bismarck Archipelago and Solomon Islands to Fiji, 36 species in total, of which 14 in New Guinea.

NOTES. *Heterospathe* species are under- and mid-storey palms, some stemless, usually with fibrous, open leaf sheaths and almost always lacking a crownshaft. Their inflorescences appear between the leaves, although may later be presented below the leaves. Importantly, the first peduncular bract is longer than and projects from the prophyll. Most *Heterospathe* species are found in submontane and montane rainforest up to 2,400 m, but some occur in the lowlands. The taxonomic history of the genus is complex and confused with the genera *Dransfieldia* and *Rhopaloblaste* (summarised and clarified by Norup *et al.* 2006). *Dransfieldia* is distinguished from *Heterospathe* by its slender, clustering habit and crownshaft, whereas New Guinea species of *Rhopaloblaste* are distinguished by their robust, solitary habit, crownshaft, dark hairs on the leaf rachis, and distinct inflorescence morphology.

In preparation for this work, the *Heterospathe elegans* complex was reviewed in detail (Trudgen & Baker 2008) and a separate revision of the species in New Guinea completed (Petoe & Baker 2019).

Key to the species of *Heterospathe* in New Guinea

1. Palm lacking aerial stem 2
1. Palm with well-defined aerial stem 3

2. Palm usually multi-stemmed; peduncle 50–140 cm long, peduncular bract attached in distal quarter of peduncle; flowers green . . . . **4b. *H. elegans* subsp. *humilis***
2. Palm single-stemmed; peduncle 45–75 cm long, peduncular bract attached in proximal half of peduncle; flowers purple . . . . **14. *H. sphaerocarpa***

3. Palm rheophytic and multi-stemmed; inflorescence 50–65 cm long including 17–22 cm peduncle . . . . **7. *H. macgregorii***
3. Palm not rheophytic and normally single-stemmed; inflorescence 50–160 cm long including 22–140 cm peduncle 4

4. Slender palm, stem 1–5 cm diam. (rarely as wide as 7.5 cm diam. in *H. lepidota*); peduncular bract inserted in proximal or distal half of peduncle 5
4. Moderately slender to robust palm, stem 6–23 cm diam.; peduncular bract inserted in proximal half of peduncle 8

5. Inflorescence with 2 or 3 orders of branching; peduncular bract inserted in proximal half of peduncle; fruit ca. 20 mm long (when dry), with 6–7 distinctive longitudinal ridges . . . **11. *H. porcata***
5. Inflorescence with 1 or 2 orders of branching; peduncular bract inserted in distal half of peduncle; fruit 10–18 mm long (when dry), not prominently ridged 6

6. Peduncular bract inserted in distal quarter of peduncle; fruit 10–15 mm long . . . . **4a. *H. elegans* subsp. *elegans***
6. Peduncular bract inserted in distal third quarter of peduncle; fruit 15–18 mm long 7

7. Leaflets 10–15 each side of rachis, the basalmost pair multi-fold; flowers green with glabrous sepals. . . . . **13. *H. pullenii***
7. Leaflets 21–35 each side of rachis, the basalmost pair single-fold; flowers yellowish with scaly sepals. . . . . **6. *H. lepidota***

8. Robust palm; leaf usually 3–5 m long including petiole (rarely as short as 1.7 m); leaflets 61–89 each side of rachis (rarely fewer); inflorescence branched to 3 or 4 orders . . . . **3. *H. elata***
8. Moderately slender to robust palm; leaf 1.2–4.1 m long including petiole; leaflets 22–56 each side of rachis; inflorescence branched to 1–3 orders. 9

9. First-order branches of inflorescence to 70 cm long; flowers reddish-brown; fruit globose with more-or-less apical stigmatic remains . . . . **10. *H. parviflora***
9. First-order branches of inflorescence to 45 cm long; flowers not reddish-brown; fruit globose to broadly ellipsoid, with subapical stigmatic remains 10

10. Crownshaft well-defined . . . . **2. *H. barfodii***
10. Crownshaft absent 11

11. Leaflets 22–30 each side of rachis; inflorescence branched to 1 order . . . . **1. *H. annectens***
11. Leaflets 40–56 each side of rachis; inflorescence branched to 2 or 3 orders. 12

12. Leaflets narrowly acuminate at their tips; male flower with 6 stamens; rachillae to 19 cm long; fruit 8–8.5 mm long . . . . **12. *H. pulchra***
12. Leaflets acute to acuminate at their tips; rachillae to 30 cm long; male flower with 6–20 stamens; fruit at least 9 mm long 13

13. Male flower pink, with 6 stamens . . . . **5. *H. ledermanniana***
13. Male flower green, yellow, cream or purple, with 12–20 stamens. 14

14. Male flower with 12–15 stamens; fruit 11–15 mm long . . . . **8. *H. muelleriana***
14. Male flower with 19–20 stamens; fruit 29–35 mm long . . . . **9. *H. obriensis***

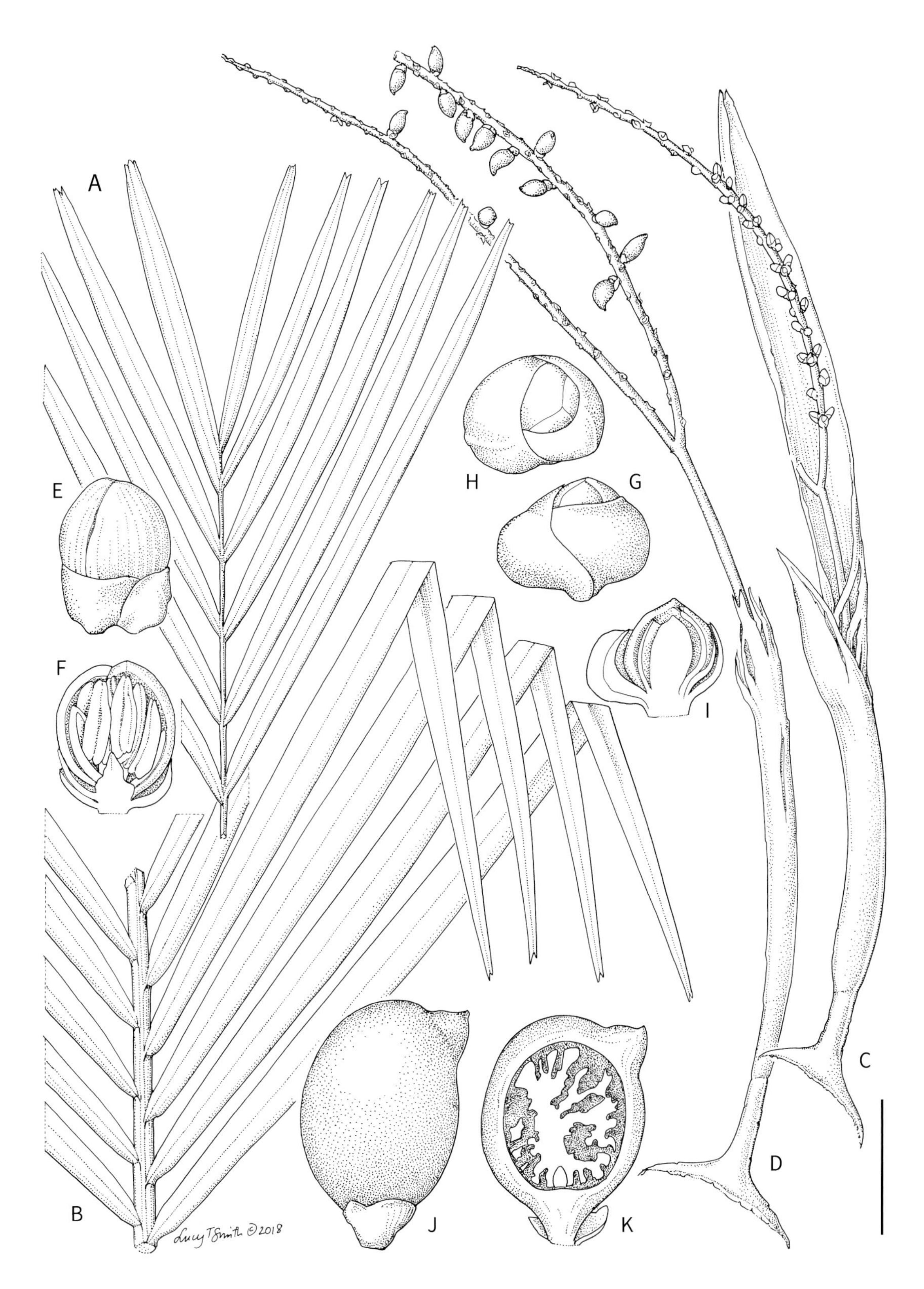

Heterospathe annectens. **A.** Leaf apex. **B.** Leaf mid-portion. **C.** Inflorescence. **D.** Inflorescence with fruit. **E**, **F.** Male flower whole and in longitudinal section. **G**, **H.** Female flower in two views. **I.** Female flower in longitudinal section. **J**, **K.** Fruit whole and in longitudinal section. Scale bar: A–D = 6 cm; E, F = 5 mm; G–I = 4 mm; J, K = 1 cm. All from *Brass 28409*. Drawn by Lucy T. Smith.

1. *Heterospathe annectens* H.E.Moore

Moderately slender, single-stemmed palm to 12 m. **Stem** ca. 8 cm diam. **Leaf** ca. 1.2 m long including petiole; sheath ca. 35 cm long, crownshaft absent; petiole ca. 54 cm long; leaflets ca. 22–30 each side of rachis, arranged regularly, single-fold, linear, acuminate at their tips. **Inflorescence** 60–70 cm long including 32–37 cm peduncle, between the leaves, branched to 1 order, colour unknown; peduncular bract ca. 30–57 × 2.5–3.5 cm, attached in proximal half of peduncle; *first-order branches 2–3, undivided*, to 34 cm long. **Male flower** *with ca. 16 stamens* and conical pistillode about half the length of filaments, cream. **Female flower** with glabrous and imbricate sepals, cream. **Fruit** ca. 18.5 × 12.5 mm (when dry and not completely mature), broadly ellipsoid, red, with subapical stigmatic remains. **Seed** ca. 10 × 10 mm (when dry), globose; endosperm ruminate.

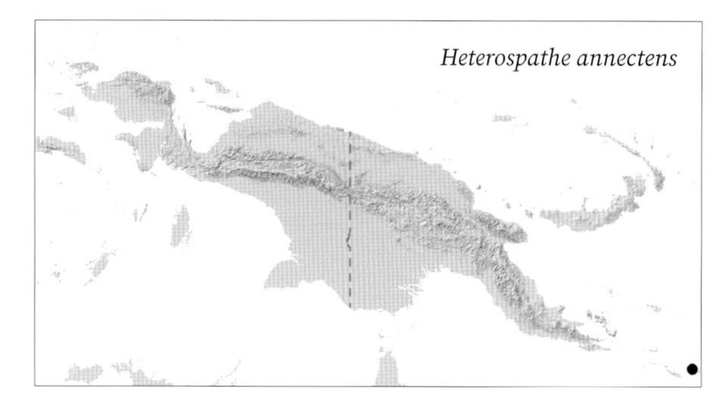

DISTRIBUTION. Rossel Island in the Louisiade Archipelago.

HABITAT. Hill forest at ca. 700 m.

LOCAL NAMES. None recorded.

USES. None recorded.

CONSERVATION STATUS. Data Deficient (IUCN 2021). This species is known from only two sites, where there is limited information on ongoing threats. Due to insufficient evidence, an extinction risk category cannot be selected.

NOTES. *Heterospathe annectens* is distinguished by its moderately slender tree habit in combination with short inflorescences with only one order of branching. Other New Guinea species with inflorescences branched to one order are either stemless or more slender. The species also has relatively large male flowers with many stamens, and large fruit.

2. *Heterospathe barfodii* L.M.Gardiner & W.J.Baker

Moderately slender to robust, single-stemmed palm to 15 m, bearing 7–8 leaves in crown. **Stem** 6–15 cm diam. **Leaf** 2.5–4 m long including petiole; sheath 50–67 cm long, *crownshaft distinct, with white, powdery indumentum*, 50–100 cm long; petiole 20–100 cm long; leaflets (38–)48–56 each side of rachis, arranged regularly, single-fold, linear, acute to acuminate at their tips. **Inflorescence** 75–150 cm long including 27–50 cm peduncle, *below the leaves*, branched to (2–)3 orders, with purple axes (at least in bud); peduncular bract ca. 55 cm long, attached in proximal half of peduncle; first-order branches 18, to 43 cm long, each with as many as 18 rachillae to 24–30 cm long. **Male flower** with 6–9 stamens and more-or-less conical pistillode about half the length of filaments, *purple*. **Female flower** with glabrous and imbricate sepals, *purple*. **Fruit** 7–10 × 7–9 mm, subglobose or globose, red, with subapical stigmatic remains. **Seed** ca. 8 × 7 mm, broadly ellipsoid; endosperm ruminate.

Heterospathe barfodii. Habit, Milne Bay (AB).

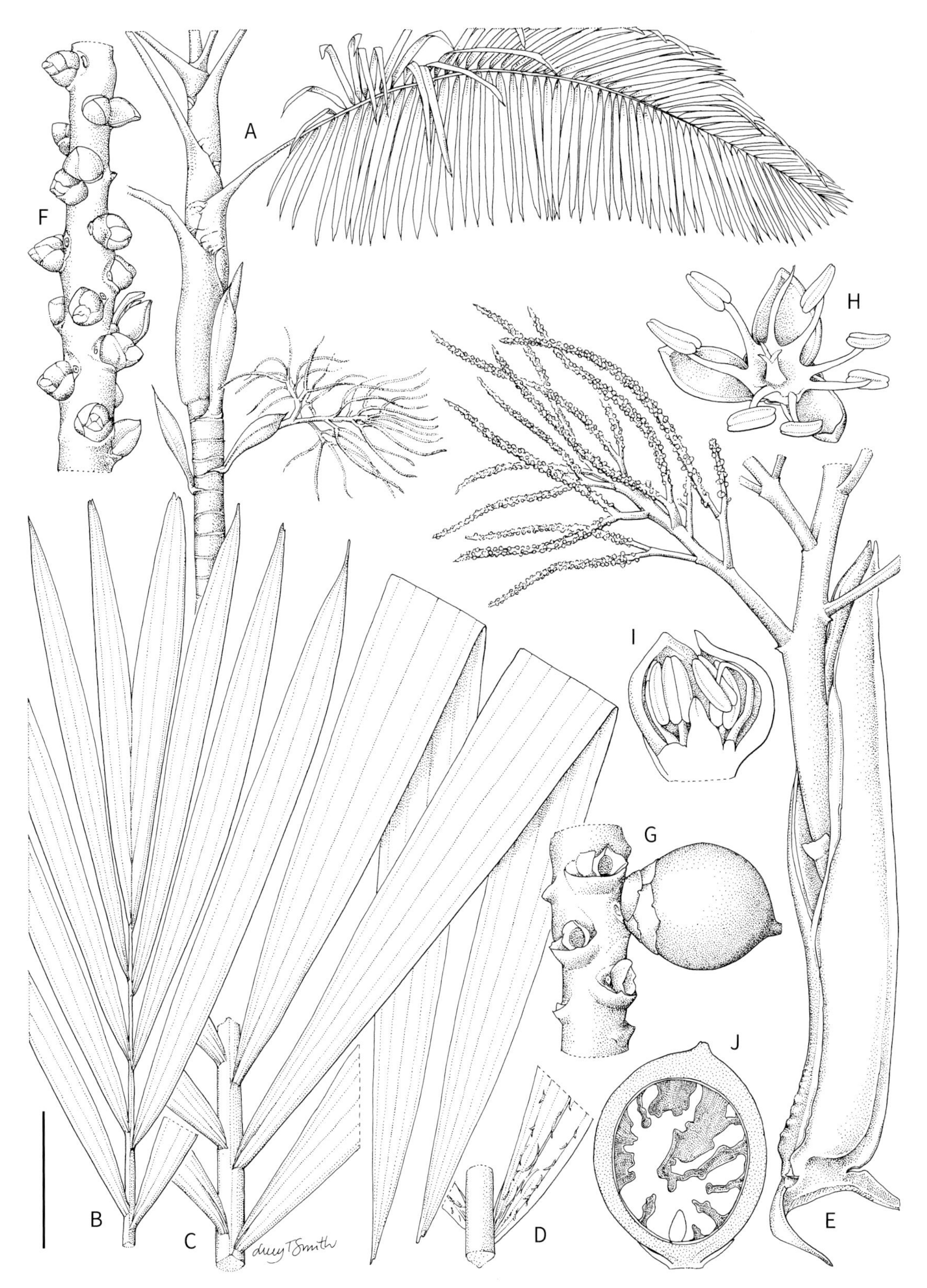

Heterospathe barfodii. **A.** Crownshaft with leaf and inflorescence. **B.** Leaf apex. **C.** Leaf mid-portion. **D.** Leaflet abaxial side with ramenta. **E.** Portion of inflorescence with prophyll. **F.** Portion of rachilla with flowers. **G.** Detail of rachilla with fruit. **H**, **I.** Male flower whole and in longitudinal section. **J.** Fruit in longitudinal section. Scale bar: A = 50 cm; B–E = 8 cm; F, G = 1 cm; H = 5 mm; I = 3 mm; J = 7 mm. All from *Marcus 1*. Drawn by Lucy T. Smith.

DISTRIBUTION. South-eastern Papua New Guinea including the D'Entrecasteaux Islands.

HABITAT. Lowland to premontane forest from sea level to 950 m.

LOCAL NAMES. *Vekiniya* (Labania), *Zagi* (unknown dialect).

USES. None recorded.

CONSERVATION STATUS. Least Concern (IUCN 2021).

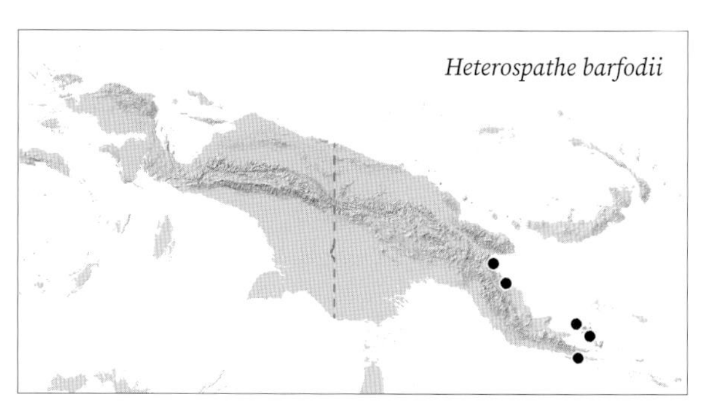

Heterospathe barfodii. ANTI-CLOCKWISE FROM TOP: inflorescences, Alotau (AB); inflorescences, Alotau (JLD); flower buds, one male flower open, cultivated Floribunda Palms and Exotics, Hawaii (JD); fruit, Alotau (JLD).

NOTES. *Heterospathe barfodii* is immediately recognisable on account of its well-defined, white crownshaft and inflorescences below the leaves (Gardiner *et al.* 2012). In many respects, it resembles a robust species of *Dransfieldia,* but DNA evidence places it firmly in *Heterospathe.*

3. *Heterospathe elata* Scheff.

Synonyms: *Heterospathe elata* var. *guamensis* Becc., *Heterospathe elata* var. *palauensis* (Becc.) Becc., *Heterospathe palauensis* Becc.

Robust, single-stemmed palm to 12(–20) m, bearing 14–21 leaves in crown. **Stem** (8–)10–23 cm diam. **Leaf** (1.7–)3–5 m long including petiole; sheath 50–85 cm long, crownshaft absent; petiole (20–)50–100 cm long; *leaflets usually 61–74(–89)* each side of rachis, arranged regularly, single-fold, linear, acuminate at their tips. **Inflorescence** 90–150 cm long including 30–55 cm peduncle, between the leaves, *branched to 3 or 4 orders,* with greenish axes; peduncular bract 55–146 × 6 cm, attached in proximal half of peduncle; first-order branches 18–25, to 60 cm long, each with as many as 83 rachillae to ca. 24 cm long. **Male flower** with *6 stamens and columnar pistillode at least as long as filaments,* cream to pinkish. **Female flower** with glabrous and imbricate sepals, cream to pinkish. **Fruit** 6.4–10 × 6.4–10 mm, globose, red, with subapical stigmatic remains, fruits sometimes containing 2 or 3 seeds (Gag Island populations), then 2-lobed or 3-lobed. **Seed** ca. 6 × 6 mm, globose; endosperm ruminate.

DISTRIBUTION. Gag Island in Indonesian New Guinea. Elsewhere in Maluku, the Philippines, Palau and Guam.

HABITAT. Limestone forest near sea level (Gag Island).

LOCAL NAMES. *Gul ways* (Gebe; Gag Island).

USES. Stems may be used for traditional house construction (Gag Island).

CONSERVATION STATUS. Least Concern (IUCN 2021).

Heterospathe elata. **A.** Habit. **B.** Leaf apex. **C.** Leaf mid-portion. **D.** Leaf base. **E.** Inflorescence. **F.** Portion of rachilla with flowers. **G**, **H.** Male flower whole and in longitudinal section. **I**, **J.** Female flower whole and in longitudinal section. **K**, **L.** Fruit whole and in longitudinal section. Scale bar: A = 2.3 m; B–D = 8 cm; E = 6 cm; F, K, L = 5 mm; G–J = 2 mm. A–D from *Heatubun 740*; E–J from *Rodin 737*; K, L from *Chapin 55*. Drawn by Lucy T. Smith.

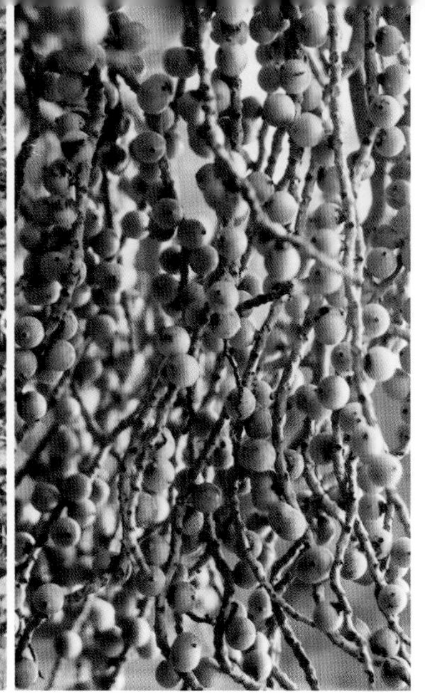

Heterospathe elata. LEFT TO RIGHT: crown; inflorescence; fruit. Gag Island (CDH).

NOTES. *Heterospathe elata* is the most widely distributed species of *Heterospathe*, although in our area it only occurs on Gag Island in the Raja Ampat Islands in Indonesian New Guinea (Heatubun *et al.* 2012b, 2014a). It is usually a robust palm with long, slightly arching leaves bearing numerous leaflets (61–74 per side), and large inflorescences branched to three orders. The population of *H. elata* from Gag Island appears somewhat distinct from other populations with individuals there being only moderately robust and displaying leaves with 74–89 leaflets on each side of the rachis and inflorescences often branched to four orders. All other species of *Heterospathe* that occur in New Guinea and adjacent islands differ in having leaves with fewer than 57 leaflets per side, and inflorescences with one to three orders of branching. The Gag population also often produces bi-lobed or tri-lobed fruit that contain two or three seeds (Heatubun *et al.* 2012b). This is highly unusual in fruits of tribe Areceae, which are generally one-seeded.

Heterospathe elata

4. *Heterospathe elegans* (Becc.) Becc.

Synonyms: *Barkerwebbia elegans* Becc., *Barkerwebbia elongata* (Becc.) Becc., *Heterospathe elongata* Becc.

Slender, single- or multi-stemmed understorey palm with or without aerial stem, bearing 7–12(–18) leaves per crown. **Stem** 1–5 cm diam. **Leaf** 1–3.3 m long including petiole; sheath poorly defined or up to 42 cm long, crownshaft absent; petiole 42–130 cm long; blade pinnate or occasionally entire-bifid; leaflets (when present) 6–43 each side of rachis, arranged regularly or rarely in pairs, single- or multi-fold with the apical leaflets comprising 1–10 folds, linear to lanceolate and sigmoid, acute to acuminate at their tips. **Inflorescence** *65–150 cm long including 50–140 cm peduncle*, between the leaves, branched to 1 or 2 orders, with green axes; peduncular bract 12.5–60 × 0.25–1.5 cm, *attached in distal quarter of peduncle* (a second, vestigial peduncular bract may be present); first-order branches 4–16, to 26 cm long, each with as many as 7 rachillae to 17 cm long. **Male flower** with 6 stamens and more-or-less columnar pistillode about the same length as filaments, green. **Female flower** with glabrous and imbricate sepals, green. **Fruit** *10–15 × 8–10 mm*, ellipsoid to globose, red, with subapical stigmatic remains. **Seed** 5–8 × 5–8 mm, globose; endosperm ruminate.

***Heterospathe elegans* subsp. *elegans*. A.** Leaf apex. **B.** Leaf bases. **C.** Inflorescence with fruit. **D.** Portion of inflorescence. **E.** Portion of rachilla with female flowers. **F, G.** Male flower whole and in longitudinal section. **H.** Fruit. Scale bar: A, C, D = 4.5 cm; B = 4 cm; E = 5 mm; F, G = 2.5 mm; H = 1 cm. A, B, D, F, G from *Baker & Kage 656*; C, H from *Baker & Kage 652*; E from *Baker et al. 634.* Drawn by Lucy T. Smith.

NOTES. *Heterospathe elegans* is one of the most widely distributed species of *Heterospathe* in New Guinea. The species is distinguished by its long, slender inflorescence, which is often exserted from the crown and has a persistent peduncular bract inserted in the distal quarter of the elongated peduncle. Two subspecies are accepted, with one being usually multi-stemmed but lacking an aerial stem (*H. elegans* subsp. *humilis*), and the other having an aerial, solitary stem reaching 2.5 m (*H. elegans* subsp. *elegans*). *Heterospathe elegans* subsp. *elegans* can be confused with *H. pullenii* and *H. lepidota* but those species have a caducous peduncular bract inserted in the distal third quarter of the peduncle, and usually larger fruit. *Heterospathe elegans* subsp. *humilis* can only be confused with *H. sphaerocarpa* on account of the absence of an aerial stem, but *H. sphaerocarpa* is single-stemmed and has purple inflorescences with the peduncular bract inserted in the proximal half of the less elongate peduncle.

4a. *Heterospathe elegans* subsp. *elegans*

Synonym: *Heterospathe versteegiana* Becc.

Single-stemmed, stem well-defined, to 2.5 m. Leaves usually divided into single-fold leaflets apart from the 2-fold apical pair, or occasionally divided into fewer, multi-fold leaflets. Inflorescence often branched to 2 orders.

DISTRIBUTION. Widespread in the eastern half of the Central Range.

HABITAT. Understorey in premontane to montane rainforest at 500–2,150 m.

LOCAL NAMES. *Dalung* (Busilmin), *Fitigit* (unknown dialect), *Tiritiri* (unknown dialect).

USES. Used for making spear heads.

CONSERVATION STATUS. Least Concern (IUCN 2021).

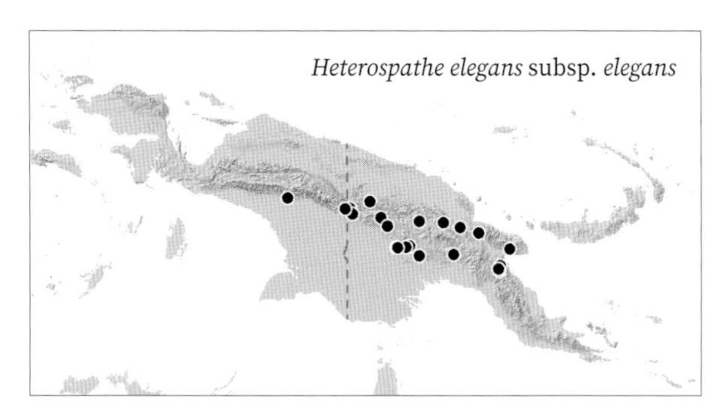

Heterospathe elegans subsp. *elegans*. LEFT TO RIGHT: habit, near Tabubil (WB); crown with inflorescence with fruit, near Tabubil (WB); crown with inflorescences, Lake Kutubu (WB).

4b. *Heterospathe elegans* subsp. *humilis* (Becc.) M.S.Trudgen & W.J.Baker

Synonyms: *Barkerwebbia humilis* (Becc.) Becc., *Heterospathe humilis* Becc., *Heterospathe pilosa* (Burret) Burret, *Rhynchocarpa pilosa* Burret

Single- or multi-stemmed, lacking an aerial stem or with a short, prostrate stem, rooting from the base. Leaves entire or divided into single- or multi-fold leaflets, or a combination of both, usually with apical and basal leaflets comprising 2–10 folds. Inflorescence branched to 1 or 2 orders.

DISTRIBUTION. Widespread in the eastern half of the Central Range and central northern New Guinea.

HABITAT. Usually premontane to montane rainforest at (50–)600–2,200 m.

LOCAL NAMES. *Sowangu* (Waskuk), *Bossowa* (Wagu), *Yawi* (Enga), *Kakanda* (unknown dialect).

USES. None recorded.

CONSERVATION STATUS. Least Concern (IUCN 2021).

Heterospathe elegans subsp. *humilis*. Habit, Nanduo (WB).

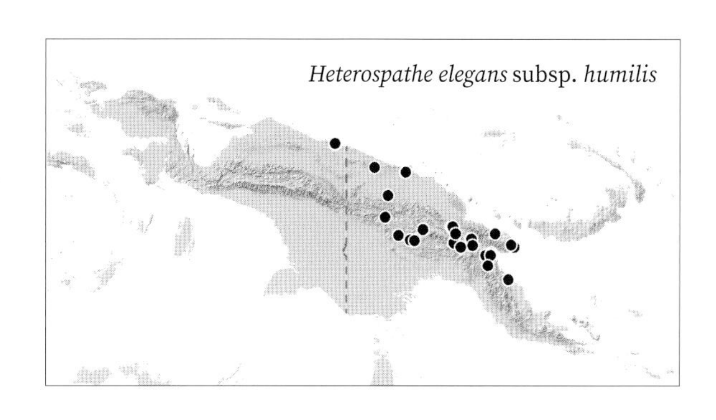

5. *Heterospathe ledermanniana* Becc.

Moderately slender, single-stemmed palm to 10 m, bearing 7–9 leaves in crown. **Stem** 6.5–7 cm diam. **Leaf** 1.7–2.5 m long including petiole; sheath ca. 30 cm long, crownshaft absent; petiole 29–37 cm long; leaflets 41–44 each side of rachis, arranged regularly, single-fold apart from the apical leaflets that may be 2-fold, linear, acuminate at their tips. **Inflorescence** ca. 100 cm long including *40–50 cm peduncle*, between the leaves, branched to 2 or 3 orders, with pink axes; peduncular bract ca. 65–75 × 5 cm, attached in proximal half of peduncle; first-order branches 18–23, to 45 cm long, each with as many as 33 rachillae to ca. 20 cm long. **Male flower** *with 6 stamens and more-or-less columnar pistillode shorter than filaments, pink*. **Female flower** with glabrous and imbricate sepals, *pink*. **Fruit** ca. 15 × 11 mm, subglobose, red, with subapical stigmatic remains. **Seed** ca. 10 × 9 mm, subglobose; endosperm ruminate.

DISTRIBUTION. Scattered records in eastern central New Guinea.

HABITAT. Montane forest on karst limestone at 1000–1,400 m.

LOCAL NAMES. None recorded.

USES. None recorded.

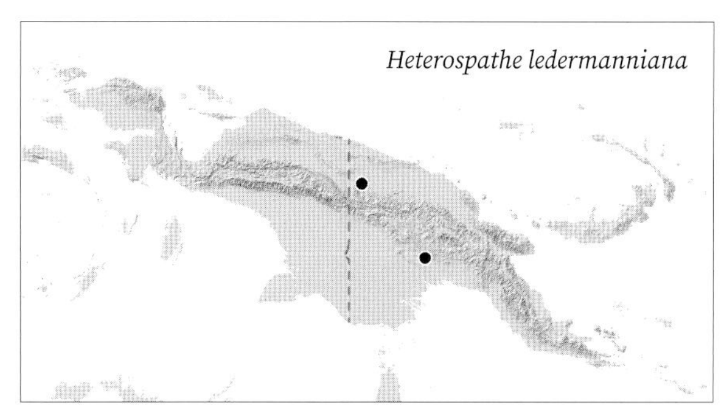

***Heterospathe ledermanniana*. A.** Habit. **B.** Leaf apex. **C.** Leaf mid-portion. **D.** Inflorescence with prophyll and peduncular bract. **E.** Portion of inflorescence with fruit. **F.** Portion of rachilla with flowers. **G**, **H.** Male flower whole and in longitudinal section. **I**, **J.** Female flower whole and in longitudinal section. **K**, **L.** Fruit whole and in longitudinal section. Scale bar: A = 1.1 m; B, C = 6 cm; D, E = 8 cm; F = 7 mm; G–J = 3.3 mm; K, L = 1 cm. All from *Baker et al. 1115*. Drawn by Lucy T. Smith.

Heterospathe ledermanniana. Habit, Southern Highlands (WB).

CONSERVATION STATUS. Data Deficient (IUCN 2021). This species is known from only two sites, where there is limited information on ongoing threats. Due to insufficient evidence, a single extinction risk category cannot be selected.

NOTES. *Heterospathe ledermanniana* is distinguished by its moderately slender tree habit in combination with pink male flowers with six stamens and a more-or-less columnar pistillode that is shorter than the filaments. The species most closely resembles *H. muelleriana*, but that species is often more robust and has green to yellow (or rarely purple) male flowers with 12–15 stamens.

6. *Heterospathe lepidota* H.E.Moore

Slender, single-stemmed palm to 5.5 m. **Stem** 2–7.5 cm diam. **Leaf** 1–2.5 m long including petiole; sheath 15–30 cm long, crownshaft absent; petiole ca. 15–40 cm long; leaflets 21–35 each side of rachis, arranged regularly, single-fold apart from the apical leaflets that may be 3–4-fold, linear, acuminate at their tips. **Inflorescence** 85–100 cm long including 57–70 cm peduncle, between the leaves, branched to 2 orders, colour unknown; peduncular bract not seen in its entirety, *attached in distal third quarter of peduncle*; first-order branches 10–15, to 45 cm long, each with

Heterospathe ledermanniana. LEFT TO RIGHT: habit; inflorescence; male flowers. Southern Highlands (WB).

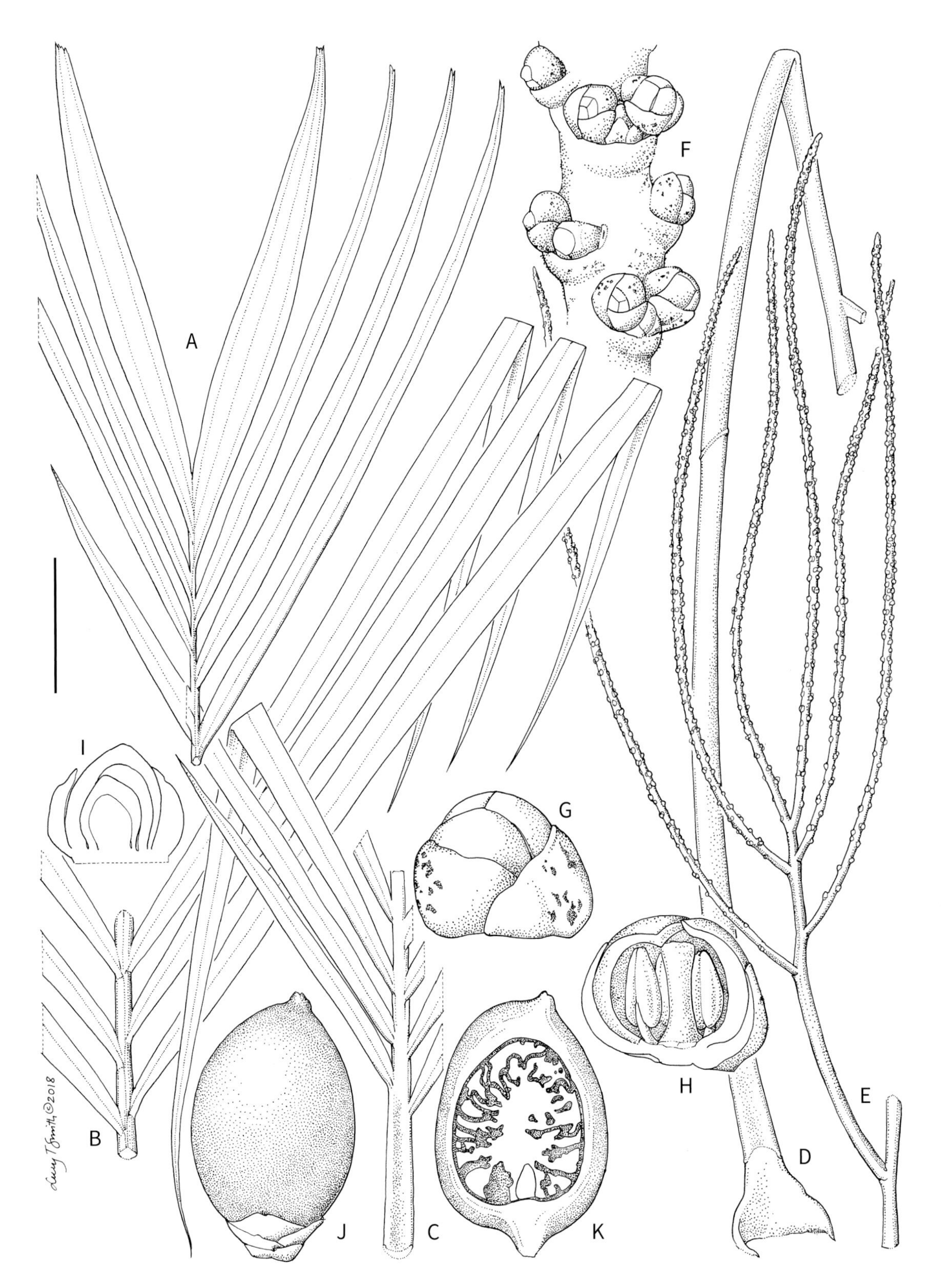

Heterospathe lepidota. **A.** Leaf apex. **B.** Leaf mid-portion. **C.** Leaf base. **D.** Inflorescence peduncle. **E.** Portion of inflorescence. **F.** Portion of rachilla with triads. **G, H.** Male flower whole and in longitudinal section. **I.** Female flower in longitudinal section. **J, K.** Fruit whole and in longitudinal section. Scale bar: A–E = 6 cm; F = 4 mm; G, H = 1.25 mm; I = 0.8 mm; J, K = 7.5 mm. All from *Hoogland & Womersley 3241*. Drawn by Lucy T. Smith.

as many as 6 rachillae to 30–35 cm long. **Male flower** with 6 stamens and columnar pistillode at least as long as filaments, *yellowish*. **Female flower** with *lepidote* and imbricate sepals, *yellowish*. **Fruit** ca. *15 × 10 mm* (when dry), broadly ellipsoid, red, with subapical stigmatic remains. **Seed** ca. 9 × 7.5 mm (when dry), ovoid; endosperm deeply ruminate.

DISTRIBUTION. South-eastern New Guinea.

HABITAT. Lowland rainforest on slopes at ca. 150–200 m.

LOCAL NAMES. *Soriki* (Orokaiva).

USES. None recorded.

CONSERVATION STATUS. Data Deficient (IUCN 2021). This species is known from only two sites, where there is limited information on ongoing threats. Due to insufficient evidence, a single extinction risk category cannot be selected.

NOTES. *Heterospathe lepidota* is a slender palm distinguished by its leaves with 21–35 leaflets on each side of the rachis, all single-fold apart from the multi-fold apical leaflet pair, a peduncular bract inserted in the distal third quarter of the peduncle, yellowish flowers with more-or-less scaly sepals, and fruit measuring at least 15 mm in length. This species most closely resembles *H. pullenii* but that species differs in having leaves with 10–15 leaflets per side, with the basal leaflet pair being multi-fold, and green flowers with glabrous sepals. *Heterospathe lepidota* can also be confused with *H. elegans* subsp. *elegans* but *H. elegans* has the peduncular bract inserted in the most distal quarter of the peduncle, occurs at higher elevation, and usually has smaller fruit.

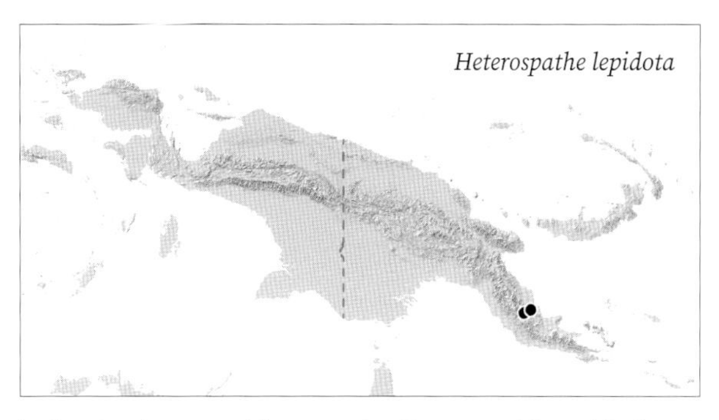

7. *Heterospathe macgregorii* (Becc.) H.E.Moore

Synonym: *Rhopaloblaste macgregorii* Becc.

Slender, *multi-stemmed, rheophytic palm* to 7 m, sometimes with adventitious roots forming cone at stem bases, bearing ca. 9 leaves per crown. **Stem** 4–7.5 cm diam. **Leaf** 1–2.5 m long including petiole; sheath 26–40 cm long, crownshaft absent; petiole 19–80 cm

Heterospathe macgregorii. Colony, showing rheophytic habit, Kikori River (OG).

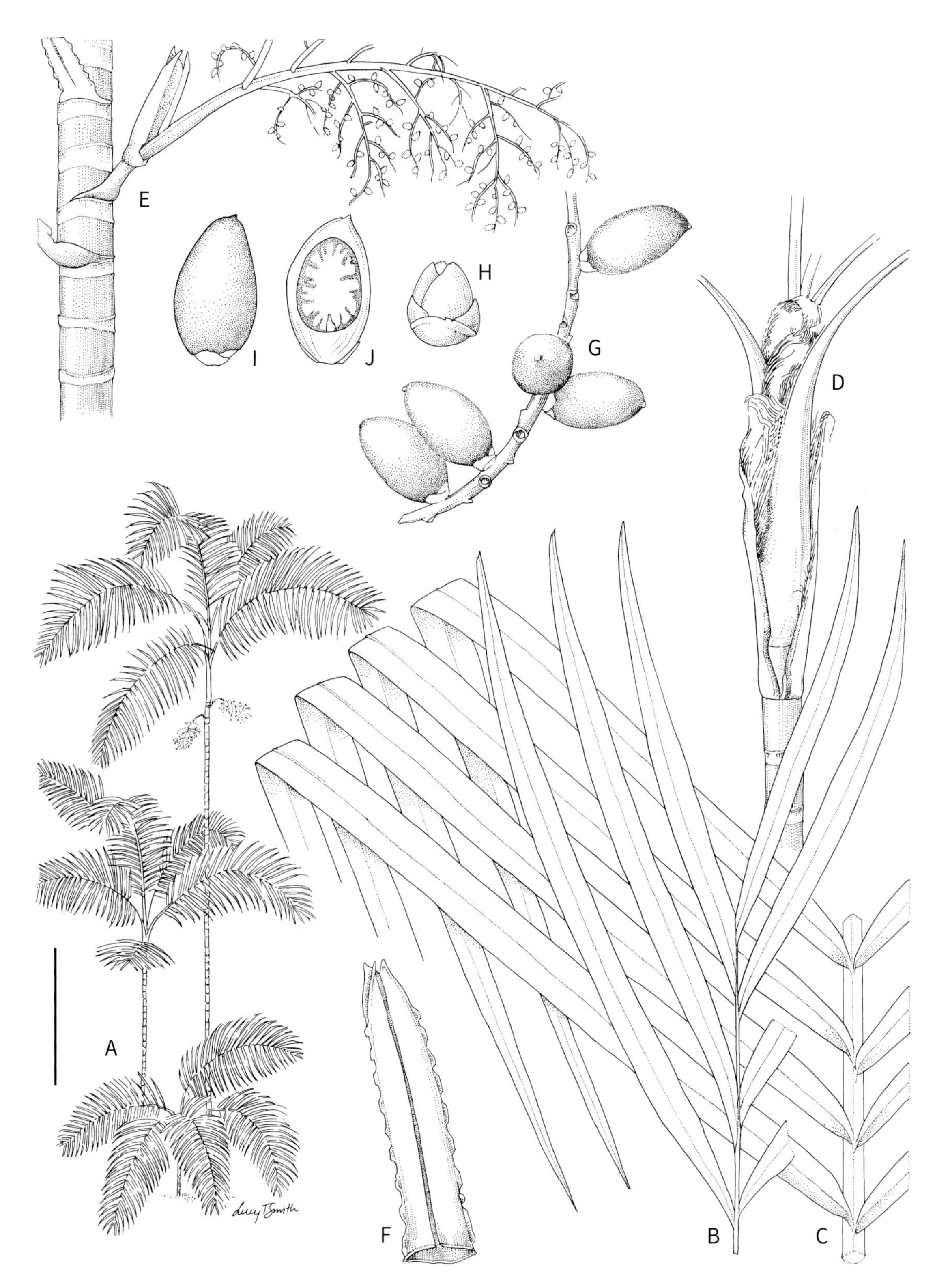

Heterospathe macgregorii. **A.** Habit. **B.** Leaf apex. **C.** Leaf mid-portion. **D.** Portion of crownshaft. **E.** Inflorescence attached to stem. **F.** Detail of prophyll. **G.** Portion of rachilla with fruit. **H.** Male flower. **I, J.** Fruit whole and in longitudinal section. Scale bar: A, D, E = unknown; B, C, F = 6 cm; G = 2.5 cm; H = 7 mm; I, J = 2 cm. A, D, E from photographs in locality of *Baker et al. 651*; B, C, F–J from *Baker et al. 651*. Drawn by Lucy T. Smith.

Heterospathe macgregorii. LEFT TO RIGHT: habit, Kikori River (OG); fruit, Kikori River (WB).

long; leaflets 32–44 each side of rachis, arranged regularly, single-fold, linear, acuminate at their tips. **Inflorescence** *50–65 cm long including 17–22 cm peduncle*, between the leaves, branched to 2 or 3 orders, with greenish axes; peduncular bract ca. 25 × 1.5 cm, attached in proximal half of peduncle; first-order branches 10–18, to 45 cm long, each with as many as 16 rachillae to 20–27 cm long. **Male flower** with 6 stamens and columnar pistillode at least as long as filaments, green. **Female flower** with glabrous and imbricate sepals, green. **Fruit** ca. 20–22 × 12 mm, *broadly ellipsoid, red*, with subapical stigmatic remains. **Seed** 12–12.5 × 8.5–9.5 mm, ellipsoid; endosperm deeply ruminate.

DISTRIBUTION. Fly, Wisaa, Kikori, Mubi, Rentoul and Strickland Rivers in eastern half of southern New Guinea.

HABITAT. Riverbanks or alluvium on karst limestone near sea level to 450 m (Baker 1997).

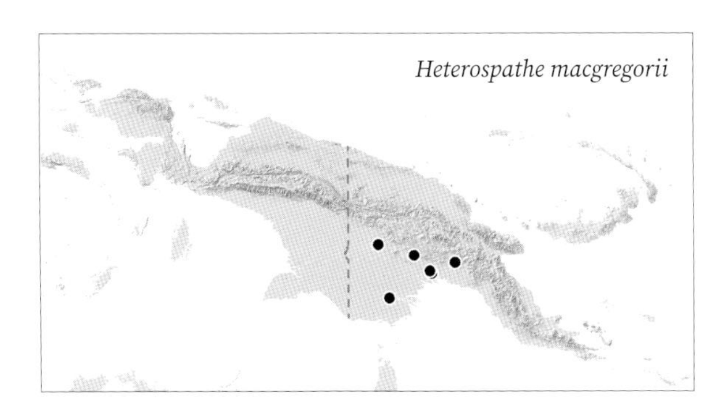

LOCAL NAMES. None recorded.

USES. None recorded.

CONSERVATION STATUS. Least Concern (IUCN 2021).

NOTES. *Heterospathe macgregorii* is an unmistakeable species. No other species of *Heterospathe* in New Guinea is rheophytic or has clustering aerial stems. The short inflorescence with relatively large, ellipsoid red fruit is also distinctive. It bears a superficial resemblance to *Hydriastele rheophytica*, but the known ranges of these species do not overlap.

8. *Heterospathe muelleriana* (Becc.) Becc.

Synonyms: *Heterospathe clemensiae* (Burret) H.E.Moore, *Heterospathe glabra* (Burret) H.E.Moore, *Ptychandra clemensiae* Burret, *Ptychandra glabra* Burret, *Ptychandra muelleriana* Becc.

Moderately slender to robust, single-stemmed palm to 15 m, bearing ca. 10 leaves in crown. **Stem** 6–16(–20) cm diam. **Leaf** (1.7–)2.3–4.1 m long including petiole; sheath 33–80 cm long, crownshaft absent; petiole (10–)25–70(–94) cm long; leaflets (40–)48–56 each side of rachis, arranged regularly, single-fold apart from the apical leaflets that may be 1–3-fold, linear, acuminate at their tips. **Inflorescence** 80–110 cm long including 40–71 cm peduncle, between the leaves, branched to 2 or 3 orders, with green axes; peduncular bract 45–100 × 3–5 cm, attached in proximal half of peduncle; first-order branches 15–21, to 35 cm long, each with as many as 14 rachillae to 25–30 cm long. **Male flower** *with 12–15 stamens, with pistillode usually conical and shorter than filaments,* green to yellow or occasionally purple. **Female flower** with glabrous and imbricate sepals, green to yellow or occasionally purple. **Fruit** ca. 11–15 × 11–15 mm, globose, red, with subapical stigmatic remains. **Seed** ca. 11 × 11 mm, globose; endosperm ruminate.

DISTRIBUTION. Widespread in the Central Range, the Cyclops Mountains, the Huon Peninsula and Karkar Island.

HABITAT. Montane rainforest on slopes and ridges or in gullies at 1,100–2,300 m.

LOCAL NAMES. *Goomnandy* (unknown dialect),

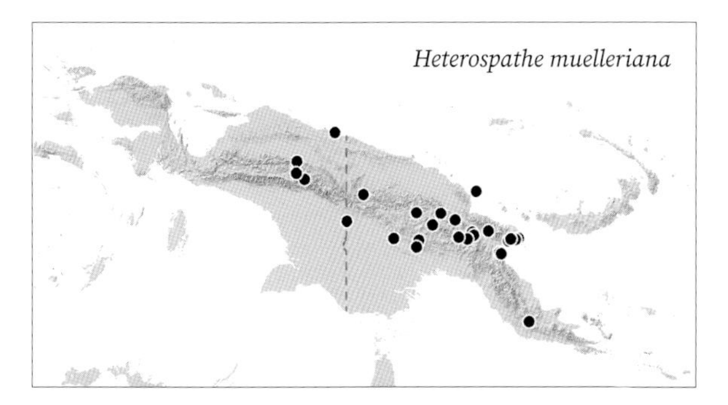

Heterospathe muelleriana. CLOCKWISE FROM TOP: habit, Southern Highlands (WB); inflorescence, Southern Highlands (WB); inflorescence, near Finschhafen (WB).

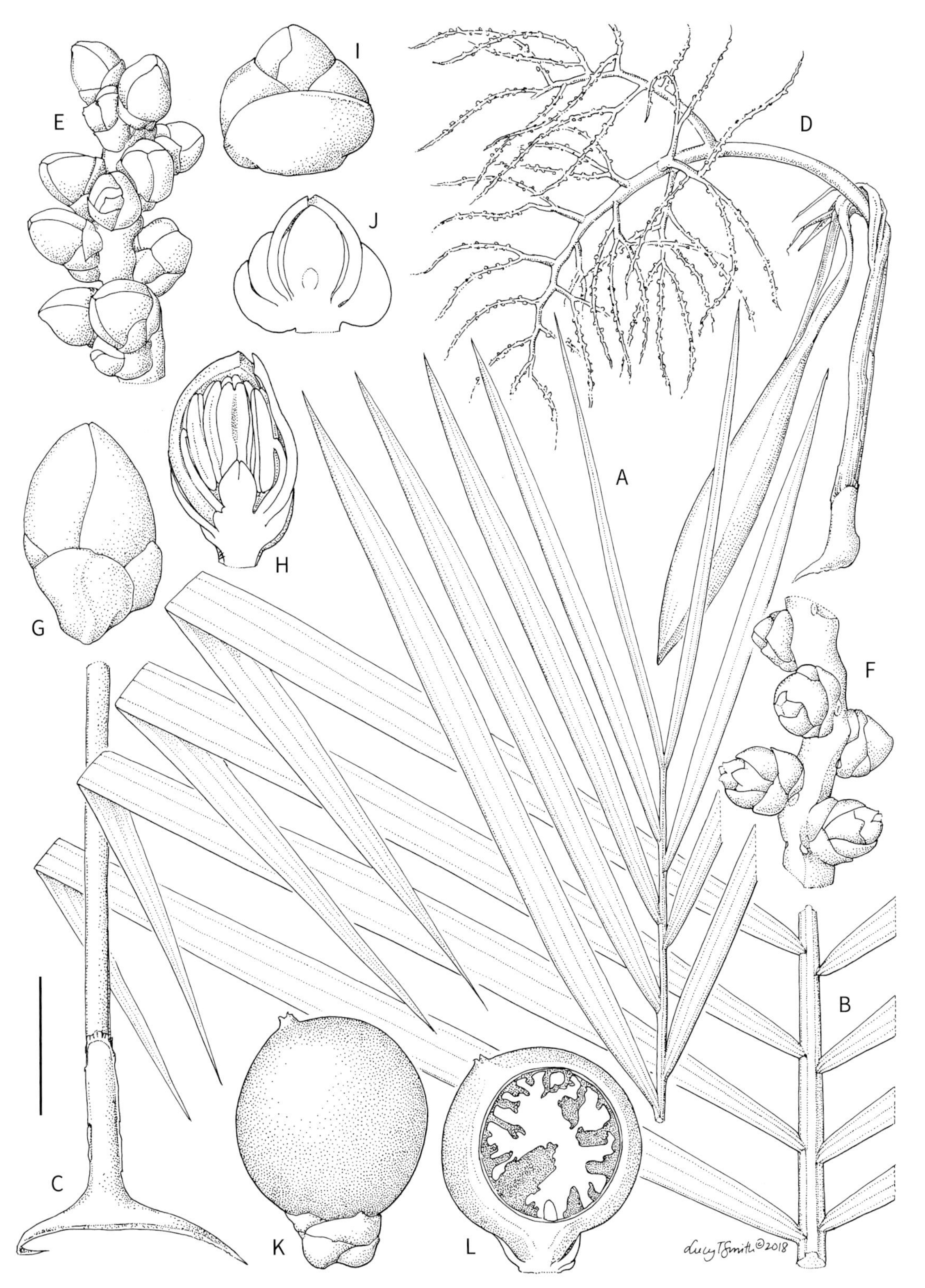

Heterospathe muelleriana. **A.** Leaf apex. **B.** Leaf mid-portion. **C.** Inflorescence peduncle. **D.** Inflorescence. **E.** Portion of rachilla with triads. **F.** Portion of rachilla with female flowers. **G**, **H.** Male flower whole and in longitudinal section. **I**, **J.** Female flower whole and in longitudinal section. **K**, **L.** Fruit whole and in longitudinal section. Scale bar: A–C = 8 cm; D = 15 cm; E, F, K, L = 1 cm; G–J = 4 mm. All from *Banka et al. 2008*. Drawn by Lucy T. Smith.

Mgamba (unknown dialect), *Yawi* (Enga), *Yali* (Wirale), *Leveng* (Hube).

USES. Stems used for bows, digging sticks, arrowheads and spears. Young fronds eaten. Armbands are made from the leaf ribs. Fruit used as a betel nut substitute.

CONSERVATION STATUS. Least Concern (IUCN 2021).

NOTES. *Heterospathe muelleriana* is a widespread and variable species with a moderately slender to robust stem. It is distinguished by its male flowers with 12 to 15 stamens, and globose fruit measuring 10 to 15 mm in diameter. The species is most similar to *H. obriensis* and *H. ledermanniana*, but the former species is often more robust, has larger male flowers with 19–20 stamens and fruit measuring 29–35 mm in length, while the latter is often more slender and has pink male flowers with only six stamens.

9. *Heterospathe obriensis* (Becc.) H.E.Moore

Synonyms: *Ptychandra montana* Burret, *Ptychandra obriensis* Becc.

Robust, single-stemmed palm to 23 m, bearing ca. 9 leaves in crown. **Stem** 10–20 cm diam. **Leaf** 1.7–3.3 m long including petiole; sheath 20–70 cm long, crownshaft absent; petiole 10–40 cm long; leaflets 40–54 each side of rachis, arranged regularly, single-fold apart from the apical leaflets that may be 1–3-fold, linear, acuminate at their tips. **Inflorescence** 70–140 cm long including 30–70 cm peduncle, between the leaves, branched to 2 or 3 orders, with green axes; peduncular bract ca. 60–80 × 4 cm, attached in proximal half of peduncle; first-order branches 9–13, to ca. 40 cm long, each with as many as 7 rachillae to 32–36 cm long. **Male flower** *with 19–20 stamens* and conical pistillode about half the length of filaments, green or cream. **Female flower** with glabrous and imbricate sepals, green or cream. **Fruit** *29–35 × 22–28 mm*, subglobose, red, with subapical stigmatic remains. **Seed** ca. 11 × 12 mm (when dry and very shrunken), subglobose; endosperm ruminate.

DISTRIBUTION. South-eastern Papua New Guinea.

HABITAT. Montane rainforest on slopes, ridges or swampy ground at 1,500–2,400 m.

LOCAL NAMES. *Koge* (Montuan).

Heterospathe obriensis. FROM TOP: habit; crown with inflorescence. Wau (WB).

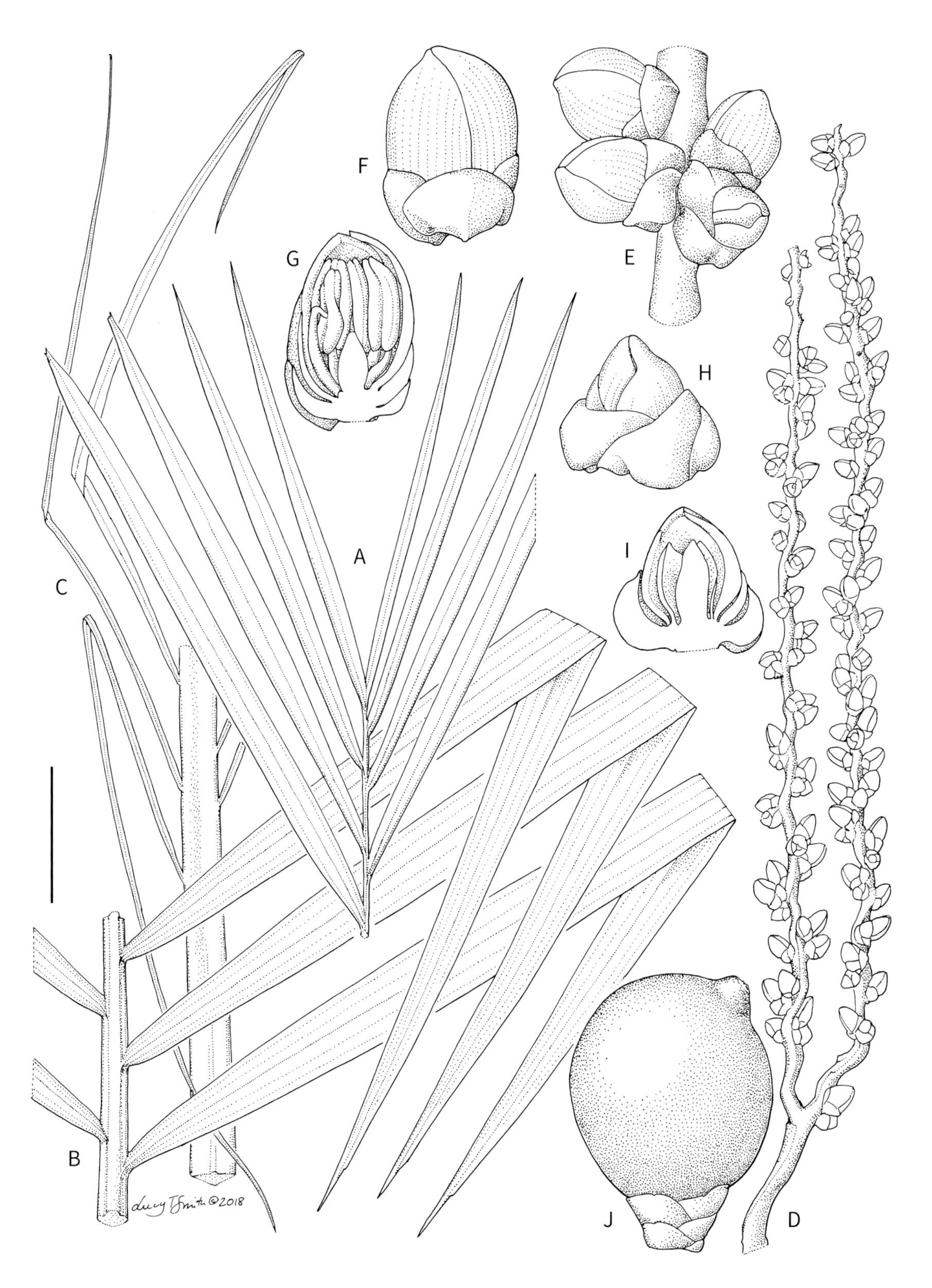

Heterospathe obriensis. **A.** Leaf apex. **B.** Leaf mid-portion. **C.** Leaf base. **D.** Portion of inflorescence. **E.** Portion of rachilla with flowers. **F**, **G.** Male flower whole and in longitudinal section. **H**, **I.** Female flower whole and in longitudinal section. **J.** Fruit. Scale bar: A–C = 8 cm; D = 4 cm; E = 1 cm; F–I = 7 mm; J = 1.5 cm. A–C from *Baker et al. 1310*; D–I from *Brass 22940*; J from *Carr 16004*. Drawn by Lucy T. Smith.

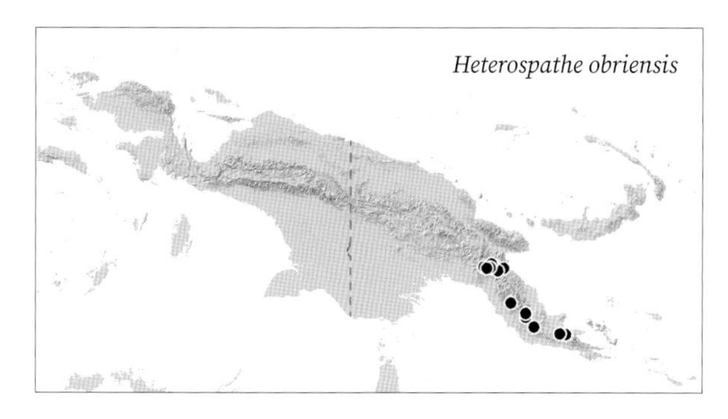

USES. None recorded.

CONSERVATION STATUS. Least Concern (IUCN 2021).

NOTES. *Heterospathe obriensis* is a robust tree palm distinguished by its male flowers with 19–20 stamens, and the large subglobose fruit measuring 29–35 mm in length. The species is most similar to *H. muelleriana* but that species is often less robust, has male flowers with 12–15 stamens, and fruit measuring only 11–15 mm in length.

10. *Heterospathe parviflora* Essig

Moderately slender to robust, single-stemmed palm to 8 m. **Stem** 7.5–10(–15) cm diam. **Leaf** (0.75–)2–3.2 m long including petiole; sheath 25–40 cm long, crownshaft absent; petiole 20–25 cm long; leaflets 26–33 each side of rachis, arranged regularly, single-fold apart from the apical leaflets that may be 2-fold, linear, acute to acuminate at their tips. **Inflorescence** ca. 150 cm long including ca. 80–90 cm peduncle, between the leaves, branched to 2 or 3 orders, with reddish grey-green

Heterospathe parviflora. CLOCKWISE FROM TOP: leaf and inflorescences; fruit; inflorescence. New Britain (FE).

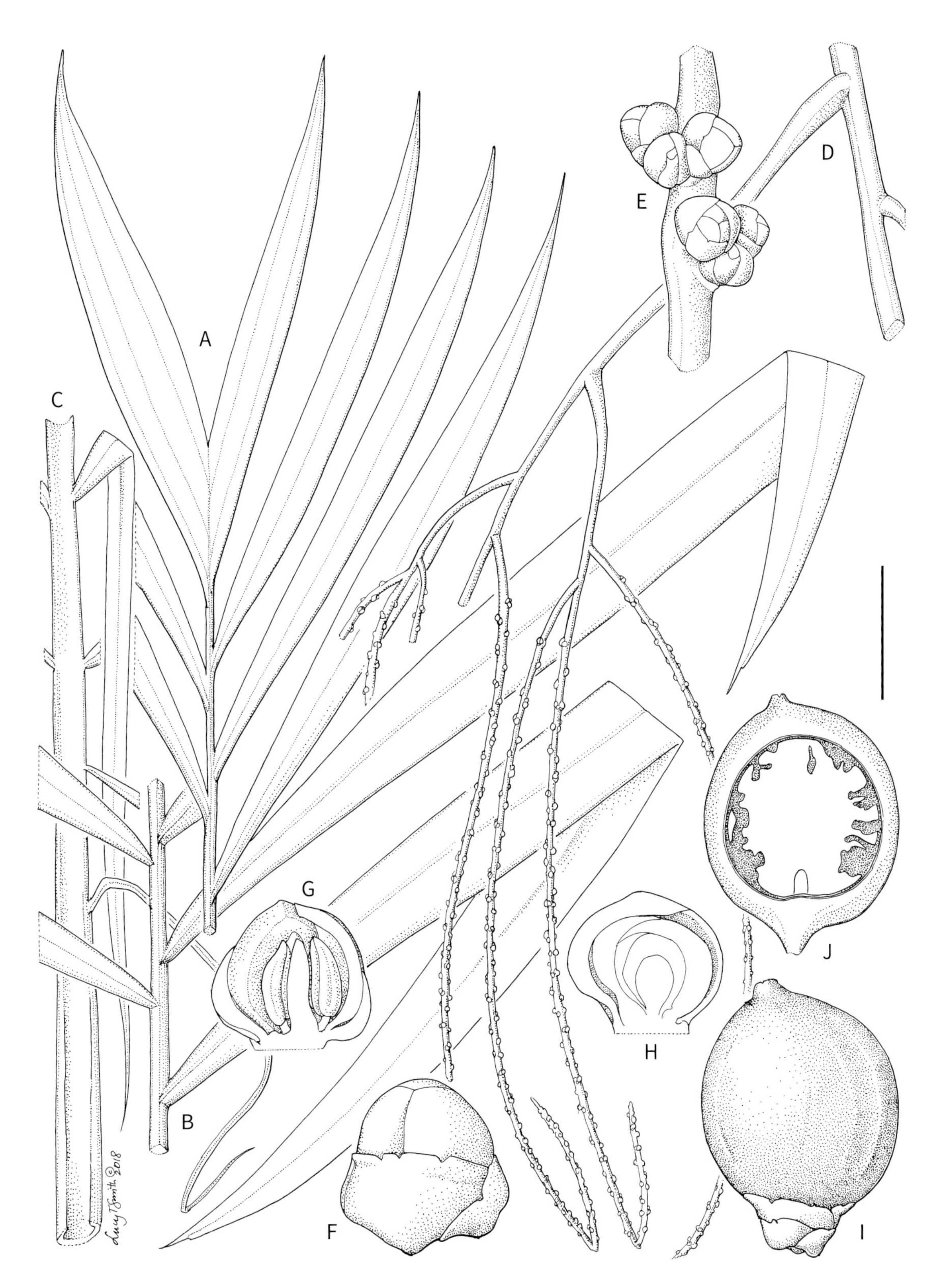

Heterospathe parviflora. **A.** Leaf apex. **B.** Leaf mid-portion. **C.** Leaf base. **D.** Portion of inflorescence. **E.** Portion of rachilla with triads. **F**, **G.** Male flower whole and in longitudinal section. **H.** Female flower in longitudinal section. **I**, **J.** Fruit whole and in longitudinal section. Scale bar: A–D = 6 cm; E = 5 mm; F–H = 2 mm; I, J = 7 mm. All from *Essig & Katik LAE 64060*. Drawn by Lucy T. Smith.

axes; peduncular bract ca. 140 cm long, attached in proximal half of peduncle; first-order branches ca. 15, *to 60–70 cm long,* each with as many as 14 rachillae to 32–37 cm long. **Male flower** with 6 stamens and *columnar pistillode at least as long as filaments, reddish-brown.* **Female flower** with glabrous and imbricate sepals, *reddish-brown.* **Fruit** ca. 12–13.5 × 10.5 mm (when dry), globose, red, with more-or-less apical stigmatic remains. **Seed** 9–10 × 8–9 mm (when dry), subglobose; endosperm deeply ruminate.

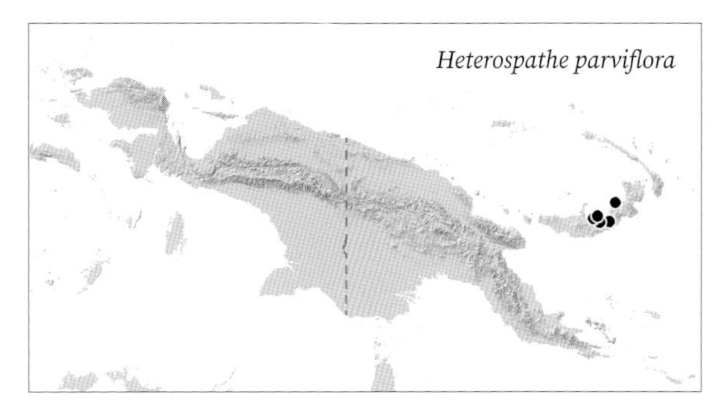

DISTRIBUTION. Central New Britain (Essig 1992).

HABITAT. Lowland or montane rainforest on volcanic soil at sea level to 1,500 m.

LOCAL NAMES. None recorded.

USES. None recorded.

CONSERVATION STATUS. Endangered (IUCN 2021). *Heterospathe parviflora* has a restricted distribution. Deforestation due to land conversion for oil palm is a major threat in its distribution range.

NOTES. *Heterospathe parviflora* is the only species of *Heterospathe* known from the Bismarck Archipelago. It is distinguished from all species in New Guinea by its large, brush-like inflorescence with long, straight branches (first-order branches reaching 70 cm in length) and its small, reddish-brown flowers and globose fruit with more-or-less apical stigmatic remains.

11. *Heterospathe porcata* W.J.Baker & Heatubun

Slender, single-stemmed understorey palm to 6 m, bearing ca. 16 leaves in crown. **Stem** ca. 3 cm diam. **Leaf** ca. 1.5 m long including petiole; entire sheath not seen, crownshaft absent; petiole ca. 50 cm long; leaflets ca. 40 each side of rachis, arranged regularly, single-fold, linear and slightly sigmoid, acuminate at their tips. **Inflorescence** 124–138 cm long including 102–110 cm peduncle, between the leaves, branched to 2 or 3 orders, colour unknown; peduncular bract ca. 68 × 0.8–1.8 cm, inserted one third to halfway along the peduncle from its base; first-order branches 9–13, to 28 cm long, each with as many as 14 rachillae to 17 cm long. **Male flower** with 6 stamens and columnar pistillode about the same length as filaments, colour unknown. **Female flower** with glabrous and imbricate sepals, colour unknown. **Fruit** ca. 20 × 9 mm (when dry), ellipsoid, *with endocarp with 6–7 distinct longitudinal ridges* running from the apex to the base (conspicuous when dry), red, with subapical stigmatic remains. **Seed** ca. 13.5 × 7 mm (when dry), elongate, conforming to contours of endocarp; endosperm ruminate.

Heterospathe porcata. Habit, Supiori Island (GP).

DISTRIBUTION. Supiori Island (Baker & Heatubun 2012).

HABITAT. Disturbed lowland rainforest on limestone near sea level.

LOCAL NAMES. None recorded.

USES. Stems and fruit are used for making bows and as a betel nut substitute, respectively.

CONSERVATION STATUS. Data Deficient (IUCN 2021). This species is known only from the type locality, where there is limited information on ongoing threats. Due to insufficient evidence, an extinction risk category cannot be selected.

***Heterospathe porcata*.** **A.** Leaf apex. **B.** Leaf mid-portion. **C.** Inflorescence. **D.** Portion of rachilla with flowers. **E.** Detail of rachilla with female flowers. **F, G.** Male flower whole and in longitudinal section. **H.** Female flower in longitudinal section. **I, J.** Fruit in two views. Scale bar: A–C = 8 cm; D, E = 7 mm; F, G = 2 mm; H = 3 mm; I, J = 1 cm. All from *Maturbongs 680*. Drawn by Lucy T. Smith.

Heterospathe porcata. LEFT TO RIGHT: crown with inflorescences; fruit. Supiori Island (GP).

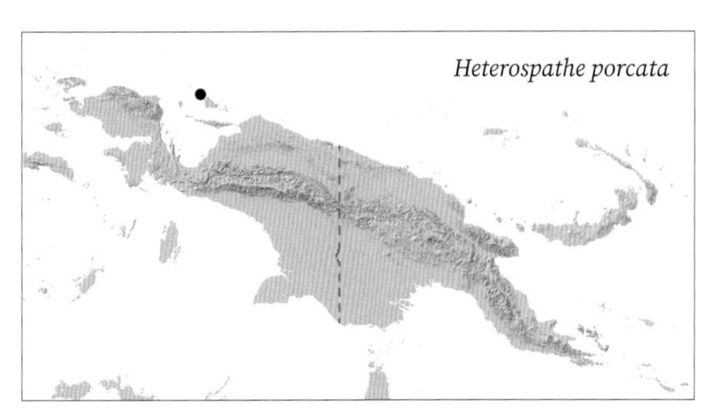

NOTES. *Heterospathe porcata* is the only member of the genus known from the Biak Islands. It is a slender palm with inflorescences branched to 2 or 3 orders and a peduncular bract inserted in the proximal half of the peduncle. This species is immediately recognisable on account of its fruit, which has an endocarp with six to seven parallel, fibrous ridges spanning its entire length. These ridges are highly distinctive and conspicuous when dry and similar features are not found in other New Guinean species of *Heterospathe*.

12. *Heterospathe pulchra* H.E.Moore

Moderately robust, single-stemmed palm to 25 m, bearing ca. 8–10 leaves in crown. **Stem** 10–15 cm diam. **Leaf** ca. 3–4 m long including petiole; sheath ca. 35 cm long, crownshaft absent; petiole ca. 55–110 cm long; leaflets (40–)53–54 each side of rachis, arranged regularly, single-fold apart from the apical leaflets that may be 1–3-fold, linear, *narrowly acuminate at their tips*. **Inflorescence** (53–)120–160 cm long including (22–)40–84 cm peduncle, between the leaves, branched to 2 or 3 orders, colour unknown; peduncular bract not seen in its entirety, attached in proximal half of peduncle; first-order branches to 37 cm long, each with as many as 22 *rachillae to 13–19 cm long*. **Male flower** *with 6 stamens and more-or-less conical pistillode shorter than filaments, cream*. **Female flower** with glabrous and imbricate sepals, *cream*. **Fruit** 8–8.5 × 8–8.5 mm (when dry), globose, red, with subapical stigmatic remains. **Seed** ca. 6 × 6 mm (when dry), globose; endosperm ruminate.

DISTRIBUTION. D'Entrecasteaux (Fergusson and Normanby Islands) and Louisiade Archipelagoes (Misima and Sudest Islands).

HABITAT. Lowland to premontane forest on slopes and ridges at 250–900 m.

LOCAL NAMES. *Bihi Bihi* (unknown dialect).

USES. Stem is used for making bows. Fruit is used as a betel nut substitute.

CONSERVATION STATUS. Least Concern (IUCN 2021). Surveys to determine population size and monitoring are required due to the restricted distribution of this species.

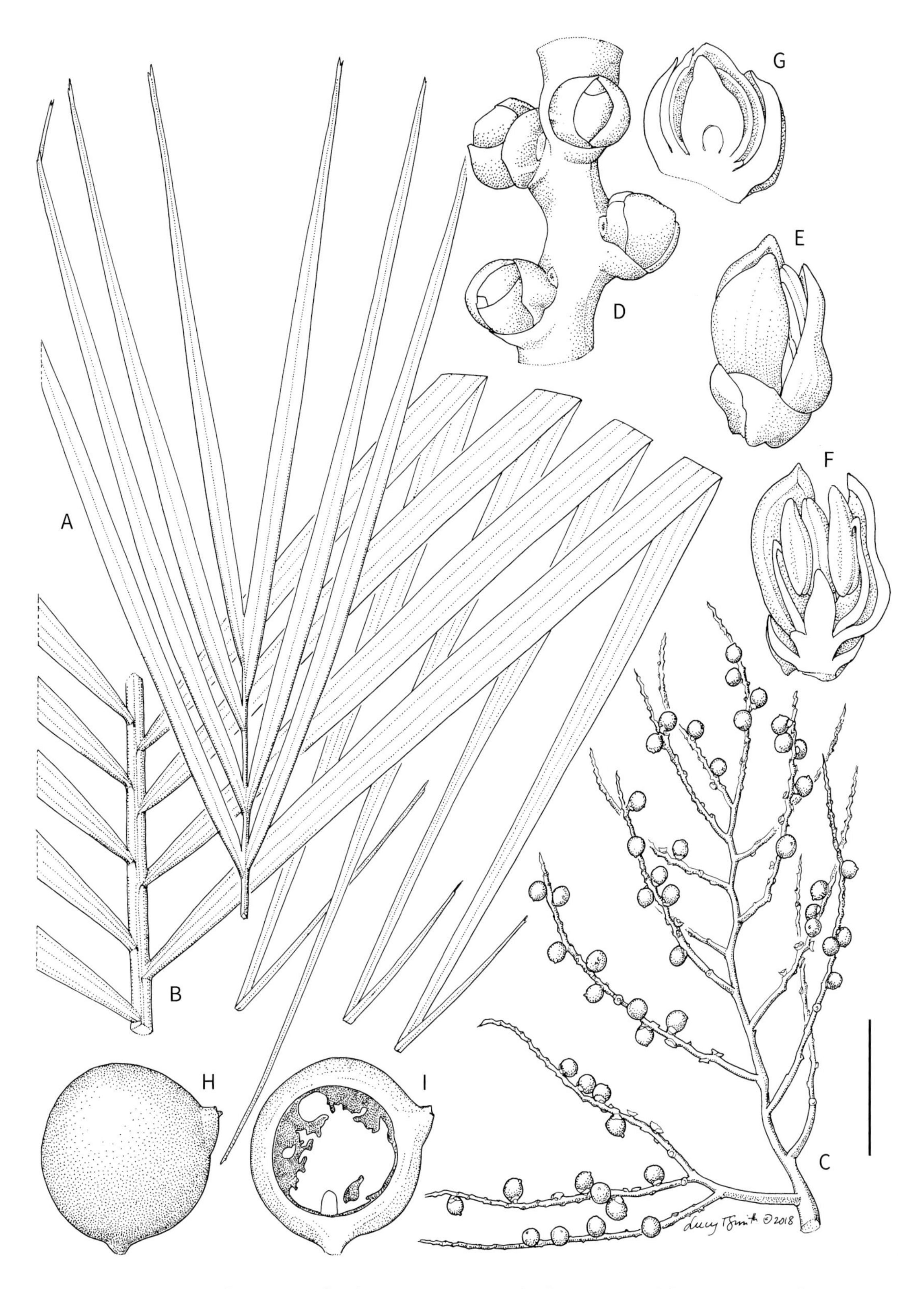

Heterospathe pulchra. **A.** Leaf apex. **B.** Leaf mid-portion. **C.** Portion of inflorescence with fruit. **D.** Portion of rachilla with female fllowers. **E**, **F.** Male flower whole and in longitudinal section. **G.** Female flower in longitudinal section. **H**, **I.** Fruit whole and in longitudinal section. Scale bar: A, B = 8 cm; C = 6 cm; D = 4 mm; E–G = 2 mm; H, I = 7 mm. All from *Brass 27116*. Drawn by Lucy T. Smith.

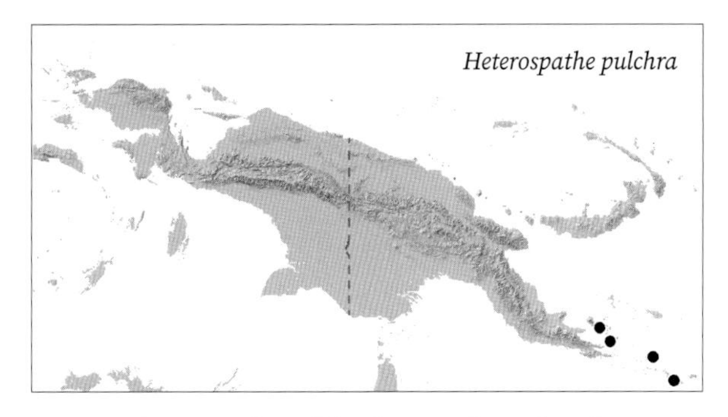

NOTES. *Heterospathe pulchra* is a moderately robust tree palm distinguished by its leaves with narrow, acuminate and prominently veined leaflets, short rachillae (as long as 19 cm), cream-coloured male flowers with six stamens, and small fruit. The species is most similar to *H. muelleriana* and *H. barfodii* but these both have less narrowly acuminate leaflets, longer rachillae and green, yellow or purple flowers. In addition, *H. muelleriana* has male flowers with at least twice as many stamens, while *H. barfodii* has a well-defined crownshaft.

13. *Heterospathe pullenii* M.S.Trudgen & W.J.Baker

Slender, single-stemmed palm to 6 m. **Stem** ca. 2–4 cm diam. **Leaf** ca. 1.2–1.5 m long including petiole; sheath 15–20 cm long, crownshaft absent; petiole 30–45 cm long; leaflets 10–15 each side of rachis, arranged regularly, *single-fold apart from the 5–8-fold basal leaflets* and 5–7-fold apical leaflets, linear, acute to acuminate at their tips. **Inflorescence** 50–120 cm long including 30–77 cm peduncle, between the leaves, branched to 1 or 2 orders, colour unknown; peduncular bract not seen in its entirety, *attached in distal third quarter of peduncle*; first-order branches 7–9, to 18–30 cm long, each with as many as 5 rachillae to 25 cm long. **Male flower** with 6 stamens and columnar pistillode about the same length as filaments, with glabrous sepals, green. **Female flower** with glabrous and imbricate sepals, green. **Fruit** ca. *18 × 10 mm* (when dry), broadly ellipsoid, red, with subapical stigmatic remains. **Seed** ca. 9 × 5.5 mm (when dry), elongate; endosperm condition unknown.

DISTRIBUTION. South-eastern New Guinea.

HABITAT. Lowland rainforest near sea level to 150 m.

LOCAL NAMES. *Saz* (Amele).

USES. None recorded.

CONSERVATION STATUS. Endangered (IUCN 2021). *Heterospathe pullenii* has a restricted distribution. Deforestation due to logging concessions is a major threat in its distribution range.

NOTES. *Heterospathe pullenii* is a slender understorey palm distinguished by its leaves with 10–15 leaflets on each side of the rachis, all single-fold apart from the multi-fold basal and apical leaflet pairs, a peduncular bract inserted in the distal third quarter of the peduncle, green flowers with glabrous sepals, and fruit measuring at least 18 mm in length. This species most closely resembles *H. lepidota* but that differs in having 21–35 leaflets per side with the basalmost pair being single-fold, and yellowish flowers with more-or-less lepidote sepals. *Heterospathe pullenii* can also be confused with *H. elegans* subsp. *elegans* but *H. elegans* has the peduncular bract inserted in the distal quarter of the peduncle, occurs at higher elevation, and has smaller fruit than *H. pullenii*.

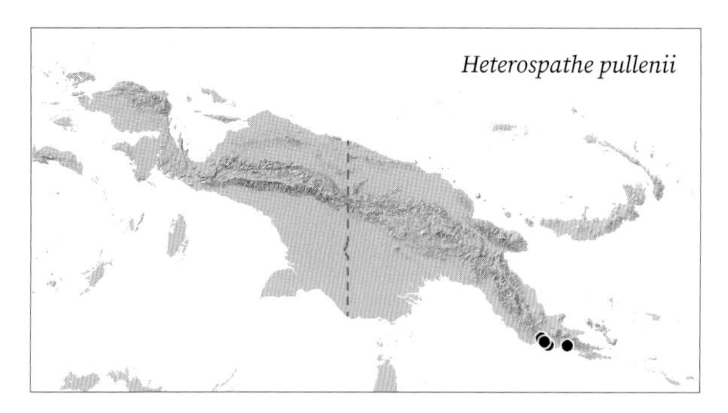

14. *Heterospathe sphaerocarpa* Burret

Synonym: *Heterospathe delicatula* H.E.Moore

Single-stemmed understorey palm bearing 6–12 leaves in crown. **Stem** *usually below ground* although rarely protruding slightly above ground. **Leaf** 1–4.5 m long including petiole; sheath distinct, if present as long as 15 cm, crownshaft absent; petiole 50–270 cm long; blade usually pinnate, rarely entire-bifid; leaflets (when present) 14–56 each side of rachis, arranged regularly, single-fold apart from the apical leaflet pair that may be 1–5-fold, linear, acute to acuminate at their tips. **Inflorescence** 56–120 cm long including 45–75 cm peduncle, between the leaves, spicate or branched to 1 or 2 orders, with purple axes (at least in bud); peduncular bract 26–60 × 1.5–2.5 cm, *attached in proximal half of peduncle*; first-order branches 0–10, to 33 cm long, each with as many as 6 rachillae to ca. 25 cm long. **Male flower** *with (6–)7–10 stamens*

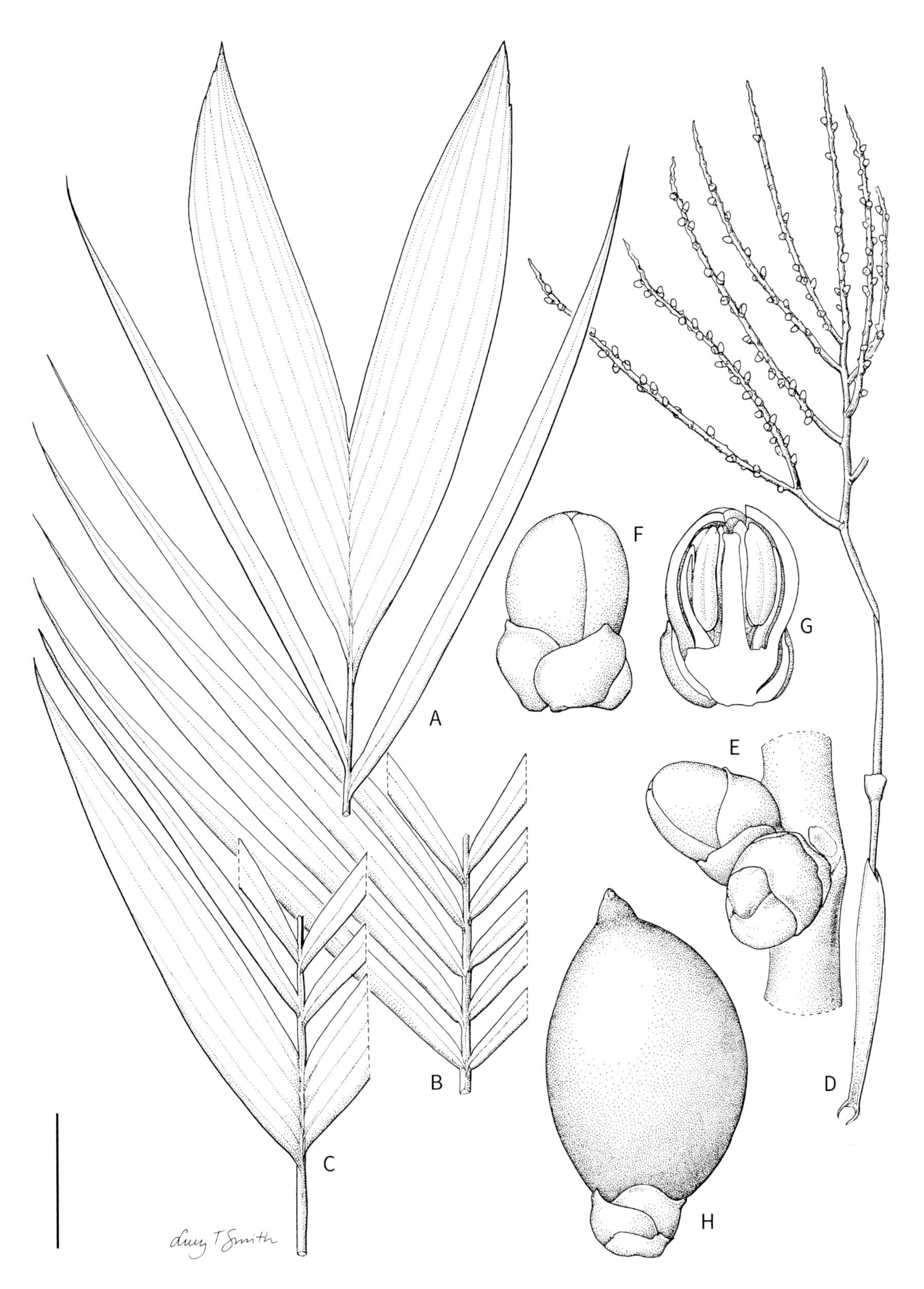

Heterospathe pullenii. **A.** Leaf apex. **B.** Leaf mid-portion. **C.** Leaf base. **D.** Inflorescence. **E.** Portion of rachilla with flowers. **F, G.** Male flower whole and in longitudinal section. **H.** Fruit. Scale bar: A–C = 8 cm; D = 6 cm; E = 2.5 mm; F, G = 2 mm; H = 7 mm. A, D from *Pullen 7640*; B, C, H from *Pullen 8116*; E, F, G from *Wiakabu LAE 70460*. Drawn by Lucy T. Smith.

Heterospathe sphaerocarpa. LEFT TO RIGHT: habit, cultivated Lyon Arboretum, Hawaii (WB); inflorescence, cultivated Lyon Arboretum, Hawaii (WB); flower buds with open male flowers, cultivated Brisbane Botanic Gardens Mount Coot-tha, Queensland (WB).

and conical pistillode about half the length of filaments, purple. **Female flower** with glabrous and imbricate sepals, *purple.* **Fruit** 9–10 × 9–10 mm, globose, red, with subapical stigmatic remains. **Seed** ca. 6.5 × 7.5 mm (when dry), subglobose; endosperm ruminate.

DISTRIBUTION. South-eastern New Guinea and the Huon Peninsula.

HABITAT. Lowland to premontane rainforest on slopes and ridges at 400–1,250 m.

LOCAL NAMES. None recorded.

USES. None recorded.

CONSERVATION STATUS. Least Concern (IUCN 2021).

NOTES. *Heterospathe sphaerocarpa* is distinguished by its single subterranean stem in combination with purple inflorescences and male flowers with 7–10 stamens (rarely 6 stamens). The lack of an aerial stem prompts comparison with *H. elegans* subsp. *humilis* but that is usually multi-stemmed and has elongate and slender, green inflorescences and male flowers with six stamens. The peduncular bract is also inserted in the distal quarter of the peduncle in *H. elegans,* whereas it is inserted in the proximal half of the peduncle in *H. sphaerocarpa.*

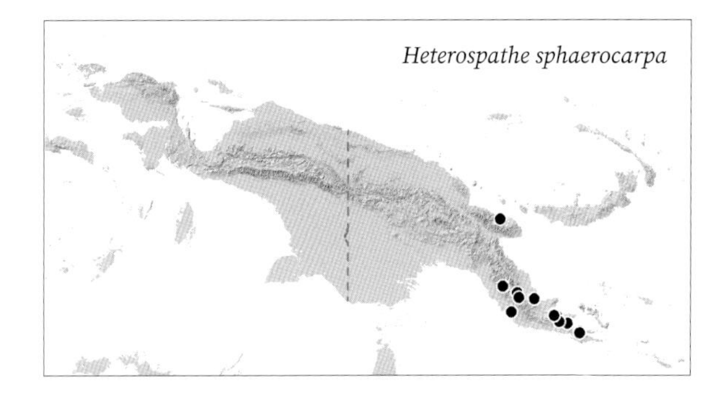

NAMES OF UNCERTAIN APPLICATION

Heterospathe arfakiana (Becc.) H.E.Moore

Synonym: *Ptychosperma arfakianum* Becc.

Beccari's original material of this species appears to be consistent with *H. elegans,* but the evidence is insufficient to justify placing *H. arfakiana* in synonymy. This is the only *Heterospathe* species known from the Bird's Head Peninsula in Indonesian New Guinea. An invalidly published name *Rhopaloblaste arfakiana* (Becc.) Becc. is associated with this species. For further discussion, see Petoe & Baker (2019).

Heterospathe compsoclada (Burret) Heatubun

Synonym: *Cyrtostachys compsoclada* Burret

The material of *H. compsoclada* (from Central Province, Papua New Guinea) is too incomplete to be interpreted accurately (Heatubun *et al.* 2009; Petoe & Baker 2019).

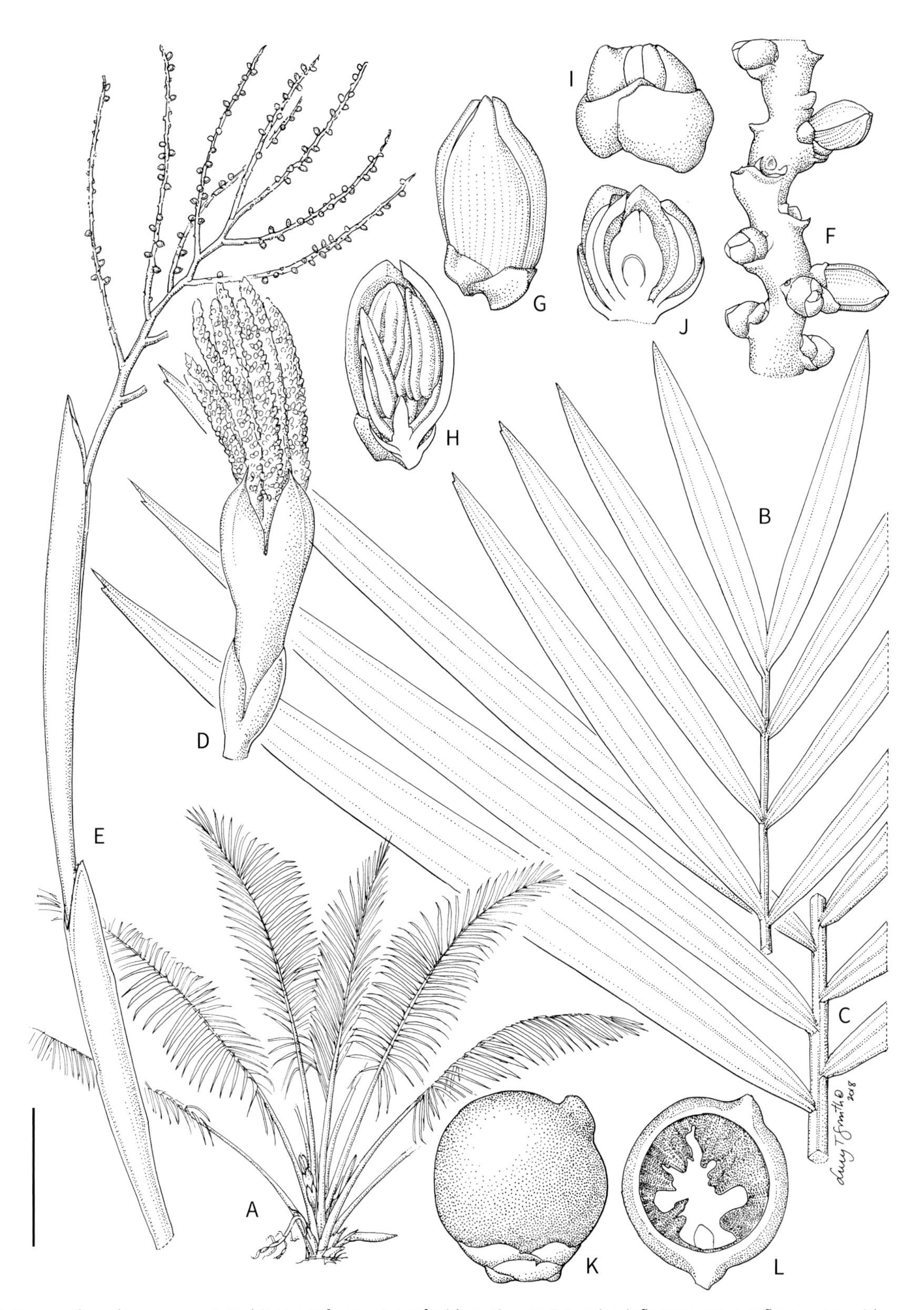

Heterospathe sphaerocarpa. **A.** Habit. **B.** Leaf apex. **C.** Leaf mid-portion. **D.** Emerging inflorescence. **E.** Inflorescence with fruit. **F.** Portion of rachilla with flowers. **G**, **H.** Male flower whole and in longitudinal section. **I**, **J.** Female flower whole and in longitudinal section. **K**, **L.** Fruit whole and in longitudinal section. Scale bar: A = 1 m; B, C = 8 cm; D = 15 cm; E = 12 cm; F, K, L = 7 mm; G–J = 3 mm. A from cultivated plant at Lyon Arboretum, Oahu, Hawaii; B, C, E from *Pullen 7784*; D from cultivated plant at Brisbane Botanic Gardens Mount Coot–tha; F–J from *Pullen 3470*; K, L from *Brass 5413*. Drawn by Lucy T. Smith.

Hydriastele biakensis. Habit, Biak Island (WB).

Hydriastele H.Wendl. & Drude

Synonyms: *Adelonenga* Hook.f, *Gronophyllum* Scheff., *Gulubia* Becc., *Gulubiopsis* Becc., *Kentia* Blume, *Leptophoenix* Becc., *Nengella* Becc., *Paragulubia* Burret, *Siphokentia* Burret

Very small to tall – leaf pinnate – crownshaft – no spines – leaflets usually jagged

Very slender to robust, single- or multi-stemmed tree palms, stem swollen in some species, crownshaft present, monoecious. **Leaf** pinnate, strongly arching in many species; sheath tubular, often with conspicuous indumentum of various kinds; petiole present (rarely absent); leaf blade occasionally not divided into leaflets (also in some juveniles); leaflets (where blade divided) few to numerous, linear to wedge-shaped, with jagged tips (rarely tips pointed), arranged regularly or irregularly, ascending, horizontal or pendulous. **Inflorescence** *below the leaves, unbranched or branched 1–4 orders, branches swept forward and resembling a horse's tail or brush*; prophyll and peduncular bract similar, enclosing inflorescence in bud, dropping off as inflorescence expands; peduncle shorter than inflorescence rachis; rachillae slender and straight (rarely somewhat sinuous) or arching. **Flowers** in triads throughout the length of the rachilla, not developing in pits, *male flowers larger than female flowers, asymmetrical and more conspicuous, with pink or white pointed petals.* **Fruit** red, brown or black, narrowly to broadly ellipsoid (rarely globose, rarely curved), stigmatic remains apical, flesh thin, endocarp thin, closely adhering to seed. **Seed** 1, ellipsoid or globose, endosperm homogeneous or ruminate.

DISTRIBUTION. Thirty-nine species distributed from Sulawesi and Palau to northern Australia and Fiji, 24 species in New Guinea.

NOTES. *Hydriastele* species range from slender understorey palms to canopy emergents. Almost all have jagged leaflet apices and bear inflorescences below the leaves, these resembling a horse's tail, unless unbranched (spicate). The fruit are small, red, brown or black (0.5–2 cm long), sometimes curved and bearing apical stigmatic remains. *Hydriastele* occurs from the lowlands to submontane rainforest, sea level to 2,200 m, and includes one of the commonest tree palms on the island, *H. costata*. It is perhaps most readily confused with *Clinostigma*, though this genus does not occur in mainland New Guinea, bears pointed not jagged leaflet tips and has asymmetric fruit with the stigma to one side of the apex. It might also be confused with *Cyrtostachys* or *Rhopaloblaste*, though these are immediately distinguished by their widely spreading inflorescences and pointed leaflet tips.

More detailed information about the species included here can be found in a monograph of the genus (Petoe *et al.* 2018a), as well as in earlier taxonomic accounts of *Gulubia* (Essig 1982), *Siphokentia* (Baker *et al.* 2000b), *Gronophyllum* and *Nengella* (Essig & Young 1985), which are now treated as synonyms of *Hydriastele* (Baker & Loo 2004), and of the Wendlandiana group (Petoe *et al.* 2018b) and the Nengella group (Heatubun *et al.* 2018). The last two group names are part of a recently established informal classification within the genus (Petoe *et al.* 2018a).

Key to the species of *Hydriastele* in New Guinea

1. Single- or multi-stemmed, slender to moderate, understorey to midstorey palms; stem diam. <10 cm (rarely up to 15 cm); leaf <2.5 m long, leaflets <40 each side of rachis 2
1. Single-stemmed, moderate to robust, subcanopy to emergent palms; stem diam. >10 cm; leaf >1.5 m long (usually >2.5 m), leaflets usually >40 each side of rachis 15

2. Inflorescence protandrous, spicate or branched to 1 or 2 orders; female flowers with petals pointed at their tips 3
2. Inflorescence protogynous, branched to 1 or 2 orders; female flowers with petals rounded at their tips 11

3. Palm up to 25 m; leaflets 27–40 each side of rachis (rarely as few as 21 per side); inflorescence branched to 2 orders .... **14. *H. lurida***
3. Palm up to 12 m; leaf variously divided into 2–23 leaflets each side of rachis or blade entire; inflorescence spicate or branched to 1 order 4

4. Moderate palm; leaf typically with 3 broad, multi-fold leaflets each side of rachis interspersed with a few single-fold leaflets; female flower with sepals fused in a cup and petals fused basally .... **7. *H. dransfieldii***
4. Slender palm; leaf not consisting of broad, multi-fold leaflets interspersed with single-fold leaflets; female flower with sepals and petals only fused briefly at the very base 5

5. Inflorescence axes often a vivid pink or reddish; endosperm ruminate 6
5. Inflorescence axes normally yellow to green; endosperm homogeneous 9

6. Leaf entire .... **21. *H. splendida***
6. Leaf dissected 7

7. Multi-stemmed palm restricted to riverbanks; stem flexible and often leaning; leaflets 14–16 each side of rachis, regularly arranged, linear .... **20. *H. simbiakii***
7. Single- or multi-stemmed palm not restricted to riverbanks; stem rigid and erect; leaflets 4–13 each side of rachis, normally irregularly arranged, linear to wedge-shaped 8

8. Multi-stemmed palm; stem diam. 0.9–1.7 cm; basal and middle leaflets narrowly linear and widely spreading; inflorescence 8–15 cm long, spicate or bifid .... **6. *H. divaricata***
8. Single- or multi-stemmed palm; stem diam. usually 1.5–7.5 cm; leaflets variable, normally broadly wedge-shaped; inflorescence usually 18–30 cm long, with 2–6 rachillae .... **17. *H. pinangoides***

9. Leaflets 15–23 each side of rachis, concave and jagged at their tips; inflorescence with 4–6 rachillae; triads decussately arranged .... **2. *H. aprica***
9. Leaflets 2–11 each side of rachis, not concavely jagged at their tips; inflorescence spicate or with 2 or 3 rachillae; triads spirally arranged 10

10. Stem 5–8 mm diam.; leaf with 9–11 narrowly linear leaflets each side of rachis .... **16. *H. montana***
10. Stem 1–2 cm diam.; leaf entire, with two pairs of leaflets, or with 3–6 wedge-shaped leaflets each side of rachis .... **8. *H. flabellata***

11. Multi-stemmed palm restricted to riverbanks; stem flexible and usually leaning; leaflets regularly arranged, linear, thin and soft; apical leaflets normally with 2 or 3 folds .... **19. *H. rheophytica***
11. Single- or multi-stemmed palm not restricted to riverbanks; stem rigid and erect; leaflets regularly or irregularly arranged, linear or narrowly wedge-shaped, relatively stiff and papery; apical leaflets usually with 3–16 folds 12

12. Juvenile leaf entire; petiole lacking in adult leaves .... **1. *H. apetiolata***
12. Juvenile leaf dissected; petiole well-defined 13

13. Multi-stemmed palm; leaflets regularly to subregularly arranged; basal leaflets pointed or obliquely jagged at their tips .... **22. *H. variabilis***
13. Single- or multi-stemmed palm; leaflets usually irregularly arranged with one or two multi-fold pairs in the middle of the leaf; basal leaflets truncately jagged at their tips 14

14. Slender, multi-stemmed palm; leaf sheath 15–30 cm long; leaflets 6–13 each side of rachis; endosperm ruminate .... **10. *H. kasesa***
14. Slender to moderate, single- or multi-stemmed palm; leaf sheath 40–73 cm long; leaflets 12–30 each side of rachis; endosperm homogeneous or ruminate .... **23. *H. wendlandiana***

15. Leaves straight or a little drooping; leaflets pendulous; fruit with light, longitudinal stripes .... **5. *H. costata***
15. Leaves straight or arching; leaflets ascending or pendulous; fruit without stripes 16

16. Inflorescence branched to 4 orders; male flower with 6 stamens that are congenitally exposed (i.e. not covered completely by the petals even in bud) .... **3. *H. biakensis***
16. Inflorescence branched to 2 or 3 orders; male flower with 6–24 stamens that are completely covered by the petals in bud 17

17. Leaves straight or a little drooping; leaflets pendulous; apical leaflets single-fold, pointed or shallowly notched at the tip 18
17. Leaves arching; leaflets ascending (at least at their bases); apical leaflets multi-fold, truncate and jagged at the tip 19

18. Rachillae ca. 2–3 mm diam.; male flower with 6 stamens; fruit 10–15 × 6–7 mm when dry .... **18. *H. procera***
18. Rachillae ca. 3.5 mm diam.; male flower with 12 stamens; fruit 7.5–9.5 × 5–5.5 mm when dry .... **24. *H. wosimiensis***

19. Stem <15 cm diam., not swollen; petiole channelled on upper surface 20
19. Stem >15 cm diam., may be strongly swollen; petiole flattened on upper surface 22

20. Leaf ca. 1.5–2 m long including ca. 10 cm petiole; peduncle elongated, at least 20 cm long .... **15. *H. manusii***
20. Leaf 2.4–2.7 m long including 32–60 cm petiole; peduncle as long as 12 cm 21

21. Leaf sheath with very thick and fluffy woolly indumentum; peduncle 11–12 cm long . **11. *H. lanata***
21. Leaf sheath with thin and smooth woolly indumentum; peduncle 3–5 cm long ..... **4. *H. calcicola***

22. Stem strongly swollen; inflorescence 55–60 cm long including 4–5 cm peduncle . . . **9. *H. gibbsiana***

22. Stem even or rarely slightly swollen; inflorescence 60–120 cm long (rarely as short as 50 cm) including 6–15 cm peduncle 23

23. Leaflets 57–70 each side of rachis (rarely as few as 48 per side); inflorescence apparently protogynous; female flowers with petals rounded at their tips; fruit subglobose to broadly ellipsoid . . . . **13. *H. longispatha***

23. Leaflets 38–51 each side of rachis; inflorescence protandrous; female flowers with petals pointed at their tips; fruit broadly ellipsoid . . . . **12. *H. ledermanniana***

1. *Hydriastele apetiolata* Petoe & W.J.Baker

Slender, multi-stemmed palm to 6 m, bearing 7–10 leaves per crown. **Stem** 4.5–6 cm diam. **Leaf** to 125 cm long; sheath ca. 60 cm long; *petiole lacking*; leaflets 23–26 each side of rachis, regularly arranged; basal leaflets single-fold, linear and grouped, obliquely or truncately jagged at their tips; apical leaflets comprising 4–6 folds, wedge-shaped, truncately jagged at the tip; *juvenile leaves entire-bifid*, with or without petiole, with blade 100–130 cm long. **Inflorescence** 25–35 cm long including 3.5–5.5 cm peduncle, branched to 2 orders, *protogynous*; triads decussately arranged. **Male flower** with 6–8 stamens. **Female flower** with free sepals and free, rounded, low petals. **Fruit** 7.4–9.4 × 6.6–8.2 mm when ripe, subglobose, red. **Seed** 5.1–7.2 × 5–6.8 mm, globose to subglobose; endosperm ruminate.

DISTRIBUTION. Known from two widely separated areas of southern New Guinea.

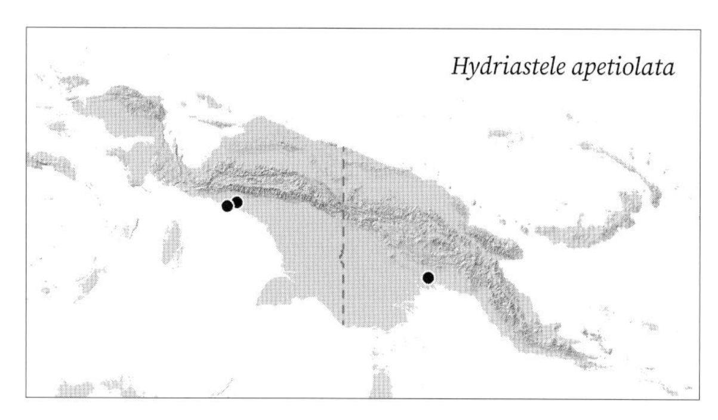

HABITAT. Lowland rainforest, near sea level.

LOCAL NAMES. None recorded.

USES. None recorded.

CONSERVATION STATUS. Data Deficient (IUCN 2019). There is limited information on ongoing threats in the area of distribution. Due to insufficient evidence, an extinction risk category cannot be selected.

Hydriastele apetiolata. Habit, near Timika (WB).

***Hydriastele apetiolata*. A.** Leaf apex. **B.** Leaf mid-portion. **C.** Leaf base with sheath. **D.** Juvenile leaf. **E.** Inflorescence with fruit. **F.** Portion of rachilla with triads. **G.** Portion of rachilla with fruit. **H.** Male flower in longitudinal section. **I, J.** Female flower whole and in longitudinal section. **K.** Fruit in longitudinal section. Scale bar: A–C, E = 6 cm; D = 12 cm; F = 7 mm; G = 1.5 cm; H = 5 mm; I, J = 3.3 cm; K = 1 cm. All from *Baker et al. 884*. Drawn by Lucy T. Smith.

Hydriastele apetiolata. LEFT TO RIGHT: crowns; entire juvenile leaves; inflorescence with fruit. Near Timika (WB).

NOTES. *Hydriastele apetiolata* is a midstorey palm with a slender stem, distinguished by its regularly pinnate, adult leaves lacking petiole and entire-bifid juvenile leaves (petiolate or apetiolate in the juvenile). The combination of entire juvenile leaves and pinnate adult leaves is not known in any other understorey or midstorey species of *Hydriastele*.

HABITAT. Montane forest on sun-exposed limestone ridges at 300–1,200 m (Young 1985).

LOCAL NAMES. None recorded.

2. *Hydriastele aprica* (B.E.Young) W.J.Baker & Loo

Synonym: *Gronophyllum apricum* B.E.Young

Slender, single-stemmed palm to 5 m, bearing ca. 7 leaves in crown. **Stem** ca. 3 cm diam. **Leaf** 48–80 cm long including petiole; sheath 21–23 cm long; petiole 12–18 cm long; leaflets 15–23 each side of rachis, irregularly arranged, drooping, wedge-shaped, *concavely jagged at their tips*. **Inflorescence** 15–26 cm long including 3.2–5.5 cm peduncle, branched to 1 order, *protandrous; rachillae 4–6; triads decussately arranged*. **Male flower** with 6 stamens, cream with purplish tips. **Female flower** with free sepals and petals with long and thin, valvate tips, dark purple. **Fruit** ca. 8 × 7 mm, subglobose, red. **Seed** ca. 5 × 4 mm; *endosperm homogeneous*.

DISTRIBUTION. Northern central New Guinea, in the vicinity of Telefomin.

Hydriastele aprica. Habit, Frieda River (FE).

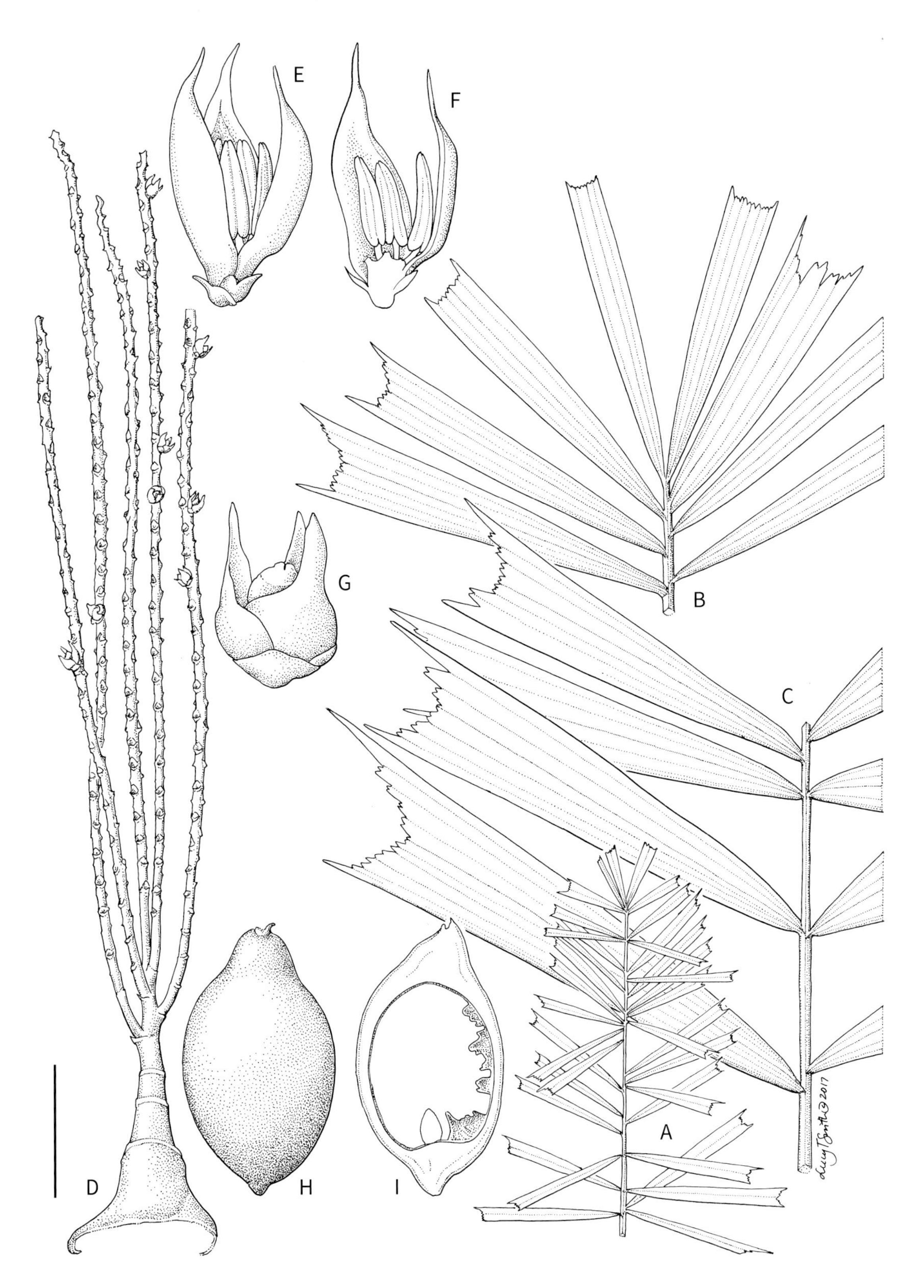

Hydriastele aprica. **A.** Whole leaf diagram. **B.** Leaf apex. **C.** Leaf mid-portion. **D.** Inflorescence. **E**, **F.** Male flower whole and in longitudinal section. **G.** Female flower. **H**, **I.** Fruit whole and in longitudinal section. Scale bar: A = 18 cm; B, C = 4 cm; D = 3 cm; E–I = 4 mm. A, G–I from *Essig & Young LAE 74049*; B–F from *Essig & Young LAE 74072*. Drawn by Lucy T. Smith.

Hydriastele aprica. LEFT TO RIGHT: leaf with inflorescence bud; inflorescence; inflorescences with fruit. Frieda River (FE).

USES. None recorded.

CONSERVATION STATUS. Data Deficient (IUCN 2019). This species is known from a restricted area, where there is limited information on ongoing threats. Due to insufficient evidence, an extinction risk category cannot be selected.

NOTES. *Hydriastele aprica* is a slender palm distinguished by its drooping leaflets with concavely jagged tips and the protandrous inflorescence in combination with decussately arranged triads. These features are not displayed by other members of the genus.

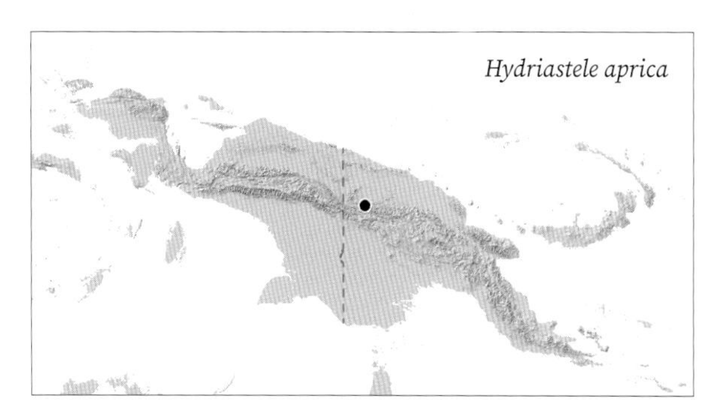

3. *Hydriastele biakensis* W.J.Baker & Heatubun

Robust, single-stemmed palm to ca. 15 m, bearing 18–24 leaves in crown. **Stem** ca. 30 cm diam. **Leaf** ca. 3.3–3.6 m long including petiole, arching; sheath ca. 170 cm long; petiole ca. 47–50 cm long; leaflets ca. 65 each side of rachis, regularly arranged, single-fold, ascending and sometimes drooping at their tips, linear, pointed or briefly notched to obliquely jagged at their tips. **Inflorescence** 95–100 cm long including 19–21 cm peduncle, *branched to 4 orders, apparently protogynous, with prominent "shoulders" formed by the abrupt constriction of the peduncle right above the prophyll scar*; prophyll often s-shaped, with pithy keels; rachillae sinuous, especially towards the tip; triads decussately arranged. **Male flower** with 6 *stamens exposed in bud*, white. **Female flower** with free sepals and free, low and rounded petals. **Fruit** 9.5–12 × 5–6 mm when ripe, *narrowly ellipsoid*, red. **Seed** 7.5–8.2 × 4–4.3 mm, cylindrical; endosperm homogeneous.

DISTRIBUTION. Biak and Numfor Islands.

HABITAT. Coastal forest near beach or on cliff edges

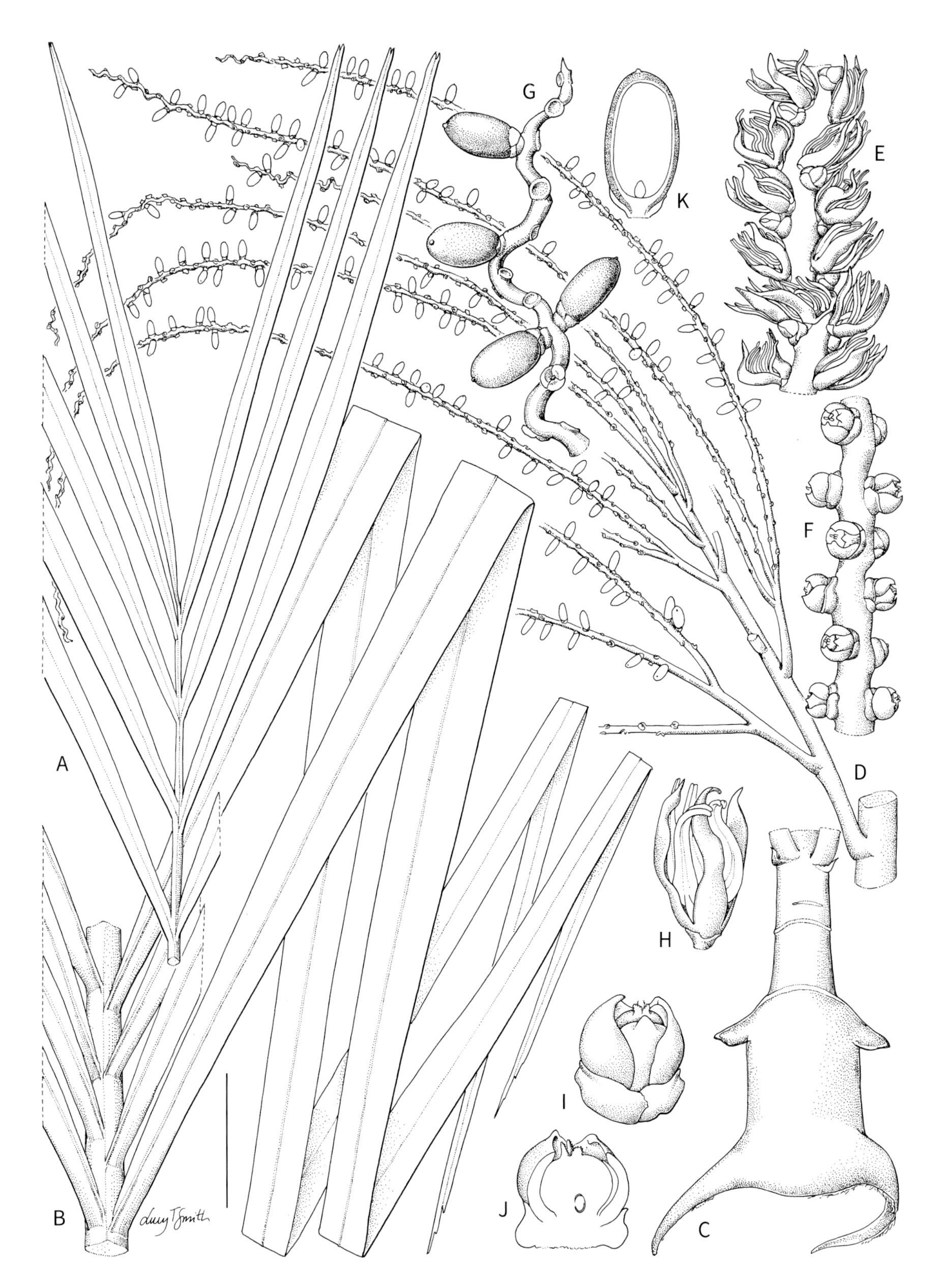

Hydriastele biakensis. **A.** Leaf apex. **B.** Leaf mid-portion. **C.** Peduncle. **D.** Portion of inflorescence with fruit. **E.** Portion of rachilla with triads. **F.** Portion of rachilla with female flowers. **G.** Portion of rachilla with fruit. **H.** Male flower. **I**, **J.** Female flower whole and in longitudinal section. **K.** Fruit in longitudinal section. Scale bar: A, B, D = 6 cm; C = 8 cm; E, G = 1.5 cm; F, K = 1 cm; H = 5 mm; I, J = 3 mm. A–D from *Baker et al. 1342*; E–K from *Heatubun et al. 970*. Drawn by Lucy T. Smith.

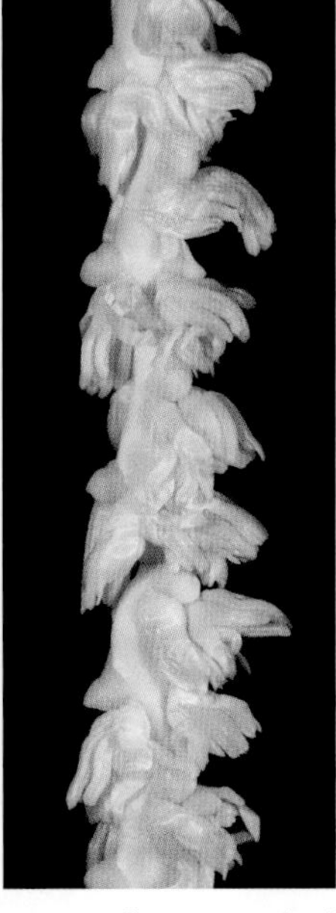

Hydriastele biakensis. LEFT TO RIGHT: inflorescence bud, showing typical sinuous shape; male flowers, congenitally open; inflorescence with fruit; fruit. Biak Island (WB).

near sea level, on rugged limestone with very thin or no topsoil (Baker & Heatubun 2012).

LOCAL NAMES. *Arwaf* (Biak).

USES. Stems and leaf sheaths used for flooring and baskets, respectively.

CONSERVATION STATUS. Endangered (IUCN 2019). *Hydriastele biakensis* has a restricted distribution and deforestation due to agriculture, logging and wood harvesting is a major threat in its distribution range.

NOTES. *Hydriastele biakensis* is endemic to the Biak Islands and is the only canopy species of the genus recorded from this area other than *H. costata*. Unlike *H. costata*, *H. biakensis* has arching leaves, and the species differs from all other New Guinean members of *Hydriastele* in the inflorescence with four orders of branching and its male flowers, which are congenitally open in bud. Congenitally open male flowers are also present in *H. palauensis* from Palau, which is closely related to *H. biakensis* (Baker & Heatubun 2012).

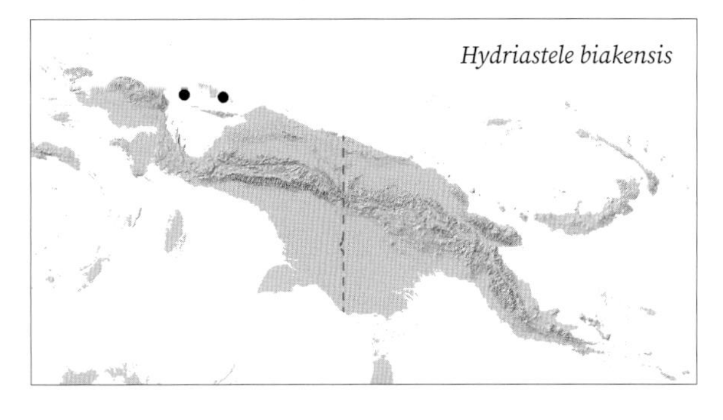

4. *Hydriastele calcicola* W.J.Baker & Petoe

Moderately robust, single-stemmed and *gregarious palm to 30 m*, bearing 15–20 leaves in crown. **Stem** *ca. 15 cm diam., inner wood very soft and pithy*. **Leaf** ca. 2.7 m long including petiole, arching; sheath ca. 100 cm long, indumentum not seen, crownshaft ca. 130 × 14 cm; petiole ca. 60 cm long, upper surface deeply channelled; leaflets ca. 46 each side of rachis, regularly arranged, ascending, linear; apical leaflets comprising ca. 2 folds, truncately jagged at the tip. **Inflorescence** ca. 70 cm long *including 3–5 cm peduncle*, branched to 2 orders, *protandrous*; rachillae ca. 26; triads decussately arranged. **Male flower** with ca. 13 stamens, cream. **Female flower** with free sepals and free petals with conspicuous, triangular tips, cream. **Fruit** ca. 15 × 9 mm when ripe, obovoid to broadly ellipsoid, bright red. **Seed** ca. 10–11 × 7.5 mm, obovoid to ellipsoid; endosperm homogeneous.

DISTRIBUTION. The Kikori and Mubi River catchments in southern New Guinea.

HABITAT. Lowland rainforest river margins on karst limestone, 50–500 m.

LOCAL NAMES. None recorded.

USES. None recorded.

CONSERVATION STATUS. Data Deficient (IUCN 2019). Due to limited information on ongoing threats, a single extinction risk category cannot be selected.

NOTES. *Hydriastele calcicola* occurs in large numbers in

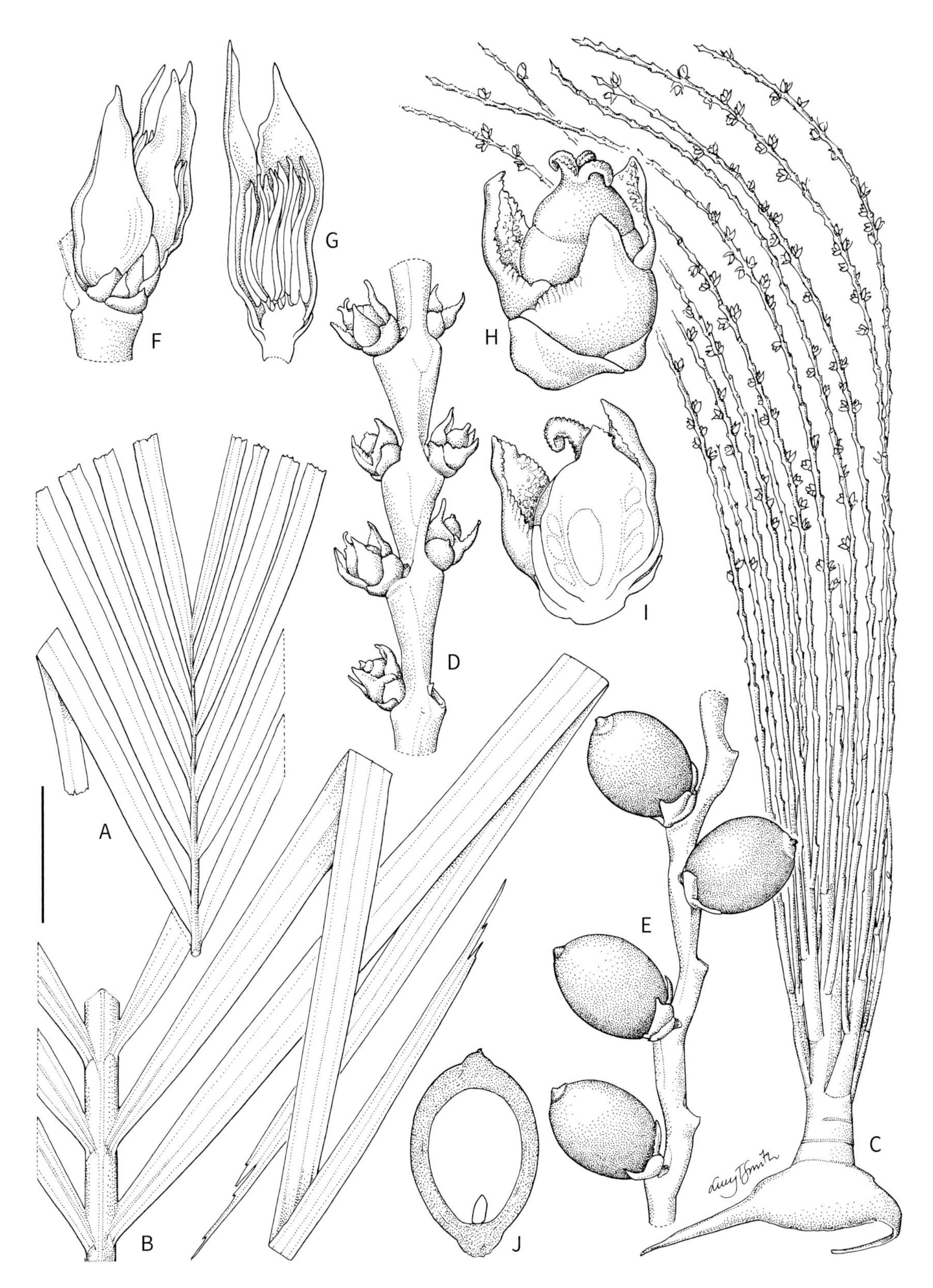

Hydriastele calcicola. **A.** Leaf apex. **B.** Leaf mid-portion. **C.** Inflorescence. **D.** Portion of rachilla with female flowers. **E.** Portion of rachilla with fruit. **F.** Triad. **G.** Male flower in longitudinal section. **H**, **I.** Female flower whole and in longitudinal section. **J.** Fruit in longitudinal section. Scale bar: A, B = 8 cm; C = 6 cm; D, E = 1.5 cm; F, J = 1 cm; G = 7 mm; H, I = 4 mm. All from *Baker et al. 1096*. Drawn by Lucy T. Smith.

Hydriastele calcicola. CLOCKWISE FROM TOP LEFT: colony (also with *Hydriastele costata*) showing gregarious habit on karst limestone, Kikori River (ZE); habit, Sirebi River (WB); floral triads with open male flowers and female flower buds, Kikori River (WB); Fruit, Kikori River (WB).

the Kikori River catchment where it grows gregariously on limestone outcrops (Baker 1997). The species occurs at relatively low elevation compared to similar species and is distinguished by its moderately robust, pithy and non-ventricose stem, by its arching leaves with deeply channelled petioles and multi-fold basal leaflets, and by its inflorescences with short peduncles. The species is most similar to *H. gibbsiana* and *H. ledermanniana*, but the former has a strongly swollen trunk while the latter is more robust.

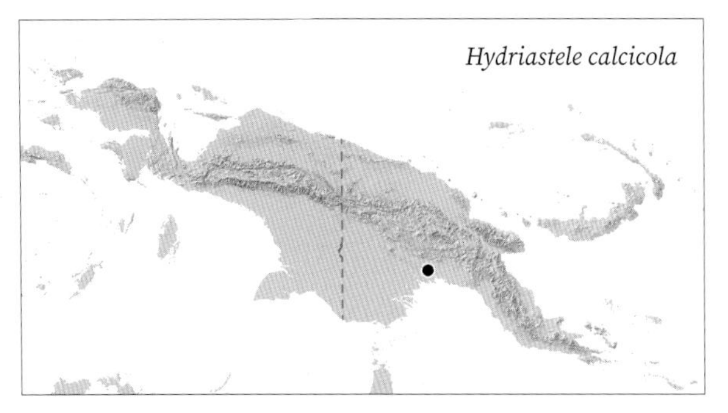

5. *Hydriastele costata* F.M.Bailey

Synonyms: *Gulubia affinis* Becc., *Gulubia costata* (Becc.) Becc., *Gulubia costata* var. *gracilior* Burret, *Gulubia costata* var. *minor* Becc., *Gulubia costata* var. *pisiformis* Becc., *Kentia costata* Becc., *Kentia costata* var. *microcarpa* Lauterb. & K.Schum. ex K.Schum. & Lauterb., *Kentia microcarpa* Warb. ex K.Schum. & Lauterb., *Pinanga pisiformis* Teijsm. ex Becc.

Robust, single-stemmed palm to 35 m, bearing 12–25 leaves in *the distinctly spherical crown.* **Stem** 15–35 cm diam. **Leaf** 2.4–5.5 m long including petiole, *straight or slightly drooping*; sheath 50–180 cm long; petiole 10–60 cm long; leaflets 58–75 each side of rachis, regularly arranged, single-fold, pendulous, linear, pointed or briefly notched at their tips. **Inflorescence** 58–100 cm long including 5–25 cm peduncle, branched to 2 or 3 orders, *apparently protogynous*; triads decussately arranged. **Male flower** with 6 stamens, white to brownish. **Female flower** with free sepals and free, low and rounded petals, cream. **Fruit** ca. 8–10 × 6 mm when ripe, ellipsoid, ripening through reddish brown

Hydriastele costata. **A.** Habit. **B.** Leaf apex. **C.** Inflorescence with fruit. **D.** Portion of rachilla with flowers. **E.** Portion of rachilla with fruit. **F, G.** Male flower whole and in longitudinal section. **H.** Female flower. **I–K.** Fruit whole, in longitudinal section and in transverse section. Scale bar: A = 1.8 m; B = 6 cm; C = 9 cm; D = 1 cm; E = 2 cm; F, G = 7 mm; H–K = 5 mm. A from photograph; B–K from *Baker et al. 836*. Drawn by Lucy T. Smith.

to dark purple or blackish maroon, *with conspicuous longitudinal white-greyish stripes*. **Seed** ca. 5.5 × 4.5 mm, ellipsoid, ridged; endosperm homogeneous.

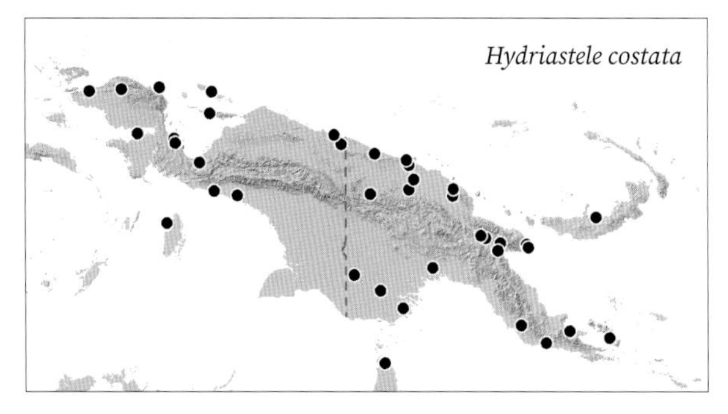

DISTRIBUTION. Widely distributed throughout lowland New Guinea and surrounding islands, including New Britain; elsewhere in Australia.

HABITAT. Rainforest, often on swampy ground, sea level to 700 m.

LOCAL NAMES. *Afos* (Miyah), *Bay* (Marap), *Kaparo* (Wandamen), *Korr* (Jal), *Mabla* (Orne), *Oratare* (Sumuri), *Poi* (Wapi), *Tab* (Timbunke), *Tabavo Nyi* (unknown dialect, North Cyclops Mountains area), *Tabuh* (Maprik), *Yawa* (Ambakanja).

USES. Stems and leaves used for flooring, house construction, and thatch. Leaf sheaths used as basins and water buckets, carrying baskets, sago containers and plates. Leaf hairs and scales used as firelighter. Old inflorescences used as brushes and brooms. Palm heart is consumed. Stem is crushed and consumed to treat shortness of breath (East Sepik). Commonly planted near villages.

CONSERVATION STATUS. Least Concern (IUCN 2019).

NOTES. *Hydriastele costata* is the commonest tree palm of New Guinea's lowland rainforests. It is immediately recognisable as a canopy emergent with a distinctive, spherical crown of more-or-less straight leaves with

Hydriastele costata. LEFT TO RIGHT: habit, Biak Island (WB); crown with inflorescences, cultivated National Botanic Garden, Lae (WB).

Hydriastele costata. LEFT TO RIGHT: floral triads with female flowers at anthesis, Frieda River (FE); male flowers at anthesis, Wandamen Peninsula (WB); female flowers after anthesis, Wandamen Peninsula (WB); fruit, Wandamen Peninsula (WB).

pendulous leaflets. Its highly distinctive longitudinally striped fruit with ridged (costate) seeds are not seen in any other member of the genus. *Hydriastele costata* has a similar habit to *H. procera* and *H. wosimiensis* but they both have protandrous inflorescences whereas *H. costata* is protogynous.

6. *Hydriastele divaricata* Heatubun, Petoe & W.J.Baker

Very slender, multi-stemmed palm to 4.5 m. **Stem** ca. 9–17 mm diam. **Leaf** ca. 80 cm long including petiole; sheath ca. 25 cm long; petiole ca. 27–30 cm long; leaflets 8–12 each side of rachis, irregularly to subregularly arranged, *widely spreading in the basal and middle portion of the leaf*, single- or 2-fold, narrowly linear, jagged at their tips. **Inflorescence** 10–12 cm long including 2–2.5 cm peduncle, *spicate or branched to 1 order, protandrous*; *rachillae 1–2*; triads spirally arranged. **Male flower** not seen. **Female flower** with free sepals and free petals with conspicuous, triangular tips. **Fruit** ca. 12 × 3 mm when ripe, cylindrical, pale green to pinkish. **Seed** ca. 8 × 2 mm, elongate to cylindrical; *endosperm shallowly ruminate*.

DISTRIBUTION. The lower slopes of the Mount Jaya region (Dransfield *et al.* 2000b) near Timika.

HABITAT. The transition zone from lowland alluvial rainforest to heath forest, ca. 100 m.

Hydriastele divaricata. Habit, Timika-Tembagapura road (WB).

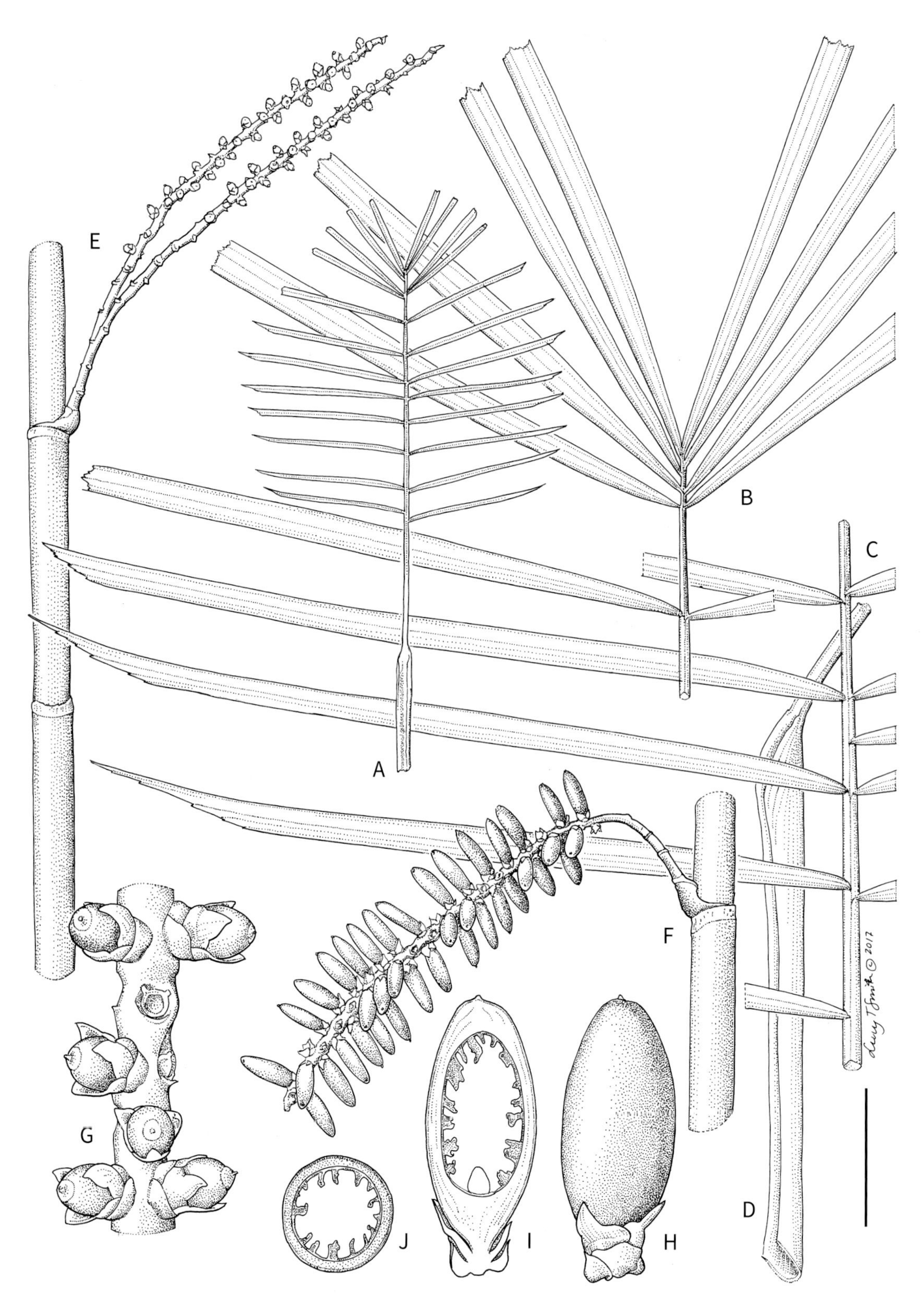

Hydriastele divaricata**. A.** Whole leaf diagram. **B.** Leaf apex. **C.** Leaf mid-portion. **D.** Leaf base with sheath. **E.** Inflorescence attached to stem. **F.** Inflorescence with fruit attached to stem. **G.** Portion of rachilla with immature fruit. **H–J.** Fruit whole, in longitudinal section and in transeverse section. Scale bar: A = 30 cm; B–D = 6 cm; E, F = 3 cm; G–J = 7 cm. All from *Baker et al. 876*. Drawn by Lucy T. Smith.

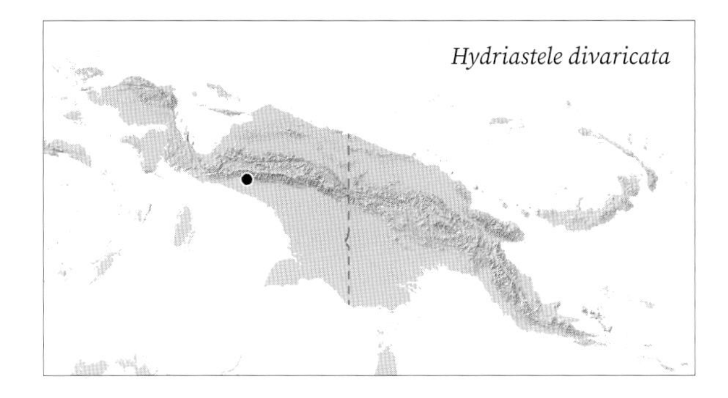

LOCAL NAMES. None recorded.

USES. None recorded.

CONSERVATION STATUS. Data Deficient (IUCN 2019). This species is known only from one site, where there is limited information on ongoing threats. Due to insufficient evidence, an extinction risk category cannot be selected.

NOTES. *Hydriastele divaricata* is a very slender understorey palm distinguished by its narrowly linear leaflets, which are widely spreading in the basal and middle section of the leaf. The inflorescence consists of just one (spicate) or two rachillae, prompting comparison with *H. flabellata* and *H. montana*, but these two species have homogeneous endosperm whereas *H. divaricata* has ruminate endosperm.

7. *Hydriastele dransfieldii* (Hambali, Maturb., Wanggai & W.J.Baker) W.J.Baker & Loo

Synonym: *Siphokentia dransfieldii* Hambali, Maturb., Wanggai & W.J.Baker

Moderate, single-stemmed palm to 12 m, bearing 6–10 leaves in crown. **Stem** 5–12 cm diam. **Leaf** 1.3–3 m long including petiole; sheath 35–80 cm long; petiole 10–65 cm long; *blade typically with 3 broad, multi-fold leaflets each side of rachis interspersed with a few single-fold leaflets*; leaflets oblong to linear, truncately jagged at their tips. **Inflorescence** 37–52 cm long including 2–5 cm peduncle, branched to 1 order, *protandrous*; rachillae 4–11, with yellowish-green axes; triads largely decussately arranged. **Male flower** with (9–)13–16 stamens, green to white. **Female flower** *with fused, greenish-white sepals forming a cylindrical tube, and white petals fused in lower half* with conspicuous, triangular tips. **Fruit** ca. 18 × 12 mm when ripe, obovoid, red. **Seed** ca. 8 × 7 mm, subglobose; endosperm ruminate.

DISTRIBUTION. Biak, Supiori and Numfor islands.

Hydriastele dransfieldii. Habit, Biak Island (WB).

HABITAT. Lowland rainforest on rocky limestone ridges and slopes at sea level to 300 m.

LOCAL NAMES. *Ombrush* (Biak).

USES. Leaves used for wrapping food items and seeds used as a betel nut substitute.

CONSERVATION STATUS. Least Concern (IUCN 2019). *Hydriastele dransfieldii* has a restricted distribution but has been observed to be locally common.

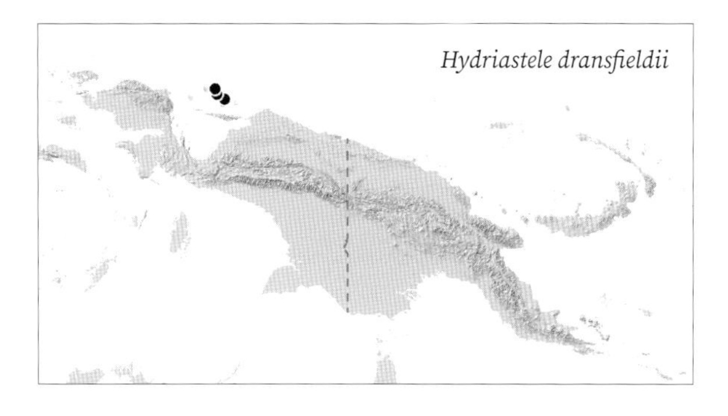

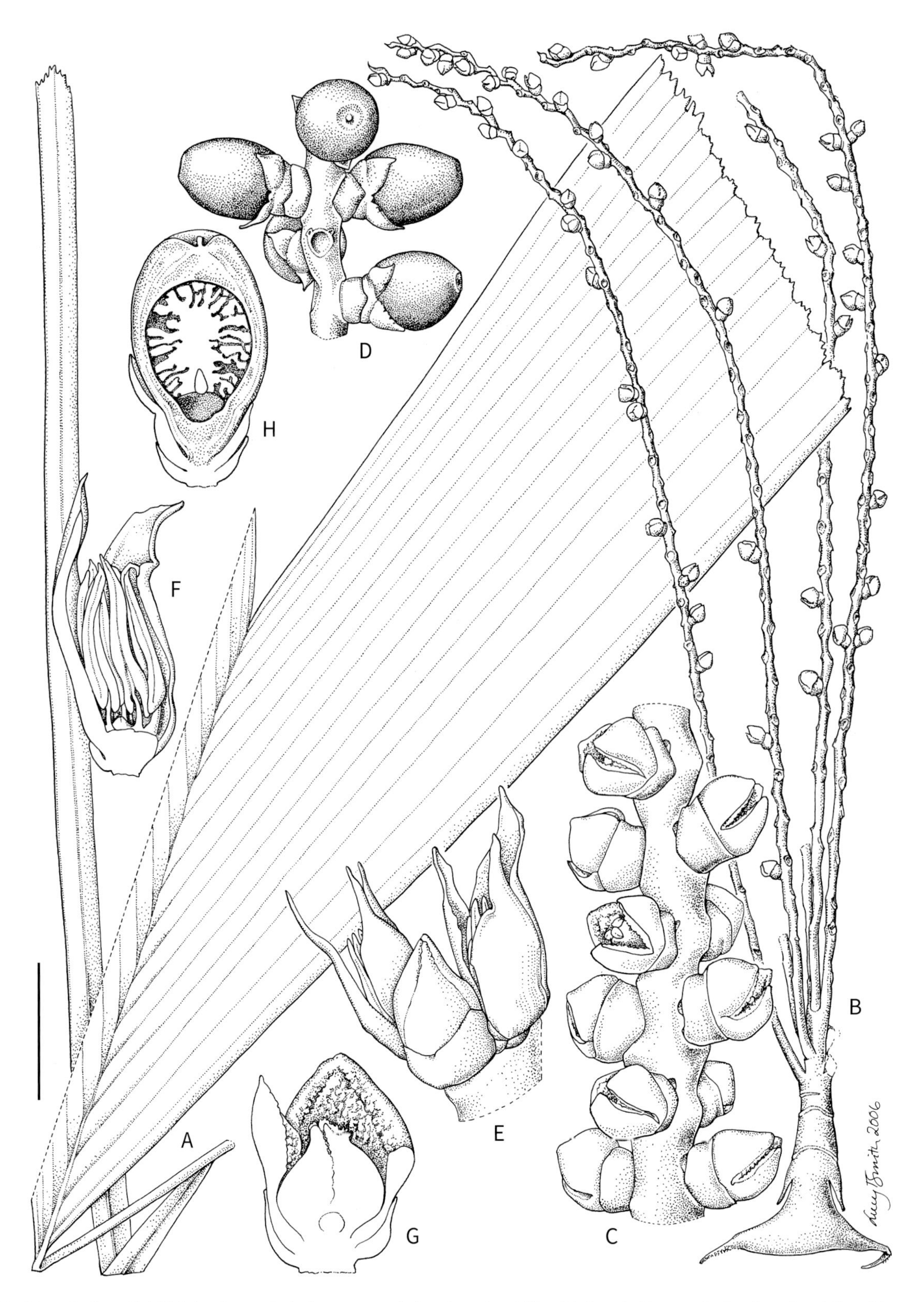

***Hydriastele dransfieldii*.** **A.** Leaf apex. **B.** Inflorescence. **C.** Portion of rachilla with female flowers. **D.** Portion of rachilla with fruit. **E.** Triad. **F.** Male flower in longitudinal section. **G.** Female flower in longitudinal section. **H.** Fruit in longitudinal section. Scale bar: A = 8 cm; B = 4 cm; C, H = 1 cm; D = 2 cm; E = 7 mm; F, G = 5 mm. A, B from *Maturbongs 555*; C, E–G from *Zona 826*; D, H from *Hambali s.n.* Drawn by Lucy T. Smith.

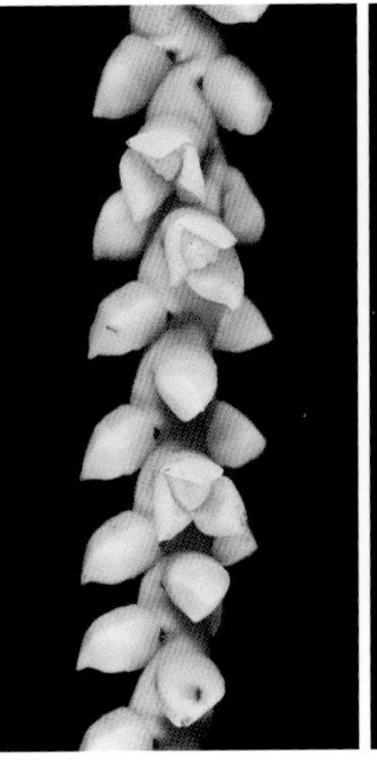

Hydriastele dransfieldii. LEFT TO RIGHT: habit, Biak Island (WB); inflorescences with fruit, cultivated Bogor, Indonesia (WB); female flowers, Biak Island (WB); fruit, Biak Island (WB).

NOTES. *Hydriastele dransfieldii* is a moderately slender midstorey palm that is endemic to Biak and adjacent islands in Indonesian New Guinea. The species is highly distinctive vegetatively as well as in its reproductive structures. Its leaves usually consist of three broad, multi-fold leaflets interspersed with a few single-fold leaflets, and the female flower has sepals that are fused entirely, forming a conspicuous cup, and petals that are fused at their bases. These features are not found in other New Guinean members of the genus. Female flowers of this type are found elsewhere only in *H. beguinii* from Maluku.

8. *Hydriastele flabellata* (Becc.) W.J.Baker & Loo

Synonyms: *Gronophyllum cariosum* Dowe & M.D.Ferrero, *Gronophyllum flabellatum* (Becc.) Essig & B.E.Young, *Gronophyllum gracile* (Burret) Essig & B.E.Young, *Gronophyllum pleurocarpum* (Burret) Essig & B.E.Young, *Gronophyllum rhopalocarpum* (Becc.) Essig & B.E.Young, *Hydriastele cariosa* (Dowe & M.D.Ferrero) W.J.Baker & Loo, *Hydriastele gracilis* (Burret) W.J.Baker & Loo, *Hydriastele pleurocarpa* (Burret) W.J.Baker & Loo, *Hydriastele rhopalocarpa* (Becc.) W.J.Baker & Loo, *Nengella calophylla* var. *montana* Becc., *Nengella calophylla* var. *rhopalocarpa* Becc., *Nengella flabellata* Becc., *Nengella gracilis* Burret, *Nengella pleurocarpa* Burret, *Nengella rhopalocarpa* (Becc.) Burret

Very slender, multi-stemmed palm to 4 m, bearing 4–8 leaves per crown. **Stem** 1–2 cm diam. **Leaf** 35–90 cm long including petiole; sheath 8–23 cm long; petiole 10–30 cm; blade 20–65 cm long, *entire-bifid or dissected with up to 6 leaflets each side of rachis*; leaflets (when present) regularly or irregularly arranged, single- or multi-fold, wedge-shaped, jagged at their tips. **Inflorescence** 8–15 cm long including 1–3 cm peduncle, spicate or branched to 1 order, *protandrous; rachillae 1–2* (rarely 3), with green to yellow axes; triads spirally arranged. **Male flower** with 6 stamens, cream to violet. **Female flower** with free sepals and free petals with conspicuous, triangular tips, cream to light green or purplish. **Fruit** 12–15 × 5–8 mm when ripe, ellipsoid to spindle-shaped, red. **Seed** 10–12 × 4–7 mm, ellipsoid; *endosperm homogeneous.*

DISTRIBUTION. Widespread in mainland New Guinea.

HABITAT. Rainforest on slopes and ridges from sea level to 700 m.

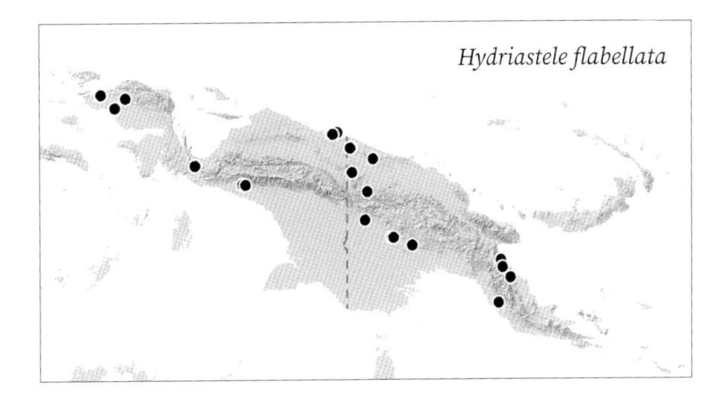

Hydriastele flabellata. CLOCKWISE FROM LEFT: habit, Bewani (AB); Habit, near Timika (WB); habit (metallic leaf form), cultivated Floribunda Palms and Exotics, Hawaii (WB); leaf (metallic leaf form), cultivated Floribunda Palms and Exotics, Hawaii (WB).

LOCAL NAMES. *Mplemponik* (Sayal), *Filiawoi Yamu* (Bewani).

USES. Leaves used as food wrapping and stems used for making arrow shafts and fish spear handles.

CONSERVATION STATUS. Least Concern (IUCN 2019).

NOTES. *Hydriastele flabellata* is an understorey palm with variable leaf morphology ranging from entire-bifid to pinnate with up to six wedge-shaped leaflets on each side of the rachis. The species is distinguished by its inflorescences, which are 8–15 cm long and typically consist of one (spicate) or two rachillae, and by its fruit with homogeneous endosperm. *Hydriastele flabellata* may be confused with *H. montana* or *H. pinangoides* but the former species has leaves with 9–11 narrowly linear leaflets on each side of the rachis while *H. pinangoides*, which can have similar foliage, is usually taller and displays 18–30 cm long inflorescences with 2–6 rachillae and ruminate endosperm.

Hydriastele flabellata. LEFT TO RIGHT: inflorescence with mature male flowers, near Timika (WB); inflorescence with female flowers, Bewani (JLD); inflorescence with fruit, near Timika (TU).

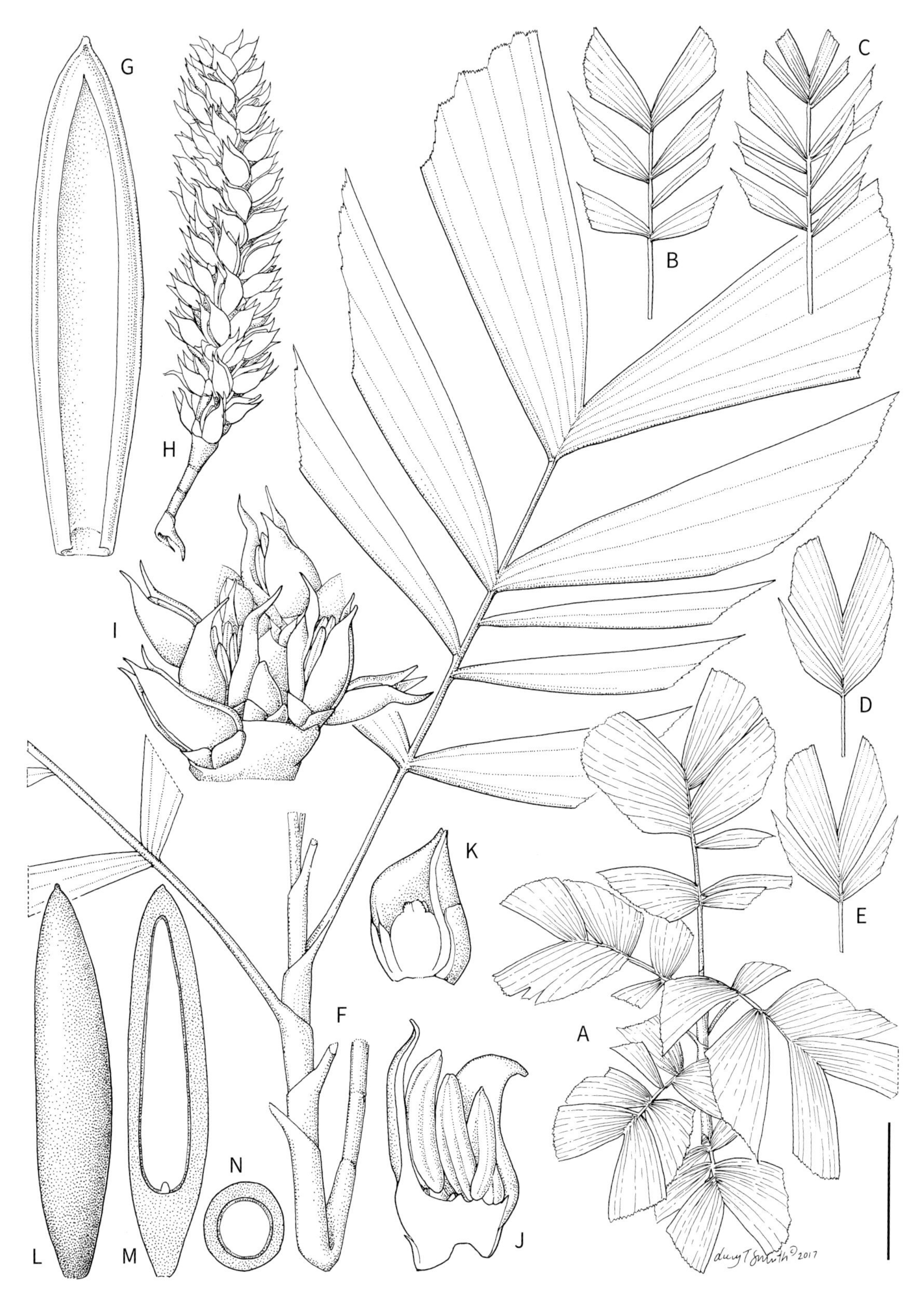

***Hydriastele flabellata*.** **A.** Habit. B–**E.** Leaf diagrams. **F.** Portion of crownshaft with leaf. **G.** Prophyll. **H.** Inflorescence with triads. **I.** Portion of rachilla with triads. **J.** Male flower in longitudinal section. **K.** Female flower in longitudinal section. **L–N.** Fruit whole, in longitudinal section and in transverse section. Scale bar: A = 20 cm; B–E = 18 cm; F = 6 cm; G, H = 2 cm; I = 5 mm; J = 4 mm; K = 3 mm; L–N = 7 mm. A, B, F–K from *Baker et al. 879*; C from *Schlechter 17466*; D, E from *Baker et al. 643*; L–N from *Heatubun 406*. Drawn by Lucy T. Smith.

9. *Hydriastele gibbsiana* (Becc.) W.J.Baker & Loo

Synonyms: *Gronophyllum gibbsianum* (Becc.) H.E.Moore, *Kentia gibbsiana* Becc.

Moderately robust, single-stemmed emergent palm to 30 m, bearing 13–15 leaves in crown. **Stem** *strongly swollen, to 30 cm diam.* (but as little as 10 cm diam. adjacent to swelling), inner wood conspicuously soft and pithy. **Leaf** 2–2.15 m long including petiole, arching; sheath 60–82 cm long, covered with thin, smooth and woolly indumentum, crownshaft ca. 100 × 15 cm; petiole ca. 30 cm long, upper surface flattened; leaflets 50–53 each side of rachis, regularly arranged, ascending and drooping at their tips, linear; apical leaflets comprising 1–5 folds, truncately jagged at the tip. **Inflorescence** *55–60 cm long* including 4–5 cm peduncle, branched to 2 orders, *protandrous;* rachillae 22–28; triads decussately arranged. **Male flower** with 6–9 stamens, cream. **Female flower** with free sepals and free petals with conspicuous, triangular tips, cream. **Fruit** ca. 13.5 × 8 mm when ripe, broadly ellipsoid, red. **Seed** ca. 9 × 6.5 mm, broadly ellipsoid; endosperm homogeneous.

DISTRIBUTION. Tamrau Mountains and Arfak Mountains of the Bird's Head Peninsula.

HABITAT. Montane rainforest on slopes and ridge tops at 950–2450 m.

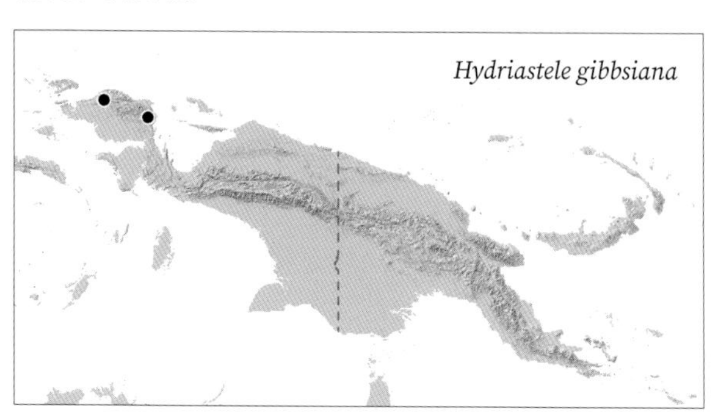

Hydriastele gibbsiana. LEFT TO RIGHT: habit showing swollen stems, Arfak Mountains (AS); crown, Tamrau Mountains (WB).

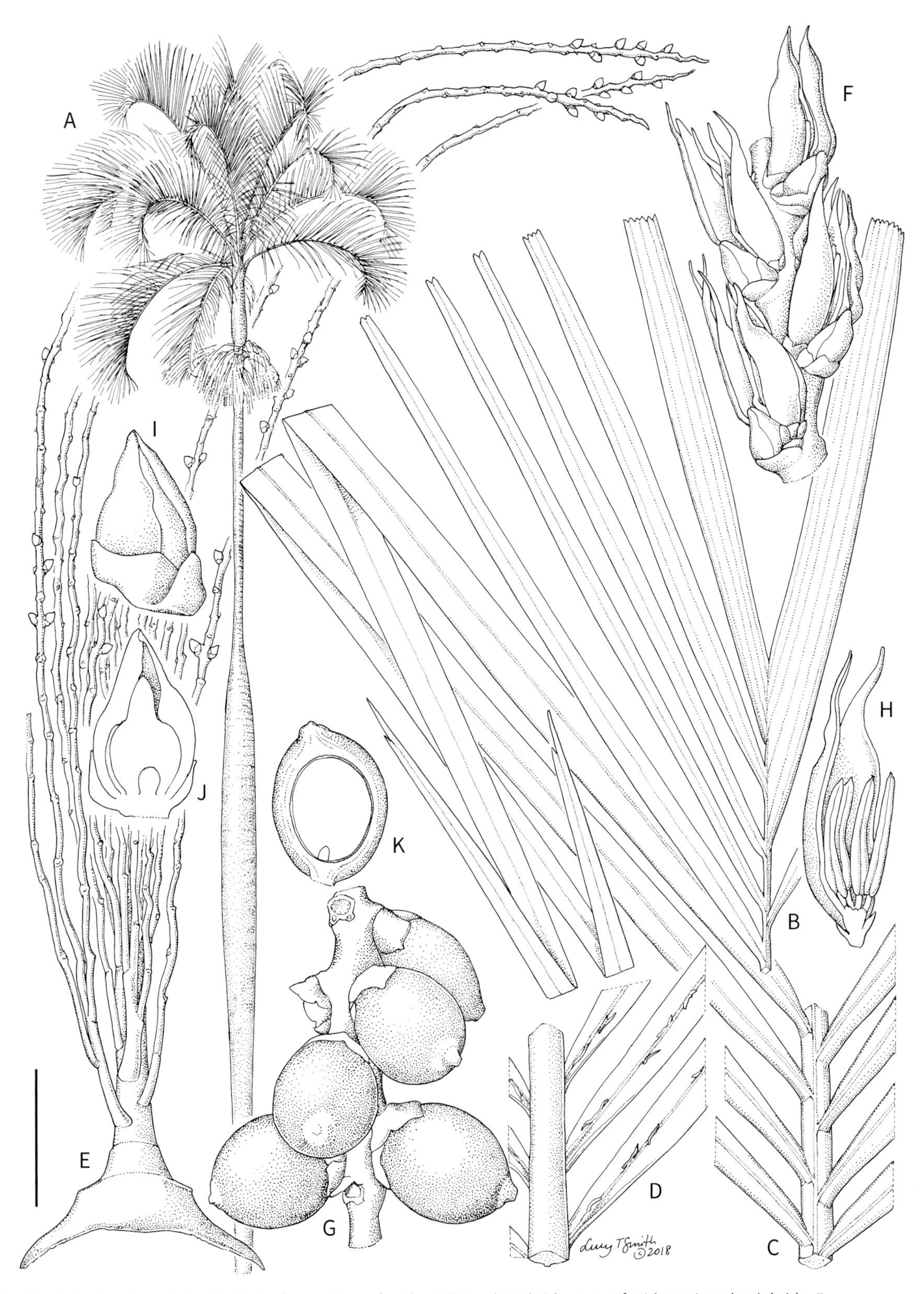

***Hydriastele gibbsiana*.** **A.** Habit. **B.** Leaf apex. **C.** Leaf mid-portion adaxial side. **D.** Leaf mid-portion abaxial side. **E.** Inflorescence. **F.** Portion of rachilla with triads. **G.** Portion of rachilla with fruit. **H.** Male flower in longitudinal section. **I**, **J.** Female flower whole and in longitudinal section. **K.** Fruit in longitudinal section. Scale bar: A = 1.5 m; B–D = 6 cm; E = 4 cm; F = 1 cm; G, K = 1.5 cm; H = 7 mm; I, J = 4 mm. All from *Baker et al. 1379*. Drawn by Lucy T. Smith.

Hydriastele gibbsiana. LEFT TO RIGHT FROM TOP: inflorescences; flower buds; fruit. Tamrau Mountains (WB).

LOCAL NAMES. *Syah* (Madik).

USES. Stem and leaves used for flooring and thatch, respectively (Tamrau Mountains). The very light and pithy trunks were, at least until the 1990s, used for making rafts in the Anggi Lakes (Petoe *et al.* 2019).

CONSERVATION STATUS. Data Deficient (IUCN 2019). However, as deforestation is a major threat in its distribution range, there is a case for this species to be placed in a threatened category.

NOTES. *Hydriastele gibbsiana* is a tall palm with a crown of arching leaves, ascending leaflets and a strikingly swollen stem that makes the species stand out among all other New Guinea palms. Two other New Guinean species of *Hydriastele, H. procera and H. wosimiensis,* have conspicuously ventricose stems but to a lesser extent, and both differ from *H. gibbsiana* in their more-or-less straight leaves with pendulous leaflets.

10. *Hydriastele kasesa* (Lauterb.) Burret

Synonyms: *Adelonenga kasesa* (Lauterb.) Becc., *Ptychosperma kasesa* Lauterb.

Slender, single- or multi-stemmed palm to 6 m, bearing 6–8 leaves per crown. **Stem** 1.5–3.8 cm diam. **Leaf** 70–150 cm long including petiole; *sheath 15–30 cm long*; petiole 15–40 cm; *leaflets 6–13 each side of rachis, irregularly arranged with a group of closely spaced leaflets in different planes in the middle of the leaf, wedge-shaped, truncately jagged at their tips*. **Inflorescence** 17–30 cm long including 2–4 cm peduncle, branched to 1 or 2 orders, *protogynous*; triads decussately arranged. **Male flower** with 6 stamens. **Female flower** with free sepals and free, rounded, low petals. **Fruit** ca. 12 × 9 mm when ripe, ovoid to subglobose, red. **Seed** ca. 6.7 × 5.4 mm (when dry), ovoid; endosperm ruminate.

DISTRIBUTION. New Britain and New Ireland in the Bismarck Archipelago.

HABITAT. Rainforest, at sea level to 800 m.

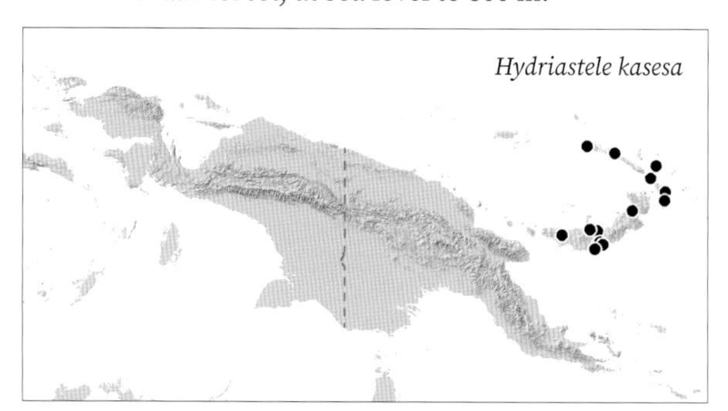

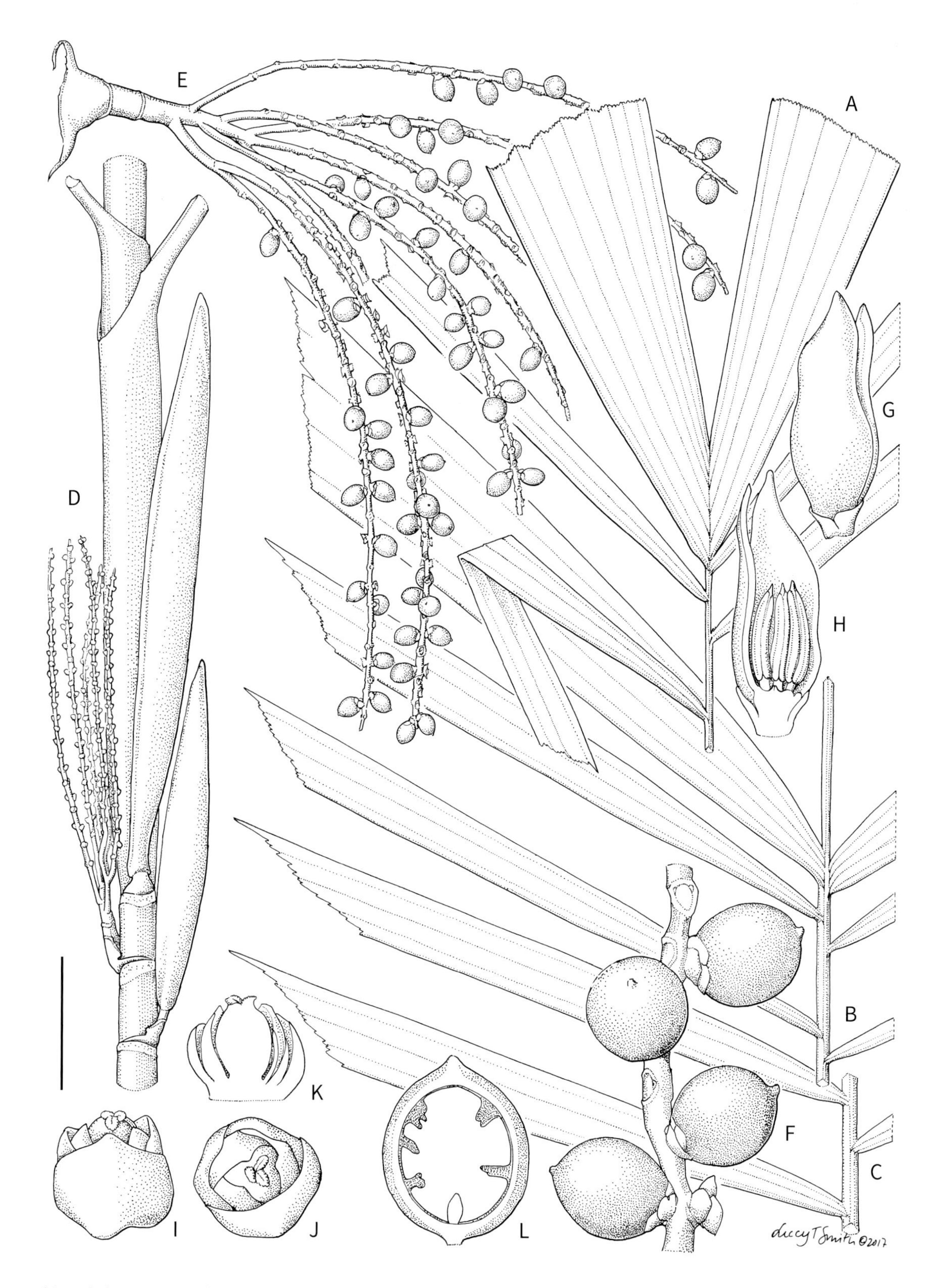

Hydriastele kasesa. **A.** Leaf apex. **B.** Leaf mid-portion. **C.** Leaf base. **D.** Leaf sheath with inflorescences. **E.** Inflorescence with fruit. **F.** Portion of rachilla with fruit. **G**, **H.** Male flower whole and in longitudinal section. **I**, **J.** Female flower in two views. **K.** Female flower in longitudinal section. **L.** Fruit in longitudinal section. Scale bar: A–C = 8 cm; D = 6 cm; E = 4 cm; F = 1 cm; G, H = 5 mm; I–K = 2.2 mm; L = 7 mm. All from *Takeuchi 9902*. Drawn by Lucy T. Smith.

LOCAL NAMES. None recorded.

USES. Stem used for flooring.

CONSERVATION STATUS. Least Concern (IUCN 2019). However, close monitoring is required due to heavy logging of the forests within the distribution of this island species.

NOTES. *Hydriastele kasesa* is a slender understorey palm distinguished by its leaves with 6–13 irregularly arranged leaflets on each side of the rachis and leaf sheaths up to 30 cm long. *Hydriastele kasesa* is most similar to *H. wendlandiana* but that species is usually taller and less slender and has leaves with 12–40 leaflets per side and leaf sheaths longer than 40 cm.

11. *Hydriastele lanata* W.J.Baker & Petoe

Moderately robust, single-stemmed palm to 10 m, bearing 10–12 leaves in crown. **Stem** *ca. 10 cm diam.* **Leaf** ca. 2.4 m long including petiole, arching; sheath ca. 120 cm long, *covered with a thick and dense layer of woolly and very fluffy indumentum,* crownshaft ca. 216 × 10 cm; petiole ca. 32 cm long, upper surface channelled; leaflets ca. 43 each side of rachis, regularly arranged, ascending and drooping at their tips, linear; apical leaflets comprising ca. 4 folds, truncately jagged at the tip. **Inflorescence** 65–70 cm long including *11–12 cm peduncle,* branched to 2 orders, *protandrous;* rachillae 17–23; triads decussately arranged. **Male flower** with 7–10 stamens, white. **Female flower** with free sepals and free petals with conspicuous, triangular tips, white. **Fruit** not seen. **Seed** not seen.

DISTRIBUTION. Known only from a site adjacent to the Ok Tedi Copper Mine, central New Guinea.

HABITAT. Montane and somewhat degraded rainforest on steep slopes at ca. 1,400 m.

LOCAL NAMES. None recorded.

USES. None recorded.

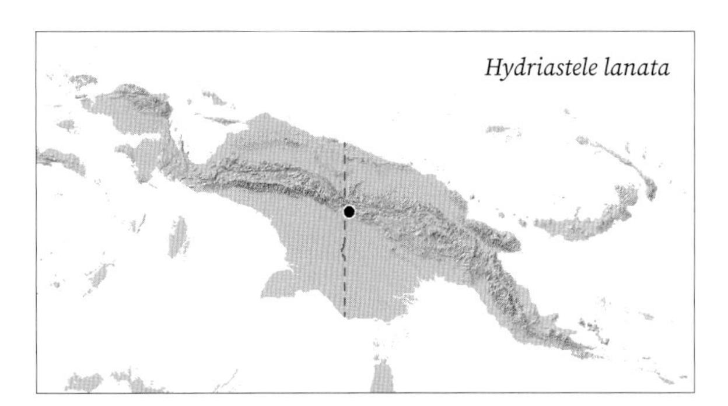

CONSERVATION STATUS. Data Deficient (IUCN 2019). Due to insufficient evidence, an extinction risk category cannot be selected. However, expansion of the mining site adjacent to the only known site for this species is likely to pose a significant threat.

NOTES. *Hydriastele lanata* is a moderately robust tree palm with arching leaves, immediately recognisable on account of the pale brown, woolly and very fluffy indumentum on the leaf sheath, which is unique in *Hydriastele.*

12. *Hydriastele ledermanniana* (Becc.) W.J.Baker & Loo

Synonyms: *Gronophyllum chaunostachys* (Burret) H.E.Moore, *Gronophyllum ledermannianum* (Becc.) H.E.Moore, *Gronophyllum mayrii* (Burret) H.E.Moore, *Hydriastele chaunostachys* (Burret) W.J.Baker & Loo, *Hydriastele mayrii* (Burret) W.J.Baker & Loo, *Kentia chaunostachys* Burret, *Kentia ledermanniana* Becc., *Kentia mayrii* Burret

Robust, single-stemmed emergent palm to 30 m, bearing 15–19 leaves in crown. **Stem** *20–35 cm diam.*

Hydriastele ledermanniana. Crown with inflorescences, Wau (WB).

Hydriastele lanata. **A.** Habit. **B.** Leaf apex. **C.** Leaf mid-portion. **D.** Inflorescence. **E.** Portion of rachilla with female flowers. **F.** Triad. **G.** Male flower in longitudinal section. **H.** Female flower in longitudinal section. Scale bar: A = 1.8 m; B, C = 8 cm; D = 6 cm; E = 1 cm; F = 7 mm; G = 5 mm; H = 3.3 mm. All from *Baker et al. 1135*. Drawn by Lucy T. Smith.

Leaf 2–3.5 m long including petiole, arching; sheath 58–200 cm long, covered with a thin layer of woolly indumentum, crownshaft 90–250 × 15–40 cm; petiole 30–80 cm long, upper surface flattened; leaflets 38–51 each side of rachis, regularly arranged, ascending and drooping at their tips, linear; apical leaflets comprising 2–4 folds, truncately jagged at the tip. **Inflorescence** *(50–)65–120 cm long including 6–15 cm peduncle*, branched to 2(–3) orders, *protandrous*; rachillae 23–66; triads decussately arranged. **Male flower** with 9–13 stamens, cream. **Female flower** with free sepals and free petals with conspicuous, triangular tips, cream. **Fruit** ca. 15–15.5 × 10–10.5 mm when ripe, broadly ellipsoid, brown to red. **Seed** ca. 10 × 8 mm, broadly ellipsoid; endosperm homogeneous.

DISTRIBUTION. Highland areas from the Cyclops Mountains to far south-eastern New Guinea.

HABITAT. Premontane to montane rainforest at 700–1,950 m.

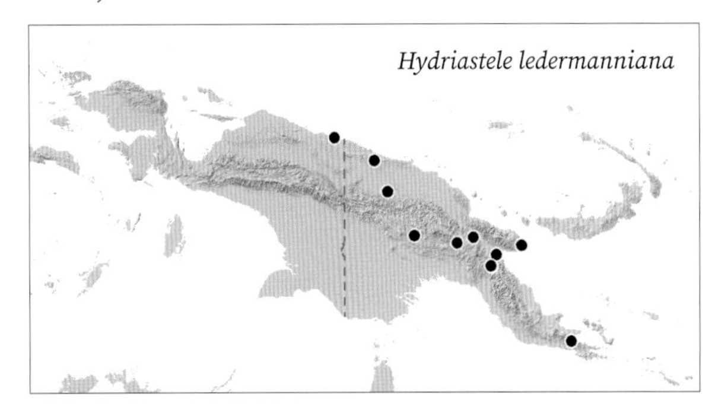

Hydriastele ledermanniana. Inflorescence, Jivewaneng (WB).

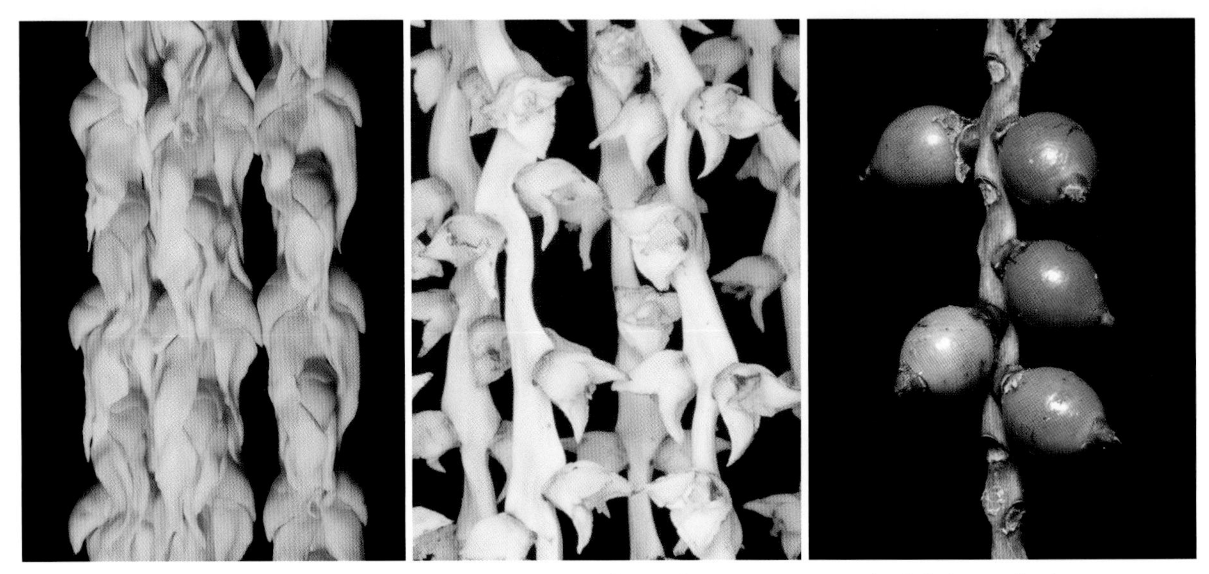

Hydriastele ledermanniana. LEFT TO RIGHT: flower buds, Wau (WB); female flowers, Jivewaneng (WB); fruit, Wau (WB).

Hydriastele ledermanniana. **A.** Leaf apex. **B.** Leaf mid-portion. **C.** Portion of inflorescence. **D.** Portion of rachilla with female flowers. **E.** Portion of rachilla with fruit. **F.** Triad. **G.** Male flower. **H**, **I.** Female flower whole and in longitudinal section. **J.** Fruit in longitudinal section. Scale bar: A, B = 8 cm; C = 6 cm; D = 1.5 cm; E = 2 cm; F, G, J = 1 cm; H, I = 7 mm. All from *Banka et al. 2000*. Drawn by Lucy T. Smith.

LOCAL NAMES. *Aidjaka* (Tari), *Limbom* (Papua New Guinea Tok Pisin), *Kawoisch* (Mendi), *Pipi* (Kotte), *Tuwenpeh* (Wapi), *Uwo* (Wagu), *Yauwi* (Kagua), *Gamu* (unknown dialect).

USES. Leaves used for thatch. Sheathing leaf bases are used as vessels to carry food. Young shoot (heart-of-palm) is edible. Split trunk used for flooring. Inflorescences used for brooms. Fruit used as a betel nut substitute.

CONSERVATION STATUS. Least Concern (IUCN 2019).

NOTES. *Hydriastele ledermanniana* is a robust, single-stemmed, canopy emergent with arching leaves and protandrous inflorescences with long (6–15 cm) peduncles. The species is most similar to *H. calcicola* and *H. gibbsiana* but those species have short peduncles (3–5 cm long) and the former is also less robust while the latter is strongly ventricose.

13. *Hydriastele longispatha* (Becc.) W.J.Baker & Loo

Synonyms: *Gulubia brassii* Burret, *Gulubia crenata* Becc., *Gulubia longispatha* Becc., *Gulubia obscura* Becc., *Gulubia valida* Essig, *Hydriastele valida* (Essig) W.J.Baker & Loo

Robust, single-stemmed emergent palm to 35 m, bearing 15–17 leaves in crown. **Stem** 15–30 cm diam. **Leaf** 2.5–3.6 m long including petiole, *arching*; sheath 1–1.5 m long, covered with a thin layer of woolly indumentum, crownshaft ca. 110–200 × 30 cm; petiole 10–40 cm long, flattened to slightly channelled on upper surface; leaflets (48–)57–70 each side of rachis, regularly arranged, ascending and drooping at their tips, linear; apical leaflets comprising 2 or 3 folds, truncately jagged at the tip. **Inflorescence** 60–90 cm long including 10–15 cm peduncle, *branched to 2 or 3 orders, apparently protogynous*; rachillae ca. 28 (rarely as few as 8–11); triads decussately arranged. **Male flower** with 8–24 stamens, cream. **Female flower** with free sepals and free, low and rounded petals, cream to brownish. **Fruit** 7–10 × 6–10 mm when ripe, subglobose to broadly ellipsoid, red. **Seed** ca. 6.5 × 5.5 mm, subglobose; endosperm homogeneous.

DISTRIBUTION. Widespread in highland areas from Yapen Island, through the Central Range and Torricelli Mountains to Goodenough Island.

HABITAT. Usually premontane and montane forest at 300–1,450 m.

LOCAL NAMES. None recorded.

Hydriastele longispatha. FROM TOP: habit; inflorescences. Near Timika (WB).

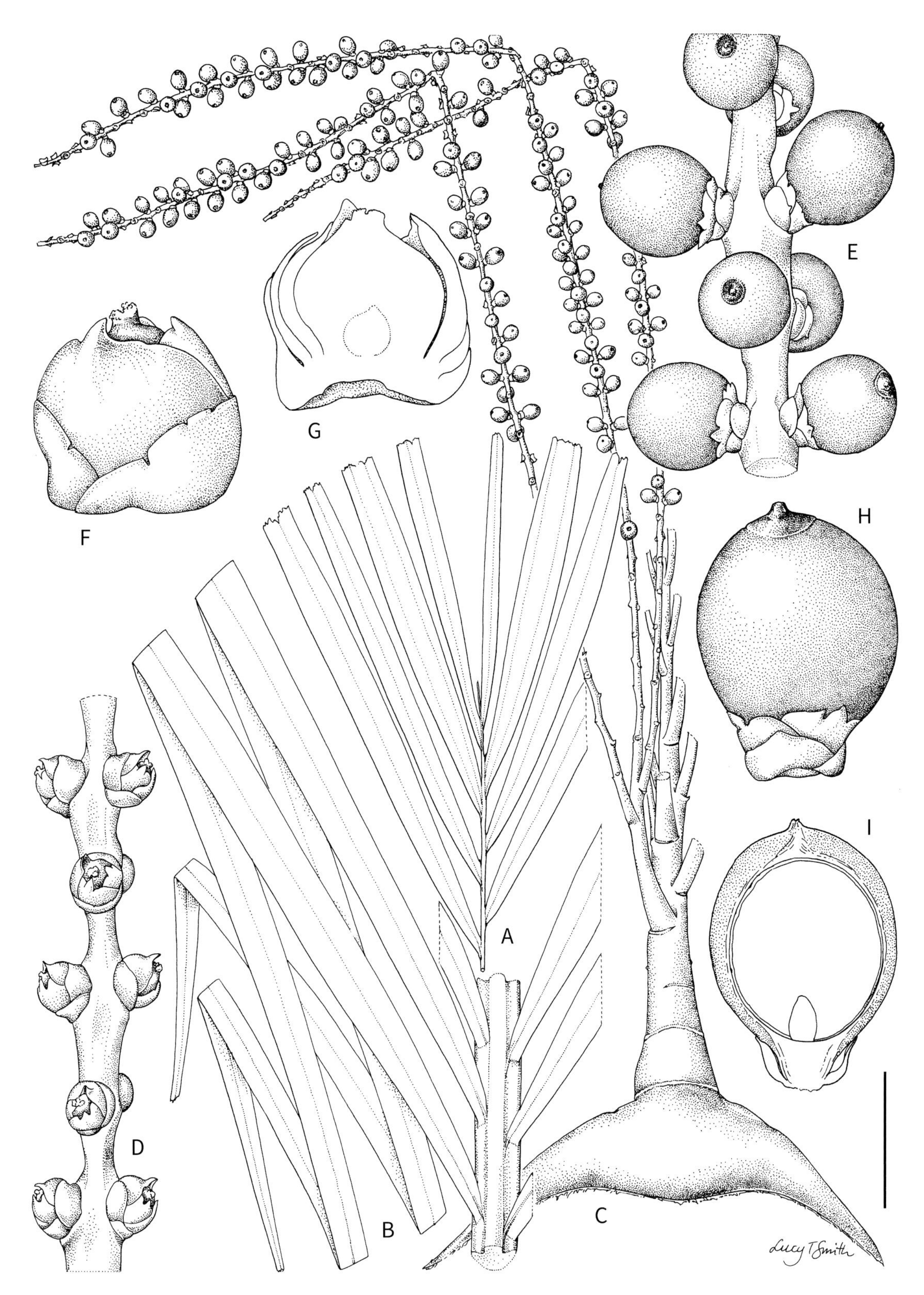

Hydriastele longispatha. **A.** Leaf apex. **B.** Leaf mid-portion. **C.** Portion of inflorescence with fruit. **D.** Portion of rachilla with female flowers. **E.** Portion of rachilla with fruit. **F**, **G.** Female flower whole and in longitudinal section. **H**, **I.** Fruit whole and in longitudinal section. Scale bar: A, B = 8 cm; C = 6 cm; D, E = 1.5 cm; F, G = 2.5 mm; H, I = 7 mm. All from *Baker et al. 882*. Drawn by Lucy T. Smith.

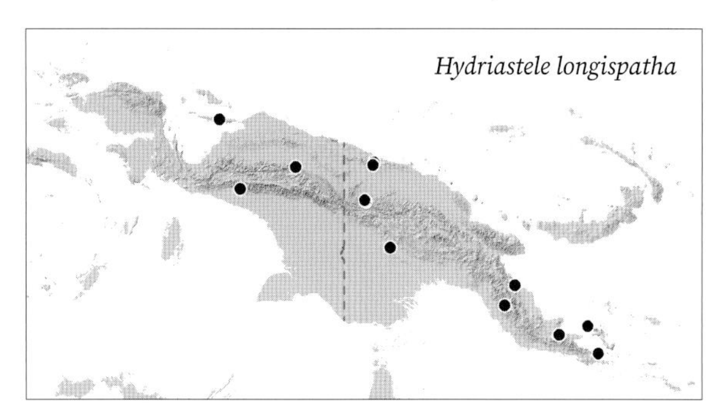

USES. Stem used in house construction (especially for flooring) and for bows. Young shoot (heart-of-palm) is edible. Fruit are edible and eaten by cassowaries.

CONSERVATION STATUS. Least Concern (IUCN 2019).

NOTES. *Hydriastele longispatha* is a robust, emergent palm distinguished by its arching leaves in combination with protogynous inflorescences branched to two or three orders and more-or-less globose to broadly ellipsoid fruit. The species can be confused with *H. costata* and *H. biakensis*, which are both protogynous, but the former has straight to drooping leaves and striped fruit while the latter displays inflorescences with four orders of branching and bullet-shaped fruit. *Hydriastele longispatha* has a similar habit to *H. ledermanniana* but that species has protandrous inflorescences.

14. *Hydriastele lurida* (Becc.) W.J.Baker & Loo

Synonyms: *Gronophyllum brassii* Burret, *Gronophyllum luridum* Becc., *Hydriastele brassii* (Burret) W.J.Baker & Loo

Single-stemmed, *moderate subcanopy palm to 25 m*, bearing 8–12 leaves in crown. **Stem** (4.5–)6–10(–15) cm diam. **Leaf** 1.3–2 m long including petiole; sheath 47–105 cm long, *conspicuously fibrous at the sheath mouth*; petiole 25–100 cm long; leaflets (21–)27–40 each side of rachis, irregularly arranged with one or more groups present, closely spaced and held in different planes, single- or multi-fold, linear to narrowly wedge-shaped, truncately jagged at their tips, brittle when dry. **Inflorescence** 42–70 cm long including 3–6 cm peduncle, *branched to 2 orders, protandrous*; rachillae (10–)15–27, with yellowish axes; *triads spirally arranged*. **Male flower** with 6(–7) stamens. **Female flower** with free sepals and free petals with conspicuous, triangular tips. **Fruit** 5–7 × 5–7 mm when ripe, globose, red. **Seed** 4–5 × 4–5 mm, globose; endosperm ruminate.

DISTRIBUTION. Widespread in west New Guinea, including Biak Islands.

Hydriastele lurida. Crown, Biak Island (WB).

HABITAT. Lowland swamps and heath forest at sea level to 500 m.

LOCAL NAMES. *Ansan* (Soon), *Gulbotom* (Gebe), *Halibou* (Mianmin), *Omdar* (Biak), *Sirata* (Sayal).

USES. Stems used for making harpoons, arrowheads, floors and beds, leaves for thatching. Shoot apices consumed.

CONSERVATION STATUS. Least Concern (IUCN 2019).

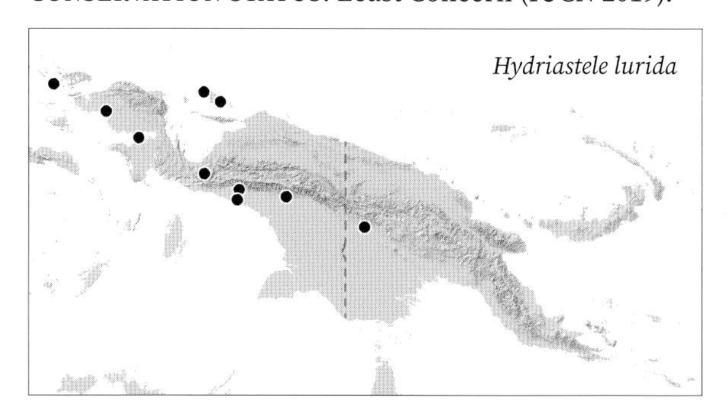

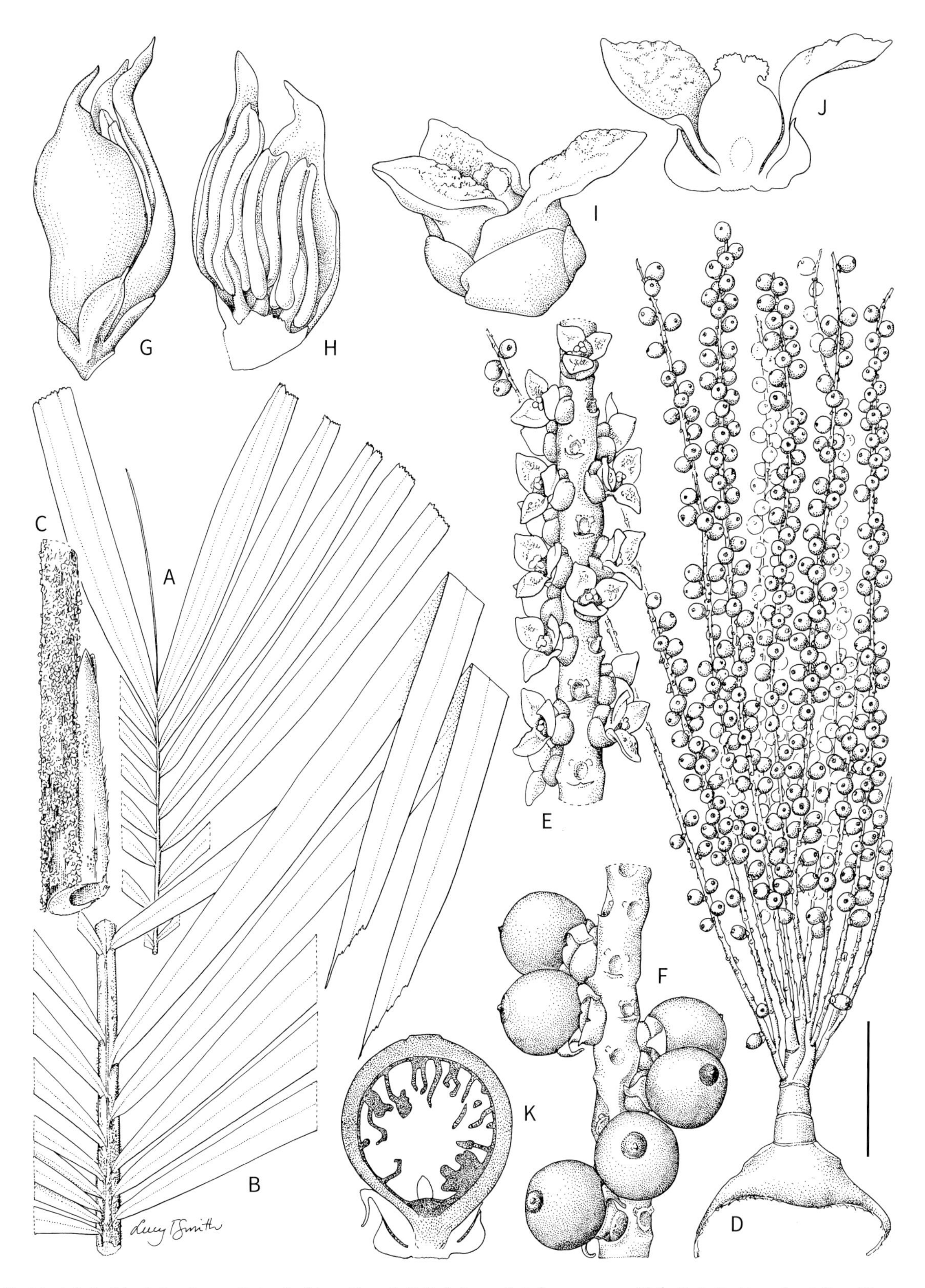

Hydriastele lurida. **A.** Leaf apex. **B.** Leaf mid-portion. **C.** Petiole base. **D.** Inflorescence with fruit. **E.** Portion of rachilla with female flowers. **F.** Portion of rachilla with fruit. **G**, **H.** Male flower whole and in longitudinal section. **I**, **J.** Female flower whole and in longitudinal section. **K.** Fruit in longitudinal section. Scale bar: A, B = 8 cm; C = 3 cm; D = 6 cm; E, F = 1 cm; G–J = 3 mm; K = 5 mm. A–D, F, K from *Baker et al. 823*; E, G–J from *Dransfield et al. JD 7682*. Drawn by Lucy T. Smith.

Hydriastele lurida. LEFT TO RIGHT: leaf bases showing petiole indumentum and fibrous sheath mouths; inflorescences with fruit. Biak Island (WB).

NOTES. *Hydriastele lurida* is a moderately slender subcanopy palm with conspicuously fibrous leaf sheaths and leaves with irregularly arranged leaflets. The species is distinguished by its protandrous inflorescences branched to 2 orders and spirally arranged triads, a combination of characters not displayed by other members of the genus in New Guinea. In habit, it may superficially resemble *H. wendlandiana,* but that species is protogynous, does not have conspicuously fibrous leaf sheaths, and normally has less numerous leaflets..

15. *Hydriastele manusii* (Essig) W.J.Baker & Loo

Synonym: *Gronophyllum manusii* Essig

Moderately robust, single-stemmed palm to 20 m, bearing 9–13 leaves in crown. **Stem** 10–15 cm diam. **Leaf** *ca. 1.5–2 m long including petiole, arching*; sheath ca. 100 cm long, with white bloom, crownshaft ca. 150 × 15 cm; petiole ca. 10 cm long, upper surface channelled; leaflets ca. 40 each side of rachis, regularly arranged, ascending and drooping at their tips, linear; apical leaflets comprising 2–4 folds, truncately jagged at the tip. **Inflorescence** 60–80 cm long including *20–22 cm peduncle,* branched to 2 orders, *protandrous; rachillae 12–15*; triads decussately arranged. **Male flower** with 6 stamens. **Female flower** with free sepals and free petals with conspicuous, triangular tips. **Fruit** ca. 12 × 4 mm (when dry), elongate and *curved* (at least when dry), red. **Seed** ca. 8.3 × 3.5 mm (when dry), curved; endosperm homogeneous.

DISTRIBUTION. Mount Dremsel, Manus Island.

HABITAT. Premontane rainforest on steep slopes at 600–750 m.

LOCAL NAMES. None recorded.

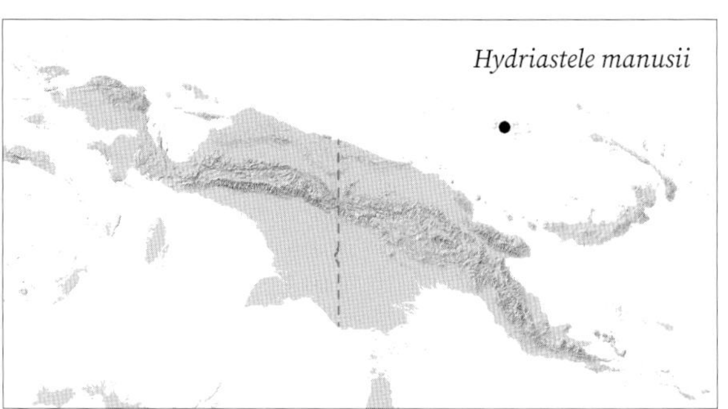

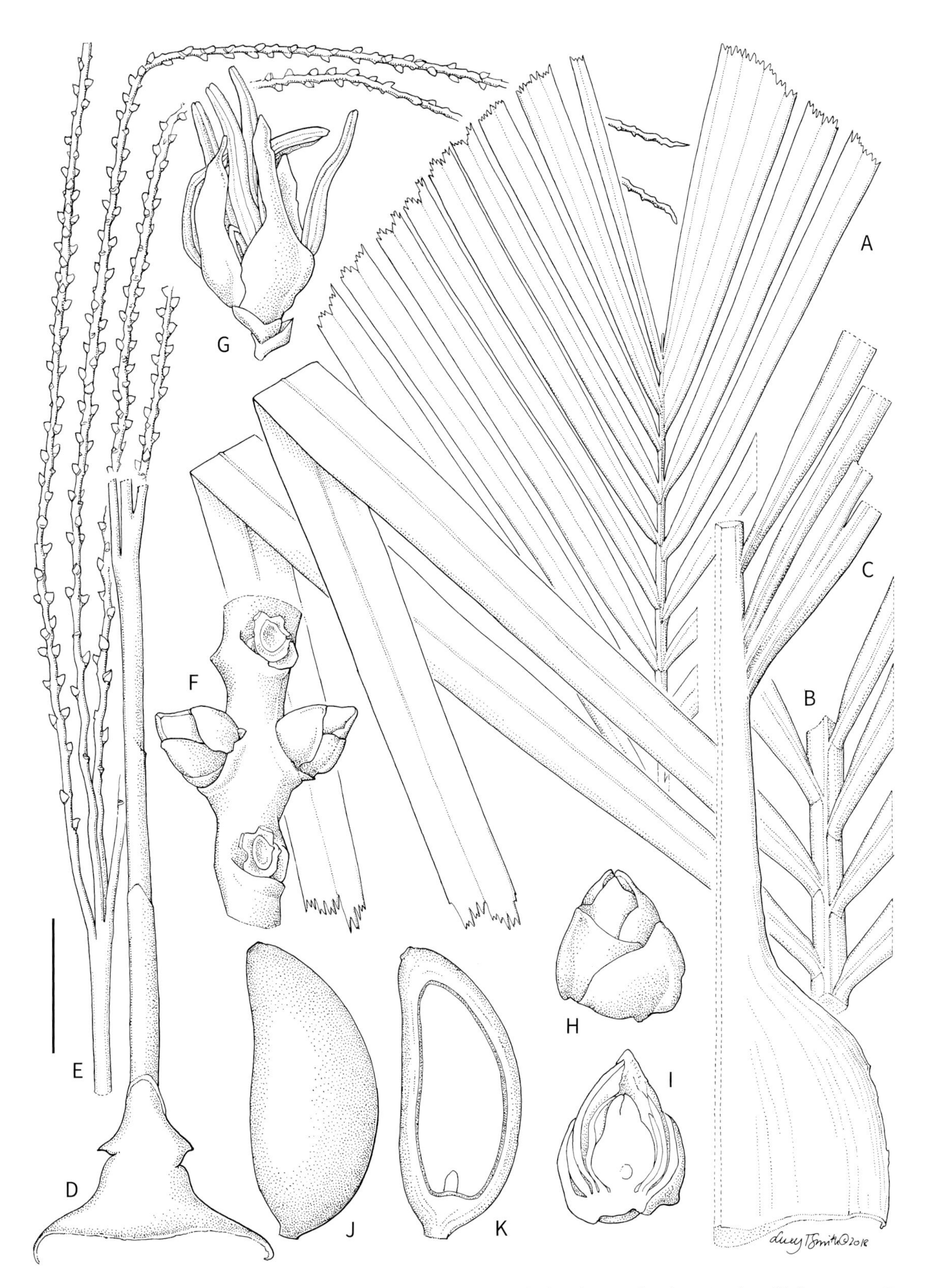

Hydriastele manusii. **A.** Leaf apex. **B.** Leaf mid-portion. **C.** Leaf base with sheath. **D.** Peduncle. **E.** Portion of inflorescence. **F.** Portion of rachilla with female flowers. **G.** Male flower. **H**, **I.** Female flower whole and in longitudinal section. **J**, **K.** Fruit whole and in longitudinal section. Scale bar: A–C = 6 cm; D, E = 4 cm; F, J, K = 5 mm; G = 4 mm; H, I = 3.3 mm. All from *Sands et al. 2880*. Drawn by Lucy T. Smith.

Hydriastele manusii. LEFT TO RIGHT: habit; juvenile individuals. Manus Island (AP).

USES. None recorded

CONSERVATION STATUS. Data Deficient (IUCN 2019). This species is known from only one site, where there is limited information on ongoing threats. Due to insufficient evidence, an extinction risk category cannot be selected.

NOTES. *Hydriastele manusii* is endemic to Manus Island. It is a moderately robust canopy palm distinguished by its short, arching leaves and inflorescences with very long (20–22 cm) peduncles. The species is unlikely to be confused with other palms.

16. *Hydriastele montana* (Becc.) W.J.Baker & Loo

Synonyms: *Gronophyllum montanum* (Becc.) Essig & B.E.Young, *Kentia beccarii* F.Muell., *Nengella montana* Becc.

Very slender, multi-stemmed and colony-forming palm to 1.5 m, bearing 4–5 leaves per crown, producing multiple running rhizomes. **Stem** 5–8 mm diam. **Leaf** 25–50 cm long including petiole; sheath 10–15 cm long; petiole 3–8 cm long; *leaflets ca. 9–11 each side of rachis, regularly arranged, single- or 2-fold, narrowly linear,* jagged or pointed at their tips. **Inflorescence** 7–9 cm long including 1–1.5 cm peduncle, *spicate, protandrous;* spike with yellowish to green axis; triads spirally arranged. **Male flower** with 6 stamens. **Female flower** with free sepals and free petals with conspicuous, triangular tips. **Fruit** ca. 12–19 × 4 mm when ripe (and dry), spindle-shaped, red. **Seed** ca. 12–14 × 3 mm (when dry), elongate to top-shaped; *endosperm homogeneous.*

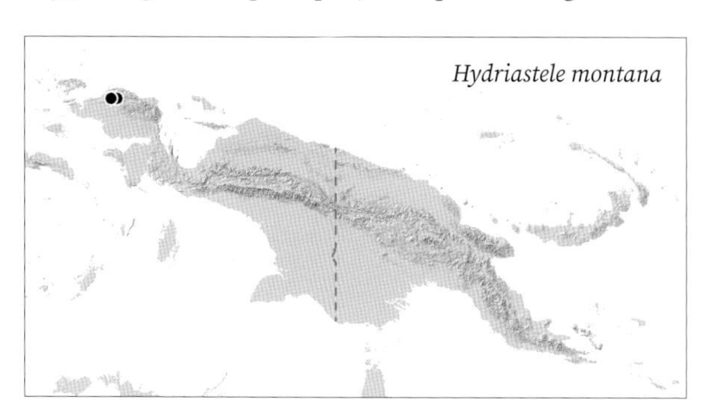

DISTRIBUTION. Tamrau Mountains, Bird's Head Peninsula.

HABITAT. Montane rainforest at 950–1,500 m.

LOCAL NAMES. None recorded.

USES. None recorded.

CONSERVATION STATUS. Data Deficient (IUCN 2019). However, as deforestation due to road building and logging is a major threat in its distribution range, a

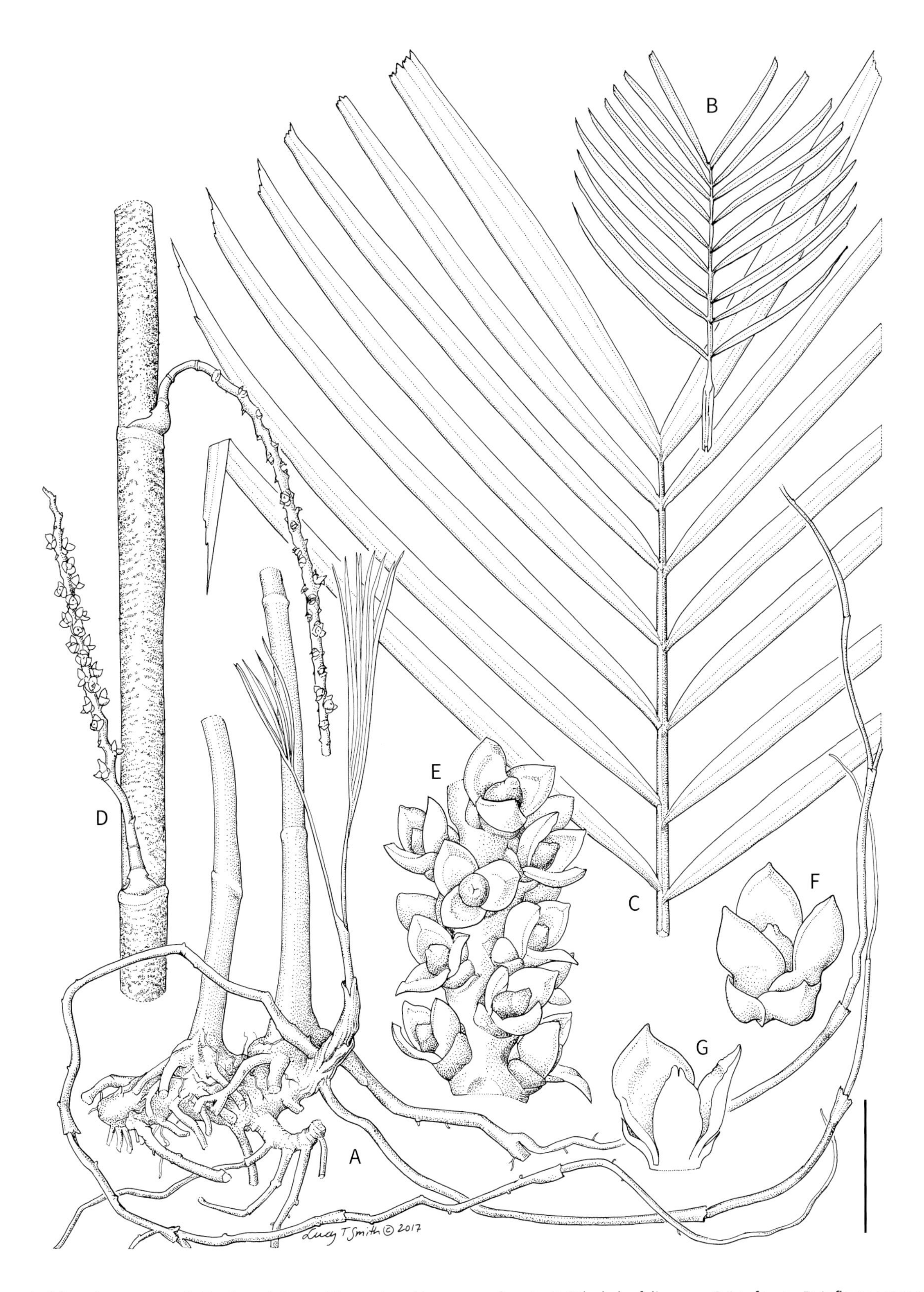

Hydriastele montana. **A.** Portion of stem with running rhizomes and roots. **B.** Whole leaf diagram. **C.** Leaf apex. **D.** Inflorescences attached to stem. **E.** Portion of rachilla with female flowers. **F**, **G.** Female flower whole and in longitudinal section. Scale bar: A, C = 4 cm; B = 18 cm; D = 2 cm; E = 7 mm; F, G = 4 mm. All from *Baker et al. 1380*. Drawn by Lucy T. Smith.

Hydriastele montana. LEFT TO RIGHT FROM TOP: colony showing gregarious habit, Aifat (CDH); crown, Tamrau Mountains (WB); inflorescence, Tamrau Mountains (WB); fruit, Aifat (CDH).

case could be made for this species to be placed in a threatened category.

NOTES. *Hydriastele montana* is a slender, colony-forming understorey palm. The species is distinguished by its leaves with 9–11 narrowly linear leaflets on each side of the rachis in combination with a spicate inflorescence and homogeneous endosperm. *Hydriastele montana* is most similar to *H. divaricata* and *H. flabellata*, but the former species has leaves with irregularly to subregularly arranged, divaricate leaflets and ruminate endosperm while the latter displays an entire-bifid leaf blade or a blade that is pinnate with up to six wedge-shaped leaflets on each side of the rachis.

17. *Hydriastele pinangoides* (Becc.) W.J.Baker & Loo

Synonyms: *Gronophyllum affine* (Becc.) Essig & B.E.Young, *Gronophyllum cyclopense* Essig & B.E.Young, *Gronophyllum densiflorum* Ridl., *Gronophyllum leonardii* Essig & B.E.Young, *Gronophyllum micranthum* (Burret) Essig & B.E.Young, *Gronophyllum pinangoides* (Becc.) Essig & B.E.Young, *Hydriastele affinis* (Becc.) W.J.Baker & Loo, *Hydriastele cyclopensis* (Essig & B.E.Young) W.J.Baker & Loo, *Hydriastele micrantha* (Burret) W.J.Baker & Loo, *Leptophoenix affinis* (Becc.) Becc., *Leptophoenix brassii* Burret, *Leptophoenix densiflora* (Ridl.) Burret, *Leptophoenix incompta* Becc., *Leptophoenix macrocarpa* Burret, *Leptophoenix mayrii* Burret, *Leptophoenix micrantha* Burret, *Leptophoenix microcarpa* Burret, *Leptophoenix minor* Becc., *Leptophoenix pinangoides*

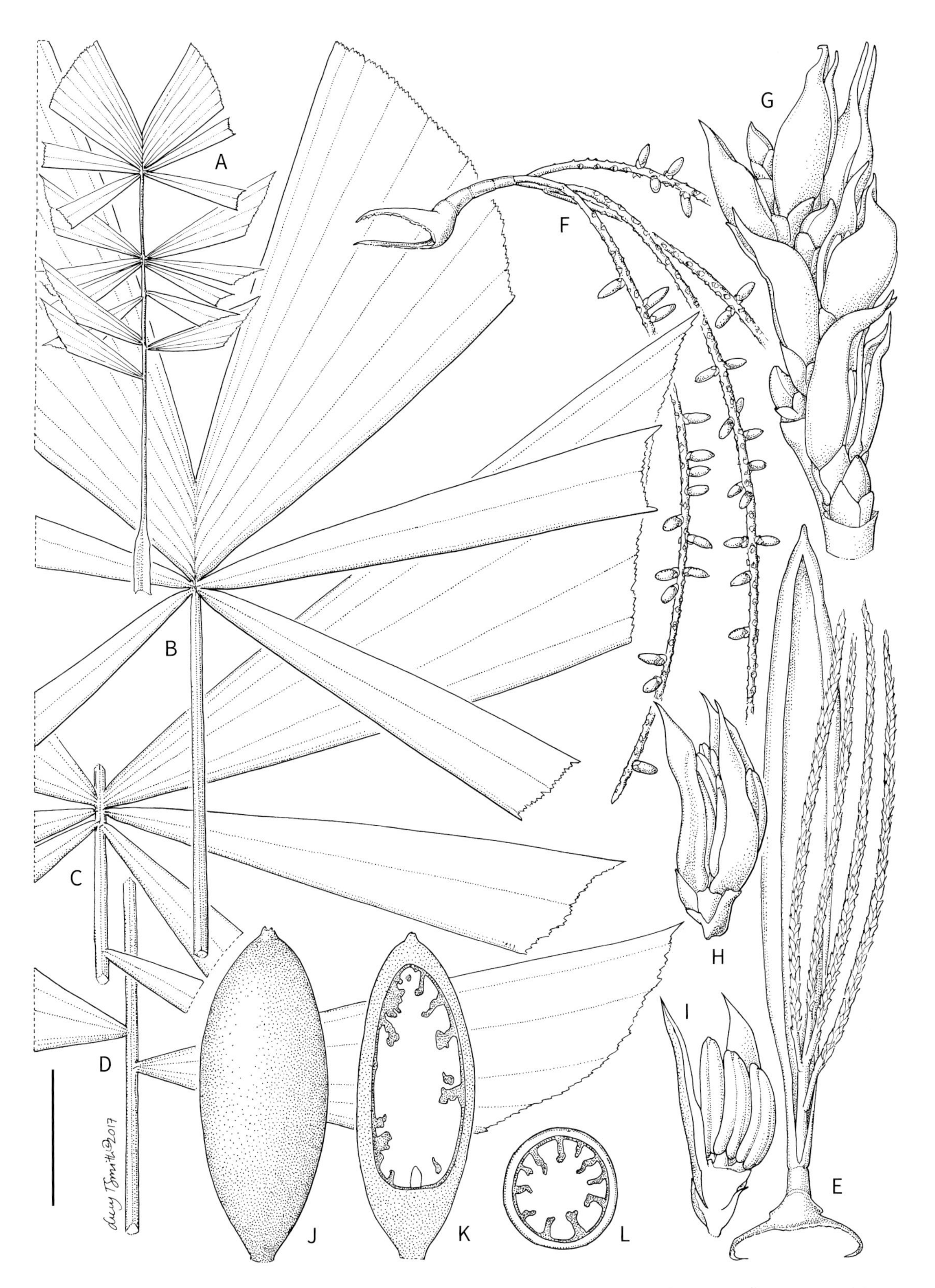

Hydriastele pinangoides. **A.** Whole leaf diagram. **B.** Leaf apex. **C.** Leaf mid-portion. **D.** Leaf base. **E.** Inflorescence with peduncular bract. **F.** Inflorescence with fruit. **G.** Portion of rachilla with triads. **H**, **I.** Male flower whole and in longitudinal section. **J–L.** Fruit whole, in longitudinal section and in transverse section. Scale bar: A = 24 cm; B–D = 8 cm; E, F = 4 cm; G–I = 5 mm; J–L = 7 mm. A–I from *Pullen 5767*; J, K from *Baker et al. 1057*. Drawn by Lucy T. Smith.

(Becc.) Becc., *Leptophoenix pterophylla* Becc., *Leptophoenix yulensis* Becc., *Nenga affinis* Becc., *Nenga calophylla* K.Schum. & Lauterb., *Nenga pinangoides* Becc., *Nengella affinis* (Becc.) Burret, *Nengella brassii* (Burret) Burret, *Nengella calophylla* (K.Schum. & Lauterb.) Becc., *Nengella densiflora* (Ridl.) Burret, *Nengella incompta* (Becc.) Burret, *Nengella macrocarpa* (Burret) Burret, *Nengella mayrii* (Burret) Burret, *Nengella micrantha* (Burret) Burret, *Nengella microcarpa* (Burret) Burret, *Nengella minor* (Becc.) Burret, *Nengella pinangoides* (Becc.) Burret, *Nengella pterophylla* (Becc.) Burret, *Nengella rhomboidea* Burret, *Nengella yulensis* (Becc.) Burret

Slender, single- or multi-stemmed palm to 7(–10) m, bearing 5–10 leaves per crown. **Stem** (0.8–)1.5–7.5 cm diam. **Leaf** 50–157 cm long including petiole; sheath 15–52 cm long; petiole 15–54 cm long; *leaflets 5–10(–13) each side of rachis, variable in size and shape, irregularly arranged usually in 2 or 3 widely spaced groups* (very rarely leaflets regularly arranged), *single- or multi-fold, broadly* (rarely narrowly) *wedge-shaped*, jagged at their tips. **Inflorescence** (13 –)18–30 cm long including 1.5–3 cm peduncle, branched to 1 order, *protandrous*; rachillae 2–5(–6), with pink axes; triads spirally arranged. **Male flower** with 6 stamens, pink to reddish. **Female flower** with free sepals and free petals with conspicuous, triangular tips, pink to reddish. **Fruit** 10–16 × 4–8 mm when ripe, ellipsoid or cylindrical to spindle-shaped, pink, red, purple or blackish. **Seed** 6–10 × 2–4 mm, ellipsoid; *endosperm ruminate.*

Hydriastele pinangoides. Habit, cultivated Royal Botanic Gardens, Kew (WB).

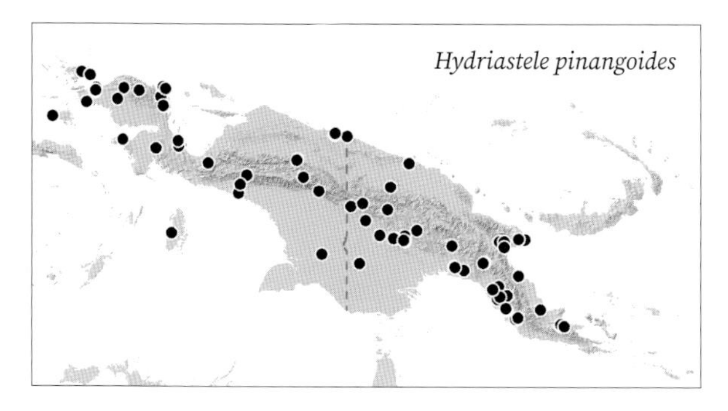

DISTRIBUTION. New Guinea, including Raja Ampat Islands, and the Aru islands.

HABITAT. Lowland to montane rainforest at sea level to 1350 m.

LOCAL NAMES. *Tapolo* (Pawaian), *Gilaia* (Waskuk), *Kobu* (Wagu), *Ugarreh* (Daga), *Mara* (Ambakanjah), *Kupal* (Gal), *Tegradri* (Irarutu), *Bim, Kabim* (Matbaat), *Biyal* (Yali), *Seraach* (Maibrat), *Sêméngbrè* (Hattam), *Manggam* (Maprik, unknown dialect), *Tooma* (Western Province, unknown dialect).

USES. Stems used for making spears, spearheads, bows, arrowheads and musical instruments and for sewing thatch. Young shoot (heart-of-palm) edible. Leaves used as roofing material. Palm used for magic and also planted locally as an ornamental.

CONSERVATION STATUS. Least Concern (IUCN 2019).

NOTES. *Hydriastele pinangoides* is a widespread, slender understorey to midstorey palm. It is distinguished by its variable leaves with 5–10 broadly wedge-shaped leaflets on each side of the rachis arranged in two or three distinct groups, in combination with 18–30 cm long inflorescences (rarely shorter) comprising 2–6 rachillae, and ruminate endosperm. Rarely, the leaflets can be more-or-less regularly arranged and narrowly wedge-shaped prompting comparison with *Hydriastele divaricata* and *H. simbiakii*, but *H. divaricata* has divaricate leaflets and inflorescences with 1–2 rachillae, while *H. simbiakii* has flexible, usually leaning stems and

Hydriastele pinangoides. LEFT TO RIGHT: flower buds, Tamrau Mountains (WB); inflorescence with immature fruit, Lake Kutubu (WB); fruit, Tamrau Mountains (WB).

14–16 leaflets on each side of the rachis. *Hydriastele pinangoides* can be confused with some pinnate-leaved forms of *H. flabellata,* although this species tends to have a shorter, spicate or bifid inflorescence, and a consistently homogeneous endosperm.

18. *Hydriastele procera* (Blume) W.J.Baker & Loo

Synonyms: *Areca procera* (Blume) Zipp. ex Blume, *Gronophyllum procerum* (Blume) H.E.Moore, *Kentia procera* Blume

Robust, single-stemmed palm to 30 m, bearing 10–15 leaves in the distinctively spherical crown. **Stem** 20–30 cm diam., moderately swollen. **Leaf** (1–)2–4 m long including petiole, *straight or slightly drooping*; sheath ca. 100 cm long, with some white bloom, crownshaft measurements unknown; petiole 15–20 cm long, upper surface flattened; leaflets 40–61 each side of rachis, regularly arranged, single-fold, pendulous, linear; apical leaflets pointed or briefly notched at the tip. **Inflorescence** 55–90 cm long including 7–12 cm peduncle, branched to 2 or 3 orders, *protandrous*; rachillae 23–40, ca. 2–3 mm diam.; triads decussately arranged; ca. 6 inflorescences present. **Male flower** with 6 stamens. **Female flower** with free sepals and free petals with conspicuous, triangular tips. **Fruit** *10–15 × 6–7 mm when ripe (and dry),* ellipsoid. **Seed** ca. 9.5 × 5 (when dry), ellipsoid; endosperm homogeneous.

DISTRIBUTION. Raja Ampat Islands (Waigeo, Gam), Kaimana and Triton Bay

HABITAT. Coastal limestone cliffs and crags at sea level to 200 m.

LOCAL NAMES. None recorded.

USES. Flooring.

CONSERVATION STATUS. Data Deficient (IUCN 2019). However, as deforestation is a major threat in its distribution range, a case could be made for this species to be placed in a threatened category.

NOTES. *Hydriastele procera* is a tall and robust palm that often occurs gregariously on limestone. The species is distinguished by its spherical crown of more-or-less straight leaves in combination with protandrous inflorescences with relatively slender rachillae (ca.

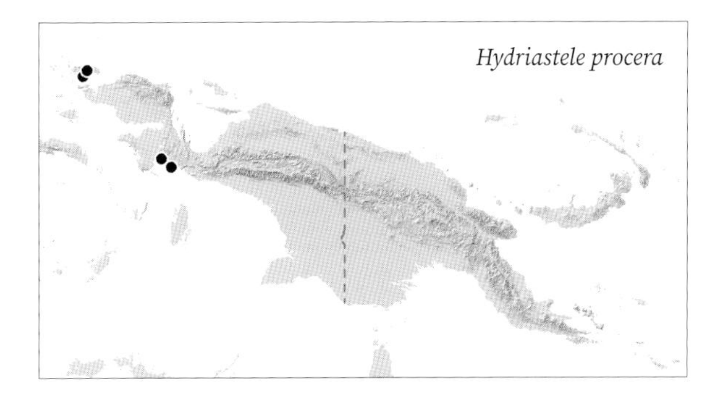

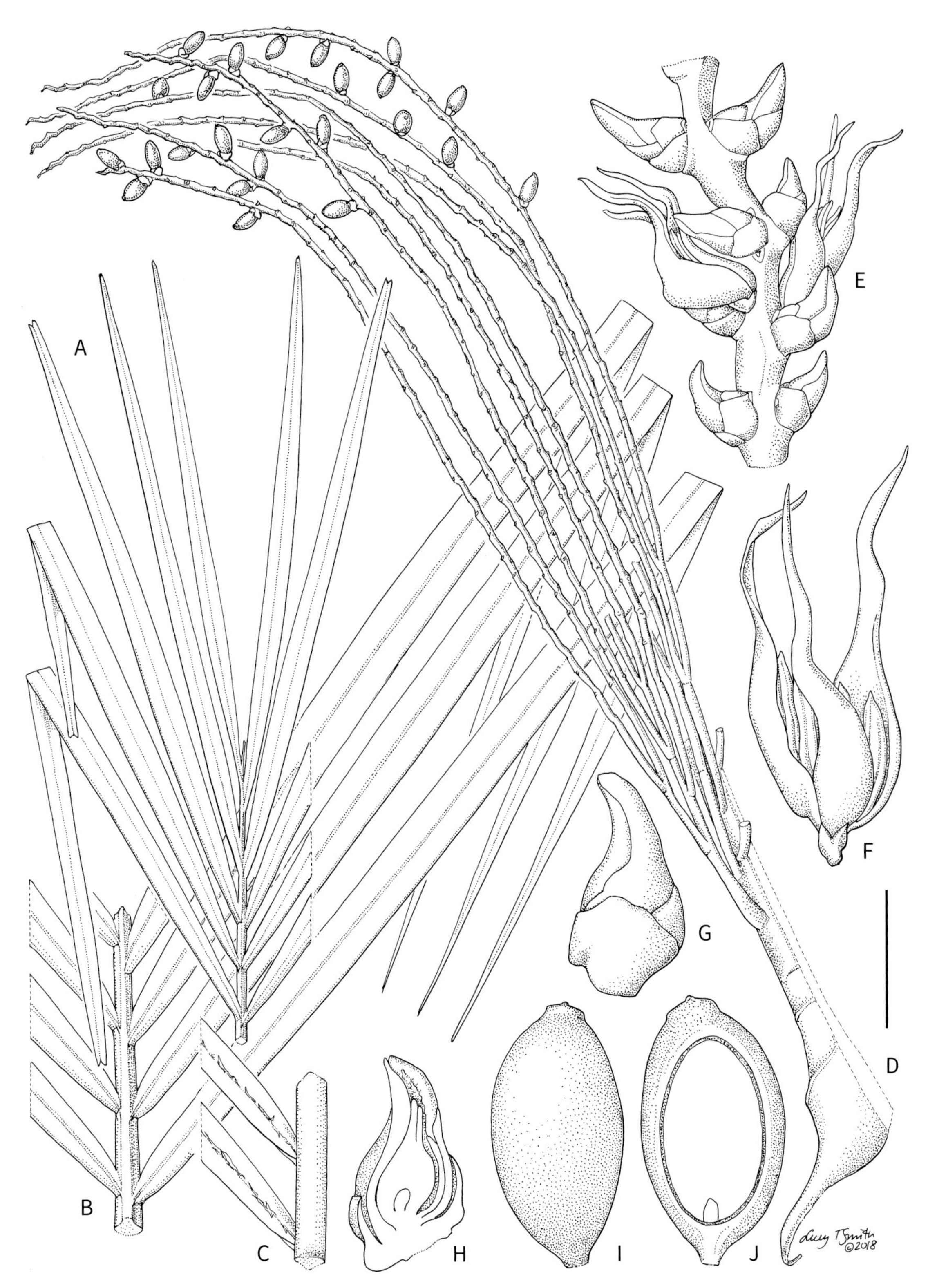

Hydriastele procera. **A.** Leaf apex. **B.** Leaf mid-portion adaxial side. **C.** Leaf mid-portion abaxial side. **D.** Portion of inflorescence with fruit. **E.** Portion of rachilla with triads. **F.** Male flower. **G**, **H.** Female flower whole and in longitudinal section. **I**, **J.** Fruit whole and in longitudinal section. Scale bar: A–D = 8 cm; E, I, J = 7.5 mm; F = 5 mm; G, H = 3.3 mm. A–D, I, J from *Maturbongs 532*; E–H from *Heatubun et al. 1133*. Drawn by Lucy T. Smith.

Hydriastele procera. FROM TOP: habit; colony growing gregariously on limestone. Triton Bay (CH).

2–3 mm diam.), male flowers with six stamens, and relatively large fruit (10–15 × 6–7 mm when dry). *Hydriastele procera* is most similar to *H. wosimiensis*, but that species has thicker rachillae (ca. 3.5 mm diam.), male flowers with 12 stamens and smaller fruit (7.5–9.5 × 5–5.5 mm when dry). *Hydriastele procera* has a similar habit to *H. costata* but the latter has protogynous inflorescences and its stem is not swollen.

19. *Hydriastele rheophytica* Dowe & M.D.Ferrero

Slender, multi-stemmed, *rheophytic palm* to 6 m, bearing 4–12 leaves per crown. **Stem** 2–2.5 cm diam., *flexible and mostly leaning*. **Leaf** ca. 95–120 cm long including petiole; sheath 40–45 cm long, petiole 20–30 cm long, flexible; *leaflets 18–32 each side of rachis, regularly arranged, linear, thin and soft*; basal leaflets single-fold, *obliquely jagged at their tips; apical leaflets comprising 2 or 3 folds*, truncately jagged at the tip. **Inflorescence** 16–30 cm long including 2.5–7 cm peduncle, branched to 1 or 2 orders, *protogynous*; triads decussately arranged. **Male flower** with 6 stamens. **Female flowers** with free sepals and free, rounded, low petals. **Fruit** ca. 7 mm long, globose to broadly ellipsoid. **Seed** globose; endosperm shallowly ruminate.

DISTRIBUTION. Idenburg and Frieda Rivers, central New Guinea (Dowe & Ferrero 2000).

HABITAT. Stream and river margins in the flood zone at ca. 850 m.

Hydriastele rheophytica. Habit, Omasai Creek (MF).

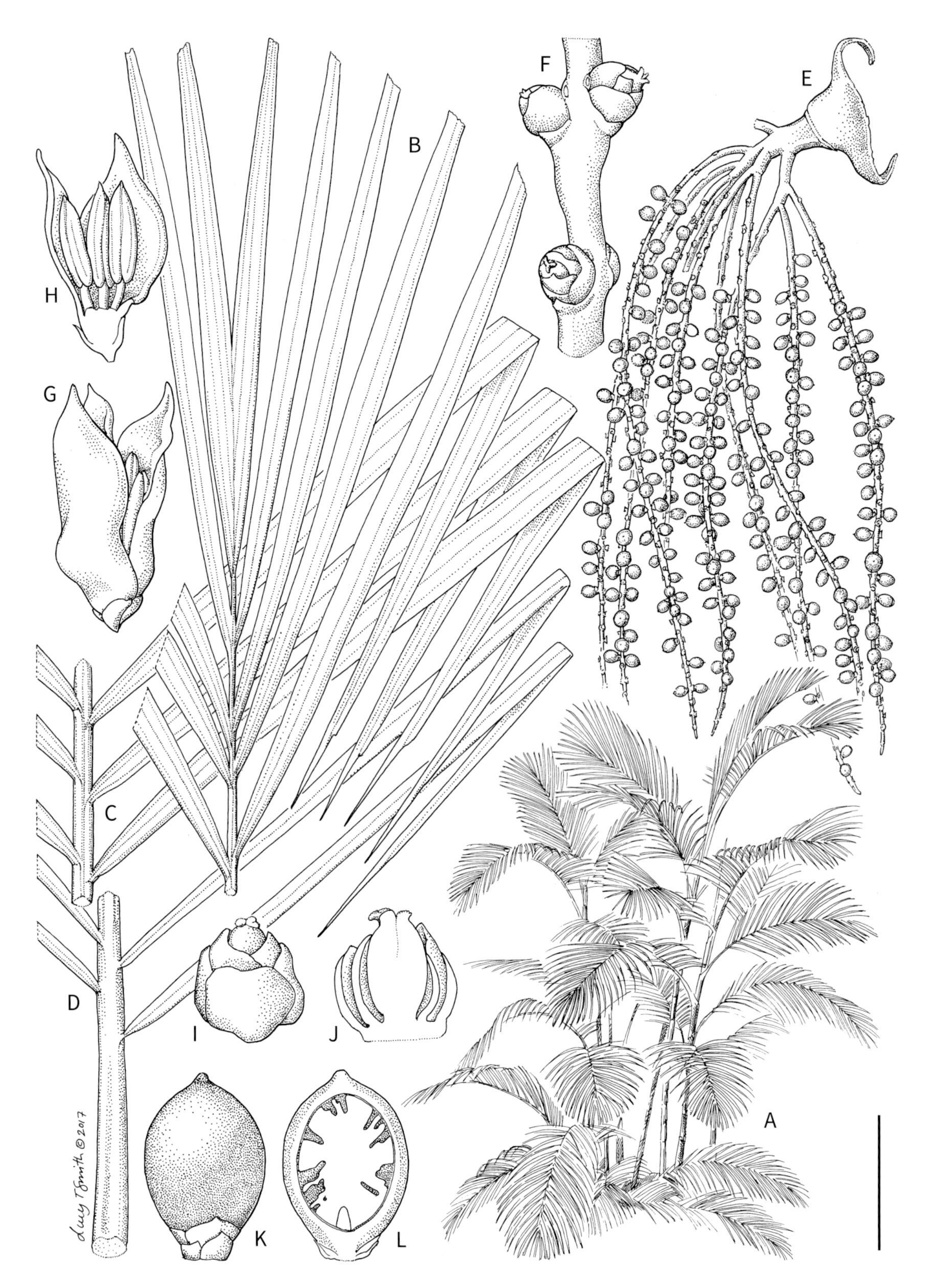

Hydriastele rheophytica. **A.** Habit. **B.** Leaf apex. **C.** Leaf mid-portion. **D.** Leaf base. **E.** Inflorescence with fruit. **F.** Portion of rachilla with female flowers. **G, H.** Male flower whole and in longitudinal section. **I, J.** Female flower whole and in longitudinal section. **K, L.** Fruit whole and in longitudinal section. Scale bar: A = 70 cm; B–D = 6 cm; E = 4 cm; F–H, K, L = 5 mm; I, J = 2.5 mm. A, K, L from *Dowe 536*; B–J from *Brass 13700*. Drawn by Lucy T. Smith.

Hydriastele rheophytica. LEFT TO RIGHT FROM TOP: habit; stem showing mottled markings; inflorescences; inflorescence with fruit. Cultivated Floribunda Palms and Exotics, Hawaii (WB).

LOCAL NAMES. None recorded.

USES. None recorded.

CONSERVATION STATUS. Data Deficient (IUCN 2019). This species is known from only two sites, where there is limited information on ongoing threats. Due to insufficient evidence, an extinction risk category cannot be selected.

NOTES. *Hydriastele rheophytica* is restricted to riverbanks where it forms large clumps of flexible, somewhat leaning stems. The species is distinguished by its leaves with numerous thin and soft, linear leaflets, the basal ones of which are obliquely jagged at their tips, and a terminal leaflet pair comprising two or three folds. This species is most similar to *H. variabilis*, but that species is not rheophytic and has erect, non-flexible stems and apical leaflets comprising 4–11 folds. See below for a comparison to the second rheophytic species of *Hydriastele*, *H. simbiakii*. The species also resembles the unrelated rheophytic *Heterospathe macgregorii*, although they are not known to co-occur.

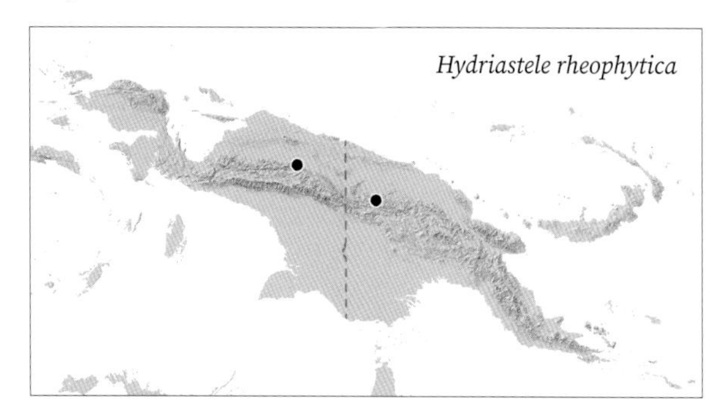

20. *Hydriastele simbiakii* Heatubun, Petoe & W.J. Baker

Slender, multi-stemmed, *rheophytic palm* to 5–6 m, bearing ca. 6–7 leaves per crown. **Stem** 1.5–2.5 cm diam., *flexible, shorter stems erect, longer stems leaning.* **Leaf** 64–104 cm long including petiole; sheath 26–36 cm long; petiole 12–25 cm long; *leaflets 14–16 each side of rachis, regularly arranged,* single- or multi-fold, linear, jagged at their tips. **Inflorescence** 13–17 cm long including 1.5–3 cm peduncle, branched to 1 order, *protandrous*; rachillae 3–4, with red axes; triads spirally arranged. **Male flower** with 6 stamens, pink. **Female flower** with free sepals and free petals with conspicuous, triangular tips, white to pinkish. **Fruit** 8–12 × 4–5 mm when ripe, ellipsoid, red. **Seed** 7–8 × 4 mm, ellipsoid; *endosperm ruminate.*

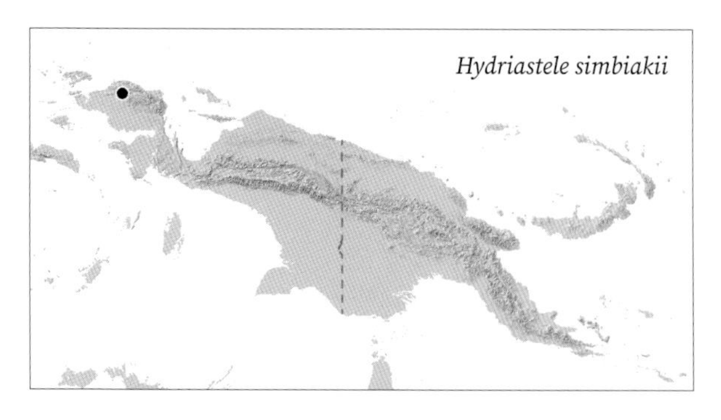

DISTRIBUTION. Tamrau Mountains, Bird's Head Peninsula.

HABITAT. Rocky river margins in the flood zone at ca. 400 m.

LOCAL NAMES. None recorded.

USES. Stems used for making spears.

CONSERVATION STATUS. Data Deficient (IUCN 2019). However, as deforestation due to road building and

Hydriastele simbiakii. Colony showing rheophytic habit, Tamrau Mountains (WB).

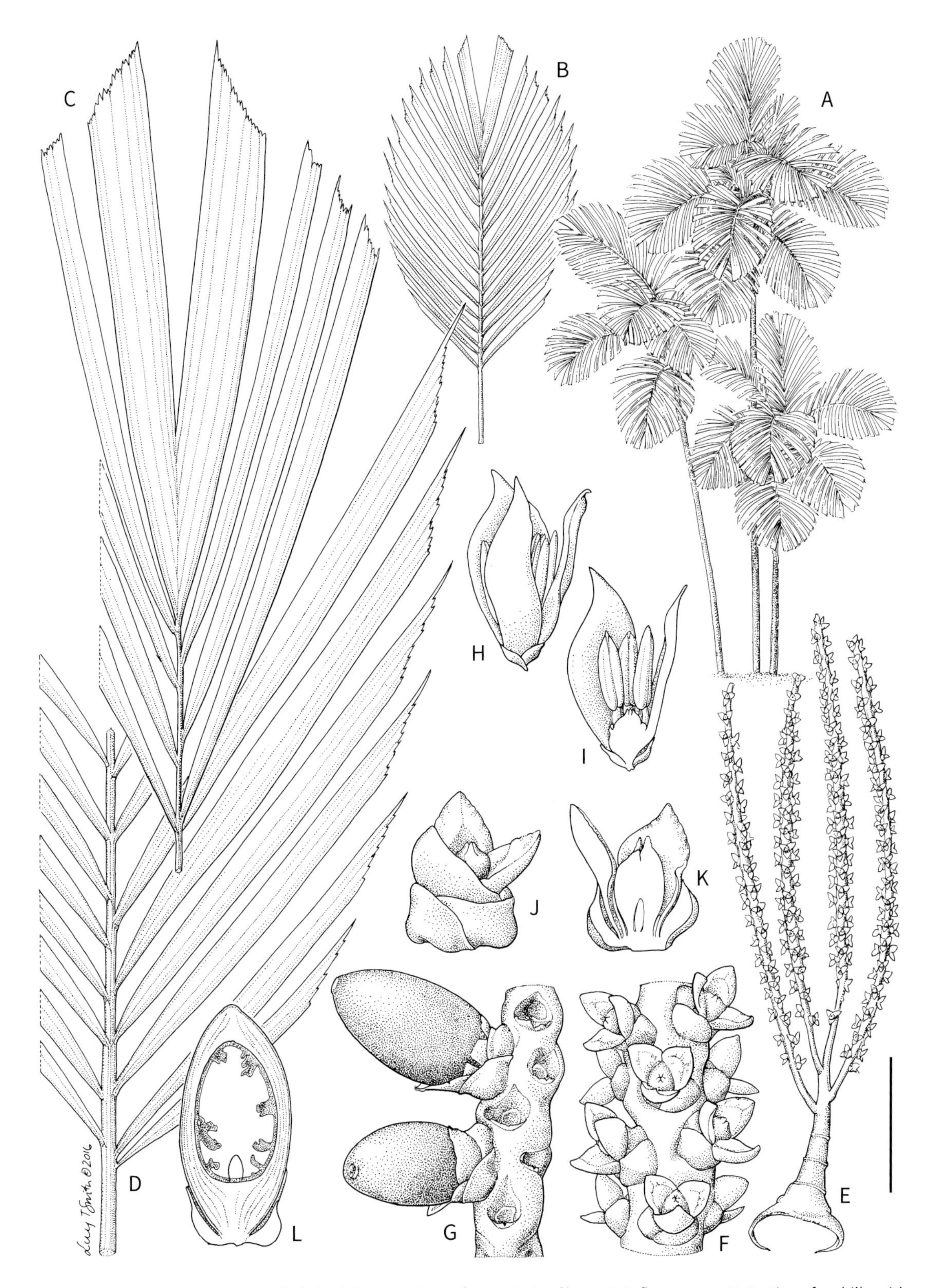

***Hydriastele simbiakii*.** **A.** Habit. **B.** Whole leaf diagram. **C.** Leaf apex. **D.** Leaf base. **E.** Inflorescence. **F.** Portion of rachilla with female flowers. **G.** Portion of rachilla with fruit. **H**, **I.** Male flower whole and in longitudinal section. **J**, **K.** Female flower whole and in longitudinal section. **L.** Fruit in longitudinal section. Scale bar: A = 60 cm; B = 24 cm; C, D = 6 cm; E = 4 cm; F–I = 8 mm; J, K = 4 mm; L = 7 mm. All from *Baker et al. 1365*. Drawn by Lucy T. Smith.

Hydriastele simbiakii. LEFT TO RIGHT FROM TOP: habit; inflorescences; female flowers. Tamrau Mountains (WB).

logging is a major threat in its distribution range, a case could be made for this species to be placed in a threatened category.

NOTES. *Hydriastele simbiakii* is one of two strictly rheophytic species known in the genus, the other being *H. rheophytica*. *Hydriastele simbiakii* forms distinctive clumps of somewhat leaning, flexible stems and has leaves with 14–16 regularly arranged, linear leaflets on each side of the rachis, and ruminate endosperm. The species is most easily confused with rare forms of *H. pinangoides* with regularly pinnate leaves and similar inflorescences, but *H. simbiakii* is distinguished from these forms by its ecological preference and leaves with more numerous leaflets.

21. *Hydriastele splendida* Heatubun, Petoe & W.J.Baker

Slender, single- or multi-stemmed palm to 4 m, bearing ca. 4–5 leaves per crown. **Stem** 1.5–2 cm diam. **Leaf** ca. 100 cm long including petiole; sheath ca. 30 cm long; petiole ca. 20 cm long; *blade ca. 80 × 40 cm, entire-bifid,* obovate, distal half of margin rounded and jagged. **Inflorescence** ca. 18 cm long including ca. 2 cm peduncle, branched to 1 order, *protandrous;* rachillae 3–6, with reddish axes; triads spirally arranged. **Male flower** not seen. **Female flower** with free sepals and free petals with conspicuous, triangular tips, pink. **Fruit** ca. 1 cm long when ripe, cylindrical, pink. **Seed** ca. 7 × 1.5 mm (when dry), top-shaped; *endosperm ruminate.*

DISTRIBUTION. Foothils below Mount Jaya (Dransfield *et al.* 2000b), near Timika.

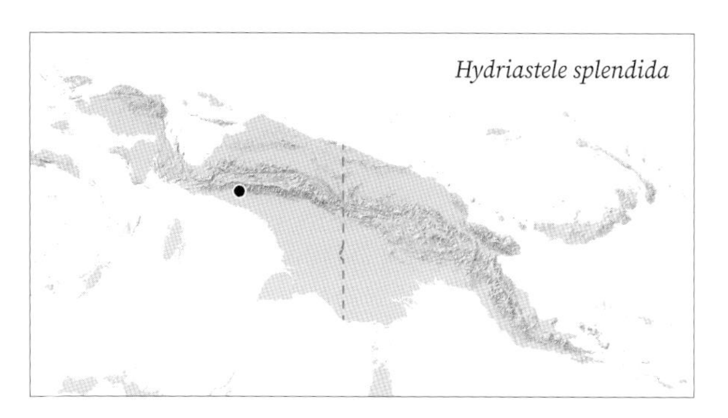

HABITAT. Heath forest on out-washed sands and gravels on steep and mossy slopes at ca. 450 m.

LOCAL NAMES. None recorded.

USES. None recorded.

CONSERVATION STATUS. Data Deficient (IUCN 2019). This species is known only from a single site, where

Hydriastele splendida. LEFT TO RIGHT: habit, Timika-Tembagapura road (WB); emerging leaf, cultivated Floribunda Palms and Exotics, Hawaii (WB).

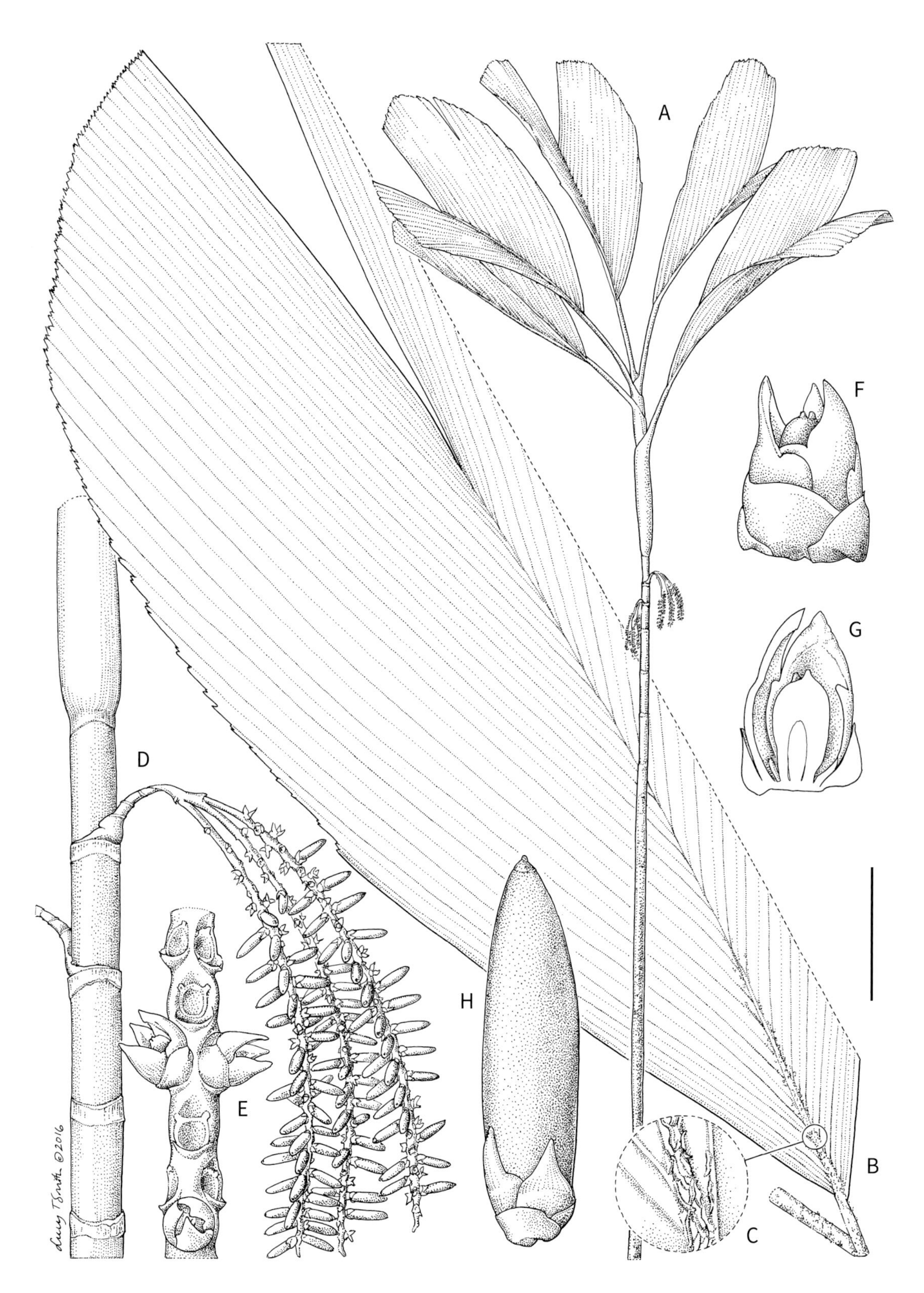

***Hydriastele splendida*. A.** Habit. **B.** Leaf. **C.** Detail of leaf indumentum. **D.** Inflorescence with fruit attached to stem. **E.** Portion of rachilla with female flowers. **F, G.** Female flower whole and in longitudinal section. **H.** Fruit. Scale bar: A = 40 cm; B = 8 cm; C = 1.5 cm; D = 4 cm; E = 7 mm; F, G = 3.3 mm; H = 5 mm. All from *Baker et al. 820*. Drawn by Lucy T. Smith.

Hydriastele splendida.
LEFT TO RIGHT: inflorescence with fruit; female flowers. Cultivated Floribunda Palms and Exotics, Hawaii (WB).

there is limited information on ongoing threats. Due to insufficient evidence, an extinction risk category cannot be selected.

NOTES. *Hydriastele splendida* is a slender understorey palm, immediately recognisable on account of its entire-bifid leaves in combination with inflorescences comprising 3–4 rachillae and ruminate endosperm. The species is similar to *H. pinangoides* in habit and reproductive morphology but is easily distinguished from this by its entire leaves. *Hydriastele splendida* is more readily confused with entire-leaved forms of *H. flabellata*, but *H. flabellata* differs in its inflorescences with typically one (spicate) or two rachillae, and homogeneous endosperm.

22. *Hydriastele variabilis* (Becc.) Burret

Synonyms: *Adelonenga variabilis* (Becc.) Becc., *Hydriastele variabilis* var. *sphaerocarpa* (Becc.) Burret, *Nenga variabilis* Becc., *Nenga variabilis* var. *sphaerocarpa* Becc.

Slender, multi-stemmed palm to 8 m, bearing 4–7 leaves per crown. **Stem** 2–4 cm diam. **Leaf** 95–140 cm long including petiole; sheath 27–45 cm long; petiole 9–30 cm long; *leaflets 11–23 each side of rachis, regularly to subregularly arranged, linear*; basal leaflets single-fold, *pointed or obliquely jagged at their tips; apical leaflets comprising 4–11 folds,* truncately jagged at the tip. **Inflorescence** 10–25 cm long including 2–2.5 cm peduncle, branched to 1 or 2 orders, *protogynous*; triads decussately arranged. **Male flower** not seen. **Female**

Hydriastele variabilis. Habit, Aifat (CDH).

Hydriastele variabilis. **A.** Habit. **B.** Leaf apex. **C.** Leaf mid-portion. **D.** Inflorescences. **E.** Portion of rachilla with female flowers. **F.** Portion of rachilla with fruit. **G**, **H.** Female flower whole and in longitudinal section. **I**, **J.** Fruit in longitudinal and transverse section. Scale bar: A = 40 cm; B, C = 4 cm; D = 6 cm; E = 1 cm; F = 1.5 cm; G, H = 3 mm; I, J = 7 mm. A, D, E, G, H from *Gardiner 424*; B, C, F, I, J from *Baker et al. 1369*. Drawn by Lucy T. Smith.

Hydriastele variabilis. LEFT TO RIGHT: crown, Tamrau Mountains (WB); inflorescences with female flowers, Tamrau Mountains (WB); inflorescence with fruit, Aifat (CDH).

flower with free sepals and free, rounded, low petals. **Fruit** 10–12 × 6.3–7.8 mm when ripe, ellipsoid to ovoid with ends tapering when dry, red. **Seed** 7–8.2 × 5–6 mm, ovoid; endosperm deeply ruminate.

DISTRIBUTION. Bird's Head and Bomberai Peninsulas.

HABITAT. Lowland or premontane rainforest at sea level to 1,200 m.

LOCAL NAMES. *Sagarofa* (Sumuri), *Pinang Oetan* (Malay).

USES. None recorded.

CONSERVATION STATUS. Least Concern (IUCN 2019). However, monitoring is required due to deforestation. Some localities have already been replaced by oil palm plantations.

NOTES. *Hydriastele variabilis* is a slender understorey to midstorey palm distinguished by its leaves, which have a well-defined petiole and regularly to subregularly arranged, linear leaflets and a terminal leaflet pair comprising 4–11 folds. The species is most similar to *H. apetiolata* and *H. rheophytica,* which both have regularly pinnate leaves, but the former species is less slender and its adult leaves lack a petiole, while the latter species is a rheophyte with flexible stems and apical leaflets comprising 2 or 3 folds.

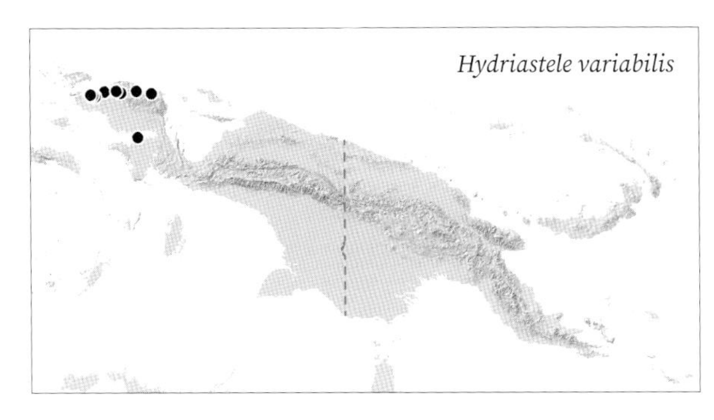

23. *Hydriastele wendlandiana* (F.Muell.) H.Wendl. & Drude

Synonyms: *Adelonenga geelvinkiana* (Becc.) Becc., *Adelonenga microspadix* (Warb. ex K.Schum. & Lauterb.) Becc., *Hydriastele beccariana* Burret, *Hydriastele carrii* Burret, *Hydriastele douglasiana* F.M.Bailey, *Hydriastele geelvinkiana* (Becc.) Burret, *Hydriastele lepidota* Burret, *Hydriastele microspadix* (Warb. ex K.Schum. & Lauterb.) Burret, *Hydriastele rostrata* Burret, *Hydriastele wendlandiana* var. *microcarpa* H.Wendl. & Drude, *Kentia microspadix* Warb. ex K.Schum. & Lauterb., *Kentia wendlandiana* F.Muell., *Nenga geelvinkiana* Becc., *Ptychosperma beccarianum* Warb. ex K.Schum. & Lauterb.

Slender to moderate, single- or multi-stemmed palm to 17 m, bearing 5–12 leaves per crown. **Stem** 2–10 cm diam. **Leaf** 1–2.5 m long including petiole; sheath 40–73 cm long; petiole (2–)10–80 cm long; *leaflets 12–30 each side of rachis, usually irregularly arranged with a group of closely spaced leaflets in different planes in the middle* *of the leaf, wedge-shaped, truncately jagged at their tips.* **Inflorescence** (16–)21–50 cm long including 2.5–7 cm peduncle, with (1–)2(–3) orders of branching, *protogynous*; triads decussately arranged. **Male flower** with (5–)6(–8) stamens. **Female flower** with free sepals and free, rounded, low petals. **Fruit** 7–9(–11) × 6–8 mm when ripe, globose to ovoid, purple to reddish. **Seed** 5–7 × 5–6.2 mm, globose to ovoid; endosperm homogeneous to ruminate.

DISTRIBUTION. Widespread throughout New Guinea and surrounding islands. Also in northern Australia.

HABITAT. Rainforest at sea level to 1,000 m.

LOCAL NAMES. *Kantrabel, Inpsal* (Kanum), *Honggomi, Patani, Sanggum* (Wondama), *Sal* (Amele), *Kitat* (Daga), *Kenege* (Kutubu), *Upo* (Meko), *Lai* (Matbat), *Sirata* (Sayal), *Kelkal* (Aru Islands), *Bil* (Mianmin), *Koeyauw* (Yei), *Befer* (Marap), *Kava Kava* (Patep, Buangs), *Sapuh* (Maprik), *Morr* (Gal), *Salvaik* (Sempi), *Fabu* (Ambakanja), *Ndzip* (Timbunke), *Kanyaweni* (Konti-unai), *Kaikinei* (Woi).

USES. Used for flooring, roofing, bed construction, arrows, bows, spears, harpoons and chicken coops. The young shoots are consumed. Locally planted as an ornamental.

CONSERVATION STATUS. Least Concern (IUCN 2019).

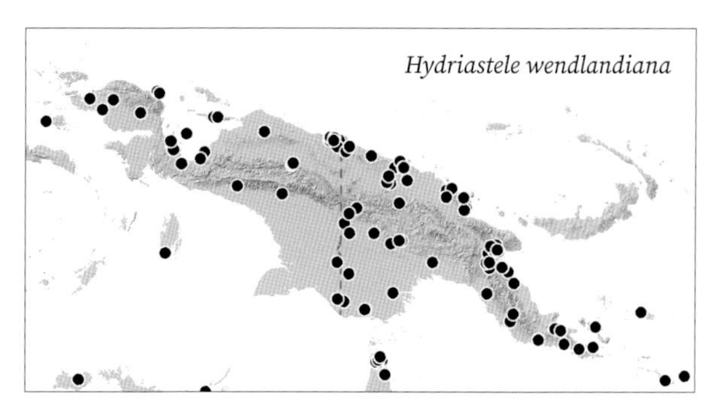

NOTES. *Hydriastele wendlandiana* is a variable, widespread and common understorey to midstorey palm distinguished by its leaves with 12–30 irregularly arranged, truncately jagged leaflets on each side of the rachis and leaf sheaths longer than 40 cm. *Hydriastele wendlandiana* is most similar to *H. kasesa*, but that species differs in being of a generally shorter stature, having leaves with 6–13 leaflets on each side of the rachis and leaf sheaths that do not exceed 30 cm in length. Rarely, leaflets of *H. wendlandiana* are regularly arranged prompting comparison with *H. variabilis* and *H. rheophytica*, but *H. variabilis* usually has shorter inflorescences (10–25 cm long as opposed to normally 21–50 cm in *H. wendlandiana*), and *H. rheophytica* is a rheophyte with flexible stems and leaves with numerous, flexible leaflets.

Hydriastele wendlandiana. LEFT TO RIGHT: crown, Nadzab (WB); inflorescences, cultivated Lae (WB); fruit, near Kikori (WB).

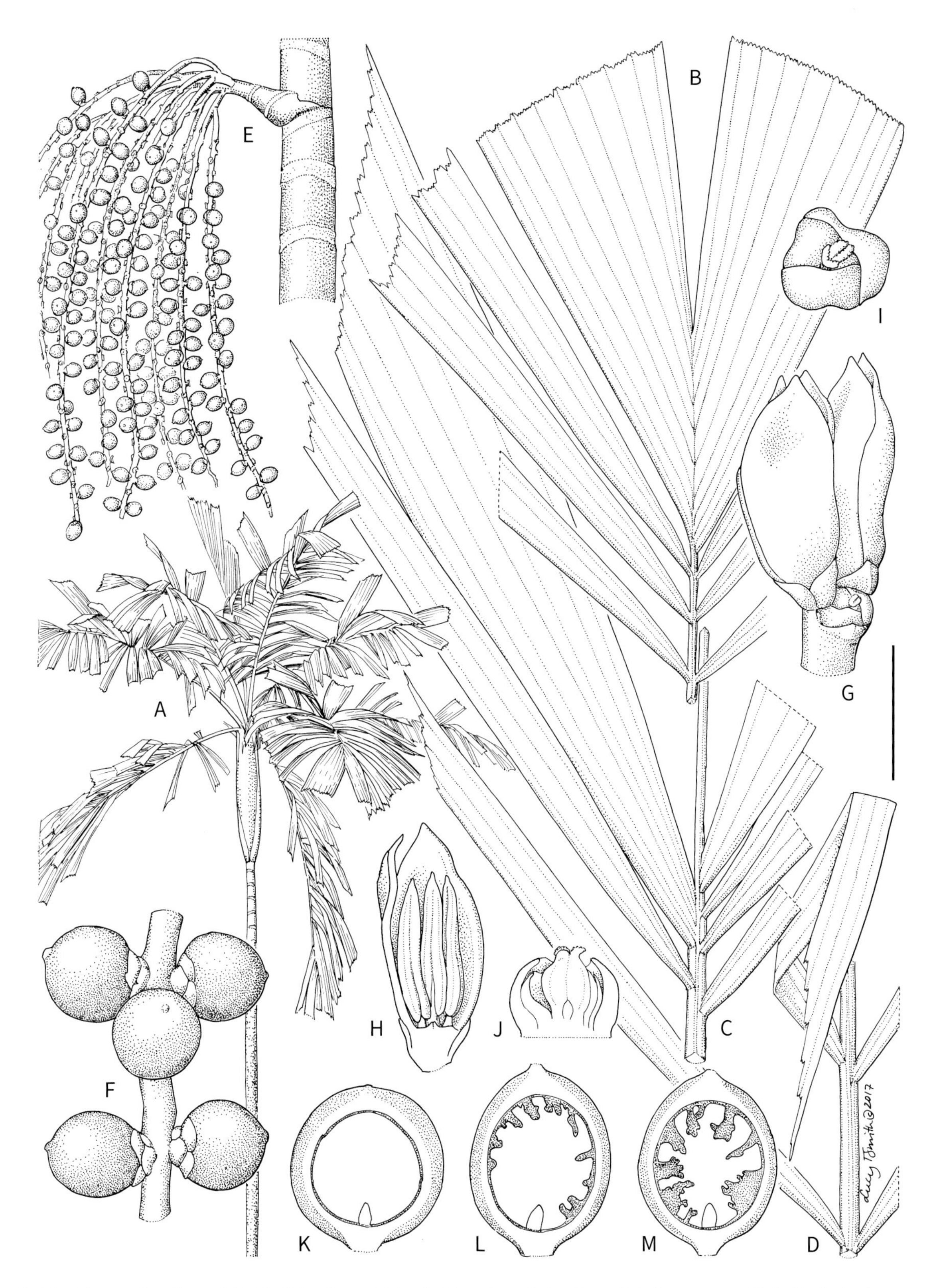

Hydriastele wendlandiana. **A.** Habit. **B.** Leaf apex. **C.** Leaf mid-portion. **D.** Leaf base. **E.** Inflorescence with fruit attached to stem. **F.** Portion of rachilla with fruit. **G.** Triad. **H.** Male flower in longitudinal section. **I, J.** Female flower whole and in longitudinal section. **K.** Homogeneous endosperm. **L.** Shallowly ruminate endosperm. **M.** Ruminate endosperm. Scale bar: A = 70 cm; B–D = 8 cm; E = 6 cm; F = 1.5 cm; G, H = 5 mm; I, J = 2.5 mm; K–M = 7 mm. A, G–J from *Baker et al. 1106*; B–E, M from *Klappa 151*; F, K from *Baker et al. 1065*; L from *Baker et al. 573*. Drawn by Lucy T. Smith.

24. *Hydriastele wosimiensis* W.J.Baker & Petoe

Robust, single-stemmed canopy palm to 30 m, bearing ca. 20 leaves in the (?spherical) crown. **Stem** ca. 25 cm diam., moderately swollen. **Leaf** 2.5–3 m long including petiole, *slightly drooping*; sheath ca. 100 cm long, covered with copious white indumentum, crownshaft ca. 110 × 25 cm; petiole ca. 40 cm long, upper surface flattened; leaflets ca. 50 each side of rachis, regularly arranged, single-fold, pendulous, linear; apical leaflets briefly notched at the tip. **Inflorescence** ca. 80 cm long including 12 cm peduncle, branched to 2 orders, *protandrous*; rachillae ca. 25, ca. 3.5 mm diam.; triads decussately arranged; ca. 26 inflorescences present (at various stages). **Male flower** with 12 stamens, cream. **Female flower** with free sepals and free petals with conspicuous, triangular tips, cream. **Fruit** *9.5–10 × 6–8 mm when ripe (7.5–9.5 × 5–5.5 mm when dried)*, ellipsoid, red. **Seed** 5.6–6.7 × 5–5.5 mm, broadly ellipsoid; endosperm homogeneous.

DISTRIBUTION. Wosimi River area, south of the Wandamen Peninsula.

HABITAT. Lowland rainforest on shallow slopes at ca. 200 m.

LOCAL NAMES. *Kaparo* (Wandama).

USES. Stems used for floor boards.

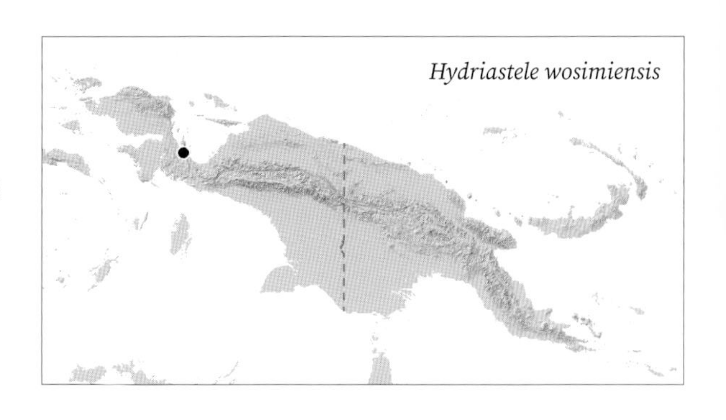

CONSERVATION STATUS. Data Deficient (IUCN 2019). However, as deforestation due to oil palm plantations and logging concessions is a major threat in its distribution range, a case could be made for this species to be placed in a threatened category.

NOTES. *Hydriastele wosimiensis* is a robust palm distinguished by its slightly drooping leaves with pendulous leaflets in combination with protandrous inflorescences with relatively thick rachillae (ca. 3.5 mm diam.), male flowers with 12 stamens, and small fruit (7.5–9.5 × 5–5.5 mm when dry). *Hydriastele wosimiensis* is most similar to *H. procera* although that species has more slender rachillae (ca. 2–3 mm diam.), male flowers with six stamens and larger fruit (10–15 × 6–7 mm when dry).

Hydriastele wosimiensis.
LEFT TO RIGHT: inflorescences; fruit. Wandamen Peninsula (WB).

Hydriastele wosimiensis. **A.** Portion of stem. **B.** Leaf apex. **C.** Leaf mid-portion. **D.** Inflorescence with fruit. **E.** Portion of rachilla with flowers. **F.** Portion of rachilla with fruit. **G**. Triad. **H.** Male flower in longitudinal section. **I**, **J.** Female flower whole and in longitudinal section. **K.** Fruit in longitudinal section. Scale bar: A = 3 cm; B, C = 8 cm; D = 6 cm; E = 1 cm; F = 1.5 cm; G, K = 7 mm; H = 5 mm; I, J = 3.3 mm. All from *Baker et al. 1068*. Drawn by Lucy T. Smith.

Rhopaloblaste ceramica.
Habit, cultivated National Botanic Garden, Lae (WB).

Rhopaloblaste Scheff.

Synonym: *Ptychoraphis* Becc.

Tall — leaf pinnate — crownshaft — no spines — leaflets pointed

Robust, single-stemmed tree palms, crownshaft present, monoecious. **Leaf** pinnate, straight, *dark hairs on the leaf rachis* (*most easily observed in recently expanded leaves*); sheath tubular; petiole usually short; leaflets numerous, with pointed tips, arranged regularly, *pendulous; seedling leaf finely pinnate.* **Inflorescence** *below the leaves,* branched to 2–5 orders, *branches widely spreading;* prophyll and peduncular bract similar, enclosing inflorescence in bud, dropping off as inflorescence expands; peduncle shorter than inflorescence rachis; rachillae slender and curving, *contorted in the inflorescence bud and resembling intestines.* **Flowers** in triads nearly throughout the length of the rachilla, *not developing in pits.* **Fruit** yellow to red, ellipsoid to subglobose, stigmatic remains apical or slightly to one side of the apex, flesh rather thin and firm, endocarp thin, closely adhering to seed. **Seed** 1, ellipsoid, endosperm strongly ruminate.

DISTRIBUTION. Six species disjunctly distributed in the Nicobar Islands, Malay Peninsula, Maluku, New Guinea, New Ireland and the Solomon Islands. Three species in the New Guinea region.

NOTES. *Rhopaloblaste* in New Guinea is a robust, single-stemmed, canopy or subcanopy palm with pendulous leaflets. The leaf has dark hairs on the leaf rachis and the seedling leaf is finely pinnate. The inflorescence is borne below the leaves, and has widely spreading branches and flowers that do not develop within pits. In bud, the inflorescence branches are contorted and resemble intestines. The genus mainly occurs in lowland forest to 550 m. It is most readily confused with *Cyrtostachys*, which has flowers usually developing in pits and very small (0.8–1.2 cm), black, ellipsoid fruit, or *Heterospathe*, which almost always lacks a crownshaft comprising usually open, fibrous leaf sheaths and a persistent prophyll.

The complex taxonomic history of *Rhopaloblaste*, which is entangled with the genera *Dransfieldia* and *Heterospathe* was clarified by Norup *et al.* (2006). A full monograph of the genus was completed in preparation for this book (Banka & Baker 2004).

Key to the species of *Rhopaloblaste* in New Guinea

1. Inflorescences branched to 4–5 orders; rachillae 2–2.9 mm diam. . . . . **3. *R. ledermanniana***
1. Inflorescences branched to 2–3 orders; rachillae greater than ca. 3 mm diam. **2**

2. Inflorescences branched to 2 orders; fruit with cupule of persistent perianth 8.5–9 mm long . . . . **2. *R. gideonii***
2. Inflorescences branched to 3 orders; fruit with cupule of persistent perianth 11–12 mm long . . . . **1. *R. ceramica***

1. *Rhopaloblaste ceramica* (Miq.) Burret

Synonyms: *Bentinckia ceramica* Miq., *Cyrtostachys ceramica* (Miq.) H.Wendl., *Rhopaloblaste hexandra* Scheff., ?*Rhopaloblaste dyscrita* H.E.Moore, ?*Rhopaloblaste micrantha* Burret (not *Rhopaloblaste micrantha* (Becc.) Hook.f. ex Salomon)

Robust, single-stemmed palm bearing up to 17 leaves in the crown to 20 m. **Stem** 15–29(–35) cm diam., internodes 12–14 cm basally, decreasing to 1 cm towards apex. **Leaf** to 4 m long including petiole; sheath 1.2–1.5 m long; petiole 3.5–4.5 cm long; *leaflets 111–120 each side of rachis*, linear; mid-leaf leaflet 100–112 × 2.3–2.5 cm. **Inflorescence** 55–130 cm long including 8–10 cm peduncle, widely spreading with basal branches strongly recurved, *branched to 3 orders*; primary branches ca. 16, to 75 cm long; rachillae 45–75 cm long, *4.9–7.3 mm diam.* **Male flower** 6.5–7 × 6.5–6.8 mm at anthesis, greenish; stamens 6. **Female flower** 4.3–4.7 × 7.6–7.9 mm. **Fruit** *30–35 × 16–18 mm*, ellipsoid-ovoid, asymmetric, yellow to red. **Seed** 21–31 mm × 14–16 mm.

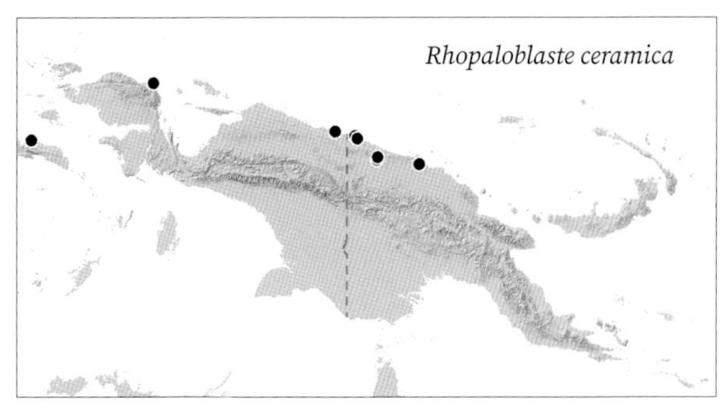

DISTRIBUTION. Widespread from the Bird's Head Peninsula to north-eastern New Guinea. Elsewhere distributed from Halmahera and Buru to Ceram.

HABITAT. Primary and secondary forest at 35–550 m (reported at up to 900 m outside New Guinea).

LOCAL NAMES. *Ansan* (Nuni).

Rhopaloblaste ceramica. LEFT TO RIGHT: habit, near Jayapura (GP); inflorescence bud, showing contorted, intestine-like branches, Lumi (FE).

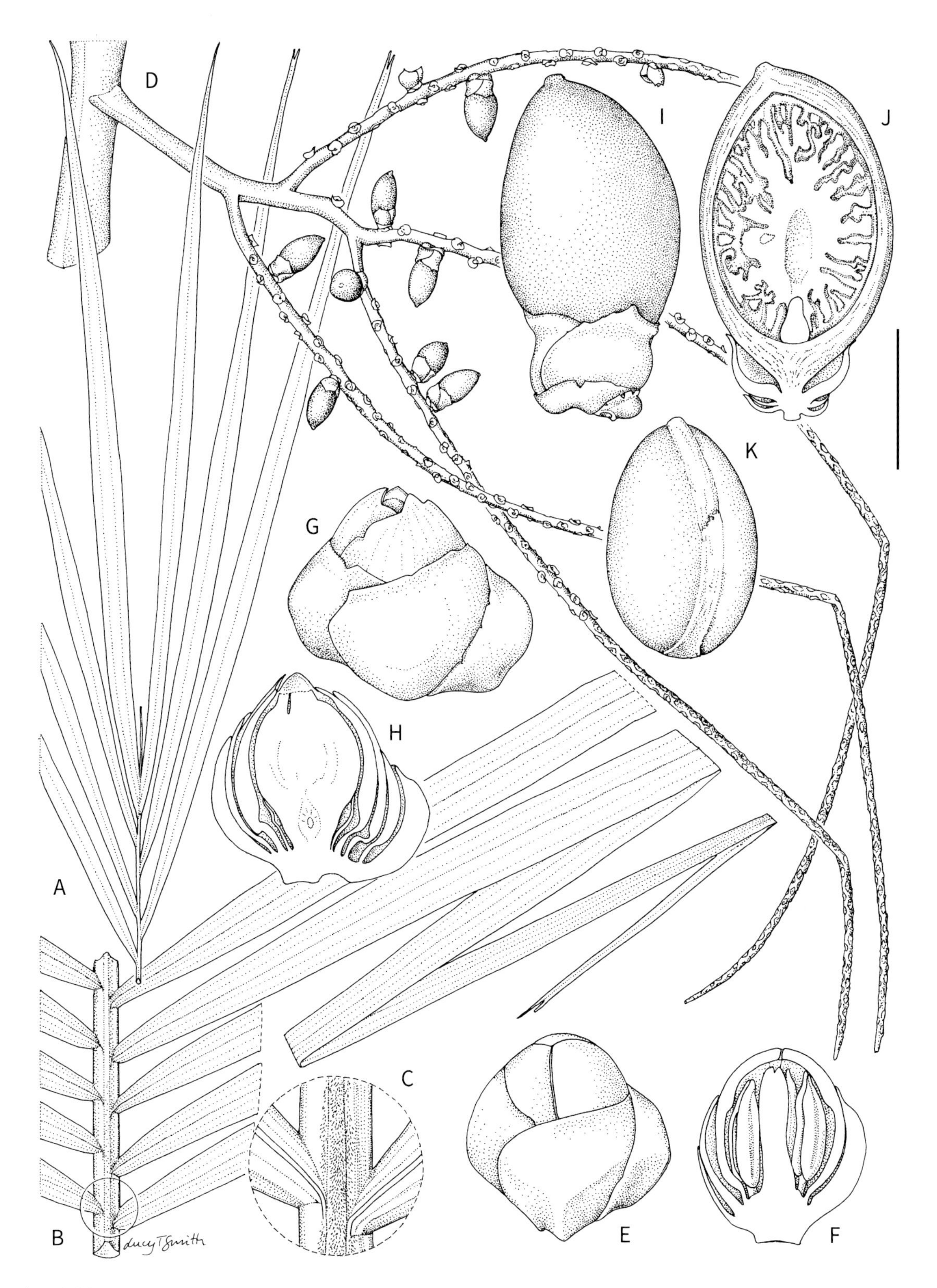

***Rhopaloblaste ceramica*.** **A.** Leaf apex. **B.** Leaf mid-portion. **C.** Detail leaf rachis indumentum. **D.** Portion of inflorescence with fruit. **E**, **F**. Male flower whole and in longitudinal section. **G**, **H**. Female flower whole and in longitudinal section. **I**, **J**. Fruit whole and in longitudinal section. **K**. Seed. Scale bar: A, B = 8 cm; C = 2.5 cm; D = 6 cm; E, F = 3 mm; G, H = 5 mm; I–K = 1.5 cm. A from *Barfod et al. 384*; B, C from *Barfod & Damborg 379*; D, G, H from *Desianto 17*; E, F from *Essig LAE 55074*; I–K from *Dransfield et al. JD 7582*. Drawn by Lucy T. Smith.

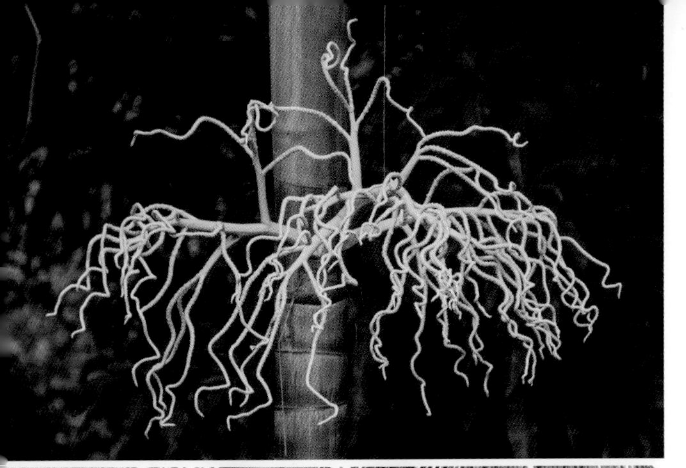

Rhopaloblaste ceramica. FROM TOP: expanding inflorescence, cultivated Merwin Conservancy, Hawaii (WB); inflorescences, cultivated Lae (WB).

USES. The shoot apex is edible. The wood is used for walking sticks, arrowheads, spear handles, and floorboards for houses. Cultivated as an ornamental.

CONSERVATION STATUS. Least Concern.

NOTES. *Rhopaloblaste ceramica* is the largest of all the species in the genus. It is easily distinguished by its large asymmetrical fruits with a substantial cupule of persistent perianth, and the inflorescence branched to three orders with very robust rachillae.

In Papua New Guinea, *R. ceramica* is known only with certainty from Sandaun and East Sepik Provinces. The dubious synonym *R. dyscrita* originates from Morobe Province (see Banka & Baker 2004), which would represent a significant range extension, but the only surviving type material (*Clemens 7987*, Harvard University Herbaria [A]) is too incomplete to confirm its identity with confidence. Further exploration is required to clarify the occurrence of *Rhopaloblaste* in eastern New Guinea.

2. *Rhopaloblaste gideonii* R.Banka

Moderately robust, single-stemmed palm to 25 m. Stem ca. 20 cm diam. **Leaf** to 4 m long including petiole; sheath 55–65 cm long; petiole 8–10 cm long; *leaflets 100–103 each side of rachis,* linear-elliptic, lower surface with few ramenta on midrib; mid-leaf leaflet 72–80 × 1.5–2.5 cm. **Inflorescence** 58–65 cm long including 1–2 cm peduncle, widely spreading with basal branches strongly recurved, *branched to 2 orders;* primary branches 12–13, to 60 cm long, rachillae 42–54 cm long, *6–6.5 mm diam.* **Male flower** 7–7.5 × 3.8–4.1 mm at anthesis. **Female flower** 5–5.5 × 5.2–5.4 mm. **Fruit** 16–18 mm × 6–8 mm (only immature fruits seen), ellipsoid-ovoid. **Seed** not seen.

DISTRIBUTION. Known only from the type locality in the Hans Meyer Range, New Ireland.

HABITAT. Lower montane forest at ca. 800 m.

LOCAL NAMES. None recorded.

USES. None recorded.

CONSERVATION STATUS. Data deficient (IUCN 2021). This species is insufficiently known to allow completion of an assessment. It is known only from one site, but this appears to be in pristine forest, so could be assessed as either Least Concern or Critically Endangered. The assessment reflects this uncertainty.

NOTES. This poorly known species appears to be similar to *Rhopaloblaste ceramica,* but is easily distinguished by its inflorescences that are branched to two orders only and by the smaller cupule of persistent perianth on the fruit. The type locality is also substantially disjunct from the known distribution of the other two New Guinea species. It is also distinct from the nearest species to the east (*R. elegans* from the Solomon Islands) in fruit shape (ellipsoid-ovoid, rather than spherical) and the more robust rachillae.

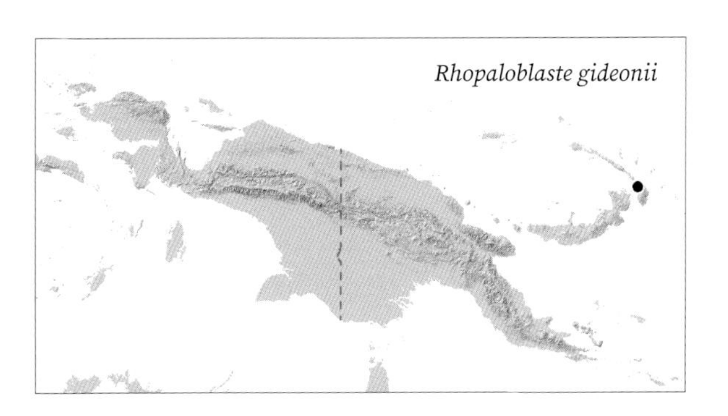

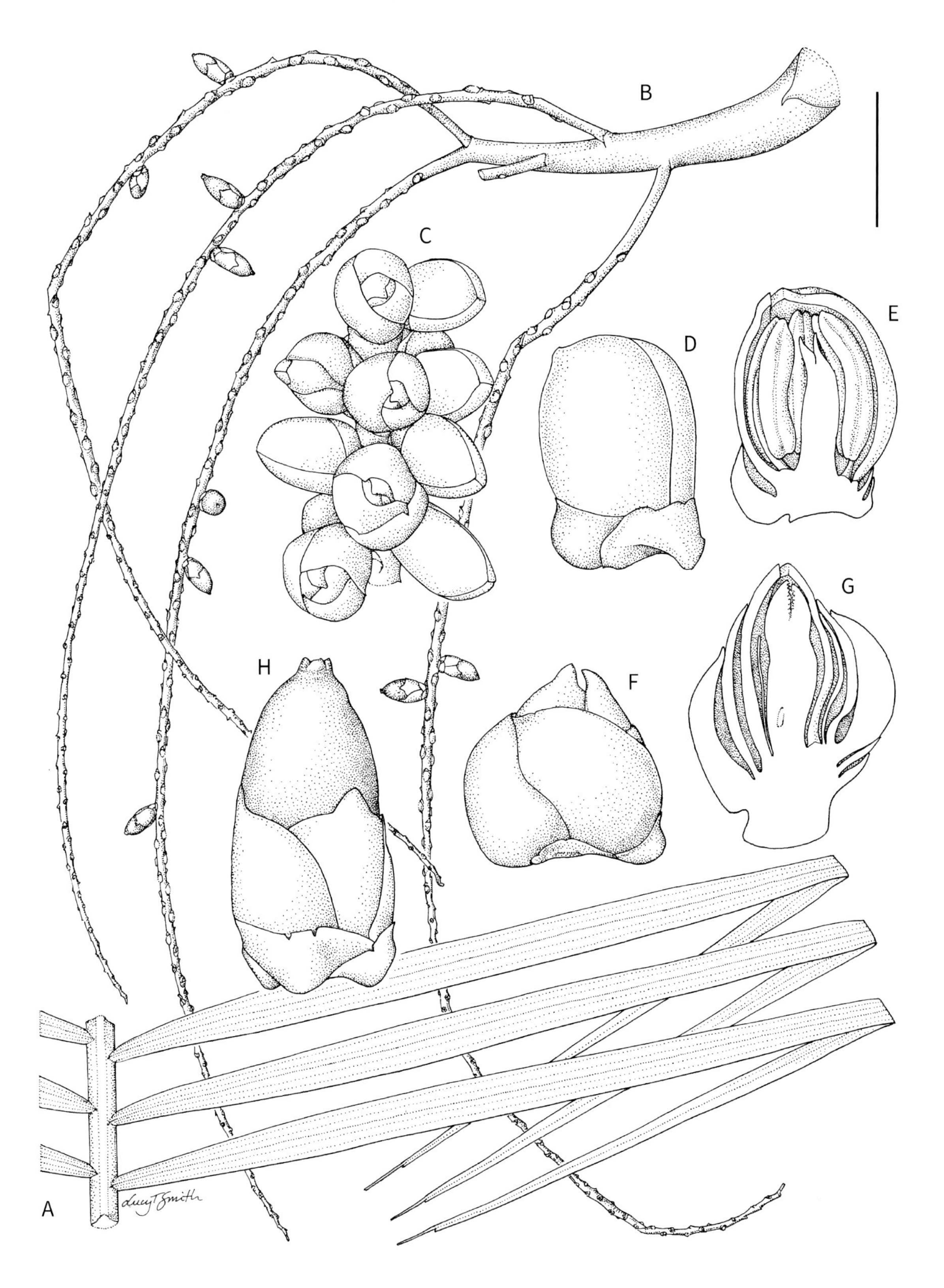

Rhopaloblaste gideonii. **A**. Leaf mid-portion. **B**. Portion of inflorescence with fruit. **C**. Portion of rachilla with flowers. **D**, **E**. Male flower whole and in longitudinal section. **F**, **G**. Female flower whole and in longitudinal section. **H**. Fruit. Scale bar: A = 8 cm; B = 4 cm; C = 5 mm; D–G = 3.3 mm; H = 7 mm. All from *Gideon et al. LAE 57194*. Drawn by Lucy T. Smith.

3. *Rhopaloblaste ledermanniana* Becc.

Synonym: *Rhopaloblaste brassii* H. E. Moore

Moderately robust, single-stemmed palm to 15 m, bearing up to 11 leaves in crown. **Stem** 8–11(–15) cm diam., internodes 6–8 cm. **Leaf** 2.2–4 m long including petiole; sheath 62–70 cm long; petiole 15–20 cm long; leaflets *59–90 each side of rachis,* linear, lower surface with twisted ramenta on midrib; mid-leaf leaflet 64–75 × 2.3–2.5 cm wide. **Inflorescence** 64–95 cm long including 1.5–4 cm peduncle, widely spreading with basal branches strongly recurved, *branched to 4 (rarely 5) orders*; primary branches ca. 18, to 70 cm long; rachillae 17–36 cm long, *2–2.9 mm diam.* **Male flower** 5–6 × 3–4 mm at anthesis; stamens 6. **Female flower** 2–3 × 3–3.5 mm. **Fruit** *15–21 mm × 9–12 mm,* ellipsoid-ovoid, yellow to orange-red. **Seed** 10–12 mm × 8–10 mm.

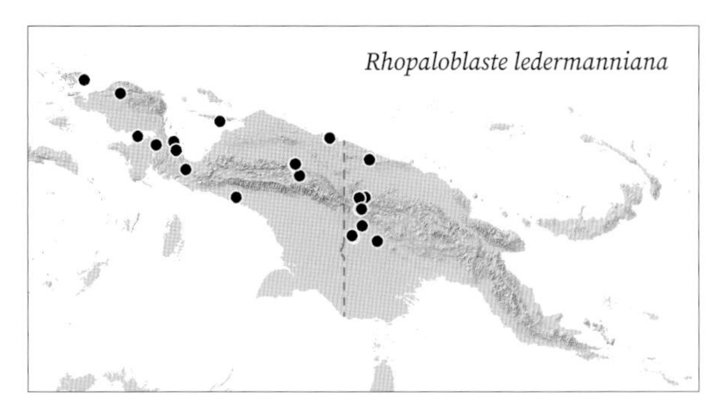

DISTRIBUTION. Widespread from north-western to central New Guinea, including the Raja Ampat Islands.

Rhopaloblaste ledermanniana. LEFT TO RIGHT: habit, Wandamen Peninsula (WB); seedlings, Tamrau Mountains (WB).

Rhopaloblaste ledermanniana. **A**. Leaf apex. **B**. Detail of rachis indumentum. **C**. Leaf mid-portion. **D**. Inflorescence with fruit base. **E**. Portion of inflorescence with fruit. **F**. Portion of rachilla with female flowers. **G**, **H**. Female flower whole and in longitudinal section. **I**, **J**. Fruit whole and in longitudinal section. K. Seed. Scale bar: A, C = 6 cm; B, I–K = 1 cm; D, E = 4 cm; F = 5 mm; G, H = 3 mm. A–C, F–H from *Barrow et al. 128*; D, E, I–K from *Maturbongs et al. 650*. Drawn by Lucy T. Smith.

HABITAT. Primary forest from sea level to 1,000 m.

LOCAL NAMES. *Black Palm* (Mian), *Flim saku* (Mian), *Imbetor* (Wandamen), *Kasira* (Wandamen), *Koah* (Nomad), Kofa (Meyah), *Kuwehleh* (Orne), *Mimini* (Kamoro), *Res* (Biak), *Saku* (Mian).

USES. The hard, outer part of the stem is used for making floorboards, walls, bows and arrow tips. The shoot apex is eaten in Mianmin and Wasior. The fruit are eaten by the Mianmin people.

CONSERVATION STATUS. Least Concern (IUCN 2021).

NOTES. *Rhopaloblaste ledermanniana* can be distinguished by its small ovoid-ellipsoid fruits, by its slender rachillae and by the inflorescence branching to four or sometimes five orders. It is the most widely encountered species of *Rhopaloblaste* in New Guinea, being widespread though rarely abundant.

Rhopaloblaste ledermanniana. LEFT TO RIGHT FROM TOP: leaf rachis, showing dark hairs, Tamrau Mountains (WB); inflorescence, Tamrau Mountains (WB); inflorescence bud, showing contorted, intestine-like branches, near Timika (WB); female flowers, Wandamen Peninsula (WB); fruit, Wandamen Peninsula (WB).

APPENDIX 1

Checklist of the palms of the New Guinea region

Accepted species are listed – see taxonomic accounts for synonyms. Extinction risk assessments combine both published assessments accessible through the IUCN Red List (IUCN 2023) and preliminary assessments prepared for this book according to the IUCN Red List Categories and Criteria (IUCN 2012). Distribution status is determined with respect to the New Guinea region, as defined in this book.

SPECIES	EXTINCTION RISK	DISTRIBUTION STATUS
Actinorhytis calapparia (Blume) H.Wendl. & Drude ex Scheff.	LC	Native non-endemic
Areca catechu L.	LC	Unknown
Areca jokowi Heatubun	CR	Native endemic
Areca macrocalyx Zipp. ex Becc.	LC	Native non-endemic
Areca mandacanii Heatubun	CR	Native endemic
Areca novohibernica (Lauterb.) Becc.	EN	Native non-endemic
Areca unipa Heatubun	CR	Native endemic
Arenga microcarpa Becc.	LC	Native non-endemic
Arenga pinnata (Wurmb.) Merr.	LC	Introduced
Borassus heineanus Becc.	NT	Native endemic
Brassiophoenix drymophloeoides Burret	LC	Native endemic
Brassiophoenix schumannii (Becc.) Essig	LC	Native endemic
Calamus altiscandens Burret	DD	Native endemic
Calamus anomalus Burret	EN	Native endemic
Calamus aruensis Becc.	LC	Native non-endemic
Calamus badius J.Dransf. & W.J.Baker	VU	Native endemic
Calamus baiyerensis W.J.Baker & J.Dransf.	CR	Native endemic
Calamus bankae W.J.Baker & J.Dransf.	CR	Native endemic
Calamus barbatus Zipp. ex Blume	LC	Native endemic
Calamus barfodii W.J.Baker & J.Dransf.	EN	Native endemic
Calamus bulubabi W.J.Baker & J.Dransf.	LC	Native endemic
Calamus calapparius Mart.	LC	Native non-endemic
Calamus capillosus W.J.Baker & J.Dransf.	CR	Native endemic
Calamus cheirophyllus J.Dransf. & W.J.Baker	DD	Native endemic

SPECIES	EXTINCTION RISK	DISTRIBUTION STATUS
Calamus croftii J.Dransf. & W.J.Baker	CR	Native endemic
Calamus cuthbertsonii Becc.	VU	Native endemic
Calamus dasyacanthus W.J.Baker, Bayton, J.Dransf. & Maturb.	LC	Native endemic
Calamus depauperatus Ridl.	LC	Native endemic
Calamus distentus Burret	DD	Native endemic
Calamus erythrocarpus W.J.Baker & J.Dransf.	CR	Native endemic
Calamus essigii W.J.Baker	CR	Native endemic
Calamus fertilis Becc.	LC	Native endemic
Calamus heatubunii W.J.Baker & J.Dransf.	EN	Native endemic
Calamus heteracanthus Zipp. ex Blume	LC	Native non-endemic
Calamus interruptus Becc.	EN	Native endemic
Calamus jacobsii W.J.Baker & J.Dransf.	CR	Native endemic
Calamus johnsii W.J.Baker & J.Dransf.	LC	Native endemic
Calamus katikii W.J.Baker & J.Dransf.	CR	Native endemic
Calamus kebariensis Maturb., J.Dransf. & W.J.Baker	CR	Native endemic
Calamus klossii Ridl.	LC	Native endemic
Calamus kostermansii W.J.Baker & J.Dransf.	CR	Native endemic
Calamus lauterbachii Becc.	LC	Native endemic
Calamus longipinna K.Schum. & Lauterb.	LC	Native non-endemic
Calamus lucysmithiae W.J.Baker & J.Dransf.	EN	Native endemic
Calamus macrochlamys Becc.	LC	Native endemic
Calamus maturbongsii W.J.Baker & J.Dransf.	CR	Native endemic
Calamus moszkowskianus Becc.	CR	Native endemic
Calamus nanduensis W.J.Baker & J.Dransf.	DD	Native endemic
Calamus nannostachys Burret	EN	Native endemic
Calamus nudus W.J.Baker & S.Venter	NT	Native endemic
Calamus oresbius W.J.Baker & J.Dransf.	LC	Native endemic
Calamus pachypus W.J.Baker, Bayton, J.Dransf. & Maturb.	LC	Native endemic
Calamus papuanus Becc.	LC	Native endemic
Calamus pholidostachys J.Dransf. & W.J.Baker	VU	Native endemic
Calamus pilosissimus Becc.	LC	Native endemic
Calamus pintaudii W.J.Baker & J.Dransf.	VU	Native endemic
Calamus polycladus Burret	EN	Native endemic

SPECIES	EXTINCTION RISK	DISTRIBUTION STATUS
Calamus pseudozebrinus Burret	CR	Native endemic
Calamus reticulatus Burret	LC	Native endemic
Calamus retroflexus J.Dransf. & W.J.Baker	LC	Native endemic
Calamus sashae J.Dransf. & W.J.Baker	CR	Native endemic
Calamus scabrispathus Becc.	CR	Native endemic
Calamus schlechterianus Becc.	LC	Native endemic
Calamus serrulatus Becc.	LC	Native endemic
Calamus spanostachys W.J.Baker & J.Dransf.	DD	Native endemic
Calamus spiculifer J.Dransf. & W.J.Baker	DD	Native endemic
Calamus superciliatus W.J.Baker & J.Dransf.	CR	Native endemic
Calamus vestitus Becc.	LC	Native endemic
Calamus vitiensis Warb. ex Becc.	LC	Native non-endemic
Calamus wanggaii W.J.Baker & J.Dransf.	CR	Native endemic
Calamus warburgii K.Schum.	LC	Native non-endemic
Calamus womersleyi J.Dransf. & W.J.Baker	EN	Native endemic
Calamus zebrinus Becc.	LC	Native endemic
Calamus zieckii Fernando	LC	Native endemic
Calamus zollingeri Becc. subsp. *zollingeri*	LC	Native non-endemic
Calyptrocalyx albertisianus Becc.	LC	Native endemic
Calyptrocalyx amoenus Dowe & M.D.Ferrero	CR	Native endemic
Calyptrocalyx arfakianus (Becc.) Dowe & M.D. Ferrero	CR	Native endemic
Calyptrocalyx awa Dowe & M.D.Ferrero	CR	Native endemic
Calyptrocalyx calcicola J.Dransf & L.T.Sm.	CR	Native endemic
Calyptrocalyx caudiculatus (Becc.) Dowe & M.D.Ferrero	NT	Native endemic
Calyptrocalyx doxanthus Dowe & M.D.Ferrero	DD	Native endemic
Calyptrocalyx elegans Becc.	LC	Native endemic
Calyptrocalyx flabellatus (Becc.) Dowe & M.D.Ferrero	LC	Native endemic
Calyptrocalyx forbesii (Ridl.) Dowe & M.D.Ferrero	LC	Native endemic
Calyptrocalyx geonomiformis (Becc.) Dowe & M.D.Ferrero	EN	Native endemic
Calyptrocalyx hollrungii (Becc.) Dowe & M.D.Ferrero	LC	Native endemic
Calyptrocalyx julianettii (Becc.) Dowe & M.D.Ferrero	LC	Native endemic
Calyptrocalyx lauterbachianus Warb. ex Becc.	LC	Native endemic
Calyptrocalyx laxiflorus Becc.	CR	Native endemic

SPECIES	EXTINCTION RISK	DISTRIBUTION STATUS
Calyptrocalyx lepidotus (Burret) Dowe & M.D.Ferrero	NT	Native endemic
Calyptrocalyx leptostachys Becc.	CR	Native endemic
Calyptrocalyx merrillianus (Burret) Dowe & M.D.Ferrero	DD	Native endemic
Calyptrocalyx micholitzii (Ridl.) Dowe & M.D.Ferrero	NT	Native endemic
Calyptrocalyx multifidus Becc.	DD	Native endemic
Calyptrocalyx pachystachys Becc.	LC	Native endemic
Calyptrocalyx pauciflorus Becc.	CR	Native endemic
Calyptrocalyx polyphyllus Becc.	NT	Native endemic
Calyptrocalyx pusillus (Becc.) Dowe & M.D.Ferrero	EN	Native endemic
Calyptrocalyx reflexus J.Dransf. & J.Marcus	DD	Native endemic
Calyptrocalyx sessiliflorus Dowe & M.D.Ferrero	VU	Native endemic
Calyptrocalyx spicatus (Lam.) Blume	LC	Native non-endemic
Calyptrocalyx yamutumene Dowe & M.D.Ferrero	CR	Native endemic
Caryota rumphiana Mart.	LC	Native non-endemic
Caryota zebrina Hambali, Maturb., Heatubun & J.Dransf.	EN	Native endemic
Clinostigma collegarum J.Dransf.	DD	Native endemic
Cocos nucifera L.	LC	Unknown
Corypha utan Lam.	LC	Native non-endemic
Cyrtostachys bakeri Heatubun	CR	Native endemic
Cyrtostachys barbata Heatubun	DD	Native endemic
Cyrtostachys elegans Burret	NT	Native endemic
Cyrtostachys excelsa Heatubun	EN	Native endemic
Cyrtostachys glauca H.E.Moore	LC	Native endemic
Cyrtostachys loriae Becc.	LC	Native non-endemic
Dransfieldia micrantha (Becc.) W.J.Baker & Zona	LC	Native endemic
Drymophloeus litigiosus (Becc.) H.E.Moore	LC	Native non-endemic
Drymophloeus oliviformis (Giseke) Mart.	LC	Native non-endemic
Heterospathe annectens H.E.Moore	DD	Native endemic
Heterospathe barfodii L.M.Gardiner & W.J.Baker	LC	Native endemic
Heterospathe elata Scheff.	LC	Native non-endemic
Heterospathe elegans (Becc.) Becc. subsp. *elegans*	LC	Native endemic
Heterospathe elegans (Becc.) Becc. subsp. *humilis* (Becc.) M.S.Trudgen & W.J.Baker	LC	Native endemic

SPECIES	EXTINCTION RISK	DISTRIBUTION STATUS
Heterospathe ledermanniana Becc.	DD	Native endemic
Heterospathe lepidota H.E.Moore	DD	Native endemic
Heterospathe macgregorii (Becc.) H.E.Moore	LC	Native endemic
Heterospathe muelleriana (Becc.) Becc.	LC	Native endemic
Heterospathe obriensis (Becc.) H.E.Moore	LC	Native endemic
Heterospathe parviflora Essig	EN	Native endemic
Heterospathe porcata W.J.Baker & Heatubun	DD	Native endemic
Heterospathe pulchra H.E.Moore	LC	Native endemic
Heterospathe pullenii M.S.Trudgen & W.J.Baker	EN	Native endemic
Heterospathe sphaerocarpa Burret	LC	Native endemic
Hydriastele apetiolata Petoe & W.J.Baker	DD	Native endemic
Hydriastele aprica (B.E.Young) W.J.Baker & Loo	DD	Native endemic
Hydriastele biakensis W.J.Baker & Heatubun	EN	Native endemic
Hydriastele calcicola W.J.Baker & Petoe	DD	Native endemic
Hydriastele costata F.M.Bailey	LC	Native non-endemic
Hydriastele divaricata Heatubun, Petoe & W.J.Baker	DD	Native endemic
Hydriastele dransfieldii (Hambali, Maturb., Wanggai & W.J.Baker) W.J.Baker & Loo	LC	Native endemic
Hydriastele flabellata (Becc.) W.J.Baker & Loo	LC	Native endemic
Hydriastele gibbsiana (Becc.) W.J.Baker & Loo	DD	Native endemic
Hydriastele kasesa (Lauterb.) Burret	LC	Native endemic
Hydriastele lanata W.J.Baker & Petoe	DD	Native endemic
Hydriastele ledermanniana (Becc.) W.J.Baker & Loo	LC	Native endemic
Hydriastele longispatha (Becc.) W.J.Baker & Loo	LC	Native endemic
Hydriastele lurida (Becc.) W.J.Baker & Loo	LC	Native endemic
Hydriastele manusii (Essig) W.J.Baker & Loo	DD	Native endemic
Hydriastele montana (Becc.) W.J.Baker & Loo	DD	Native endemic
Hydriastele pinangoides (Becc.) W.J.Baker & Loo	LC	Native non-endemic
Hydriastele procera (Blume) W.J.Baker & Loo	DD	Native endemic
Hydriastele rheophytica Dowe & M.D.Ferrero	DD	Native endemic
Hydriastele simbiakii Heatubun, Petoe & W.J.Baker	DD	Native endemic
Hydriastele splendida Heatubun, Petoe & W.J.Baker	DD	Native endemic
Hydriastele variabilis (Becc.) Burret	LC	Native endemic

SPECIES	EXTINCTION RISK	DISTRIBUTION STATUS
Hydriastele wendlandiana (F.Muell.) H.Wendl. & Drude	LC	Native non-endemic
Hydriastele wosimiensis W.J.Baker & Petoe	DD	Native endemic
Korthalsia zippelii Blume	LC	Native non-endemic
Licuala anomala Becc.	NT	Native endemic
Licuala bacularia Becc.	LC	Native endemic
Licuala bakeri Heatubun & Barfod	NT	Native endemic
Licuala bankae Heatubun & Barfod	CR	Native endemic
Licuala bellatula Becc.	NT	Native endemic
Licuala bifida Heatubun & Barfod	EN	Native endemic
Licuala brevicalyx Becc.	NT	Native endemic
Licuala coccinisedes Heatubun & Barfod	VU	Native endemic
Licuala essigii Heatubun & Barfod	CR	Native endemic
Licuala flexuosa Burret	LC	Native non-endemic
Licuala graminifolia Heatubun & Barfod	DD	Native endemic
Licuala grandiflora Ridl.	EN	Native endemic
Licuala heatubunii Barfod & W.J.Baker	CR	Native endemic
Licuala insignis Becc.	LC	Native endemic
Licuala lauterbachii Dammer & K.Schum. subsp. *lauterbachii*	LC	Native endemic
Licuala lauterbachii Dammer & K.Schum. subsp. *peekelii* (Lauterb.) Barfod	LC	Native non-endemic
Licuala longispadix Banka & Barfod	CR	Native endemic
Licuala montana Dammer & K.Schum.	LC	Native endemic
Licuala multibracteata Heatubun & Barfod	EN	Native endemic
Licuala parviflora Dammer ex Becc.	LC	Native endemic
Licuala penduliflora (Blume) Zipp. ex Blume	LC	Native non-endemic
Licuala sandsiana Heatubun & Barfod	EN	Native endemic
Licuala simplex (Lauterb. & K.Schum.) Becc.	LC	Native endemic
Licuala suprafolia Heatubun & Barfod	CR	Native endemic
Licuala telifera Becc.	LC	Native endemic
Licuala urciflora Barfod & Heatubun	DD	Native endemic
Linospadix albertisianus (Becc.) Burret	LC	Native endemic
Livistona benthamii F.M.Bailey	LC	Native non-endemic
Livistona muelleri F.M.Bailey	LC	Native non-endemic

SPECIES	EXTINCTION RISK	DISTRIBUTION STATUS
Manjekia maturbongsii (W.J.Baker & Heatubun) W.J.Baker & Heatubun	EN	Native endemic
Metroxylon sagu Rottb.	LC	Native non-endemic
Nypa fruticans Wurmb	LC	Native non-endemic
Orania archboldiana Burret	LC	Native endemic
Orania bakeri A.P.Keim & J.Dransf.	CR	Native endemic
Orania dafonsoroensis A.P.Keim & J.Dransf.	EN	Native endemic
Orania deflexa A.P.Keim & J.Dransf.	CR	Native endemic
Orania disticha Burret	LC	Native endemic
Orania ferruginea A.P.Keim & J.Dransf.	CR	Native endemic
Orania gagavu Essig	CR	Native endemic
Orania glauca Essig	DD	Native endemic
Orania grandiflora A.P.Keim & J.Dransf.	CR	Native endemic
Orania lauterbachiana Becc.	LC	Native endemic
Orania littoralis A.P.Keim & J.Dransf.	CR	Native endemic
Orania longistaminodia A.P.Keim & J.Dransf.	CR	Native endemic
Orania macropetala K.Schum. & Lauterb.	LC	Native endemic
Orania micrantha Becc.	VU	Native endemic
Orania oreophila Essig	EN	Native endemic
Orania palindan (Blanco) Merr.	LC	Native non-endemic
Orania parva Essig	DD	Native endemic
Orania regalis Zipp.	LC	Native non-endemic
Orania subdisticha A.P.Keim & J.Dransf.	VU	Native endemic
Orania tabubilensis A.P.Keim & J.Dransf.	CR	Native endemic
Orania timikae A.P.Keim & J.Dransf.	CR	Native endemic
Orania zonae A.P.Keim & J.Dransf.	CR	Native endemic
Physokentia avia H.E.Moore	EN	Native endemic
Pigafetta filaris (Giseke) Becc.	LC	Native non-endemic
Pinanga rumphiana (Mart.) J.Dransf. & Govaerts	LC	Native non-endemic
Ponapea hentyi (Essig) C.Lewis & Zona	EN	Native endemic
Ptychococcus lepidotus H.E.Moore	LC	Native endemic
Ptychococcus paradoxus (Scheff.) Becc	LC	Native non-endemic
Ptychosperma ambiguum (Becc.) Becc. ex Martelli	CR	Native endemic

SPECIES	EXTINCTION RISK	DISTRIBUTION STATUS
Ptychosperma buabe Essig	CR	Native endemic
Ptychosperma caryotoides Ridl.	LC	Native endemic
Ptychosperma cuneatum (Burret) Burret	EN	Native endemic
Ptychosperma furcatum (Becc.) Becc. ex Martelli	LC	Native endemic
Ptychosperma gracile Labill.	LC	Native non-endemic
Ptychosperma keiense (Becc.) Becc. ex Martelli	LC	Native non-endemic
Ptychosperma lauterbachii Becc.	LC	Native endemic
Ptychosperma lineare (Burret) Burret	VU	Native endemic
Ptychosperma mambare (F.M.Bailey) Becc. ex Martelli	CR	Native endemic
Ptychosperma microcarpum (Burret) Burret	EN	Native endemic
Ptychosperma mooreanum Essig	LC	Native endemic
Ptychosperma nicolai (Sander ex André) Burret	DD	Native endemic
Ptychosperma propinquum (Becc.) Becc. ex Martelli	LC	Native non-endemic
Ptychosperma pullenii Essig	CR	Native endemic
Ptychosperma ramosissimum Essig	EN	Native endemic
Ptychosperma rosselense Essig	NT	Native endemic
Ptychosperma sanderianum Ridl.	EW	Native endemic
Ptychosperma schefferi Becc. ex Martelli	CR	Native endemic
Ptychosperma tagulense Essig	DD	Native endemic
Ptychosperma vestitum Essig	DD	Native endemic
Ptychosperma waitianum Essig	NT	Native endemic
Rhopaloblaste ceramica (Miq.) Burret	LC	Native non-endemic
Rhopaloblaste gideonii R.Banka	DD	Native endemic
Rhopaloblaste ledermanniana Becc.	LC	Native endemic
Saribus brevifolius (Dowe & Mogea) Bacon & W.J.Baker	LC	Native endemic
Saribus chocolatinus (Dowe) Bacon & W.J.Baker	NT	Native endemic
Saribus papuanus (Becc.) Kuntze	NT	Native endemic
Saribus pendulinus Dowe & S.Venter	VU	Native endemic
Saribus rotundifolius (Lam.) Blume	LC	Native non-endemic
Saribus surru (Dowe & Barfod) Bacon & W.J.Baker	EN	Native endemic
Saribus tothur (Dowe & Barfod) Bacon & W.J.Baker	EN	Native endemic
Saribus woodfordii (Ridl.) Bacon & W.J.Baker	LC	Native non-endemic
Sommieria leucophylla Becc.	LC	Native endemic
Wallaceodoxa raja-ampat Heatubun & W.J.Baker	CR	Native endemic

APPENDIX 2

Checklists of the palms of subdivisions of the New Guinea region

Indonesian New Guinea

Actinorhytis calapparia
Areca catechu
Areca jokowi
Areca macrocalyx
Areca mandacanii
Areca unipa
Arenga microcarpa
Arenga pinnata
Borassus heineanus
Calamus aruensis
Calamus badius
Calamus barbatus
Calamus bulubabi
Calamus calapparius
Calamus capillosus
Calamus dasyacanthus
Calamus depauperatus
Calamus fertilis
Calamus heatubunii
Calamus heteracanthus
Calamus interruptus
Calamus kebariensis
Calamus klossii
Calamus kostermansii
Calamus lauterbachii
Calamus longipinna
Calamus macrochlamys
Calamus maturbongsii
Calamus moszkowskianus
Calamus nudus
Calamus pachypus
Calamus papuanus
Calamus pilosissimus
Calamus retroflexus
Calamus sashae
Calamus serrulatus
Calamus spanostachys
Calamus spiculifer
Calamus superciliatus
Calamus vestitus
Calamus vitiensis
Calamus wanggaii
Calamus warburgii
Calamus zebrinus
Calamus zieckii
Calamus zollingeri subsp. *zollingeri*
Calyptrocalyx albertisianus
Calyptrocalyx arfakianus
Calyptrocalyx caudiculatus
Calyptrocalyx doxanthus
Calyptrocalyx elegans
Calyptrocalyx flabellatus
Calyptrocalyx geonomiformis
Calyptrocalyx hollrungii
Calyptrocalyx julianettii
Calyptrocalyx micholitzii
Calyptrocalyx multifidus
Calyptrocalyx sessiliflorus
Calyptrocalyx spicatus
Caryota rumphiana
Caryota zebrina
Cocos nucifera
Corypha utan
Cyrtostachys barbata
Cyrtostachys elegans
Cyrtostachys excelsa
Cyrtostachys loriae
Dransfieldia micrantha
Drymophloeus litigiosus
Drymophloeus oliviformis
Heterospathe elata
Heterospathe elegans subsp. *elegans*
Heterospathe elegans subsp. *humilis*
Heterospathe muelleriana
Heterospathe porcata
Hydriastele apetiolata
Hydriastele biakensis
Hydriastele costata
Hydriastele divaricata
Hydriastele dransfieldii
Hydriastele flabellata
Hydriastele gibbsiana
Hydriastele ledermanniana
Hydriastele longispatha
Hydriastele lurida
Hydriastele montana
Hydriastele pinangoides
Hydriastele procera
Hydriastele rheophytica
Hydriastele simbiakii
Hydriastele splendida
Hydriastele variabilis
Hydriastele wendlandiana
Hydriastele wosimiensis
Korthalsia zippelii
Licuala anomala
Licuala bacularia
Licuala bakeri
Licuala bellatula

Licuala bifida
Licuala brevicalyx
Licuala coccinisedes
Licuala flexuosa
Licuala graminifolia
Licuala grandiflora
Licuala heatubunii
Licuala insignis
Licuala lauterbachii subsp. *lauterbachii*
Licuala montana
Licuala parviflora
Licuala penduliflora
Licuala simplex
Licuala telifera
Licuala urciflora
Linospadix albertisianus
Livistona benthamii
Livistona muelleri
Manjekia maturbongsii
Metroxylon sagu
Nypa fruticans
Orania archboldiana
Orania dafonsoroensis
Orania disticha
Orania ferruginea
Orania grandiflora
Orania lauterbachiana
Orania macropetala
Orania palindan
Orania regalis
Orania tabubilensis
Orania timikae
Orania zonae
Pigafetta filaris
Pinanga rumphiana
Ptychococcus lepidotus
Ptychococcus paradoxus
Ptychosperma ambiguum
Ptychosperma cuneatum
Ptychosperma furcatum
Ptychosperma keiense
Ptychosperma lauterbachii
Ptychosperma nicolai
Ptychosperma propinquum
Ptychosperma schefferi
Rhopaloblaste ceramica
Rhopaloblaste ledermanniana
Saribus brevifolius
Saribus papuanus
Saribus rotundifolius
Sommieria leucophylla
Wallaceodoxa raja-ampat

Papua New Guinea

Actinorhytis calapparia
Areca catechu
Areca macrocalyx
Areca novohibernica
Arenga microcarpa
Borassus heineanus
Brassiophoenix drymophloeoides
Brassiophoenix schumannii
Calamus altiscandens
Calamus anomalus
Calamus aruensis
Calamus baiyerensis
Calamus bankae
Calamus barbatus
Calamus barfodii
Calamus bulubabi
Calamus cheirophyllus
Calamus croftii
Calamus cuthbertsonii
Calamus dasyacanthus
Calamus distentus
Calamus erythrocarpus
Calamus essigii
Calamus fertilis
Calamus heteracanthus
Calamus jacobsii
Calamus johnsii
Calamus katikii
Calamus klossii
Calamus lauterbachii
Calamus longipinna
Calamus lucysmithiae
Calamus macrochlamys
Calamus nanduensis
Calamus nannostachys
Calamus nudus
Calamus oresbius
Calamus pachypus
Calamus pholidostachys
Calamus pilosissimus
Calamus pintaudii
Calamus polycladus
Calamus pseudozebrinus
Calamus reticulatus
Calamus retroflexus
Calamus scabrispathus
Calamus schlechterianus
Calamus serrulatus
Calamus spiculifer
Calamus vestitus
Calamus vitiensis
Calamus warburgii
Calamus womersleyi
Calamus zebrinus
Calamus zieckii
Calyptrocalyx albertisianus
Calyptrocalyx amoenus
Calyptrocalyx awa
Calyptrocalyx calcicola
Calyptrocalyx elegans

Calyptrocalyx forbesii
Calyptrocalyx hollrungii
Calyptrocalyx julianettii
Calyptrocalyx lauterbachianus
Calyptrocalyx laxiflorus
Calyptrocalyx lepidotus
Calyptrocalyx leptostachys
Calyptrocalyx merrillianus
Calyptrocalyx pachystachys
Calyptrocalyx pauciflorus
Calyptrocalyx polyphyllus
Calyptrocalyx pusillus
Calyptrocalyx reflexus
Calyptrocalyx sessiliflorus
Calyptrocalyx yamutumene
Caryota rumphiana
Caryota zebrina
Clinostigma collegarum
Cocos nucifera
Corypha utan
Cyrtostachys bakeri
Cyrtostachys glauca
Cyrtostachys loriae
Heterospathe annectens
Heterospathe barfodii
Heterospathe elegans subsp. *elegans*
Heterospathe elegans subsp. *humilis*
Heterospathe ledermanniana
Heterospathe lepidota
Heterospathe macgregorii
Heterospathe muelleriana
Heterospathe parviflora
Heterospathe pulchra
Heterospathe pullenii
Heterospathe sphaerocarpa
Hydriastele apetiolata
Hydriastele aprica
Hydriastele calcicola
Hydriastele costata
Hydriastele flabellata
Hydriastele kasesa
Hydriastele lanata
Hydriastele ledermanniana
Hydriastele longispatha
Hydriastele lurida
Hydriastele manusii
Hydriastele pinangoides
Hydriastele rheophytica
Hydriastele wendlandiana
Korthalsia zippelii
Licuala bankae
Licuala bellatula
Licuala brevicalyx
Licuala coccinisedes
Licuala essigii
Licuala insignis
Licuala lauterbachii subsp. *lauterbachii*
Licuala lauterbachii subsp. *peekelii*
Licuala longispadix
Licuala montana
Licuala multibracteata
Licuala parviflora
Licuala penduliflora
Licuala sandsiana
Licuala simplex
Licuala suprafolia
Licuala telifera
Linospadix albertisianus
Livistona benthamii
Livistona muelleri
Metroxylon sagu
Nypa fruticans
Orania archboldiana
Orania bakeri
Orania deflexa
Orania disticha
Orania gagavu
Orania glauca
Orania lauterbachiana
Orania littoralis
Orania longistaminodia
Orania macropetala
Orania micrantha
Orania oreophila
Orania palindan
Orania parva
Orania subdisticha
Orania tabubilensis
Physokentia avia
Pigafetta filaris
Pinanga rumphiana
Ponapea hentyi
Ptychococcus lepidotus
Ptychococcus paradoxus
Ptychosperma buabe
Ptychosperma caryotoides
Ptychosperma cuneatum
Ptychosperma furcatum
Ptychosperma gracile
Ptychosperma lauterbachii
Ptychosperma lineare
Ptychosperma mambare
Ptychosperma microcarpum
Ptychosperma mooreanum
Ptychosperma propinquum
Ptychosperma pullenii
Ptychosperma ramosissimum
Ptychosperma rosselense
Ptychosperma sanderianum
Ptychosperma schefferi
Ptychosperma tagulense
Ptychosperma vestitum
Ptychosperma waitianum
Rhopaloblaste ceramica
Rhopaloblaste gideonii
Rhopaloblaste ledermanniana
Saribus chocolatinus
Saribus pendulinus
Saribus surru
Saribus tothur
Saribus woodfordii
Sommieria leucophylla

Mainland New Guinea

Actinorhytis calapparia
Areca catechu
Areca jokowi
Areca macrocalyx
Areca mandacanii
Areca unipa
Arenga microcarpa
Arenga pinnata
Borassus heineanus
Brassiophoenix drymophloeoides
Brassiophoenix schumannii
Calamus altiscandens
Calamus anomalus
Calamus aruensis
Calamus badius
Calamus baiyerensis
Calamus bankae
Calamus barbatus
Calamus barfodii
Calamus bulubabi
Calamus calapparius
Calamus capillosus
Calamus cheirophyllus
Calamus croftii
Calamus cuthbertsonii
Calamus dasyacanthus
Calamus depauperatus
Calamus distentus
Calamus erythrocarpus
Calamus essigii
Calamus fertilis
Calamus heatubunii
Calamus heteracanthus
Calamus interruptus
Calamus jacobsii
Calamus johnsii
Calamus katikii
Calamus kebariensis
Calamus klossii
Calamus kostermansii
Calamus lauterbachii
Calamus longipinna
Calamus lucysmithiae
Calamus macrochlamys
Calamus maturbongsii
Calamus moszkowskianus
Calamus nanduensis
Calamus nannostachys
Calamus nudus
Calamus oresbius
Calamus pachypus
Calamus papuanus
Calamus pholidostachys
Calamus pilosissimus
Calamus pintaudii
Calamus polycladus
Calamus pseudozebrinus
Calamus reticulatus
Calamus retroflexus
Calamus sashae
Calamus scabrispathus
Calamus schlechterianus
Calamus serrulatus
Calamus spanostachys
Calamus spiculifer
Calamus superciliatus
Calamus vestitus
Calamus vitiensis
Calamus wanggaii
Calamus warburgii
Calamus womersleyi
Calamus zebrinus
Calamus zieckii
Calyptrocalyx albertisianus
Calyptrocalyx amoenus
Calyptrocalyx arfakianus
Calyptrocalyx awa
Calyptrocalyx calcicola
Calyptrocalyx caudiculatus
Calyptrocalyx doxanthus
Calyptrocalyx elegans
Calyptrocalyx flabellatus
Calyptrocalyx forbesii
Calyptrocalyx geonomiformis
Calyptrocalyx hollrungii
Calyptrocalyx julianettii
Calyptrocalyx lauterbachianus
Calyptrocalyx laxiflorus
Calyptrocalyx lepidotus
Calyptrocalyx leptostachys
Calyptrocalyx merrillianus
Calyptrocalyx micholitzii
Calyptrocalyx multifidus
Calyptrocalyx pachystachys
Calyptrocalyx pauciflorus
Calyptrocalyx polyphyllus
Calyptrocalyx pusillus
Calyptrocalyx reflexus
Calyptrocalyx sessiliflorus
Calyptrocalyx yamutumene
Caryota rumphiana
Caryota zebrina
Cocos nucifera
Corypha utan
Cyrtostachys bakeri
Cyrtostachys barbata
Cyrtostachys elegans
Cyrtostachys excelsa
Cyrtostachys glauca
Cyrtostachys loriae
Dransfieldia micrantha
Drymophloeus litigiosus
Drymophloeus oliviformis
Heterospathe barfodii

Heterospathe elegans subsp. *elegans*
Heterospathe elegans subsp. *humilis*
Heterospathe ledermanniana
Heterospathe lepidota
Heterospathe macgregorii
Heterospathe muelleriana
Heterospathe obriensis
Heterospathe pullenii
Heterospathe sphaerocarpa
Hydriastele apetiolata
Hydriastele biakensis
Hydriastele calcicola
Hydriastele costata
Hydriastele divaricata
Hydriastele flabellata
Hydriastele gibbsiana
Hydriastele lanata
Hydriastele ledermanniana
Hydriastele longispatha
Hydriastele lurida
Hydriastele montana
Hydriastele pinangoides
Hydriastele procera
Hydriastele rheophytica
Hydriastele simbiakii
Hydriastele splendida
Hydriastele variabilis
Hydriastele wendlandiana
Hydriastele wosimiensis
Korthalsia zippelii
Licuala anomala
Licuala bacularia
Licuala bakeri
Licuala bankae
Licuala bellatula
Licuala bifida
Licuala brevicalyx
Licuala coccinisedes
Licuala essigii
Licuala graminifolia
Licuala grandiflora
Licuala heatubunii
Licuala insignis
Licuala lauterbachii subsp. *lauterbachii*
Licuala longispadix
Licuala montana
Licuala multibracteata
Licuala parviflora
Licuala penduliflora
Licuala simplex
Licuala suprafolia
Licuala telifera
Linospadix albertisianus
Livistona benthamii
Livistona muelleri
Metroxylon sagu
Nypa fruticans
Orania archboldiana
Orania bakeri
Orania dafonsoroensis
Orania deflexa
Orania disticha
Orania ferruginea
Orania gagavu
Orania glauca
Orania grandiflora
Orania lauterbachiana
Orania littoralis
Orania longistaminodia
Orania macropetala
Orania micrantha
Orania oreophila
Orania palindan
Orania parva
Orania regalis
Orania subdisticha
Orania tabubilensis
Orania timikae
Orania zonae
Pigafetta filaris
Pinanga rumphiana
Ptychococcus lepidotus
Ptychococcus paradoxus
Ptychosperma ambiguum
Ptychosperma buabe
Ptychosperma caryotoides
Ptychosperma cuneatum
Ptychosperma furcatum
Ptychosperma keiense
Ptychosperma lauterbachii
Ptychosperma lineare
Ptychosperma mambare
Ptychosperma microcarpum
Ptychosperma mooreanum
Ptychosperma nicolai
Ptychosperma propinquum
Ptychosperma pullenii
Ptychosperma sanderianum
Ptychosperma schefferi
Ptychosperma vestitum
Ptychosperma waitianum
Rhopaloblaste ceramica
Rhopaloblaste ledermanniana
Saribus chocolatinus
Saribus papuanus
Saribus pendulinus
Saribus surru
Saribus tothur
Saribus woodfordii
Sommieria leucophylla

Aru Islands

Areca catechu
Areca macrocalyx
Calamus aruensis
Calamus heteracanthus
Cocos nucifera
Hydriastele costata
Hydriastele pinangoides
Hydriastele wendlandiana
Korthalsia zippelii
Licuala penduliflora
Metroxylon sagu
Nypa fruticans
Orania regalis
Ptychosperma keiense
Ptychosperma propinquum

Raja Ampat Islands

Areca catechu
Areca macrocalyx
Calamus aruensis
Calamus capillosus
Calamus heatubunii
Calamus heteracanthus
Calamus papuanus
Calamus zebrinus
Calamus zollingeri subsp. *zollingeri*
Calyptrocalyx arfakianus
Calyptrocalyx spicatus
Caryota rumphiana
Cocos nucifera
Dransfieldia micrantha
Drymophloeus litigiosus
Heterospathe elata
Hydriastele lurida
Hydriastele pinangoides
Hydriastele procera
Hydriastele wendlandiana
Korthalsia zippelii
Licuala flexuosa
Licuala penduliflora
Licuala telifera
Licuala urciflora
Metroxylon sagu
Nypa fruticans
Orania regalis
Pigafetta filaris
Ptychosperma propinquum
Rhopaloblaste ledermanniana
Saribus brevifolius
Saribus papuanus
Saribus rotundifolius
Sommieria leucophylla
Wallaceodoxa raja-ampat

Cenderawasih Bay Islands (Biak, Supiori, Numfor, Yapen)

Areca catechu
Areca macrocalyx
Arenga microcarpa
Calamus aruensis
Calamus dasyacanthus
Calamus heteracanthus
Calamus lauterbachii
Calamus serrulatus
Calamus vitiensis
Calamus zebrinus
Calyptrocalyx sessiliflorus
Cocos nucifera
Heterospathe porcata
Hydriastele biakensis
Hydriastele costata
Hydriastele dransfieldii
Hydriastele longispatha
Hydriastele lurida
Hydriastele wendlandiana
Korthalsia zippelii
Licuala bakeri
Licuala lauterbachii subsp. *lauterbachii*
Licuala telifera
Manjekia maturbongsii
Metroxylon sagu
Nypa fruticans
Pigafetta filaris
Pinanga rumphiana
Rhopaloblaste ledermanniana
Saribus papuanus

Bismarck Archipelago

Actinorhytis calapparia
Areca catechu
Areca macrocalyx
Areca novohibernica
Calamus aruensis
Calamus fertilis
Calamus longipinna
Calamus pachypus
Calamus vitiensis
Calamus warburgii
Calamus zebrinus
Calamus zieckii
Calyptrocalyx albertisianus
Caryota rumphiana
Clinostigma collegarum
Cocos nucifera
Cyrtostachys loriae
Heterospathe parviflora
Hydriastele costata
Hydriastele kasesa
Hydriastele manusii
Korthalsia zippelii
Licuala lauterbachii subsp. *peekelii*
Licuala sandsiana
Metroxylon sagu
Nypa fruticans
Physokentia avia
Ponapea hentyi
Ptychococcus paradoxus
Ptychosperma gracile
Rhopaloblaste gideonii

Milne Bay Islands
(D'Entrecasteaux Islands, Trobriand Islands, Woodlark Island, Louisiade Archipelago)

Areca catechu
Calamus johnsii
Calyptrocalyx albertisianus
Calyptrocalyx forbesii
Caryota rumphiana
Cocos nucifera
Cyrtostachys glauca
Heterospathe annectens
Heterospathe barfodii
Heterospathe pulchra
Hydriastele costata
Hydriastele longispatha
Hydriastele wendlandiana
Metroxylon sagu
Nypa fruticans
Orania lauterbachiana
Ptychosperma gracile
Ptychosperma lauterbachii
Ptychosperma mooreanum
Ptychosperma ramosissimum
Ptychosperma rosselense
Ptychosperma tagulense
Ptychosperma waitianum
Saribus woodfordii

APPENDIX 3

Useful palm species in the New Guinea region

The number of use reports per use category is shown for each species, based on a review of the literature and herbarium specimens. For a description of use categories, see Macía *et al.* (2011).

	USE CATEGORY			
SPECIES	ANIMAL FOOD	CONSTRUCTION	CULTURAL	ENVIRONMENTAL
Actinorhytis calapparia	1	1	6	1
Areca catechu		11	46	1
Areca jokowi			1	
Areca macrocalyx	2	4	33	
Areca mandacanii		1	1	
Areca novohibernica			2	
Areca unipa			2	
Arenga microcarpa		9		2
Borassus heineanus	1	1		
Brassiophoenix drymophloeoides		1		
Brassiophoenix schumannii		1	1	
Calamus anomalus			1	1
Calamus aruensis	1	6	5	
Calamus barbatus				
Calamus bulubabi				
Calamus calapparius		4		
Calamus depauperatus		4	3	1
Calamus heteracanthus		1		
Calamus klossii			1	
Calamus lauterbachii				
Calamus longipinna				
Calamus nannostachys				
Calamus oresbius		1	1	1
Calamus pachypus		1	1	
Calamus papuanus				
Calamus pintaudii		2		1
Calamus retroflexus		1		
Calamus schlechterianus				
Calamus serrulatus		3		
Calamus vestitus		2		
Calamus vitiensis		1	2	

USE CATEGORY

FUEL	HUMAN FOOD	MEDICINAL AND VETERINARY	TOXIC	UTENSILS AND TOOLS	OTHER
	2			1	
1	11	70	1	9	2
	8	1		2	
	13	2	2	4	1
	1				
				2	
	4	1		22	1
	1			3	
				1	
				1	
				13	1
	1			5	
				2	
	1			5	
				3	
				1	
				1	
1				3	
				1	
				1	
				1	
	1	2		5	
		2		2	

USE CATEGORY				
SPECIES	ANIMAL FOOD	CONSTRUCTION	CULTURAL	ENVIRONMENTAL
Calamus warburgii				
Calamus zebrinus	1	1	2	
Calamus zollingeri		1		
Calyptrocalyx albertisianus	1		2	
Calyptrocalyx amoenus		1		
Calyptrocalyx awa				
Calyptrocalyx doxanthus		1		
Calyptrocalyx elegans				
Calyptrocalyx hollrungii				
Calyptrocalyx lauterbachianus			3	
Calyptrocalyx merrillianus				
Calyptrocalyx pachystachys				
Calyptrocalyx pauciflorus				
Calyptrocalyx polyphyllus				
Calyptrocalyx yamutumene				
Caryota rumphiana	1	20	3	5
Cocos nucifera	5	29	29	4
Corypha utan	3	4	8	1
Cyrtostachys excelsa		2		
Cyrtostachys glauca		1		
Cyrtostachys loriae		8		
Dransfieldia micrantha		4		
Drymophloeus litigiosus				
Drymophloeus oliviformis		1		
Heterospathe elata		2		
Heterospathe elegans				
Heterospathe muelleriana			2	
Heterospathe porcata			1	
Heterospathe pulchra			1	
Hydriastele biakensis		3		
Hydriastele costata	1	8		
Hydriastele dransfieldii				
Hydriastele flabellata				
Hydriastele gibbsiana		2		
Hydriastele kasesa		1		
Hydriastele ledermanniana		2		
Hydriastele longispatha		2		
Hydriastele lurida		2		
Hydriastele pinangoides		3	2	1
Hydriastele procera		1		
Hydriastele simbiakii				
Hydriastele wendlandiana		3		1

USE CATEGORY

FUEL	HUMAN FOOD	MEDICINAL AND VETERINARY	TOXIC	UTENSILS AND TOOLS	OTHER
	1			5	
1				8	
				1	
				1	
				1	
				2	
				2	1
				1	
	3			1	
				2	
				1	
				1	
				1	
	1				
5	33	3	3	22	3
12	81	67		64	7
	19	14	2	12	2
	2			1	
				1	
				3	
				3	
	2			4	
	2			5	
				1	
	1			2	
				3	
1	1	1		7	2
				1	
				7	
	3			3	
	2			1	
	1			1	
	1		1	10	
				1	
	1			6	

	USE CATEGORY			
SPECIES	ANIMAL FOOD	CONSTRUCTION	CULTURAL	ENVIRONMENTAL
Hydriastele wosimiensis		1		
Korthalsia zippelii		4		2
Licuala anomala		1	1	
Licuala bacularia				
Licuala bakeri				
Licuala brevicalyx		2		
Licuala insignis		2		
Licuala lauterbachii	1	1	1	
Licuala penduliflora		1		
Licuala simplex		1	1	
Licuala telifera				
Licuala urciflora				
Linospadix albertisianus				
Livistona muelleri			1	
Manjekia maturbongsii		2		
Metroxylon sagu	5	53	21	5
Nypa fruticans	2	28	20	
Orania disticha		1		
Orania lauterbachiana		2		
Orania longistaminodia				
Orania macropetala	1			
Orania palindan	1	2		
Orania regalis		1		
Orania zonae				
Pigafetta filaris		13		1
Pinanga rumphiana		3	2	3
Ptychococcus lepidotus		1		
Ptychococcus paradoxus	1	2	6	
Ptychosperma buabe				
Ptychosperma keiense				
Ptychosperma lauterbachii				
Ptychosperma microcarpum				
Ptychosperma propinquum				1
Rhopaloblaste ceramica		3		1
Rhopaloblaste ledermanniana		2		
Saribus rotundifolius		4	1	1
Saribus surru		5	2	
Saribus tothur		1	2	
Sommieria leucophylla		2	2	
Wallaceodoxa raja-ampat		2	1	

USE CATEGORY					
FUEL	HUMAN FOOD	MEDICINAL AND VETERINARY	TOXIC	UTENSILS AND TOOLS	OTHER
	1	2		8	
				1	
				1	
				1	
				1	
				2	
				1	
	1				
	1				
				2	
				1	
				1	
	2				
5	83	19		26	16
2	56	15		20	
		1			
			2	2	
				1	
1	1			2	1
	1			4	
				4	
	1			12	
				1	
	1				
				1	
				1	
	1			1	
	2			4	
	9			4	
	3			3	
				5	
	2			2	
				2	

BIBLIOGRAPHY

Alamgir, M., Sloan, S., Campbell, M.J., Engert, J., Kiele, R., Porolak, G., Mutton, T., Brenier, A., Ibisch, P.L. & Laurance, W.F. (2019) Infrastructure expansion challenges sustainable development in Papua New Guinea. *PLoS ONE* 14: e0219408.

Alapetite, E., Baker, W.J. & Nadot, S. (2014) Evolution of stamen number in Ptychospermatinae (Arecaceae): insights from a new molecular phylogeny of the subtribe. *Molecular Phylogenetics and Evolution* 76: 227–240.

André, É.-F. (1898) Les palmiers nouveaux a l'exposition quinquennale de Gand. *Revue Horticole (Paris)* 70: 260–264.

Applequist, W.L. 2016. Report of the Nomenclature Committee for Vascular Plants: 67. *Taxon* 65:169–182.

Bachman, S., Moat, J., Hill, A.W., de Torre, J. & Scott, B. (2011) Supporting Red List threat assessments with GeoCAT: geospatial conservation assessment tool. *ZooKeys*: 117–126.

Bacon, C.D. & Baker, W.J. (2011) *Saribus* resurrected. *Palms* 55: 109–116.

Bacon, C.D., Baker, W.J. & Simmons, M.P. (2012) Miocene dispersal drives island radiations in the palm tribe Trachycarpeae (Arecaceae). *Systematic Biology* 61: 426–442.

Bacon, C.D., Michonneau, F., Henderson, A.J., McKenna, M.J., Milroy, A.M. & Simmons, M.P. (2013) Geographic and taxonomic disparities in species diversity: dispersal and diversification rates across Wallace's Line. *Evolution* 67: 2058–2071.

Bachman, S., Baker, W.J., Brummitt, N., Dransfield, J. & Moat, J. (2004) Elevational gradients, area and tropical island diversity: an example from the palms of New Guinea. *Ecography* 27: 299–310.

Baker, W.J. (1997) Rattans and rheophytes — palms of the Mubi River. *Principes* 41: 148–157.

Baker, W.J. (2000) The palms of New Guinea project. *Palms* 44: 160, 165.

Baker, W.J. (2001a) The palms of New Guinea project. *Palms and Cycads* 71: 3–7.

Baker, W.J. (2001b) The palms of New Guinea project. *New Guinea Tropical Ecology and Biodiversity Digest* 11: 6–7.

Baker, W.J. (2002a) The palms of New Guinea project. *Flora Malesiana Bulletin* 13: 35–37.

Baker, W.J. (2002b) Two unusual *Calamus* species from New Guinea. *Kew Bulletin* 57: 719–724.

Baker, W.J. (2015) A revised delimitation of the rattan genus *Calamus* (Arecaceae). *Phytotaxa* 197: 139–152.

Baker, W.J. & Couvreur, T.L.P. (2012) Biogeography and distribution patterns of Southeast Asian palms. *In:* Gower, D., Johnson, K., Richardson, J.E., Rosen, B., Rüber, L. & Williams, S. (eds) *Biotic evolution and environmental change in Southeast Asia*. Cambridge University Press, Cambridge, pp. 164–190.

Baker, W.J. & Couvreur, T.L.P. (2013a) Global biogeography and diversification of palms sheds light on the evolution of tropical lineages. I. Historical biogeography. *Journal of Biogeography* 40: 274–285.

Baker, W.J. & Couvreur, T.L.P. (2013b) Global biogeography and diversification of palms sheds light on the evolution of tropical lineages. II. Diversification history and origin of regional assemblages. *Journal of Biogeography* 40: 286–298.

Baker, W.J. & Dransfield, J. (2002a) *Calamus longipinna* (Arecaceae: Calamoideae) and its relatives in New Guinea. *Kew Bulletin* 57: 853–866.

Baker, W.J. & Dransfield, J. (2002b) *Calamus maturbongsii*, an unusual new rattan species from New Guinea. *Kew Bulletin* 57: 725–728.

Baker, W.J. & Dransfield, J. (2006a) *Field Guide to the Palms of New Guinea.* Royal Botanic Gardens, Kew, Richmond, 108 pp.

Baker, W.J. & Dransfield, J. (2006b) *Sebuah Panduan Lapangan untuk Palem New Guinea. Translation by Ary P. Keim.* Royal Botanic Gardens, Kew, Richmond, 108 pp.

Baker, W.J. & Dransfield, J. (2007) Arecaceae of Papua. *In:* Marshall, A.J. & Beehler, B.M. (eds) *The Ecology of Papua. Part 1.* Periplus Editions, Singapore, pp. 359–370.

Baker, W.J. & Dransfield, J. (2014) New rattans from New Guinea (*Calamus*, Arecaceae). *Phytotaxa* 163: 181–215.

Baker, W.J. & Dransfield, J. (2016) Beyond *Genera Palmarum*: progress and prospects in palm systematics. *Botanical Journal of the Linnean Society* 182: 207–233.

Baker, W.J. & Dransfield, J. (2017) More new rattans from New Guinea and the Solomon Islands (*Calamus*, Arecaceae). *Phytotaxa* 305: 61–86.

Baker, W.J. & Heatubun, C.D. (2012) New palms from Biak and Supiori, western New Guinea. *Palms* 56: 131–150.

Baker, W.J. & Loo, A.H.B. (2004) A synopsis of the genus *Hydriastele*. *Kew Bulletin* 59: 61–68.

Baker, W.J. & Venter, S. (2019) *Calamus nudus* – an exceptional new rattan from New Guinea. *Palms* 63: 189–196.

Baker, W.J. & Zona, S. (2006) *Dransfieldia* deciphered. *Palms* 50: 71–75.

Baker, W.J., Barfod, A.S., Dowe, J.L., Heatubun, C.D., Petoe, P., Zona, S. & Dransfield, J. (in prep.) A checklist of the palms of New Guinea. *Kew Bulletin.*

Baker, W.J., Bayton, R.P., Dransfield, J. & Maturbongs, R.A. (2003) A revision of the *Calamus aruensis* (Arecaceae) complex in New Guinea and the Pacific. *Kew Bulletin* 58: 351–370.

Baker, W.J., Heatubun, C.D. & Dransfield, J. (2021) Arecales. *In:* Utteridge, T.M.A. & Jennings, L.V.S. (eds) *Trees of New Guinea*. Royal Botanic Gardens, Kew, Richmond, pp. 98–122.

Baker, W.J., Heatubun, C.D. & Petoe, P. (2018) New finds in New Guinea *Hydriastele*. *Palms* 62: 145–154.

Baker, W.J., Keim, A.P. & Heatubun, C.D. (2000a) *Orania regalis* rediscovered. *Palms* 44: 166–169.

Baker, W.J., Maturbongs, R.A., Wanggai, J. & Hambali, G. (2000b) *Siphokentia*. *Palms* 44: 175–181.

Baker, W.J., Savolainen, V., Asmussen-Lange, C.B., Chase, M.W., Dransfield, J., Forest, F., Harley, M.M., Uhl, N.W. & Wilkinson, M. (2009) Complete generic-level phylogenetic analyses of palms (Arecaceae) with comparisons of supertree and supermatrix approaches. *Systematic Biology* 58: 240–256.

Baker, W.J, Zona, S., Heatubun, C.D., Lewis, C.E., Maturbongs, R. & Norup, M. (2006) *Dransfieldia* (Arecaceae) — A new palm genus from western New Guinea. *Systematic Botany* 31: 61–69.

Banka, R. & Baker, W.J. (2004) A monograph of the genus *Rhopaloblaste* (Arecaceae). *Kew Bulletin* 59: 47–60.

Banka, R. & Barfod, A.S. (2004) A spectacular new species of *Licuala* (Arecaceae, Coryphoideae) from New Guinea. *Kew Bulletin* 59: 73–75.

Barfod, A.S. (2000) A new species of *Licuala* from New Guinea. *Palms* 44:198–201.

Barfod, A.S. & Baker, W.J. (2022) A new, large-flowered *Licuala* from New Guinea. *Palms* 62: 69–71.

Barfod, A.S. & Heatubun, C.D. (2009) Two new species of *Licuala* Thunb. (Arecaceae: Coryphoideae) from North Moluccas and Western New Guinea. *Kew Bulletin* 64: 553–554.

Barfod, A.S. & Heatubun, C.D. (2022) Seven new species of *Licuala* (Arecaceae; Livistoninae) from Papua New Guinea. *Phytotaxa* 555: 1–16.

Barfod, A.S., Banka, R. & Dowe, J.L. (2001) *Field Guide to Palms in Papua New Guinea — with a multi-access key and notes on the genera. AAU Reports 40*. Aarhus University Press, Aarhus, 79 pp.

Barfod, A.S., Hagen, M. & Borchsenius, F. (2011) Twenty-five years of progress in understanding pollination mechanisms in palms (Arecaceae). *Annals of Botany* 108: 1503–1516.

Barstow, M., Jimbo, T. & Davies, K. (2023) Extinction risk to the endemic trees of Papua New Guinea. *Plants, People, Planet* 5: 508–509.

Bayton, R.P. (2007) A revision of *Borassus* L. (Arecaceae). *Kew Bulletin* 62: 561–586.

Beccari, O. (1877a) Della disseminazione delle palme. *Bullettino della R. Società Toscana di Orticultura* 2: 167–173.

Beccari, O. (1877b) Le specie di palme raccolte alla Nuova Guinea da O. Beccari e dal medesimo adesso descritte, con note sulle specie dei paesi circonvicini. *Malesia* 1: 9–102.

Beccari, O. (1886) Nuovi studi sulle palme Asiatiche. *Malesia* 3: 58–149.

Beccari, O. (1908) Asiatic palms - Lepidocaryeae. Part 1. The species of *Calamus*. *Annals of the Royal Botanic Garden, Calcutta* 11: 1–518.

Beccari, O. (1923) Neue Palmen Papuasiens II. *Botanische Jahrbücher für Systematik, Pflanzengeschichte und Pflanzengeographie* 58: 441–462.

Beccari, O. (1931). *Asiatic palms–Corypheae* (ed. U. Martelli) *Annals of the Royal Botanic Garden, Calcutta* 13: 1–356.

Beehler, B.M. (2007) Introduction to Papua. *In:* Marshall, A.J. & Beehler, B.M. (eds) *The Ecology of Papua. Part 1*. Periplus Editions, Singapore, pp. 3–13.

Beehler, B.M., Pratt, T.K. & Zimmerman, D.A. (1986) *Birds of New Guinea*. Princeton University Press, Princeton, 295 pp.

Beentje, H. (2016) *The Kew Plant Glossary - an illustrated dictionary of plant terms*. Royal Botanic Gardens, Kew, Richmond, 184 pp.

Bellot, S., Bayton, R.P., Couvreur, T.L.P., Dodsworth, S., Eiserhardt, W.L., Guignard, M.S., Pritchard, H.W., Roberts, L., Toorop, P.E. & Baker, W.J. (2020b) On the origin of giant seeds: the macroevolution of the double coconut (*Lodoicea maldivica*) and its relatives (Borasseae, Arecaceae). *New Phytologist* 228: 1134–1148.

Bellot, S., Lu, Y., Antonelli, A., Baker, W.J., Dransfield, J., Forest, F., Kissling, W.D., Leitch, I.J., Nic Lughadha, E., Ondo, I., Pironon, S., Walker, B.E., Cámara-Leret, R. & Bachman, S.P. (2022) The likely extinction of hundreds of palm species threatens their contributions to people and ecosystems. *Nature Ecology & Evolution* 6: 1710–1722.

Bellot, S., Odufuwa, P., Dransfield, J., Eiserhardt, W.L., Perez-Escobar, O.A., Petoe, P., Usher, E. & Baker, W.J. (2020a) Why and how to develop DNA barcoding for palms? A case study of *Pinanga*. *Palms* 64: 109–120.

Bintoro, M.H., Nurulhaq, M.I., Pratama, A.J., Ahmad, F. & Ayulia, L. (2018) Growing area of sago palm and its environment. *In:* Ehara, H., Toyoda, Y. & Johnson, D.V. (eds) *Sago Palm — multiple contributions to food security and sustainable livelihoods*. Springer, Singapore, pp. 17–29.

Blume, C.L. (1839) De quibusdam generibus e tribu Arecinearum. *Rumphia* 2: 63–99.

Bourke, R.M. & Vlassak, V. (2004). *Estimates of food crop production in Papua New Guinea*. Land Management Group, The Australian National University, Canberra.

Bradford, M.G. & Westcott, D.A. (2011) Predation of cassowary dispersed seeds: is the cassowary an effective disperser? *Integrative Zoology* 6: 168–177.

Brooks, T.M., Mittermeier, R.A., da Fonseca, G.A.B., Gerlach, J., Hoffmann, M., Lamoreux, J.F., Mittermeier, C.G., Pilgrim, J.D. & Rodrigues, A.S.L. (2006) Global Biodiversity Conservation Priorities. *Science* 313: 58–61.

Bryan, J.E. & Shearman, P.L. (eds) (2015) *The State of the Forests of Papua New Guinea 2014: measuring change over the period 2002–2014*. University of Papua New Guinea, Port Moresby, 209 pp.

Burret, M. (1933) Neue Palmen aus Neu-Guinea. *Notizblatt des Botanischen Gartens und Museums zu Berlin-Dahlem* 11: 704–713.

Burret, M. (1935) Neue Palmen aus Neuguinea II. *Notizblatt des Botanischen Gartens und Museums zu Berlin-Dahlem* 12: 309–348.

Burret, M. (1936) Neue Palmen aus Neuguinea IV. *Notizblatt des Botanischen Gartens und Museums zu Berlin-Dahlem* 13: 317–332.

Burret, M. (1941). Beiträge zur Palmengattung *Licuala* Wurmb. *Notizblatt des Botanischen Gartens und Museums zu Berlin-Dahlem* 15: 327–336.

Cámara-Leret, R. & Dennehy, Z. (2019a). Quantitative ethnobotany of palms (Arecaceae) in New Guinea. *Gardens' Bulletin Singapore* 71: 321–364.

Cámara-Leret, R. & Dennehy, Z. (2019b). Information gaps in indigenous and local knowledge for science-policy assessments. *Nature Sustainability* 2: 736–741.

Cámara-Leret, R., Frodin, D.G., Adema, F., Anderson, C., Appelhans, M.S., Argent, G., Arias Guerrero, S., Ashton, P., Baker, W.J., Barfod, A.S., Barrington, D., Borosova, R., Bramley, G.L.C., Briggs, M., Buerki, S., Cahen, D., Callmander, M.W., Cheek, M., Chen, C.-W., Conn, B.J., Coode, M.J.E., Darbyshire, I., Dawson, S., Dransfield, J., Drinkell, C., Duyfjes, B., Ebihara, A., Ezedin, Z., Fu, L.-F., Gideon, O., Girmansyah, D., Govaerts, R., Fortune-Hopkins, H., Hassemer, G., Hay, A., Heatubun, C.D., Hind, D.J.N., Hoch, P., Homot, P., Hovenkamp, P., Hughes, M., Jebb, M., Jennings, L., Jimbo, T., Kessler, M., Kiew, R., Knapp, S., Lamei, P., Lehnert, M., Lewis, G.P., Linder, H.P., Lindsay, S., Low, Y.W., Lucas, E., Mancera, J.P., Monro, A.K., Moore, A., Middleton, D.J., Nagamasu, H., Newman, M.F., Nic Lughadha, E., Melo, P.H.A., Ohlsen, D.J., Pannell, C.M., Parris, B., Pearce, L., Penneys, D.S., Perrie, L.R., Petoe, P., Poulsen, A.D., Prance, G.T., Quakenbush, J.P., Raes, N., Rodda, M., Rogers, Z.S., Schuiteman, A., Schwartsburd, P., Scotland, R.W., Simmons, M.P., Simpson, D.A., Stevens, P., Sundue, M., Testo, W., Trias-Blasi, A., Turner, I., Utteridge, T., Walsingham, L., Webber, B.L., Wei, R., Weiblen, G.D., Weigend, M., Weston, P., de Wilde, W., Wilkie, P., Wilmot-Dear, C.M., Wilson, H.P., Wood, J.R.I., Zhang, L.-B. & van Welzen, P.C. (2020) New Guinea has the world's richest island flora. *Nature* 584: 579–583.

Cámara-Leret, R., Schuiteman, A., Utteridge, T., Bramley, G., Deverell, R., Fisher, L.A., McLeod, J., Hannah, L., Roehrdanz, P., Laman, T.G., Scholes, E., de Fretes, Y. & Heatubun, C. (2019) The Manokwari Declaration: challenges ahead in conserving 70% of Tanah Papua's forests. *Forest and Society* 3: 148–151.

Cookson, M. (2000) The Archbold Expeditions to New Guinea: A preliminary survey of archival materials held at the American Museum of Natural History, New York City. *Journal of Pacific History* 35: 313–318.

Couvreur, T.L., Kissling, W.D., Condamine, F.L., Svenning, J.C., Rowe, N.P. & Baker, W.J. (2014) Global diversification of a tropical plant growth form: environmental correlates and historical contingencies in climbing palms. *Frontiers in Genetics* 5: 452.

Crayn, D.M., Costion, C. & Harrington, M.G. (2015) The Sahul–Sunda floristic exchange: dated molecular phylogenies document Cenozoic intercontinental dispersal dynamics. *Journal of Biogeography* 42: 11–24.

Crisp, M.D., Isagi, Y., Kato, Y., Cook, L.G. & Bowman, D.M.J.S. (2010) *Livistona* palms in Australia: ancient relics or opportunistic immigrants? *Molecular Phylogenetics and Evolution* 54: 512–523.

Currie, D.J. & Kerr, J.T. (2008) Tests of the mid-domain hypothesis: a review of the evidence. *Ecological Monographs* 78: 3–18.

Darbyshire, I., Anderson, S., Asatryan, A., Byfield, A., Cheek, M., Clubbe, C., Ghrabi, Z., Harris, T., Heatubun, C.D., Kalema, J., Magassouba, S., McCarthy, B., Milliken, W., de Montmollin, B., Lughadha, E.N., Onana, J.-M., Saïdou, D., Sârbu, A., Shrestha, K. & Radford, E.A. (2017) Important Plant Areas: revised selection criteria for a global approach to plant conservation. *Biodiversity and Conservation* 26: 1767–1800.

Dickie, J.B., Balick, M.J. & Linington, I.M. (1993) Studies on the practicality of *ex situ* preservation of palm seeds. *Palms* 37: 94–98.

Dirzo, R., Young, H.S., Galetti, M., Ceballos, G., Isaac, N.J.B. & Collen, B. (2014) Defaunation in the Anthropocene. *Science* 345: 401–406.

Dowe, J.L. (2007) *Ptychosperma macarthurii*: discovery, horticulture and taxonomy. *Palms* 51: 85–96.

Dowe, J.L. (2009) A taxonomic account of *Livistona* (Arecaceae). *Gardens' Bulletin Singapore* 60: 185–344.

Dowe, J.L. (2010) *Australian Palms: Biogeography, Ecology and Systematics.* CSIRO Publishing, Collingwood, 304 pp.

Dowe, J.L. & Ferrero, M.D. (2000) A new species of rheophytic palm from New Guinea. *Palms* 44: 194–197.

Dowe, J.L. & Ferrero, M.D. (2001) Revision of *Calyptrocalyx* and the New Guinea species of *Linospadix* (Linospadicinae: Arecoideae: Arecaceae). *Blumea* 46: 207–251.

Dowe, J.L. & Latifah, D. (2020) Oranges or kings? The cryptic etymology of Zippelius's *Orania regalis*. *Palms* 64: 121–130.

Dowe, J.L. & Venter, S. (2023) *Saribus pendulinus,* a new fan palm from New Guinea. *Palms* 67: 13–25.

Dransfield, J. (1981) A synopsis of the genus *Korthalsia* (Palmae: Lepidocaryoideae). *Kew Bulletin* 36: 163–194.

Dransfield, J. (1982) *Clinostigma* in New Ireland. *Principes* 26: 73–76.

Dransfield, J. (1987) Bicentric distributions in Malesia as exemplified by palms. *In:* Whitmore, T.C. (ed.) *Biogeographical evolution of the Malay archipelago*. Clarendon Press, Oxford, pp. 60–72.

Dransfield, J. (1998). *Pigafetta*. *Principes* 42: 34–40.

Dransfield, J. & Baker, W.J. (2003) An account of the Papuasian species of *Calamus* (Arecaceae) with paired fruit. *Kew Bulletin* 58: 371–387.

Dransfield, J. & Beentje, H. (1995) *The Palms of Madagascar*. Royal Botanic Gardens, Kew, Richmond & International Palm Society, 475 pp.

Dransfield, J. & Marcus, J. (2020) *Calyptrocalyx reflexus,* a new species described from cultivation. *Palms* 64: 43–48.

Dransfield, J., Baker, W.J., Heatubun, C.D. & Witono, J. (2000b) The palms of Mount Jaya. *Palms* 44: 202–208.

Dransfield, J., Hambali, G.G., Maturbongs, R.A. & Heatubun, C.D. (2000a) *Caryota zebrina*. *Palms* 44: 170–174.

Dransfield, J., Uhl, N.W., Asmussen, C.B., Baker, W.J., Harley, M.M. & Lewis, C.E. (2008) *Genera Palmarum — the evolution and classification of palms*. Royal Botanic Gardens, Kew, Richmond, 732 pp.

Ehara, H., Toyoda, Y. & Johnson, D.V. (eds) (2018) *Sago Palm — multiple contributions to food security and sustainable livelihoods*. Springer, Singapore, 330 pages.

Eiserhardt, W.L., Bellot, S., Cowan, R.S., Dransfield, J., Hansen, L.E.S.F., Heyduk, K., Rabarijaona, R.N., Rakotoarinivo, M. & Baker, W.J. (2022) Phylogenomics and generic limits of Dypsidinae (Arecaceae), the largest palm radiation in Madagascar. *Taxon* 71: 1170–1195.

Essig, F.B. (1972) Palms in the botanic garden at Lae, Papua New Guinea. *Principes* 16: 119–127.

Essig, F.B. (1973) Pollination in some New Guinea palms. *Principes* 17: 75–83.

Essig, F.B. (1977) A preliminary analysis of the palm flora of New Guinea and the Bismarck Archipelago. *Papua New Guinea Botany Bulletin* 9: 1–39.
Essig, F.B. (1978) A revision of the genus *Ptychosperma* Labill. (Arecaceae). *Allertonia* 1: 415–478.
Essig, F.B. (1980) The genus *Orania* Zipp. (Arecaceae) in New Guinea. *Lyonia* 1: 211–233.
Essig, F.B. (1982) A synopsis of the genus *Gulubia*. *Principes* 26: 159–173.
Essig, F.B. (1992) A new species of *Heterospathe* (Palmae) from New Britain. *Principes* 36: 4–6.
Essig, F.B. (1995) A checklist and analysis of the palms of the Bismarck Archipelago. *Principes* 39: 123–129.
Essig, F.B. & Young, B.E. (1980) Palm collecting in Papua New Guinea. I. The Northeast. *Principes* 24: 14–28.
Essig, F.B & Young, B.E. (1981a) Palm collecting in Papua New Guinea. II. The Sepik and the North Coast. *Principes* 25: 3–15.
Essig, F.B & Young, B.E. (1981b) Palm collecting in Papua New Guinea. III. Papua. *Principes* 25: 16–28.
Essig, F.B. & Young, B.E. (1985) A reconsideration of *Gronophyllum* and *Nengella* (Arecoideae). *Principes* 29: 129–137.
Fairchild, D. (1944). *Garden Islands of the Great East*. Scribners, New York, 239 pp.
Fernando, E.S. (2014) Three new species in *Calamus* sect. *Podocephalus* (Arecaceae: Calamoideae) from the Philippines, Indonesia, and Papua New Guinea. *Phytotaxa* 166: 69–76.
Ferrero, M.D. (1996) *Ptychococcus lepidotus*: in from the cold? A promising palm from the highlands of New Guinea. *Palms & Cycads* 52/53: 48–54.
Ferrero, M.D. (1997) A checklist of Palmae for New Guinea. *Palms and Cycads* 55/56: 2–39.
Ferrero, M.D. & Dowe, J. L. (1996) Notes on *Brassiophoenix* (Arecaceae) — the bat-wing palm from Papua New Guinea. *Mooreana* 6: 18–22.
Flach, M. (1997) *Sago palm:* Metroxylon sagu *Rottb. Promoting the conservation and use of underutilized and neglected crops. 13.* International Plant Genetic Resources Institute, Rome, 76 pp.
Flynn T. (2004) *Morphological variation and species limits in the genus* Areca *(Palmae) in New Guinea and the Solomon Islands*. MSc thesis, University of Wales, Bangor.
Forster, P.I. (1997) Len Brass and his contribution to palm discoveries in New Guinea and the Solomon Islands. *Principes* 41: 158–162.
Frazier, S. (2007) Threats to biodiversity. *In:* Marshall, A.J. & Beehler, B.M. (eds) *The Ecology of Papua. Part 2.* Periplus Editions, Singapore, pp. 1199–1229.
Fullagar, R., Field, J., Denham, T. & Lentfer, C. (2006) Early and mid-Holocene tool-use and processing of taro (*Colocasia esculenta*), yam (*Dioscorea* sp.) and other plants at Kuk Swamp in the highlands of Papua New Guinea. *Journal of Archaeological Science* 33: 595–614.
Furtado, C.X. (1955) Palmae Malesicae — XVIII. Two new calamoid genera of Malaysia. *Gardens' Bulletin Singapore* 14: 517–529.
Furtado, C.X. (1956) Palmae Malesicae — IX. The genus *Calamus* in the Malayan Peninsula. *Gardens' Bulletin Singapore* 15: 32–262.
Galetti, M., Bovendorp, R.S. & Guevara, R. (2015) Defaunation of large mammals leads to an increase in seed predation in the Atlantic forests. *Global Ecology and Conservation* 3: 824–830.
Gamoga, G., Turia, R., Abe, H., Haraguchi, M. & Iuda, O. (2021) The forest extent in 2015 and the drivers of forest change between 2000 and 2015 in Papua New Guinea: deforestation and forest degradation in Papua New Guinea. *Case Studies in the Environment* 5: 1442018.
Gardiner, L.M., Dransfield, J., Marcus, J. & Baker, W.J. (2012) *Heterospathe barfodii,* a new species from Papua New Guinea. *Palms* 56: 91–100.

Gaveau, D.L.A., Santos, L., Locatelli, B., Salim, M.A., Husnayaen, H., Meijaard, E., Heatubun, C. & Sheil, D. (2021) Forest loss in Indonesian New Guinea (2001–2019): trends, drivers and outlook. *Biological Conservation* 261: 109225.

GEBCO Compilation Group (2022) *GEBCO 2022 Grid.* Available from: https://download.gebco.net/. doi:10.5285/e0f0bb80-ab44-2739-e053-6c86abc0289c

Global Forest Watch (2023) *Global Forest Watch.* Available at: https://www.globalforestwatch.org/

Gold, D.P., Casas-Gallego, M., Holm, R., Webb, M. & White, L.T. (2020) New tectonic reconstructions of New Guinea derived from biostratigraphy and geochronology. *In: Proceedings, Indonesian Petroleum Association*, pp. PA20 G 61.

Govaerts, R., Nic Lughadha, E., Black, N., Turner, R. & Paton, A. (2021) The World Checklist of Vascular Plants, a continuously updated resource for exploring global plant diversity. *Scientific Data* 8: 215.

Greenhill, A.R., Shipton, W.A., Blaney, B.J., & Warner, J.M. 2007. Fungal colonization of sago starch in Papua New Guinea. *International Journal of Food Microbiology* 119: 284–290.

Gunn, B.F., Baudoin, L. & Olsen, K.M. (2011) Independent origins of cultivated coconut (*Cocos nucifera* L.) in the Old World tropics. *PLoS ONE* 6: e21143.

Gupta, P.C. & Warnakulasuriya, S. (2002) Global epidemiology of areca nut usage. *Addiction Biology* 7: 77–83.

Hall, R. (2009) Southeast Asia's changing palaeogeography. *Blumea* 54: 148–161.

Hall, R. (2012) Sundaland and Wallacea: geology, plate tectonics and palaeogeography. *In:* Gower, D., Johnson, K., Richardson, J.E., Rosen, B., Rüber, L. & Williams, S. (eds) *Biotic Evolution and Environmental Change in Southeast Asia*. Cambridge University Press, Cambridge, pp. 32–78.

Hall, R. (2017) Southeast Asia: new views of the geology of the Malay Archipelago. *Annual Review of Earth and Planetary Sciences* 45: 331–358.

Hansen, L.E.S.F., Baker, W.J., Tietje, M. & Eiserhardt, W.L. (2021) Testing tropical biogeographical regions using the palm family as a model clade. *Journal of Biogeography* 48: 2502–2511.

Hay, A. (1984) Part 3. Palmae. *In:* Johns, R.J. & Hay, A. (eds) *A Guide to the Monocotyledons of Papua New Guinea.* Papua New Guinea University of Technology, Lae, 195–319 pp.

Heatubun, C.D. (2000) In search of *Caryota zebrina* — a palm expedition to the Cyclops Mountains. *Palms* 44: 187–193.

Heatubun, C.D. (2002) A monograph of *Sommieria*. *Kew Bulletin* 57: 599–611.

Heatubun, C.D. (2008) A new *Areca* from western New Guinea. *Palms* 52: 198–202.

Heatubun, C.D. (2016). *Areca jokowi*: a new species of betel nut palm (Arecaceae) from western New Guinea. *Phytotaxa* 288: 175–180.

Heatubun, C.D. (ed.) (2022) *Papua Barat — menuju pembangunan berkelanjutan.* PT Micepro, Indonesia, 254 pp.

Heatubun, C.D. & Barfod, A. S. (2008). Two new species of *Licuala* (Arecaceae: Coryphoideae) from western New Guinea. *Blumea* 53: 429–434.

Heatubun, C.D., Baker, W.J., Mogea, J.P., Harley, M.M., Tjitrosoedirdjo, S.S. & Dransfield, J. (2009) A monograph of *Cyrtostachys* (Arecaceae). *Kew Bulletin* 64: 67–94.

Heatubun, C.D., Dransfield, J., Flynn, T., Tjitrosoedirdjo, S.S., Mogea, J.P. & Baker, W.J. (2012a) A monograph of the betel nut palms (*Areca*: Arecaceae) of East Malesia. *Botanical Journal of the Linnean Society* 168: 147–173.

Heatubun, C.D., Dransfield, J., Govaerts, R. & Baker, W.J. (2014b) (2279) Proposal to reject the name *Areca glandiformis* (Arecaceae). *Taxon* 63: 434–435.

Heatubun, C.D., Gardiner, L. M. & Baker, W.J. (2012b) *Heterospathe elata*, a new record for the New Guinea

islands. *Palms* 56: 61–64.
Heatubun, C.D, Iwanggin, M. & Simbiak, V.I. (2013) A new species of betel nut palm (*Areca*: Arecaceae) from western New Guinea. *Phytotaxa* 154: 59–64.
Heatubun, C.D., Lekitoo, K. & Matani, O.P. (2014a) Palms on the nickel island: an expedition to Gag Island, western New Guinea. *Palms* 58: 115–134.
Heatubun, C.D., Petoe, P. & Baker, W.J. (2018) A monograph of the *Nengella* group of *Hydriastele* (Arecaceae). *Kew Bulletin* 73: 18.
Heatubun, C.D., Zona, S. & Baker, W.J. (2014c) Three new genera of arecoid palm (Arecaceae) from eastern Malesia. *Kew Bulletin* 69: 9525.
Heatubun, C.D., Zona, S. & Baker, W.J. (2014d) Three new palm genera from Indonesia. *Palms* 58: 197–202.
Heinen, J.H., Florens, F.B.V., Baider, C., Hume, J.P., Kissling, W.D., Whittaker, R.J., Rahbek, C. & Borregaard, M.K. (2023) Novel plant–frugivore network on Mauritius is unlikely to compensate for the extinction of seed dispersers. *Nature Communications* 14: 1019.
Henderson, A. (2020) A revision of *Calamus* (Arecaceae, Calamoideae, Calameae, Calaminae). *Phytotaxa* 445: 1–656.
Henderson, A.J. & Bacon, C.D. (2011) *Lanonia* (Arecaceae: Palmae), a new genus from Asia, with a revision of the species. *Systematic Botany* 36: 883–895.
Hirashi, T. (2008) Effectiveness of coastal forests in mitigating tsunami hazards. *In*: Chan, H. T. & Ong, J. E. (eds) *Proceedings of the Meeting and Workshop on Guidelines for the Rehabilitation of Mangroves and other Coastal Forests Damaged by Tsunamis and other Natural Hazards in the Asia-Pacific Region*. International Society for Mangrove Ecosystems, Okinawa, pp. 65–73.
Hodel, D.R., Baker, W.J., Bellot, S., Pérez-Calle, V., Cumberledge, A. & Barrett, C.F. (2021) Reassessment of the Archontophoenicinae of New Caledonia and description of a new species. *Palms* 65: 109–131.
Hodel, D.R., Butaud, J.F., Barrett, C.F., Grayum, M.H., Komen, J., Lorence, D.H., Marcus, J. & Falchetto, A. (2019). Reassessment of *Pelagodoxa*. *Palms* 63: 113–152.
Hope, G.S. (2007) The history of human impact on New Guinea. *In:* Marshall, A.J. & Beehler, B.M. (eds) *The Ecology of Papua. Part 2*. Periplus Editions, Singapore, pp. 1087–1097.
IUCN (1998–2023) *The IUCN Red List of Threatened Species*. Available at: https://www.iucnredlist.org.
IUCN (2012) *IUCN Red List Categories and Criteria: Version 3.1. Second edition*. IUCN, Gland, Switzerland & Cambridge, 32 pp.
Jeanson, M.L. (2011). *Systematics of tribe Caryoteae (Arecaceae)*. PhD thesis, Muséum National d'Histoire Naturelle, Paris.
Jimbo, T. & Kipiro, W. (2021) *Cyrtostachys bakeri. The IUCN Red List of Threatened Species* 2021: e.T189125838A189758475. https://dx.doi.org/10.2305/IUCN.UK.2021-3.RLTS.T189125838A189758475.en
Jimbo, T., Saulei, S., Moses, J., Lawong, B., Kaina, G., Kiapranis, R., Hitofumi, A., Novotny, V., Attorre, F., Testolin, R. & Cicuzza, D. (2023) Beyond the trees: a comparison of nonwoody species, and their ecology, in Papua New Guinea elevational gradient forest. *Case Studies in the Environment* 7: 1831407.
Johns, R.J. & Taurereko, R. (1989a) *A Preliminary Checklist of the Collections of* Calamus *and* Daemonorops *from the Papuasian Region. Rattan Research Report 1989/2*. Forestry Department and Papua New Guinea University of Technology, Lae, 67 pp.
Johns, R.J. & Taurereko, R. (1989b) *A Guide to the Collection and Field Description of* Calamus *(Palmae) from Papuasia. Rattan Research Report 1989/3*. Forestry Department and Papua New Guinea University of Technology, Lae, 34 pp.

Johns, R.J. & Zibe, S. (1989) *A Checklist of the Species of* Calamus *and* Korthalsia *in Papuasia. Rattan Research Report 1989/1*. Forestry Department and Papua New Guinea University of Technology, Lae, 13 pp.

Johnson, N.W., Warnakulasuriya, S., Gupta, P.C., Dimba, E., Chindia, M., Otoh, E.C., Sankaranarayanan, R., Califano, J. & Kowalski, L. (2011). Global oral health inequalities in incidence and outcomes for oral cancer: causes and solutions. *Advances in Dental Research,* 23: 237–246.

Jong, F.S. (1995) *Research for the development of sago palm (*Metroxylon sagu *Rottb.) cultivation in Sarawak, Malaysia.* PhD thesis, Wageningen Agricultural University.

Joyce, E.M., Thiele, K.R., Slik, J.W.F. & Crayn, D.M. (2020) Plants will cross the lines: climate and available land mass are the major determinants of phytogeographical patterns in the Sunda–Sahul Convergence Zone. *Biological Journal of the Linnean Society* 132: 374–387.

Keim, A. & Dransfield, J. (2012) A monograph of the genus *Orania* (Arecaceae: Oranieae). *Kew Bulletin* 67: 127–190.

Kissling, W.D., Baker, W.J., Balslev, H., Barfod, A.S., Borchsenius, F., Dransfield, J., Govaerts, R. & Svenning, J.-C. (2012b) Quaternary and pre-Quaternary historical legacies in the global distribution of a major tropical plant lineage. *Global Ecology and Biogeography* 21: 909–921.

Kissling, W.D., Eiserhardt, W.L., Baker, W.J., Borchsenius, F., Couvreur, T.L.P., Balslev, H. & Svenning, J.-C. (2012a) Cenozoic imprints on the phylogenetic structure of palm species assemblages worldwide. *Proceedings of the National Academy of Sciences of the United States of America* 109: 7379–7384.

Kjaer, A., Barfod, A.S., Asmussen, C.B. & Seberg, O. (2004) Investigation of genetic and morphological variation in the sago palm (*Metroxylon sagu*; Arecaceae) in Papua New Guinea. *Annals of Botany* 94: 109–117.

Kooyman, R.M., Morley, R.J., Crayn, D.M., Joyce, E.M., Rossetto, M., Slik, J.W.F., Strijk, J.S., Su, T., Yap, J.-Y.S. & Wilf, P. (2019) Origins and assembly of Malesian rainforests. *Annual Review of Ecology, Evolution, and Systematics* 50: 119–143.

Kuhnhäuser, B.G. (2021) *Phylogenomics and Biogeography of the Calamoid Palms*. PhD thesis, University of Oxford.

Kuhnhäuser, B.G., Bellot, S., Couvreur, T.L.P., Dransfield, J., Henderson, A., Schley, R., Chomicki, G., Eiserhardt, W.L., Hiscock, S.J. & Baker, W.J. (2021) A robust phylogenomic framework for the calamoid palms. *Molecular Phylogenetics and Evolution* 157: 107067.

Labillardière, J.J.H.d. (1809) Mémoire sur un nouveau genre de palmier. *Mémoires de la classe des sciences mathématiques et physiques de l'Institut National de France* 9: 251–256.

Lepš, J., Novotný, V., Čížek, L., Molem, K., Isua, B., William, B., Kutil, R., Auga, J., Kasbal, M., Manumbor, M. & Hiuk, S. (2002) Successful invasion of the neotropical species *Piper aduncum* in rain forests in Papua New Guinea. *Applied Vegetation Science* 5: 255–262.

Lim, J.Y., Svenning, J.-C., Göldel, B., Faurby, S. & Kissling, W.D. (2020) Frugivore-fruit size relationships between palms and mammals reveal past and future defaunation impacts. *Nature Communications* 11: 4904.

Loo, A., Dransfield, J., Chase, M. & Baker, W. (2006) Low-copy nuclear DNA, phylogeny and the evolution of dichogamy in the betel nut palms and their relatives (Arecinae; Arecaceae). *Molecular Phylogenetics and Evolution* 39: 598–618.

Macía, M.J., Armesilla, P.J., Cámara-Leret, R., Paniagua-Zambrana, N., Villalba, S., Balslev, H. & Pardo-de-Santayana, M. (2011). Palm uses in northwestern South America: a quantitative review. *The Botanical Review* 77: 462–570.

Mantiquilla, J.A., Abad, R.G., Barro, K.M.G., Basilio, J.A.M., Rivero, G.C. & Silvosa, C.S.C. (2016) Potential pollinators of nipa palm (*Nypa fruticans* Wurmb.). *Asia Life Science* 25: 1–22.

Marsh, S.T., Brummitt, N.A., de Kok, R.P.J. & Utteridge, T.M.A. (2009) Large-scale patterns of plant diversity and conservation priorities in South East Asia. *Blumea* 54: 103–108.

Marshall, A.J. & Beehler, B.M. (eds) (2007) *The Ecology of Papua (Parts 1 & 2)*. Periplus Editions, Singapore, 1467 pp.

Masters, M.T. (1898) New or noteworthy plants. *Gardeners' Chronicle, series 3* 24: 241–243.

Maturbongs, R.A., Dransfield, J. & Baker, W.J. (2014) *Calamus kebariensis* (Arecaceae) — a new montane rattan from New Guinea. *Phytotaxa* 163: 235–238.

Maturbongs, R.A., Dransfield, J. & Mogea, J.P. (2015) *Daemonorops komsaryi* (Arecaceae) — a new rattan from the Bird's Head Peninsula, Indonesian New Guinea. *Phytotaxa* 195: 297–300.

McClatchey, W.C. (1996) A revision of the genus *Metroxylon* section *Coelococcus* (Arecaceae). PhD dissertation. University of Florida.

Merklinger, F.F., Baker, W.J. & Rudall, P.J. (2014) Comparative development of the rattan ocrea, a structural innovation that facilitates ant-plant mutualism. *Plant Systematics and Evolution* 300: 1973–1983.

Middleton, D.J., Armstrong, K., Baba, Y., Balslev, H., Chayamarit, K., Chung, R.C.K., Conn, B.J., Fernando, E.S., Fujikawa, K., Kiew, R., Luu, H.T., Mu Mu, A., Newman, M.F., Tagane, S., Tanaka, N., Thomas, D.C., Tran, T.B., Utteridge, T.M.A., Welzen, P.C.v., Widyatmoko, D., Yahara, T. & Wong, K.M. (2019) Progress on Southeast Asia's flora projects. *Gardens' Bulletin (Singapore)* 71: 267–319.

Miettinen, J., Shi, C. & Liew, S.C. (2011) Deforestation rates in insular Southeast Asia between 2000 and 2010. *Global Change Biology* 17: 2261–2270.

Miklouho-Maclay, N., De (1885). List of plants in use by the natives of the Maclay coast, New Guinea. *Proceedings of the Linnean Society of New South Wales* 10: 346–358.

Mittermeier, R.A., Mittermeier, C.G., Brooks, T.M., Pilgrim, J.D., Konstant, W.R., da Fonseca, G.A.B. & Kormos, C. (2003) Wilderness and biodiversity conservation. *Proceedings of the National Academy of Sciences of the United States of America* 100: 10309–10313.

Mogea, J.P. (1991). Revisi marga *Arenga*. PhD thesis, Universitas Indonesia.

Moore, H.E. (1966) Palm hunting around the World. V. New Guinea to the New Hebrides. *Principes* 10: 64–85.

Moore, H.E., Jr. (1977) New palms from the Pacific, IV. *Principes* 21: 86–88.

Moore, H.E., Jr. & Fosberg, F.R. (1956) The palms of Micronesia and the Bonin Islands. *Gentes Herbarum* 8:423–478.

Muscarella, R., Emilio, T., Phillips, O.L., Lewis, S.L., Slik, F., Baker, W.J., Couvreur, T.L.P., Eiserhardt, W.L., Svenning, J.-C., Affum-Baffoe, K., Aiba, S.-I., de Almeida, E.C., de Almeida, S.S., de Oliveira, E.A., Álvarez-Dávila, E., Alves, L.F., Alvez-Valles, C.M., Carvalho, F.A., Guarin, F.A., Andrade, A., Aragão, L.E.O.C., Murakami, A.A., Arroyo, L., Ashton, P.S., Corredor, G.A.A., Baker, T.R., de Camargo, P.B., Barlow, J., Bastin, J.-F., Bengone, N.N., Berenguer, E., Berry, N., Blanc, L., Böhning-Gaese, K., Bonal, D., Bongers, F., Bradford, M., Brambach, F., Brearley, F.Q., Brewer, S.W., Camargo, J.L.C., Campbell, D.G., Castilho, C.V., Castro, W., Catchpole, D., Cerón Martínez, C.E., Chen, S., Chhang, P., Cho, P., Chutipong, W., Clark, C., Collins, M., Comiskey, J.A., Medina, M.N.C., Costa, F.R.C., Culmsee, H., David-Higuita, H., Davidar, P., del Aguila-Pasquel, J., Derroire, G., Di Fiore, A., Van Do, T., Doucet, J.-L., Dourdain, A., Drake, D.R., Ensslin, A., Erwin, T., Ewango, C.E.N., Ewers, R.M., Fauset, S., Feldpausch, T.R., Ferreira, J., Ferreira, L.V., Fischer, M., Franklin, J., Fredriksson, G.M., Gillespie, T.W., Gilpin, M., Gonmadje, C., Gunatilleke, A.U.N., Hakeem, K.R., Hall, J.S., Hamer, K.C., Harris,

D.J., Harrison, R.D., Hector, A., Hemp, A., Herault, B., Pizango, C.G.H., Coronado, E.N.H., Hubau, W., Hussain, M.S., Ibrahim, F.-H., Imai, N., Joly, C.A., Joseph, S., Anitha,K., Kartawinata, K., Kassi, J., Killeen, T.J., Kitayama, K., Klitgård, B.B., Kooyman, R., Labrière, N., Larney, E., Laumonier, Y., Laurance, S.G., Laurance, W.F., Lawes, M.J., Levesley, A., Lisingo, J., Lovejoy, T., Lovett, J.C., Lu, X., Lykke, A.M., Magnusson, W.E., Mahayani, N.P.D., Malhi, Y., Mansor, A., Peña, J.L.M., Marimon-Junior, B.H., Marshall, A.J., Melgaco, K., Bautista, C.M., Mihindou, V., Millet, J., Milliken, W., Mohandass, D., Mendoza, A.L.M., Mugerwa, B., Nagamasu, H., Nagy, L., Seuaturien, N., Nascimento, M.T., Neill, D.A., Neto, L.M., Nilus, R., Vargas, M.P.N., Nurtjahya, E., de Araújo, R.N.O., Onrizal, O., Palacios, W.A., Palacios-Ramos, S., Parren, M., Paudel, E., Morandi, P.S., Pennington, R.T., Pickavance, G., Pipoly III, J.J., Pitman, N.C.A., Poedjirahajoe, E., Poorter, L., Poulsen, J.R., Rama Chandra Prasad, P., Prieto, A., Puyravaud, J.-P., Qie, L., Quesada, C.A., Ramírez-Angulo, H., Razafimahaimodison, J.C., Reitsma, J.M., Requena-Rojas, E.J., Restrepo Correa, R., Rodriguez, C.R., Roopsind, A., Rovero, F., Rozak, A., Lleras, A.R., Rutishauser, E., Rutten, G., Punchi-Manage, R., Salomão, R.P., Van Sam, H., Sarker, S.K., Satdichanh, M., Schietti, J., Schmitt, C.B., Marimon, B.S., Senbeta, F., Nath Sharma, L., Sheil, D., Sierra, R., Silva-Espejo, J.E., Silveira, M., Sonké, B., Steininger, M.K., Steinmetz, R., Stévart, T., Sukumar, R., Sultana, A., Sunderland, T.C.H., Suresh, H.S., Tang, J., Tanner, E., ter Steege, H., Terborgh, J.W., Theilade, I., Timberlake, J., Torres-Lezama, A., Umunay, P., Uriarte, M., Gamarra, L.V., van de Bult, M., van der Hout, P., Martinez, R.V., Vieira, I.C.G., Vieira, S.A., Vilanova, E., Cayo, J.V., Wang, O., Webb, C.O., Webb, E.L., White, L., Whitfeld, T.J.S., Wich, S., Willcock, S., Wiser, S.K., Young, K.R., Zakaria, R., Zang, R., Zartman, C.E., Zo-Bi, I.C. & Balslev, H. (2020) The global abundance of tree palms. *Global Ecology and Biogeography* 29: 1495–1514.

Myers, N., Mittermeier, R.A., Mittermeier, C.G., da Fonseca, G.A.B. & Kent, J. (2000) Biodiversity hotspots for conservation priorities. *Nature* 403: 853–858.

Nelson, P.N., Gabriel, J., Filer, C., Banabas, M., Sayer, J.A., Curry, G.N., Koczberski, G. & Venter, O. (2014) Oil palm and deforestation in Papua New Guinea. *Conservation Letters* 7: 188–195.

Nic Lughadha, E., Bachman, S.P., Leão, T.C.C., Forest, F., Halley, J.M., Moat, J., Acedo, C., Bacon, K.L., Brewer, R.F.A., Gâteblé, G., Gonçalves, S.C., Govaerts, R., Hollingsworth, P.M., Krisai-Greilhuber, I., de Lirio, E.J., Moore, P.G.P., Negrão, R., Onana, J.M., Rajaovelona, L.R., Razanajatovo, H., Reich, P.B., Richards, S.L., Rivers, M.C., Cooper, A., Iganci, J., Lewis, G.P., Smidt, E.C., Antonelli, A., Mueller, G.M. & Walker, B.E. (2020) Extinction risk and threats to plants and fungi. *Plants, People, Planet* 2: 389–408.

Norton, S.A. (1998) Betel: consumption and consequences. *Journal of American Academy of Dermatology* 38: 81–88.

Norup, M.V. (2005) *Alsmithia* subsumed in *Heterospathe* (Arecaceae, Arecoideae). *Novon* 15: 455–457.

Norup, M.V., Dransfield, J., Chase, M.W., Barfod, A.S., Fernando, E.S. & Baker, W.J. (2006) Homoplasious character combinations and generic delimitation: a case study from the Indo-Pacific arecoid palms (Arecaceae: Areceae). *American Journal of Botany* 93: 1065–1080.

Ohtsuka, R. (1983) *Oriomo Papuans: Ecology of Sago-Eaters in Lowland Papua*. University of Tokyo Press, 197 pp.

Paijmans, K. (ed.) (1976) *New Guinea Vegetation*. Australian National University Press, Canberra, 213 pp.

Pangau-Adam, M. & Mühlenberg, M. (2014) Palm species in the diet of the northern cassowary (*Casuarius unappendiculatus*) in Jayapura region, Papua, Indonesia. *Palms* 58: 19–26.

Pemerintah Provinsi Papua Barat (2021) *Laporan hasil evaluasi perizinan perkebunan kelapa sawit provinsi Papua Barat (Report on the Results of the Evaluation of Permits for Oil Palm Plantations in West Papua Province)*. Yayasan EcoNusa, Jakarta, 91 pp.

Petoe, P. & Baker, W.J. (2019) A monograph of *Heterospathe* (Areceae, Arecaceae) in New Guinea. *Phytotaxa* 413: 71–116.

Petoe, P., Cámara-Leret, R. & Baker, W.J. (2018b) A monograph of the *Hydriastele wendlandiana* group (Arecaceae: *Hydriastele*). *Kew Bulletin* 73: 17.

Petoe, P., Heatubun, C.D. & Baker, W.J. (2018a) A monograph of *Hydriastele* (Areceae, Arecaceae) in New Guinea and Australia. *Phytotaxa* 370: 1–92.

Petoe, P., Schuiteman, A. & Baker, W.J. (2019) *Hydriastele gibbsiana,* a remarkable belly palm from New Guinea. *Palms* 63: 5–10.

Polhemus, D.A. (2007) Tectonic Geology of Papua. *In:* Marshall, A.J. & Beehler, B.M. (eds) *The Ecology of Papua. Part 1.* Periplus Editions, Singapore, pp. 137–164.

POWO (2023) *Plants of the World Online.* Facilitated by the Royal Botanic Gardens, Kew. Available at: http://www.plantsoftheworldonline.org/ (accessed 3 April 2023).

Pratt, T.K. (1983) Diet of the Dwarf Cassowary *Casuarius bennetti picticollis* at Wau, Papua New Guinea. *Emu Austral Ornithology* 82: 283–285.

Prentice, M.L. & Hope, G.S. (2007) Climate of Papua. *In:* Marshall, A.J. & Beehler, B.M. (eds) *The Ecology of Papua. Part 1.* Periplus Editions, Singapore, pp. 177–195.

Raes, N. & Van Welzen, P.C. (2009) The demarcation and internal division of Flora Malesiana: 1857–present. *Blumea* 54: 6–8.

Rakotoarinivo, M., Dransfield, J., Bachman, S.P., Moat, J. & Baker, W.J. (2014) Comprehensive Red List assessment reveals exceptionally high extinction risk to Madagascar palms. *PLoS ONE* 9: e103684.

Rauwerdink, J.B. (1985) An essay on *Metroxylon,* the sago palm. *Principes* 30: 165–180.

Richardson, J.E., Costion, C.M. & Muellner, A.N. (2012) The Malesian floristic interchange: plant migration patterns across Wallace's Line. *In:* Gower, D., Johnson, K., Richardson, J.E., Rosen, B., Rüber, L. & Williams, S. (eds) *Biotic Evolution and Environmental Change in Southeast Asia.* Cambridge University Press, Cambridge, pp. 138–163.

Roscoe, P. (2002) The hunters and gatherers of New Guinea. *Current Anthropology* 43: 153–162.

Sâm, L.N., Baker, W.J., Bellot, S., Dransfield, J., Eiserhardt, W.L. & Henderson, A. (2023) *Truongsonia* (Arecaceae: Arecoideae: Truongsonieae) — a new palm genus and tribe from Vietnam. *Phytotaxa* 613: 201–212.

Schrödl, M. (2020) *Colonization and Diversification History of Madagascan Palms with New Phylogenomic Evidence from the Genus* Orania *(Arecaceae).* Masters thesis, Aarhus University.

Schumann, K. (1898) Die Flora von Neu-Pommern. *Notizblatt des Königl. Botanischen Gartens und Museums zu Berlin* 2: 59–158.

Schumann, K. & Hollrung, M. (1889) *Die Flora von Kaiser Wilhelms Land.* Asher & Co., Berlin, 137 pp.

Schumann, K. & Lauterbach, K. (1900) *Die Flora der Deutschen Schutzgebiete in der Südsee.* Verlag von Gebrüder Borntraeger, Leipzig, 613 pp.

Secretan, B., Straif, K., Baan, R., Grosse, Y., El Ghissassi, F., Bouvard, V., Benbrahim-Tallaa, L., Guha, N., Freeman, C., Galichet, L. & Cogliano, V. (2009) A review of human carcinogens — Part E: tobacco, areca nut, alcohol, coal smoke, and salted fish. *The Lancet Oncology* 10: 1033–1034.

Shahimi, S., Conejero, M., Prychid, C.J., Rudall, P.J., Hawkins, J.A. & Baker, W.J. (2019) A taxonomic revision of the myrmecophilous species of the rattan genus *Korthalsia* (Arecaceae). *Kew Bulletin* 74: 69.

Shearman, P. & Bryan, J. (2011) A bioregional analysis of the distribution of rainforest cover, deforestation and degradation in Papua New Guinea. *Austral Ecology* 36: 9–24.

Shearman, P.L., Ash, J., Mackey, B., Bryan, J.E. & Lokes, B. (2009) Forest conversion and degradation in Papua New Guinea 1972–2002. *Biotropica* 41: 379–390.

Shearman, P.L., Bryan, J.E., Ash, J., Hunnam, P., Mackey, B. & Lokes, B. (2008) *The State of the Forests of Papua New Guinea. Mapping the Extent and Condition of Forest Cover and Measuring the Drivers of Forest Change in the Period 1972–2002.* University of Papua New Guinea, Port Moresby, 148 pp.

Shee, Z.Q., Frodin, D.G., Cámara-Leret, R. & Pokorny, L. (2020) Reconstructing the complex evolutionary history of the Papuasian *Schefflera* radiation through herbariomics. *Frontiers in Plant Science* 11: 258.

Simons, G.F. & Fennig, C.D. (eds) (2018) *Ethnologue: Languages of the World, 21st edn.* SIL International. Available at: http://www.ethnologue.com.

Slik, J.W.F., Franklin, J., Arroyo-Rodríguez, V., Field, R., Aguilar, S., Aguirre, N., Ahumada, J., Aiba, S.-I., Alves, L.F., Anitha, K., Avella, A., Mora, F., Aymard C. G.A., Báez, S., Balvanera, P., Bastian, M.L., Bastin, J.-F., Bellingham, P.J., van den Berg, E., da Conceição Bispo, P., Boeckx, P., Boehning-Gaese, K., Bongers, F., Boyle, B., Brambach, F., Brearley, F.Q., Brown, S., Chai, S.-L., Chazdon, R.L., Chen, S., Chhang, P., Chuyong, G., Ewango, C., Coronado, I.M., Cristóbal-Azkarate, J., Culmsee, H., Damas, K., Dattaraja, H.S., Davidar, P., DeWalt, S.J., Din, H., Drake, D.R., Duque, A., Durigan, G., Eichhorn, K., Eler, E.S., Enoki, T., Ensslin, A., Fandohan, A.B., Farwig, N., Feeley, K.J., Fischer, M., Forshed, O., Garcia, Q.S., Garkoti, S.C., Gillespie, T.W., Gillet, J.-F., Gonmadje, C., Granzow-de la Cerda, I., Griffith, D.M., Grogan, J., Hakeem, K.R., Harris, D.J., Harrison, R.D., Hector, A., Hemp, A., Homeier, J., Hussain, M.S., Ibarra-Manríquez, G., Hanum, I.F., Imai, N., Jansen, P.A., Joly, C.A., Joseph, S., Kartawinata, K., Kearsley, E., Kelly, D.L., Kessler, M., Killeen, T.J., Kooyman, R.M., Laumonier, Y., Laurance, S.G., Laurance, W.F., Lawes, M.J., Letcher, S.G., Lindsell, J., Lovett, J., Lozada, J., Lu, X., Lykke, A.M., Mahmud, K.B., Mahayani, N.P.D., Mansor, A., Marshall, A.J., Martin, E.H., Calderado Leal Matos, D., Meave, J.A., Melo, F.P.L., Mendoza, Z.H.A., Metali, F., Medjibe, V.P., Metzger, J.P., Metzker, T., Mohandass, D., Munguía-Rosas, M.A., Muñoz, R., Nurtjahy, E., de Oliveira, E.L., Onrizal, Parolin, P., Parren, M., Parthasarathy, N., Paudel, E., Perez, R., Pérez-García, E.A., Pommer, U., Poorter, L., Qie, L., Piedade, M.T.F., Pinto, J.R.R., Poulsen, A.D., Poulsen, J.R., Powers, J.S., Prasad, R.C., Puyravaud, J.-P., Rangel, O., Reitsma, J., Rocha, D.S.B., Rolim, S., Rovero, F., Rozak, A., Ruokolainen, K., Rutishauser, E., Rutten, G., Mohd Said, M.N., Saiter, F.Z., Saner, P., Santos, B., dos Santos, J.R., Sarker, S.K., Schmitt, C.B., Schoengart, J., Schulze, M., Sheil, D., Sist, P., Souza, A.F., Spironello, W.R., Sposito, T., Steinmetz, R., Stevart, T., Suganuma, M.S., Sukri, R., Sultana, A., Sukumar, R., Sunderland, T., Supriyadi, Suresh, H.S., Suzuki, E., Tabarelli, M., Tang, J., Tanner, E.V.J., Targhetta, N., Theilade, I., Thomas, D., Timberlake, J., de Morisson Valeriano, M., van Valkenburg, J., Van Do, T., Van Sam, H., Vandermeer, J.H., Verbeeck, H., Vetaas, O.R., Adekunle, V., Vieira, S.A., Webb, C.O., Webb, E.L., Whitfeld, T., Wich, S., Williams, J., Wiser, S., Wittmann, F., Yang, X., Adou Yao, C.Y., Yap, S.L., Zahawi, R.A., Zakaria, R. & Zang, R. (2018) Phylogenetic classification of the world's tropical forests. *Proceedings of the National Academy of Sciences of the United States of America* 115: 1837–1842.

Sloss, C.R., Nothdurft, L., Hua, Q., O'Connor, S.G., Moss, P.T., Rosendahl, D., Petherick, L.M., Nanson, R.A., Mackenzie, L.L., Sternes, A., Jacobsen, G.E. & Ulm, S. (2018) Holocene sea-level change and coastal landscape evolution in the southern Gulf of Carpentaria, Australia. *Holocene* 28: 1411–1430.

Sneed, M.W. (1985) *Gronophyllum procerum*, people and places, pictured in their stone age habitat. *Principes* 29: 47–54.

Stauffer, F., Baker, W.J., Dransfield, J. & Endress, P. (2004) Comparative floral structure and systematics of *Pelagodoxa* and *Sommieria* (Arecaceae). *Botanical Journal of the Linnean Society* 146: 27–39.

Straarup, M., Hoppe, L.E., Pooma, R. & Barfod, A.S. (2018) The role of beetles in the pollination of the mangrove palm *Nypa fruticans*. *Nordic Journal of Botany* 36: e01967.

Sunderland, T.C.H. (2012) A taxonomic revision of the rattans of Africa (Arecaceae: Calamoideae). *Phytotaxa* 51: 1–76.

Takeuchi, W.N. (2000). A floristic and ethnobotanical account of the Josephstaal Forest Management Agreement Area, Papua New Guinea. *SIDA, Contributions to Botany* 19: 1–63.

Takeuchi, W.N. (2007) Introduction to the Flora of Papua. *In:* Marshall, A.J. & Beehler, B.M. (eds) *The Ecology of Papua. Part 1.* Periplus Editions, Singapore, pp. 269–302.

Terborgh, J. & Diamond, J.M. (1970) Niche overlap in feeding assemblages of New Guinea birds. *The Wilson Bulletin* 82: 29–52.

Thiers, B.M. (2023) *Index Herbariorum.* Available at: https://sweetgum.nybg.org/science/ih/ (continuously updated).

Thomson, B.H. (1889) New Guinea: narrative of an exploring expedition to the Louisiade and D'Entrecasteaux Islands. *Proceedings of the Royal Geographical Society of London, new series* 11: 525–542.

Tomlinson, P.B. (2006) The uniqueness of palms. *Botanical Journal of the Linnean Society* 151: 5–14.

Trudgen, M.S. & Baker, W.J. (2008) A revision of the *Heterospathe elegans* (Arecaceae) complex in New Guinea. *Kew Bulletin* 63: 639–647.

Turner, J., Petoe, P. & Baker, W.J. (2021) *Heterospathe parviflora. The IUCN Red List of Threatened Species* 2021: e.T171871838A171872085. https://dx.doi.org/10.2305/IUCN.UK.2021-3.RLTS.T171871838A171872085.en.

Uhl, N.W. & Dransfield, J. (1987) *Genera Palmarum, a classification of palms based on the work of Harold E. Moore Jr.* Lawrence, Kansas: L.H. Bailey Hortorium and the International Palm Society.

Utteridge, T.M.A. & Jennings, L.V.S. (eds) (2021) *Trees of New Guinea*. Royal Botanic Gardens, Kew, Richmond, pp. 648.

Van Balgooy, M. (1976) Phytogeography. *In:* Paijmans, K. (eds) *New Guinea Vegetation*. Australian National University Press, Canberra, pp. 1–22.

Van Steenis, C.G.G.J. (1950) The delimitation of Malaysia and its main plant geographical divisions. *In:* Van Steenis, C.G.G.J. (eds) *Flora Malesiana Volume 1*. Noordhof-Kolff N.V., Jakarta, pp. LXX–LXXV.

Van Steenis-Kruseman, M.J. (1950) Cyclopaedia of collectors. *In:* Van Steenis, C.G.G.J. (eds) *Flora Malesiana Volume 1*. Noordhof-Kolff N.V., Jakarta, pp. 1–639.

Van Welzen, P.C. & Slik, J.W.F. (2009) Patterns in species richness and composition of plant families in the Malay Archipelago. *Blumea* 54: 166–171.

Veron, J.E.N., Devantier, L.M., Turak, E., Green, A.L., Kininmonth, S., Stafford-Smith, M. & Peterson, N. (2009) Delineating the Coral Triangle. *Galaxea, Journal of Coral Reef Studies* 11: 91–100.

Vorontsova, M.S., Clark, L.G., Dransfield, J., Govaerts, R. & Baker, W.J. (2016) *World Checklist of Bamboos and Rattans. INBAR Technical Report No. 37*. International Network of Bamboo and Rattan, Beijing, 454 pp.

Wallace, A.R. (1860) On the zoological geography of the Malay Archipelago. *Journal of the Proceedings of the Linnean Society: Zoology* 4: 172–184.

Warburg, O. (1891) Beiträge zur Kenntnis der papuanischen Flora. *Botanische Jahrbücher fur Systematik, Pflanzengeschichte und Pflanzengeographie* 13: 130–455.
Westphal, E. & Jansen, P.C.M. (eds) (1989) *Plant resources of South-East Asia: a selection*. Pudoc/Prosea, Wageningen, 322 pp.
Young, B.E. (1985) A new species of *Gronophyllum* (Palmae) from Papua New Guinea. *Principes* 29: 138–141.
Zieck, J.F.U. (1972) *Minor Forest Products – rattans, etc. in some parts of the Eastern/Western Highlands, Chimbu districts and Jimi valley (W.H.D.) (U11/167-1-6)*. Forest Product Research Centre, Port Moresby, 7 pp.
Zona, S. (1999) Revision of *Drymophloeus* (Arecaceae: Arecoideae). *Blumea* 44: 1–24.
Zona, S. (2000) A personal history of *Drymophloeus* in New Guinea. *Palms* 44: 184–186.
Zona, S. (2003) Endosperm condition and the paradox of *Ptychococcus paradoxus*. *Telopea* 10: 179–185.
Zona, S. (2005) A revision of *Ptychococcus* (Arecaceae). *Systematic Botany* 30: 520–529.
Zona, S. & Essig, F.B. (1999) How many species of *Brassiophoenix*? *Palms* 43: 45–48.
Zona, S. & Henderson, A. (1989) A review of animal-mediated seed dispersal of palms. *Selbyana* 11: 6–21.
Zona, S., Francisco-Ortega, J., Jestrow, B., Baker, W.J. & Lewis C.E. (2011) Molecular phylogenetics of the palm subtribe Ptychospermatinae (Arecaceae). *American Journal of Botany* 98: 1716–1726.
Zumbroich, T.J. (2008) The origin and diffusion of the betel chewing: a synthesis of evidence from South Asia, Southeast Asia and beyond. *Electronic Journal of Indian Medicine* 1: 63–116.

INDEX OF PLANT NAMES

Main genus and species entries indicated by bold page number. Synonyms and names of uncertain application are in *italic* text.

M

N

O

P

R

PACIFIC
Pulau Waigeo
almahera Sea
Kepulauan Raja Ampat
Sorong
Tamrau Mountains
Bird's Head Peninsula
Manokwari
Arfak Mountains
Gunung Umsini (2970 m)
Pulau Biak
Pulau Yapen
Pulau Misool
Seram Sea
Gunung Binaiya (3027 m)
Pulau Seram
Fakfak
Bomberai Peninsula
Teluk Cenderwasih
Nabire
Weremai
Sarmi
Sungei Mamberamo
Gautier Range
Van Rees Range
Jayapura
Vanimo
Torricelli Mountain
Sungei Taritatu
Maoke Mountains
Puncak Jaya (4884 m)
Sudiman Range
Jayawijaya Mountains
Timika
Star Mountains
Ok Tedi
INDONESIA
Banda Sea
Agats
NEW GUI
Kepulauan Aru
Sungai Digul
Fly
Pulau Yamdena
Merauke
Arafura Sea
Torres
Gulf of Carpentaria
AUSTRAL
NORTHERN TERRITORY